ADVANCES IN NANOSTRUCTURED COMPOSITES

Volume 1: Carbon Nanotube and Graphene Composites

Editor

Mahmood Aliofkhazraei
Department of Materials Engineering
Tarbiat Modares University
Tehran, Iran

CRC Press
Taylor & Francis Group
Boca Raton London New York

CRC Press is an imprint of the
Taylor & Francis Group, an **informa** business

A SCIENCE PUBLISHERS BOOK

Cover credit: Tyler Boyes/Shutterstock.com

CRC Press
Taylor & Francis Group
6000 Broken Sound Parkway NW, Suite 300
Boca Raton, FL 33487-2742

First issued in paperback 2020

© 2019 by Taylor & Francis Group, LLC
CRC Press is an imprint of Taylor & Francis Group, an Informa business

No claim to original U.S. Government works

ISBN-13: 978-1-4822-3663-7 (hbk)
ISBN-13: 978-0-367-77998-6 (pbk)

Library of Congress Cataloging-in-Publication Data

Names: Aliofkhazraei, Mahmood, editor.
Title: Carbon nanotube and graphene composites / editor Mahmood
   Aliofkhazraei, Department of Materials Engineering, Tarbiat Modares
   University, Tehran, Iran.
Description: Boca ERaton, FL : CRC Press, Taylor & Francis Group, 2018. |
   Series: A science publishers book | Series: Advances in nanostructured
   composites ; volume 1 | Includes bibliographical references and index.
Identifiers: LCCN 2018032343 | ISBN 9781482236637 (hardback)
Subjects: LCSH: Nanocomposites (Materials) | Carbon nanotubes. | Graphene.
Classification: LCC TA418.9.C6 C3228 2018 | DDC 620.1/18--dc23
LC record available at https://lccn.loc.gov/2018032343

Visit the Taylor & Francis Web site at
http://www.taylorandfrancis.com

and the CRC Press Web site at
http://www.crcpress.com

Preface

Composites and nanocomposites are used in cases where long durability and strength of components are required, i.e., where high stress levels, erosion processes and multiphase environments are present, including the parts under collision and impact, the parts under rotating motion and erosion (like excavation drills in oil and gas wells). The first volume of this book aims to provide a guide for fabrication of new nanocomposites mainly based on carbon nanotubes and graphene. The main topics of this volume are: Application of Nano-powders for Formation of Metal Matrix of Composites, Conjugated Polymer Nanocomposites, Biopolymer Nanocomposites, Dental Nanocomposites, Graphene-based Nanocomposites for Electrochemical Energy Storage, Polymer/Filler Composites for Optical Diffuse Reflectors, Synthesis and Applications of LDH-Based Nanocomposites, Rubber—CNT Nanocomposites, Nanocomposite Fibers with Carbon Nanotubes, Fabrications of Graphene Based Nanocomposites for Electrochemical Sensing of Drug Molecules, Recent Advances in Graphene Metal Oxide Based Nanocomposites.

Editor

Contents

SECTION II: CNT AND GRAPHENE NANOCOMPOSITES

Section I

Introduction to Nanocomposites Fabrication

1

Application of Nano-powders for Formation of Metal Matrix of Composites

Vladimir A. Popov

Introduction

Metal matrix composites have improved service characteristics, i.e., reduced thermal-expansion coefficient, improved wear resistance or lubrication characteristics, improved material properties, particularly stiffness and strength, either providing increased component durability or permitting more extreme service conditions, while an aluminum matrix has the highest possible strength-to-weight ratio which is impossible to obtain in regular metals or alloys. This is the subject of great interest in metal matrix composites development including composites with aluminum matrix (Sijo and Jayadevan 2016, Rajan et al. 2010, Mazahery and Ostad Shabani 2012, Carreño-Gallardo et al. 2014, Manigandan et al. 2012, Kollo et al. 2010, Aigbodion and Hassan 2007, Eslamian et al. 2008, Ureña et al. 2007, Kollo et al. 2011, Kalkanlı and Yılmaz 2008, Tham et al. 2001). For this particular reason, the concern regarding this class of materials has recently increased, especially on the part of the manufacturers in automotive and electronics technologies. However, existing methods of manufacturing metal matrix composites do not allow the effective obtaining of a material with reinforcements less than 1 micron by size. On the other hand, there is already a demand in materials with nano-size reinforcements in the industry (Kollo et al. 2011). The basic problems of such methods are poor wettability (sometimes full absence of wettability) between matrix material melt and reinforcements for casting methods of obtaining MMC or inadequate strength due to loose cohesion on the matrix-reinforcing particles interface of components for powder metallurgy methods of obtaining MMC.

There might be various solutions to this problem. This paper offers a method to improve binding strength between the matrix and reinforcing particles, based on the use of nanomaterials for matrix forming.

National University of Science and Technology "MISIS", Department of Physical Metallurgy, Leninsky prospect, 4, 119049 Moscow, Russia.
Email: popov58@inbox.ru

Materials, research methods and experimental equipment

The research has been carried out in development of metal-matrix composites with aluminum and copper matrices and reinforcing particles of silicon carbide. Base materials were commercially available powders of silicon carbide and wire of commercial-purity aluminum and copper. MMC was produced according to the following process flow scheme: production of metal nanopowders by means of electrical explosion of wire method (Burtsev et al. 1990, Kotov et al. 1997, Gromov and Teipel 2014, Il'in et al. 2013, Nazarenko 2005) (the equipment used has been developed in the Institute of Electrophysics of the Ural Division of the Russian Academy of Sciences, Yekaterinburg, Russia); after production, the powders were exposed to conservation in hexane, mixing nanopowders with reinforcing particles, dynamic pressing (magnetic-pulse treatment (Kaygorodov et al. 2007, Ivanov et al. 1995, Nikonov et al. 2006, Chae et al. 2011, Li et al. 2010, Ivanov et al. 2006, Bhuiyan et al. 2013), the equipment and methods used have been developed in the Institute of Electrophysics of the Ural Division of the Russian Academy of Sciences, Yekaterinburg, Russia).

Preparation of nanopowders for metal-matrix composites by means of electrical detonation

In order to obtain Al and Cu nanopowders, the method of electrical detonation of a wire (EDW) (Kollo et al. 2010), has been utilized. The method is as follows: through the wire of a desired material, a current pulse of high density ($10^4\ldots10^6$ A per mm^2), received usually at discharge of a capacitor bank, is passed. Due to inertia of metal thermal expansion and very short time of energy input ($10^{-8}\ldots10^{-5}$ sec), it is possible to overheat the material essentially up to its destruction. Before termination of detonation, it is possible to enter into a wire a lot of energy (up to 3 times of the material bonding energy). As a result of overheat, the explosive destruction of the metal followed with a light exposure, shock wave and disintegration of the wire metal to nano-particles may be observed. The nano-particles obtained have spherical shape. Varying the diameter of blasted out wire, metal overheat and pressure of ambient gas, it is possible to receive particles of the size of up to tens to hundreds of nanometers. Due to high activity of nano-particles obtained by EDW, indifferent gases should be used as working gases while manufacturing nanopowders of metals and alloys. In the same gas, packing and conservation of the powder have been carried out.

For this study, some lots of aluminum and copper nanopowder were obtained and conserved in hexane:

1) 8 lots of aluminum nanopowder with mean values of specific surface in a lot from S = 18.5 m^2/gram (d = 120 nm) up to S = 14 m^2/gram (d = 160 nm) (Figure 1).

2) 3 lots of copper nanopowder with mean values of specific surface in a lot from S = 10.3 m^2/gram (d = 63 нм) up to S = 9.5 m^2/gram (d = 70 nm). The SEM-picture is shown in Figure 2.

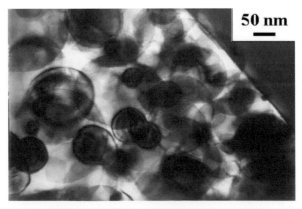

Figure 1. TEM image of aluminum nanopowder.

Silicon carbide powder of two types has been used. At first, the commercial SiC powder with specific surface S = 2.54 m²/g has been applied. Secondly, the part of silicon carbide powder has been treated in a planetary grinding mill; this dust was certificated for specific surface, S = 11.7 m²/g, and, then, its sedimentation analysis was conducted. It has been shown that at tenfold serial ultrasonic processing and consequent washing out, one manages to extract from the powder about 30 weight percentage of particles with specific surface S = 20.5 m²/g. As the transmission electron microscopy (Figure 3) has shown, at a small-sized fraction (S = 20.5 m²/g), there are agglomerates of more small-sized particles. These agglomerates are not separated by ultrasonic processing and have size down to 300 nm.

As it was already mentioned, the developed method includes preparation of a mixture from nanopowders and its consequent compacting. The mixture has been prepared directly prior to compacting. At first, the disintegration of aluminum powder has been carried out in hexane during one hour. Then the necessary amount of SiC powder, also disintegrated in hexane, was added into it. This mixture was dispersed with simultaneous stirring and evaporation of hexane within 1.5 hours. The creamy mass obtained was transmitted for compaction, which represented either magnetic impulse pressurizing (MIP) or combination of MIP with processing by explosion.

Figure 2. SEM of initial copper nanopowder.

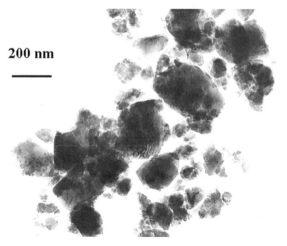

Figure 3. TEM image of silicon carbide nanopowder.

The method of magnetic impulse pressurizing of powder mixtures of MMC

The single-axis magnetic impulse press was used as the baseline equipment for pressurizing of MMC powder mixtures. The impulse force contracting the powder in the mold is generated as a result of interaction of an impulse magnetic field of the inductor and conducting surface of the concentrator facing it. The powerful current pulses for feeding the inductor are generated by discharge of a capacitive energy store (2400 μf, 5 KV). The amplitude of current pulses in operational modes varied within the limits of 5…60 kA. The duration of the first half-wave of a current pulse depends on operational mode of the inductor, and lays within the limits of 300…600 μsec. The amplitude of impulse pressing force on the installation can reach 800 kN.

The unit of degassing is intended for removal of gases and adsorbed matters from molds filled with powder. It is implemented by the means of continuous vacuum spilling of the pressurized chamber with the mold inside capable of heating and cooling off the powder within the temperature range of 20…500°C. The chamber is supplied with the adapter transmitting gain of pressing from an atmosphere of air to vacuum, that ensures pressing powder in vacuum.

Two types of pressurizing tools were applied. At first, sectional molds made of high-speed steels were used. The die with the diameter of 15 mm shaping the channel of pressing consists of three identical quadrants which makes it convenient to extract pressurized samples. The materials of dies and punches allow the making of powder degassing at temperatures up to 500°C without deterioration of mechanical characteristics of the tool, and then pressing at pressures up to 1.6 GPa. The similar modes of pressurizing tool working regimes limit its reusable operation.

Secondly, the one-time mold containing non-reusable solid cylindrical die and two punches has been developed with the use of the available equipment. The set of such tools was produced in order to carry out a separate experiment on pressurizing. This method allowed pressing powders at higher pressures reaching 2.7 GPa. The utmost pressure of pressurizing is limited to plastic deformation of punches and their jamming in dies.

Study of the structure and properties was held by means of scanning and transmission electron microscopy.

Theoretical estimation of the processes while using casting methods of obtaining MMC

By means of existing casting methods, it is impossible to obtain MMC with the size of reinforcements less than 10 microns. New methods should be developed in this case but would that be practical? To resolve this problem, it is necessary to simulate the process of obtaining MMC.

Substantial necessity in qualitative forecasting properties of the materials obtained depending on the various factors generates a set of mathematical models and approaches describing the process of obtaining MMC and their properties. In the papers (Popov 2017, Popov et al. 2005a), the analysis of the process of grains nucleation of solute on reinforcing particles in a melt are submitted depending on such parameters of the reinforcement as temperature, volumetric fraction and chemical composition defining surface energy.

These papers show that the reduction of the reinforcing particle size results in sharp increase of chemical heterogeneity of the alloy which is due to significant deterioration of material properties. For example, it was shown that already at the size of the particle of 0.5 microns, the concentration of the melt in layer around the particle will be more than that of a remaining melt by 9%. The further decrease of the size of reinforcing particles results in even greater heterogeneity of chemical composition of the melt.

So, the theoretical estimation of processes of obtaining metal-matrix composites with formation of the fluid phase has shown that at particle sizes less than 0.5 microns, the application of casting methods in production of metal-matrix composites results in appearance of a considerable heterogeneity of distribution of alloying components in the matrix, which is undesirable, as it will result in non-uniformity of their properties.

Two new methods of manufacturing metal-matrix composites without formation of fluid phases were, therefore, offered. The first method is based on application of nano-materials for obtaining the matrix. Second one is based on application of mechanical alloying.

Development of methods of manufacturing MMC on the basis of application of nano-materials

The main idea of this method consists in using high activity of particles at nano-scale state (Popov et al. 2002, Popov et al. 2005b). Firstly, research for checking the conjecture about positive influence of using nano-dust of aluminum for the processes of wetting of silicon carbide by aluminum melt was conducted. For this purpose, the following experiment has been carried out. Aluminum nano-dust at ambient temperature was mixed with silicon carbide dust, then the mixture was a little bit heated, put in aluminum melt and the melt was quickly cooled off to solidification. On the samples of the material obtained, the interface of aluminum melt with reinforcing particles of silicon carbide plated by particles of aluminum nanopowder has been studied.

The interface between aluminum melt and silicon carbide particles plated by aluminum nano-particles was investigated with the help of scanning electron microscope. In Figure 4, the general view of this boundary and adjoining sites is exhibited. The picture shows, that the silicon carbide particle plated by nano-particles of aluminum is clearly visible. Nano-particles are uniformly distributed on a surface of the silicon carbide particle (position 1 in Figure 4), that allows an aluminum melt to effectively wet the silicon carbide particle. This is represented by the flowing property of an aluminum melt (position 2 in Figure 4) on a surface of the particle of silicon carbide plated by aluminum nano-particles. In order to obtain more detailed information about interface surface between aluminum nano-particles and silicon carbide particles, the research was conducted by means of a transmission electron microscope.

The Figure 5 demonstrates a view of aluminum nano-particles on the surface of silicon carbide particle. No pretreatment of the surface of silicon carbide particle was applied.

The shapes of aluminum nano-particles as well as the value of wetting angle speak about high wettability between aluminum and silicon carbide in the case of application of nano-size aluminum powder. This figure helps to understand that the size of silicon carbide particles practically does not influence the processes of wetting them by aluminum, therefore, it is possible to apply silicon carbide particles of submicron sizes.

This study has allowed for the development of a method of manufacturing MMC on the base of applications of nano-materials to form matrix. The method developed includes preparation of nanopowders, mixtures for them and their consequent compacting.

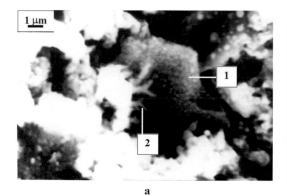 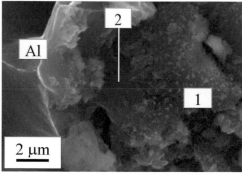

a   b

Figure 4. Fracture surface of the specimen with small quantity of aluminum nanopowder after putting in aluminum melt: 1 – SiC-particle covered by aluminum nanopowder; 2 – aluminum spreading on surface of SiC-particle covered by aluminum nanopowder (SEM).

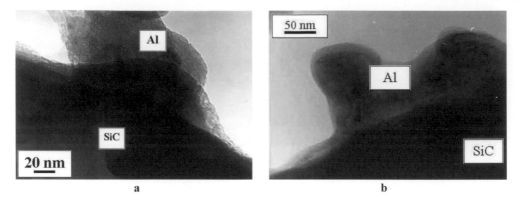

a   b

Figure 5. Aluminum nano-particles on the surface of silicon carbide particle.

Magnetic impulse pressurizing of powder mixtures of MMC

The above described method has been used for studying magnetic impulse pressurizing of powder mixtures of aluminum nanopowders with micro-size SiC powder in three weight proportions, as stated above: Al+0.2SiC, Al+0.4SiC, Al+0.6SiC. In this case, the temperature of pressurizing varied within the limits of 20…300°C, along with gauging impulse pressure in the process of powder compression. Experiments on pressurizing of a composite mixture of copper nanopowder with silicon carbide powder (Cu+0.07SiC) were also conducted. The powders were conserved in hexane. A part of samples after MIP was treated for compaction by an explosive method.

Investigation of structure and properties of metal-matrix composites obtained with application of nano-materials

Metal-matrix composites with aluminum matrix

In order to estimate the quality of obtained samples, the investigation of their structure was conducted by means of optical, scanning electron and transmission electronic microscopy. In order to identify the basic features of structure formation when using nano-materials, investigation of MMC with reinforcing particles by the size 10 microns was conducted at the beginning.

This study has shown that due to the method developed, the dense contact between matrix and silicon carbide particles after magnetic impulse pressurizing is reached. The processing by explosion improves quality of the material under certain conditions. So, processing, by explosion of MMC with large contents (60%) of silicon carbide reinforcement and with the SiC particle size larger than 10 microns, results in cracks initiation. Reduction of silicon carbide reinforcement contents to 40% considerably decreases probability of cracks initiation, but the defects are still present. At the reduction of silicon carbide reinforcement contents to 20%, no defects are visible (Figure 6). Full elimination of defects also results in decrease of the silicon carbide particle size.

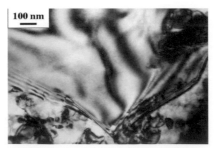

Figure 6. No defects in metal matrix composite structure.

It was earlier shown that the surface activity of a nanopowder positively influences the process of wetting. Thus, nanopowder immerses into anything, even the narrowest gaps between particles (Figure 7). The major concern represents the state of an interface between matrix and reinforcing particles, as it determines the strength of an aggregate. The research has shown that this interface does not contain any inclusions, defects, etc., only dense contact between components is visible (Figure 8).

During the treatment, the matrix structure is developing. The view of initial state of aluminum nanopowder is shown in Figure 1. The applied load induces deformation of nano-particles, thus, numerous defects arise in them (Figure 9a).

The following stage of structure development may be assumed as the first stage of recrystallization. Figure 9b shows nanopowder particles fitting closely to each other becoming nano-grains of the dense matrix. The density of defects is reducing. Nanopowder particles have not yet lost their individuality. The next stage of structure development is grain growth by collective re-crystallization type (Figure 9c).

The state of "matrix – particle" interface, thus (Figure 10), is not worsened.

The method developed enables to obtain MMC with uniform distribution of reinforcing particles through the volume (Figure 11).

Furthermore, according to the developed method, MMC with reinforcing particles of size less than 150 nm were obtained. All basic patterns of structure development were preserved. "Matrix – particle" interface has no defects either. Figure 12 presents the view of small-size reinforcing particle in an aluminum matrix.

The conducted material processing results in the occurrence of defects in silicon carbide nano-particles (Figure 13). It is difficult to reveal the time of these defects appearance, whether they have appeared during manufacturing and preparation of the powder or during dynamic compacting. However, the tendency of decreasing quantity of defects with reduction of particle size is stable (compare Figure 13 and Figure 12).

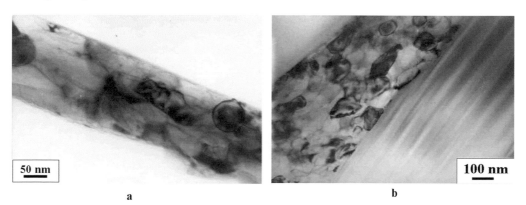

Figure 7. Matrix material nano-particles penetrate into narrow gaps (100–200 nm) between SiC particles.

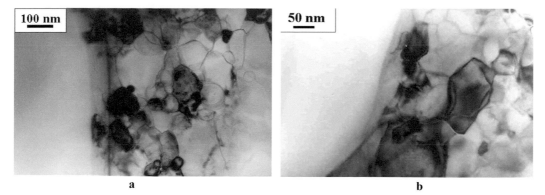

Figure 8. The "matrix-reinforcement" interface.

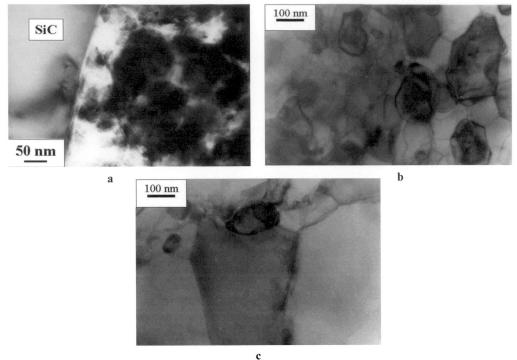

c

Figure 9. Stages of structure development.

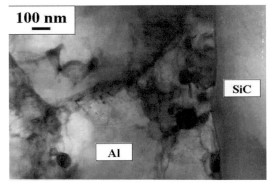

Figure 10. Interface of "matrix-reinforcement" during grain growth.

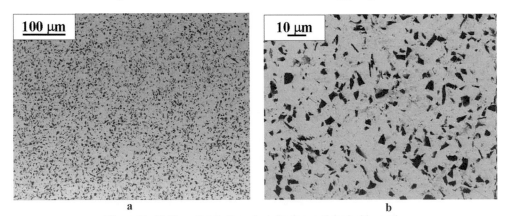

Figure 11. Uniform distribution of reinforcing particles inside matrix.

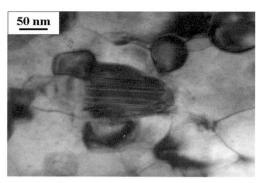

Figure 12. A fine reinforcing particle in an aluminum matrix.

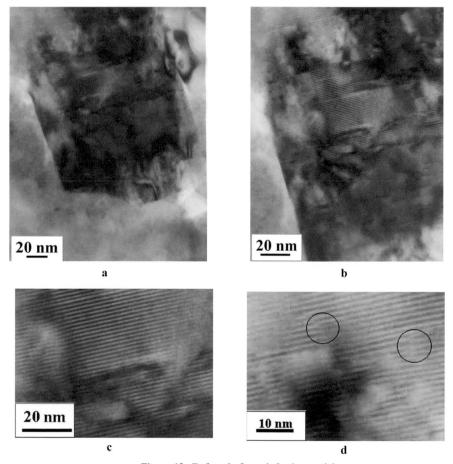

Figure 13. Defects in fine reinforcing particles.

In order to estimate the properties of the obtained material, its hardness and micro-hardness were measured. The results of measuring are presented in the Tables 1 and 2, where it is visible that the increase in percentage of silicon carbide results in increase of hardness and micro-hardness of the material. The increase of pressurizing temperature results in some decrease of hardness.

Table 1. Micro-hardness of aluminum-based metal-matrix composites.

Composition	Temperature of processing, °C	Micro-hardness, MPa
Al-20%SiC	300	2,890
Al-40%SiC	300	4,170
Al-60%SiC	300	4,310

Table 2. Hardness of aluminum-based metal-matrix composites.

Composition	Temperature of processing, °C	Vickers hardness, HV
Al+0,2SiC	300	155±4
Al+0,4SiC	300	181±4
Al+0,4SiC	200	195±3

Metal-matrix composites with copper matrix

There is a fair quantity of matrix-reinforcement versions. Among them, the considerable concerns represent MMC on the basis of copper and copper-alloys (Almomani et al. 2016, Prosviryakov 2015, Bagheri 2016, Dinaharan et al. 2016, Modling and Grong 1995, Coupard et al. 1997). The goal of this research stage is to study the combination of copper and silicon carbide reinforcement. The method described above was applied to obtain such combination.

The research has shown that each technological operation is relevant to obtain solid material. Nanopowder creates agglomerates. Before mixing, it is necessary to treat it in the powder disperser; mixing itself should also be conducted in the powder disperser. Insufficient period of treatment results in formation of zones with the lack of silicon carbide particles in the material (Figure 14). When following the developed modes, one can obtain the material with uniform distribution of reinforcing particles and without any defects (Figure 15). TEM studies have shown that between matrix and silicon carbide reinforcing particles, there is a dense contact without availability of any defects that should ensure the high level of material properties (Figure 16).

The structure of matrix undergoes the following changes. In the initial state, copper nano-particles have the spherical shape (Figure 2). After compacting, nano-grains lose the spherical shape (Figure 17a). Furthermore, grain growth, according to the mechanism of collective recrystallization, takes place. The grain size increases up to 500 nm (Figure 17b).

Table 3 shows the values of hardness of the obtained material.

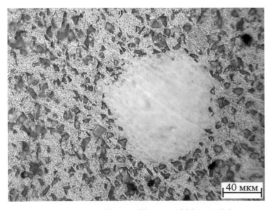

Figure 14. Zones without silicon carbide particles as a result of weak treatment in disperser.

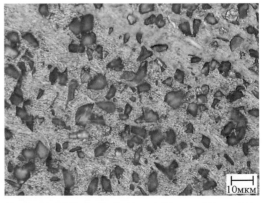

Figure 15. Uniform distribution of silicon carbide particles inside copper matrix.

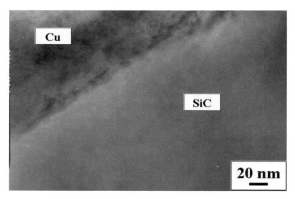

Figure 16. Interface of "copper-silicon carbide particle".

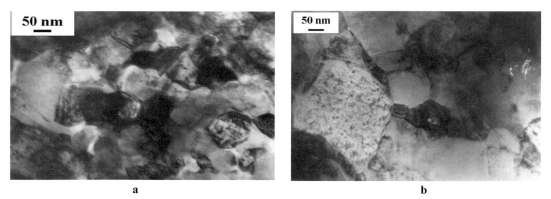

Figure 17. Structure development of copper matrix.

Table 3. Hardness of copper-based matrix composites.

Composition	Temperature of processing, °C	Vickers hardness, HV
Cu+0.07SiC	300	223±3
Cu+0.07SiC	300	222±4

Conclusion

The present study shows that the use of nanopowders for composite material matrix forming allows for considerable increase of wetting between components of the composite wherein this is typical of various materials. This technological approach allows for improvement of the contact between particles of the composite and upgrade of its properties. Study of the structure has shown absence of any defects, both on the surface of the "matrix-reinforcing particle" interface and within the body of the matrix material.

Acknowledgments

The experimental part of this work was partially done on the equipment of the Institute of Electrophysics of the Ural Division of the Russian Academy of Sciences (Yekaterinburg, Russia). Author is grateful to I.I. Khodos, V.V. Ivanov for help in the study of specimens. The research leading to these results has received funding from the Ministry of Education and Science of Russian Federation under the project number 14.587.21.0030 (identifier RFMEFI58716X0030).

References

Aigbodion, V.S. and S.B. Hassan. 2007. Effect of silicon carbide reinforcement on microstructure and properties of cast Al-Si-Fe/SiC particulate composites. Materials Science and Engineering: A 447(1-2): 355–360.

Almomani, M.A., Wa'il Radwan Tyfour and Mohammed Hani Nemrat. 2016. Effect of silicon carbide addition on the corrosion behavior of powder metallurgy Cu–30Zn brass in a 3.5 wt% NaCl solution. Journal of Alloys and Compounds 679: 104–114.

Bagheri, G.H.A. 2016. The effect of reinforcement percentages on properties of copper matrix composites reinforced with TiC particles. Journal of Alloys and Compounds 676: 120–126.

Bhuiyan, M.H., Yukihiro Isoda, Taek-Soo Kim and Soon-Jik Hong. 2013. Thermoelectric properties of n-type 95%Bi2Te3–5%Bi2Se3 materials fabricated by magnetic pulse compaction. Intermetallics 34: 49–55.

Burtsev, V.A., N.V. Kalinin and F.V. Luchinsky. 1990. Electric explosion of conductors and its application in the electrical installations. Energoatomizdat, Moscow (in Russian).

Carreño-Gallardo, C., I. Estrada-Guel, C. López-Meléndez and R. Martínez-Sánchez. 2014. Dispersion of silicon carbide nanoparticles in a AA2024 aluminum alloy by a high-energy ball mill. Journal of Alloys and Compounds 586: S68–S72.

Chae, H.J., Young Do Kim and Taek-Soo Kim. 2011. Microstructure and mechanical properties of rapidly solidified Mg alloy powders compacted by magnetic pulsed compaction (MPC) method. Journal of Alloys and Compounds 509: S250–S253.

Coupard, D., M.C. Castro, J. Coleto, Ana Garcia, Javier Goni and J.K. Palacios. 1997. Wear behavior of copper matrix composites. Key Engineering Materials 127-131: 1009–1016.

Dinaharan, I., Ramasamy Sathiskumar and Nadarajan Murugan. 2016. Effect of ceramic particulate type on microstructure and properties of copper matrix composites synthesized by friction stir processing. J. Mater Res. Technol. 5: 302–316.

Eslamian, M., Joel Rak and Nasser Ashgriz. 2008. Preparation of aluminum/silicon carbide metal matrix composites using centrifugal atomization. Powder Technology 184: 11–20.

Gromov, A. and U. Teipel. 2014. Metal Nanopowders: Production, Characterization, and Energetic Applications. Wiley-VCH Verlag GmbH & Co., KGaA, Weinheim, Germany, XXI.

Il'in, A.P., O.B. Nazarenko and D.V. Tikhonov. 2013. Peculiarities of nanopowders in terms of electrical explosion of conductors: a monograph. National Research Tomsk Polytechnic University. Izdatelstvo TPU, Tomsk (in Russian).

Ivanov, V., Y.A. Kotov, O.H. Samatov, R. Böhme, H.U. Karow and G. Schumacher. 1995. Synthesis and dynamic compaction of ceramic nano powders by techniques based on electric pulsed power. Nanostructured Materials 6(1-4): 287–290.

Ivanov, V.V., A.S. Lipilin, Yu.A. Kotov, V.R. Khrustov, S.N. Shkerin, S.N. Paranin et al. 2006. Formation of a thin-layer electrolyte for SOFC by magnetic pulse compaction of tapes cast of nanopowders. Journal of Power Sources 159(1): 605–612.

Kalkanlı, A. and S. Yılmaz. 2008. Synthesis and characterization of aluminum alloy 7074 reinforced with silicon carbide particulates. Materials & Design 29(4): 775–780.

Kaygorodov, A.S., V.V. Ivanov, V.R. Khrustov, Yu.A. Kotov, A.I. Medvedev, V.V. Osipov et al. 2007. Fabrication of Nd: Y2O3 transparent ceramics by pulsed compaction and sintering of weakly agglomerated nanopowders. Journal of the European Ceramic Society 27(2-3): 1165–1169.

Kollo, L., M. Leparoux, C.R. Bradbury, C. Jäggi, E. Carreño-Morelli and M. Rodríguez-Arbaizar. 2010. Investigation of planetary milling for nano-silicon carbide reinforced aluminum metal matrix composites. Journal of Alloys and Compounds 489(2): 394–400.

Kollo, L., C.R. Bradbury, R. Veinthal, C. Jäggi, E. Carreño-Morelli and M. Leparoux. 2011. Nano-silicon carbide reinforced aluminium produced by high-energy milling and hot consolidation. Materials Science and Engineering: A 528: 6606–6615.

Kotov, Yu., E.I. Azarkevich, I.V. Beketov, T.M. Demina, A.M. Murzakaev and O.M. Samatov. 1997. Producing Al and Al2O3 nanopowders by electrical explosion of wire. Key Engineering Materials 132-136: 173–176.

Li, M., Hai-ping Yu and Chun-feng Li. 2010. Microstructure and mechanical properties of Ti6Al4V powder compacts prepared by magnetic pulse compaction. Transactions of Nonferrous Metals Society of China 20(4): 553–558.

Manigandan, K., T.S. Srivatsan and T. Quick. 2012. Influence of silicon carbide particulates on tensile fracture behavior of an aluminum alloy. Materials Science and Engineering: A 534: 711–715.

Mazahery, A. and Mohsen Ostad Shabani. 2012. Nano-sized silicon carbide reinforced commercial casting aluminum alloy matrix: Experimental and novel modeling evaluation. Powder Technology 217: 558–565.

Modling, O.T. and Ø. Grong. 1995. Processing and properties of particle reinforced Al-SiC MMCs. Key Engineering Materials 104-107: 329–354.

Nazarenko, O.B. 2005. Electroexplosion nanopowders: production, properties, applications. Izdatelstvo TPU, Tomsk (in Russian).

Nikonov, A.V., E.M. Kelder, J. Schoonman, V.V. Ivanov and N.M. Pivkin. 2006. Characteristic changes under pulsed pressure action in electrode materials based on LiMn2O4 and Li4Ti5O12 spinels. Solid State Ionics 177(26-32): 2779–2785.

Popov, V.A., D.R. Lesuer, I.A. Kotov et al. 2002. On the Development of Microstructure in a MMC using Nano-materials. TMS Annual Meeting, February, Seattle, USA.

Popov, V.A., A.V. Marmulev and M. Yu. Kondratenkov. 2005a. Theoretical evaluation of the possibility of obtaining metal-matrix composites with small strengthening particle dimension. Russian Journal of Non-Ferrous Metals 46(2): 24–29.

Popov, V.A., Yu.A. Kotov and V.V. Ivanov. 2005b. New methods of production of metal matrix composites based on using nanomaterials. Non-ferrous Metals 1: 92–95 (in Russian).

Popov, V.A. 2017. Theoretical estimation of the processes while using casting methods of obtaining MMC. Proc. Int. Conf. ICMAEM-2017-IOP-061, India, Accept. In Press.

Prosviryakov, A.S. 2015. SiC content effect on the properties of Cu-SiC composites produced by mechanical alloying. Journal of Alloys and Compounds 632: 707–710.

Rajan, T.P.D., R.M. Pillai and B.C. Pai. 2010. Characterization of centrifugal cast functionally graded aluminum-silicon carbide metal matrix composites. Materials Characterization 61(10): 923–928.

Sijo, M.T. and K.R. Jayadevan. 2016. Analysis of stir cast aluminium silicon carbide metal matrix composite: a comprehensive review. Procedia Technology 24: 379–385.

Tham, L.M., M. Gupta and L. Cheng. 2001. Effect of limited matrix-reinforcement interfacial reaction on enhancing the mechanical properties of aluminium-silicon carbide composites. Acta Materialia 49(16): 3243–3253.

Ureña, A., M.V. Utrilla, J. Rams, P. Rodrigo and M. Ferrer. 2007. Electroless multilayer coatings on aluminium-silicon carbide composites for electronic packaging. Journal of the European Ceramic Society 27(13-15): 3983–3986.

2

Nanocomposites: Recent Trends, Developments and Applications

Ravindra Pratap Singh

Introduction

The material which is made up of more than one component is known as composite whereas nanocomposites are materials that combine plastic and nanosized nanofibers. They have many advantages over synthetic plastic, and are utilized in car parts; for example, Toyota started using nanocomposites to make bumpers for their cars, General Motors (GM) made nanocomposite step-assists external running boards that help people get into and out of cars. Nanocomposites are lighter, stiffer, less brittle, and more dent- and scratch-resistant than conventional plastics. Some nanocomposites are also more recyclable, more flame retardant, less porous, better conductors of electricity, and can be painted more easily. Small additions, i.e., nanofillers and big changes are the advantages of nanocomposites, long lasting, more resistant to wear-and-tear and breakage without affecting the surface quality or the transparency. Most nanocomposites use nanoparticles of smectite, a pure type of clay, and more advanced carbon nanotubes as the filler. General Motor uses a specific type of nanocomposite in its cars called a thermoplastic olefin (TPO) based nanocomposite. Thermoplastic is a material that melts when heated whereas olefin is a chemical which has a molecular structure with two carbon atoms doubly bonded together. Nanocomposites have a very important advantage: they are lighter and use less gasoline, which means that they are less expensive. The high heat resistance and low flammability of some nanocomposites also make them good choices to use as insulators and wire coverings. Another important property of nanocomposites is that they are less porous than regular plastics, making them ideal to use in the packaging of foods and drinks, vacuum packs, and to protect medical instruments, film, and other products from outside contamination. The properties of nanocomposites, such as increased durability, strength, or recyclability have been utilized in movies and TV (Caseri 2003).

Nanocomposites have become a central domain of scientific and technical activity to help humanity. The advances in nanostructured composites and their trends, developmental applications in various fields including in water treatment, green energy generation, anticorrosive, hard coatings, antiballistic, optoelectronic devices, solar cells, bio-sensors, and nanodevices, has been reported by several investigators tremendously. All these developments are possible due to the unique properties

Indira Gandhi National Tribal University, Department of Biotechnology, Lalpur, Amarkantak-484887, M.P., India.
Email: rpsnpl69@gmail.com

of nanostructures composites at nanoscale level which can change dramatically by a reduction in size, shape, and dimension and exhibit newer properties including reactivity, electrical conductivity, insulating behavior, elasticity, strength, and color. Nanocomposite materials are combinations of two or more components, typically consists of a few matrix containing one or more fillers made up of particles, sheets and fibers with dimensions less than 100 nm, having a higher surface to volume ratio. Silver/poly aniline (Ag-Pani) nanocomposites were prepared via *in situ* reduction of silver in aniline by photolysis at 265 nm for not only promising approach for electro catalytic hydrazine oxidation but utilization in other biosensing applications. The nanocomposites are utilized in new technology development like nanobiosensor based bioanalysis and biodetection applications and for business to all sectors of industry, with ecofriendly manner (Singh et al. 2011a, Singh 2016).

Nanocomposite is an interdisciplinary research and development area to help humanity by using a wide range of building blocks with nano dimensions to design and create new materials with flexibility. The nanocomposites are related to the theory, preparation, and characterization of cellulose and metal nanoparticles, polymer/clay, polymer/Carbon and polymer-graphene nanocomposites and several other exciting topics while it introduces the various applications of nanocomposites in water treatment, supercapacitors, green energy generation, anticorrosive and antistatic applications, hard coatings, antiballistic and electroconductive scaffolds. Multifunctional nanocomposite is an exciting and fast-evolving field that has better adsorption capacity, selectivity, and stability. Therefore, they have immense potential for heavy metal detection, pollutants, toxicants and adulterants from the contaminated water. This chapter presents a comprehensive discussion on applications of nanocomposites. It covers a vast background of the current literature based on applications of nanocomposites in various domains which is highly useful to researchers involved in this topic of an interdisciplinary approach and also throws light on the recent trends and advances in the field (Singh and Pandey 2011, Singh et al. 2011b, Singh and Choi 2010, Shukla et al. 2010).

The nanocomposites having additives as well as filler counterparts showed substantial property improvements such as mechanical properties, decreased permeability of gases, water and hydrocarbons, thermal stability, flame retardancy and reduced smoke emissions, chemical resistance, surface appearance, electrical conductivity and optical property (Gilmann 1999, Zhu et al. 2001, Lee et al. 1997c, Morgan et al. 2002, Bourbigot et al. 2002).

Applications of nanocomposites

Nanocomposite materials are used in general automotive and industrial applications including mirror housings on various vehicle types, door handles, engine covers and belt covers, used as impellers and blades for vacuum cleaners, power tool housings, covers for portable electronic equipment such as mobile phones, pagers, etc. Nanotechnology, with the help of nanocomposites, is revolutionizing the world of materials. It has very high impact in developing a new generation of composites with enhanced functionality and a wide range of applications. The data on processing, characterization and applications helps researchers in understanding and utilizing the special chemical and material principles underlying these cutting edge polymer nanocomposites. Although nanocomposites have many key applications in numerous industrial fields but a number of key technical and economic barriers exist to widespread commercialization. These include impact performance, the complex formulation relationships and routes to achieving and measuring nanofiller dispersion and exfoliation in the polymer matrix. Investment in state-of-the-art equipment and the enlargement of core research teams are some hurdles to bring out innovative technologies on nanocomposites. Future trends include the extension of this nanotechnology to additional types of polymer system, where the development of new compatibility strategies would be adopted. A range of nanostructural nanocomposites, i.e., functional nanomaterials are used in the construction of biosensors and these nanobiosensors are used in biological, chemical and environmental monitoring (Singh et al. 2014, Singh et al. 2012a, Singh et al. 2012b, Tiwari et al. 2012a, Yadav et al. 2012, Tiwari et al. 2012b, Singh et al. 2012c, Tiwari et al. 2012c, Singh et al. 2010).

The gas barrier property improvements shown by nanoclay as nanofiller in nanocomposites have been established. Such excellent barrier characteristics have resulted in food packaging applications for processed meats, cheese, confectionery, cereals, fruit juice, dairy products, beer and carbonated drinks bottles. They have ability to enhance the shelf life of many foods. Honeywell has a combined active/ passive oxygen barrier system in nature. Artificial polyamide-6 with nanoclay particles incorporation may show an oxygen scavenging species as an artificial combined active/passive system. Nanoclay polymer composites are currently using food packaging materials which do not require refrigeration and are capable of maintaining food freshness for long time. It has excellent gaseous barrier. It may also be possible to develop films for artificial intestines in future. A nanoclay filler material in nanocomposites based fuel tank for cars would reduce solvent leakage and also reduce the cost. The presence of nanofiller in nanocomposites may have significant effects on the transparency and haze characteristics when targeted to make films used in windows of car. Nanoclay may enhance transparency and reduce haze. Nano modified polymers are used as coating polymeric transparency materials which enhance toughness and hardness without obstructing light transmission (Garces et al. 2000, Caroline 2002).

In addition to these, they have an ability to resist high velocity impact and enhanced abrasion resistance. Nanoclay incorporation in polymeric nanocomposite can reduce the extent of water absorption in atmosphere. The ability of nanoclay incorporation to reduce the flammability of polymeric materials was established. Nanocomposites are producing batteries with greater power output and faster charging or discharging of power, for example, anodes for lithium ion batteries (Si/CNTs). Recently, nanocomposites have gained much interest due to not only the properties of their individual parents but also because their morphology and interfacial characteristics. Optimized fabrication process and controlled nanosized materials have enhanced thermal stability and mechanical properties such as adhesion resistance, flexural strength, toughness and hardness with nanodispersion. The nanodispersion property is helpful for producing materials with tailored physical and electronic applications from drug delivery to corrosion prevention to electronic/automotive parts to industrial equipment, etc. The nanocomposite material is fillers like particles, fibers, or fragments which is surrounded and bound together in the matrix with a wide range of materials, 3D metal matrix, 2D lamellar, 1D nanowires and 0D core shells. The physical, chemical and biological properties of nanomaterials differ from the properties of individual atoms and molecules or bulk matter. By creating nanoparticles, it is possible to control the fundamental properties of materials, such as their melting temperature, magnetic properties, charge capacity and even their color without changing the material's chemical compositions. Nanoparticles and nanolayers have very high surface to volume and aspect ratios and this makes them ideal for use in polymeric materials. Such structures combine the best properties of each component to possess enhanced mechanical and superconducting properties for advanced applications. The inorganic components can be three dimensional framework systems such as zeolites, two-dimensional layered materials such as clays, metal oxides, metal phosphates, chalcogenides, and even one-dimensional and zero-dimensional materials such as $(Mo^{+3}Se^{-3})_n$ chains and clusters. Thus, nanocomposites promise new applications in many fields such as battery cathodes, nanowires, sensors and other systems. Polymer nanocomposites are a polymeric material with a blend of nanoparticles. A few of the most commonly used nanoparticles include montmorillonite organoclays, carbon nanofibers, polyhedral oligomeric silsesquioxane, carbon nanotubes, nanosilica, nanoaluminum oxide and nanotitanium oxide. In addition, there are a few thermosets and thermoplastics which are used as matrices for making nanocomposites which include nylons, polyolefin, e.g., polypropylene, polystyrene, ethylenevinyl acetate copolymer, epoxy resins, polyurethanes, polyimides, and poly ethylene terephthalate (Camargo et al. 2009).

Nanocomposites are a high performance material with unique property and design chances. Being environmental friendly, applications of nanocomposites offer new technology and business opportunities for aerospace, automotive, electronics and biotechnology industries. Nanocomposites are composites with at least one of the component at dimensions of nanoscale range and are reported to be the materials of 21st century and offer new technology and business opportunities for all sectors of industry. Nanocomposite materials may be classified on the basis of their matrix materials. Few different types of nanocomposites are metal matrix nanocomposites (Fe-Cr/Al_2O_3, Ni/Al_2O_3, Co/Cr, Fe/MgO, Al/CNT, and Mg/CNT), ceramic matrix nanocomposites (Al_2O_3/SiO_2, SiO_2/Ni, Al_2O_3/TiO_2, and Al_2O_3/SiC) and polymer matrix nanocomposites (thermoplastic/thermoset polymer/layered silicates, polyester/TiO_2, Polymer/CNT, etc.). Nanocomposites systems have feature sizes for significant changes which may be expected in their properties such as catalytic property at < 5 nm, making hard magnetic materials soft at < 20 nm, producing referective index changes < 50 nm, producing supermagnetism and others electromagnetic phenomenon at < 100 nm, producing strengthening and toughening < 100 nm, and modifying hardness and plasticity < 100 nm (Kamigaito 1991).

Nanocomposites provide a lot of benefits. Tribological coating of tools for hard and dry cuts operation such as drilling, turning and milling are possible efficiently by nanocomposites emergence (Lim et al. 2002, Voevodin and Zabinski 2005a). Carbon Bucky fibers based polymer nanocomposites provide light weight bodies of cars. Aerospace, electronic and military domains metal and ceramic based nanocomposites are highly utilized in battery cathodes, microelectronics, sensors, catalysts, structural materials, electronic, optical, magnetic, and energy conversion devices, etc. Few metal nanocomposites are Fe/MgO for catalysts and magnetic devices, Ni/TiO_2 for photoelectrical applications, Al/SiC for aerospace, naval and automotive structures, Cu/Al_2O_3 for electronic packaging and Au/Ag for optical devices and light energy conversion (Sternitzk 1997, Choa et al. 2003). CNT based ceramic nanocomposites which are utilized in aerospace and sports goods, composite mirrors and automotive spares are also useful for flat panel displays, gas storage devices, toxic gas sensors, Li$^+$ batteries, and conducting paints, e.g., Al_2O_3/CNT composite. Also, these are highly useful in engineering and biomedical applications. However, these are industrially used in limit mainly because of challenges in processing and their cost (Andrews and Weisenberger 2004, Peigney et al. 2000, Alexandre and Dubois 2000). Polymer based nanocomposites having insulating, semiconducting or metallic nanoparticles, are highly used in electronic and food packaging industries, e.g., Nylon-6, polypropylene for packaging and injection molded articles (Nylon-6/surface-modified montmorillonite). They are also useful in tyres, fuel systems, seat textiles, mirror housings, door handles, engine covers, timing belt covers and pollution filters. Polymer/inorganic nanocomposites showed improved conductivity, permeability, and water management property to develop water nanofilter (Gangopadhyay and Amitabha 2000, Giannelis 1996, Fischer 2003, Pandey et al. 2005a).

The developments of ecofriendly nanocomposites for better packaging materials have been reported this can reduce the extent of solid waste to users. For example, the use of nanoclay particles into thermoplastic resins improves barrier properties and package survivability. CNT polymer composites are to be used for data storage media, photovoltaic cells and photo diodes, optical limiting devices, drums for printers, etc. (Dresselhaus et al. 2001, Choi and Awaji 2005, Ajayan et al. 2003). Nanocomposites speed healing process for broken bones, for example, CNTs/polymer nanocomposite. This nanocomposite conducts electricity and its property makes stress sensor, which is used in windmill blades to alarm the excessive damage and then shut down to save blade. Epoxy/CNTs composite makes a windmill blade which is a strong lightweight blade. Graphene/epoxy composites are stronger/stiffer because graphene bonds better to the polymers in the epoxy in comparison to CNT. This property allows the manufacturing of windmill blades or aircraft components. In the biomedical field, nanocomposite blend, with magnetic and fluorescent nanoparticles, makes tumor cells more visible during an MRI which helps the oncologist for surgery purposes (Breuer and Sunderraj 2004, Ke and Bai 2005, Presting and Konig 2003, Swearingen et al. 2003).

Nanocomposites for water treatment

Water is one of the universal solvent responsible for life on earth as two thirds of earth's surface is covered by water. 97 percent of the world's water is found in oceans. 2.5% of the world's water is non-saline fresh water. Although 75% of all the fresh water is bound up in glaciers and ice caps, remaining 25% fresh water is found in lakes, ponds, rivers and 24% is present as ground water. Water is the essential resource for living system, industry, agriculture and domestic use. An adequate supply of safe drinking water is a pre-requisite for a healthy life. Pollution occurs when a product added to natural environment adversely affects nature's ability to dispose it off. A pollutant adversely interferes with health, comfort, property or environment of the people. Most pollutants are introduced in the environment as sewage, agricultural waste, domestic waste, industrial waste, accidental discharge and as compounds used to protect plants and animals. As a result of the increasing demand for water and shortage of supply, it is necessary to increase the rate of water development in the world and to ensure that the water is used more efficiently by treatment and by management (Itodo and Itodo 2010, Duncan 2003).

DNA/PPy-PVS nanocomposite films have been developed for estimation of o-chlorophenal and 2-aminoanthracine which is present in water and waste water samples (Arora et al. 2006). PANi/ClO$_4$ films as nanocomposites have been developed (Singh et al. 2010a) and immobilize DNA as biosensing platform to detect sanguinarine, an alkaloid present in adulterated edible mustard oil. dsDNA/Cyt c/GSH-SAM/Au was developed as nanocomposite (Singh et al. 2011c) for the detection of various toxicants like 2-amino anthracine (2-AA), a poly aromatic compound as a pesticide and potent carcinogen, 3-bromobenzanthrone (3-BBA), a dye intermediate which causes hepatic injury, Bisphenol A (BPhA) which is an intermediate, and xenoestrogen, a well known pollutant present in water.

Nanoclay is considered to be potential nanofiller for the manufacturing of natural fiber nanocomposites. The hydrophilic nature of natural fibers affects negatively its adhesion to hydrophobic polymer matrix. Propionic anhydride (PA) treated jute were used for the manufacturing of jute/polyethylene/nanoclay nanocomposites (Wypych and Satyanarayana 2004). Ag/Fe$_3$O$_4$ nanocomposites were synthesized by a facile and cost-effective method using starch, which is not only a biocompatible capping agent but also acts as a reducing agent for the reduction of silver ions in an alkaline medium. It has superparamagnetic property with high-antibacterial activity against *Escherichia coli* (Seyedeh and Seyed 2016). Fe$_2$O$_3$/CuFe$_2$O$_4$/chitosan nanocomposites have been synthesized via a new sol-gel auto-combustion route utilizing onion as a green reductant for the first time. In this nanocomposite, chitosan was used to functionalize and modify the nanostructures and also to improve surface properties (Fatemeh et al. 2016). TiO$_2$ nanocomposites doped with Al$_2$O$_3$, Bi$_2$O$_3$, CuO and ZrO$_2$ which were synthesized by sonochemical method and showed photocatalytic activity to minimize activity of organic pollutants like methyl orange, methylene blue and rhodamine B dyes and antibacterial activity of these nanocomposites against gram negative bacteria (*Escherichia coli*) had been proved (Magesan et al. 2016).

The wide application of reduced graphene oxide (rGO) in wastewater treatment was reported but suffered from high cost and reusage. To address these issues, a series of zinc ferrite-reduced graphene oxide (ZF-rGO) nanocomposites with enhanced magnetic properties were successfully prepared using graphene oxide (GO), Fe^{3+}, Zn^{2+} and ethylene glycol (EG) via a one-pot solvothermal method and the adsorption property was evaluated by the adsorption amount of methylene blue (MB) solution on the samples. Thus, the as-prepared nanocomposite material can be a potential excellent absorbent for removing dye pollutants (Peng et al. 2016b). The silica-containing nanocomposites with a controlled carbon phase based on high quality reduced graphene oxide (rGO) were prepared via a cheap and facile sol-gel method, followed by pyrolysis in inert gas atmosphere. The resulting nanocomposite materials have an intimate bonding between rGO of low defect density and a partially crystalline silica matrix (Cornelia et al. 2016). The tethering graphene nanoplatelets (GNPs) with Fe$_3$O$_4$ nanoparticles to enable their alignment in an epoxy using a weak magnetic field have been established. The GNPs are first stabilized in water using polyvinylpyrrolidone (PVP) and Fe$_3$O$_4$ nanoparticles and then co-precipitation. The Fe$_3$O$_4$/PVP-GNPs nanohybrids are superparamagnetic (Wu et al. 2016c). The nanosized titanium dioxide (TiO$_2$) has been extensively studied due to its unique properties and broad applicability, being very promising for application in photocatalysis and photoprotection (Alice et al. 2016). Chitosan-based magnetite nanocomposites were synthesized using a versatile ultrasound assisted *in situ* method involving one quick step. The procedure will permit a further diversity of applications into nanocomposite materials engineering (Freire et al. 2016). Alumina-coated silica nanoparticles (NPs) grafted with phosphonic acids of different hydrophobicity were used as filler in poly(ethylacrylate) nanocomposites. Preliminary evidence for an improved compatibility of grafted with respect to bare NPs is found, as opposed to their aqueous precursor suspensions where some pre-aggregation is induced by grafting (Pauly et al. 2016). The photocatalytic activities of mesoporous TiO$_2$ modified by the addition of carbon nanotubes (CNTs) and Cu have been reported. Nanocomposites of carbon nanotubes (CNTs) containing varying amounts of Cu were formed by treatment with Cu^{2+}, and were then reduced to Cu^0 using NaBH$_4$ as the reducing agent. The degradation efficiency for MO was a synergistic effect of photodegradation of TiO$_2$ and may be due to improvement of the electrical conductivity of the system by the presence of the CNT/Cu networks, since the photodegradation of MO and the photocatalytic activity of the photoactive systems increased with increasing copper content (Amirabbas et al. 2016). A new method was developed for the preparation of polyethylene magnetic nanocomposites by *in situ* polymerization. Carbon nanotubes (CNTs), synthesized by the chemical vapor deposition method using ferrocene as the precursor and catalyst and silica (SiO_2) as the support, were used as fillers and were compared with a commercial CNT. The synthesized nanofillers had iron magnetic particles encapsulated in CNTs. The thermal properties showed that the polymeric matrix did not change their properties significantly (Muhammad et al. 2016). Ibuprofen (IBU) estimation is a persistent organic pollutant (POPs) which can cause severe adverse effects in humans and wildlife. If removing IBU from water which is a worldwide necessity, then superparamagnetic Bi_2O_4/Fe_3O_4 nanocomposites have been prepared by an *in situ* growth method and utilized for photocatalytic removal of IBU. Bi_2O_4/Fe_3O_4 (1:2.5) was demonstrated by the efficient photocatalytic degradation of IBU in actual drinking water (Xia and Lo 2016). The PFNCs synthesis, characterization and performance in adsorption processes as well as the potential environmental risks and perspectives have been reported. Pollution by metal and metalloid ions is one of the most widespread environmental concerns. They are non-biodegradable and generally present high water solubility, facilitating their environmental mobilisation, interacting with abiotic and biotic components such as adsorption onto natural colloids or even accumulation by living organisms, thus, threatening human health and ecosystems. There is a high demand for effective removal treatments of heavy metals, making the application of adsorption materials such as polymer-functionalized nanocomposites (PFNCs), increasingly attractive. PFNCs retain the inherent remarkable surface properties of nanoparticles, while the polymeric support materials provide high stability and processability. These nanoparticle-matrix materials are of great interest for metals and metalloids removal, thanks to the functional groups of the polymeric matrixes that provide specific bindings to target pollutants (Giusy et al. 2016). The photocatalytic paper sheets which were prepared by addition of different ratios of TiO_2/Sodium alginate (TSA) nanocomposite. The high photocatalytic activity and anti-bacterial effect against *Salmonella typhimurium* was higher than standard antibiotic, beside other microorganisms such as *Candida albicans*. The results obtained confirm the possible utilization of the modified paper in both hygienic and food packaging applications (Rehim et al. 2016). The delivery and therapeutic actions of galantamine drug (GAL) against Alzheimer's disease (AD) in rat brain through attaching GAL to ceria-containing hydroxyapatite (GAL/Ce/HAp) as well ceria-containing carboxymethyl chitosan-coated hydroxyapatite (GAL/Ce/HAp/CMC) nanocomposites have been established (Sanaa et al. 2016). Intercalated polyaniline (PANI) into Montmorillonite (MMT) clay layers have been reported and showed that the clay acts as a thermal barrier and precludes the thermal decomposition of PANI (Nascimento and Pradie 2016). A facile and controllable method for *in situ* synthesis of the magnetic carbon nanotubes nanocomposites (Fe/CNTs) in water-ethylene glycol (EG) mixed solvents is reported by the deposition-precipitation method following annealing. The as-prepared Fe/CNTs nanocomposites display superparamagnetic property at room temperature and the water/EG ratio determines the magnetization of the sample (Liu et al. 2016c). The dual-responsive shape-memory polymer/clay nanocomposite is successfully prepared using *in situ* polymerization approach with the presence of exfoliated clay platelets. Incorporation of clays into polymer matrix effectively improved the hydrophilic performance of materials, thus the water-induced shape recovery process of the nanocomposites can be finished within 6 s with excellent recovery ratio. Therefore, the preparation of dual responsive shape memory polymer/clay nanocomposites pave a new way to design and fabricate high mechanical shape memory materials with dual triggering mechanism (Xianqi et al. 2016). The magnetically separable $rGO/MgFe_2O_4$ nanocomposites have developed and showed photocatalytic activity, which were prepared by one step hydrothermal method. The photocatalytic studies were carried out by taking methylene blue as model organic compound under visible light irradiation. The photo-electrochemical (PEC) water splitting properties of the $rGO/MgFe_2O_4$ nanocomposite exhibited higher PEC performance compared to $MgFe_2O_4$ nanoparticles owing to the suppression of charge recombination. The results showed that nanocomposite can be used as photocatalysts for degradation of organic compounds in the polluted water (Ain et al. 2016). The elastomeric nanocomposites (NCs) based flame retardant (FR) materials have been established and are being explored to give new direction for the development of the FR materials, which would be more accessible to the emerging field of materials science (Khobragade et al. 2016). The dispersions of silver nanoparticles (Ag NPs) in a tetrafunctional thiol were prepared by *in situ* reduction of silver nitrate with 2, 6-di-tert-butyl-para-cresol. The Ag NPs were maintained in a stable colloidal state for more than eleven months at ambient temperature. The thiol monomer acts as both stabilizing agent and reactive solvent. The thiol-methacrylate matrix of the nanocomposites was characterized by measuring the glass transition temperature, the flexural modulus and the compressive strength. The low toxicity of the reactants used in the synthesis in combination with antimicrobial activity of Ag NPs makes the nanocomposite materials synthesized in this study very attractive for the preparation of biomaterials and coating with improved biocompatibility (Silvana and Vallo 2016). An antibacterial nanocomposites (NC-1/NC-4) based on cellulose acetate have been developed, which were prepared by dispersing ZnO nanofillers in the cellulose acetate matrix. Further, the selectivity of anti-bacterial nanocomposites was investigated toward different metal ions, including Zn^{2+}, Cd^{2+}, Pb^{2+}, Mn^{2+}, Ni^{2+}, Fe^{2+}, Al^{3+}, Sb^{3+}, and Sr^{3+}. NC1 was subjected to water permeability to explore the role of anti-bacterial nanocomposite as membrane for water purification. The results suggest that these materials are possibly appropriate for water treatments (Khan et al. 2015). A new quaternary magnetic Fe_3O_4/ZnO/Ag_3VO_4/AgI nanocomposite have been developed and showed an excellent visible-light-driven photocatalysts, fabricated via preparation of Fe_3O_4/ZnO/Ag_3VO_4 nanocomposite followed by coupling it with silver iodide through ultrasonic irradiation. Magnetic properties of the nanocomposite provide a convenient route for separation of the photocatalyst from the reaction mixture using an external magnet (Aziz et al. 2016). The silver nanowire (AgNW)-carbon fiber cloth (CC) nanocomposites have been developed which were synthesized by a rapid and facile method. Acting as filter in an electrical gravity filtration device, the AgNW/CC nanocomposites were applied to electrochemical point-of-use water disinfection. Their disinfection performance toward *Escherichia coli* and bacteriophage MS-2 was evaluated by inhibition zone tests, optical density growth curve tests, and flow tests. They exhibited excellent intrinsic antibacterial activities against *E. coli*. The concentration of AgNWs and UV adhesive controlled the released silver and hence, led to the change in antibacterial activity. The convenient synthesis and outstanding disinfection performance offer AgNW/CC nanocomposites opportunities in the application of electrochemical point-of-use drinking water disinfection (Hong et al. 2016). For the first time, the silver (Ag)-doped $MWCNT/TiO_2$ (multi-walled carbon nanotubes@titanium dioxide) core-shell nanocomposites with enhanced visible-light-responsive properties have been established for the application in carbon dioxide (CO_2) photoreduction (Gui et al. 2015). The mechanisms of sulfamethazine (SMT) sorption on a novel carbonaceous nanocomposite have established the effects of harsh aging on SMT sorption in the presence and absence of soil and before as well as after aging. The carbonaceous nanocomposites were synthesized by dip-coating straw biomass in carboxyl functionalized multi-walled carbon nanotubes solution and then pyrolyzed at 300°C and 600°C in the absence of air. The carbonaceous nanocomposites have potential in removal of SMT and possibly other persistent organic pollutants from wastewater (Zhang et al. 2016c). The chitosan nanocomposites have been developed to optimize the minimum amount and contact time required to achieving the complete inactivation of bacteria in water. Gram-negative *Escherichia coli* and Gram-positive *Enterococcus faecalis* bacteria were used to test the antibacterial activity of chitosan cross-linked with glutaraldehyde and chitosan nanocomposites in water. The silver and zinc oxide nanoparticles supported on bentonite were synthesized using microwave-assisted synthesis method. The resulting bentonite-supported silver and zinc oxide nanoparticles were dispersed in a chitosan biopolymer to prepare bentonite chitosan nanocomposites. Finally, leaching tests demonstrated that bentonite chitosan nanocomposites were stable and, consequently, could be effectively used as antibacterial materials for water disinfection (Sarah et al. 2015).

The nanophoto active cellulosic fabric has been developed which was prepared through *in situ* phytosynthesis of star-like Ag/ZnO nanocomposites, using the ashes of Seidlitzia rosmarinus plants so-called Keliab in alkali media as a vital condition for synthesis of nanocomposites, further increasing the reduceability of cellulosic chains by activation of hydroxyl groups. The intermolecular dehydrolysis of intermediate ions under thermal and alkaline conditions leads to formation of Ag/ZnO heterostructure. The treated cotton samples exhibited self-cleaning activities through methylene blue degradation under day-light exposure along with improved wettability and whiteness. The prepared sample in optimized conditions showed good antibacterial activities against *Staphylococcus aureus* and *Escherichia coli* with enhanced fabric tensile strength (Aladpoosh and Montazer 2016). Polypyrrole wrapped oxidized multiwalled carbon nanotubes nanocomposites (PPy/OMWCNTs/NCs) have been developed which were prepared via *in situ* chemical polymerization of pyrrole (Py) monomer in the presence of OMWCNTs using $FeCl_3$ as oxidant for the effective removal of hexavalent chromium [Cr (VI)]. Characterization results suggested that PPy was uniformly covered on the OMWCNTs surface and resulted in enhanced specific surface area. Adsorption experiments were carried out in batch sorption mode to investigate the effect of pH, dose of adsorbent, contact time, concentration of Cr (VI) and temperature (Bhaumik et al. 2016). The various montmorillonite (Mt) nanocomposite adsorbents have been developed which were prepared with Al^{+3} cations, dodecyl trimethyl ammonium chloride (DTAC) or dodecyl amine (DA). The adsorption of hexavalent chromium, Cr (VI) onto various Mt nanocomposites as a function of adsorbent dosage, initial Cr (VI) concentration, and contact time and solution pH was investigated. The removal efficiency of Cr (VI) ions increased with increasing the adsorbent dosage and contact time, but decreased with increasing initial Cr (VI) concentration, as expected. The adsorption results indicated that among all the adsorbents used in this experiment, the dodecyl amine and Al^{+3} cations composited with Mt (DA-Al-Mt) was the most effective for removing Cr (VI) from wastewater (Wang et al. 2016b). The metal/ZnO nanoparticles have been developed which were synthesized by a solvothermal method. The influence of reaction time, noble metal presence or kind of noble metal (Ag or Pt) was evaluated. The photocatalytic behavior of the synthesized systems was studied as well by the removing reaction of methylene blue (MB) in water solution. It was verified that metal/ZnO nanocomposites synthesized by solvothermal method for usage in environmental applications (Fernandez et al. 2016). SnO_2 quantum dots (QDs) decorated reduced graphene oxide (RGO) nanocomposite type adsorbents have been reported which is used for the removal of different organic dyes (methylene blue (MB), methyl orange (MO) and rhodamine B (RhB)), toxic metal ions (Co^{2+}, Ni^{2+}, Cu^{2+}, Cd^{2+}, Cr^{3+}, Pb^{2+}, Hg^{2+} and As^{3+}), and pathogenic bacteria (*Escherichia coli*) from wastewater. It exhibits good removal capacity and fast adsorption rate for cationic dye MB. QDs-RGO showed higher antibacterial activity towards *E. coli* than the individually constituted materials. QDs-RGO is an ideal aspirant for the removal of positively charged organic dyes, toxic heavy metal ions and bacteria pathogen from wastewater (Dutta et al. 2016). The castor oil-based waterborne polyurethane/silver-halloysite (WPU/Ag/HNT) antibacterial nanocomposites have been developed which were prepared by incorporating silver-halloysite composite (Ag-HNT) into WPU via *in situ* polymerization. Ag/HNT was synthesized by modification of halloysite nanotubes (HNTs) with 3-(2-Aminoethylamino) propyldimethoxymethylsilane (AEAPTMS) and chitosan, and then mixed with silver nitrate for combining silver ions. Finally, the silver nanoparticles were loaded on the surface of HNTs by reducing silver ions with $NaBH_4$. The antibacterial test indicated that the nanocomposites exhibited excellent antibacterial activity against *Staphylococcus aureus* and *Escherichia coli*. The obtained nanocomposites will have promising applications in high performance antibacterial coatings (Fu et al. 2016). Ni-P-TiN nanocomposites have been developed which were deposited onto 45 steel sheets. The microstructure, microhardness and corrosion behavior of the nanocomposites. Electrochemical test results illustrate that the Ni/P/TiN nanocomposites heat-treated at 500°C for 10 min has the most optimum corrosion resistance among all as-plated and heat-treated specimens (Ma et al. 2016b). Adsorption becomes widely used for the removal of inorganic and organic micropollutants from aqueous solution which require numerous adsorbents for wastewater treatment. Clay-polymer nanocomposite (CPN) adsorbents treat water by adsorption and flocculation of both inorganic and organic micropollutants from aqueous solutions. Some of these CPNs, when modified with biocides, also have the ability to efficiently remove microorganisms such as *Escherichia coli*, *Pseudomonas aeruginosa*, *Staphylococcus aureus* and *Candida albicans* from water. CPNs therefore show an excellent potential as highly efficient water and waste treatment agents. Recently, CPNs have been prepared and used as an adsorbent in the removal of micropollutants (inorganic, organic and biological) from aqueous solutions. CPN is used as water treatment agent that has not yet realized their full potential (Emmanuel and Taubert 2016).

The rod-like chitin whiskers (CHWs) have been developed by acid hydrolysis and then surface-modified with l-lactide to obtain grafted CHWs (g-CHWs) via solution casting. The resulting nanocomposites were fully characterized in terms of crystallization behavior, thermal transitions, morphology, mechanical properties and cytocompatibility and suitable for cell adhesion and differentiation than those of the neat PLLA and CHW/PLLA nanocomposites (Li et al. 2016a). The manganese oxide incorporated ferric oxide nanocomposite (MIFN) nanocomposite have been developed as a novel adsorbent and synthesized, characterized and explored for the removal of Cr (VI) from contaminated water. Adsorption of Cr (VI) was strongly inhibited by phosphate and sulphate, whereas fluoride, carbonate, bicarbonate and silicate have no significant interference. Adsorption efficiency of the spent adsorbent (MIFN) could be rejuvenated around 78–80% by 1.0 M NaOH and subsequently be reused. MIFN could be an efficient adsorbent for scavenging Cr (VI) from contaminated water because of its high adsorption capacity and reusability (Gosh et al. 2015). The titanium dioxide (TiO_2) nanoparticles/ halloysite nanotube (HNT) nanocomposites have been developed via one-step solvothermal method which is used as nanofillers in the preparation of thin film nanocomposite (TFN) membranes for forward osmosis (FO) application. The unique tubular structure of HNTs coupled with the excellent anti-fouling features of TiO_2 have made TiO_2/HNTs a reliable material with a bright perspective in improving the antifouling affinity of conventional thin film composite membranes for FO applications (Ghanbari et al. 2015). The nanocrystalline TiO_2 and MWCNTs are important functional materials; the composites of TiO_2 nanoparticles with varying amount of functionalized MWCNTs were prepared by using a solution-based method. The physico-chemical properties of these composites were studied in connection with their antibacterial activity. TiO_2 particles have non-spherical shape with size in the range of 8–15 nm. The experimental findings suggested that TiO_2-MWCNTs nanocomposites have been developed as efficient antibacterial agents against a wide range of microorganisms to prevent and control the persistence and spreading of bacterial infections (Koli et al. 2016). Ni/Fe-Fe_3O_4 nanocomposites have been synthesized for dechlorination of 2, 4-dichlorophenol (2,4-DCP). The nanocomposites were easily separated from the solution by an applied magnetic field. Ni/Fe/Fe_3O_4 nanocomposites can be considered as a potentially effective tool for remediation of pollution by 2, 4-DCP (Ray and Sunanda 2016, Xu et al. 2016a). The synthesis of graphene oxide (GO) based nanocomposite materials have been developed as an environment-friendly nanostructure suitable for novel applications. A series of GO have been fabricated utilizing the modified Hummers' method. Graphene oxide was not only being used as an effective solid stabilizer for Pickering emulsion, but also dispersed homogeneously into the production of polystyrene as highly effective nanofiller (Liu et al. 2016d). A multifunctional Ag-TiO_2/reduced graphene oxide (rGO) nanocomposite was prepared by a simple one-step hydrothermal reaction using TiO_2 NP (Degussa P25), $AgNO_3$, and graphene oxide (GO) without reducing agents. Growth of the Ag/TiO_2 NPs over GO and reduction of GO were simultaneously carried out under hydrothermal condition. This method provides the notable advantages of a single-step reaction without employing toxic solvents or reducing agents, thereby providing a novel green synthetic route to produce the nanocomposite of Ag/TiO_2/rGO and enhances the optical response of the nanocomposites and contributed for the absorption of light in the visible spectrum through localized surface plasmon resonance effects. The superior photocatalytic activities, antibacterial properties, and the easy recovery make the Ag/TiO_2/rGO nanocomposite a promising candidate for wastewater treatment (Pant et al. 2016). A novel organic-inorganic heterostructured photocatalyst, porous graphitic carbon nitride (g-C_3N_4) hybrid with copper sulfide (CuS) has been synthesized via a precipitation-deposition method at low temperature for the first time. Due to the high stability, the porous g-C_3N_4/CuS could be applied in the field of environmental remediation and also provides a facile way to improve the photocatalytic activity (Chen et al. 2016a). Au-Fe_3O_4 nanocomposites were synthesized and modified with 1-(2-mercaptoethyl)-1,2,3,4,5,6-hexanhydro-s-triazine-2,4,6-trione (MTT) for the bimodal detection of melamine. The bimodal detection is based on the aggregation of Au-Fe_3O_4/MTT nanoparticles (NPs) from the dispersed state upon the formation of special triple hydrogen bonds between melamine and MTT. This phenomenon causes the red shift of the absorption peak and the increase of the spin-spin relaxation time (T2) of the water protons upon addition of melamine at different concentrations to an aqueous solution of Au-Fe_3O_4/MTT NPs. Furthermore, the dual-modal method exhibits a clear selectivity toward melamine (Shen et al. 2016).

Novel CdS/$BiVO_4$ nanocomposites were synthesized by simple solvothermal method. In the nanocomposites, CdS particles were deposited on the surface of the $BiVO_4$. The photocatalytic tests showed that the CdS/$BiVO_4$ nanocomposites possessed a higher rate for degradation of malachite green (MG) than the pure $BiVO_4$ under visible light irradiation. The results showed that the nanocomposite construction between CdS and $BiVO_4$ played a very important role in their photocatalytic properties, which has a potential application in solving environmental pollution issues utilizing solar energy effectively (Fang et al. 2016). The nanocomposite was produced via phenolic resin infiltrating into a carbon nanotube (CNT) buckypaper preform containing B_4C fillers and amorphous Si particles followed by an *in situ* reaction between resin-derived carbon and Si to form SiC matrix. The buckypaper preform combined with the *in situ* reaction avoided the phase segregation and significantly increased the volume fraction of CNTs. These nanocomposites became much more electrically conductive with high loading level of CNTs (Cai et al. 2016).

Magnetic graphene oxide based composites of the nano-particle size of < 10 nm were developed, synthesized, and characterized. The adsorption of Cs (I), Co (II), Ni (II), Cu (II) and Pb (II) to nano-composites was studied in a wide range of initial concentrations contaminated solution (Lujaniene et al. 2016). SnS_2/TiO_2 nanocomposites which have reported were synthesized via microwave assisted hydrothermal treatment of tetrabutyl titanate in the presence of SnS_2 nanoplates in the solvent of ethanol at 160°C for 1 h. The experimental results showed that the SnS_2/TiO_2 nanocomposites exhibited excellent reduction efficiency of Cr (VI) (~ 87%) than that of pure TiO_2 and SnS_2. The SnS_2/TiO_2 nanocomposites were expected to be a promising candidate as effective photocatalysts in the treatment of Cr (VI) wastewater (Deng et al. 2016). A facile two-step strategy to fabricate magnetically separable 3-dimensional (3D) hierarchical carbon-coated nickel (Ni/C) nanocomposites has been developed by calcinating nickel based metal organic framework $(Ni_3(OH)_2(C_8H_4O_4)_2(H_2O)_4)$. The specific surface area of the nanocomposites at room temperature magnetic measurement indicates the nanocomposites show soft magnetism property, which endows the nanocomposites with an ideal fast magnetic separable property for rhodamine B. Nanocomposites also exhibit a high adsorption capacity for heavy metal ions. The adsorbent can be very easily separated from the solution by using a common magnet without exterior energy. The Ni/C nanocomposites can be applied in waste water treatment on a large-scale as a new adsorbent with high efficiency and excellent recyclability (Song et al. 2015a). Bicomponent ZnO/$ZnCo_2O_4$ nanocomposites have been developed via a facile and scalable synthesis method by controlling the ratio of Zn to Co in the synthesis stage. The ZnO/$ZnCo_2O_4$ nanocomposites showed interconnecting porous nanosheets possessing loose porous nanostructures with abundant open space and electroactive surface sites used as an anode material for lithium ion batteries, its electrode exhibits high capacity, good cycling stability and excellent rate capability (Xu et al. 2016b). Active visible light photocatalysts $BiOBr/TiO_2$ nanocomposites have been developed by a facile one-pot solvothermal approach. The material shows excellent photocatalytic performance towards photodegradation of Rhodamine B under visible light irradiation. The high photocatalytic activity of $BiOBr/TiO_2$ nanocomposites under visible light showed the large surface area, mesoporous structure, appropriate band-gap, as well as synergistic effect between TiO_2 and BiOBr. The $BiOBr/TiO_2$ photocatalyst is very promising for water purification as well as other environmental applications (Wei et al. 2013).

Carbon nanotubes (CNTs), a carbonaceous material, which depend on morphology, size and diameter and have highly unique properties in mechanical strength, thermal stability electrical conductivity, catalytic and adsorption, have been discussed over the past decades; combination of carbon nanotube with metal oxide is an effective way to build hybrid carbon architectures with fascinating new properties. Combination of CNT with metal oxides such as aluminum dioxide, titanium dioxide, zinc oxide, iron oxide, etc. have been reported and applied toward novel device fabrication with remarkable properties for a wide range of applications such as sensor, supercapacitors, absorbent, photocatalytic, photovoltaic, etc. (Mallakpour and Khadem 2016). Magnetic carbon nanotube adsorbents, CNTs/Fe_3O_4 nanocomposites have been developed in which carbon nanotubes (CNTs) were grafted to magnetic Fe_3O_4 particles by a facile hydrothermal method. The adsorption behaviors of the CNTs/Fe_3O_4 nanocomposites were evaluated for the removal of bisphenol A (BPA) in aqueous solution. The results demonstrated a recyclable adsorbent (Li et al. 2015b). Green methods for synthesis of hierarchical ZnO-reduced graphene oxide (ZnO-RGO) nanocomposites have been developed via a hydrothermal method as an efficient photocatalyst for the photodegradation of azure B dye. The synergistic effect between surface adsorption characteristic and photocatalytic potential of the nanocomposites on the photodegradation of azure B was studied (Rabieh et al. 2016). Cu_2O/Carbon Nanotubes (CNTs) hierarchical chrysanthemum-like nanocomposite have been developed by a facile wet chemical method for the first time. Chrysanthemum-like structure has the capability to enlarge the range of absorbed light and enhance the absorption intensity of light. Due to this ability, degradation of phenol under visible light irradiation is possible. Photocatalytic decomposition of phenol is possible due to their unique 3D ordered nanostructures, which is also beneficial to the reagent diffusion and mass transportation (Yang et al. 2016a). Graphitic-phase C_3N_4 (g-C_3N_4) loaded with metal Ag nanoparticles (Ag NPs) using grape seed extract as reducing agent and stabilizing agent has been developed for a new type of antimicrobials without drug-resistence which showed better antibacterial activity than the green synthesized bare Ag NPs (Liu et al. 2016b). ZnO/Ag_2O nanocomposites have been developed by combination of thermal decomposition and precipitation technique. The photocatalytic performance of the photocatalysts was evaluated towards the degradation of a methyl orange (MO) under UV and visible light. ZnO/Ag_2O (4:2) nanocomposite showed excellent stability towards the photodegradation of MO under visible light (Kadam et al. 2016). $CoFe_2O_4$/$BiVO_4$ nanocomposites have been reported by a facile hydrothermal method. The photocatalytic activities of nanocomposites were established to degrade methylene blue (MB) under visible light irradiation (Duangjam et al. 2016). Nano silicon carbide (SiC) designed chitosan nanocomposites have been developed by solution technique. The oxygen permeability of chitosan/SiC nanocomposites was reduced when compared to the virgin chitosan. The chemical resistance properties of chitosan were enhanced due to the incorporation of nano SiC. The tensile strength of chitosan/SiC nanocomposites was increased with increasing percentage of SiC. With a substantial reduction in oxygen barrier properties in combination with increased thermal stability, tensile strength and chemical resistance properties, the synthesized nanocomposite may be suitable for packaging applications (Pradhan et al. 2015). Magnetic α-Fe_2O_3/TiO_2 nanocomposites have been developed and showed photocatalytic activity and degraded methyl orange. It can be transferred easily by a magnet and shows potential for application for wastewater treatment (Liu et al. 2016c). Green synthesis of Cu/eggshell, Fe_3O_4/eggshell and Cu/Fe_3O_4/eggshell nanocomposites via an environmental and economical method using aqueous extract of the leaves of *Orchis mascula* without any stabilizer or surfactant have been developed. The nanocomposites showed high catalytic activity in the reduction of a variety of dyes like 4-nitrophenol (4-NP), Methyl orange (MO), Congo red (CR), Methylene blue (MB) and Rhodamine B (RhB) in water at room temperature. The catalysts could be easily removed from the reaction media by mild centrifugation or an external magnet and reused several times without any negative effect on the initial results (Nasrollahzadeh et al. 2016).

Copper (II)-halloysite nanotube (Cu^{2+}-Hal) nanocomposites were used as adsorbents for the removal of 2,4,6-trichlorophenol (2,4,6-TCP) in water at ambient temperature (30°C). The nanocomposites show higher adsorption capacity when compared to the pristine Hal and was attributed to the enhancement in specific surface area and pore volume in the Cu^{2+}-Hal nanocomposites which provide more adsorption sites as compared to the pristine Hal (Zango et al. 2016). Hybrid montmorillonite/cellulose nanowhiskers (MMT/CNW) reinforced polylactic acid (PLA) nanocomposites have been developed which were produced through solution casting. The tensile, thermal, morphological and biodegradability properties of PLA hybrid nanocomposites were achieved (Arjmandi et al. 2015). Ag_2S/TiO_2 nanocomposites have been developed by a simple thermal decomposition approach. The Ag_2S-TiO_2 nanocomposites act as good catalyst for the photodegradation of Rhodamine B in aqueous solutions in the presence of sunlight (Yadav and Jeevananadam 2015). Palygorskite (Pal), which is a naturally available one-dimensional (1D) nanomaterial, rod crystals exhibit remarkable improvement of the colloidal, adsorptive, mechanical, thermal, and surface properties which reflect its key role to develop various functional nanocomposites (Wang and Wang 2016). Multiwall-carbon nanotubes and TiO_2/SiO_2 nanocomposites (MWCNT-TiO_2-SiO_2) for the removal of both pollutants, i.e., bisphenol A (BPA) and carbamazepine (CBZ) from water solution have been reported. Toxicity to *Vibrio fischeri* and *Daphnia magna* was reduced, indicating formation of non-toxic products of photooxidation of tested contaminants (Czech and Waldemar 2015).

Ceramic-polymer nanocomposites have been used in biomedical field as synthetic scaffolds to overcome the limitations of autografts and allografts. Polymers give flexibility and resorbability whereas bioactive ceramic particles provide strength and osteoconductivity. Ceramic-polymer nanocomposites are useful in the bone tissue regeneration field and also provide suggestions for future research and development.

Imogolite (naturally occurring aluminosilicate nanotube), consisting of a single-walled with a composition of $(OH)_3Al_2O_3SiOH$, with Al-OH and Si-OH groups distributed on the external and internal surfaces of the tube wall, have been developed. The surface modification of imogolite utilizing specific interaction between Al-OH and phosphonic acid is reported (Ma et al. 2016a). A simple *in situ* method to synthesize polypyrrole (PPy)/TiO_2 nanocomposites has been developed. It tested photocatalytic activity under simulated solar light and tested Rhodamine B degradation and production of methanol from CO_2. Widespread concerns about the impacts of exposure to chemical substances with textile dyeing are well known (Gao et al. 2016). Graphene/C_3N_4/semiconductor (CNS) nanocomposites have been developed as a catalyst in the photocatalytic destruction of industrial effluents, i.e., dyestuff effluents. The efficacy of these nanocomposites in removing of selected dye as pollutants is established. The g-C_3N_4-based nanocomposites would be the possible enhancement of environmental conservation (Lam et al. 2016). A facile one-step approach for the preparation of carbon nanotubes-silver (including single-walled carbon nanotubes-silver (SWCNTs-Ag), multi-walled carbon nanotubes-silver (MWCNTs-Ag)) and graphene oxide-silver (GO/Ag) nanoparticles have been reported. Out of which GO/AgNPs with excellent disinfection efficiency against *E. coli* and *S. aureus* was established (Chang et al. 2016). Tribological performances of developed Si_3N_4-based nanocomposites have been reported which were enhanced by either Si_3N_4 nanowhiskers or TiN nanoparticles. A chemical resistance of the Si_3N_4-TiN nanocomposite acids and bases at elevated temperatures were established (Lozynskyy et al. 2015). Magnetic iron oxide/carbon nanocomposites were synthesized (Istratie et al. 2016) as adsorbents for the removal of methyl orange (MO) and phenol (Ph) from aqueous solutions. A significant increase in the removal efficiency, both for MO and phenol, with the increase in the carbon content of the magnetic nanopowder has been reported. The ternary AgI/FG/TiO_2 nanocomposites have been developed via fast microwave technique. The functionalize graphene (FG) sheets were used for the attachment of AgI and TiO_2 nanoparticles to form a ternary composites. The photocatalytic activity for hydrogen evolution was affected by FG and ternary component in the system (Ullah et al. 2015).

Poly(vinyl alcohol) (PVA) nanocomposite films have been prepared using different concentrations of functionalized graphene to study the permeability of oxygen gas in the PVA nanocomposites. Graphene layers were covalently functionalized with tris (hydroxymethyl) aminomethane (TRIS) and used as nanofiller in the PVA matrix. The PVA/TRIS-functionalized graphene nanosheets (TRIS-GO) films have water absorbent ability. The oxygen permeability of the nanocomposites was increased by the presence of TRIS-GO due to an increase in hydrophilic volume (Mallakpour et al. 2016). Bentonite/nitrile butadiene rubber nanocomposites have been developed and their effects of different silane coupling agents on the mechanical strength, tribological properties, thermal behaviors, swelling properties and rheological properties were explored (Ge et al. 2015). Magnetic hollow Fe_3O_4 nanoparticles (MNPs) have been developed by two-step process, which can be fast and efficiently separate oils from water surface under a magnetic field. The magnetic Fe_3O_4 nanoparticles (MNPs) coated with a polystyrene layer acts as water-repellent and have oil-absorbing surfaces, which absorb lubricating oil. The oils could be readily removed from the surfaces of nanocomposites. Due to highly hydrophobic and superoleophilic characteristics, nanocomposites have an excellent recyclability in the oil absorbent capacity. The removal of such organic contaminants (oil spills from water surface) is of great technological importance for environmental protection. In addition, it will open broad application in wastewater treatment (Chen et al. 2013b). Poly (lactic acid) (PLA) composites with titanium oxide (TiO_2) as PLA/TiO_2 nanocomposites have been developed which showed biocidal properties, and significantly reduced *Escherichia coli* colonies without UVA irradiation when compared with pure PLA. The biocidal characteristic can be increased under UVA irradiation with these nanocomposites (Fonseca et al. 2015).

Halloysite nanotubes (HNTs) have been developed which showed natural nanotubular structures with exciting applications of HNTs which was due to their abundant deposit, nanoscale lumens, high length-to-diameter ratios, and relatively low surface hydroxyl group density. It was used as ideal templates for conveniently immobilizing nanoparticles, and for the construction of designed nano-architectures as a heterogeneous cataly utilized in the fuel cells (Zhang et al. 2016a). The textural properties and phosphate adsorption capability of modified-biochar containing Mg-Al assembled nanocomposites have been established which were prepared by an effective electro-assisted modification method with

MgCl$_2$ as an electrolyte. Structure and chemical analyses of the modified-biochar showed that nano-sized stonelike or flowerlike Mg-Al assembled composites, MgO, spinel MgAl$_2$O$_4$, AlOOH, and Al$_2$O$_3$, were densely grown and uniformly dispersed on the biochar surface. It was developed that the biochar/Mg-Al assembled nanocomposites provide an excellent adsorbent that can effectively remove phosphate from aqueous solutions (Jung et al. 2015). CdS/ZnS sandwich and core-shell nanocomposites have been developed and synthesized by a simple and modified chemical precipitation method under ambient conditions. Photocatalytic properties of the as-synthesized composites were reported using Acid Blue dye and p-chlorophenol under visible light irradiation and results showed the formation of stable core-shell nanocomposites and their efficient photocatalytic properties (Qutub et al. 2015). Graphene oxide (GO) sheets have been synthesized by modified Hummers and Offeman's method as ZnSe/graphene nanocomposites via hydrothermal conditions (180°C; 12 hr) which act as photocatalysts under visible light tested with methylene blue (MB) dye which were significantly degraded in aqueous phase (Hsieh et al. 2015). A new magnetically responsive three-component nanocomposite NiFe$_2$O$_4$/PAMA/Ag have been established and act as an antibacterial activities, avoiding the contamination of the environment (Allafchian et al. 2015).

Nanocomposites for biosensors and nanodevices

ChOx/FNAB/ODT/Au nanocomposite film was developed by Arya et al. (2006) which has been utilized for estimation of cholesterol in solution using SPR technique. Now-a-days, monitoring of toxicants, contaminants or pollutants in the air, water and soil is very important to save human health and ecosystems from the risk posed to them. In these contexts, recent advances in the development and application of biosensor arrays using aptamers, aptaymes for environmental detection are highlighted (Singh et al. 2008). Figure 1 shows the basic concept of biosensor/nanobiosensor. Au/APTES/GDA/GST nanocomposite film was developed (Singh et al. 2009a) for the electrochemical detection of captan in contaminated water. Captan is a known harmful chemical and potential carcinogen to a water ecosystem, and its continual use poses an environmental pollution. Biosensors based on conjugated conducting polymers such as polyaniline, polypyrrole, polythiophene blend with metal, ceramic to form nanocomposites for wide applications to detect toxicants in environmental samples (Singh and Choi 2009). CAT/PANi/ITO nanocomposite film was developed (Singh et al. 2009b) for the detection of hydrogen peroxide

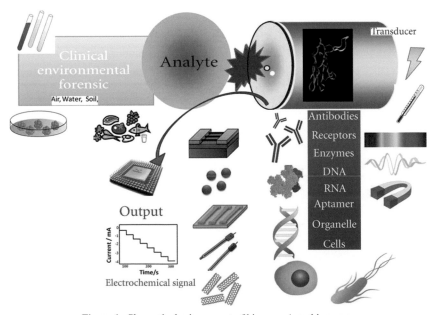

Figure 1. Shows the basic concept of biosensor/nanobiosensor.

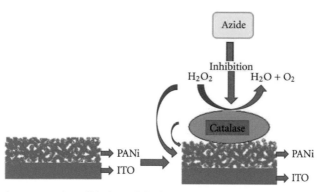

Figure 2. Shows schematic representation of bioelectrode in the development of biosensor for the H_2O_2 determination under optimal conditions.

and azide. Figure 2 shows schematic representation of bioelectrode in the development of biosensor for the H_2O_2 determination under optimal conditions. Further, Figure 3 shows the morphology of the polyaniline-silver nanocomposite. It indicates that nano-sized PANi possesses nanofibers morphology and strongly influences the composite morphology. Since the silver nanoparticles were synthesized in the polyaniline solution, the nanoparticles were embedded into the polyaniline matrix. Hydrogen peroxide causes oxidative damage when in high concentration whereas azide is toxic chemical widely used in agriculture for pest control. Progress and development towards biosensor development will focus upon the technology of nanocomposites for the real time detection of pesticides, antibiotics, pathogen, toxins and biomolecules in food, soil and water. The current trends and challenges with nanocomposites for various applications will have focused not only biosensor development but also on nanobiosensor (Singh 2011). The current trends and challenges with smart nanomaterials for the various applications pertain to development of biosensor, nanotechnology, and nanobiotechnology. All these growing areas will have a remarkable influence on the development of new ultra biosensing devices (nanobiosensor) to resolve the severe pollution problems in the future that not only challenge the human health but also adversely affect various other areas of living entities (Singh et al. 2012d).

Alginate/chitosan nanocomposite particles (GSNO-acNCPs), i.e., S-nitrosoglutathione (GSNO) loaded polymeric nanoparticles incorporated into an alginate and chitosan matrix have been developed and were able to increase the effective GSNO loading capacity, a nitric oxide (NO) donor, and to sustain its release from the intestine following oral administration. In conclusion, GSNO-acNCPs enhance GSNO intestinal absorption and promote the formation of releasable NO stores into the rat aorta. GSNO-

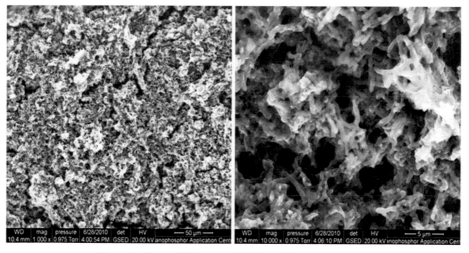

Figure 3. Shows SEM image of PANi-AgNPs nanocomposite.

acNCPs are promising carriers for chronic oral application devoted to the treatment of cardiovascular diseases (Wu et al. 2016b). Biofunctionalized carbon nanocomposites, mainly bioreceptor-functionalized nanocomposites, have been established which are able to generate sensitive detection systems in clinical/environmental monitoring (Sabherwal et al. 2016). Anisotropic yolk/shell or Janus inorganic/polystyrene nanocomposites have been developed which were prepared by combining miniemulsion polymerization and sol-gel reaction. These nanocomposites showed stable and strong fluorescence by introducing quantum dots as the co-seeds. These composites with unique anisotropic properties will have high potential in biomedical applications, particularly in biodetection (Han et al. 2016). Graphene oxide (GO) based polymer nanocomposites have attracted extensive research interest for their outstanding physicochemical properties and potential applications. The resultant products exhibit high water disperisibility, excellent biocompatibility and high efficient drug loading capability, making these PEGylated GO nanocomposites promising candidates for biomedical applications (Peng et al. 2016b). The four different morphologies of the CdSe semiconductor nanograss have been developed and grown on graphene oxide (GO) sheets via hydrothermal method at 220°C for 12 hr and showed that GO-CdSe nanocomposites nanorods have best field emission (FE) properties (Xie et al. 2016).

Porous Co nanobeads/rGO nanocomposites derived from rGO/Co-metal organic frameworks have been established for glucose sensing and have become one of the functional materials in sensor, catalysis, energy conversion, etc. The porous and hierarchical superstructures, good electrical conductivity and large specific surface area of Co nanobeads/rGO nanocomposites contributed to the good performance of the sensor towards the detection of glucose (Song et al. 2015b). Magnetic nanocomposites (MNCs) consisting of polymer matrix and encapsulated functional nanoparticles have been reported to represent innovative stimuli-responsive materials which are currently used in progressive technologies and biomedicine. They have designed the novel therapeutic systems for controlled drug release, and the biodegradable composites based on poly(3-hydroxybutyrate), chitosan, with encapsulated magnetic nanoparticles (MNPs), and the modeling drugs (dipyridamole and rifampicin) were tested (Iordanskii et al. 2016). Cassava starch acetate (CSA)-polyethylene glycol (PEG)-gelatin (G) nanocomposites have been established as controlled drug delivery systems which can be used for the controlled release of an anticancer drug. The findings revealed that the cross linked CSA-PEG-G nanocomposites can be a potential polymeric carrier for the controlled delivery of drug (Raj and Prabha 2015). Luminol electrochemiluminescence (ECL) strategy based on TiO_2/CNTs nanocomposites for detection of glucose has been prepared by a sol-gel method and modified on the glass carbon electrode. TiO_2/CNTs nanocomposites have offered a novel material for the signal enhancement in electrochemiluminescence sensor (Wang et al. 2015f). Nanocomposites containing two or more functional constituents have synthesized multifunctional Ag/Fe_3O_4-CS nanocomposites using chitosan as a stabilizing and cross-linking agent. Fe_3O_4 and Ag NPs were uniformly dispersed in the chitosan matrix. The hybrid NPs exhibited strong antibacterial property against Pseudomonas aeruginosa. Ag/Fe_3O_4-CS composite showed that the multicomponent hybrid nanostructures appeared and were used for local hyperthermia treatment of cancers (Nguyen et al. 2015). Cobalt ferrite/graphene oxide $(CoFe_2O_4/GO)$ nanocomposites have been developed and synthesized by a facile sonochemical method. The $CoFe_2O_4$/GO-based ferrofluid exhibited typical MR effect with increasing viscosity, shear stress and yield stress depending on the applied magnetic field strength (Wang et al. 2015e). Nickel disulphide decorated carbon nanotube nanocomposites (NiS_2-CNT NCs) have been established by the wet-chemical method in alkaline medium. Toxic chemical 4-methoxyphenol (4-MP) was used as a target analyte for efficient phenolic sensor development in environment and healthcare (Mohammed et al. 2016).

Eu^{3+} doped SiO_2/PMMA hybrid nanocomposites have been developed by a sol-gel process and to study the effects of thermal treatment on the structure and luminescent properties of the material. Hybrid nanocomposites are multifunctional materials and their properties are the consequence of molecular interaction between inorganic and organic phases. These materials may be used in future applications as thermal sensors (Filipe et al. 2016). The dispersions of silver nanoparticles (Ag NPs) in a tetrafunctional thiol were reported and prepared by *in situ* reduction of silver nitrate with 2, 6-di-tert-butyl-para-cresol. The Ag NPs were maintained in a stable colloidal state for more than eleven months at ambient temperature. The thiol monomer acts as both stabilizing agent and reactive solvent. The thiol-methacrylate matrix of the nanocomposites was characterized by measuring the glass transition temperature, the flexural modulus and the compressive strength. The low toxicity of the reactants used in the synthesis in combination with antimicrobial activity of Ag NPs makes the nanocomposite materials synthesized in this study very attractive for the preparation of biomaterials and coating with improved biocompatibility (Silvana and Vallo 2016). An innovative and effective approach has been developed which is introduced to functionalize multi-walled carbon nanotubes (f-MWCNTs) by *in situ* chemical precipitation of hydroxyapatite (HA) to improve their magnetic properties. The HA/f-MWCNTs nanocomposites are obtained by pressureless sintering in vacuum atmosphere. This nanocomposite has the potential to be used as a biomaterial for hyperthermia treatment of bone cancer and other biomedical applications (Afroze et al. 2016). Modified tri-axial electrospinning processes for the generation of a new type of pH-sensitive polymer/lipid nanocomposite have been reported. The systems produced are able to promote both dissolution and permeation of a poorly modeled water-soluble drug. The new tri-axial electrospinning process was developed which provides a platform to fabricate structural nanomaterials, and the core-shell polymer-PL nanocomposites which have significant potential applications for oral colon-targeted drug delivery. The strategy of a combined usage of polymeric excipients and phospholipid in a core-shell format should provide new possibilities of developing novel drug delivery systems for efficacious oral administration of poorly-water soluble drugs (Yang et al. 2016b).

Magnetic Resonance Imaging (MRI) and X-ray imaging of biological tissues with magnetic nanocomposites (MNC) have been reported for the first time for the cancer treatments. These novel phantoms show a long-term stability of over several months up to years. The presented phantoms have been found to be suitable to act as a body tissue substitute for XCT imaging as well as an acceptable T2 phantom of biological tissue enriched with magnetic nanoparticles for MRI (Rahn et al. 2016). Anti-inflammatory effect of triamcinolone acetonide-loaded hydroxyapatite (TA-loaded HAp) nanocomposites in the arthritic rat model has been reported. The HAp nanocomposites were synthesized through a chemical precipitation method and the drug was subsequently incorporated into the nanocomposites using an impregnation method. TA-loaded HAp nanocomposites are potentially suggested for treatment of rheumatoid arthritis after further required evaluations (Jafari et al. 2016). The developing multifunctional theranostic platforms with complementary roles have drawn considerable attention in recent years. In this study, superparamagnetic cobalt ferrite/graphene oxide ($CoFe_2O_4$/GO) nanocomposites with integrated characteristics of magnetic resonance imaging and controlled drug delivery were prepared by sonochemical method. Furthermore, the $CoFe_2O_4$/GO showed negligible cytotoxicity even at a high concentration after being treated for 96 h. Doxorubicin hydrochloride (DOX), as an anti-tumor model drug, was loaded on $CoFe_2O_4$/GO. The as-prepared $CoFe_2O_4$/GO showed great potential as an effective multifunctional nanoplatform for magnetic resonance imaging and controlled drug delivery for simultaneous cancer diagnosis and chemotherapy (Wang et al. 2016e).

Au-Fe_3O_4 nanocomposites have been synthesized and modified with l-(2-mercaptoethyl)-1,2,3,4,5,6-hexanhydro-s-triazine-2,4,6-trione (MTT) for the bimodal detection of melamine. The bimodal detection is based on the aggregation of Au-Fe_3O_4@MTT nanoparticles (NPs) from the dispersed state upon the formation of special triple hydrogen bonds between melamine and MTT. This phenomenon causes the red shift of the absorption peak and the increase of the spin-spin relaxation time (T2) of the water protons upon addition of melamine at different concentrations to an aqueous solution of Au-Fe_3O_4@MTT NPs. Furthermore, the dual-modal method exhibits a clear selectivity toward melamine (Shen et al. 2016). 5-Fluorouracil encapsulated chitosan/silver and chitosan/silver/multiwalled carbon nanotubes have been synthesized to comparatively study the release profile and cytotoxicity of the systems towards MCF-7 cell line. The release profile showed a prolonged release for 5-Fluorouracil encapsulated Chitosan/silver/ multiwalled carbon nanotube and a better cytotoxicity with IC_{50} of 50 μg/ml was observed for the same (Nivethaa et al. 2016). MWCNTs/$NaGdF_4$:Yb^{3+}, Er^{3+} nanocomposites were successfully synthesized by a simple liquid method using PVP as a significant dispersant agent. The as-prepared nanocomposites have nice paramagnetic, up-conversion luminescent, photothermal properties and excellent biocompatibility simultaneously. After combining with MWCNTs, the upconversion luminescence process of $NaGdF_4$:Yb^{3+}:Er^{3+} has not been affected, the nanocomposites still emit green light mainly. Therefore, the multifunctional MWCNTs/$NaGdF_4$:Yb^{3+}, Er^{3+} nanocomposites have potential applications in the fields of biomarkers, drugs targeting, and the diagnosis and treatment of cancers (Wang et al. 2016d).

Hexagonal ring structured PbO/CdO/ZnO and spherical shaped PbS/CdS/ZnS nanocomposites have been synthesized by chemical method in air. The combined electrochemical and optical experiments indicated that PbO/CdO/ZnO and PbS/CdS/ZnS nanocomposites can be considered as efficient candidates for optical and electrochemical applications (Murugadoss et al. 2016). Prussian blue (PB) and gold modified polystyrene (PS) nanocomposites (PS-Au-PB) have been reported and showed peroxidase-like and catalase-like activities which indicated that the PS-Au-PB nanocomposites possess strong affinity with the substrates and high catalytic activity. A sensitive method for glucose detection was developed using glucose oxidase (GOx) and PS-Au-PB nanocomposites. Also, PS-Au-PB nanocomposites can serve as a useful reagent in some biological detection (Zhang et al. 2016b). Manganese ferrite/graphene oxide ($MnFe_2O_4$/GO) nanocomposites have been reported as controlled targeted drug deliveries which were prepared by a facile sonochemical method. The *in vitro* cytotoxicity testing exhibited negligible cytotoxicity of as-prepared $MnFe_2O_4$/GO, even at the concentration as high as 150 µg/mL. Doxorubicin hydrochloride (DOX) is an anti-tumor model drug which was utilized to explore the application potential of $MnFe_2O_4$/GO for controlled drug delivery (Wang et al. 2016c). Poly (lactic acid) (PLA) composites with titanium oxide (TiO_2) as PLA/TiO_2 nanocomposites have showed biocidal properties, significantly reducing *Escherichia coli* colonies without UVA irradiation when compared with pure PLA. The biocidal characteristic can be increased under UVA irradiation with these nanocomposites (Fonseca et al. 2015). Graphene and its derivatives have unique physical and chemical properties that make them promising vehicles for photothermal therapy (PTT)-based cancer treatment. Graphene-based nanosystems exhibit multifunctional properties that are useful for PTT applications including enhancement of multimodalities, guided imaging, enhanced chemotherapy and low-power efficient PTT for optimum therapeutic efficiency (Chen et al. 2016c).

Green nanocomposite

Green materials are those which have biodegradable and renewable property. In similar fashion, if polymer is said to be green, it means it possesses environmentally favorable properties such as renewability and degradability. Green chemistry is the design of chemical products and processes that reduce or eliminate the use or generation of substances hazardous to humans, animals, plants, and the environment. Thus, green chemistry seeks to reduce and prevent pollution at its source. Natural polymers are almost green like biodegradable polymers (e.g., cellulose, chitin, starch, polyhydroxyalkanoates, polylactide, polycaprolactone, collagen and polypeptides). However, few microorganisms and enzymes capable of degrading green polymers have been identified (Kaplan et al. 1993, Chandra and Rustogi 1998). Figure 4 shows scheme of intercalated and exfoliated nanocomposite structure.

These nanocomposites are environmental friendly and sustainable bio-reinforced composites used in automotive, construction, packaging and medical fields. These have unique properties including strength, elastic modulus, dimensional stability, permeability towards gases, water, and hydrocarbons, thermal stability, heat distortion temperature, smoke emissions, chemical resistance, surface appearance, weight and electrical conductivity at nanoscale level which possess superior properties. Green nanocomposites

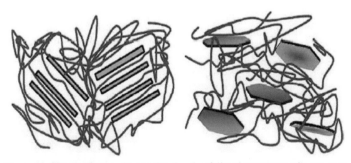

Figure 4. Shows scheme of intercalated and exfoliated nanocomposite structure.

are biodegradable, exibit biological degradability and reduction in the volume of carbon dioxide in the atmosphere which have attracted tremendous research interests (Leza and Lewandowicz 2010).

To design, develop, and characterize films, blends, coatings, and nanocomposite formulations which have shown antimicrobial and/or antioxidant properties and they have to improve food safety and food shelf-life. Active biomolecules have a significant role in the food application. Chitosan is a biopolymer that shows great potential as an ingredient for the preparation of new edible films and multifunctional formulations for different food applications. It has showed high antimicrobial activity against a wide variety of pathogenic and spoilage microorganisms, including fungi, and gram-positive and gram-negative bacteria. The chitosan-based edible films, blends, and multifunctional nanocomposites were used as package materials and potential antimicrobial activity. Materials are said to be 'green' when they are biodegradable and renewable. The major attractions about green composites are that they are environmental friendly, fully degradable and sustainable in every way. At the end of their service life, they can be easily disposed of or composted without harming the environment. The challenge of green composite basically involves the challenge of obtaining 'green' polymers that are used as matrix for the production of the composites. Polymer is said to be green when it possesses environmentally favorable properties such as renewability and degradability. Biodegradation implies degradation of a polymer in natural environment that includes changes in the chemical structure, loss of mechanical and structural properties and changing into other compounds that are beneficial to the environment (Jamshidian et al. 2010, Njuguna et al. 2008, Chandra and Rustgi 1998, Mohanty et al. 2000). Polymers from natural sources (such as starch, lignin, cellulose acetate, poly-lactic acid (PLA), polyhydroxylalkanoates (PHA), polyhydroxylbutyrate (PHB), etc.) and some synthetic sources (aliphatic and aromatic polyesters, polyvinyl alcohol, modified polyolefins, etc.) that are degradable are classified as biopolymers (John and Thomas 2008). However, those from synthetic sources are not renewable and therefore do not conform wholly to the concept of renewability and degradability (Siracusa et al. 2008). A number of natural and other degradable polymers have been used in green composites and some of these are discussed as follows.

Thermoplastic starch based nanocomposites

Thermoplastic starch properties also appear to benefit from silica addition and have reported that inclusion of dry powder SiO_2 particles in starch PVOH films increased tensile strength at break and improved water barrier properties (Tang et al. 2008). The improved mechanical properties, transmittance, and water resistance of starch films containing nano-SiO_2 particles have been reported (Xiong et al. 2008). Thermoplastic starches are commercially viable, biodegradable and compostable starch polymers. They have starch content greater than 70% to have biodegradable or compostable action. The properties of a nanocomposite depend on the volume fraction of the filler, its shape, size and interfacial adhesion. Starch Based Resin 11C is a biodegradable and compostable resin based on a blend of thermoplastic starch for the manufacturing of film type products. It can be directly used in the film blowing process (Billmeyer 2008).

The use of thermoplastic starch for the production of composites by melt intercalation in twin screw extruder has been reported. The composites were prepared with regular cornstarch plasticized with glycerin and reinforced with hydrated kaolin. The sequence of addition of plasticizers to determine the effect of plasticizers on the mechanical and structural properties using solution method was established. Thermal stability, mechanical properties and water absorption studies were conducted to measure the material properties whence it was deduced that the sequence of addition of components (starch/plasticizer (glycerol)/clay) had a significant effect on the nature of composites formed and consequently, the properties were altered with no well established sequence to determine appropriate process method for the composites (De Carvalho et al. 2001, Pandey and Singh 2005). Starch modified with acetate and produced biocomposites with cellulose fibers have been reported (Guan and Hanna 2006).

Polylactic acid (PLA) based nanocomposites

The polylactic acid (PLA) has been proving a viable alternative biopolymer to petrochemical based plastics for many applications. PLA is produced from lactic acid that is itself derived from the fermentation of corn or sugar beet and due to its biodegradation ability; PLA presents a major advantage to enter in the natural cycle implying its return to the biomass. Studies on the thermal, mechanical and morphological properties of PLA-based composites have been reported (Lee et al. 2008b). Nanocomposite of PLA with a compatibilizer and cellulose fibrils have been developed (Qu et al. 2010).

These composites were prepared by melt compounding and injection molding. Thermal degradation, thermal transition, morphological, and mechanical properties of the composites were evaluated. More recently, nanocomposites of PLA with a compatibilizer and cellulose fibrils have been developed by Qu et al. (2010). Organically modified PLA clay blends were developed (Ogata et al. 1997). PLA have unique properties such as high mechanical strength, low toxicity and good barrier properties but limited in applications due to its low glass transition temperature, weak thermal stability, low ductility and toughness and low modulus above the glass transition temperature (Harada et al. 2007). The commercial PLA have reported the matrix with cellulose whiskers treated with anionic surfactant as reinforcement (Bondeson and Oksman 2007). Studies on the thermal, mechanical and morphological properties of PLA-based composites have been reported (Lee et al. 2008b).

Cellulose based nanocomposites

Cellulose from agricultural products is a source of biopolymer which can replace petroleum polymers. Green nanocomposites have been successfully produced from cellulose acetate (CA), triethyl citrate (TEC) plasticizer and organically modified clay via melt compounding (Mishra et al. 2004). A cellulose based nanocomposite material has been developed as a flexible humidity and temperature sensor (Mahadeva et al. 2011). Cellulose was obtained from cotton pulp; its nanocomposites were prepared by polymerization induced adsorption process. An active antimicrobial packaging material had been developed using methyl cellulose (MC) as the base material with montmorillonite (MMT) as reinforcement (Tunc et al. 2011). Cellulose based nanocomposite with hydroxylapatite for medical applications have been developed (Zimmermann et al. 2011, Zadegan et al. 2011).

Plant oil based nanocomposites

Biobased nanocomposites have been developed from epoxidized soybean oil, diglycidyl ether of bisphenol-A (DGEBA) and organically modified montmorillonite (OMM) (Sithique et al. 2010). The composite produced is not completely from green source because of the use of organophillic montmorillonite clay as reinforcement. The use of mango puree-based edible films for nanocomposites has reported potential packaging applications (Azeredo et al. 2009). Soy based polyurethane can be used as a matrix for the production of biobased nanocomposites (Tate et al. 2010).

Polymer-polymer blends based nanocomposites

The blending of two or more polymers has to achieve a nanocomposite polymer. These polymers have to be used as degradable polymers for composite applications. Such polymer blends as Starch/PLA blends, poly butylenes succinate/cellulose acetate blends, starch/modified polyester blends, polycarprolactone/poly vinyl alchohol blends and thermoplastic starch/polyesteramide blends have been reported by various investigators (Ke and Sun 2001, Uesaka et al. 2000, Kesel et al. 1997, Averous et al. 2000, Willet et al. 2002, Martin and Averus 2001). Recent use of binary and ternary blends of polylactide (PLA), polycaprolactone (PCL) and thermoplastic starch (TPS) as composites has been reported by Sarazin et al. (2008). Natural rubber was blended with starch to form nanocomposites (Majdzadeh-Ardakani and Sadeghi-Ardakani 2010). Polyhydroxyl butyrate (PHB) is a natural occurring polyester produced by numerous bacteria in nature. Gelatin is a biopolymer obtained by thermal denaturation of collagen isolated from animal skin and bones and Chitosan is a natural polymer found in exoskeletons of crustaceans and insects and in the cell wall of fungi and microorganisms. Mechanical and water vapor barrier properties of chitosan based nanocomposites were improved by addition of cellulose nanofiber whereas chitosan filled with hydroxyapatite, tricalcium phosphate, carbon fiber and montmorillonite did not show substantial improvement in mechanical properties (Maiti et al. 2003, Zheng et al. 2002, Takegawa et al. 2010).

Cellulose is the most abundant renewable fiber in nature. Moreover, they can be produced from agricultural waste which makes it more economically viable than any other source of fibers currently in use. Cellulose fiber offers an outstanding balance of excellent mechanical properties, low density, safer handling and working conditions compared to synthetic fibers. The usages of sisal, kenaf, hemp, jute and coir fibers as reinforcement in polypropylene based composites were investigated by several investigators (Wamba et al. 2003). Also, chemically modified banana fiber reinforced in polyester resin was developed (Pothan and Thomus 2003). The study on the potential of wheat straw fibers prepared by mechanical and chemical processes as reinforcing additives for thermoplastics was carried out (Panthapulakkal et al. 2006). A study of the synthesis and mechanical properties of new series of green composites involving Hibiscus sabdariffa fiber as a reinforcing material in urea-formaldehyde (UF) resin based polymer matrix has been reported (Singha and Thakur 2008). The challenge in the making of nanocomposites is compatibility of the hydrophobic (water repelling) polymer matrix and hydrophilic (water absorbing) fibers. To improve the affinity and adhesion between fibers and thermoplastic matrices in composites production, additives are required but no significant result is obtained. True green polymers with good mechanical properties are used as matrix material for better result. Biopolymers like starch have poor water resistance, inferior tensile properties and are highly brittle due to their large particle size which necessitates the use of plasticizers such as glycerol in the presence of heat and pressure (George et al. 1994).

In conclusion, PLA is the most promising of all the biopolymers currently in use and require much attention. Other biopolymer sources such as cellulose, gelatin, chitosan and plant-based oils are more scarce sources and involve a more tedious and costly production process. It is difficult to take the orientation of nanosize fibers into consideration but in macromechanics, it has been observed that the orientation of the fibers have an overiding effect on the mechanical properties of the composite material to achieve proper orientation of particles in the matrix. Nanocomposites have been used in several applications such as mirror housings on various vehicle types, door handles, door panels, trunk liners, instrument panels, parcel shelves, head rests, roofs, upholstery and engine covers and intake manifolds and timing belt covers. Other applications currently being considered include impellers and blades for vacuum cleaners, power tool housings, mower hoods and covers for portable electronic equipment such as mobile phones, pagers, etc. Its excellent barrier properties, chemical resistance and surface appearance make it an excellent material for packaging applications such as in beer and carbonated drinks bottles and paperboard for fruit juice and dairy. Nanocomposites also have future in aerospace applications because of their light weight. Natural fiber reinforced composites using biodegradable polymer as matrix are considered as the most environmental friendly material having renewability and degradability. These are thermoplastic starch and its blends, PLA and its modifications, cellulose, gelatin, chitosan, etc. Natural fibers are preferred over synthetic fibers for environmental reasons such as using kenaf, jute, hemp, flax, banana, bamboo, sisal and coconut coir, etc. There is a wide range of possible applications of nanocomposites with few problems such as poor adhesion of matrix and fiber, difficulty of fiber orientation, the achievement of nanoscale sizes and the evolution of truly green polymers (Ashori 2008, Kim et al. 2006, Teixeira et al. 2009).

Generally, nanocomposites have been used in several applications such as mirror housings on various vehicle types, door handles, door panels, trunk liners, instrument panels, parcel shelves, head rests, roofs, upholstery and engine covers and intake manifolds and timing belt covers, impellers and blades for vacuum cleaners, power tool housings, mower hoods and covers for portable electronic equipment such as mobile phones, pagers, etc. Honeywell developed commercial clay/nylon-6 nanocomposite products for drink packaging applications. Starch Based Resin 11C is a biodegradable and compostable resin based on a blend of thermoplastic starch to make compostable bags (shopping bags/check-out bags, green bin liners) and meat liners (overwrap packaging, mulch film, breathable film). The innovation and developments in technologies for the green polymer nanocomposites using biodegradable polymer as matrix have shown properties such as renewability and degradability. A series of interesting polymers have been realized through multiple research activities ranging from thermoplastic starch and its blends,

PLA and its modifications, cellulose, gelatine, chitosan, etc. Natural fibers have preferred over synthetic fibers for environmental reasons. They have been synthesized from agricultural sources such as kenaf, jute, hemp, flax, banana, bamboo, sisal and coconut coir, etc. There is thus a wide range of possible applications of nanocomposites (Pandey et al. 2007b, Gao 2004).

Nanocomposites as anticorrosive

The degradation of metal surfaces due to atmospheric corrosion is a major problem for many exposed metallic structures, such as bridges, pipelines and storage tanks. Seawater has an ability to increase in the degree of corrosion. ZnO nanoparticles/polypyrrole (ZnO-NPs/PPy) has been reported as hybrid nanocomposites and used them as additives in an epoxy paint to protect SAE 1020 carbon steel from corrosion (Valenca et al. 2015). Corrosion costs are very high. The corrosive action of the atmosphere depends primarily on factors such as relative humidity, pollutants, temperature and residence time of electrolyte solutions on the metal surface. The physicochemical characteristics that may interfere with the corrosive action of the environment are the presence of water, salts, gases, differences in pH and electrical conductivity. The use of zinc as an anticorrosive agent in paints is often adopted as a way to improve the corrosion resistance of metallic components, and these types of coatings can act as a physical barrier (Hammouda et al. 2011, Shi et al. 2011). Treatment with chromium and phosphate can be used to prolong the lifetime of zinc coatings (Hihara et al. 2013, Tomachuk et al. 2010). However, the leaching of such compounds in coatings may cause serious ecological damage, and that encourages the development of corrosion inhibitor compounds that could be less impacting for the environment. Conductive polymers, such as polyaniline (PANI) and polypyrrole (PPy), are organic compounds that usually have a good corrosion stability when in contact with solution or/and in the dry state. They also are a possible answer to the demand for "green" corrosion inhibitors to minimize the health risk to humans and damage to the environment (Riccardis and Martina 2014, Sitaram et al. 1997, Branzoi et al. 2008, Pan 2013). ZnO/PPy hybrid materials have been reported with its variety of applications in optoelectronic devices and anti-corrosion properties. The zinc oxide has high refractive index and thermal stability, UV protection, good transparency and presents high electron mobility. These properties are used in emerging applications where polypyrrole is added as an anticorrosive additive for steel coating (Batool et al. 2012, Lehr and Saidram 2013, Wang 2004).

Our interest aims the development of new anticorrosive additives to protect carbon steel; more specifically, we want to investigate the efficiency of the addition of new (ZnO/Polypyrrole) hybrid nanoparticles to an organic matrix upon the protection against the corrosion of carbon steel substrates (Jagtap et al. 2008). ZnO-NPs/PPy nanocomposites have to be used as anticorrosive pigments. The synthesis of ZnO-NPs/PPy nanocomposites is obtained by polymerization of pyrrole in the presence of SDS and ZnO-NPs. Coatings with higher percentages of nanocomposites have presented lower protection efficiency, and are attributed to the increased porosity of the coating due to particles agglomeration.

Nanocomposites for supercapacitors

Supercapacitors or ultracapacitors or electrochemical capacitors (ECs) are energy storage devices that store energy in which charging-discharging has occurred on the electrode surface of supercapacitor and possess excellent cycling ability. It is because of this unique feature that it is regarded as one of the most promising energy storage devices. There are two types of supercapacitors: electrochemical double layer capacitors (EDLCs) and pseudocapacitors. In EDLCs, the energy is stored electrostatically at the electrode-electrolyte interface in the double layer, while in pseudocapacitors; charge storage occurs via fast redox reactions on the electrode surface. There are three major types of electrode materials for supercapacitors: carbon-based materials, metal oxides/hydroxides and conducting polymers. Carbon-based materials such as activated carbon, mesoporous carbon, carbon nanotubes, graphene and carbon fibres are used as electrode active materials in EDLCs, while conducting polymers such as polyaniline, polypyrrole and polythiophene or metal oxides such as MnO_2, V_2O_5, and RuO_2 are used for pseudocapacitors. EDLCs depend only on the surface area of the carbon-based materials to storage charge, therefore, often exhibit very higher power output and better cycling ability. However, EDLCs have lower energy density values than pseudocapacitors since pseudocapacitors involve redox active materials to store charge both on the surface as well as in sub-surface layer (Yang 2011).

Although carbon-based materials, metal oxides/hydroxides and conducting polymers are the most common electroactive materials for supercapacitor, each type of material has its own unique advantages and disadvantages, for example, carbon-based materials can provide high power density and long life cycle but its small specific capacitance (mainly double layer capacitance) limits its application for high energy density devices. Metal oxides/hydroxides possess pseudocapacitance in addition to double layer capacitance and have wide charge/discharge potential range; however, they have relatively small surface area and poor cycle life. Conducting polymers have the advantages of high capacitance, good conductivity, low cost and ease of fabrication but they have relatively low mechanical stability and cycle life. Coupling the unique advantages of these nano-scale dissimilar capacitive materials to form nanocomposite electroactive materials is an important approach to control, develop and optimize the structures and properties of electrode material for enhancing their performance for supercapacitors (Wu et al. 2010a).

The properties of nanocomposite electrodes depend not only upon the individual components used but also on the morphology and the interfacial characteristics. Recently, considerable efforts have been placed to develop all kinds of nanocomposite capacitive materials, such as mixed metal oxides, conducting polymers mixed with metal oxides, carbon nanotubes mixed with conducting polymers, or metal oxides, and graphene mixed with metal oxides or conducting polymers. Design and fabrication of nanocomposite electroactive materials for supercapacitors applications needs the consideration of many factors, such as material selection, synthesis methods, fabrication process parameters, interfacial characteristics, electrical conductivity, nanocrystallite size, and surface area, etc. Although significant progress has been made to develop nanocomposite electroactive materials for supercapacitor applications, there are still a lot of challenges to be overcome. Nanocomposite electroactive materials that have been developed so far have demonstrated huge potential for supercapacitor applications. Different types of nanocomposite electroactive materials, such as mixed metal oxides, polymers mixed with metal oxides, carbon nanotubes mixed with polymers, or metal oxides, and graphene mixed with metal oxides or polymers, can be fabricated by various processes such as solid state reactions, mechanical mixing, chemical coprecipitation, electrochemical anodic deposition, sol-gel, *in situ* polymerization and other wet chemical synthesis. It has been shown that significant improvement in terms of specific surface area, electrical and ionic conductivities, specific capacitance, cyclic stability, and energy and power density, of supercapacitors can be achieved by using nanocomposite electroactive materials (Lee et al. 2005a).

Dielectric nanocomposites

The nanostructures and nanocomposites based on nanoparticles of semiconductor materials exhibit a wide range of nonlinear properties and can be used in various applied fields. The dielectric nanostructures are the heterogeneous medium formed by liquid or solid dielectric matrices such as polymer glasses and oils and nanoparticles of dielectrics such as Al_2O_3, SiO_2, and MgO, etc. Nanostructures have nonlinear optical properties. They are used to develop and create new optoelectronic and fibre-optic devices to control process and transmit the information like an electrical circuit. The study of nonlinear optical properties of dielectric nanostructures containing nanoscale objects of different chemical natures, shapes and sizes has existence of a low threshold optical response due to a number of conditions. The experimental study of changes of optical characteristics of the dielectric nanostructures based on Al_2O_3, SiO_2, TiO_2, ZnO nanoparticles and theoretical description of these characteristics allows estimating the conditions of observing the low-threshold optical nonlinearity under low-intensity optical fields. The ability to observe this nonlinearity is directly connected with the peculiarities of the energy spectrum of nanoparticle charge carriers. Because of the wide band gap of the bulk dielectric material, it is not possible to excite electron transitions to the conduction band by a visible light. The energy spectrum of nanoparticle electrons is of a different structure: the band gap has defect levels containing a lot of electrons due to a high density of crystal defects on the nanoparticle's surface; the small size and shape of nanoparticle leads to strong broadening of the band of high-density exciton states from the bottom of the conduction band up to defect levels. The existence of an absorption band in visible light spectrum is observed only for nanoparticles of broad-band dielectrics (Al_2O_3, SiO_2). The absorption band in the energy spectrum of electrons of narrow-band dielectric (TiO_2, ZnO) nanoparticles is not manifested in a visible light spectrum; however, it can be manifested within the infrared region and adjoins the bottom of the conduction band. The qualitative agreement between experimental and theoretical results was also obtained and the proposed theory model of optical nonlinearity can be applied to explain a number of phenomena in physics of nanoscale dielectrics, e.g., proteins and blood bodies (Dzyuba et al. 2011, Ho et al. 2011).

Nanocomposites as hard coatings

Nanostructured and nanocomposite films are a new generation of nanomaterials and are in nanocrystalline and/or amorphous structure. The unique physical and functional properties of these films are the main driving force stimulating the huge development of hard and superhard films. The nanostructured films with controlled size of grains in the range from 1 to 10 nm as a new advanced coatings have showed unique physical and functional properties, such as nanocrystallization from amorphous phase, electronic charge transfer between nanograins with different chemical composition and different Fermi energies again with the aim to produce films with new functional properties, development of new nanostructured and nanocomposite coatings based on oxides and mixture of oxides and nitrides or carbides or other compounds development of protective coatings with high oxidation resistance exceeding 2000°C, low temperature deposition of crystalline nanostructured functional films on heated sensitive substrates such as polymer foils and polycarbonate, and development of new PVD systems for the production of nanostructured coatings under new physical conditions and with high deposition rate. The geometry of the grains is responsible not only for the enhanced hardness of hard nanocomposite films but also for other enhanced properties of nanostructured films, for instance, enhanced magnetic or catalytic properties. Unique properties of the nanocomposite films are due to their nanostructure with a metastable phase. The nanostructure converts into large grains or a new crystalline phase, and determines the thermal stability of the nanocomposite. High-temperature (high-T) oxidation resistance is a very desirable property of the hard nanocomposite coatings in many applications (Musil 2000, Voevodin and Zabinski 2000b).

Perspectives of nanocomposites

Due to the outstanding potential application of nanocomposites, they have generated a great impact on world economy and business, locally as well as globally. Polymer/clay nanocomposites have been commercialized by many leading industries for packaging, coating and automotive use. High demands due to high performance systems using nanofillers at low costs make a bright future for a wide range of applications, for example, consumer products like electroconductive polymers, nanosmart switches and sensors for automotives. Further power production and storage devices like hydrogen storage, fuel cells, supercapacitors and batteries which are utilizing metal/ceramic/polymer nanocomposites with low cost and having good impact by improving fire retardancy in interior parts as well as weather condition improvement in exterior parts of cars. Nanoclays and CNTs have utilized in automotive and beverage/food packaging. CNT based nanocomposites have utilized in electrical and electronic industries. Polymer based nanocomposites have utilized in for engineered plastics. However, with few challenges and issues pertaining to nanocomposites, future R&D would be required for ecofriendly nanocomposites for good health and cleaner environment.

Nanocomposite materials are an emerging technology and creating macroscopic engineered materials as product at nanoscale level. Nanocomposites would benefit many domains of our society, such as electronics, chemical, space, general automobiles, medicine, health care and environmental protection. Due to these, nanocomposites will be able to improve our quality of life in the near future. However, social implications of nanocomposites as an emerging technology are a serious cause of concern because of their large surface area, crystalline structure as well as reactivity, which may facilitate their easy transport into the environment which, in turn, interact with cell bioconstituents and elicit many harmful effects. Thus, utilization of nanocomposites and their release into the environment is a major health and safety issue. In these contexts, there is an increasing need for R&D to devlop nanotoxicological profiles of nanocomposites before their utilization and applications. U.S. Environmental Protection Agency (EPA) is looking and evaluating the potential impacts of nanocomposites on human health and safeguard of the environment to not only minimize their use but also check the toxicity on human beings.

Conclusions

An emerging field requires not only starting materials but also potential ingredients. These have unique/ versatile properties which improves their performance when compared with the conventional, existing one. Due to scientific and technological advances of nanocomposites, these are the suitable materials to meet the demands. A lot of applications of nanomaterials are known and also open new applications for future because of their unique properties. Thus, nanocomposites provide opportunities and rewards for creating new fields.

Acknowledgement

Dr. Ravindra Pratap Singh (Singh R.P.) thanks IGN Tribal University, a central University, Amarkantak, M.P., India for providing facilities to prepare this book chapter.

References

Afroze, J.D., M.J. Abden, M.S. Alam, N.M. Bahadur and M.A. Gafur. 2016. Development of functionalized carbon nanotube reinforced hydroxyapatite magnetic nanocomposites. Mat. Lett. 169: 24–27.

Ain, N., W. Shaheen, B. Bashir, N.M. Abdelsalam, M.F. Warsi, M.A. Khan et al. 2016. Electrical, magnetic and photoelectrochemical activity of rGO/MgFe$_2$O$_4$ nanocomposites under visible light irradiation. Ceram. Internat. 42: 10, 12401–12408.

Ajayan, P.M., L. Schadler and P.V. Braun. 2003. Nanocomposite science and technology. Weinheim: Wiley-VCH, Verlag Gmbh & Co. KgaA.

Aladpoosh, R. and M. Montazer. 2016. Nano-photo active cellulosic fabric through *in situ* phytosynthesis of star-like Ag/ZnO nanocomposites: Investigation and optimization of attributes associated with photocatalytic activity. Carbohyd. Polym. 141: 116–125.

Alexandre, M. and P. Dubois. 2000. Polymer-layered silicate nanocomposites: preparation, properties and uses of a new class of materials. Mater. Sc. Engin. 28: 1–63.

Alice, A.M.L.F. Jardim, Rebeca Bacani, Fernanda F. Camilo, Márcia C.A. Fantini and T.S. Martins. 2016. SBA-15:TiO$_2$ nanocomposites. I. Synthesis with ionic liquids and properties. Microporous and Mesoporous Materials 228: 37–44.

Allafchian, A., H. Bahramian, S.A.H. Jalali and H. Ahmadvand. 2015. Synthesis, characterization and antibacterial effect of new magnetically core-shell nanocomposites. J. Magnet. Mag. Mat. 394: 318–324.

Amirabbas, N., S. Abbaspour, M. Masood, S.N. Mirsattari, A.Vahedi and K.J.D. Mackenzie. 2016. Photocatalytic properties of mesoporous TiO$_2$ nanocomposites modified with carbon nanotubes and copper. Cer. Intern. 42(10): 11901–11906.

Andrews, R. and M.C. Weisenberger. 2004. Carbon nanotube polymer composites. Curr. Opin. Solid State and Mater. Sci. 8(1): 31–37.

Arora, K., A. Chaubey, R. Singhal, R.P. Singh, M.K. Pandey, S.B. Samanta et al. 2006. Application of electrochemically prepared polypyrrole-polyvinylsulphonate films to DNA biosensor. Biosen. Bioelect. 21(9): 1777–1783.

Arjmandi, R., A. Hassan, M.K. Mohamad Haafiz and Z. Zakaria. 2015. Partial replacement effect of montmorillonite with cellulose nanowhiskers on polylactic acid nanocomposites. Int. J. Biol. Macromol. 81: 91 99.

Arya, S.K., P.R. Solanki, R.P. Singh, M.K. Pandey, M. Datta and B.D. Malhotra. 2006. Application of octadecanethiol self assembled monolayer to cholesterol biosensor based on surface Plasmon Resonance technique. Talanta 69(4): 918–926.

Ashori, A. 2008. Wood-plastic composites as promising green-composites for automotive industries! Bioresource Biotechnol. 99: 4661–4667.

Averous, L., N. Fauconnier and L. Moro. 2000. Fringant blends of thermoplastic starch and polyesteramide: Processing and properties. J. Appl. Polym. Sci. 76: 1117–1128.

Azeredo, H.M.C., L.H.C. Mattoso, D. Wood, T.G. Williams, R.J. Avena-Bustillos and T.H. Mchugh. 2009. Nanocomposite edible films from mango puree reinforced with cellulose nanofibers. J. Food Sci. 74: 31–35.

Aziz, H.Y. and M.S. Gohari. 2016. Fe$_3$O$_4$/ZnO/Ag$_3$VO$_4$/AgI nanocomposites: quaternary magnetic photocatalysts with excellent activity in degradation of water pollutants under visible light. Sep. Pur. Tech. 166: 63–72.

Batool, A., F. Kanwal, M. Imran, T. Jamil and S.A. Siddiqi. 2012. Synthesis of polypyrrole/zinc oxide composites and study of their structural, thermal and electrical properties. Synth. Met. 161(23-24): 2753–2758.

Bhaumik, M., S. Agarwal, V.K. Gupta and A. Maity. 2016. Enhanced removal of Cr(VI) from aqueous solutions using polypyrrole wrapped oxidized MWCNTs nanocomposites adsorbent. J. Colloid Interf. Sc. 470: 257–267.

Bondeson, D. and K. Oksman. 2007. Dispersion and characteristics of surfactant modified cellulose whiskers nanocomposites. Compos. Interface 14: 617–630.

Billmeyer, F.W.J. 1984. Textbook of Polymer Science. 3 Ed. John Wiley & Sons. Inc.

Bourbigot, S., E. Devaux and X. Flambard. 2002. Flammability of polyamide-6/clay hybrid nanocomposite textiles. Polym. Degrad. Stab. 75: 397–402.

Branzoi, V., A. Pruna and F. Branzoi. 2013. Inhibition effects of some organic compounds on Zinc corrosion in 35% NaCl. Revista de Chimie 59(5): 540–554.

Breuer, O. and U. Sunderraj. 2004. Big returns from small fibers: a review of polymer/carbon nanotube composites. Polym. Compos. 25(6): 630–645.

Cai, Y., C. Lingqi, Y. Hongjiang, G. Jihua, C. Laifei, Y. Xiaowei et al. 2016. Mechanical and electrical properties of carbon nanotube buckypaper reinforced silicon carbide nanocomposites. Ceram. Internat. 42: 4984–4992.

Chandra, R. and R. Rustogi. 1998. Biodegradable polymers. Prog. Polym. Sci. 23(7): 1273–1335.

Camargo, P.H.C., K.G. Satyanarayana and F. Wypych. 2009. Nanocomposites: synthesis, structure, properties and new application opportunities. Mat. Res. 12(1): 1–39.

Caroline, E. 2002. Auto applications drive commercialization of nanocomposites. Plast. Add. Compound. 4(11): 30–33.

Caseri, W. 2003. Nanocomposites. In: Yang, P. (ed.). The Chemistry of Nanostructured Materials. World Scientific, Singapore, 359.

Carvalho, A.J.F., A.A.S. Curvelo and J.A.M.A. Agnelli. 2001. First insight on composites of thermoplastic starch and kaolin. Carbohyd. Polym. 45: 189–194.

Chang, Y.N., J.L. Gong, G.M. Zeng, X.M. Ou, B. Song, M. Guo et al. 2016. Antimicrobial behavior comparison and antimicrobial mechanism of silver coated carbon nanocomposites. Proc. Saf. Environ. Prot. 102: 596–605.

Chen, X., L. Huankun, W. Yuxin, W. Hanshuo, W. Laidi, T. Pengfei et al. 2016a. Facile fabrication of novel porous graphitic carbon nitride/copper sulfide nanocomposites with enhanced visible light driven photocatalytic performance. J. Coll. Interf. Sc. 476: 132–143.

Chen, M., W. Jiang, F. Wang, P. Shen, P. Ma, J. Gu et al. 2013b. Synthesis of highly hydrophobic floating magnetic polymer nanocomposites for the removal of oils from water surface. Appl. Surf. Sc. 286: 249–256.

Chen, Y.W., Y.L. Su, S.H. Hu and S.Y. Chen. 2016c. Functionalized graphene nanocomposites for enhancing photothermal therapy in tumor treatment. Adv. Drug Del. Rev. 105(Pt B): 190–204.

Choa, Y.H., J.K. Yang, B.H. Kim, Y.K. Jeong, J.S. Lee and T. Nakayama. 2003. Preparation and characterization of metal: ceramic nanoporous nanocomposite powders. J. Magnet. Magn. Mat. 266(1-2): 12–19.

Choi, S.M. and H. Awaji. 2005. Nanocomposites: a new material design concept. Sc. Technol. Adv. Mat. 6(1): 2–10.

Cornelia, H., M. Koji, R. Ralf, L. Emanuel and M. Gabriela. 2016. Facile sol–gel synthesis of reduced graphene oxide/silica nanocomposites. J. Europ. Ceram. Soc. 36(12): 2923–2930.

Czech, B. and B. Waldemar. 2015. Photocatalytic treatment of pharmaceutical wastewater using new multiwall-carbon nanotubes/TiO$_2$/SiO$_2$ nanocomposites. Environ. Res. 137: 176–184.

Deng, L., L. Hui, G. Xiaoyi, S. Xing and Z. Zhenfeng. 2016. SnS$_2$/TiO$_2$ nanocomposites with enhanced visible light-driven photoreduction of aqueous Cr (VI). Ceram. Internat. 42: 3808–3815.

Dresselhaus, M.S., G. Dresslhaus and P. Avouris. 2001. Carbon Nanotubes: Synthesis, Structure, Properties and Applications. Berlin: Springer Verlag.

Duangjam, S., K. Wetchakun, S. Phanichphant and N. Wetchakun. 2016. Hydrothermal synthesis of novel CoFe$_2$O$_4$/BiVO$_4$ nanocomposites with enhanced visible-light-driven photocatalytic activities. Mat. Lett. 181: 86–91.

Dzyuba, V., V. Milichko and K. Yurii. 2011. Non typical photoinduced optical nonlinearity of dielectric nanostructures. Journal of Nanophotonics 5(053528): 1–13.

Dutta, D., S. Thiyagarajan and D. Bahadur. 2016. SnO$_2$ quantum dots decorated reduced graphene oxide nanocomposites for efficient water remediation. Chem. Engin. J. 297: 55–65.

Duncan, M. 2003. Domestic Water Treatment in Developing Countries. Duncan Mara. Cromwell Press, U.K.

Emmanuel, I.U. and A. Taubert. 2016. Clay-polymer nanocomposites (CPNs): Adsorbents of the future for water treatment. Appl. Clay Sc. 99: 83–92.

Fang, S., S. Xue, W. Can, W. Guanqiu, W. Xi, L. Qian et al. 2016. Fabrication and characterization of CdS/BiVO4 nanocomposites with efficient visible light driven photocatalytic activities. Ceram. Internat. 42: 4421–4428.

Fatemeh, A., S. Azam and A.S. Masoud. 2016. Green synthesis of magnetic chitosan nanocomposites by a new sol-gel auto-combustion method. J. Magnet. Magnet. Mat. 410: 27–33.

Fernandez, L.M., A.S. Fernandez, O. Milošević and M.E. Rabanal. 2016. Solvothermal synthesis of Ag/ZnO and Pt/ZnO nanocomposites and comparison of their photocatalytic behaviors on dyes degradation. Adv. Powd. Techn. 27: 983–993.

Filipe, A.d.J., S.T.S. Santos, J.M.A. Caiut and V.H.V. Sarmento. 2016. Effects of thermal treatment on the structure and luminescent properties of Eu^{3+} doped SiO$_2$-PMMA hybrid nanocomposites prepared by a sol-gel process. J. Luminesc. 170(Part 2): 588–593.

Fischer, H. 2003. Polymer nanocomposites: from fundamental research to specific applications. Mat. Sc. Engin.: C. 23(6-8): 763–772.

Fonseca, C., A. Ochoa, M.T. Ulloa, E. Alvarez, D. Canales and P.A. Zapata. 2015. Poly (lactic acid)/TiO$_2$ nanocomposites as alternative biocidal and antifungal materials. Mat. Sc. Engin.: C. 57: 314–320.

Fu, H., W. Yin, L. Xiaoya and C. Weifeng. 2016. Synthesis of vegetable oil-based waterborne polyurethane/silver-halloysite antibacterial nanocomposites. Comp. Sc. Techn. 126: 86–93.

Freire, T.M., L.M.U. Dutra, D.C. Queiroz, N.M.P.S. Ricardo, K. Barreto, J.C. Denardin et al. 2016. Fast ultrasound assisted synthesis of chitosan-based magnetite nanocomposites as a modified electrode sensor. Carbohyd. Polym. 151: 760–769.

Gangopadhyay, R. and D. Amitabha. 2000. Conducting polymer nanocomposites: a brief overview. Chem. Mat. 12(7): 608–622.

Gao, F. 2004. Clay/polymer composites: the story. Materials Today 7(11): 50–55.

Gao, F., X. Hou, A. Wang, G. Chu, W. Wu, J. Chen et al. 2016. Preparation of polypyrrole/TiO$_2$ nanocomposites with enhanced photocatalytic performance. Particuol. 26: 73–78.

Garces, J.M., D.J. Moll, J. Bicerano, R. Fibiger and D.G. McLeod. 2000. Polymeric nanocomposites for automotive applications. Adv. Mat. 12(23): 1835–1839.

Ge, X., M.C. Li, X.X. Li and U.R. Cho. 2015. Effects of silane coupling agents on the properties of bentonite/nitrile butadiene rubber nanocomposites synthesized by a novel green method. Appl. Clay Sc. 118: 265–275.

George, E.R., T.M. Sullivan and E.H. Park. 1994. Preparation of high moisture content thermoplastic polyester starch. Polym. Eng. Sci. 34: 17–24.

Ghosh, A., P. Madhubonti, B. Krishna, C.G. Uday and M. Biswaranjan. 2015. Manganese oxide incorporated ferric oxide nanocomposites (MIFN): A novel adsorbent for effective removal of Cr (VI) from contaminated water. J. Wat. Proc. Engin. 7: 176–186.

Ghanbari, M., D. Emadzadeh, W.J. Lau, T. Matsuura, M. Davoody and A.F. Ismail. 2015. Super hydrophilic TiO$_2$/HNT nanocomposites as a new approach for fabrication of high performance thin film nanocomposite membranes for FO application. Desalin. 371: 104–114.

Giannelis, E.P. 1996. Polymer layered silicate nanocomposites. Adv. Mat. 8(1): 29–35.

Gilmann, J.W. 1999. Flammability and thermal stability studies of polymer-layered-silicate (clay) nanocomposites. Appl. Clay Sc. 15: 31–49.

Giusy, L.M.C., L. Giovanni, F.D. Rute, M. Arjen, D. Luciana, K.G. Ravindra et al. 2016. Polymer functionalized nanocomposites for metals removal from water and wastewater: an overview. Wat. Res. 92: 22–37.

Gui, M.M., W.M.P. Wong, S.P. Chai and A.R. Mohamed. 2015. One-pot synthesis of Ag-MWCNT@TiO$_2$ core-shell nanocomposites for photocatalytic reduction of CO$_2$ with water under visible light irradiation. Chem. Engin. J. 278: 272–278.

Guan, J. and M.A. Hanna. 2006. Selected morphological and functional properties of extruded acetylated starch-cellulose foams. Bioresource Technol. 97: 17161726.

Han, X., S. Huang, Y. Wang and D. Shi. 2016. Design and development of anisotropic inorganic/polystyrene nanocomposites by surface modification of zinc oxide nanoparticles. Materials Science and Engineering: C. 64: 87–92.

Harada, M., T. Ohya, K. Iida, H. Hayashi, K. Hirano and H. Fukuda. 2007. Increased impact strength of biodegradable poly(lactic acid)/poly(butylenes succinate) blend composites by using isocyanate as a reactive processing agent. J. Appl. Polym. Sci. 106: 1813–1820.

Hammouda, N., H. Chadli, G. Guillemot and K. Belmokre. 2011. The corrosion protection behaviour of zinc rich epoxy paint in 3% NaCl solution. Adv. Chem. Engin. Sc. 1(2): 51–60.

Hihara, L.H., R.P.I. Adler and R.M. Latanision (eds.). 2013. Environmental Degradation of Advanced and Traditional Engineering Materials. Boca Raton: CRC Press, 719 p.

Ho, C.H., C.H. Chan, L.C. Tien and Y.S. Huang. 2011. Direct optical observation of band-edge excitons, band gap, and fermi level in degenerate semiconducting oxide nanowires In$_2$O$_3$. Journal of Physical Chemistry C 115: 25088–25096.

Hong, X., J. Wen, X. Xiong and Y. Hu. 2016. Silver nanowire-carbon fiber cloth nanocomposites synthesized by UV curing adhesive for electrochemical point-of-use water disinfection. Chemosph. 154: 537–545.

Hsieh, S.H., W.J. Chen and T.H. Yeh. 2015. Effect of various amounts of graphene oxide on the degradation characteristics of the ZnSe/graphene nanocomposites. Appl. Surf. Sc. 358: 63–69.

Iordanskii, A.L., A.V. Bychkova, K.Z. Gumargalieva and A.A. Berlin. 2016. Magnetoanisotropic biodegradable nanocomposites for controlled drug release. Nanobiomat. Drug Del. 171–196, Chapter 6.

Istratie, R., M. Stoia, C. Păcurariu and C. Locovei. 2016. Single and simultaneous adsorption of methyl orange and phenol onto magnetic iron oxide/carbon nanocomposites. Arab. J. Chem. In Press.

Itodo, A.U. and H.U. Itodo. 2010. Quantitative specification of potentially toxic metals in expired canned tomatoes found in village markets. Nature Sci. 8(4): 54–59.

Jafari, S., M.D. Nasrin, B. Jaleh, B.J. Mohammad, R. Maryam and A. Khosro. 2016. Physicochemical characterization and *in vivo* evaluation of triamcinolone acetonide-loaded hydroxyapatite nanocomposites for treatment of rheumatoid arthritis. Coll. Surf. B: Biointerf. 140: 223–232.

Jagtap, R.N., P.P. Patil and S.Z. Hassan. 2008. Effect of zinc oxide in combating corrosion in zinc-rich primer. Prog. Org. Coat. 63(4): 389–394.

Jamshidian, M., E.A. Tehrany, M. Imran, M. Jacquot and S. Desobry. 2010. Poly-lactic acid: Production, applications, nanocomposites, and release studies. Compr. Rev. Food Sci. Food Saf. 9: 552–571.

John, M.J. and S. Thomas. 2008. Biofibres and biocomposites. Carbohyd. Polym. 71: 343–364.

Jung, K.W., T.U. Jeong, M.J. Hwang, K. Kim and K.H. Ahn. 2015. Phosphate adsorption ability of biochar/Mg–Al assembled nanocomposites prepared by aluminum-electrode based electro-assisted modification method with MgCl$_2$ as electrolyte. Biores. Tech. 198: 603–610.

Kadam, A., R. Dhabbe, A. Gophane, T. Sathe and K. Garadkar. 2016. Template free synthesis of ZnO/Ag$_2$O nanocomposites as a highly efficient visible active photocatalyst for detoxification of methyl orange. J. Photochem. Photobiol. B: Biol. 154: 24–33.

Kamigaito, O. 1991. What can be improved by nanometer composites? J. Jap. Soc. Powd. Metalur. 38: 315–321.

Kaplan, D.L., J.M. Mayer, D. Ball, J. McCassie, A.L. Allen and P. Stenhouse. 1993. Fundamentals of biodegradable polymers. pp. 1–42. *In*: Ching, C., D.L. Kaplan and E.L. Thomas (eds.). Biodegradable Polymers and Packaging. Technomic Pub. Co, Lancaster.

Ke, Z. and Y.P. Bai. 2005. Improve the gas barrier property of pet film with montmorillonite by *in situ* interlayer polymerization. Mat. Lett. 59(27): 3348–3351.

Ke, T.Y. and X.Z. Sun. 2001. Effects of moisture content and heat treatment on the physical properties of starch and poly(lactic acid) blends. J. Appl. Polym. Sci. 81: 3069–82.

Kesel, C.D., C.V. Wauven and C. David. 1997. Biodegradation of polycaprolactone and its blends with poly(vinylalcohol) by micro-organisms from a compost of household refuse. Polym. Degrad. Stab. 55: 107–113.

Khan, S.B., K.A. Alamry, E.N. Bifari, A.M. Asiri, M. Yasir, L. Gzara et al. 2015. Assessment of antibacterial cellulose nanocomposites for water permeability and salt rejection. J. Indust. Engin. Chem. 24: 266–275.

Khobragade, P.S., D.P. Hansora, J.B. Naik and A. Chatterjee. 2016. Flame retarding performance of elastomeric nanocomposites: a review. Polym. Degrad. Stab. In Press, Available online 7 June 2016.

Kim, J.P., T.H. Yoon, S.P. Mun, J.M. Rhee and J.S. Lee. 2006. Wood-polyethylene composites using ethylene-vinyl alcohol copolymer as adhesion promoter. Bioresource Biotechnol. 97: 494–499.

Koli, V.B., G.D. Ananta, V.R. Abhinav, D.T. Nanasaheb, H.P. Shivaji and D.D. Sagar. 2016. Visible light photo-induced antibacterial activity of TiO$_2$-MWCNTs nanocomposites with varying the contents of MWCNTs. J. Photochem. Photobiol. A: Chem. 328: 50–58.

Lam, S.M., J.C. Sin and A.R. Mohamed. 2016. A review on photocatalytic application of g-C$_3$N$_4$/semiconductor (CNS) nanocomposites towards the erasure of dyeing wastewater. Mat. Sc. Semicond. Proc. 47: 62–84.

Leja, K. and G. Lewandowicz. 2010. Polymer biodegradation and biodegradable polymers—a review. Polish. J. Environ. Stud. 19: 255–266.

Lee, J.Y., K. Liang, K.H. An and Y. Lee. 2005a. Nickel oxide/carbon nanotubes nanocomposite for electrochemical capacitance. Synthetic Metals 150–153.

Lee, S., I. Kang, G. Doh, H. Yoon, B. Park and Q. Wu. 2008b. Thermal and mechanical properties of wood flour/talc-filled polylactic acid composites: effect of filler content and coupling treatment. J. Thermoplast. Compos. Mater. 21: 209223.

Lee, J., T. Takekoshi and E.P. Giannelis. 1997c. Fire retardant polyetherimide nanocomposites. Mat. Res. Soc. Symp. Proc. 457: 513–518.

Lehr, I.L. and S.B. Saidman. 2013. Anticorrosive properties of polypyrrole films modified with Zinc onto SAE 4140 steel. Prog. Org. Coat. 76(11): 1586–1593.

Li, C., H. Liu, B. Luo, W. Wen, L. He, M. Liu et al. 2016a. Nanocomposites of poly(l-lactide) and surface-modified chitin whiskers with improved mechanical properties and cytocompatibility. Europ. Polym. J. 81: 266–283.

Li, S., Y. Gong, Y. Yang, C. He, L. Hu, L. Zhu et al. 2015b. Recyclable CNTs/Fe$_3$O$_4$ magnetic nanocomposites as adsorbents to remove bisphenol A from water and their regeneration. Chem. Engin. J. 260: 231–239.

Lim, D.S., J.W. An and H.J. Lee. 2002. Effect of carbon nanotube addition on the tribological behavior of carbon/carbon composites. Wear. 252: 512–517.

Liu, R., Y. Qiao, Y. Xu, X. Ma and Z. Li. 2016a. A facile controlled *in situ* synthesis of monodisperse magnetic carbon nanotubes nanocomposites using water-ethylene glycol mixed solvents. J. All. Comp. 657: 138–143.

Liu, C., L. Wang, H. Xu, S. Wang, S. Gao, X. Ji et al. 2016b. "One pot" green synthesis and the antibacterial activity of g-C$_3$N$_4$/Ag nanocomposites. Mat. Lett. 164: 567–570.

Liŭ, D., Z. Li, W. Wang and G. Wang. 2016c. Hematite doped magnetic TiO$_2$ nanocomposites with improved photocatalytic activity. All. Comp. 654: 491–497.

Liu, Y., Z. Yuhong, D. Lanlan, Z. Weili, S. Mingji, S. Zhengguang et al. 2016d. Polystyrene/graphene oxide nanocomposites synthesized via Pickering polymerization. Prog. Org. Coat. 99: 23–31.

Lozynskyy, O.Z., V. Varchenko, N. Tischenko, A. Ragulya, M. Andrzejczuk and A. Polotai. 2015. Tribological behaviour of Si$_3$N$_4$-based nanocomposites. Trib. Intern. 91: 85–93.

Lujaniene, G., S. Semcuk, A. Lecinskyte, I. Kulakauskaite, K. Mazeika, D. Valiulis et al. 2016. Magnetic graphene oxide based nano-composites for removal of radionuclides and metals from contaminated solutions. J. Environ. Radioact 02.014.

Ma, W., Y. Higaki and A. Takahara. 2016a. Imogolite polymer nanocomposites. Develop. Clay Sc. 7: 628–671, Chap 24.

Ma, C., X. Guo, J. Leang and F. Xia. 2016b. Synthesis and characterization of Ni–P–TiN nanocomposites fabricated by magnetic electrodeposition technology. Ceram. Internat. 42: 10428–10432.

Magesan, P.G. and M.J. Umapathy. 2016. Ultrasonic-assisted synthesis of doped TiO$_2$ nanocomposites: Characterization and evaluation of photocatalytic and antimicrobial activity. Opt. Internat. J. Light Elect. Opt. 127: 5171–5180.

Mahadeva, S.K., S. Yun and J. Kim. 2011. Flexible humidity and temperature sensor based on cellulose-polypyrrole nanocomposite. Sensor. Actuator. A Phys. 165: 194199.

Majdzadeh-Ardakani, K. and S. Sadeghi-Ardakani. 2010. Experimental investigation of mechanical properties of starch/ natural rubber/clay nanocomposites. Digest J. Nanomater. Biostruct. 5: 307–316.

Maiti, P., C.A. Batt and E.P. Giannelis. 2003. Renewable plastics: synthesis and properties of PHB nanocomposites. Polym. Mater. Sci. Eng. 88: 58–59.

Mallakpour, S. and E. Khadem. 2016. Carbon nanotube–metal oxide nanocomposites: Fabrication, properties and applications. Chem. Engin. J. 302: 344–367.

Mallakpour, S., A. Abdolmaleki, Z. khalesi and S. Borandeh. 2016. Surface functionalization of GO, preparation and characterization of PVA/TRIS-GO nanocomposites. Polym. 81: 140–150.

Martin, O. and L. Averous. 2001. Poly(lactic acid): Plasticization and properties of biodegradable multiphase systems. Polym. 42: 6209–6219.

Misra, M., H. Park, A.K. Mohanty and L.T. Drzal. 2004. Injection molded 'Green' nanocomposite materials from renewable resources. Presented at the Global Plastics Environmental Conference, Detroit, MI, USA, 2004.

Mohammed, M.R., J. Ahmed, A.M. Asiri, I.A. Siddiquey and M.A. Hasnat. 2016. Development of 4-methoxyphenol chemical sensor based on NiS_2-CNT nanocomposites. J. Taiw. Inst. Chem. Engin. 64: 157–165.

Mohanty, A.K., M. Misra and G. Hinrichsen. 2000. Biofibres, biodegradable polymers and biocomposites: an overview. Macrmol. Mater Eng. 276(277): 1–24.

Morgan, A.B., R.H. Harris, T. Kashiwagi, L.J. Chyall and J.W. Gilman. 2002. Flammability of polystyrene layered silicate (Clay) nanocomposites: carbonaceous char formation. Fir. Mat. 26: 247–253.

Murugadoss, G., J. Ramasamy, T. Rangasamy and R.K. Manavalan. 2016. PbO/CdO/ZnO and PbS/CdS/ZnS nanocomposites: Studies on optical, electrochemical and thermal properties. J. Luminesc. 170: 78–89.

Muhammad, N., C. Bergmann, J. Geshev, R. Quijada and G.B. Galland. 2016. An efficient approach to the preparation of polyethylene magnetic nanocomposites. Polym. 97: 131–137.

Musil, J. 2000. Surface Coatings Technology 125, 322.

Nasrollahzadeh, M., S.M. Sajadi and H. Arezo. 2016. Waste chicken eggshell as a natural valuable resource and environmentally benign support for biosynthesis of catalytically active Cu/eggshell, Fe_3O_4/eggshell and Cu/Fe_3O_4/eggshell nanocomposites. Appl. Cat. B: Environ. 191: 209–227.

Nivethaa, E.A.K., S. Dhanavel, A. Rebekah, V. Narayanan and A. Stephen. 2016. A comparative study of 5-Fluorouracil release from chitosan/silver and chitosan/silver/MWCNT nanocomposites and their cytotoxicity towards MCF-7. Mat. Sc. Engin.: C 66: 244–250.

Njuguna, J., K. Pielichowski and S. Desai. 2008. Nanofiller reinforced polymer nanocomposites. Polym. Adv. Technol. 19: 947–959.

Nguyen, N.T., D.L. Tran, D.C. Nguyen, T.L. Nguyen, T.C. Ba, B.H. Nguyen et al. 2015. Facile synthesis of multifunctional Ag/Fe_3O_4-CS nanocomposites for antibacterial and hyperthermic applications. Curr. Appl. Phys. 15: 1482–1487.

Nascimento, G.M. and A.N. Pradie. 2016. Deprotonation, Raman dispersion and thermal behavior of polyaniline–montmorillonite nanocomposites. Synth. Met. 217: 109–116.

Ochi, S. 2008. Mechanical properties of Kenaf fibers and Kenaf/PLA composites. Mech. Mater. 40: 446–452.

Ogata, N., G. Jimenez, H. Kawai and T. Ogihara. 1997. Structure and thermal/mechanical properties of poly(L-lactide)-clay blend. J. Polym. Sci. Part B: Polym. Phys. 35: 389–96.

Pandey, J.K., K.R. Reddy, A.P. Kumar and R.P. Singh. 2005a. An overview on the degradability of polymer nanocomposites. Poly. Degrad. Stab. 88(2): 234–250.

Pandey, J.K. and R.P. Singh. 2005. Green nanocomposites from renewable resources: Effect of plasticizer on the structure and material properties of clay-filled starch. Starch/Starke 57: 8–15.

Pandey, J.K., W.S. Chu, C.S. Lee and S.H. Ahn. 2007b. Preparation characterization and performance evaluation of nanocomposites from natural fiber reinforced biodegradable polymer matrix for automotive applications. Presented at the International Symposium on Polymers and the Environment: Emerging Technology and Science, Bioenvironmental Polymer Society (BEPS), Vancouver, WA, USA.

Pant, B., S.S. Prem, P. Mira, J.P. Soo and Y.K. Hak. 2016. General one-pot strategy to prepare Ag-TiO_2 decorated reduced graphene oxide nanocomposites for chemical and biological disinfectant. J. All. Comp. 671: 51–59.

Pan, T. 2013. Intrinsically conducting polymer-based heavy-duty and environmentally friendly coating system for corrosion protection of structural steels. Spectros. Lett. 46(4): 268–276.

Panthapulakka, S., A. Zereshkian and M. Sain. 2006. Preparation and characterization of wheat straw for reinforcing application in injection molded thermoplastic composites. Biores. Biotechnol. 97: 265–272.

Pauly, C., S.A.C. Genix, J.G. Alauzun, J. Jestin, M. Sztucki, P.H. Mutin et al. 2016. Structure of alumina-silica nanoparticles grafted with alkylphosphonic acids in poly(ethylacrylate) nanocomposites. Polym. 97: 138–146.

Peigney, A., C.H. Laurent, E. Flahaut and A. Rousset. 2000. Carbon nanotubes in novel ceramic matrix nanocomposites. Ceramic International 26(6): 677–683.

Peng, F., W. Qiang, Z. Ming and S. Bitao. 2016a. Preparation and adsorption properties of enhanced magnetic zinc ferrite-reduced graphene oxide nanocomposites via a facile one-pot solvothermal method. J. All. Comp. 685: 411–417.

Peng, G., L. Meiying, D. Jianwen, D. Fengjie, W. Ke, X. Dazhuang et al. 2016b. Improving the drug delivery characteristics of graphene oxide based polymer nanocomposites through the "one-pot" synthetic approach of single-electron-transfer living radical polymerization. Appl. Surf. Sc. 378: 22–29.

Pothan, L.A. and S. Thomas. 2003. Polarity parameters and dynamic mechanical behavior of chemically modified banana fiber reinforced polyester composites. Compos. Sci. Technol. 63: 1231–1240.

Pradhan, G.C., S. Dash and S.K. Swain. 2015. Barrier properties of nano silicon carbide designed chitosan nanocomposites. Carb. Poly. 134: 60–65.

Presting, H. and U. Konig. 2003. Future nanotechnology developments for automotive applications. Mat. Sc. Engin.: C. 23(6-8): 737–741.

Qutub, N., B.M. Pirzada, K. Umar, O. Mehraj, M. Muneer and S. Sabir. 2015. Synthesis, characterization and visible-light driven photocatalysis by differently structured CdS/ZnS sandwich and core–shell nanocomposites. Phys. E: Low-dimen. Syst. Nanostr. 74: 74–86.

Qu, P., Y. Gao, G. Wu and L. Zhang. 2010. Nanocomposites of poly(lactic acid) reinforced with cellulose nanofibrils. Bio Resources 5: 1811–1823.

Rabieh, S., K. Nassimi and M. Bagheri. 2016. Synthesis of hierarchical ZnO-reduced graphene oxide nanocomposites with enhanced adsorption–photocatalytic performance. Mat. Lett. 162: 28–31.

Raj, V. and G. Prabha. 2015. Synthesis, characterization and *in vitro* drug release of cisplatin loaded Cassava starch acetate–PEG/gelatin nanocomposites. J. Assoc. Arab Univ. Bas. Appl. Sc. In Press, Available online 19 November 2015.

Rahn, H., W. Robert, H. Michael, E. Diana, F. Kirk, D. Silvio et al. 2016. Calibration standard of body tissue with magnetic nanocomposites for MRI and X-ray imaging. J. Magnet. Magnetic Mat. 405: 78–87.

Ray, D. and S. Sunanda. 2016. *In situ* processing of cellulose nanocomposites. Compos. Part A: Appl. Sc. Manufact. 83: 19–37.

Rehim, M.H.A., M.A. El-Samahy, A.A. Badawy and M.E. Mohram. 2016. Photocatalytic activity and antimicrobial properties of paper sheets modified with TiO$_2$/Sodium alginate nanocomposites. Carbohyd. Polym. 148: 194–199.

Riccardis, M.F. and V. Martina. 2014. Hybrid conducting nanocomposites coatings for corrosion protection. pp. 271–317. *In*: M. Aliofkhazraei (ed.). Developments in Corrosion Protection. Rijeka: InTech.

Sabherwal, P., R. Mutreja and C.R. Suri. 2016. Biofunctionalized carbon nanocomposites: New-generation diagnostic tools. TrAC Tr. Anal. Chem. 82: 12–21.

Sana, M.R.W., S.D. Atef and M.K. Sara. 2016. Ceria-containing uncoated and coated hydroxyapatite-based galantamine nanocomposites for formidable treatment of Alzheimer's disease in ovariectomized albino-rat model. Mat. Sc. Engin.: C. 65: 151–163.

Sarazin, P., G. Li, W.J. Orts and B.D. Favis. 2008. Binary and ternary blends of polylactide, polycaprolactone and thermoplastic starch. Polym. 49: 599609.

Sarah, C.M., S.S. Ray, M.S. Onyango and M.N.B. Momba. 2015. Preparation and antibacterial activity of chitosan-based nanocomposites containing bentonite-supported silver and zinc oxide nanoparticles for water disinfection. Appl. Clay Sc. 114: 330–339.

Seyedeh, M.G. and A.S. Seyed. 2016. Evaluation of the antibacterial activity of Ag/Fe$_3$O$_4$ nanocomposites synthesized using starch. Carbohyd. Polym. 144: 454–463.

Shen, J., Y. Yan, Z. Yang, Y. Hong, Z. Zhiguo and Y. Shiping. 2016. Functionalized Au-Fe$_3$O$_4$ nanocomposites as a magnetic and colorimetric bimodal sensor for melamine. Sens. Actuat. B: Chem. 226: 512–517.

Shi, H., F. Liu and E.H. Han. 2011. The corrosion behavior of zinc-rich paints on steel: Influence of simulated salts deposition in an offshore atmosphere at the steel/paint interface. Surf. Coat. Techn. 205(19): 4532–4539.

Shukla, V.K., R.P. Singh and A.C. Pandey. 2010. Black pepper assisted biomimetic synthesis of silver nanoparticles. J. Alloy Comp. 507(1): L13–L16.

Silvana, V.A. and C.I. Vallo. 2016 Facile preparation of silver-based nanocomposites via thiol-methacrylate 'click' photopolymerization. Europ. Polym. J. 79: 163–175.

Singha, A.S. and V.K. Thakur. 2008. Mechanical properties of natural fiber reinforced polymer composites. Bull. Mater. Sci. 31: 791–799.

Singh, R.P., B.K. Oh, K.K. Koo, J.Y. Jyoung, S. Jeong and J.W. Choi. 2008. Biosensor arrays for environmental pollutants detection. Bioch. J. 2: 4, 223–234.

Singh, R.P., Y.J. Kim, B.K. Oh and J.W. Choi. 2009a. Glutathione-s-transferase based electrochemical biosensor for the detection of captan. Electrochem. Commun. 11: 181–185.

Singh, R.P. and J.W. Choi. 2009. Biosensors development based on potential target of conducting polymers. Sen. Transd. J. 104(5): 1–18.

Singh, R.P., D.Y. Kang, B.K. Oh and J.W. Choi. 2009b. Polyaniline based catalase biosensor for the detection of hydrogen peroxide and Azide. Biotech. Biopr. Engin. 14(4): 443–449.

Singh, R.P., D.Y. Kang and J.W. Choi. 2010a. Electrochemical DNA biosensor for the detection of sanguinarine in adulterated mustard oil. Adv. Mat. Lett. 1(1): 48–54.

Singh, R.P., B.K. Oh and J.W. Choi. 2010b. Application of peptide nucleic acid towards development of nanobiosensor arrays. Bioelectrochem. 79(2): 153–161.

Singh, R.P. and J.W. Choi. 2010. Bio-nanomaterials for versatile bio-molecules detection technology. Letter to Editors. Adv. Mat. Lett. 1(1): 83–84.

Singh, R.P. 2011. Prospects of nanobiomaterials for biosensing. Internat. J. Electrochem. Publisher SAGE-Hindawi journal collection. 2011, Vol. 2011, Review article ID 125487, 30 pages, doi:10.4061/2011/125487.

Singh, R.P., A. Tiwari and A.C. Pandey. 2011a. Silver/polyaniline nanocomposite for the electrocatalytic hydrazine oxidation. J. Inorg. Organomet. Polym. Mat. 21: 788–792.

Singh, R.P. and A.C. Pandey. 2011. Silver nano-sieve using 1, 2-benzenedicarboxylic acid as a sensor for detecting hydrogen peroxide. Anal. Meth., RSC. 3: 586–592.

Singh, R.P., V.K. Shukla, R.S. Yadav, P.K. Sharma, P.K. Singh and A.C. Pandey. 2011b. Biological approach of zinc oxide nanoparticles formation and its characterization. Adv. Mat. Lett. 2(4): 313–317.

Singh, R.P., D.Y. Kang and J.W. Choi. 2011c. Nanofabrication of bio-self assembled monolayer and its electrochemical property for toxicant detection. J. Nanosc. Nanotech. 11: 408–412.

Singh, R.P., J.W. Choi, A. Tiwari and A.C. Pandey. 2012a. Utility and potential application of nanomaterials in medicine. *In*: A. Tiwari, M. Ramalingam, H. Kobayashi and A.P.F. Turner (eds.). Biomedical Materials and Diagnostic Devices. John Wiley & Sons, Inc., Hoboken, NJ, USA. doi:10.1002/9781118523025.ch7.

Singh, R.P., K. Kumar, R. Rai, A. Tiwari, J.W. Choi and A.C. Pandey. 2012b. Synthesis, characterization of Metal oxide based nanomaterials and its application in Biosensing, to the upcoming book Synthesis, characterization and application of Smart material. Nova Science Publishers, Inc USA. 2012. Chapter 11, pp. 225–238.

Singh, R.P., J.W. Choi, A. Tiwari and A.C. Pandey. 2012c. Biomimetic materials toward application of nanobiodevices. pp. 741–782, *In*: A. Tiwari, A.K. Mishra, H. Kobayashi and A.P. Turner (eds.). Intelligent Nanomaterials: Processes, Properties, and Applications. John Wiley & Sons, Inc., Hoboken, NJ, USA. Chapter 20. Published Online: 21 Feb 2012.

Singh, R.P., J.W. Choi and A.C. Pandey. 2012d. Smart nanomaterials for biosensors, biochips and molecular bioelectronics. Bentham Science Publisher (USA). pp. 3–41. *In*: S. Li, Y. Ge and H. Li (eds.). Electronic Book 'Smart Nanomaterials for Sensor Application' Editors: Songjun Li, Jiangsu University, China, Yi Ge, Cranfield University, UK, He Li, University of Jinan, China, Chapter 1.

Singh, R.P., J.W. Choi, A. Tiwari and A.C. Pandey. 2014. Functional nanomaterials for multifarious nanomedicine. *In*: A. Tiwari and A.P.F. Turner (eds.). Biosensors Nanotechnology. John Wiley & Sons, Inc., Hoboken, NJ, USA.

Singh, R.P. 2016. Nanobiosensors: Potentiality towards bioanalysis. J. Bioanal. Biomed. 8: e143. doi:10.4172/1948–593X.1000e143.

Sitaram, S.P., J.O. Stoffer and T.J. O'Keefe. 1997. Application of conducting polymers in corrosion protection. J. Coat. Tech. 69(866): 65–69.

Sithique, M.A. and M. Alagar. 2010. Preparation and properties of bio-based nanocomposites from epoxidized soy bean oil and layered silicate. Malaysian Polym. J. 5: 151–161.

Siracusa, V., P. Rocculi, S. Romani and M.D. Rosa. 2008. Biodegradable polymers for food packaging: a review. Trends Food Sci. Technol. 19: 634–643.

Sternitzke, M. 1997. Review: structural ceramic nanocomposites. J. Europ. Ceram. Soc. 17(9): 1061–1082.

Song, Y., T. Qiang, M. Ye, Q. Ma and Z. Fang. 2015a. Metal organic framework derived magnetically separable 3-dimensional hierarchical Ni@C nanocomposites: Synthesis and adsorption properties. Appl. Surf. Sc. 359: 834–840.

Song, Y., C. Wei, J. He, X. Li, X. Lu and L. Wang. 2015b. Porous Co nanobeads/rGO nanocomposites derived from rGO/Co-metal organic frameworks for glucose sensing. Sens. Actuat. B: Chem. 220: 1056–1063.

Swearingen, C., S. Macha and A. Fitch. 2003. Leashed ferrocenes at clay surfaces: potential applications for environmental catalysis. J. Mole. Catal. A - Chemical 199(1-2): 149–160.

Takegawa, A., M. Murakami, Y. Kaneko and J. Kadokawa. 2010. Preparation of chitin/cellulose composite gels and films with ionic liquids. Carbohyd. Polym. 79: 85–90.

Tang, S.Z.P., H. Xiong and H. Tang. 2008. Effects of nano-SiO_2 on the performance of starch/polyvinyl alcohol blend films. Carbohydrate Polymers 72: 521–526.

Tate, J.S., A.T. Akinola and D. Kabakov. 2010. Bio-based nanocomposites: an alternative to traditional composites. J. Technol. Stud. 1: 25–32.

Teixeira, E., D. Pasquini, A.S. Antonio, C.E. Corradini, M.N. Belgacem and A. Dufresne. 2009. Cassava baggasse cellulose nanofibrils reinforced thermoplastic cassava starch. Carbohydrate Polymers 78: 422–431.

Tiwari, A., D. Terada, H. Kobayashi, R.P. Singh and R. Rai. 2012a. Bionanomaterials for emerging biosensors technology. pp. 137–154. *In*: Radheshyam Rai (ed.). Synthesis, Characterization and Application of Smart Materials. Nova Publishers, Hauppauge, New York, USA. Chapter 7.

Tiwari, A., R.P. Singh and R. Rai. 2012b. Vinyls modified guar gum biodegradable plastics. pp. 125–136. *In*: Radheshyam Rai (ed.). Synthesis, Characterization and Application of Smart Materials. Nova Publishers, Hauppauge, New York, USA. Chapter 6.

Tiwari, A., A. Tiwari and R.P. Singh. 2012c. Bionanocomposite matrices in electrochemical biosensors. *In*: A. Tiwari, M. Ramalingam, H. Kobayashi and A.P.F. Turner (eds.). Biomedical Materials and Diagnostic Devices. John Wiley & Sons, Inc., Hoboken, NJ, USA. doi:10.1002/9781118523025.ch10.

Tomachuk, C.R., A.R. Sarli and C.L. Elsner. 2010. Anti-corrosion performance of Cr+6-free passivating layers applied on electrogalvanized. Mat. Sc. Appl. 1(4): 202–209.

Tunc, S. and O. Duman. 2011. Preparation of active antimicrobial methyl cellulose/carvacrol/montmorillonite nanocomposite films and investigation of carvacrol release. Food Sci. Technol. 44: 465–472.

Uesaka, T., K. Nakane, S. Maeda, T. Ogihara and N. Ogata. 2000. Structure and physical properties of poly(butylene succinate)/cellulose acetate blends. Polymer 41: 8449–54.

Ullah, K., A. Ullah, A. Aldalbahi, J. Chung and W.C. Oh. 2015. Enhanced visible light photocatalytic activity and hydrogen evolution through novel heterostructure AgI–FG–TiO_2 nanocomposites. J. Mol. Cat. A: Chemical 410: 242–252.

Valença, D.P., K.G.B. Alves, C.P. Melo and N. Bouchonneau. 2015. Study of the efficiency of polypyrrole/ZnO nanocomposites as additives in anticorrosion coatings. Mat. Res. 18: 2–10.

Voevodin, A.A. and J.S. Zabinski. 2005a. Nanocomposite and nanostructured tribological materials for space applications. Compos. Sc. Technol. 65: 741–748.

Voevodin, A.A. and J.S. Zabinski. 2000b. Thin Solid Films 37: 223.

Wambua, P., J. Ivens and I. Verpoest. 2003b. Natural fibers: Can they replace glass in fiber reinforced plastics? Compos. Sci. Technol. 63: 1259–1264.

Wang, D., L. Guixia, D. Xiangting and W. Jinxian. 2016d. Magnetic-optical-thermal properties assembled into MWCNTs/ NaGdF$_4$:Yb^{3+}, Er^{3+} multifunctional nanocomposites. Coll. Surf. A: Physicochem. Engin. Aspects 490: 283–290.

Wang, G., Y. Ma, X. Dong, Y. Tong, L. Zhang, J. Mu et al. 2015e. Facile synthesis and magnetorheological properties of superparamagnetic CoFe$_2$O$_4$/GO nanocomposites. Appl. Surf. Sc. 357: 2131–2135.

Wang, G., Y. Ma, L. Zhang, J. Mu, Z. Zhang, X. Zhang et al. 2016a. Facile synthesis of manganese ferrite/graphene oxide nanocomposites for controlled targeted drug delivery. J. Magnet. Magnet. Mat. 401: 647–650.

Wang, G., Y. Hua, X. Su, S. Komarneni, S. Ma and Y. Wang. 2016b. Cr(VI) adsorption by montmorillonite nanocomposites. Appl. Clay Sc. 124-125: 111–118.

Wang, G., M. Yingying, W. Zhiyong and Q. Min. 2016c. Development of multifunctional cobalt ferrite/graphene oxide nanocomposites for magnetic resonance imaging and controlled drug delivery. Chem. Engin. J. 289: 150–160.

Wang, Z.L. 2004. Nanostructures of zinc oxide. Mat. Tod. 7(6): 26–33.

Wang, W. and A. Wang. 2016. Recent progress in dispersion of palygorskite crystal bundles for nanocomposites. Appl. Clay Sc. 119(Part 1): 18–30.

Wang, Y.H., F.L. Li, Y.Q. Wang, S. Wu, X.X. He and K.M. Wang. 2015f. A TiO$_2$/CNTs nanocomposites enhanced luminol electrochemiluminescence assay for glucose detection. Chin. J. Anal. Chem. 43: 1682–1687.

Wei, X.X., H. Cui, S. Guo, L. Zhao and W. Li. 2013. Hybrid BiOBr-TiO$_2$ nanocomposites with high visible light photocatalytic activity for water treatment. J. Hazard. Mat. 263: 650–658.

Willett, J.L. and R.L. Shogren. 2002. Processing and properties of extruded starch/polymer foams. Polym. 43: 5935–5947.

Wu, Q., Y. Xu, Z. Yao and A. Liu. 2010a. Supercapacitors based on flexible graphene/polyaniline nanofiber composite films. ACS Nano. 4(4): 1963–1970.

Wu, W.C.P.S., H. Ming, I. Lartaud, P. Maincent, X.M. Hu, A.S.Minet et al. 2016b. Polymer nanocomposites enhance S-nitrosoglutathione intestinal absorption and promote the formation of releasable nitric oxide stores in rat aorta. Nanomed. Nanotech. Biol. Med. 12(7): 1795–1803.

Wu, S., J. Zhang, B. Raj, Ladani, G.K. Ghorbani, A.P. Mouritz et al. 2016c. A novel route for tethering graphene with iron oxide and its magnetic field alignment in polymer nanocomposites. Polym. 97: 273–284.

Wypych, F. 2004. *In*: Wypych, F. and K.G. Satyanarayana (eds.). Clay Surfaces: Fundamentals, Applications. Amsterdam: Academic Press.

Xia, D. and I.M.C. Lo. 2016. Synthesis of magnetically separable Bi$_2$O$_4$/Fe$_3$O$_4$ hybrid nanocomposites with enhanced photocatalytic removal of ibuprofen under visible light irradiation. Wat. Res. 100: 393–404.

Xiong, H.T.S., H. Tang and P. Zou. 2008. The structure and properties of a starch-based biodegradable film. Carbohydrate Polymers pp. 263–268.

Xianqi, F., G. Zhang, S. Zhuo, H. Jiang, J. Shi, F. Li et al. 2016. Dual responsive shape memory polymer/clay nanocomposites. Composit. Sc. Technol. 129: 53–60.

Xie, P., S. Xue, J. Wei, J. Han, W. Zhou and R. Zou. 2016. Morphology-controlled synthesis of grass-like GO-CdSe nanocomposites with excellent optical properties and field emission properties. J. Solid State Chem. 234: 63–71.

Xu, C., L. Rui, C. Lvjun and T. Jialu. 2016a. Enhanced dechlorination of 2,4-dichlorophenol by recoverable Ni/Fe-Fe$_3$O$_4$ nanocomposites. J. Environ. Sc. 48: 92–101.

Xu, J.L., H.Y. Wang, C. Zhang and Y. Zhang. 2016b. Preparation of bi-component ZnO/ZnCo$_2$O$_4$ nanocomposites with improved electrochemical performance as anode materials for lithium-ion batteries. Electrochim. Acta 191: 417–425.

Yang, L., D. Chu, L. Wang, X. Wu and J. Luo. 2016a. Synthesis and photocatalytic activity of chrysanthemum-like Cu$_2$O/ Carbon Nanotubes nanocomposites. Ceram. Internat. 42: Issue 2, Part A, 2502–2509.

Yang, C., G.Y. Deng, P. Deng, K.L. Xin, W. Xia, A.S.W. Bligh et al. 2016b. Electrospun pH-sensitive core-shell polymer nanocomposites fabricated using a tri-axial process. Acta Biomaterialia 35: 77–86.

Yadav, S.K. and P. Jeevanandam. 2015. Synthesis of Ag$_2$S-TiO$_2$ nanocomposites and their catalytic activity towards rhodamine B photodegradation. J. All. Comp. 649: 483–490.

Yadav, R.S., R.P. Singh, P. Verma, A. Tiwari and A.C. Pandey. 2012. Smart nanomaterials for space and energy applications. pp. 213–250. *In*: A. Tiwari, A.K. Mishra, H. Kobayashi and A.P. Turner (eds.). Intelligent Nanomaterials: Processes, Properties, and Applications. John Wiley & Sons, Inc., Hoboken, NJ, USA. Chapter 6.

Yang, D. 2011. Pulsed laser deposition of manganese oxide thin films for supercapacitor applications. Journal of Power Sources 196–8843.

Zabihzadeh, S.M. 2010. Water uptake and flexural properties of natural Filler/HDPE composites. BioResour. 5: 316–323.

Zadegan, S., M. Hosainalipour, H.R. Rezaie, H. Ghassai and M.A. Shokrgozar. 2011. Synthesis and biocompatibility evaluation of cellulose/hydroxyapatite nanocomposite scaffold in 1-n-allyl-3-methylimidazolium chloride. Mater. Sci. Eng. 31: 954–961.

Zango, Z.U., N.G. Zaharaddeen, N.H.H.A. Bakar, W.L. Tan and M.A. Bakar. 2016. Adsorption studies of Cu^{2+}–Hal nanocomposites for the removal of 2,4, 6-trichlorophenol. Appl. Clay Sc. In Press, Available online 31 May 2016.

Zhang, Y., A. Tang, H. Yang and J. Ouyang. 2016a. Applications and interfaces of halloysite nanocomposites. Appl. Clay Sc. 119: 8–17.

Zhang, X.Z., Y. Zhou, W. Zhang, Y. Zhang and N. Gu. 2016b. Polystyrene@Au@prussian blue nanocomposites with enzyme-like activity and their application in glucose detection. Coll. Surf. A: Physicochem. Engin. Asp. 490: 291–299.

Zhang, C., C. Lai, G. Zeng, D. Huang, C. Yang, Y. Wang et al. 2016c. Efficacy of carbonaceous nanocomposites for sorbing ionizable antibiotic sulfamethazine from aqueous solution. Wat. Res. 95: 103–112.

Zheng, J.P., P. Li, Y.L. Ma and K.D. Yao. 2002. Gelatine/montmorillonite hybrid nanocomposite. I. Preparation and properties. J. Appl. Polym. Sci. 86: 1189–1194.

Zimmermann, K.A., J.M. LeBlanc, K.T. Sheets, R.W. Fox and P. Gatenholm. 2011. Biomimetic design of a bacterial cellulose/hydroxyapatite nanocomposite for bone healing applications. Mater. Sci. Eng. 31: 43–49.

Zhu, J., A.B. Morgan, F.J. Lamelas and C.A. Wilkie. 2001. Fire properties of polystyrene-clay nanocomposites. Chem. Mat. 13: 3774–3780.

3

Conjugated Polymer Nanocomposites

Pradip Kar

Introduction

Polymers are macromolecules, which are formed by the repetitive union (mer unit or repeating unit) of a large number of a reactive small molecule in a regular sequence. The simplest example is polyethylene, in which ethylene moiety is the 'mer unit'. The polymers are the most widely used versatile materials in the globe due to having some serious advantages over other materials, viz. flexibility, tailorability, processability, environmental stability, low cost, light weight (Gowariker et al. 2005), etc. Even till the 5th decade of the last century, the polymers with structures of saturated long chain framework were used as insulating cover on the electrical wire, insulating gloves, electrical switches, insulating coating on electronic circuit board, etc. (Gowariker et al. 2005). So called insulating polymers having the molecular orbital band gap, i.e., the energy required to excite electrons from the highest occupied molecular orbital (HOMO) to lowest unoccupied molecular orbital (LUMO) (Potember et al. 1987) greater than 10 eV generally show a surface resistivity higher than 10^{12} ohm-cm. The versatility of the polymer materials is extended, as electronic conducting behavior has been included into the characteristics of some of the polymers. The most exciting development in this area is related to the discovery of intrinsically conductive polymers (ICP) and the interest in this field is increasing day by day after the Noble Prize was awarded, in the year 2000, to three scientists, Prof. Alan J. Heeger, Prof. Alan G. MacDiarmid and Prof. Hideki Shirakawa (Pickup 1990). Polymers can be made intrinsically conducting by introducing π-electron conjugation in their chain structure, i.e., alternative double bond and single bond structure. However, in neutral (undoped) state, these polymers having band gap within the range 4–8 eV behave like weak conductors or semiconductors. Those polymers are converted to electronic conductors or semiconductors by doping through injection of electrons or holes into the super orbital to reduce the band gap up to 2–6 eV (Bidan 1992). The conjugated polymers, having a wide range of conductivity (10^{-5}–10^3 S/cm), interesting optical and mechanical properties, serious advantages like their architecture flexibility, tailorability, versatility, light weight, environmental stability, etc., cover a broad spectrum of applications from the solid-state technology to biotechnology as a substitute of inorganic electronic, optoelectronic, semiconducting materials. Up-until today, the conjugated polymers are not highly successful in the marketplace in place of inorganic materials as their applications in various fields have not been widely recognized, especially by trade and business. This is because the conducting polymers are suffering from fundamental drawbacks like problems in synthesis, problems in reproducibility, problems in doping,

Department of Chemistry, Birla Institute of Technology, Mesra, Ranchi, Jharkhand, India, 835215.
    Email: pradipkgp@gmail.com; pkar@bitmesra.ac.in

problems in processability, problems in mechanical and chemical stability, etc. In order to find out the reliable solutions for the above problems, the following approaches have been adopted: structure modification and surface modification or composition modification. Due to structural rigidity, most of the conjugated polymers are insoluble and infusible within the degradation temperature and therefore the processing of the synthesized conducting polymers is a real challenge today. In view of that, the stuctural modifications of the conjugated polymers have been achieved by two ways: by making polymers from substituted monomers and through copolymerization. Many researchers have tried to solve this problem by chemical group substitution, especially long carbon chain substitution in the monomer. In addition, co-polymerization between un-substituted and/or substituted monomers is also one of the most effective methods to get the desired properties through structural modification. However, the conducting polymers with substituted side chain often suffer from the problems like less product yield, low molecular weight of synthesized polymers, poor mechanical strength, poor film forming properties, poor conductivity, etc. due to introduction of steric hindrance through substitution. Moreover, the structure modification by co-polymerization through chemical or electrochemical technique is not an easy task as the reactivity and polymerization conditions should not be very similar for those substituted monomers in comparison with the parent homopolymer.

The other one, which is the easiest route for the surface modification or composition modification without too much influencing the intrinsic properties of the polymers, is the composite formation. Composites are made from two or more constituent materials with significantly different physical or chemical properties as compared to their individual components which remain separate and distinct at the macroscopic or microscopic scale. In addition, it is well established that the properties of materials change as their size approaches towards the nanoscale and as the percentage of atoms at the surface of the material becomes significant. Therefore, recent development in this field has been directed towards the polymer based composites with various nanomaterials. This is because the composites of organic conjugated polymers with the nanomaterials of inorganic metals or metal compounds have shown synergistic and hybrid properties derived from the wide range of components. The conjugated polymer composites combine the properties of two or more different materials with the possibility of introduction of novel mechanical, electronic or chemical behavior. Such nanocomposites not only bridge the world of nanomaterials with that of the macromolecules but also solve some of the fundamental problems associated with conjugated polymers. The resulting conjugated polymer based nanocomposites have found successful applications with better performance than the pristine conjugated polymers in versatile fields viz. battery cathodes, microelectronics, nonlinear optics, sensors, etc. However, the novel properties of nanocomposites can be derived from the successful selection and combination of the characteristics of parent constituents into a single material. Exploring the present state-of-art of preparation, properties of conjugated polymer nanocomposites to be used in various applications with better performance than the pristine conjugated polymers is the primary aim. The concepts to be covered in this chapter are schematically shown in the Figure 1.

Conjugated polymers

Electronic structure

Polymers can be made intrinsically conducting by introducing π-electron conjugation in their chain structure and all the carbon atoms or heteroatoms centered throughout the conjugated chain are sp^2 hybridized. That means in each atom, one p-orbital (unhybridized) remains perpendicular to the plane of the polymer chain and all those p-orbitals remain parallel to each other. Each p-orbital can laterally overlap to form π-orbital with either side of the nearest p-orbital and ultimately the p-orbitals are delocalized throughout the polymer chain. So the energy of delocalized highest occupied molecular orbital (HOMO) is increased while that of the lowest unoccupied molecular orbital (LUMO) is decreased for those types of long chain conjugated polymers. As a result, energy gap between the highest occupied molecular orbital (HOMO) and the lowest unoccupied molecular orbital is decreased. Like the other molecules, the HOMO of the conjugated polymer is generally filled by electrons and LUMO remains vacant. The

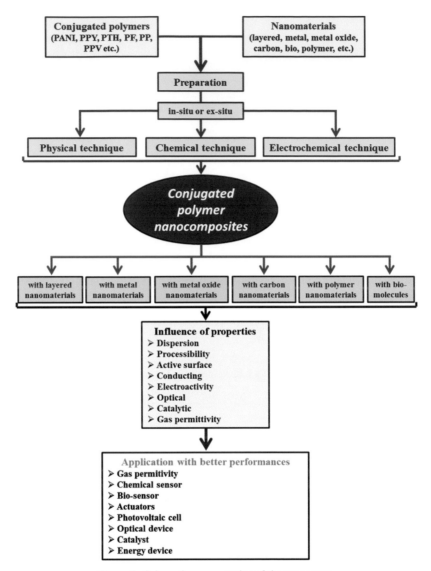

Figure 1. Schematic representation of chapter content.

'band gap energy', i.e., energy required to promote electron from valance band to conduction band is responsible for the electrical conduction. In the neutral stable (undoped) state, the conjugated polymers conduct electricity only in lower semiconducting region due to having higher band gap energy. In general, the conductivity of those undoped conjugated polymers is in the range of 10^{-7}–10^{-11} S cm^{-1}, while for the application of conjugated polymers as the substitutes of inorganic or traditional semiconductors, some higher conductivity range is required. Interestingly, when the conjugated polymers are 'doped', the electrical conductivity of the polymers is increased to several folds by decreasing the band gap energy. The term 'doping' is employed from that in the inorganic or traditional semiconductors. The doped polymers are considered as polymeric organic salts as the counter ions are simultaneously inserted into the conjugated polymer matrix. The oxidizing or reducing agents, which convert polymer to polymer salt, are known as 'doping agents' or 'dopants'. So the doping in conjugated polymer is meant by the addition of donor or acceptor molecule to the polymer. Due to doping in the conjugated polymers, the defects are incorporated into the polymer matrix and the conductivity of the polymers is increased by several folds. The name of the species having such defects in the conjugated polymer matrix is soliton, polaron and bipolaron. The first conjugated inorganic polymer, polythiazyl (SN)$_x$, was discovered in 1975 (Saxena and Malhotra 2003). It was reported to show good conductivity and become super conductive at 0–29 K (Greene et al. 1975). However, the idea of using conjugated polymers as electrically conducting one actually emerged in 1977 with the findings of Shirakawa et al. (1977), according to which, the iodine doped trans-polyacetylene (CH)$_x$, exhibits conductivity of 10^3 S cm^{-1}. Since then, the interests for the development of other organic conjugated polymers possessing conducting property have been initiated. Some of the examples of such other conducting polymers are: polyaniline (PANI), polypyrrole (PPY), polythiophene (PTH), polyfuran (PF), poly(p-phenylene) and polycarbazole. The molecular structures of a few conjugated conducting polymers are shown in Figure 2.

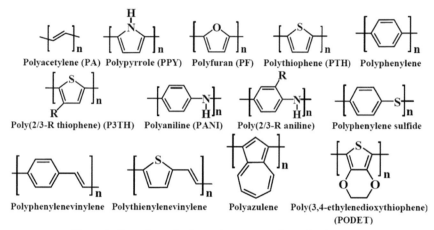

Figure 2. Repeating unit of some important conjugated polymers.

Polymer synthesis

Presently, the conjugated polymers such as polypyrrole, polythiophene, polyfuran, polyisothionapthalene, polyindole, polyaniline, polycarbazole, polyazulene, poly-*p*-phenylene, poly *p*-phenylene vinylene, etc., are used in various device applications. The conjugated polymers are usually synthesized following either of the two popular methods, chemical or electrochemical oxidation of the corresponding monomers like thiophene, furan, carbazole, aniline, indole, azulene, and their derivatives. The solubility, conductivity, materials property, yield of the conjugated polymers greatly depends on the conditions used during the polymerization.

Chemical synthesis

The majority of conjugated polymers may be synthesized by chemical polymerization of corresponding monomer through radical initiated oxidative coupling pathway. In a typical procedure, the monomer is dissolved first in a suitable solvent, followed by addition of radical oxidative initiator for the polymerization. According to the polymerization medium used like aqueous, organic or aqueous/ organic mixture (miscible or immiscible), the chemical polymerization may be classified as (Figure 3): (a) Aqueous chemical polymerization, (b) Dispersion chemical polymerization, (c) Emulsion or Inverse emulsion chemical polymerization and (d) Interfacial chemical polymerization. In aqueous chemical polymerization, acidic aqueous medium is used for the homogenous polymerization of monomer by adding oxidant solution in water. Following this method, the polymers of aniline, pyrroles, thiophenes, indoles, azines, and their derivatives, etc., have been successfully synthesized and a comprehensive picture may be obtained from some comprehensive review reports (MacDiarmid and Epstein 1989, Syed and Dinesan 1991, Martin et al. 1993, Toshima and Hara 1995, Feast et al. 1996, Smith 1998). For aqueous chemical polymerization of monomer, aniline and its derivatives are taken in aqueous acid solution or monomer salts with inorganic acids like hydrochloric, sulfuric acid, etc. are taken in water solution. The water

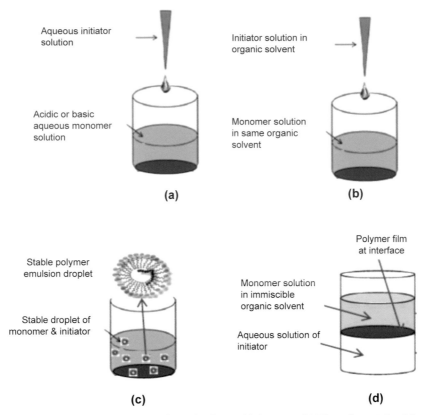

Figure 3. Types of chemical polymerization of conjugated polymer: (a) Aqueous, (b) Dispersion, (c) Emulsion or Inverse emulsion and (d) Interfacial chemical polymerization.

soluble acid may be inorganic acids like hydrochloric acid, sulfuric acid, phosphoric acid, perchloric acid, nitric acid, etc. or organic acids like oxalic, malic, succinic, formic, acetic, p-toluene sulfonic, sullphonic, camphor sulfonic acid, etc. The examples of commonly used oxidative initiators are ammonium persulfate, sodium persulfate, potassium persulfate, ferric chloride, cupric chloride, hydrogen peroxide, potassium dichromate, cerium sulfate, sodium vanadate, potassium ferricyanide, potassium iodate, etc. The chemical polymerization methods, which are suitable for producing bulk quantities of polymers, are very easy and simple method. The monomer, acid and oxidant are dispersed in the organic solvent for the chemical synthesis of conjugated polymer in dispersion chemical polymerization. For example, aniline and acid were dispersed in nonaqueous dimethyl sulfoxide solvent followed by addition of perdisulfate initiator to produce PANI as a stable dispersion (Ghosh et al. 2001). The dispersion polymerization of aniline may be also carried out in miscible solvent mixture like chloroform (Gul et al. 2013) and 2-butanol or dimethyl formamide and water, adding the suitable oxidant solution (Sengupta et al. 2011). Emulsion polymerization process entails polymerization of emulsion particles consisting of monomer and oxidant in a small portion of miscible organic medium with a water-insoluble organic acid. For example, PANI was synthesized by emulsion polymerization in mixture of small amount of 2-butoxyethanol with water using water insoluble dinonylnaphthalenesulfonic acid and ammonium persulfate initiator (Kinlen et al. 1998). In inverse emulsion polymerization, an aqueous solution of the monomer is emulsified in an immiscible nonpolar organic solvent and the polymerization is initiated by adding an oil soluble initiator. The PANI was synthesized using benzoyl peroxide initiator by inverted emulsion process in a mixture of water with non-polar toluene-isooctane solvent in the presence of protonic acids and an emulsifier (sodium lauryl sulphate) (Rao et al. 2002). Typically, the interfacial polymerization is performed in the interface between the immiscible organic solution phase of monomer and the aqueous solution phase of oxidant or/and oxidant with suitable acid dopant. For example, aniline dissolved in the organic phase such as hexane, benzene, toluene, xylene, diethyl ether, carbon disulfide, carbon tetrachloride, chloroform, o-dichlorobenzene, or methylene chloride, was polymerized at the interface of aqueous solution of ammonium persulfate oxidant and dopant acids viz., hydrochloric, sulfuric, nitric, phosphoric, perchloric, acetic, formic, tartaric, camphorsulfonic, methylsulfonic, ethylsulfonic, or 4-toluenesulfonic acid (Huang and Kaner 2004).

Electrochemical synthesis

The electrochemically deposited film of conjugated polymers on the electrode surface is one of the solutions for the problem of solution processability or film casting. If possible, the polymer-deposited electrodes are used directly or the polymers deposited on the electrode surface are peeled off as self-standing films for the particular applications. For example, the conjugated polymers such as long-chain alkyl-substituted PTH (Sato et al. 1986), PPY (MacDiarmid and Epstein 1989, Syed and Dinesan 1991, Martin et al. 1993), PANI (Miasik et al. 1986, Pei and Inganas 1993) were synthesized by electrochemical polymerization method. In general, a monomers solution is oxidized or reduced electrochemically to an activated form that leads to form a polymer film directly on the electrode surface made of materials like platinum, stainless steel, gold, indium tin oxide (ITO), glass, etc. Three electrodes system having a working electrode, control or counter electrode and reference electrode is preferred over the two electrodes having only working electrode and control or counter electrode system as additional electrochemical characterization is possible to study in three electrodes set-up (Kar 2013). As shown in Figure 4 (Balint et al. 2014), typical two and three electrode set-up with different principles, viz., galvanostatic, potentiostatic, and potentiodynamic methods are followed to synthesize conjugated polymers electrochemically (Toshima and Hara 1995, Smith 1998, Feast et al. 1996). In potentiostatic principle, the potential of the electrodes is controlled, while the current varies in galvanostatic polymerization. During potentiodynamic electrochemical polymerization, the potential is swept between a low and high potential limit in cycles. The structure, morphology and properties of the PEDOT were compared for the electrochemical synthesis of the polymer following those three methods (Patra et al. 2008). In order to dope, the oxidation state of the synthesized conjugated polymer can also be varied electrochemically during or after the polymerization by cycling the potential between oxidized, conducting, and the neutral, insulating state, or by using suitable redox compounds. However, the method is a sophisticated one and fine set-up is required for the purpose.

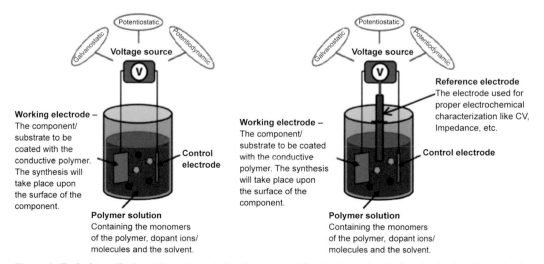

Figure 4. Typical two (first) and three electrode (next) set-up used for the electrochemical synthesis of conjugated polymer (Reproduced with permission from Balint, R., N.J. Cassidy and S.H. Cartmell. 2014. Conductive polymers: Towards a smart biomaterial for tissue engineering. Acta Biomaterialia 10: 2341–2353, Copyright© Elsevier Ltd.).

About nanocomposites

Nanomaterials

Over the past decade, development of nano-structured materials has been the subject of outstanding research interest in chemistry, biochemistry and materials science for wide range of applications in industrial, biomedical, and electronic field. The nanomaterial is the material in which at least one of the dimensions of the material is the order of 1–100 nanometers (nm). Nanomaterials are not simply miniaturization of macro or micro materials, but the nano-world lies midway between the scale of atomic and quantum phenomena, as the transition from macro or micro-materials to nanomaterials yields dramatic changes in physical properties due to having a large surface area for a given volume, i.e., high aspect ratios. This is because many important chemical and physical properties as well as interactions with other materials are governed by surface area and surface properties. In general, these materials are classified into four types, according to their geometries and respective surface area-to-volume ratios, as shown in Figure 5 (Thostenson et al. 2005). The particulate or zero dimensional nanomaterials, which is the first type are the particle or sphere with diameter < 100 nm, having all the three dimension near about zero such as carbon black, silica, metallic nanoparticles or quantum dots, etc. (Schmidt et al. 2002). Second, the fibrous or one dimensional nanomaterials are the fibers, wires or tubes having diameter less than 100 nm and aspect ratio more than 100 in one dimension towards the length. Carbon nanofibers, carbon nanotubes, metallic nano-wires, etc., are examples of fibrous nanomaterials (Schmidt et al. 2002). The two dimensional or layered nanomaterials, the third type, are the layered or plate-like nanofillers with thickness typically in the order of 1 nm and aspect ratio in the other two dimensions is in the range of 30–1000. For example, the layered nanomaterials are nanoclay, organosilicate, MoO_3, etc. (Alexandre and Dubois 2000). In addition, the fourth type is the polyhedral nanomaterials or nano structured nanomaterials or three dimensional nanomaterials having all the tree dimensions in nano meter range and these include bulk materials composed of the individual blocks which are in the nanometer scale (1–100 nm) (Figure 5) (Otero 2009, Thostenson et al. 2005, Luo and Daniel 2003). The compact or consolidated (bulk) polycrystal with nanosized grains is an important type of three-dimensional nano structured materials.

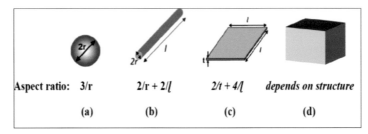

Figure 5. Types of nanomaterials (a) Particulate nanomaterials (0 dimensional), (b) Fibrous nanomaterials (1 dimensional), (c) Layered nanomaterials (2 dimensional) and (d) Polyhedral nanomaterials (nano structured nanomaterials or three dimensional nanomaterials).

Nanocomposites

Composites can be defined as materials consisting of two or more chemically and physically different phases separated by a distinct interface and having bulk properties significantly different from those of any of the constituents. The well-known composites used in our day-to-day life are wood, concrete, ceramics, and so on. The basic difference between blends and composites is that the constituents in the composites are easily recognizable as separated by distinct boundary, while in blends these are not recognizable due to their miscibility. In the composite, one of the material is called the reinforcing phase, is in the form of fibers, sheets, or particles, and is embedded in the other material called the matrix phase. In contrast, the nanocomposites are composite in which some nanomaterials are mixed inside a continuous matrix like polymer. Here, the reinforcing nanomaterials are dispersed within the continuous polymer matrix in the conjugated polymers based nanocomposites. In general, the properties of the nanocomposites are very much close to that of the matrix phase as it contains higher percentage than that of the reinforcing phase. The less percentage of reinforcing phase positively influences one or more properties of the derived nanocomposites. For example, the thermal stability of PANI/Na+-MMT nanocomposites was more improved than that of pristine PANI by inclusion of Na^+-MMT nanomaterials due to reinforcing effect (Lee and Char 2002).

Types of these nanocomposites

The conjugated polymer is a generic name given to a wide variety of polymers with conjugated long chain structure. Those conjugated polymers can have different chemical structure, physical properties, doping ability, electronic properties, thermal behavior, filed of applications, etc. Typically, vast varieties of available nano-materials have been introduced for the development of nanocomposite with conjugated polymers. On the basis of matrix phase, composites can be classified into: (1) metal matrix composites (MMCs), (2) ceramic matrix composites (CMCs) and (3) polymer matrix composites (PMCs). Among the various classifications of composites, the PMCs are the most widely studied as more than 95% of used composite products to date falls into this category. The other reinforcing phase like metals, metal oxides, carbon nanomaterials, etc., are well distributed within the continuous polymer phase in their composites. In those composites, polymers having chief advantages like low cost, easy processability, good chemical resistance and low specific gravity, etc., are generally used as matrix material or continuous phase. However, only PMC is the sole type of nanocomposite possible for conjugated polymer nanocomposite as almost in all the cases, the conjugated polymers are used as matrix phase. Still, the fundamental materials available for PMC fabrication with conjugated polymers should be organic or inorganic according to their chemical nature. In general, the organic materials are made from (or extracted from) plants or animals having mainly the compounds of carbon along with other elements and the inorganic materials are made from rocks and minerals. Therefore, the fundamental classifications of conjugated polymer nanocomposite should be: conjugated polymer/organic nanocomposites and conjugated polymer/inorganic nanocomposites. The most common organic materials for preparation of nanocomposites with conjugated polymers are used as polymers or the bio-molecules. In other words, the classification can be done according to the major types of inorganic nanomaterials used, e.g., layered materials, metal, metal oxides and carbon materials for the preparation of nanocomposites with conjugated polymers. The special properties of conjugated polymers, various reinforcing nanomaterials and influencing properties in nanocomposites for proposed applications are shown in Table 1. In order to discuss the preparations, properties and applications of conjugated polymer based nanocomposites accordingly, the final classifications can be made as:

1. Conjugated polymer nanocomposites with layered nanomaterials
2. Conjugated polymer nanocomposites with metallic nanomaterials
3. Conjugated polymer nanocomposites with metal oxide nanomaterials
4. Conjugated polymer nanocomposites with carbon nanomaterials
5. Conjugated polymer nanocomposites with polymer materials
6. Conjugated polymer nanomaterials with bio-molecules

Other than the nature of the components used, the properties of a nanocomposite are greatly influenced by the size and scale of its component phases, the degree of mixing between the two phases and effective interaction between them. Novel properties of conjugated polymers based nanocomposites with various nanomaterials can be derived from the successful combination of the characteristics of parent constituents into a single material by following any of the suitable methods shown in Figure 6. Significant differences in properties of composite may be obtained even for same and/or similar type of nanomaterials depending on the method of preparation.

Table 1. Properties of conjugated polymers, various reinforcing nanomaterials and their nanocomposites for important applications.

Special properties of conjugated polymer matrix	Reinforcing nanomaterials	Unique properties	Important influencing properties in their nanocomposite	Important applications proposed for the nanocomposite
Easy chemical and electrochemical syntheses, solution processability, mechanical flexibility, architecture tailorability, reversible redox switching and doping, electroactivity and semiconducting properties, electrochromic properties	Clays	Chief, easy handling, colloidal size of particles, moderate layer charge, high degree of layer stacking disorder, easy surface modification, cation and anion exchange capacity	Mechanical properties, dispersion property, barrier property, anticorrosion property, gas permeability, electrochemical property	Anticorrosion protection on metal
	Metal	Electrical properties, optical properties, catalytic properties, magnetic properties	Improved conductivity and enhanced stability, chemical receptor for analyte, catalytic influence	Chemical sensing application and energy device
	Metal oxide	Semiconductors, catalytic effect	Improved conductivity and enhanced stability, chemical receptor for analyte, catalytic influence	Chemical sensing application, energy device, solar cell, actuators
	Carbon	High mechanical strength and modulus, good electrical conductivity and excellent chemical stability, possibility of modification	Improved electrical and mechanical properties, interaction site for analyte	Chemical sensing application, energy device, solar cell, actuators
	Polymers	Mechanical properties, miscibility, solution processability, easy synthesis	Electrical, redox, or optical properties, mechanical properties	Chemical sensing application, energy device
	Bio-molecules	Biologically activity, biological recognition components, enzyme catalyst	High loading of the biologically specific component with minimal, enzymes and the biologically active material leaching, electrochemical mediator	Bio-sensing, bio-medical

Composites with layered nanomaterials

In the past decades, the nanocomposites of layered materials with various synthetic polymers have attracted great research interest (Loo and Gleason 2004, Kim et al. 2005b, Uthirakumar et al. 2005) due to the potential of these materials to enhance the various properties, significantly greater than that of the other conventional fillers (Dean et al. 2005, Mishra et al. 2004). Recently, the nanocomposites of different layered materials and conjugated conducting polymers (Yeh et al. 2002, Anuar et al. 2004, Yeh et al. 2001) have found to exhibit novel electronic, chemical, structural and mechanical properties obtained from successful combination of the two dissimilar chemical components at the molecular level (Lee et al. 2003, Chang et al. 2006). Among such types of layered solids like graphite, clay minerals, transition metal dichalcogenides, metal phosphates, phosphonates and layered double hydroxides, etc., the clay nanomaterials have been used widely for their unique structure and properties. The clay nanomaterials are classified according to the magnitude of net layer charge per formula unit as 1:1, or 2:1, and according to

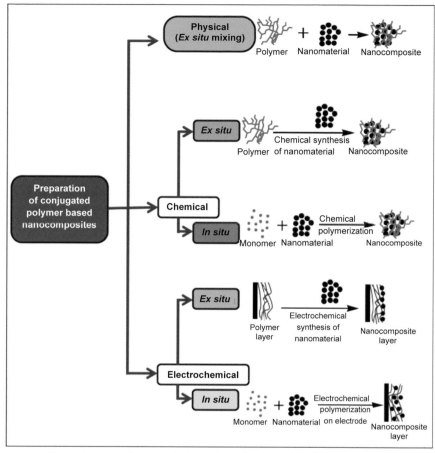

Figure 6. Preparation methods for conjugated polymers based nanocomposites.

the type of interlayer species as (1) kaolin-serpentine, (2) pyrophyllite-talc, (3) smectite, (4) vermiculite, (5) mica, (6) chlorite, and (7) interstratified clay minerals (Martin et al. 1991). Another classification can be made according to their occurrence: natural and synthesized. To date, the most widely used nanomaterials are the type of natural clays, e.g., montmorillonite, hectorite and saponite, etc. (Zeng et al. 2005). Therefore, the nanocomposites of conjugated polymers with the nanoclay materials are only considered here for a brief discussion. In order to achieve the enhanced properties of nanocomposites, the approaches are directed towards the incorporation of polymer matrix into the interlayer space of layered inorganic solids with considerable interfacial interactions. These types of interactions in nanocomposites of conjugated polymers with nanoclays have been demonstrated as the weak van der Waals interactions, which hold the layer together having conjugated polymers in-between. Three different types of nanocomposites of layered silicates with conjugated polymers are generally formed (Figure 7) in such immiscible or partially miscible systems, namely (1) intercalated nanocomposites, (2) flocculated nanocomposites and (3) exfoliated nanocomposites (Ray and Okamoto 2003). Normally, insertion of conjugated polymer matrix into the layered silicate structure in a crystallographically regular fashion is defined as intercalated nanocomposites. The layer structure is somewhat broken in flocculated structure and almost homogenous dispersion is obtained in the conjugated polymer matrix by complete breaking of layer structure in exfoliated nanocomposites. The exfoliated morphology is to be considered as the best morphology through successful combination of two components in the best dispersion state in terms of achieved properties. However, the phase separated mixed nano-morphology with little intercalated or exfoliating statures is generally obtained for the composite of nanoclay with conjugated polymer. This is because the intrinsic incompatibility of hydrophilic nanoclays with that of the hydrophobic conjugated

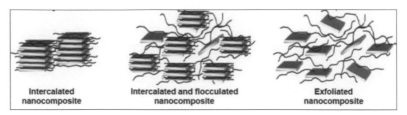

Figure 7. Three types of morphological structure of layered nanomaterials composites with conjugated polymers.

polymers matrices hinders the exfoliation and preparation of well dispersed stable nanocomposite with improved properties. Moreover, the face to face stacking in agglomerated tactoids structure of pristine nanoclay makes them very difficult to disperse into a conjugated polymer matrix. The surface modification of nanoclay layers by chemical methods increases the hydrophobicity by reducing surface energy of clay layers to make it organophilic in nature by matching their surface polarity with polymer polarity. Therefore, the syntheses, surface modifications of nanolayer and their successful combination with different conjugated polymers by right preparation for various applications according to their achieved properties have been discussed briefly as follows.

Synthesis of nanomaterials

As described in the previous section, the classification of nanoclays can be made according to their occurrence: natural and synthesized. In general, the clay minerals are a class of phyllosilicates and the naturally occurring or natural nanoclay are usually formed from the various silicate minerals at the surface of the earth as a result of environmental process. Three typical environmental processes under which the clay nanomaterials are obtained are chemical weathering, hydrothermal alterations, and sedimentary rock. Several applications of layered clay minerals have been invented by the human beings since prehistoric civilization for their widespread distribution and a great diversity in nature (Vaccari 1998, Carretero and Lagaly 2007). Those easily available and stable naturally occurring nanomaterials are directly used to fabricate great numbers of nanocomposites by incorporating the various species (conjugated polymers should be considered here) into the interlayer space for various applications (Yariv and Cross 2002, Zhou et al. 2004, Tong et al. 2009).

The variations in naturally occurring minerals, which suffer from several drawbacks like the existence of impurities, random mineralogical composition in natural deposits, available insufficient quantities (Bergaya and Lagaly 2006, Zhang et al. 2010), may cause the limiting of their potential use. Here is the need to obtain the material in pure phase and homogeneous state through controllable synthesis, which also allows the controlling of isomorphous substitutions in the synthetic solids. Therefore, the synthetic clay minerals with well-designed structure and composition provide advanced functional materials for new applications (Zhang et al. 2009, Zhou 2010). Inspiring by the natural process, the laboratory synthesis can be performed in two ways: (i) Synthesis at low temperature close to early earth surroundings, and (ii) Transformations from natural minerals (Zhang et al. 2010). The clay nanomaterials are synthesized by the most popular, low temperature, hydrothermal process from primary naturally occurring minerals such as feldspars, albite and muscovite, etc. (Parham 1969, Murray 1988). The synthesis conditions may influence the morphology of the nanomaterials, such as platy, lath-like, rod-like, and spherical due to the composition and structure of the starting gels (Miyawaki et al. 1993). In general, the formation of pyrophyllite through dehydroxylation has been promoted by increasing the temperature in hydrothermal process (Zhang et al. 2010). The other individual various chemical methods or little modified methods have been also adopted for the successful syntheses of different types of nanoclays (Zhang et al. 2010).

Modification of nanomaterials

The hydrophobic or organophilic property of the clay minerals has been introduced by surface modification to achieve the improved property of the derived nanocomposites by making it compatible with the polymer matrix. The well-known routes employed for the surface modifications of clays are adsorption, ion exchange, binding with inorganic and organic anions, grafting, reaction with acids, pillaring by different types of poly(hydroxo metal) cations, intra-particle and inter-particle polymerization, dehydroxylation and calcination, delamination and reaggregation of smectites, lyophilisation, ultrasound process, and plasma process (Bergaya and Lagaly 2001, dePaiva et al. 2008). In the nanoclays, the layers are stacked with a regular structure through Van Deer Waals forces between excess positive charges of alkali or alkaline earth cations. These cations are usually replaced by the isomorphic substitution, e.g., Al^{3+}, Mg^{2+}, Fe^{2+}, Mg^{2+}, Li^+, etc., in inorganic cation modification process (Figure 8). However, the organofillicity of these types of modified clays is only slightly improved. In order to make these organophilic, the cations of the interlayer can be replaced by organic cations such as alkylammonium or alkylphosphonium (onium). The types of structures are shown in the Figure 8, which are obtained by the modification of clays using inorganic or organic cation exchange reaction (Hung et al. 2011). About four types of structures, namely, unilayer, bilayer, pseudo-trilayer or slit (Figure 8), are obtained during organic cation modification. Moreover, the distribution of the modified clays within the conjugated polymers matrix greatly depends on those structures obtained during modification. For example, the clays are modified by the quaternary ammonium salts in 80% of cases. The other types of compounds used for modifications are polymeric quaternary alkylammonium salts and copolymers, crown-ethers, crypt and, 2-aminopyrimidine, poly(ethylene glycol), alkyl imadazolium salts, alcohols, aldehydes, n-alkyl pyrrolidones, maleic anhydride, pentaerythritol, phosphonium salts, silanes and biomolecules (dePaiva et al. 2008). All those methods are briefly reviewed in the literature (Bergaya and Lagaly 2001, dePaiva et al. 2008). Among those methods, the ion exchange with alkylammonium ions is well-known, easy, reliable and preferential method for the modification of organoclays (dePaiva et al. 2008).

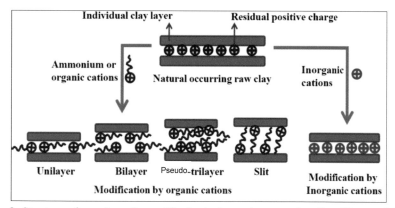

Figure 8. Structures of nanoclays after modification by inorganic or organic cations (Hung et al. 2011).

Preparation of nanocomposites

Physical technique

The conjugated polymer nanocomposites can be prepared through physical, chemical or electrochemical process following *in situ* or *ex situ* route for each process, as described in Figure 6. In the so-called *ex situ* physical preparation technique, the clay nanomaterials can be distributed within the matrix of chemically or electrochemically pre-synthesized host conjugated polymer by means of physical mixing. The easy and effective solvent mixing method is generally used to prepare the nanocomposites of conjugated polymers with nano clays. For example, PPY/laponite clay nanocomposite was prepared by mixing the dispersion of both the components in water (Mohamed and Mohammed 2015). In most of the cases, the insolubility of conjugated polymer or nanoclay in common organic solvent or water and poor disparity inside the conjugated polymer matrix is the major problem for preparation of their nanocomposite following this method. Moreover, the physical mixing of a polymer and layered silicate may form a blend rather than nanocomposite formation due to the segregation or aggregation of phase, which leads to detrimental weakening of the mechanical, optical and electrical properties of the final nanocomposite material. To achieve the homogenous dispersion of nano clay filler inside the conjugated polymer matrix, the surface modified organophilic clay minerals are preferred over the pristine one. In general, the properties of the nanocomposites prepared by this method exhibit poor dispersion in the matrix due to very weak interaction between the conjugated polymer and nanoclay.

Chemical technique

Among the *in situ* or *ex situ* chemical pathway, the *in situ* chemical preparation method is generally followed for the preparation of conjugated polymer nanocomposites with nanoclays. In *in situ* chemical preparation method, the nanoclays are incorporated during the chemical polymerization of the corresponding monomer. For example, the nanocomposites of montmorillonite nanoclay (MMT) with various conjugated polymers were prepared by following the *in situ* chemical synthesis methods (Ray and Okamoto 2003, Yeh et al. 2002, Ray and Biswas 1999, Oriakhi and Lerner 1999, Boukerma et al. 2006, Anuar et al. 2004). In fact, *in situ* chemical preparation of nanocomposites of conjugated polymer with pristine or modified nanoclays is the reliable, easy and the most commonly used one. For example, the exfoliated morphology having a well distribution of the nanoclays inside the polymer matrix was obtained for PPY-organically modified montmorillonite clay nanocomposite prepared via *in situ* polymerization in ethanol water mixture (Setshedi et al. 2013). As shown in Figure 9, the exfoliated structure was obtained due to the achievement of better dispersion of organically modified montmorillonite clays in ethanol water mixture as well as within the PPY matrix (Setshedi et al. 2013). The hydrophilic nanoclays have exhibited better dispersion within conjugated polymer matrix by aqueous emulsion polymerization than that of organophilic organo-MMT clay (Chang et al. 2006). Novel *in situ* intercalative emulsion polymerization of aniline in aqueous dispersion of clays, which was acting as template, was carried out in presence of bifunctional amphiphilic dopant, 3-pentadecyl phenol-4-sulphonic acid to prepare nano/microstructured conducting PANI/clay nanocomposite (Sudha et al. 2007). Similarly, the PANI/Na⁺MMT nanocomposite was prepared by *in situ* chemical technique by emulsion polymerization using dodecylbenzenesulfonic acid (DBSA) as emulsifier and dopant as well (Kim et al. 2001). The formation of PANI/Na⁺MMT nanocomposite having intercalated morphology into nanoclay gallery was explained by the micro emulsion polymerization using sodium lauryl sulfate and pentanol, as shown in Figure 10 (Song et al. 2008). In order to synthesize PPY/clay nanocomposite, an inverted emulsion pathway polymerization method was introduced using DBSA as both an emulsifier and a dopant (Kim et al. 2003). Nanocomposite of PPY and layered montmorillonite (MMT) clay platelets were prepared via an *in situ* oxidative polymerization using dodecylbenzene sulfonic acid (DBSA) as dopant (Yeh et al. 2003).

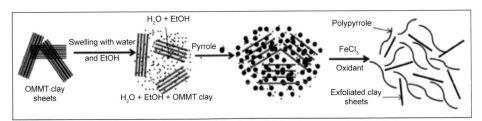

Figure 9. Schematic presentation for the formation of exfoliated polypyrrole-organically modified montmorillonite clay nanocomposite via *in situ* chemical polymerization (Reproduced with permission from Setshedi, K.Z., M. Bhaumik, S. Songwane, M.S. Onyango and A. Maity. 2013. Exfoliated polypyrrole-organically modified montmorillonite clay nanocomposite as a potential adsorbent for Cr(VI) removal. Chem. Eng. J. 222: 186–197, Copyright© Elsevier B.V.).

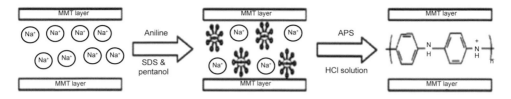

Figure 10. Schematic presentation of intercalated structure of polyaniline/Na⁺MMT nanocomposite within the clay gallery by microemulsion polymerization (Reproduced with permission from Song, D.H., H.M. Lee, K.H. Lee and H.J. Choi. 2008. Intercalated conducting polyaniline-clay nanocomposites and their electrical characteristics. J. Phys. Chem. Solids 69: 1383–1385, Copyright© Elsevier Ltd.).

Electrochemical technique

The *in situ* or *ex situ* electrochemical preparations of nanocomposites are rarely exploited due to some disadvantages like complicated set-up, difficulty in film deposition on the electrode, etc. (Singh and Prakash 2012). In *in situ* electrochemical methods, the conjugated polymer was electrochemically deposited on the electrode surface from the corresponding solution of monomer with nanoclay dispersion. The examples of such type of *in situ* electrochemical method are the electrochemical deposition of nanocomposite films of pristine or modified MMT with polycarbazole under ambient conditions (Singh and Prakash 2012, Singh et al. 2009a, Gupta et al. 2010). Similarly, electrochemical preparation of PANI-MMT was reported by electropolymerization of aniline on a clay modified electrode surface intercalated by dipping the electrode in aniline followed by drying in air (Inoue and Yoneyama 1987). The mechanism could be understood from the Figure 11 (doNascimento et al. 2004). According to the author (doNascimento et al. 2004), the protonated aniline monomer (An⁺) was intercalated between the nanoclays layers in aqueous solution and PANI-MMT nanocomposite was formed during the polymerization of intercalated An⁺ in aqueous suspension. In order to obtain homogeneous PANI-MMT nanocomposite, the aniline was

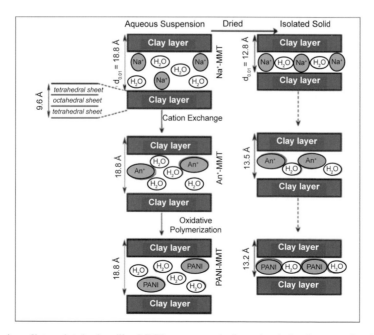

Figure 11. Mechanism of intercalated polyaniline-MMT nanocomposite formation during the *in situ* chemical polymerization (Reproduced with permission from doNascimento, G.M., V.R.L. Constantino, R. Landers and M.L.A. Temperini. 2004. Aniline polymerization into montmorillonite clay: A spectroscopic investigation of the intercalated conducting polymer. Macromolecules 37: 9373–9385, Copyright© American Chemical Society).

electropolymerized from a pre-treated mixture of aniline-MMT in aqueous hydrochloric acid solution (Feng et al. 2001).

Influence of properties

Influence of processability, gas permeability, optical properties, mechanical properties, anticorrosion properties, electrical conductivity on the composition of conjugated polymer nanocomposite with nanoclays were investigated widely. The effects of the material composition on the anticorrosion, gas barrier, thermal stability, flammability, mechanical strength, and electrical conductivity properties of the poly(3-hexylthiophene) and poly(3-hexylthiophene)/MMT clay nanocomposite were studied briefly (Yu et al. 2004). It is well established that the nanocomposites of nanoclays with various conjugated polymers usually exhibit superior physical properties, especially optical properties, heat resistance, corrosion resistance, better processability, better mechanical strength, thermal stability, etc. (Alexandre and Dubois 2000, Ke and Strove 2005, Walls et al. 2003). However, the properties of conjugated polymer nanocomposite greatly depend on the degree of the miscibility of the two phases. For example, the formation of an immiscible nanocomposite of Cloisite 30B nanoclay in PANI matrix observed in SEM image and obtained poor crystallinity in XRD pattern was considered as the reason for the poor properties of the nanocomposite (Figure 12) (Akbarinezhad et al. 2011). As explained, the white portions in the Figure 12a were assumed as the nanoclay layers, which were separated from the black PANI matrix. The conjugated polymers are inherently insoluble in nature or very sparingly soluble in water or even in organic solvent due to rigid π-conjugated long chain structure. In order to improve the processability, the initial objectives behind the synthesis of nanocomposite between conjugated polymer and nanoclays were to keep the conducting polymer in solution/dispersion within water/organic solvent for versatile applications. For example, PANI was little soluble in dimethyl sulfoxide or dimethyl formamide, while PANI-MMT nanocomposite was found to form stable colloid, even in water solvent (Biswas and Ray 2000). In most of the cases, the composite of doped conjugated polymer with nanoclays show less conductivity than that of doped conjugated polymer itself. The poor conductivity for this type of non-interacting as well as insulating nanoparticles may result due to interruption of doping interaction for the doped conjugated polymer. As for example, the conductivity was decreased considerably, when nanoclays were added to dodecyl benzene sulfonic acid doped PPY due to interruption of effective doping (Geng et al. 2007). Similarly, the intercalated clay layers interrupt the effective doping of PANI matrix resulting in lower conductivity of the nanocomposite (Kim et al. 2001). For this reason, dispersed nanolayers of MMT clays in poly(3-hexylthiophene) matrix led to a significant decrease of electrical conductivity (Yu et al. 2004).

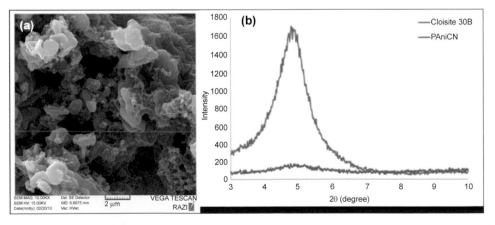

Figure 12. (a) SEM image and (b) XRD patter of polyaniline/clay (PAniCN) nanocomposite (Reproduced with permission from Akbarinezhad, E., M. Ebrahimi, F. Sharif, M.M. Attar and H.R. Faridi. 2011. Synthesis and evaluating corrosion protection effects of emeraldine base PAniCN/clay nanocomposite as a barrier pigment in zinc-rich ethyl silicate primer. Prog. Org. Coatings 70: 39–44, Copyright© Elsevier B.V.).

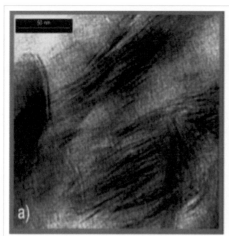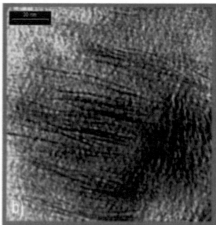

Figure 13. TEM micrographs of polypyrrole/MMT nanocomposite having intercalated structure of between the clay gallery spaces of the clay (Reproduced with permission from Peighambardoust, S.J. and B. Pourabbas. 2007. Synthesis and characterization of conductive polypyrrole/montmorillonite nanocomposites via one-pot emulsion polymerization. Macromol. Symp. 247: 99–109, Copyright© WILEY-VCH Verlag GmbH & Co.).

However, the higher conductivity for the nanocomposite like PPY/MMT clay might be observed due to having PPY rich-surface with intercalated structure between the clay gallery spaces of the layer, as shown in the TEM figure (Figure 13) (Peighambardoust and Pourabbas 2007). A higher electrical conductivity was observed for the clay nanocomposite implying that conjugated polymer was essentially located at the surface. In contrast, the nanocomposite of conducting polymer within the galleries of the clay, i.e., clay rich surface should have a rather low conductivity. For example, lower conductivity of MMT/PPY was recorded as montmorillonite on the surface of nanocomposite and a rather high conductivity was observed for organically modified MMT/PPY nanocomposite having a PPY-rich surface (Mravcakova et al. 2006). The molecular weights of poly(3-hexylthiophene) in its nanocomposite with MMT materials were found to be slightly lower than those of bulk pristine polymer. For the reason, the decomposition temperature of the nanocomposite was slightly shifted to lower temperature range than that of pristine one (Yu et al. 2004). The incorporation of nanolayers of MMT clay in the poly(3-hexylthiophene) matrix led to a decrease in mechanical strength indicated by less storage modulus or higher glass transition temperature in dynamic mechanical analysis. This might be attributed to the significantly decreased crystallinity and molecular weight of poly(3-hexylthiophene) formed in nanocomposite with MMT clay (Yu et al. 2004). Therefore, it is understood that the aim for preparation of conjugated polymers nanocomposites with various nanoclays is not to enhance the conductivity, rather to improve the thermal stability, mechanical strength, fire resistance, corrosion resistance and molecular barrier properties. However, the poor physical interactions between the nanoclays and immiscible conjugated polymers lead to poor mechanical and thermal properties. In general, the enhancement of material properties of those nanocomposites have been linked to the improved interfacial interaction by introducing suitable functional group modified nanoclay in conjugated polymer matrix.

Important applications

Like conducting polymers, its nanocomposites with clay nanomaterials have been investigated as potential material for various applications such as corrosion inhibitor, biosensors, solid-state batteries, smart windows and other electrochemical devices (Zeng et al. 2005). The important application of the conjugated polymer nanocomposite with layered nanomaterials has been reported as corrosion inhibitor with other rare applications such as gas permission layer and fire retardant.

Corrosion inhibitor

The corrosion protection of electroactive conjugated polymer nanocomposites with nanoclays has attracted extensive research interest in the recent decade over the other applications. The enhanced corrosion inhibitor on the metallic substances can be expected from the interaction of electroactive conjugated polymer and pellet like structure nanoclays at the molecular level in their nanocomposite. For instance, the nanocomposites of MMT nano-layers with pristine conducting polymers, viz., PANI, poly(o-methoxyaniline), poly(o-ethoxyaniline), poly(3-alkylthiophene), PPY, etc. have been used for the corrosion protection applications (Qiang et al. 2008, Yeh and Chin 2003, Yu et al. 2004, Ding et al. 2002). The coatings of pristine conjugated polymer protect metal surface either by creating a barrier for oxygen and moisture or by forming a passive oxide film through an oxidation–reduction process (Twite and Bierwagen 1998, Kinlen et al. 1997). The increase of anticorrosion property of PPY/MMT nanocomposite over the pristine PPY was attributed to the dramatic improvement in barrier properties of polymers due to dispersing silicate nanolayers of clay in the PPY matrix. Usually, the plate-like materials like nanoclays are employed to improve the corrosion resistance by increasing the length of the diffusion pathways and by decreasing the permeability of the coating for both oxygen and water (Figure 14). Therefore, the presence of layered nanoclays improves the corrosion resistance properties of the conjugated polymer coatings by decreasing the coating porosity, decreasing the permeability and stabilizing their electronic structure. The nanocomposite of poly(3-hexylthiophene) with low MMT clay exhibited better anticorrosion property as compared to bulk poly(3-hexylthiophene) (Yu et al. 2004).As

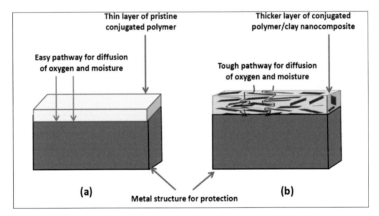

Figure 14. Anti-corrosion mechanism through layer of (a) pristine conjugated polymer and (b) conjugated polymer/clay nanocomposite.

shown in the Figure 14, the gas diffusion was decreased for the nanocomposite layer through the tough pathway (Olad et al. 2013). The corrosion protection effect of PANI nanocomposites with either pristine or modified MMT clay on metal surface was attributed to the combination of redox catalytic property of PANI matrix and barrier effect of MMT clay platelets (Chang et al. 2006).

Other applications

Also, to be used in a rare application like the exfoliated PPY-organically modified montmorillonite clay, nanocomposite was found to have better adsorption property for toxic Cr(VI) from aqueous solution (Setshedi et al. 2013). The poly(3-hexylthiophene)/MMT clay nanocomposite was found to be superior in the gas barrier over those of bulk poly(3-hexylthiophene) based on oxygen gas-permeability measurements (Yu et al. 2004). This indicated that the incorporation of MMT into the polymer matrix efficiently increased the flame retardancy of the polymers (Yu et al. 2004). Similarly, the PPY/MMT nanocomposite was introduced as fire retardant due to having a lower thermal decomposition rate than that of the pristine PPY (Peighambardoust and Pourabbas 2007).

Composites with metallic nanomaterials

Metal nanoparticles itself exhibit novel electronic, optical and chemical properties, which are far different from the bulk metal for quantum size effects. Thus, they have attracted continuous attention for promising applications in optical, electronic, sensing, biomedical and energy devices. In addition, the conjugated polymers are also widely used in various applications due to having distinct properties like conductivity, optical and electroactive properties. The conjugated polymer nanocomposites with metal nanoparticles have been attacked interests as important class of materials due to improving derive properties by synergetic pertaining of the properties of two components. The efforts are directed towards the preparation of nanocomposites of different metals with different types of conjugated polymers for the wide variety of electronic and optoelectronic applications. However, the properties as well as performances of those nanocomposites are strongly dependent on particle size, morphology, composition, crystallinity, distribution within polymer matrix, surface structures, etc.

Synthesis of nanomaterials

The synthesis of stable nanoparticles from metal is a very difficult task and it opens up a new dimension for the researchers worldwide. Like the other nanomaterials, the metal nanomaterials can be made using different approaches in both top-down and bottom-up techniques (Figure 15) (Iravani 2011). The name of few such type of approaches are homogeneous nucleation from liquid or gas phase, heterogeneous nucleation on a substrate, phase separation through annealing of certain solids or micelle synthesis, thermodynamic equilibrium approach, kinetic approach, etc. During the synthesis of metal nanoparticles, the main challenges that are faced are the poor control over the particle size, the stability of those nanoparticles and the difference in their structural, electronic, spectral, magnetic, and chemical characteristics according to their size. However, the variation of properties of various metal nanoparticles by controlling the size during synthesis have been reported elsewhere. In this section, important synthesis procedure with some example of metal nanoparticles is considered. In typical "one-pot" chemical approach, which is discussed briefly in the next section, the nanocomposite is directly obtained through

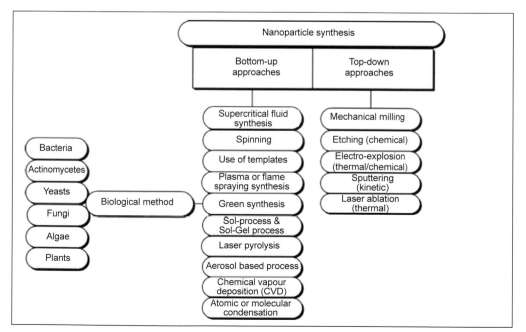

Figure 15. Some important methods for the syntheses of metal or metal oxide nanomaterials (Reproduced with permission from Iravani, S. 2011. Green synthesis of metal nanoparticles using plants. Green Chem. 13: 2638–2650, Copyright© Royal Society of Chemistry).

the reduction of metal salt to metal nanoparticles by the oxidation of the conjugated polymer matrix or the corresponding monomer to polymer. The major chemical routes that have been followed to synthesize the nanoparticles are, reduction of metal salt in solution by using a suitable reducing agent, e.g., sodium borohydride, hydrazine, ammonia, etc. without or with stabilizer like dopant, monomer, oligomer, etc. For example, the nanocomposites of conjugated polymer/gold or silver metal nanoparticles were prepared by reduction of corresponding salt with sodium borohydride (Choudhury 2009, Choudhury et al. 2009, Zhai and McCullough 2004). The syntheses of metal nanoparticles have been developed from corresponding metal salt or metal complex by electrochemical reduction either on pre-deposited polymers or in growing polymer films. In a typical method, the pre-deposited conjugated polymer by electro-oxidation on an electrode was dipped in a solution containing metal salts of Pt^{6+}, Ag^+ or Cu^{2+} followed by electrochemical reduction, yielding Pt, Ag or Cu nano-clusters embedded in the conjugated polymer (Kost et al. 1988, Holdcroft and Funt 1988, Tourillon et al. 1986). Alternatively, the pre-deposited conjugated polymer film on electrode was immersed in an aqueous solution of metal anions (Pd, Ru or Cu) for the formation of anion exchange metal complex to facilitate metal nucleation of nanoparticles during the electrochemical reduction (Sigaud et al. 2004, Coche and Moutet 1987, Zouaoui et al. 1999). In order to get metal nanoparticles, the photoreduction of the metal salt, e.g., copper, silver or gold salts might be carried out with simulations photopolymerization of monomer (Breimer et al. 2001, Zhou et al. 1996a). Following the method, palladium nanoparticles were synthesized by the γ-radiolysis from the heated solution containing palladium chloride and substituted aniline monomers (Athawale et al. 2003).

Preparation of nanocomposites

Several chemical and electrochemical approaches have been developed to prepare nanocomposites of conjugated polymers with metal nanoparticles. However, the so-called *ex situ* physical preparation method (Kar et al. 2015), where the nanomaterials are mixed up within the pre-synthesized conjugated polymer matrix, has been rarely demonstrated. Such type of example is the preparation of PANI/copper nanocomposite by simple mechanical stirring for 12 hours continuously in water medium (Shanmugapriya and Velraj 2014).

Chemical technique

The incorporation of metal nanomaterials into the conjugated polymer matrix is achieved through *in situ, ex situ* or 'one pot' chemical pathway. In typical, *in situ* method the chemical polymerization of suitable monomers was carried out in the dispersion of pre-synthesized metal nanomaterials. Following the methodology, the preparation of PANI nanocomposites with metal nanoparticles like copper, silver, gold, palladium, etc. has been reported (Sharma et al. 2002, Do and Chang 2004, Athawale et al. 2006, Li et al. 2009, Choudhury et al. 2009). In *ex situ* chemical preparation method, the metal nanoparticles might be synthesized chemically within the pre-synthesized conjugated polymer. For example, the nanocomposites of poly(m-aminophenol) with silver nanoparticles were prepared by easy thermal decomposition of silver nitrate ammonia complex during the film casting of the polymer (Kar et al. 2011a). Similarly, the PANI/Pt nanocomposite was prepared on the surface of pre-synthesized PANI by reducing a Pt salt with ethylene glycol (Chen et al. 2006). As shown in the Figure 16, the Pd, Au, Pt metal nanoparticles were synthesized through reduction of pre-synthesized poly(dithiafulvene) in DMSO medium by simple stirring (Zhou et al. 2002). In order to prepare nanocomposite following "one-pot" approach, the metal nanoparticles were synthesized from reduction of corresponding metal salt either by the oxidation of undoped conjugated polymer to doped one or by the oxidation of monomer to the respective polymer (Sih and Wolf 2005). As the conjugated polymers exhibited reversible oxidation/reduction (i.e., redox) properties, those polymers could be oxidized by strong oxidants such as $HAuCl_4$, H_2PdCl_4, H_2PtCl_6, $AgNO_3$, etc., and the salts were reduced to the corresponding metal nanoparticles, rusting the metal embedded conjugated polymer nanocomposite. As for example, the PANI/silver nanocomposite was prepared by the reduction of silver nitrate to silver nanoparticles during the oxidation of PANI (Huang et al. 2004, Stejskal et al. 2009). Following the same method, the PANI/Pd or PPY/metal (Ag and Au) nanocomposites were prepared

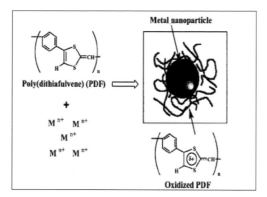

Figure 16. Schematic presentation of formation of poly(dithiafulvene)-metal (Pd, Au, Pt) nanocomposite (Reproduced with permission from Zhou, Y., H. Itoh, T. Uemura, K. Naka and Y. Chujo. 2002. Synthesis of novel stable nanometer-sized metal (M = Pd, Au, Pt) colloids protected by a π-conjugated polymer. Langmuir 18: 277–283, Copyright© American Chemical Society).

(Tseng et al. 2005, Gallon et al. 2007, Zhang and Manohar 2005, Xu et al. 2009). Nanocomposites of PANI derivatives were prepared following the second approach, i.e., the gold salt was reduced to gold nanoparticles during the oxidation of monomer such as 2-methoxyaniline or o-anisidine to corresponding polymer (Tan et al. 2003, Dai et al. 2002). In a modified method, one reagent like hydrogen peroxide was used both for reduction of gold salt and oxidative polymerization of monomer, aniline to prepare PANI/gold nanocomposite (Sarma et al. 2002, Sarma and Chattopadhyay 2004).

Electrochemical technique

Electrochemical synthesis, following either *in situ* or *ex situ* pathway, is also a well-established and efficient technique for the preparation of metal nanoparticles incorporated conjugated polymer nanocomposite. Initially, *ex situ* attempts were made to prepare nanocomposite in two-step process, i.e., the metal salt was electrochemically reduced onto the surface of pre-deposited conjugated polymer on electrode. First, a conjugated polymer such as PANI, PPY or poly(methylthiophene) deposited on electrode was dipped in a solution containing metal salts of Pt^{6+}, Ag^+ or Cu^{2+}, followed by electrochemical reduction of metal ions to prepare corresponding nanocomposites (Kost et al. 1988, Holdcroft and Funt 1988, Tourillon et al. 1986). Such type of two-step synthesis is not preferred as the metal nanoparticles are mostly deposited at the polymer/electrolyte interface, i.e., on the surface of the polymer (Figure 6). In an alternate *in situ* approach, metal nanoparticles are incorporated into a conjugated polymer matrix during the electropolymerization of the monomer dissolved in the colloidal dispersion of pre-synthesized nanomaterials (Bose and Rajeshwar 1992, Katz et al. 2004). This method is preferred over the other method, as a homogeneous distribution of metal nanoparticles inside the conjugated matrix is possible to achieve (Figure 6). In one-pot electrochemical co-deposition method, the simultaneous electrochemical oxidative polymerization of aniline and reduction of silver ions to silver nanowires was performed (Drury et al. 2007).

Influence of properties

Conjugated polymer/metal nanocomposite offers a convenient way to combine the advantages as well as properties of organic polymer materials with that of the inorganic metal nanoparticles. However, the properties of the nanocomposites may be tuned via controlling shape, size, inter-particle spacing, distribution or changing the various types of materials selected, synthetic process, and processing methods. Incorporation of metal nano-species, an excellent conducting one, into the conjugated polymer matrices can enhance the electron transfer through a direct or mediated mechanism to improve the conductivity of the nanocomposite. The increasing of conductivity of nanocomposite might be a result of having some induced doping effect of metal nanoparticles on the undoped polymer. In general, the conductivity of the nanocomposite is increased gradually with the increasing concentration of metal nanoparticles. However, the nanocomposite only shows the highest conductivity at an optimum concentration of the metal nanoparticles into the conjugated polymer matrix for well-dispersed doping by metal. For example, the conductivity of poly(m-aminophenol) nanocomposite with only 3.2% concentration of silver nanoparticles was recorded as 10^{-6} S/cm order, though the undoped pristine polymer was found to have the conductivity less than 10^{-11} S/cm (Kar et al. 2011a,b). Good linearity was also observed in the V-I characteristics plots of the nanocomposite, indicating Ohmic conduction in the nanocomposite, as shown in Figure 17 (Kar et al. 2011a). Similarly, the conductivity of nanocomposites of conjugated polymers like PPY, PANI and polythiophene with gold nanoparticles was found to increase by approximately two orders of magnitude compared to pristine polymers (Breimer et al. 2001, Sarma et al. 2002, Zhai and McCullough 2004). In nanocomposites, the coupling between surface plasmon absorption of individual metal nanoparticles with the relatively weak optical properties of conjugated polymer could lead to improve the optical properties of their nanocomposites. The examples of such effects are either the shift of surface plasmon of gold nanoparticles embedded in poly(3,4-ethylenethiophene) and poly(dithiafulvalene) (Li et al. 2002a, Zhou et al. 2002) or a red-shift in the π–π* absorption of poly(1,6-di(N-carbazolyl)-2,4-hexadiyne) and poly(diacetylene) surrounded by silver nanoparticles (Zhou et al. 1996a,b). Some of the metal nanoparticles itself possess good catalytic properties due to having active surface and metallic bonding ability for adsorption of chemical species. The metal nanoparticles generally impart the catalytic properties to the conjugated polymer and the derived nanocomposite can have catalytic properties by improving the chemical adsorption. For example, PANI/metal nanocomposites have received great attention over the pure PANI for catalytic capabilities of metal nanoparticles to improve the adsorption of chemical analyte (Kitani et al. 2001, Granot et al. 2005, Choudhury 2009, Choudhury et al. 2009). The influence of magnetic properties of PANI nanocomposites with magnetic metal nanoparticles such as Fe, Co, Ni, etc. have also been extensively studied (Cao et al. 2001a,b,c).

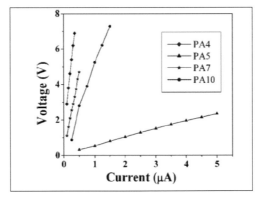

Figure 17. Voltage vs. Current plot of poly(m-aminophenol)/silver nanocomposites having 2.30 (PA4), 3.20 (PA5), 4.43 (PA7) and 6.30 (PA10) wt% silver nanoparticles (Reproduced with permission from Kar, P., N.C. Pradhan and B. Adhikari. 2011a. Doping of processable conducting poly(m-aminophenol) with silver nanoparticles. Polym. Adv. Technol. 22: 1060–1066, Copyright© John Wiley & Sons, Ltd.).

Important applications

The unique properties of conjugated polymer/metal nanocomposites obtained by combining the properties of individual component have led to several possible applications with improved performances than that of pure conjugated polymers. Among those two important applications are energy device applications like light-emitting diodes, solar cells, capacitor, field effect transistor, rechargeable battery, etc. and sensor device for chemical or bio analytes. Some other applications have also been reported such as memory device, catalyst, etc.

Energy device

To date, the photocatalytic or optoelectronic behavior of conjugated polymer/metal nanocomposites was most extensively explored for the possible applications in light-emitting diodes and solar cells. The blue polymer light emitting diodes of poly(9,9'-dioctylfluorene)/gold nanocomposites were found to have improved luminescence stability due to quenching of the triplet state (Park et al. 2004a). The electroactivity of the metal nanoparticles has been attributed to the increase of the redox activity of the nanocomposites with conjugated polymers. For the application as cathode material in improved lithium rechargeable batteries, the increasing redox activity of such types of nanocomposites leads to better charge–discharge performance. 2,5-dimercapto-1,3,4-thiadiazole-PANI nanocomposite with silver or palladium nanoparticles have been reported as a better redox active material with improved charge–discharge performance compared to pristine material (Park et al. 2002, Park et al. 2004b). In a typical comparison of galvanostatic charge/discharge curves of the metal/PPY hollow nanocomposites (Figure 18), it was found that the discharge time for the Mn/PPY hollow nanocomposite cell was longer than that for the Ni/PPY hollow nanocomposite recorded at a current density of 0.1 A g^{-1} (Ahn et al. 2015). The Scotty diode and field effect transistor (FET) devices could also be fabricated by the nanocomposites of conjugated polymers with metal nanocomponents. For example, room temperature diode was fabricated by using Au–PPY–Cd–Au four segment nanocomposite (Park et al. 2004c) or Co–PPY–Co nanocomposite might be used for FET application (Chung et al. 2005).

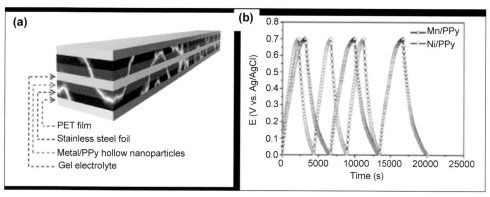

Figure 18. Flexible solid-state cells based on metal/PPY nanocomposite electrodes (a) Construction of the cell and (b) Galvanostatic charge/discharge curves recorded at a current density of 0.1 A g^{-1} (Reproduced with permission from Ahn, K.-J., Y. Lee, H. Choi, M.-S. Kim, K. Im, S. Noh and H. Yoon. 2015. Surfactant-templated synthesis of polypyrrole nanocages as redox mediators for efficient energy storage. Sci. Rep. 5: 14097–14106, Copyright© Nature Publishing Group).

Sensor device

The nanocomposite sensor for chemical or bio analytes have been found to have better response, better recovery, good reproducibility and good reversibility than that of the conjugated polymer sensor. This is due to increasing the specific surface area and improving catalytic efficiency by incorporating the metal nanoparticles into the conjugated polymers matrices (Yang et al. 2010, Shirsat et al. 2009, Srivastava et al. 2012, Li et al. 2009). For this reason, the PANI/silver nanocomposite exhibited fastest response, good reproducibility and long-term stability towards ethanol or acetone vapor in comparison with pure PANI (Choudhury 2009, Choudhury et al. 2009). Some of the inert volatile organic compounds vapors do not interact or weakly interact with the common conjugated polymers. However, by introducing metal nanoparticle into the conjugated polymer matrix, those analyte vapors could be successfully and selectively analyzed due to interacting effectively. For example, the chloroform, methanol or acetone/ethanol vapor which was weakly detected by PANI, was successfully detected by the PANI/Cu (Sharma et al. 2002), PANI/Pd (Athawale et al. 2006) or PANI/Ag (Choudhury 2009, Choudhury et al. 2009) nanocomposite sensing layer. The problem of incomplete recovery of the conjugated polymer based sensor is solved effectively by introducing the metal nanoparticles into the polymer matrix. Typically, ammonia gas sensing by hydrochloric acid doped poly(m-aminophenol)/Ag nanocomposite was reported without having the problem of incomplete recovery for ammonia sensing by HCl doped PANI (Kar et al. 2011b). Moreover, the average relative response 86.75 ± 2.9% was observed for varying concentration of ammonia and only the response time was decreased with increasing concentration of ammonia vapor in air (Figure 19). Use of conducting polymer/metal nanocomposite for sensing of bio analyte like H_2O_2, glucose, enzyme, antibody, etc., has been examined. For example, PANI/Pt/Pd nanocomposite immobilized with glucose oxidase was used for the successful sensing of glucose (Guo et al. 2009). The conjugated polymer nanocomposite with metal nanoparticles can be precisely deposited on electrodes for electrochemical biosensors with more effective transduction mechanism through easy electron transfer at the interface. The Au–Ag nanoparticles adsorbed onto the PPY film was applied to the detection of human immunodeficiency virus (HIV) sequences by electrochemical impedance spectroscopy (Fu et al. 2006).

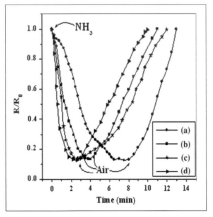

Figure 19. Sensing characteristics of HCl-doped PmAP–silver nanocomposite at varying concentration, viz. (a) 100 (b) 250 (c) 500 (d) 1,000 ppm of ammonia in air (Reproduced with permission from Kar, P., N.C. Pradhan and B. Adhikari. 2011b. Ammonia sensing by hydrochloric acid doped poly(m-aminophenol)–silver nanocomposite. J. Mater. Sci. 46: 2905–2913, Copyright© Springer Science+Business Media, LLC).

Other device

By incorporating the metal nanomaterials, the electroactive as well as semiconducting properties were enhanced for the high speed and high density nonvolatile memory devices. Such an example is PANI/Au composite, which was demonstrated to fabricate the memory device for processing of more and more data information in a short period of time (Tseng et al. 2005, Tseng et al. 2007). Recently, conjugated polymer/metal nanocomposites, viz. P3HT/gold nanoparticle have been used to fabricate electric bistable memory device, as the conductivity of the nanocomposite film was increased sharply by several orders of magnitude due to the presence of metal nanoparticles when a certain potential was applied on it (turn on) (Chen et al. 2007a). The catalytic properties of the metal nanoparticles might be imparted to its nanocomposites with conjugated polymers. For example, the catalytic efficiency of PANI/Pd nanocomposite has been studied for Suzuki coupling reactions (Gallon et al. 2007) and for the reduction of p-nitroaniline by hydrazine hydroxide in ethanol (Kong et al. 2009). Conjugated polymers/metal nanocomposite electrodes might be used with better catalytic behavior for electro-oxidation of methanol in fuel cell, since a 3D catalyst array was formed by the trapped metal particles into the conjugated polymer matrices. For example, PPY/Pt and PANI/Pt nanocomposites were used as the catalytic electrodes for electro-oxidation of methanol (Rajesh et al. 2003, Li and Lin 2007).

Composites with metal oxide nanomaterials

A number of different types of metal oxide nanoparticles have been dispersed within the conducting polymers matrices giving rise to lots of nanocomposites. Like conjugated polymer-metal nanocomposites, the metal oxide nanocomposites are a class of material that successfully combines the properties of both organic polymer material and inorganic semiconductor oxides. Due to such synergetic behavior, those nanocomposites possess unique physical, chemical, optical, mechanical, magnetic and electrical properties, which are unavailable in both the component materials. The nanocomposites are employed in wide range of device applications. However, the wide varieties of properties of nanocomposites depend upon the individual polymer or metal oxide used, the morphology obtained and the interfacial interactions between them. Such types of metal oxides like NiO, Cu_2O, CuO, TiO_2, WO_3, ZnO, Fe_2O_3, Fe_3O_4, MnO_2, SnO_2, V_2O_5, RuO_2, etc., have been used to prepare the nanocomposites with conjugated polymer.

Synthesis of nanomaterials

The synthesis of nanostuctured metal oxide with controllable sizes, shape and distribution is a real challenge. For the identification and control of structural features in the form of nanocrystals, nanorods, nanowires, or even nanoparticles to achieve the desirable electronic, optical, or magnetic properties, tremendous research efforts are being put worldwide for right formulation procedure. As shown in Figure 15 (Iravani 2011), the same procedure might be followed to synthesize the metal oxide nanomaterials. Here, the descriptions of all those synthetic procedures are not included as it is the other dimension of this broad field and so many literatures are available on this. In general, the synthesis methods for nanostructured metal oxide can be classified broadly into solution based and vapor phase methods. Important solution phase methods include hydrothermal, sol-gel, electrochemical deposition, precipitation methods, thermolysis, flame spray pyrolysis, etc., while vapor phase methods include physical vapor deposition, chemical vapor deposition method, arc discharge, laser pyrolysis, etc. Structures and properties of the nanostructured metal oxides are strongly dependent on synthesis techniques and synthesis conditions. However, solution phase chemical synthesis methods are always preferred due to cost efficiency and practicality. As for example, tin oxide nanoparticles were synthesized from hydrolysis of $SnCl_4 \cdot 5H_2O$ at pH $\leq$ 4 using dilute HCl, followed by addition of hydrogen peroxide for oxidation of tin ions to tin oxide suspension (Deshpande et al. 2009).

Preparation of nanocomposites

Conjugated polymers based nanocomposites with metal oxide nanomaterials can easily be prepared by *ex situ* physical processes and *ex situ* or *in situ* pathway following both the chemical or electrochemical process. The techniques are discussed briefly in the following sections.

Physical technique

The preparation of nanocomposites by *ex situ* physical process such as vacuum thermal evaporation and sputter deposition or coating has been reported. For example, poly(o-phenylenediamine)/WO_3 nanocomposite was prepared by vacuum thermal evaporation of mixed powder of both the pre-synthesized components (Tiwari and Li 2009, Tiwari et al. 2010). As shown in the Figure 20, first aligned SnO_2 nanosheets on a substrate was prepared by hydrothermal treatment of electrospun nanofibers of poly(vinyl butyral) containing $SnCl_2$ and then PANI/SnO_2 nanocomposite was prepared by dip coating with water-processable PANI (Li et al. 2016). In other method, PEDOT-PSS/TiO_2 nanocomposite was prepared by simple dip-coating PEDOT-PSS on fibers. From the AFM micrograph, it was confirmed that the PEDOT-PSS coating on TiO_2 was formed as PEDOT-PSS/TiO_2-tween rather than nanocomposite (Zampetti et al. 2013).

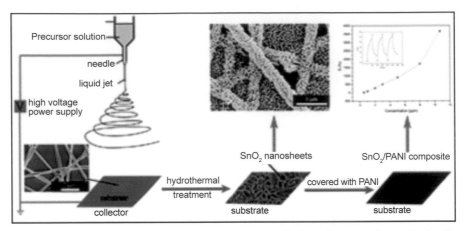

Figure 20. *Ex situ* physical preparation of polyaniline/SnO₂ nanocomposite layer for gas sensing application (Reproduced with permission from Li, Y., H. Ban, M. Jiaoa and M. Yanga. 2016. *In situ* growth of SnO₂ nanosheets on a substrate via hydrothermal synthesis assisted by electrospinning and the gas sensing properties of SnO₂/polyaniline nanocomposites. RSC Adv. 6: 74944–74956, Copyright© Royal Society of Chemistry).

Chemical technique

The *in situ* chemical route involving the incorporation of pre-synthesized nanomaterials during the chemical polymerization of suitable monomers to prepare polymer-metal oxide nanocomposite films was reported. In a typical method, tin oxide-intercalated PANI nanocomposite was deposited on glass substrate by dipping in the synthesis medium of polyaniline with suspension of pre-synthesized tin oxide nanoparticles (Deshpande et al. 2009). Alternatively, the nanomaterials are chemically synthesized from the corresponding precursor in the polymerization medium during chemical polymerization of monomer. Following the *in situ* chemical methods, the nanocomposites of various conjugated polymers like PANI, PPY, pothiophene, etc., were prepared with different nanostructured metals oxides such as SnO_2, TiO_2,

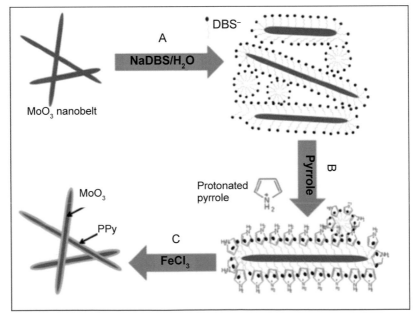

Figure 21. *In situ* chemical preparation of PPY/MoO₃ nanocomposite (Reproduced with permission from Liu, Y., B. Zhang, Y. Yang, Z. Chang, Z. Wen and Y. Wu. 2013. Polypyrrole-coated a-MoO₃ nanobelts with good electrochemical performance as anode materials for aqueous supercapacitors. J. Mater. Chem. A 1: 13582–13587, Copyright© Royal Society of Chemistry).

In$_2$O$_3$, WO$_3$, etc. The metal oxide nanomaterial templates have been attracting tremendous interest to develop nano-structured polymer composites with that nanomaterial. For example, the core-shell structured silica–PANI nanocomposite has been synthesized chemically using templates of silica cores (Jang et al. 2006). As shown in Figure 21, the composites of PPY coated on MoO$_3$ nanobelt was prepared through the *in situ* chemical polymerization of pyrrole on the surface of the MoO$_3$ in presence of sodium dodecylbenzenesulfonate (Liu et al. 2013). The *ex situ* chemical preparation method was very rarely used or not used at all.

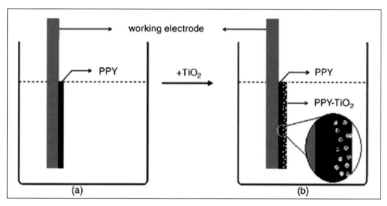

Figure 22. Scheme of (a) electrochemical deposition of PPY and (b) successive *in situ* electrochemical preparation of PPY-TiO$_2$ nanocomposite (Reproduced with permission from He, X.M. and G.Q. Shi. 2006. Electrochemical actuator based on monolithic polypyrrole-TiO$_2$ nanoparticle composite film. Sens. Actuators B: Chem. 115: 488–493, Copyright© Elsevier B.V.).

Electrochemical technique

The conjugated polymer was deposited electrochemically onto the electrode pre-deposited by metal oxide nanomaterials in *ex situ* electrochemical preparation of conjugated polymer-metal oxide nanocomposites. For example, PPY was successfully deposited by electrochemical polymerization on to the electrode containing pre-deposited TiO$_2$ nanotubes (Mi et al. 2009). Like the metal nanomaterials, the metal oxide nanomaterials can also be introduced into the conjugated polymer matrix by *in situ* electrochemical polymerization of the monomer from its solution with colloidal dispersion of metal oxide nanomaterials. In order to incorporate metal oxide nanoparticles into the conjugated polymer matrix following this type of electrochemical co-deposition method, negatively charged surface of nanoparticles is required. Following the method, the WO$_3$, SiO$_2$, SnO$_2$, and Ta$_2$O$_5$ nanoparticles were successfully dispersed in electrochemically growing PPY film to prepare their nanocomposite (Yoneyama et al. 1989). Similar to that of chemical pathway, the nanostructured conjugated polymer based nanocomposites can be fabricated electrochemically in the presence of metal oxide template. For example, the nanocomposites of conducting PPY, PANI and PTH derivative were synthesized by the electrochemical polymerization method, using Al$_2$O$_3$ nanoporous templates (Kim et al. 2005a). In a typical method (He and Shi 2006), a pure PPY layer was electrochemically deposited on the working electrode (Figure 22a), followed by successive electrochemical deposition of nanocomposite layer from aqueous solution of pyrrole with TiO$_2$ dispersion (Figure 22b).

Influence of properties

The properties of the metal oxides materials as well as their derived nanocomposites with conjugated polymer are sensibly dependant on their sizes in different morphologies like nanorods, nanowires, nanotubes, etc. In general, the electrical behavior of metal oxide nanomaterials can vary from electrically insulating (MgO, and Al$_2$O$_3$), wide-band semiconductor (TiO$_2$, SnO$_2$, ZnO, Ti$_2$O$_3$) to metal-like (V$_2$O$_3$, ReO$_3$, RuO$_2$) properties. The conductivity of various conjugated polymer/metal oxide nanocomposites was pointed out to be a lack of any consistent order due to depending on the nature of polymer and metal oxide counterpart. In general, the electrical conductivity of conjugated polymer nanocomposites with metal oxide would decrease due to incorporating a low conducting to insulating component like metal oxide. The reason might be attributed to the partial blockage of conductive pathway by those insulating components. For example, the electrical conductivity of β-naphthalene sulfonic acid doped PANI nanotubes having the value 0.154 S/cm, was decreased to 0.13 and 0.045 S/cm for the same acid doped PANI nanocomposites with 6 and 20 wt% Fe_3O_4 nanoparticles (Long et al. 2005). However, the metal oxide nanomaterials increase the localization of charge carriers in their nanocomposites with conjugated polymers and for this reason, the electroactive properties of nanocomposites may be improved. Thus the conjugated polymers nanocomposites with metal oxides such as ruthenium oxide, MnO_2, NiO, etc., were found to have high specific capacitance (Zheng et al. 1995, Nam et al. 2002, Hu and Wang 2003, Chang and Tsai 2003). The metal oxides with good optical properties can be introduced to conjugated polymers in order to be developed as photosensitive materials in solar cell application. The metal oxide nanomaterials with magnetic properties might be incorporated in conjugated polymer matrix and the derived nanocomposite have been examined for the applications like electro-chromic devices, electromagnetic interference shielding, and nonlinear optical systems. The magnetic property of PANI was improved in its nanocomposite with magnetic metal oxide nanomaterials like Fe_3O_4 composite (Zhang et al. 2005b, Lu et al. 2006, Long et al. 2005). The conjugated polymer like PANI with visible light absorbing chromophores was used as sensitizers to enhance the photocatalytic efficiency of less absorbing TiO_2 in their nanocomposite (Xiong et al. 2004). The unique photovoltaic properties of metal oxides were combined with conducting polymers having good light absorbing and hole transporting properties. For example, the P3HT/ZnO nanocomposite was found to have better photovoltaic efficiency than that of the constituent components (Olson et al. 2006).

Important applications

Metal oxide nanomaterials dispersed in conjugated polymer matrices have shown better gas sensing features than that of the pristine metal oxide or conjugated polymers. Thus, the sensor applications are to be considered as the main important applications of these types of nanocomposites. Other than sensor application, the nanocomposites are also investigated for the application as solar cell, capacitor, super capacitor, rechargeable battery, actuators, etc.

Gas sensor device

The metal oxides having disadvantages like poor selectivity, high operating temperature (300–500°C), poor mechanical property, etc., have shown good recognition characteristics for the gaseous chemical analytes. On the other hand, the conjugated polymers having distinguishable electrical properties, mechanical flexibility, and relative ease of processing are regarded as promising materials for chemical gaseous analyte with poor selectivity and sensitivity. Thus, conjugated polymer matrices embedded with metal oxide nanomaterials are considered as a new class of nanocomposites with better gas sensing features. Unlike pristine metal oxide nanomaterials, the nanocomposites responded well at room temperature due to the formation of metal oxide-conjugated polymer junction. For example, the n-type SnO_2 or ZnO forms a pn-hetero junction with p-type conjugated polymer like PANI or PPY in their nanocomposite. The PANI/SnO_2 nanocomposite was demonstrated to have ~ 3700% response towards 10.7 ppm of NH_3 with ~ 46 ppb detection limit at room temperature due to formation of p/n heterojunction at the interface (Li et al. 2016). The introduction of metal oxide nanoparticles into the conjugated polymers could further improve the interaction with analyte by modifying the physical and chemical properties of their nanocomposite to detect the analyte selectively. For example, PANI/In_2O_3 nanocomposite was used for the sensing of H_2, CO and NO_2 at room temperature over the In_2O_3, which required high operating temperature for the sensing of those gases (Sadek et al. 2006). The PANI (PANI)/titanium dioxide (TiO_2) nanocomposite layer deposited on a quartz crystal microbalance (QCM) chip was exhibited for the sensing of different gases. As shown in Figure 23, the layer was found to have high sensitivity toward 10 ppm of ammonia, hydrogen sulfide

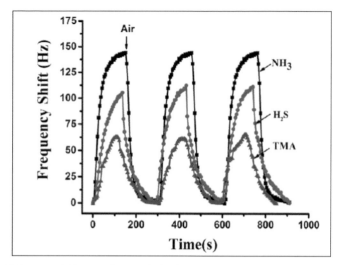

Figure 23. QCM sensitivity toward 10 ppm of ammonia, hydrogen sulfide and trimethylamine gases by polyaniline/TiO$_2$ nanocomposite layer (Reproduced with permission from Cui, S., J. Wang and X. Wang. 2015. Fabrication and design of a toxic gas sensor based on polyaniline/titanium dioxide nanocomposite film by layer-by-layer self-assembly. RSC Adv. 5: 58211–58219, Copyright© Royal Society of Chemistry).

and trimethylamine gases with an evident frequency shift, fast response, good recovery time, excellent reversibility, long-term stability as well as good selectivity towards ammonia (Cui et al. 2015).

Other device

The electrodes coated with thin film of conjugated polymer/metal oxide nanocomposites are shown to have great potential for solar cell application. This is because of significant absorption of dye on the active surface and great optical properties of dispersed nanomaterials in the conjugated polymer matrix. For example, the photoanodes of nanocomposites of conjugated polymers with TiO$_2$, SnO$_2$, Nb$_2$O$_5$, ZnO, etc. metal oxides have been employed for the conversion of maximum solar energy to electrical energy. The PPY/MoO$_3$ nanocomposite as the anode exhibited good rate capability as well as excellent cycling performance for supercapacitor application in 0.5 M K$_2$SO$_4$ aqueous solution as the electrolyte with activated carbon (Liu et al. 2013). The nanocomposites of conjugated polymers with metal nanomaterials have been extensively explored for potential application in nanophotonic devices due to having their unique tunable optical properties. The nanocomposite of vanadium oxide/PANI was used as cathode materials in Li-ion rechargeable batteries (Malta et al. 2003). MnO$_2$/PEDOT coaxial nanowires prepared by a coelectro-deposition approach using an AAO template were used for the supercapacitor application (Liu and Lee 2008). An actuator fabricated using bi-layers of pure PPY and PPY/TiO$_2$ nanocomposite films exhibited a large bending angle, a high response rate and a long lifetime due to having non-delamination, asymmetric structure and high conductivity of the nanocomposite (He and Shi 2006).

Composites with carbon nanomaterials

Until the mid-1980s, pure solid carbon was thought to exist in only two physical forms with different properties called allotropes, diamond and graphite. After that, the Buckminsterfullerene, often referred to as buckyball, fullerene or simply C$_{60}$, other related molecules (C$_{36}$, C$_{70}$, C$_{76}$ and C$_{84}$) composed of only carbon atoms were also discovered and they were recognized as allotropes of carbon (Figure 24). Another new form of carbon is known as graphene formed by a layer of carbon atoms in a hexagonal (honeycomb) mesh. Then the graphene layer is wrapped in the shape of a cylinder and bonded together to form a carbon nanotube (CNT). These molecules are shaped like a tube which can be considered as a sheet of graphite (a hexagonal lattice of carbon) rolled into a cylinder or chicken wire rolled into a tube having average diameter of 1–2 nm. To make things more interesting, besides having a single cylindrical wall (SWNTs), nanotubes can have multiple walls (MWNTs), i.e., cylinders inside the other cylinders. As shown in Figure 24 below, the carbon nanotubes can have a single outer wall of carbon, or they can be made of multiple walls (cylinders inside other cylinders of carbon). Some of the extraordinary properties of graphene or CNT include:

a) Extraordinary electrical conductivity, heat conductivity, and mechanical properties.
b) They are probably the best electron field emitter known, largely due to their high length-to-diameter ratios.
c) As pure carbon polymers, they can be manipulated using the well-known and the tremendously rich chemistry of that element.

These extraordinary characteristics of graphene or CNTs are employed in numerous applications like field emitters/emission, conductive or reinforced plastics, molecular electronics, non-volatile RAM, transistors, energy storage, fibers and fabrics, aerospace ceramics, biomedical applications, etc. The nanocomposites of conjugated polymers with carbon nanomaterials like carbon black (CB), carbon nanotubes (CNT), carbon nanofibre (CNF) and graphene have been prepared for the above applications with improved performances. Though carbon black nanoparticles are one of the simplest carbon materials, but carbon nanotubes (CNT), carbon nanofibres (CNF) and graphenes are preferred to prepare nanocomposites with conjugated polymers. This is because of having a higher surface area, and better electronic properties of graphene, CNTs, CNFs than that of the carbon black.

Synthesis of nanomaterials

The activated carbons were generally produced from different precursors, e.g., wood, coal, or nutshell (Beguin and Frackowiak 2010). The Arc Discharge method, which is the most common and easiest way, has been reported to prepare fullerenes as well as carbon nanotubes. In a typical method, nanotubes are produced through arc-vaporization of two carbon rods in inert gas such as helium or argon environment. During the above synthesis, the SWNTs was produced using metal catalyst such as Fe, Co, Ni, Y, or Mo and the MWNTs was produced with the use of pure graphite without catalyst. In Laser Ablation method, a pulsed or continuous laser is used to synthesize SWCNT or MWCNT by vaporizing a graphite target at a very high temperature in the presence of helium or argon gas. The other methods like Chemical Vapors Deposition method, Flame Synthesis method, Silane Solution Method, etc. have been also used to prepare the CNTs. In order to purify the nanotubes containing a large amount of impurities such as metal particles, amorphous carbon, and multishell, etc., different processes like air oxidation, acid or strong acid refluxing, surfactant aided sonication, filtration, annealing (Peter and Harris 1999), etc. are used.

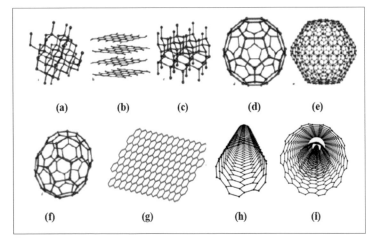

Figure 24. Eight allotropes of carbon: (a) diamond, (b) graphite, (c) lonsdaleite, (d) C60, (e) C540, (f) C70, (g) Graphene sheet (h) SWCNT and (i) MWCNT (Sharma et al. 2015).

Modification of nanomaterials

It is very difficult to disperse CNTs within the conjugated polymer matrix due to having very weak van der Waals type of interfacial interactions between the CNTs. Therefore, significant efforts have been directed towards the development of various methods to modify the surface of CNTs by breaking the hexagonal structures. The various physical and chemical functionalization methods, which are also conveniently same with non-covalent and covalent functionalization of CNTs, respectively,have been developed. A comparison of advantages and disadvantages of both the methods are shown in the Table 2. However, the physical method having serious disadvantages is not a very convenient one. In that respect, the chemical functionalization is based on the covalent linkage of functional entities onto carbon scaffold of CNTs through some chemical reaction like fluorination followed by amino, alkyl and hydroxyl group modification or carbene and nitrene addition, chlorination, bromination, hydrogenation, azomethine ylides, carboxylation, amination, sulfonation, etc. The most commonly used processes are shown in the Figure 25 (Yang et al. 2007). Covalent functionalization of CNTs using chemical method

Table 2. Advantages and disadvantages of various MWCNT functionalization methods (Ma et al. 2010).

Methods	Principle	Advantages	Disadvantages
Physical modification methods	Polymer wrapping by weak forces like Vander Waals force, π–π stacking, etc.	No damage to CNT structure, ease of functionalization and fabrication of composites	According to the degree of miscibility, variable weak interaction with polymer matrix, re-agglomeration is possible in polymer matrix
	Surfactant adsorption by physical adsorption		
	Endohedral method by capillary method		
Chemical modification methods	Side wall and/or end wall functionalization by breaking of double bonds and Sp^2 C centre changed to Sp^3	Strong interaction with the polymer matrix, re-agglomeration is not possible in polymer matrix	Damage to CNT structure, difficult to functionalization or fabrication of composites
	Defect transformation by introduction of suitable functional group chemically		

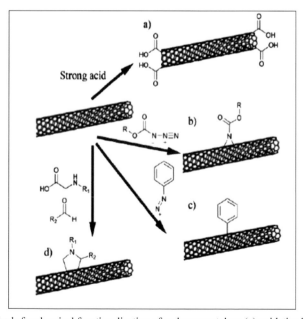

Figure 25. Common methods for chemical functionalization of carbon nanotubes: (a) oxidation by strong acids, (b) nitrene cycloaddition, (c) arylation using diazonium salts, and (d) 1,3-dipolar cycloadditions (Reproduced with permission from Yang, W., P. Thordarson, J.J. Gooding, S.P. Ringer and F. Braet. 2007. Carbon nanotubes for biological and biomedical applications. Nanotechnology 18: 412001–412013, Copyright© IOP Publishing).

is associated with insert functional groups by breaking of hybridization from sp^2 hexagonal carbon network to sp^3 carbon center. A typical covalent functionalization, predominantly at the more reactive (open) ends with carboxylic acid (–COOH) groups on carbon nanotubes, involves the refluxing of CNT with strong concentrated acid, e.g., nitric acid or nitric/sulfuric acid mixture (Yang et al. 2007). The covalent functionalization of CNTs is favored for nanocomposite preparation over the non-covalent functionalization as the covalently functionalized CNTs are better dispersible in polymerization medium or conjugated polymer matrix by improving interaction. Moreover, the chemically functionalized CNTs can be attached with many polymers by covalent bonds for strong interfacial interaction, allowing CNT-based nanocomposites to possess better reinforcing effects.

Preparation of nanocomposites

Physical technique

In an *ex situ* physical technique, the conjugated polymers can be prepared by mixing the solution of pre-synthesized conjugated polymer chemically or electrochemically in a suitable solvent with the dispersion of carbon nanomaterials in the same solvent. Then the film of nanocomposite was obtained by solvent removal trough vacuum drying. Following the technique the PANI, PPY, poly(3,4-ethylenedioxythiophene): poly(styrenesulphonate), etc., nanocomposites have been prepared. However, the technique is not much popular as poor distribution of nanomaterials resulted within the conjugated polymer matrix.

Chemical technique

The *in situ* chemical method is very popular and easy for the preparation of conjugated polymer nanocomposites with CNTs. For example, the nanocomposites of conjugated polymer such as PANI, polyanisidine, polyaminophenol, etc., with pristine CNT or carboxylic acid functionalized CNT were prepared following this method. In a typical method, first the pristine or functionalized carbon nanomaterials were dispersed in suitable polymerization medium followed by chemical polymerization of monomers within the medium using some radical oxidative initiator. Following the method, PANI or PPY coated carbon nanocoils composites was prepared (Rakhi et al. 2012). Aniline, pyrrole, thiophene, 3,4-ethylenedioxythiophene, etc., can be polymerized by radical initiator such as ammonium persulfate within acidic aqueous dispersion of graphene to prepare the nanocomposite. The only possible interaction for pristine CNT with conducting polymer is π–π stacking interactions as described in literature (Verma et al. 2015a). In order to improve the interfacial interaction between the CNTs and conjugated polymer, the functionalized CNTs was preferred. As shown in Figure 26, core–shell structure was obtained for carboxyl-functionalized multiwall carbon nanotube/poly(m-aminophenol) nanocomposite through *in situ* chemical polymerization of m-aminophenol in the presence of MWCNTs (Verma et al. 2016). In addition, surfactants have been employed with CNTs to adsorb and arrange the nanomaterials regularly and the monomers were chemically polymerized *in situ* to obtain the nanocomposite. Following the method, CNTs nanocomposites with PPY, PANI, PTH, etc., were prepared in the presence of cationic surfactants like cetyltrimethylammonium bromide, sodium dodecyl sulphate, benzenesulfonate, etc. (Zhang et al. 2004a,b, Zhang et al. 2005a, Yu et al. 2005a,b). Similar type of emulsion or microemulsion approach was introduced in the presence of suitable surfactant to prepare CNTs nanocomposites with PPY, PANI, etc. (Guo et al. 2008, Yu et al. 2005a,b). Other than those, the chemical bonding was introduced between PANI and the CNTs through functionalization by *in situ* chemical preparation method (Mekki et al. 2014). PANI nanocomposite with MWCNT was formed via surface functionalized of MWNTs with p-phenylenediamine followed by *in situ* chemical oxidative polymerization of aniline on phenylamine functional groups (Philip et al. 2005). The nanocomposite of CNTs with PANI was prepared through covalent bonding of poly(maminobenzene sulfonic acid) (Zhao et al. 2004). Similarly, the nanocomposites of polythiophene with MWCNTs was prepared by making covalent bonding between poly[3-(2-hydroxyethyl)-2,5-thienylene] and –COCl functionalized MWNTs (Philip et al. 2004).

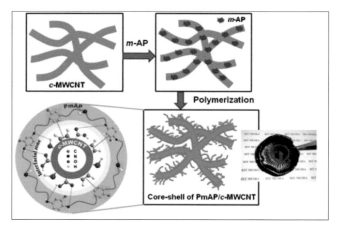

Figure 26. Core–shell structure formation of carboxyl-functionalized multiwall carbon nanotube/poly(m-aminophenol) nanocomposite by *in situ* chemical polymerization (Reproduced with permission from Verma, S.K., M. Kumar, P. Kar and A. Choudhury. 2016. Core–shell functionalized MWCNT/poly(maminophenol) nanocomposite with large dielectric permittivity and low dielectric loss. Polym. Adv. Technol. 27: 1596–1603, Copyright© John Wiley & Sons, Ltd.).

Electrochemical technique

Like *in situ* chemical method, the nanocomposites were also prepared by following *in situ* electrochemical polymerization process, i.e., the monomer was polymerized in presence of CNT in polymerization medium. For example, PANI/MWCNT or PANI/graphene nanocomposite was deposited on an electrode by *in situ* electrochemical polymerization process (Vishnu et al. 2013, Hu et al. 2012). The graphene based nanocomposites were prepared by the *in situ* electropolymerization of aniline, pyrrole, thiophene, 3,4-ethylenedioxythiophene, etc. The PANI/graphene oxide nanocomposites were prepared following a modified electrochemical polymerization, i.e., aniline in chloroform was electrochemically polymerized in the interface of acidic aqueous medium with graphene oxide dispersion (Figure 27). As shown in the Figure 27, three electrode set-up was used, consisting a platinum sheet as counter electrode, stainless steel foil as working electrode, and an Ag/AgCl electrode employed as the counter electrode (Li et al. 2015). *Ex situ* type of electrochemical method was followed to prepared PANI/CNT nanocomposite by electrochemical polymerization of aniline in sulfuric acid medium on the CNT electrode (Downs et al. 1999).

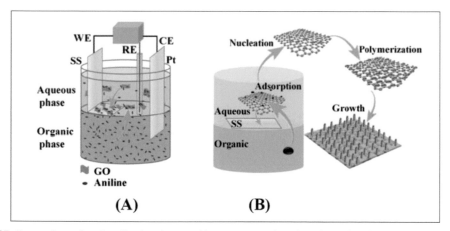

Figure 27. Preparation of polyaniline/graphene oxide nanocomposites by electrochemical interfacial polymerize (Reproduced with permission from Li, D., Y. Li, Y. Feng, W. Hu and W. Feng. 2015. Hierarchical graphene oxide/polyaniline nanocomposites prepared by interfacial electrochemical polymerization for flexible solid-state supercapacitors. J. Mater. Chem. A 3: 2135–2143, Copyright© Royal Society of Chemistry).

Influence of properties

Carbon-based nanomaterials such as CNTs, fullerene, amorphous carbon and graphene are frequently used fillers to reinforce the property of the conjugated polymer matrices in their nanocomposites. However, the properties of conjugated polymers/carbon nanomaterials composites are greatly dependent on many factors such as (i) the preparation procedure adopted, (ii) the solvent and electrolyte used, (iii) adhesion of the polymer film to surface of the nanotubes (iv) thickness of the polymer film deposited on the CNT surface, (v) roughness of the deposited film, etc. (Pieta et al. 2013). The functionalized CNT, which shows high disparity in polymerization medium and compatibility within the conjugated polymer matrix due to improvement of interactions has been studied widely to prepare the nanocomposites over the pristine CNT (Sun et al. 2001, Spitalsky et al. 2010, Bauhofer and Kovacs 2009). The carbon nanofillers are highly conducting in nature and also strongly interact with the conjugated polymers in their nanocomposites through charge transfer doping or site selective interaction. As observed, the incorporation of even small amount of pristine or functionalized carbon nanofillers into the matrices of conjugated polymers significantly improves the electrical conductivity. For example, electrical conductivity of reduced graphene oxide/poly(3,4-ethylenedioxythiophene) doped with polystyrene sulfonic acid nanocomposite was increased gradually with increasing nanofillers loading (Zhang et al. 2013). Similarly, the electrical resistance for PANI or poly(m-aminophenol) was significantly decreased upon incorporation of MWCNT or carboxylic acid group functionalized MWCNT (Verma et al. 2015a,b, Verma et al. 2016). As an example of the synergistic effect of the two components, the conductivity of the CNT/PANI nanocomposites was observed to be higher than pristine PANI. The electrical properties of conjugated polymer/CNTs nanocomposites follow power-law dependence in alignment as well as concentration. The best electrical property is observed at a fixed concentration of CNTs, i.e., at percolation threshold with fixed aspect ratio. The electrical conductivity of nanocomposite was increased sharply up to an optimum loading of carbon nanomaterials within the conjugated polymer matrix and then the conductivity increased negligibly with further loading of nanomaterials. For convenience, it was observed from Figure 28a that the electrical conductivity of PPY–CNT nanocomposites was increased with the increase in MWCNT loading. The sharp increment in DC conductivity from zero to 9.1–13.04 wt% CNT in the nanocomposite was observed and then a marginal increment in DC conductivity was observed with concentration change from 23.1 wt% CNT in the nanocomposite (Long et al. 2004). The percolation threshold/limit was well explained by the Sigmoidal curve of variation of electrical conductivity against filler concentration expressed in volume fraction (Figure 28b). According to the figure, it was considered that the uniform distribution and dispersion of filler was observed at percolation threshold/critical concentration whereas preferential

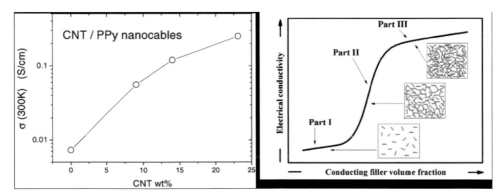

Figure 28. (a) DC conductivity (σ) vs. wt% of CNT in PPY matrix (Reproduced with permission from Long, Y., Z. Chen, X. Zhang, J. Zhang and Z. Liu. 2004. Electrical properties of multi-walled carbon nanotube/polypyrrole nanocables: percolation-dominated conductivity. J. Phys. D: Appl. Phys. 37: 1965–1976, Copyright© IOP Publishing) and (b) Sigmoidal curve of electrical conductivity against filler concentration; Part I is insulating zone, Part II is percolation zone, and Part III is conductive zone. (Reproduced with permission from Ram, R., M. Rahaman and D. Khastgir. 2015. Electrical properties of polyvinylidene fluoride (PVDF)/multi-walled carbon nanotube (MWCNT) semi-transparent composites: Modelling of DC conductivity. Composites A 69: 30–39, Copyright© Elsevier Ltd.)

distribution and dispersion of filler was accrued at lower or even higher concentration than percolation threshold (Ram et al. 2015). However, the percolation threshold depends on various factors such as nature of components, preparation methods, processing parameters, etc.

Important applications

Sensor device

The use of pristine or functionalized CNTs in chemical sensor is restricted for the lack of chemical sensitivity, selectivity which is due to having limited specific interactions with analyte molecules. Thus the utilization of pristine or functionalized CNTs as dispersed secondary material in conjugated polymer can increase the chemical sensing performances for the nanocomposites by increasing number of interacting sites with the analyte, mobility of charge in the nanocomposite, adsorption/desorption of analyte, electronic conductivity, etc. As for evidence, the PANI/carboxyl-functionalized MWCNTs nanocomposite, which favorably interacted with each other, created a pathway for strong dipole interaction with the analyte molecules (Figure 29) to enhance the chloroform sensing performances (Kar and Choudhury 2013). Similarly, the responses and selectivity of the poly(m-aminophenol) (PmAP)/carboxylic acid group which functionalized MWCNT nanocomposite sensor toward aliphatic alcohol vapor were substantially improved with good reproducibility and reversibility compared to pure PmAP due to having better site selective interaction in the presence of CNT (Verma et al. 2015b).The PPY/MWCNT or PPY/carbon nanofiber (CNF) nanocomposite was used for ammonia sensing over a wide range of concentration of ammonia with a higher sensitivity than that of the pristine polymer (Jang and Bae 2007, Chen et al. 2007b). Therefore, the conjugated polymer nanocomposites with pristine or functionalized carbon nanofillers generally show better sensitivity, selectivity, stability, detection limit over the pristine nanofillers or the conjugated polymers. For example, the sensitivity for NO_2 gas of PPY-single walled CNT nanocomposite is about ten times higher than that of pristine PPY (An et al. 2004). The P3TH/MWCNT nanocomposite was used for the detection of chloromethane gases and the nanocomposite sensor was also found to be selective towards methane gas (Santhanam et al. 2005). As shown in Figure 30, the hierarchically nanostructured reduced graphene–PANI nanocomposite film exhibited better sensing performance towards NH_3 gas than that of the individual components, rGO or PANI nanofiber (Guo et al. 2016). Recently, conjugated polymer/CNT nanocomposites have received significant interest for bio-sensing applications because the incorporation of CNTs into conjugated polymers can lead to better electrochemical performances as well. For example, the biosensor immobilized abundant CNTs

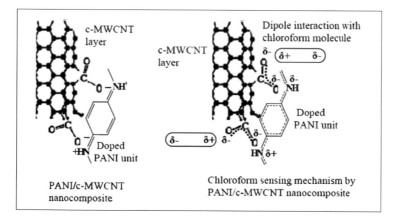

Figure 29. Mechanism of better chloroform sensing through enhanced dipole interaction by PANI/carboxylic acid group functionalized MWCNTs nanocomposite (Reproduced with permission from Kar, P. and A. Choudhury. 2013. Carboxylic acid functionalized multi-walled carbon nanotube doped polyaniline for chloroform sensors. Sens. Actuators B: Chem. 183: 25–33, Copyright© Elsevier B.V.).

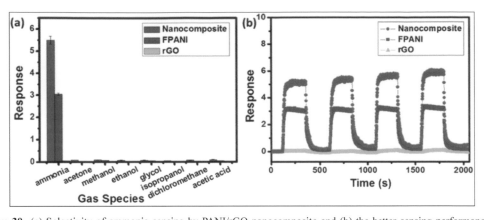

Figure 30. (a) Selectivity of ammonia sensing by PANI/rGO nanocomposite and (b) the better sensing performances of PANI/rGO nanocomposite over PANI nanofiber or rGO (Reproduced with permission from Guo, Y., T. Wang, F. Chen, X. Sun, X. Li, Z. Yu, P. Wan and X. Chen. 2016. Hierarchical graphene–polyaniline nanocomposite films for high-performance flexible electronic gas sensors. Nanoscale 8: 12073–12080, Copyright© Royal Society of Chemistry).

with polyaniline were used for a rapid response and also expanded the linear response range (Qu et al. 2005).

Electroactive electrode

The carbon nanomaterials itself have been used for various application as electrode materials and thus the nanocomposites of conjugated polymer with this good electrode materials have also been introduced in those applications with improved performances. The supercapacitors based on the conjugated polymer based nanocomposites with carbon nanomaterials such as, CNT/PPY (Ju et al. 2008), activated carbon nanofibers (ACNF)/CNT/PPY (Hughes et al. 2002), CNT/PANI and CNT/PPY (Khomenko et al. 2005), SWNT/PANI (Gupta and Miura 2006), CNT/PEDOT (Chen et al. 2009), etc., have shown better performances over the pristine components. The values obtained for specific capacitance and maximum storage energy per unit mass of the polymer/CNT composites were found to be comparable to the best reported values for well dispersed MWNT in two electrode configuration. The carbon nanomaterials can be used as an effective carbon backbone as they allow excellent dispersion of conducting polymers (Rakhi et al. 2012). The graphene-PEDOT nanocomposite as electrode material was found to have good charge/discharge property with improved specific capacitance in hydrochloric or sulfuric acid solution. The charge/discharge mechanism for the above nanocomposite is well explained in the Figure 31 with the help of electrochemical double layer capacitor mechanism (Alvi et al. 2011). As shown in the Figure 31, the specific high power capability was attributed to fast and short diffusive transport of ions through the interface of nanocomposite coated electrode. In order to enhance the electroactive properties for the EMI shielding application, the conducting polymers might be combined with other nanocomponents. For example, CNT/PANI nanocomposite was applied for shielding purposes with better microwave absorption (Saini et al. 2009). The nanocomposites of conjugated polymers with conducting carbon fillers (CNT, active carbon nanoparticles) which are acting as template for electro deposition of polymer layer are regarded as good candidates for capacitors. For example, PANI/CNT nanocomposite electrode synthesized by electrochemical polymerization on CNT was used for capacitor material (Zhang et al. 2008). The CNT nanocomposites with conjugated polymers were also investigated as electroactive materials for dielectric applications. For example, the improved dielectric permittivity has been reported for dielectric application of PmAP/carboxylic acid functionalized nanocomposite or PANI/carboxylic acid functionalized nanocomposite (Kar and Choudhury 2012, Verma et al. 2016). In a rare case, PANI (emeraldine base)-MWNT nanocomposite having luminescent property was investigated for use in optoelectronic devices with improved optical properties (Sainz et al. 2005).

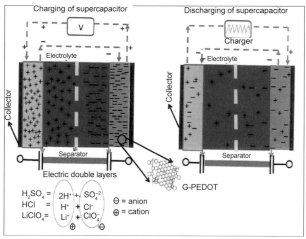

Figure 31. The electrochemical double layer capacitor mechanism for charging/discharging of graphene-PEDOT (G-PEDOT) nanocomposite (Reproduced with permission from Alvi, F., M.K. Ram, P.S. Basnayaka, E. Stefanakos, Y. Goswami and A. Kumar. 2011. Graphene-polyethylenedioxythiophene conducting polymer nanocomposite based supercapacitor. Electrochim. Acta 56: 9406–9412, Copyright© Elsevier Ltd.).

Composites with polymer nanomaterials

Composites (better known as blend) of conjugated polymer with other polymers having either of the components with nano morphology are promising material for various applications. In order to enhance the processability, mechanical and thermal stability, the nanocomposites of conducting polymers are often prepared with the other flexible insulating polymers, such as poly(vinylalcohol), poly(methyl methacrylate), poly(vinylacetate), etc. (Pud et al. 2003).

Preparation of nanocomposite

Composites of conjugated polymer with polymer nanomaterials are typically fabricated following the physical *ex situ* method by mixing the solution of both the polymers in same solvent. For example, the PANI/poly-(ethylene oxide) nanowire sensors fabricated by electrospinning method were used for ammonia gas sensing application with rapid response and recovery time (Liu et al. 2004). Following the co-evaporation electrospinning technique, the conjugated polymers/insulating polymers nanocomposites, such as PANI/polystyrene, PANI/nylon-6, PANI/polylactic acid, etc. were prepared (Zhu et al. 2006, Hong and Kang 2006, Picciani et al. 2009). In an alternative method, either of the polymers was deposited first on the surface as nanowires by electrospinning followed by deposition of other polymer. Typically, PANI/nylon-6 nanocomposite was prepared via the chemical polymerization of aniline on the surface of electrospun nylon-6 nanofibers templates (Hong et al. 2005). Following the same method, the films or layer of PPY/polymethyl methacrylate, PANI/polyvinyl alcohol PPY/polyvinyl acetate, PPY/polystyrene, PPY/polyvinyl chloride nanocomposites were fabricated for gas sensing applications (Jung et al. 2008, Liu et al. 2004, Ruangchuay et al. 2003, McGovern et al. 2005, Hosseini and Entezami 2003). The template-guided synthesis, where the surface of polymer chains was used as the template to form nanoparticles of conjugated polymer, was carried out with a slightly modified procedure. As shown in Figure 32, aniline monomer salt with camphor sulfonic acid (CSA) was formed complex on the surface of polyacrylic acid (PAA) through acid-base interaction and the polymerization of locally concentrated aniline monomer on the surface was performed to obtain PAA/PANI/CSA nanocomposite (Li et al. 2002b). These types of nanocomposite can be easily prepared chemically by the simulations or co-polymerization process as both the components are polymeric in nature. For example, nanofiber composites of PANI/poly(3,4-ethylenedioxythiophene) were fabricated by the simulations polymerization of aniline and 3,4-ethylenedioxythiophene in aqueous medium using ammonium persulfate initiator (Huh and Basavaraja 2015).

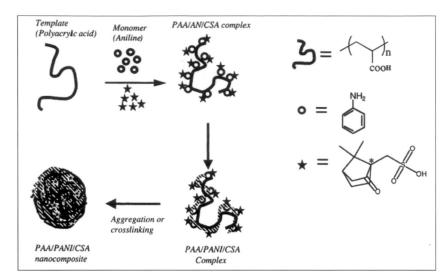

Figure 32. The schematic representation of template guided preparation of PAA/PANI/CSA nanocomposite (Reproduced with permission from Li, W., P.A. McCarthy, D. Liu, J. Huang, S.C. Yang and H.L. Wang. 2002b. Toward understanding and optimizing the template-guided synthesis of chiral polyaniline nanocomposites. Macromolecules 35: 9975–9982, Copyright© American Chemical Society).

Important applications

The sensing performances of those conjugated polymer nanocomposites greatly improved due to having better conductivity and more rapid electrochemical switching speeds. The advantages of sensitivity, spatial resolution, and rapid response associated with individual nanowires are imparted in the nanocomposites along with the material advantages associated with conjugated polymeric conductors (Innis and Wallace 2002). Moreover, those nanocomposites having better mechanical strength as well as easy fabrication of sensing layer could have improved detection limit due to having much greater exposed surface area and much greater penetration depth for nanowires relative to their bulky counterparts (Pokhodenko et al. 1999). For example, ethanol vapor sensor based on poly(3,4-ethylenedioxythiophene): poly-(styrenesulfonate) conducting polymer nanowires shows a higher and more linear resistance response than that of the bulk nanocomposite film having same thickness (Jung et al. 2008). The flexible mechanically strong film of poly(methyl methacrylate)/P3HT or PANI/poly(ethylene oxide) nanocomposite was investigated for the application as Schottky diode (Kuo et al. 2009) or field effect transistor (FET) (Pinto et al. 2003). The films of conducting nanocomposites of conjugated polymer with insulating polymer were also used as gas sensors to detect chemical or bio-analytes. For examples, poly(methyl methacrylate)/PANI coaxial composite nanofiber was used as triethylamine vapor sensor (Ji et al. 2009). In order to analyze bio-analyte, the PANI/polystyrene nanocomposite prepared by electrospinning was employed for the sensing of H_2O_2 (Aussawasathien et al. 2005). The nanocomposite film of conjugated polymer with the bio-compatible polymer should have better suitable property for the applications in drug delivery due to introducing their large surface area and better mechanical strength. For example, the poly(lactic acid-glutaric acid) nanocomposites with PEDOT or PPY synthesized via electrospinning were demonstrated as controlled drug delivery materials (Abidian et al. 2006).

Nanocomposites with bio-molecules

The nanocomposites of conjugated polymers with biomolecules like enzymes, DNA, RNA, proteins (amino acids), antibodies, antigens, etc., have been tried for a variety of applications due to their enhanced physical, chemical and especially electrochemical properties compared to the pure components. Among those applications, the main applications of conjugated polymer nanocomposites with bio-molecules are the sensor applications.

Preparation of nanocomposites

In the fabrication procedure, the problem that should be kept in mind is that most of the biomolecules are water soluble while the conjugated polymers are not at all soluble in water. Conjugated polymers nanocomposites with those biomolecules have been fabricated via several approaches (Ma et al. 2004, Nickels et al. 2004, Singh et al. 2009b, Ghanbari et al. 2008, Ko and Jang 2008, Yang et al. 2006, Dawn and Nandi 2005). In physical method, simple adsorption or entrapment technique is followed stepwise on the porous material for the synthesis of nanocomposites. For example, the urease enzyme was immobilized by a physical entrapment approach onto the pre-deposited PPY nanotubes over porous carbon paper substrate (Syu and Chang 2009). Chemical synthetic procedure involves the molecular combining of biomolecules with the monomer through electrostatic interaction followed by polymerization to get composite. For example, PANI nanowires composite with DNA was fabricated by chemically polymerized protonated aniline monomers organized along the DNA chains (Ma et al. 2004). In a modified chemical method, the biomolecule RNA can be used as a template by absorbing protonated aniline molecules on the negatively charged hybridized surface of RNA for the formation of RNA/PANI composite nanowires (Fan et al. 2007). Similarly, disaccharide and polysaccharide have been employed as the templates for the preparation of nanocomposite of saccharide with nanostructured conjugated polymer (Numata et al. 2004, Shi et al. 2006, Yu et al. 2006). Covalent functionalization of conjugated polymer with the biomolecules through crosslinking by suitable small chemical compound is also a reliable method for the fabrication of such types of nanocomposites. For example, covalent bonding of uricase was achieved with PANI using glutaraldehyde as a cross-linker (Dhand et al. 2009). As shown in the Figure 33, the example of similar types of chemically covalent functionalized nanocomposites are biotin-functionalized polythiophene, biotintylated polythiophenes, purine or pyrimidine attaching a pyrimidine or triazine to bi-, tri-, or polythiophenes (McQuade et al. 2000). Immobilization of biologically active molecules as nanocomposite with conjugated polymer during electrochemical polymerization has been deposited

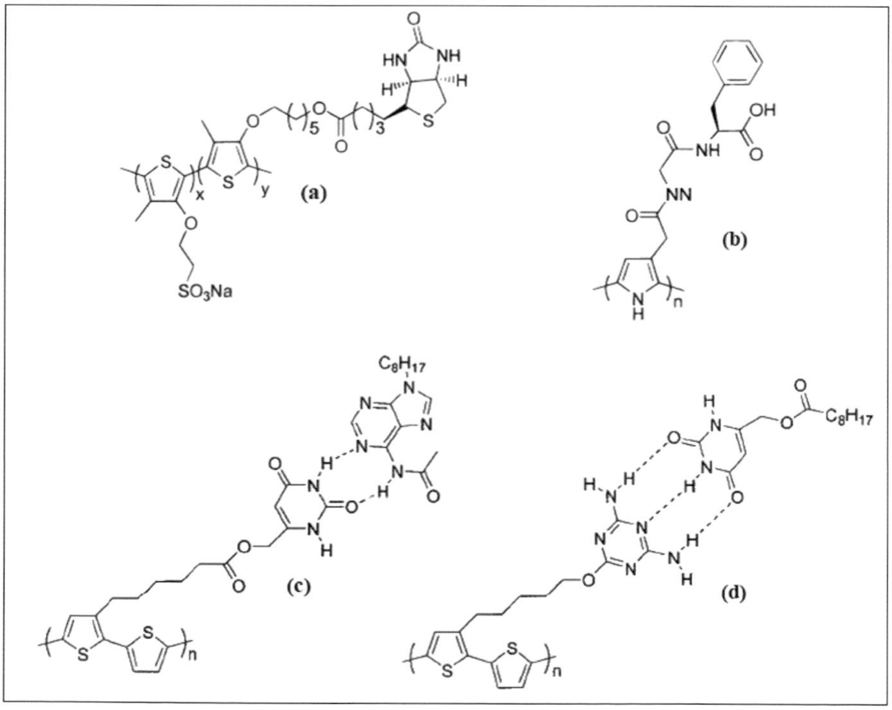

Figure 33. Chemical structure of covalent functionalized nanocomposites of (a) biotin-functionalized polythiophene, (b) biotintylated polythiophenes, (c) purine or pyrimidine attaching a pyrimidine or (d) triazine to bi-, tri-, or polythiophenes (Reproduced with permission from McQuade, D.T., A.E. Pullen and T.M. Swager. 2000. Conjugated polymer-based chemical sensors. Chem. Rev. 100: 2537–2574, Copyright© American Chemical Society).

on the electrode surface and that can be directly applied for the electrochemical sensing application. Following this method, biologically active molecules within PPY or glucose oxidase into chitosan-PANI membranes have been successfully fabricated (Ramanavicius et al. 2006, Xu et al. 2006).

Sensor applications

The conjugated polymer nanocomposites with biomaterials have been mainly investigated for sensing applications. As an electrochemical transducer, a conjugate polymer having intrinsic charge transfer property can convert the bio-response into an electric current with improved detection speed, stability, sensitivity and the response time. Even those types of nanocomposites were also successfully used to detect the chemical analyte like HCl, ammonia, etc. For example, the PANI/DNA nanocomposite on Si surfaces was used for the detection of NH_3 and HCl (Ma et al. 2004). The nanocomposites of bio-molecules with conducting polymers were mainly introduced as biosensors with enhanced speed, stability, and sensitivity. Therefore, the fabrications of conjugated polymers nanocomposites with biological materials in the form of film on electrodes have been demonstrated. A typical configuration for DNA sensors based on conjugated polymer nanocomposites with corresponding DNA probe biomolecule is shown in Figure 34 (Peng et al. 2009). The sensor performance was enhanced due to the improving transducer through the conjugated polymer layer for the signal of recognition event taking place at the conjugated polymer nanocomposite with DNA probe and electrolyte interface. For example, PPY/DNA nanocomposite prepared by electropolymerization method was used for the sensing application (Ghanbari et al. 2008). Moreover, the introduction of covalent bond between the enzyme and the conjugated polymer matrix enhanced the stability of the sensor membrane. The covalently immobilized lipase nanocomposite with PANI via glutaraldehyde reactions was found to have good sensitivity for triglyceride with good linearity (Dhand et al. 2009). Immobilization of DNA onto conducting polymers like PANI or PPY was extensively studied for detection of various DNA target sequences and microorganisms (Singh et al. 2009b, Ghanbari et al. 2008).

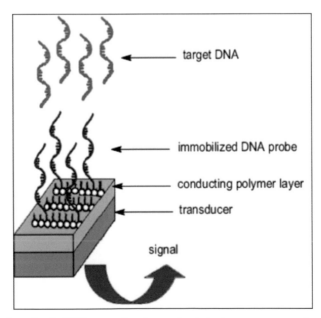

Figure 34. A typical configuration for DNA sensors based on conjugated polymer nanocomposites with corresponding DNA probe biomolecule (Reproduced with permission from Peng, H., L. Zhang, C. Soeller and J.T. Sejdic. 2009. Conducting polymers for electrochemical DNA sensing. Biomaterials 30: 2132–2148, Copyright© Elsevier Ltd).

Composites with other nanomaterials

The optical properties of CdS/PANI nanocomposite film with improved mechanical strength synthesized by electrochemical technique were significantly enhanced and thus the nanocomposite was demonstrated as a material for the solar cell (Xi et al. 2005). In order to further increase the mechanical strength of the film, the electrospun PEO/PANI/CdS three-component nanocomposite showing good optical properties was introduced for the solar cell application (Yu et al. 2008). Other than metal or metal oxide nanomaterials, the metal salts can also be incorporated into the conducting polymer matrix for gas sensing applications. For example, the PANI/CuCl$_2$ nanocomposite exhibited a high response for H$_2$S gas compared to the pristine PANI (Virji et al. 2005). GOX was incorporated into CNT/PPY nanocomposite with Fe nanoparticles on the tip for the sensing of H$_2$O$_2$ through the electrochemical redox reaction on the surface of Fe nanoparticles catalyst (Gao et al. 2003). Similarly, three component nanocomposite containing PANI, Fe$_3$O$_4$ and CNTs were used to fabricate the glucose biosensors (Liu et al. 2008). In general, the study on conjugated polymer based nanocomposites of multi-components are just the extension of bi-component system having one or more improved property incorporated by the individual components. For example, similar type of *in situ* electrochemical polymerization (Figure 35) of pyrrole in presence of CNT and graphene dispersion was carried out to prepare PPY/CNT/Graphene nanocomposite film deposited on electrode (Aphale et al. 2015). The electrode coated with PPY/CNT/ Graphene nanocomposite was examined for the supercapacitor application with improved performances (Aphale et al. 2015).

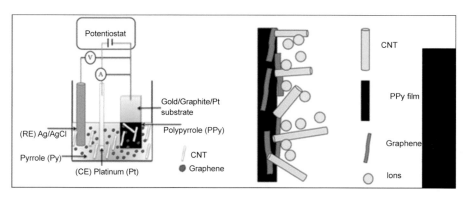

Figure 35. *In situ* electrochemical polymerization of pyrrole in presence of CNT and graphene dispersion to deposite PPY/CNT/Graphene nanocomposite film on electrode (Reproduced with permission from Aphale, A., K. Maisuria, M.K. Mahapatra, A. Santiago, P. Singh and Prabir Patra. 2015. Hybrid electrodes by *in situ* integration of graphene and carbon-nanotubes in polypyrrole for supercapacitors. Sci. Rep. 5: 14445–14452, Copyright© Nature Publishing Group).

Remarks

The global research and development on the field of conjugated polymers based nanocomposites has occurred only within the last 30–40 years. However, the achievement of those nanocomposites for prospective applications in the various fields has not been widely recognized, especially by trade and business. The problems are still associated with both the components. The conjugated polymers inherently suffer from poor processability, little reproducibility of their properties, problems in doping, problems in processability, problems in mechanical and chemical stability, etc. For example, the conjugated polymers like PANI or its derivatives synthesized from the acid medium are *in situ* doped with the corresponding acids and in order to make them processable, the salt form is reduced to base form by dipping in ammonia, phenyl hydrazine or hydrazine solution. On the other hand, the nanomaterials also have its own problems like poor control over size, shape and properties. As such, those nanocomposites suffer from the basic problems of poor solution processability, poor environmental stability, poor mechanical strength and poor conductivity, etc., like conjugated polymers and that restricts the applications. The properties of nanocomposites of conjugated polymers with nanomaterials are also associated with fundamental challenges like successful assembling of nanomaterials in their nanocomposite for applied research. Moreover, the film casting of chemically synthesized polymers is often hampered due to having poor mechanical strength or bad film forming properties of the conjugated polymer as for getting good performance, a thin film would be better than a pellet. In general, the mixing of nanomaterials as second immiscible component further reduces the mechanical strength of the nanocomposites film. The major obstacles for successful combination of those two components in single component to reinforce the property are uniform dispersion, control over size, shape, and surface chemistry of small nanomaterial as well as poor control over interfacial interactions. Therefore, the conducting polymer-based nanocomposites are novel materials with a short history of study, which makes it difficult to foresee all of or even the main fields for future applications of these materials. It can be only expected that those novel nanocomposites of conjugated polymers will progress with time to being as commonplace as the electronic materials are today.

References

Abidian, M.R., D.H. Kim and D.C. Martin. 2006. Conducting-polymer nanotubes for controlled drug release. Adv. Mater. 18: 405–409.

Ahn, K.-J., Y. Lee, H. Choi, M.-S. Kim, K. Im, S. Noh et al. 2015. Surfactant-templated synthesis of polypyrrole nanocages as redox mediators for efficient energy storage. Sci. Rep. 5: 14097–14106.

Akbarinezhad, E., M. Ebrahimi, F. Sharif, M.M. Attar and H.R. Faridi. 2011. Synthesis and evaluating corrosion protection effects of emeraldine base PAni/clay nanocomposite as a barrier pigment in zinc-rich ethyl silicate primer. Prog. Org. Coatings 70: 39–44.

Alexandre, M. and P. Dubois. 2000. Polymer layered-silicate nanocomposites: Preparation, properties and use of a new class of materials. Mater. Sci. Eng. Rep. 28: 1–63.

Alvi, F., M.K. Ram, P.S. Basnayaka, E. Stefanakos, Y. Goswami and A. Kumar. 2011. Graphene-polyethylenedioxythiophene conducting polymer nanocomposite based supercapacitor. Electrochim. Acta 56: 9406–9412.

An, K.H., S.Y. Jeong, H.R. Hwang and Y.H. Lee. 2004. Enhanced sensitivity of a gas sensor incorporating single-walled carbon nanotube–polypyrrole nanocomposites. Adv. Mater. 16: 1005–1009.

Anuar, K., S. Murali, A. Fariz and H.N.M.M. Ekramul. 2004. Polymer/clay composites: Preparation and characterization. Mater. Sci. 10: 255–258.

Aphale, A., K. Maisuria, M.K. Mahapatra, A. Santiago, P. Singh and Prabir Patra. 2015. Hybrid electrodes by *in situ* integration of graphene and carbon-nanotubes in polypyrrole for supercapacitors. Sci. Rep. 5: 14445–14452.

Athawale, A.A., S.V. Bhagwat, P.P. Katre, A.J. Chandwadkar and P. Karandikar. 2003. Aniline as a stablizer for metal nanoparticles. Mater. Lett. 57: 3889–3894.

Athawale, A.A., S.V. Bhagwat and P.P. Katre. 2006. Nanocomposite of Pd-polyaniline as a selective methanol sensor. Sens. Actuators B: Chem. 114: 263–267.

Aussawasathien, D., J.H. Dong and L. Dai. 2005. Electrospun polymer nanofibers sensors. Synth. Met. 154: 37–40.

Balint, R., N.J. Cassidy and S.H. Cartmell. 2014. Conductive polymers: Towards a smart biomaterial for tissue engineering. Acta Biomaterialia 10: 2341–2353.

Bauhofer, W. and J.Z. Kovacs. 2009. A review and analysis of electrical percolation in carbon nanotube polymer composites. Compos. Sci. Technol. 69: 1486–1498.

Beguin, F. and E. Frackowiak. 2010. Carbons for Electrochemical Energy Storage and Conversion Systems. CRC Press, Boca Raton.

Bergaya, F. and G. Lagaly. 2001. Surface modifications of clay minerals. Appl. Clay Sci. 19: 1–3.

Bergaya, F. and G. Lagaly. 2006. General introduction: Clays, clay minerals, and clay science. pp. 1–18. *In*: F. Bergaya, B.K.G. Theng and G. Lagaly (eds.). Handbook of Clay Science: Developments in Clay Science. Vol. 1. Amsterdam, Elsevier.

Bidan, G. 1992. Electroconducting conjugated polymers: new sensitive matrices to build up chemical or electrochemical sensors: a review. Sens. Actuators B: Chem. 6: 45–56.

Biswas, M. and S.S. Ray. 2000. Water-dispersible nanocomposites of polyaniline and montmorillonite. J. Appl. Polym. Sci. 77: 2948–2956.

Bose, C.S.C. and K. Rajeshwar. 1992. Efficient electrocatalyst assemblies for proton and oxygen reduction: the electrosynthesis and characterization of polypyrrole films containing nanodispersed platinum particles. J. Electroanal. Chem. 333: 235–256.

Boukerma, K., J.-Y. Piquemal, M.M. Chehimi, M. Mravcakova, M. Omastova and P. Beaunier. 2006. Synthesis and interfacial properties of montmorillonite/polypyrrole nanocomposites. Polymer 47: 569–576.

Breimer, M.A., G. Yevgeny, S. Sy and O.A. Sadik. 2001. Incorporation of metal nanoparticles in photopolymerized organic conducting polymers: a mechanistic insight. Nano Lett. 1: 305–308.

Cao, H., Z. Xu, H. Sang, D. Sheng and C. Tie. 2001a. Template synthesis and magnetic behavior of an array of cobalt nanowires encapsulated in polyaniline nanotubules. Adv. Mater. 13: 121–123.

Cao, H., Z. Xu, D. Sheng, J. Hong, H. Sang and Y. Du. 2001b. An array of iron nanowires encapsulated in polyaniline nanotubules and its magnetic behavior. J. Mater. Chem. 11: 958–960.

Cao, H., C. Tie, Z. Xu, J. Hong and H. Sang. 2001c. Array of nickel nanowires enveloped in polyaniline nanotubules and its magnetic behavior. Appl. Phys. Lett. 78: 1592–1594.

Carretero, M.I. and G. Lagaly. 2007. Clays and health: an introduction. Appl. Clay Sci. 36: 1–3.

Chang, J.K. and W.T. Tsai. 2003. Material characterization and electrochemical performance of hydrous manganese oxide electrodes for use in electrochemical pseudocapacitors. J. Electrochem. Soc. 150: A1333–A1338.

Chang, K.-C., M.-C. Lai, C.-W. Peng, Y.-T. Chen, J.-M. Yeh, C.-L. Lin et al. 2006. Comparative studies on the corrosion protection effect of DBSA-doped polyaniline prepared from *in situ* emulsion polymerization in the presence of hydrophilic Na$^+$-MMT and organophilic organo-MMT clay platelets. Electrochim. Acta 51: 5645–5653.

Chen, Z., L. Xu, W. Li, M. Waje and Y. Yan. 2006. Polyaniline nanofibre supported platinum nanoelectrocatalysts for direct methanol fuel cells. Nanotechnology 17: 5254–5259.

Chen, Q., L. Zhao, C. Li and G.Q. Shi. 2007a. Electrochemical fabrication of a memory device based on conducting polymer nanocomposites. J. Phys. Chem. C 111: 18392–18396.

Chen, Y., Y. Li, H. Wang and M. Yang. 2007b. Gas sensitivity of a composite of multi-walled carbon nanotubes and polypyrrole prepared by vapor phase polymerization. Carbon 45: 357–363.

Chen, L., C.Z. Yuan, H. Dou, B. Gao, S. Chen and X. Zhang. 2009. Synthesis and electrochemical capacitance of core-shell poly(3,4-ethylenedioxythiophene)/poly(sodium 4-styrenesulfonate)-modified multiwalled carbon nanotube nanocomposites. Electrochim. Acta 54: 2335–2341.

Choudhury, A. 2009. Polyaniline/silver nanocomposites: Dielectric properties and ethanol vapour sensitivity. Sens. Actuators B: Chem. 138: 318–325.

Choudhury, A., P. Kar, M. Mukherjee and B. Adhikari. 2009. Acetone sensing by polyaniline/silver nanocomposite. Sensor Lett. 7: 592–598.

Chung, H.J., H.H. Jung, Y.S. Cho, S. Lee, J.H. Ha, J.H. Choi et al. 2005. Cobalt–polypyrrole–cobalt nanowire field-effect transistors. Appl. Phys. Lett. 86: 213113–213116.

Coche, L. and J.C. Moutet. 1987. Electrocatalytic hydrogenation of organic compounds on carbon electrodes modified by precious metal microparticles in redox active polymer films. J. Am. Chem. Soc. 109: 6887–6889.

Cui, S., J. Wang and X. Wang. 2015. Fabrication and design of a toxic gas sensor based on polyaniline/titanium dioxide nanocomposite film by layer-by-layer self-assembly. RSC Adv. 5: 58211–58219.

Dai, X., Y. Tan and J. Xu. 2002. Formation of gold nanoparticles in the presence of o-anisidine and the dependence of the structure of poly(o-anisidine) on synthetic conditions. Langmuir 18: 9010–9016.

Dawn, A. and A.K. Nandi. 2005. Slow doping rate in DNA−poly(o-methoxyaniline) hybrid: Uncoiling of poly(o-methoxyaniline) chain on DNA template. Macromolecules 38: 10067–10073.

Dean, D., R. Walker, M. Theodore, E. Hampton and E. Nyairo. 2005. Chemorheology and properties of epoxy/layered silicate nanocomposites. Polymer 46: 3014–3021.

dePaiva, L.B., A.R. Morales and F.R.V. Diaz. 2008. Organoclays: Properties, preparation and applications. Appl. Clay Sci. 42: 8–24.

Deshpande, N.G., Y.G. Gudage, R. Sharma, J.C. Vyas, J.B. Kim and Y.P. Lee. 2009. Studies on tin oxide-intercalated polyaniline nanocomposite for ammonia gas sensing applications. Sen. Actuators B: Chem. 138: 76–84.

Dhand, C., P.R. Solanki, K.N. Sood, M. Datta and B.D. Malhotra. 2009. Polyaniline nanotubes for impedimetric triglyceride detection. Electrochem. Commun. 11: 1482–1486.

Ding, K., Z. Jia, W. Ma, R. Tong and X. Wang. 2002. Polyaniline and polyaniline-thiokol rubber composite coatings for the corrosion protection of mild steel. Mater. Chem. Phys. 76: 137–142.

Do, J.S. and W.B. Chang. 2004. Amperometric nitrogen dioxide gas sensor based on PAn/Au/Nafion® prepared by constant current and cyclic voltammetry methods. Sens. Actuators B: Chem. 101: 97.

doNascimento, G.M., V.R.L. Constantino, R. Landers and M.L.A. Temperini. 2004. Aniline polymerization into montmorillonite clay: A spectroscopic investigation of the intercalated conducting polymer. Macromolecules 37: 9373–9385.

Downs, C., J. Nugent, P.M. Ajayan, D.J. Duquette and K.S.V. Santhanam. 1999. Efficient polymerization of aniline at carbon nanotube electrodes. Adv. Mater. 11: 1028–1031.

Drury, A., S. Chaure, M. Kroell, V. Nicolosi, N. Chaure and W.J. Blau. 2007. Fabrication and characterization of silver/polyaniline composite nanowires in porous anodic alumina. Chem. Mater. 19: 4252–4258.

Fan, Y., X. Chen, A.D. Trigg, C. Tung, J. Kong and Z. Gao. 2007. Detection of micro RNAs using target-guided formation of conducting polymer nanowires in nanogaps. J. Am. Chem. Soc. 129: 5437–5443.

Feast, W.J., J. Tsibouklis, K.L. Pouwer, L. Groenendaal and E.W. Meijer. 1996. Synthesis, processing and material properties of conjugated polymers. Polymer 37: 5017–5047.

Feng, B., Y. Su, J. Song and K. Kong. 2001. Electropolymerization of polyaniline/montmorillonite nanocomposite. Mater. Sci. Let. 20: 293–294.

Fu, Y., R. Yuan, Y. Chai, L. Zhou and Y. Zhang. 2006. Coupling of a reagentless electrochemical DNA biosensor with conducting polymer film and nanocomposite as matrices for the detection of the HIV DNA sequences. Anal. Lett. 39: 467–482.

Gallon, B.J., R.W. Kojima, R.B. Kaner and P.L. Diaconescu. 2007. Palladium nanoparticles supported on polyaniline nanofibers as a semiheterogeneous catalyst in water. Angew. Chem. Int. Ed. 46: 7251–7254.

Gao, M., L. Dai and G.G. Wallace. 2003. Biosensors based on aligned carbon nanotubes coated with inherently conducting polymers. Electroanalysis 15: 1089–1094.

Geng, L., Y. Zhao, X. Huang, S. Wang, S. Zhang and S. Wu. 2007. Characterization and gas sensitivity study of polyaniline/ SnO_2 hybrid material prepared by hydrothermal route. Sens. Actuators B: Chem. 120: 568–572.

Ghanbari, K., S.Z. Bathaie and M.F. Mousavi. 2008. Electrochemically fabricated polypyrrole nanofiber-modified electrode as a new electrochemical DNA biosensor. Biosens. Bioelectronics 23: 1825–1831.

Ghosh, P., S. Siddhanta, S.R. Haque and A. Chakrabarti. 2001. Stable polyaniline dispersions prepared in nonaqueous medium: Synthesis and characterization. Synth. Met. 123: 83–89.

Gowariker, V.R., N.V. Viswanathan and J. Sreedhar. 2005. Polymer Science (1st Ed.) New Age Publication (P) Ltd., New Delhi, India, pp. 230–245.

Granot, E., E. Katz, B. Basnar and I. Willner. 2005. Enhanced bioelectrocatalysis using Au-nanoparticle/polyaniline hybrid systems in thin films and microstructured rods assembled on electrodes. Chem. Mater. 17: 4600–4609.

Greene, R.L., G.B. Street and L.J. Sutude. 1975. Superconductivity in polysulfur nitride $(SN)_x$. Phys. Rev. Lett. 34: 577–579.

Gul, S., A.H.A. Shah and S. Bilal. 2013. Synthesis and characterization of processable polyaniline salts. J. Phys. Conf. Ser. 439: 012002–012008.

Guo, H., H. Zhu, H. Lin and J. Zhang. 2008. Polypyrrole–multi-walled carbon nanotube nanocomposites synthesized in oil–water microemulsion. Colloid. Polym. Sci. 286: 587–591.

Guo, S., S. Dong and E. Wang. 2009. Polyaniline/Pt hybrid nanofibers: high efficiency nanoelectrocatalysts for electrochemical devices. Small 5: 1869–1876.

Guo, Y., T. Wang, F. Chen, X. Sun, X. Li, Z. Yu et al. 2016. Hierarchical graphene–polyaniline nanocomposite films for high-performance flexible electronic gas sensors. Nanoscale 8: 12073–12080.

Gupta, V. and N. Miura. 2006. Polyaniline/single-wall carbon nanotube (PANI/SWCNT) composites for high performance supercapacitors. Electrochim. Acta 52: 1721–1726.

Gupta, B., A.K. Singh and R. Prakash. 2010. Electrolyte effects on various properties of polycarbazole. Thin Solid Films 519: 1016–1019.

He, X.M. and G.Q. Shi. 2006. Electrochemical actuator based on monolithic polypyrrole-TiO_2 nanoparticle composite film. Sens. Actuators B: Chem. 115: 488–493.

Holdcroft, S. and B.L. Funt. 1988. Preparation and electrocatalytic properties of conducting films of polypyrrole containing platinum microparticulates. J. Electroanal. Chem. Interf. Electrochem. 240: 89–103.

Hong, K.H., K.W. Oh and T.J. Kang. 2005. Preparation of conducting nylon-6 electrospun fiber webs by the *in situ* polymerization of polyaniline. J. Appl. Polym. Sci. 96: 983–991.

Hong, K.H. and T.J. Kang. 2006. Polyaniline-nylon 6 composite nanowires prepared by emulsion polymerization and electrospinning process. J. Appl. Polym. Sci. 99: 1277–1286.

Hosseini, S.H. and A.A. Entezami. 2003. Conducting polymer blends of polypyrrole with polyvinyl acetate, polystyrene, and polyvinyl chloride based toxic gas sensors. J. Appl. Polym. Sci. 90: 49–62.

Hu, C.C. and C.C. Wang. 2003. Nanostructures and capacitive characteristics of hydrous manganese oxide prepared by electrochemical deposition. J. Electrochem. Soc. 150: A1079–A1084.

Hu, L., J. Tu, S. Jiao, J. Hou, H. Zhua and D.J. Fray. 2012. *In situ* electrochemical polymerization of a nanorod-PANI–Graphene composite in a reverse micelle electrolyte and its application in a supercapacitor. Phys. Chem. Chem. Phys. 14: 15652–15656.

Huang, J. and R.B. Kaner. 2004. A general chemical route to polyaniline nanofibers. J. Am. Chem. Soc. 126: 851–855.

Huang, J., S. Virji, B.H. Weiller and R.B. Kaner. 2004. Nanostructured polyaniline sensors. Chem. A Eur. J. 10: 1314–1323.

Hughes, M., G.Z. Chen, M.S.P. Shaffer, D.J. Fray and A.H. Windle. 2002. Electrochemical capacitance of a nanoporous composite of carbon nanotubes and polypyrrole. Chem. Mater. 14: 1610–1613.

Huh, D.S. and C. Basavaraja. 2015. Improving electron transport and mechanical properties of polyaniline-based composites. Plastic Res. Online 10.2417/spepro.006131.

Hung, W.I., K.C. Chang, Y.H. Chang and J.M. Yeh. 2011. Advanced anticorrosive coatings prepared from polymer-clay nanocomposite materials. pp. 561–582. *In*: B. Reddy (ed.). Advances in Nanocomposites—Synthesis, Characterization and Industrial Applications. Rijeka, Croatia.

Innis, P.C. and G.G. Wallace. 2002. Inherently conducting polymeric nanostructures. J. Nano Sci. Nanotechnol. 2: 441–451.

Inoue, H. and H. Yoneyama. 1987. Electropolymerization of aniline intercalated in montmorillonite. J. Electroanal. Chem. 233: 291–294.

Iravani, S. 2011. Green synthesis of metal nanoparticles using plants. Green Chem. 13: 2638–2650.

Jang, J., J. Ha and B.K. Lim. 2006. Synthesis and characterization of monodisperse silica–polyaniline core–shell nanoparticles. Chem. Commun. 1622–1624.

Jang, J. and J. Bae. 2007. Carbon nanofiber/polypyrrole nanocable as toxic gas sensor. Sens. Actuators B: Chem. 122: 7–13.

Ji, S., Y. Li and M. Yang. 2008. Gas sensing properties of a composite composed of electrospun poly(methyl methacrylate) nanofibers and *in situ* polymerized polyaniline. Sens. Actuators B: Chem. 133: 644–649.

Ju, Y.W., G.R. Choi, H.R. Jung and W.J. Lee. 2008. Electrochemical properties of electrospun PAN/MWCNT carbon nanofibers electrodes coated with polypyrrole. Electrochim. Acta 53: 5796–5803.

Jung, Y.S., W.C. Jung, H.L. Tuller and C. Ross. 2008. Nanowire conductive polymer gas sensor patterned using self-assembled block copolymer lithography. Nano Lett. 8: 3776–3780.

Kar, P., N.C. Pradhan and B. Adhikari. 2011a. Doping of processable conducting poly(m-aminophenol) with silver nanoparticles. Polym. Adv. Technol. 22: 1060–1066.

Kar, P., N.C. Pradhan and B. Adhikari. 2011b. Ammonia sensing by hydrochloric acid doped poly(m-aminophenol)–silver nanocomposite. J. Mater. Sci. 46: 2905–2913.

Kar, P. and A. Choudhury. 2012. Electrical and dielectric properties of polyaniline doped with carboxyl-functionalized multiwalled carbon nanotube. Adv. Polym. Technol. 32: E760–E770.

Kar, P. 2013. Doping in Conjugated Polymers. Wiley-Screensaver, New Jersey.

Kar, P. and A. Choudhury. 2013. Carboxylic acid functionalized multi-walled carbon nanotube doped polyaniline for chloroform sensors. Sens. Actuators B: Chem. 183: 25–33.

Kar, P., A. Choudhury and S.K. Verma. 2015. Chapter 12: Conjugated polymer nanocomposites based chemical sensors. pp. 621–685. *In:* P. Saini (ed.). Fundamentals of Conjugated Polymer Blends, Copolymers and Composites. Wiley-Scrivener, New Jersey.

Katz, E., A.N. Shipway and I. Willner. 2004. Chemically functionalized metal nanoparticles: Synthesis, properties and applications. pp. 5–78. *In:* Liz-Marzan, L.M. and P.V. Kamat (eds.). Nanoscale Materials. Springer, Boston.

Ke, Y.C. and P. Strove. 2005. Polymer Layered Silicate, Silica Nanocomposites. Elsevier, Amsterdam.

Khomenko, V., E. Frackowiak and F. Beguin. 2005. Determination of the specific capacitance of conducting polymer/nanotubes composite electrodes using different cell configurations. Electrochim. Acta 50: 2499–2506.

Kim, B.H., J.H. Jung, S.H. Hong, J.W. Kim, H.J. Choi and J. Joo. 2001. Physical characterization of emulsion intercalated polyaniline–clay nanocomposite. Curr. Appl. Phys. 1: 112–115.

Kim, J.W., F. Liu, H.J. Choi, S.H. Hong and J. Joo. 2003. Intercalated polypyrrole/Na+-montmorillonite nanocomposite via an inverted emulsion pathway method. Polymer 44: 289–293.

Kim, B.H., D.H. Park, J. Joo, S.G. Yu and S.H. Lee. 2005a. Synthesis, characteristics, and field emission of doped and de-doped polypyrrole, polyaniline, poly(3,4-ethylenedioxythiophene) nanotubes and nanowires. Synth. Met. Vol. 150: 279–284.

Kim, J.-K., C. Hu, R.S.C. Woo and M.-L. Sham. 2005b. Moisture barrier characteristics of organoclay–epoxy nanocomposites. Compos. Sci. Technol. 65: 805–813.

Kinlen, P.J., D.C. Silverman and C.R. Jeffreys. 1997. Corrosion protection using polyaniline coating formulations. Synth. Met. 85: 1327–1332.

Kinlen, P.J., J. Liu, Y. Ding, C.R. Graham and E.E. Remsen. 1998. Emulsion polymerization process for organically soluble and electrically conducting polyaniline. Macromolecules 31: 1735–1744.

Kitani, A., T. Akashi, K. Sugimoto and S. Ito. 2001. Electrocatalytic oxidation of methanol on platinum modified polyaniline electrodes. Synth. Met. 121: 1301–1302.

Ko, S. and J. Jang. 2008. Label-free target DNA recognition using oligonucleotide-functionalized polypyrrole nanotubes. Ultramicroscopy 108: 1328–1333.

Kong, L., X. Lu, E. Jin, S. Jiang, C. Wang and W. Zhang. 2009. Templated synthesis of polyaniline nanotubes with Pd nanoparticles attached onto their inner walls and its catalytic activity on the reduction of p-nitroanilinum. Composites Sci. Technol. 69: 561–566.

Kost, K.M., D.E. Bartak, B. Kazee and T. Kuwana. 1988. Electrodeposition of platinum microparticles into polyaniline films with electrocatalytic applications. Anal. Chem. 60: 2379–2384.

Kuo, C.C., C.T. Wang and W.C. Chen. 2009. Poly(3-hexylthiophene)/poly(methyl methacrylate) core-shell electrospun fibers for sensory applications. Macromol. Symp. 279: 41–47.

Lee, D. and K. Char. 2002. Thermal degradation behavior of polyaniline in polyaniline/Na+-montmorillonite nanocomposites. Polym. Degrad. Stabil. 75: 555–560.

Lee, D., K. Char, S.W. Lee and Y.W. Park. 2003. Structural changes of polyaniline/montmorillonite nanocomposites and their effects on physical properties. J. Mater. Chem. 13: 2942–2947.

Li, X., Y. Li, Y. Tan, C. Yang and Y. Li. 2002a. Self-assembly of gold nanoparticles prepared with 3,4-ethylenedioxythiophene as reductant. J. Phys. Chem. B 108: 5192–5199.

Li, W., P.A. McCarthy, D. Liu, J. Huang, S.C. Yang and H.L. Wang. 2002b. Toward understanding and optimizing the template-guided synthesis of chiral polyaniline nanocomposites. Macromolecules 35: 9975–9982.

Li, J. and X.Q. Lin. 2007. A composite of pyrrole nanowire platinum modified electrode for oxygen reduction and methanol oxidation reaction. J. Electrochem. Soc. 154: B1074–B1079.

Li, X., Y. Gao, J. Gong, L. Zhang and L. Qu. 2009. Polyaniline/Ag composite nanotubes prepared through UV rays irradiation via fiber template approach and their NH_3 gas sensitivity. J. Phys. Chem. C 113: 69–73.

Li, D., Y. Li, Y. Feng, W. Hu and W. Feng. 2015. Hierarchical graphene oxide/polyaniline nanocomposites prepared by interfacial electrochemical polymerization for flexible solid-state supercapacitors. J. Mater. Chem. A 3: 2135–2143.

Li, Y., H. Ban, M. Jiaoa and M. Yanga. 2016. *In situ* growth of SnO_2 nanosheets on a substrate via hydrothermal synthesis assisted by electrospinning and the gas sensing properties of SnO_2/polyaniline nanocomposites. RSC Adv. 6: 74944–74956.

Liu, R. and S.B. Lee. 2008. MnO_2/poly(3,4-ethylenedioxythiophene) coaxial nanowires by one-step coelectrodeposition for electrochemical energy storage. J. Am. Chem. Soc. 130: 2942–2943.

Liu, H., J. Kameoka, D.A. Czaplewski and H.G. Craighead. 2004. Polymeric nanowire chemical sensor. Nano Lett. 4: 671–675.

Liu, Z., J. Wang, D. Xie and G. Chen. 2008. Polyaniline-coated Fe_3O_4 nanoparticle–carbon-nanotube composite and its application in electrochemical biosensing. Small 4: 462–466.

Liu, Y., B. Zhang, Y. Yang, Z. Chang, Z. Wen and Y. Wu. 2013. Polypyrrole-coated a-MoO$_3$ nanobelts with good electrochemical performance as anode materials for aqueous supercapacitors. J. Mater. Chem. A 1: 13582–13587.

Long, Y., Z. Chen, X. Zhang, J. Zhang and Z. Liu. 2004. Electrical properties of multi-walled carbon nanotube/polypyrrole nanocables: percolation-dominated conductivity. J. Phys. D: Appl. Phys. 37: 1965–1976.

Long, Y., Z. Chen, J.L. Duvail, Z. Zhang and M. Wan. 2005. Electrical and magnetic properties of polyaniline/Fe$_3$O$_4$ nanostructures. Physical B 370: 121–130.

Loo, L.S. and K.K. Gleason. 2004. Investigation of polymer and nanoclay orientation distribution in nylon 6/montmorillonite nanocomposite. Polymer 45: 5933–5939.

Lu, X., H. Mao, D. Chao, W. Zhang and Y. Wei. 2006. Ultrasonic synthesis of polyaniline nanotubes containing Fe$_3$O$_4$ nanoparticles. J. Solid State Chem. 179: 2609–2615.

Luo, J.J. and I.M. Daniel. 2003. Characterization and modeling of mechanical behavior of polymer/clay nanocomposites. Compos. Sci. Technol. 63: 1607–1616.

Ma, Y., J. Zhang, G. Zhang and H. He. 2004. Polyaniline nanowires on Si surfaces fabricated with DNA templates. J. Am. Chem. Soc. 126: 7097–7101.

Ma, P.C., N.A. Siddiqui, G. Marom and J.K. Kim. 2010. Dispersion and functionalization of carbon nanotubes for polymer-based nanocomposites: a review. Composites: Part A 41: 1345–1367.

MacDiarmid, A.G. and A.J. Epstein. 1989. Polyanilines: A novel class of conducting polymers. Faraday Discuss. Chem. Soc. 88: 317–332.

Malta, M., G. Louarn, N. Errien and R.M. Torresi. 2003. Nanofibers composite vanadium oxide/polyaniline: Synthesis and characterization of an electroactive anisotropic structure. Electrochem. Commun. 5: 1011–1015.

Martin, R.T., S.W. Bailey, D.D. Eberl, D.S. Fanning, S. Guggenheim, H. Kodama et al. 1991. Report of the clay minerals society nomenclature committee: revised classification of claymaterials. Clays Clay Miner. 39: 333–335.

Martin, C.R., R. Parthasarathy and V. Menon. 1993. Template synthesis of electronically conductive polymers—a new route for achieving higher electronic conductivities. Synth. Met. 55: 1165–1170.

McGovern, S.T., G.M. Spinks and G.G. Wallace. 2005. Micro-humidity sensors based on a processible polyaniline blend. Sens. Actuators B: Chem. 107: 657–665.

McQuade, D.T., A.E. Pullen and T.M. Swager. 2000. Conjugated polymer-based chemical sensors. Chem. Rev. 100: 2537–2574.

Mekki, A., S. Samanta, A. Singh, Z. Salmi, R. Mahmoud, M.M. Chehimi et al. 2014. Core/shell, protuberance-free multiwalled carbon nanotube/polyaniline nanocomposites via interfacial chemistry of aryl diazonium salts. J. Colloid Interface Sci. 418: 185–192.

Mi, O., B. Ru, X. Yi, Z. Cheng, M.A. Chunan, W. Mang et al. 2009. Fabrication of polypyrrole/TiO$_2$ nanocomposite via electrochemical process and its photoconductivity. Trans. Nonferrous Met. Soc. China 19: 1572–1577.

Miasik, J., A. Hooper and B. Tofield. 1986. Conducting polymer gas sensors. J. Chem. Soc. Faraday Trans. 182: 1117–1127.

Mishra, J.K., K. Hwang and C. Ha. 2004. Preparation, mechanical and rheological properties of a thermoplastic polyolefin (TPO)/organoclay nanocomposite with reference to the effect of maleic anhydride modified polypropylene as a compatibilizer. Polymer 46: 1995–2002.

Miyawaki, R., S. Tomura, K. Inukai, M. Okazaki, K. Toriyama, Y. Shibasaki et al. 1993. Formation process of kaolinite from amorphous calcium silicate and aluminum chloride. Clay Sci. 9: 21–32.

Mohamed, F. and E.S. Mohammed. 2015. Nanoscale phase separation in laponite–polypyrrole nanocomposites. Application to electrodes for energy storage. RSC Adv. 5: 21550–21557.

Mravcakova, M., K. Boukerma, M. Omastova and M.M. Chehimi. 2006. Montmorillonite/polypyrrole nanocomposites. The effect of organic modification of clay on the chemical and electrical. Mater. Sci. Eng. C 26: 306–313.

Murray, H.H. 1988. Kaolin minerals; their genesis and occurrences. Rev. Mineral. Geochem. 19: 67–89.

Nam, K.W., W.S. Yoon and K.B. Kim. 2002. X-ray absorption spectroscopy studies of nickel oxide thin film electrodes for supercapacitors. Electrochim. Acta 47: 3201–3209.

Nickels, P., W.U. Dittmer, S. Beyer and J.P. Kotthaus. 2004. Polyaniline nanowire synthesis templated by DNA. Nanotechnol. 15: 1524–1529.

Numata, M., T. Hasegawa, T. Fujisawa, K. Sakurai and S. Shinkai. 2004. β-1, 3-glucan (schizophyllan) can act as a one-dimensional host for creation of novel poly(aniline) nanofiber structures. Org. Lett. 6: 4447–4450.

Olad, A. and A. Rashidzadeh. 2008. Preparation and anticorrosive properties of PANI/Na-MMT and PANI/O-MMT nanocomposites. Prog. Org. Coat. 62: 293–298.

Olad, A., A. Rashidzadeh and M. Amini. 2013. Preparation of polypyrrole nanocomposites with organophilic and hydrophilic montmorillonite and investigation of their corrosion protection on iron. Adv. Polym. Technol. 32: 21337–21347.

Olson, D.C., J. Piris, R.T. Collins, S.E. Shaheen and D.S. Ginley. 2006. Hybrid photovoltaic devices of polymer and ZnO nanofiber composites. Thin Solid Films 496: 26–29.

Oriakhi, C.O. and M.M. Lerner. 1995. Poly(pyrrole) and poly(thiophene)/clay nanocomposites via latex-colloid interaction. Mater. Res. Bull. 30: 723–729.

Otero, T.F. 2009. Soft, wet, and reactive polymers. Sensing artificial muscles and conformational energy. J. Mater. Chem. 19: 681–689.

Parham, W.E. 1969. Formation of halloysite from feldspar: low temperature. Artif. Weather. Vers. Nat. Weather. 17: 13–22.

Park, J.-E., S. Kim, S. Mihashi, O. Hatozaki and N. Oyama. 2002. Roles of metal nanoparticles on organosulfur-conducting polymer composites for lithium battery with high energy density. Macromol. Symp. 186: 35–40.

Park, J.H., Y.T. Lim, O.O. Park, J.K. Kim, J.-W. Yu and Y.C. Kim. 2004a. Polymer/gold nanoparticle nanocomposite light-emitting diodes: enhancement of electroluminescence stability and quantum efficiency of blue-light-emitting polymers. Chem. Mater. 16: 688–692.

Park, J.-E., S.-G. Park, A. Koukitu, O. Hatozaki and N. Oyama. 2004b. Effect of adding Pd nanoparticles to dimercaptan-polyaniline cathodes for lithium polymer battery. Synth. Met. 140: 121–126.

Park, S., S.W. Chung and C.A. Mirkin. 2004c. Hybrid organic-inorganic, rod-shaped nanoresistors and diodes. J. Am. Chem. Soc. 126: 11772–11775.

Patra, S., K. Barai and N. Munichandraiah. 2008. Scanning electron microscopy studies of PEDOT prepared by various electrochemical routes. Synth. Met. 158: 430–435.

Pei, Q. and O. Inganas. 1993. Conjugated polymers as smart materials, gas sensors and actuators using bending beams. Synth. Met. 57: 3730–3735.

Peighambardoust, S.J. and B. Pourabbas. 2007. Synthesis and characterization of conductive polypyrrole/montmorillonite nanocomposites via one-pot emulsion polymerization. Macromol. Symp. 247: 99–109.

Peng, H., L. Zhang, C. Soeller and J.T. Sejdic. 2009. Conducting polymers for electrochemical DNA sensing. Biomaterials 30: 2132–2148.

Peter, J. and F. Harris. 1999. Carbon Nanotube and Related Structures, New Materials for the Twenty First Century. University of Cambridge, Cambridge.

Philip, B., J. Xie, A. Chandrasekhar, J. Abraham and V.K. Varada. 2004. A novel nanocomposite from multiwalled carbon nanotubes functionalized with a conducting polymer. Smart. Mater. Struct. 13: 295–298.

Philip, B., J. Xie, J. Abraham and V.K. Varada. 2005. Polyaniline/carbon nanotube composites: starting with phenylamino functionalized carbon nanotubes. Polym. Bull. 53: 127–138.

Picciani, P.H.S., E.S. Medeiros, Z. Pan, W.J. Orts, L.H.C. Mattoso and B.G. Soares. 2009. Development of conducting polyaniline/poly(lactic acid) nanofibers by electrospinning. J. Appl. Polym. Sci. 112: 744–753.

Pickup, P.G. 1990. Alternating current impedance study of a polypyrrole based anion-exchange polymer. J. Chem. Soc. Faraday Trans. 86: 3631–3636.

Pieta, P., I. Obraztsov, F. D'Souza and W. Kutnera. 2013. Composites of conducting polymers and various carbon nanostructures for electrochemical supercapacitors. ECS J. Solid State Sci. Technol. 2: M3120–M3134.

Pinto, N.J., A.T. Johnson Jr, A.G. MacDiarmid, C.H. Mueller, N. Theofylaktos, D.C. Robinson et al. 2003. Electrospun polyaniline/polyethylene oxide nanofiber field-effect transistor. Appl. Phys. Lett. 83: 4244–4246.

Pokhodenko, V.D., V.A. Krylov, Y.I. Kurys and O.Y. Posudievsky. 1999. Nanosized effects in composites based on polyaniline and vanadium or iron oxides. Phys. Chem. Chem. Phys. 1: 905–908.

Potember, R.S., R.C. Hoffman, H.S. Hu, J.E. Cocchiaro, C.A. Viands, R.A. Murphy et al. 1987. Conducting organics and polymers for electronic and optical devices. Polymers 28: 574–580.

Pud, A., N. Ogurtsov, A. Korzhenko and G. Shapoval. 2003. Some aspects of preparation methods and properties of polyaniline blends and composites with organic polymers. Prog. Polym. Sci. 28: 1701–53.

Qiang, M., T. Chen, R.P. Yao and L. Chen. 2008. Effect of preparation condition of polyaniline-monmorillonite nanocomposite. J. Coat. Technol. Res. 5: 241–249.

Qu, F., M. Yang, J. Jiang, G. Shen and R. Yu. 2005. Amperometric biosensor for choline based on layer-by-layer assembled functionalized carbon nanotube and polyaniline multilayer film. Anal. Biochem. 344: 108–114.

Rajesh, B., K.R. Thampi, J.M. Bonard, H.J. Mathieu, N. Xanthopoulos and B. Viswanathan. 2003. Conducting polymer nanotubes as high performance methanol oxidation catalyst support. Chem. Commun. 2022–2023.

Rakhi, R.B., W. Chen and H.N. Alshareef. 2012. Conducting polymer/carbon nanocoil composite electrodes for efficient supercapacitors. J. Mater. Chem. 22: 5177–5183.

Ram, R., M. Rahaman and D. Khastgir. 2015. Electrical properties of polyvinylidene fluoride (PVDF)/multi-walled carbon nanotube (MWCNT) semi-transparent composites: Modelling of DC conductivity. Composites A 69: 30–39.

Ramanavicius, A., A. Ramanaviciene and A. Malinauskas. 2006. Electrochemical sensors based on conducting polymer—polypyrrole. Electrochim. Acta 51: 6025–6037.

Rao, P.S., S. Subrahmanya and D.N. Sathyanarayana. 2002. Inverse emulsion polymerization: A new route for the synthesis of conducting polyaniline. Synth. Met. 128: 311–316.

Ray, S.S. and M. Biswas. 1999. Preparation and evaluation of composites from montmorillonite and some heterocyclic polymers: 3. A water dispersible nanocomposite from pyrrole-montmorillonite polymerization system. Mater. Res. Bull. 34: 1187–1194.

Ray, S.S. and M. Okamoto. 2003. Polymer/layered silicate nanocomposites: a review from preparation to processing. Prog. Polym. Sci. 28: 1539–1641.

Ruangchuay, L., A. Sirivat and J. Schwank. 2003. Polypyrrole/poly(methylmethacrylate) blend as selective sensor for acetone in lacquer. Talanta 60: 25–30.

Sadek, A.Z., W. Wlodarski, K. Shin, R.B. Kaner and K. Kalantar-zadeh. 2006. A layered surface acoustic wave gas sensor based on a polyaniline/In$_2$O$_3$ nanofibre composite. Nanotechnology 17: 4488–4492.

Saini, P., V. Choudhary, B.P. Singh, R.B. Mathur and S.K. Dhawan. 2009. Polyaniline-MWCNT nanocomposites for microwave absorption and EMI shielding. Mater. Chem. Phys. 113: 919–926.

Sainz, R., A.M. Benito, M.T. Martinez, J.F. Galindo, J. Sotres, A.M. Baro et al. 2005. A soluble and highly functional polyaniline–carbon nanotube composite. Nanotechnol. 16: S150–S154.

Santhanam, K.S.V., R. Sangoi and L. Fuller. 2005. A chemical sensor for chloromethanes using a nanocomposite of multiwalled carbon nanotubes with poly(3-methylthiophene). Sens. Actuators B 106: 766–771.

Sarma, T.K., D. Chowdhury, A. Paul and A. Chattopadhyay. 2002. Synthesis of Au nanoparticle–conductive polyaniline composite using H_2O_2 as oxidising as well as reducing agent. Chem. Commun. 1048–1049.

Sarma, T.K. and A. Chattopadhyay. 2004. One pot synthesis of nanoparticles of aqueous colloidal polyaniline and its Au-nanoparticle composite from monomer vapor. J. Phys. Chem. A 108: 7837–7842.

Sato, M., S. Tanaka and K. Kacriyama. 1986. Electrochemical preparation of conducting poly(3-methylthiophene): comparison with polythiophene and poly(3-ethylthiophene). Synth. Met. 14: 279–288.

Saxena, V. and B.D. Malhotra. 2003. Prospects of conducting polymers in molecular electronics. Curr. Appl. Phys. 3: 293–305.

Schmidt, D., D. Shah and E.P. Giannelis. 2002. New advances in polymer/layered silicate nanocomposites. Curr. Opin. Solid State Mater. Sci. 6: 205–212.

Sengupta, P.P., P. Kar and B. Adhikari. 2011. Influence of dielectric constant of the polymerization medium on the processability and ammonia gas sensing properties of polyaniline. Bul. Mater. Sci. 34: 261–270.

Setshedi, K.Z., M. Bhaumik, S. Songwane, M.S. Onyango and A. Maity. 2013. Exfoliated polypyrrole-organically modified montmorillonite clay nanocomposite as a potential adsorbent for Cr(VI) removal. Chem. Eng. J. 222: 186–197.

Shanmugapriya, C. and G. Velraj. 2014. Studies on structural and electrical conducting properties of micro and nano copper doped polyaniline. Chem. Sci. Rev. Lett. 3: 506–511.

Sharma, S., C. Nirkhe, S. Pethkar and A.A. Athawale. 2002. Chloroform vapour sensor based on copper/polyaniline nanocomposite. Sens. Actuators B: Chem. 85: 131–136.

Sharma, V.K., F. Jelen and L. Trnkova. 2015. Functionalized solid electrodes for electrochemical biosensing of purine nucleobases and their analogues: a review. Sensors 15: 1564–1600.

Shi, W., D. Ge, J. Wang, Z. Jiang, R. Len and Q. Zhang. 2006. Heparin-controlled growth of polypyrrole nanowires. Macromol. Rapid Commun. 27: 926–930.

Shirakawa, H., E.J. Louis, A.G. MacDiarmid, C.K. Chiang and A.J. Heeger. 1977. Synthesis of electrically conducting organic polymers: halogen derivatives of polyacetylene, (CH)x. J. Chem. Soc. Chem. Commun. 578–580.

Shirsat, M.D., M.A. Bangar, M.A. Deshusses, N.V. Myung and A. Mulchandani. 2009. Polyaniline nanowires-gold nanoparticles hybrid network based chemiresistive hydrogen sulfide sensor. Appl. Phys. Lett. 94: 083502–083505.

Sigaud, M., M. Li, S.C. Noblat, F.J.C.S. Aires, Y.S. Olivier, J.P. Simon et al. 2004. Electrochemical preparation of nanometer sized noble metal particles into a polypyrrole functionalized by a molecular electrocatalyst precursor. J. Mater. Chem. 14: 2606–2608.

Sih, B.C. and M.O. Wolf. 2005. Metal nanoparticle—conjugated polymer nanocomposites. Chem. Commun. 3375–3384.

Singh, A.K., A.D.D. Dwivedi, P. Chakrabarti and R. Prakash. 2009a. Electronic and optical properties of electrochemically polymerized polycarbazole/aluminum Schottky diodes. J. Appl. Phys. 105: 114506–114507.

Singh, R., R. Prasad, G. Sumana, K. Arora, S. Sood, R.K. Gupta and B.D. Malhotra. 2009b. STD sensor based on nucleic acid functionalized nanostructured polyaniline. Biosens. Bioelectron. 24: 2232–2238.

Singh, A.K. and R. Prakash. 2012. Organic Schottky diode based on conducting polymer-nanoclay composite. RSC Adv. 2: 5277–5283.

Smith, J.D.S. 1998. Intrinsically electrically conducting polymers. Synthesis, characterization, and their applications. Prog. Polym. Sci. 23: 57–79.

Song, D.H., H.M. Lee, K.H. Lee and H.J. Choi. 2008. Intercalated conducting polyaniline-clay nanocomposites and their electrical characteristics. J. Phys. Chem. Solids 69: 1383–1385.

Spitalsky, Z., D. Tasis, K. Papagelis and C. Galiotis. 2010. Carbon nanotube–polymer composites: chemistry, processing, mechanical and electrical properties. Prog. Polym. Sci. 35: 357–401.

Srivastava, S., S. Kumar and Y.K. Vijay. 2012. Preparation and characterization of tantalum/polyaniline composite based chemiresistor type sensor for hydrogen gas sensing application. Int. J. Hydrogen Energy 37: 3825–3832.

Stejskal, J., M. Trchova, J. Kovarova, L. Brozova and J. Prokes. 2009. The reduction of silver nitrate with various polyaniline salts to polyaniline–silver composites. React. Funct. Polym. 69: 86–90.

Sudha, J.D., V.L. Reena and C. Pavithran. 2007. Facile green strategy for micro/nano structured conducting polyaniline-clay nanocomposite via template polymerization using amphiphilic dopant, 3-pentadecyl phenol 4-sulphonic acid. J. Polym. Sci. Part B: Polym. Phys. 45: 2664–2673.

Sun, Y.P., W.J. Huang, Y. Li, K.F. Fu, A. Kitaygorodskiy, L.A. Riddle et al. 2001. Carroll. Soluble dendron-functionalized carbon nanotubes: Preparation, characterization, and properties. Chem. Mater. 13: 2864–2869.

Syed, A.A. and M.K. Dinesan. 1991. Review: Polyaniline—A novel polymeric material. Talanta 38: 815–837.

Syu, M.J. and Y.S. Chang. 2009. Ionic effect investigation of a potentiometric sensor for urea and surface morphology observation of entrapped urease/polypyrrole matrix. Biosens. Bioelectron. 24: 2671–2677.

Tan, Y., Y. Li and D. Zhu. 2003. Synthesis of poly(2-methoxyaniline)/Au nanoparticles in aqueous solution with chlorauric acid as the oxidant. Synth. Met. 135-136: 847–848.

Tiwari, A. and S. Li. 2009. Vacuum-deposited poly(o-phenylenediamine)/$WO_3 \cdot nH_2O$ nanocomposite thin film for NO_2 gas sensor. Polym. J. 41: 726–732.

Tiwari, A., M. Prabaharan, R.R. Pandey and S. Li. 2010. Vacuum-deposited thin film of aniline-formaldehyde condensate/$WO_3 \cdot nH_2O$ nanocomposite for NO_2 Gas sensor. J. Inorg. Organomet. Polym. Mater. 20: 380–386.

Tong, D.S., H.S. Xia and C.H. Zhou. 2009. Designed preparation and catalysis of smectite clay-based catalytic materials. Chin. J. Catal. 30: 1170–1187.

Toshima, N. and S. Hara. 1995. Direct synthesis of conducting polymers from simple monomers. Prog. Polym. Sci. 20: 155–183.

Thostenson, E., C. Li and T. Chou. 2005. Review nanocomposites in context. Comps. Sci. Technol. 65: 491–516.

Tourillon, G., E. Dartyge, A. Fontaine and A. Jucha. 1986. Dispersive X-ray spectroscopy for time-resolved *in situ* observations of electrochemical inclusions of metallic clusters within a conducting polymer. Phys. Rev. Lett. 57: 603–606.

Tseng, R.J., J. Huang, J. Ouyang, R.B. Kaner and Y. Yang. 2005. Polyaniline nanofiber/gold nanoparticle nonvolatile memory. Nano. Lett. 5: 1077–1080.

Tseng, R.J., C.O. Baker, C. Shedd, J. Huang, R.B. Kaner, J. Ouyang et al. 2007. Charge transfer effect in the polyaniline-gold nanoparticle memory system. Appl. Phys. Lett. 90: 053101–053103.

Twite, R.L. and G.P. Bierwagen. 1998. Review of alternatives to chromate for corrosion protection of aluminum aerospace alloys. Prog. Org. Coat. 33: 91–100.

Uthirakumar, P., M.-K. Song, C. Nah and Y.-S. Lee. 2005. Preparation and characterization of exfoliated polystyrene/clay nanocomposites using a cationic radical initiator-MMT hybrid. Eur. Polym. J. 41: 211–217.

Vaccari, A. 1998. Preparation and catalytic properties of cationic and anionic clays. Appl. Clay Sci. 41: 53–71.

Verma, S.K., A. Choudhury and P. Kar. 2015a. Interaction of multi-walled carbon nanotube with poly(m-aminophenol) in their processable conducting nanocomposite. Phys. Stat. Solidi A 212: 2044–2052.

Verma, S.K., P. Kar, D.J. Yang and A. Choudhury. 2015b. Poly(m-aminophenol)/functionalized multi-walled carbon nanotube nanocomposite based alcohol sensors. Sens. Actuators B: Chem. 219: 199–208.

Verma, S.K., M. Kumar, P. Kar and A. Choudhury. 2016. Core–shell functionalized MWCNT/poly(maminophenol) nanocomposite with large dielectric permittivity and low dielectric loss. Polym. Adv. Technol. 27: 1596–1603.

Virji, S., J.D. Fowler, C.O. Baker, J. Huang, R.B. Kaner and B.H. Weiller. 2005. Polyaniline nanofiber composites with metal salts: chemical sensors for hydrogen sulfide. Small 1: 624–627.

Vishnu, N., A.S. Kumar and K.C. Pillai. 2013. Unusual neutral pH assisted electrochemical polymerization of aniline on a MWCNT modified electrode and its enhanced electro-analytical features. Analyst 138: 6296–6300.

Walls, H.J., M.W. Riley, R.R. Singhal, R.J. Spontak, P.S. Fedkiw and S.A. Khan. 2003. Nanocomposite electrolytes with fumed silica and hectorite clay networks: Passive versus active fillers. Adv. Funct. Mater. 13: 710–717.

Wang, L., X. Lu, S. Lei and Y. Song. 2014. Graphene-based polyaniline nanocomposites: Preparation, properties and applications. J. Mater. Chem. A 2: 4491–4509.

Xi, Y., J. Zhou, H. Guo, C. Cai and Z. Lin. 2005. Enhanced photoluminescence in core-sheath CdS-PANI coaxial nanocables: a charge transfer mechanism. Chem. Phys. Lett. 412: 60–64.

Xiong, S., Q. Wang and H. Xia. 2004. Template synthesis of polyaniline/TiO_2 bilayer microtubes. Synth. Met. 146: 37–42.

Xu, X.H., G.L. Ren, J. Cheng, Q. Liu and D.G. Li. 2006. Layer by layer self-assembly immobilization of glucose oxidase onto chitosan-graft-polyaniline polymers. J. Mater. Sci. 41: 3147–3149.

Xu, J., J. Hu, B. Quan and Z. Wei. 2009. Decorating polypyrrole nanotubes with Au nanoparticles by an *in situ* reduction process. Macromol. Rapid. Commun. 30: 936–940.

Yang, T., G. Wei, L. Niu and Z. Li. 2006. Fabrication of linear aniline-DNA complex nanowires and DNA-templated nanowires. Chem. J. Chin. Univ. 27: 1126–1130.

Yang, W., P. Thordarson, J.J. Gooding, S.P. Ringer and F. Braet. 2007. Carbon nanotubes for biological and biomedical applications. Nanotechnology 18: 412001–412013.

Yang, X., L. Li and F. Yan. 2010. Polypyrrole/silver composite nanotubes for gas sensors. Sens. Actuators B: Chem. 145: 495–500.

Yeh, J.-M., S.-J. Liou, C.-Y. Lai and P.-C. Wu. 2001. Enhancement of corrosion protection effect in polyaniline via the formation of polyaniline–clay nanocomposite materials. Chem. Mater. 13: 1131–1136.

Yeh, J.-M., C.-L. Chen, Y.-C. Chen, C.-Y. Ma, K.-R. Lee, Y. Wei et al. 2002. Enhancement of corrosion protection effect of poly (o-ethoxyaniline) via the formation of poly (o-ethoxyaniline)–clay nanocomposite materials. Polymer 43: 2729–2736.

Yeh, J.M. and C.P. Chin. 2003. Structure and properties of poly (o-methoxyaniline)-clay nanocomposite materials. J. Appl. Polym. Sci. 88: 1072–1080.

Yeh, J.-M., C.-P. Chin and S. Chang. 2003. Enhanced corrosion protection coatings prepared from soluble electronically conductive polypyrrole-clay nanocomposite materials. J. Appl. Polym. Sci. 88: 3264–3272.

Yoneyama, H., Y. Shoji and K. Kawai. 1989. Electrochemical synthesis of polypyrrole films containing metal oxide particles. Chem. Lett. 18: 1067–1070.

Yu, Y.-H., C.-C. Jen, H.-Y. Huang, P.-C. Wu, C.-C. Huang and J.-M. Yeh. 2004. Preparation and properties of heterocyclically conjugated poly (3-hexylthiophene)-clay nanocomposite. J. Appl. Polym. Sci. 91: 3438–3446.

Yu, Y., C. Ouyang, Y. Gao, Z. Si, W. Chen, Z. Wang et al. 2005a. Synthesis and characterization of carbon nanotube/polypyrrole core-shell nanocomposites via *in situ* inverse microemulsion. J. Polym. Sci. Part A Polym. Chem. 43: 6105–6115.

Yu, Y., B. Che, Z. Si, L. Li, W. Chen and G. Xue. 2005b. Carbon nanotube/polyaniline core-shell nanowires prepared by *in situ* inverse microemulsion. Synth. Met. 150: 271–277.

Yu, Y., Z. Si, S. Chen, C. Bian, W. Chen and G. Xue. 2006. Facile synthesis of polyaniline-sodium alginate nanofibers. Langmuir 22: 3899–3905.

Yu, G., X. Li, X. Cai, W. Cui, S. Zhou and J. Weng. 2008. The photoluminescence enhancement of electrospun poly(ethylene oxide) fibers with CdS and polyaniline inoculations. Acta Mater. 56: 5775–5782.

Yariv, S. and H. Cross. 2002. Organo-Clay Complexes and Interactions. Maecel Dekker, Inc., New York.

Zampetti, E., S. Pantalei, A. Muzyczuk, A. Bearzotti, F.D. Cesare, C. Spinella et al. 2013. A high sensitive NO$_2$ gas sensor based on PEDOT–PSS/TiO$_2$ nanofibres. Sens. Actuators B: Chem. 176: 390–398.

Zehhaf, A., E. Morallon and A. Benyoucef. 2013. Polyaniline/montmorillonite nanocomposites obtained by *in situ* intercalation and oxidative polymerization in cationic modified-clay (sodium, copper and iron). J. Inorg. Organometal. Polym. Mater. 23: 1485–1491.

Zeng, Q.H., A.B. Yu, G.Q. Lu and D.R. Paul. 2005. Clay based polymer nanocomposites: Research and commercial development. J. Nanosci. Nanotechnol. 5: 1574–1592.

Zhai, L. and R.D. McCullough. 2004. Regioregular polythiophene/gold nanoparticle hybrid materials. J. Mater. Chem. 14: 141–143.

Zhang, X., J. Zhang, R. Wang, T. Zhu and Z. Liu. 2004a. Surfactant-directed polypyrrole/CNT nanocables: synthesis, characterization, and enhanced electrical properties. Chem. Phys. Chem. 5: 998–1002.

Zhang, X., J. Zhang, R. Wang and Z. Liu. 2004b. Cationic surfactant directed polyaniline/CNT nanocables: synthesis, characterization, and enhanced electrical properties. Carbon 42: 1455–1461.

Zhang, X., Z. Lu, M. Wen, H. Liang, J. Zhang and Z. Liu. 2005a. Single-walled carbon nanotube-based coaxial nanowires: synthesis, characterization, and electrical properties. J. Phys. Chem. B 109: 1101–1107.

Zhang, Z., M. Wan and Y. Wei. 2005b. Electromagnetic functionalized polyaniline nanostructures. Nanotechnology 16: 2827–2832.

Zhang, X. and S.K. Manohar. 2005. Narrow pore-diameter polypyrrole nanotubes. J. Am. Chem. Soc. 127: 14156–14163.

Zhang, H., G.P. Cao, Z.Y. Wang, Y.S. Yang, Z.J. Shi and Z.N. Gu. 2008. High-rate lithium-ion battery cathodes using nanostructured polyaniline/carbon nanotube array composites. Electrochem. Solid State Lett. 11: A223–A225.

Zhang, D., D.S. Tong, H.S. Xia, G.F. Jiang, X. Xia, M. Liu et al. 2009. Hydrothermal synthesis of clay minerals. 14th Intl. Clay Conf., Micro ET Nano ScientIE Mare Magnum, Castellaneta M., Italy, pp. 432.

Zhang, D., C.-H. Zhou, C.-X. Lin, D.-S. Tong and W.-H. Yu. 2010. Synthesis of clay minerals. Appl. Clay Sci. 50: 1–11.

Zhang, K., Y. Zhang and S. Wang. 2013. Enhancing thermoelectric properties of organic composites through hierarchical nanostructures. Sci. Rep. 3: 3448–3455.

Zhao, B., H. Hu and R.C. Haddon. 2004. Synthesis and properties of a watersoluble single-walled carbon nanotube-poly(m-aminobenzene) sulfonic acid graft copolymer. Adv. Funct. Mater. 14: 71–76.

Zheng, J.P., P.J. Cygon and T.R. Jow. 1995. Hydrous ruthenium oxide as an electrode material for electrochemical capacitors. J. Electrochem. Soc. 142: 2699–2703.

Zhou, C.H. 2010. Emerging trends and challenges in synthetic clay-based materials and layered double hydroxides. Appl. Clay Sci. 48: 1–4.

Zhou, H.S., T. Wada, H. Sasabe and H. Komiyama. 1996a. Synthesis of nanometer-size silver coated polymerized diacetylene composite particles. Appl. Phys. Lett. 68: 1288–1289.

Zhou, H.S., T. Wada, H. Sasabe and H. Komiyama. 1996b. Synthesis and optical properties of nanocomposite silver—polydiacetylene. Synth. Met. 81: 129–132.

Zhou, Y., H. Itoh, T. Uemura, K. Naka and Y. Chujo. 2002. Synthesis of novel stable nanometer-sized metal (M = Pd, Au, Pt) colloids protected by a π-conjugated polymer. Langmuir 18: 277–283.

Zhou, C.H., X.N. Li, Z.H. Ge, Q.W. Li and D.S. Tong. 2004. Synthesis and acid catalysis of nanoporous silica/alumina–clay composites. Catal. Today 93: 607–613.

Zhu, Y., J. Zhang, Y. Zheng, Z. Huang, L. Feng and L. Jiang. 2006. Stable, superhydrophobic, and conductive polyaniline/polystyrene films for corrosive environments. Adv. Funct. Mater. 16: 568–574.

Zouaoui, A., O. Stephan, M. Carrier and J.-C. Moutet. 1999. Electrodeposition of copper into functionalized polypyrrole films. J. Electroanal. Chem. 474: 113–122.

4

Biopolymer Nanocomposites—From Basic Concepts to Nanotechnologies

Magdalena Aflori

Introduction

The interest in using biopolymer as an alternative to petroleum based polymers has increased in recent years due to the limited resources of crude oil, environmental awareness of consumers and global warming. Biopolymers are fully renewable, do not contribute to an accumulation of carbon dioxide in the atmosphere after their life cycle, are a part of the ecosystem and many of them are also biologically degradable, thus reducing waste management problems. The most important applications of such materials are different tissue supporting materials (scaffolds), barrier materials and drug delivery systems (Vainionpä et al. 1989, Armentano et al. 2010). Polymers are the primary materials in tissue engineering applications and many types of biodegradable polymeric materials have been already used in this field (Tian et al. 2012, Tan et al. 2013). It is known that the material's surface has the most critical influence on the biocompatibility and, in many cases, surface modification is essential for a conventional polymer to be used as a biomedical material (Ikada et al. 1994, Yoshida et al. 2013). To obtain materials with preferred properties for specific applications, a wide spectrum of techniques could be used for the modification. There are many ways to modify the polymers surface: plasma or laser treatment, wet-chemical treatment, metal coating, immobilization of bioactive compounds, etc. (Goddard et al. 2007, Rimpelová et al. 2013, Kasálková et al. 2014, Slepicka et al. 2012). In general, a cleaning procedure or an alteration of the polymer surface such as wet chemical treatment (wet etching) is preferred for appropriate surface modification (Kim et al. 2006). One way to improve the properties of biopolymers and greatly enhance their commercial potential is to incorporate nanosized reinforcement in the polymer —the reinforcing material having the dimensions in nanometric scale (Pandey et al. 2005). Due to their nanoscale dispersion, the reinforcement efficiency of nanocomposites can match that of conventional composites with 40–50% of loading with classical fillers. To enhance the mechanical properties and cellular adhesion and proliferation, the incorporation of nanoparticles (e.g., apatite component, carbon nanostructures and metal nanoparticles) has been extensively investigated.

"Petru Poni" Institute of Macromolecular Chemistry Iasi Romania, Department of Polymeric Materials Physics, 41A Grigore Ghica Voda Alley, Iasi, Romania.
Email: maflori@icmpp.ro

Very often, poly-L-lactic acid (PLLA) and polyhydroxybutyrate (PHB) are used such a material in tissue engineering (Chen et al. 2005, Chen et al. 2013). Poly-L-lactic acid (PLLA) is produced from renewable resources, is readily biodegradable and can compete with petroleum based polymers on a cost-performance basis (Vink et al. 2003). PLLA also has excellent properties comparable to many petroleum based plastics (Lunt 1998). However, some properties like melt strength, impact strength, thermal stability, gas barrier, etc. do not meet the demands for some end-use applications and has to be improved (Perego et al. 1996). PLLA can be extruded into films, injection molded into different shapes or spun to obtain fibers and has attracted considerable attention by its inherent biodegradability and possibilities to be composted in an industrial environment (Mohanty et al. 2003). PHB is an aliphatic polyester derived from microorganisms, with high melting point, and high crystallinity.

Noble metal nanoparticles (Au, Ag, Pt and Pd) are of continuing interest because of their unusual properties compared to bulk metals and their potential applications in novel electrical, optical, magnetic, catalytic, and sensing technologies. Silver nanoparticles have not only been noted to show strong broad antimicrobial activity but limited cytotoxicity towards mammalian cells (Hackenberg et al. 2011, Freire et al. 2015). Flores et al. reported that cubic Ag nanoparticles exhibited inhibitory effects towards S. aureus and P. aeruginosa at low concentrations and demonstrate low cytotoxicity towards UMR-106 cell lines (Flores et al. 2013). This finding suggests that if Ag nanoparticles loaded matrix is properly designed with the purpose of releasing nanoparticles, the nanoparticles loaded matrix could maintain a sterile environment against microorganisms without inhibiting the growth of bone cells. Travan et al. synthesized alginate–chitlac hydrogels containing silver nanoparticles and noticed strong bactericidal activity against *S. epidermis, S. aureus, E. coli* and *P. aeruginosa* while no cytotoxic effect was observed towards mouse fibroblast (NIH-3T3), human hepatocarcinoma (HepG2), and human osteosarcoma (MG63) cells (Travan et al. 2009).

Surface treatment techniques, such as plasma treatment, ion sputtering, oxidation and corona discharge affect the chemical and physical properties of the surface layer without changing the bulk material properties. The influence of plasma treatment and wet etching on the surface physico-chemical properties of PLLA and PHB was determined (Kasálková et al. 2014). The saturation and stabilization of the polymer surface was observed after 7–15 days (for PLLA) and 7–20 (for PHB) from the modification depending on the use of etching agent. Plasma treatment and subsequent wet etching with acetone leads to significant mass loss of the surface layer, changes in surface morphology and roughness. Pronounced mass loss was observed for PHB. Using acetone–water solution as etching agent leads to diffusion of the oxygen or nitrogen to the volume of the modified PLLA and PHB. Activated biopolymers may find strong potential application as carriers in tissue engineering (Kasálková et al. 2014).

In this chapter, a comprehensive state of the art highlighting the various techniques used to prepare biodegradable polymer-based nanocomposites, their physicochemical characterization, their biodegradability, processing, and future prospect will be discussed. In addition, the design of the surface proprieties of the poly-L-lactic acid (PLLA) and polyhydroxybutyrate (PHB) by a two-step treatment (plasma treatments and subsequent silver-nanoparticle solution chemical modification) is presented.

Biodegradable polymer-based nanocomposites

Biodegradable polymers are defined as those polymers that fully decompose to carbon dioxide, methane, water, biomass and inorganic compounds under aerobic or anaerobic conditions and the action of living organisms. Biodegradable polymers undergo microbial induced chain scission leading to the mineralization (a process in which an organic substance is converted to an inorganic substance, such as carbon dioxide) under certain conditions of pH, humidity, oxygenation and the presence of some metals. Different factors are involved in the process of biodegradation: a large variety of polymer structures combinations, numerous enzymes produced by microorganisms, and variable reaction conditions. The biodegradation process occurs because polymers are perceived by the living organisms (bacteria, fungi, algae) as their food: a source of organic compounds (e.g., simple monosaccharides, amino acids, etc.) and energy that sustain them. Intracellular and extracellular enzymes (endo- and exoenzymes) from the living organisms provoke chemical reactions in the polymer, which degrades by the process of scission of the polymer chain, oxidation, etc. This is the way to create increasingly smaller molecules which enter into cellular metabolic processes to generate energy and to turn into water, carbon dioxide, biomass and other basic products of biotic decomposition.

Biodegradable polymers can be classified in four categories depending on the synthesis and on the sources (Bordes et al. 2009, Averous et al. 2004, Tharanathan 2003) (Scheme 1).

The first category (1) is considered as agro-polymers, the others two (2) and (3) are called biodegradable biopolyesters (Bertan et al. 2005), while the last category (4) is obtained from non-renewable resources. The main target applications of biodegradable polymer films are food packaging, surgery and pharmaceutical uses, recent innovations being widely discussed in the literature (Guilbert et al. 1997, Krochta et al. 1997, Arvanitoyannis et al. 1997, Kester et al. 1986).

The application area is restricted by some properties of the biodegradable polymers, such as brittleness, high gas permeability, low heat distortion temperature, low melt viscosity for further processing, etc. Thus, a recent challenge for materials scientists is the modification of the biodegradable polymers through innovative technology, an effective way being the nanoreinforcement of pristine polymers in order to obtain the next generation materials: nanocomposites. On the other hand, nanocomposites are defined as composites with reinforcement in the nanometer range (< 100 nm) in at least one dimension and, depending on the geometrical shape of the nanoreinforcement, are classified into three groups (Bondeson 2007) (Scheme 2).

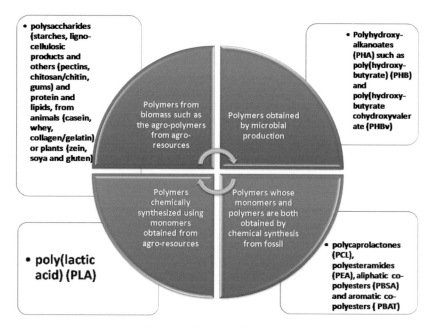

Scheme 1. Biodegradable polymers.

According to this classification, by changing the nanocomposite composition used and the nanoreinforcement, preparation to processing of biodegradable polymer-based nanocomposites, that is, green nanocomposites is the wave of the future due to the improvement of proprieties like electrical conductivity, abrasion and wear resistance, permittivity and breakdown strength, dimensional and thermal stability, and mechanical, barrier, and flammability properties (Schadler 2003). All these benefits can be achieved due to the nanometric size of the reinforcement, at a nanoreinforcement content of 5–10 wt%. The significantly enhanced properties in the composite is due to the substantial reduction of the probability of finding defects such as voids, grain boundaries, dislocations and imperfections in nanosized particles compared to micro or macro sized ones. Moreover, the interfacial area of the reinforcement increases significantly from microscale to nanoscale, and the number of possible interaction

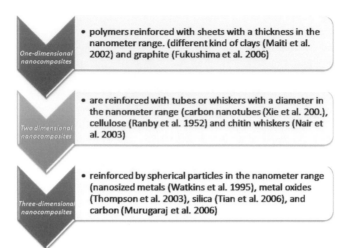

Scheme 2. Classification of nanocomposites.

sites between the reinforcement and the matrix substantially increases. The entire polymer matrix is affected by the interface of the nanoparticle, supposing the interfacial region between the polymer and the nanoparticle is of few nanometers, the entire polymer matrix presenting a different behavior than the bulk. The interaction strength between the polymer and the nanoreinforcement is direct proportional with the polymer chain mobility at the interface: by changing the interaction degree, the size of the interfacial region and hence the properties of the entire matrix can be changed. The matrix can be reinforced with relatively low nanoreinforcement content, and with quite small size particles, preserving the optical clarity or transparency. Despite the classical composites, the nanosized reinforcements do not create large stress concentrations and thus do not compromise the ductility of the polymer (Schadler 2003).

Apatite-base nanocomposites

In recent years, synthetic or natural hydroxyapatite (HA) has been intensively investigated as the major component of scaffold materials for bone tissue engineering (Knowles 2003). Around 65 wt% of human bone is made of HA, $Ca_{10}(PO_4)_6(OH)_2$, the Ca/P ratio of 1.50–1.67 being the key issue to promote bone regeneration. By changing size, composition and morphology of HA, the researchers improved the osteoconductive properties (Gay et al. 2009). In this context, the use of nano-sized HA (nHA) may have other special properties due to its small size and huge specific surface area, a significant increase in protein adsorption and osteoblast adhesion on the nanosized ceramic materials being obtained (Webster et al. 2000). Nejati et al. demonstrated the effect of the synthesis of nHA on the scaffolds morphology and mechanical properties in poly-L-lactic acid (PLLA)-based nanocomposites in maintaining a regular internal ladder-like pore structure, similar to neat PLLA scaffold with a typical morphology processed by thermally induced phase separation (Nejati et al. 2008). An increase in the mechanical properties has been observed by reducing grain size of bioceramics to nanolevels or by increase in the interfacial strength between PLLA and HA in blends (Li et al. 2008). An improvement in cell compatibility due to the favorable biocompatibility of the nHA and more uniform distribution of the surface grafted on the PLLA film surface was also demonstrated (Hong et al. 2004). The materials containing nHA have a structure that induces and promotes new bone formation. Moreover, calcium phosphate biomaterials certainly posses osteoconductive properties and may bind directly to bone under certain conditions (Nejati et al. 2008) and are suitable for the calcified tissue generation.

Bionanocomposites obtained by green synthesis

A green synthetic approach for preparing antimicrobial silver nanoparticles has been suggested by using carbohydrates such as sucrose or waxy corn starch. The carbohydrates act as both reducing and stabilizing agents and also as a template for carrying silver nanoparticles with excellent antibacterial activity. Some authors obtained a solution of bovine serum albumin containing silver nanoparticles (BSA/AgNp) and they functionalized a degradable biopolymer, i.e., PHBV, with hydroxyl functional group, immobilized with extra-cellular matrix (Type I collagen from calf skin) (Bakare et al. 2016). Then they loaded BSA/Ag NP to form a potential scaffold for tissue engineering applications and they demonstrated that the concentration of BSA/Ag NP solution, molecular weight of immobilized collagen, and pH of wash buffer solutions influenced the extent of NPs loading. The drop in the pH of the bacterial cells (*E. coli*, *S. aureus*, and *P. aeruginosa*) during incubation with Ag/BSA NPs loaded collagen immobilized PHBV film triggered the release of NPs and promoted antimicrobial activities. MTT assay result showed that MCTC3-E1 osteoblast cells adhesion on both NLMWCP and NHMWCP was similar to tissue culture polystyrene standard. A notable improvement of this study is that significantly low concentration of biocompatible Ag/BSA NPs is required to show antibacterial activity compared to the one reported by Travan et al. (Travan et al. 2009). Future studies will investigate the antimicrobial activity and cell viability of Ag/BSA nanoparticles loaded collagen immobilized PHBV matrix when osteoblast cell media is contaminated with *S. aureus*, *E. coli*, and *P. aeruginosa* (Bakare et al. 2016).

Chitosan nanoparticles loaded with various nanoparticles such as Ag^+, Cu^{2+}, and Zn^{2+} showed a significantly increased antimicrobial activity against *E. coli*, *Salmonella choleraesuis*, and *S. aureus* (Du et al. 2009). Silver nanoparticles have been incorporated or obtained into biopolymer films such as chitosan and starch, the efficiency of antimicrobial function of these polymeric nanocomposites being greatly influenced by various factors such as particle size, size distribution, degree of particle agglomeration, silver content, and interaction of silver surface with the base polymer (Kim et al. 2007). In order to fully exploit the properties of silver nanoparticles against both Gram-positive and Gram-negative bacteria, they should be well dispersed on the surface of the polymer matrix without the formation of aggregation, which otherwise dramatically reduce the antimicrobial effect of silver (Bi et al. 2011, Yoksan et al. 2011).

Carbon-base nanocomposites

Nowadays, carbon nanostructures in polymer matrix are a good alternative for biomedical scaffolds preparation (Harrison et al. 2007), the carbon networks being intensively used as a biocompatible mesh for restoring or reinforcing damaged tissues because of no cytotoxicity of the networks. The honeycomb-like matrices of multi-walled nanotube have been fabricated to be used in tissue engineering (Mwenifumbo et al. 2007). A useful tool to direct cell growth is the electrical conductivity of the nanocomposites including carbon nanostructures because they can conduct an electrical stimulus in the tissue healing process. An experiment performing proliferation of osteoblast on PLLA/multi-walled nanotube nanocomposites under alternating current stimulation (Supronowicz et al. 2002) showed an increase in the extra-cellular calcium deposition on the nanocomposites as compared with the control samples. Some authors (Armentano et al. 2008) demonstrated the influence of the functional group nature at the carbon nanotube surface in the mechanism of interaction with cells. One of the most promising materials for use in multifunctional nanocomposites for various applications (Mochalin et al. 2012) are nanodiamonds synthesized by detonation dispersed into single particles. The surface area of the nanoparticles available for interaction with the matrix depends on the quality of the nanodiamonds filler dispersion in the matrix, the nanocomposites properties like strength, toughness, and thermal stability being improved when nanodiamonds were adequately dispersed. Zhang et al. (Zhang et al. 2012) prepared multifunctional bone scaffold materials composed of PLLA and octadecylamine functionalized nanodiamonds solution casting, followed by compression molding. The results demonstrated that the addition of 10 wt% of octadecylamine functionalized nanodiamonds resulted in a 280% increase in the strain to failure and a 310% increase in fracture energy as compared to neat PLLA. Both of these parameters are crucial for bone tissue engineering and for manufacturing of orthopedic surgical fixation devices, PLLA/octadecylamine functionalized nanodiamonds being a promising material for bone surgical fixation devices and regenerative medicine.

Metal nanoparticles-based nanocomposites

Nowadays, metal nanoparticle-based nanocomposites are used in various biomedical applications. Among the numerous nanoparticles used for functionalizing polymeric materials, silver nanoparticles have been one of the most widely used for the development of innovative packaging materials as well as scaffolds in biomedical applications. This is mainly due to their unique properties such as electric, optical, catalytic, thermal stability, and particularly antimicrobial properties (Carlson et al. 2008, Dallas et al. 2011). Silver (Ag) is known to have a disinfecting effect and antibacterial property, and has found applications in traditional medicines (Rai et al. 2009). The controlled antibacterial effect of the nano-sized Ag is the effect of its high surface area. Biopolymer-embedded Ag nanoparticles have been intensively investigated by numerous authors (Lee et al. 2006) for PLLA-based nanocomposite fibers including Ag nanoparticles, an antibacterial effect longer than three weeks being shown (Xu et al. 2006).

The action of silver ions and metallic silver nanoparticles is responsible for the antimicrobial function of silver in the manner presented in Scheme 3.

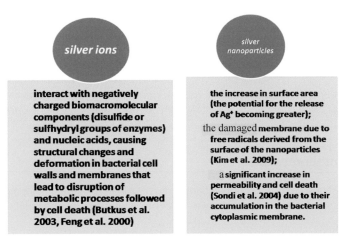

Scheme 3. Antimicrobial function of silver.

Other metal oxides that have also been exploited for the preparation of antimicrobial films are TiO_2, ZnO, and MgO due to their strong antimicrobial activity against both Gram-positive (*S. aureus*) and Gram-negative (*E. coli*) bacteria (Applerot et al. 2009), with high stability compared with organic antimicrobial agents (Zhang et al. 2010). The mechanism of their antimicrobial activity is based on the generation of highly reactive oxygen species by photocatalystis irradion with ultraviolet light (Dong et al. 2011). Among the metallic oxides, TiO_2 is inert, non-toxic, inexpensive, and environmental friendly with antimicrobial activity against a wide variety of microorganisms (Fujishima et al. 2000), but its practical applications are limited due to its low photon utilization efficiency and necessity of the ultraviolet (UV) as an excitation source. TiO_2 can be prepared as thin films on several substrates by various techniques such as chemical vapor deposition, evaporation, magnetron sputtering, ion beam technique, chemical spray pyrolysis, electro-deposition, and sol–gel method. The sol–gel process is suitable for producing composite materials of high purity without multiple steps (Bordianu et al. 2015).

Copper is not concentrated by animals and thus has few adverse effects on higher animals. Copper ions can destroy microorganisms and viruses, and is indispensable for life as a constituent of metallic enzymes.

Gold nanoparticles have been generated on a 4,4'-oxydiphthalic anhydride/4,4'-diamino-4''-hydroxytriphenylmethane-based polyimide matrix via electrospinning technique (Aflori et al. 2015), by thermally treating the gold salt-containing polyimide fibers. The average fiber diameters of the hybrid fibers decreased up to 48% as compared with the one corresponding to the pristine polyimide fibers due to the increased Coulombic repulsions within the fibers (Aflori et al. 2015).

The degradation rate of the nanocomposites is enhanced by the presence of the metal nanoparticles which changes thermal conductivity, surface wettability and roughness of the nanocomposites. It is important to note that the World Health Organization recognized and listed Ag nanoparticles as carcinogenic materials, immediate and thorough action being expected not only from the environment and human health viewpoints, but also from the perspective of socio-economic benefits.

Degradation and health implication of biopolymer nanocomposites

Biodegradable polymers have attracted considerable attention for applications in medical devices, and will play an important role in the design and function of medical devices. Biodegradability of bio-nanocomposites is one of the most interesting and controversial issues in the bionanocomposite materials. For biodegradable polymers, biodegradation may mean fragmentation, loss of mechanical properties, or sometimes degradation through the action of microorganisms such as bacteria, fungi, and algae. The biodegradation of polymers is a complex process, which can proceed via hydrolysis catalyzed by enzymes and oxidation (Paul et al. 2005).

One of the most important criteria of polymer materials used for medical devices include the degradation time, which must be appropriate to the medical purpose. In addition, the materials should not evoke toxic or immune responses, and they should be metabolized in the body after fulfilling their tasks. The degradation behavior has a crucial impact on the long-term performance of a tissue-engineered cell/polymer construct because tissue engineering aims at the regeneration of new tissues. The degradation kinetics of the biomaterials may affect a range of processes such as cell growth, tissue regeneration, and host response. The biomaterials are expected to be degradable and absorbable with a proper rate to match the speed of new tissue formation, but the bio-erosion of the material is determined by the surface hydrolysis of the polymer. Since one of the main reasons for using biopolymers for the preparation of bio-nanocomposite materials is to utilize the biodegradability of the biopolymer matrix, it is expected that the biodegradability of the resulting nanocomposites should not be sacrificed after the formation of nanocomposites. The scaffolds can lead to heterogeneous degradation, with the neutralization of carboxylic end groups located at the surface by the external buffer solution (*in vitro* or *in vivo*). These phenomena contribute to the reduction of the acidity at the surface whereas, in the bulk, degradation rate is enhanced by autocatalysis due to carboxylic end groups of aliphatic polyesters. The nanotoxicology defined as the systematic evaluation of possible toxic effects elicited by nanotechnology products on cells or biological fluids and animals rises from the development of some properties of nanoparticles (Au, carbon nanotubes, Al, Cu, and Ag) in industrial applications which might be detrimental both for biological systems and environment. Different polyesters can exhibit quite distinct degradation kinetics in aqueous solutions. For example, poly(glycolic acid) is a stronger acid and is more hydrophilic than PLA, which is hydrophobic due to its methyl groups. Among the biopolymer-based nanocomposite materials, PLA-based nanocomposites deserve a particular attention for the study of their biodegradation in the environment (Zhou et al. 2008). Some authors (Sinha et al. 2003) compared biodegradability of both PLA and PLA nanocomposite films using soil compost test at 58°C with films prepared by melt blending. They found no clear difference in biodegradability between neat PLA and PLA nanocomposite until a month of composting, but within two months, the PLA nanocomposite was completely degraded by compost. Some authors (Nieddu et al. 2009) reported similar results of enhanced biodegradation of PLA-based nanocomposites prepared with five different types of nanoclays and different level of clay content using a melt intercalation method. By measuring both the amount of lactic acid released and weight change of the sample materials during hydrolytic degradation in plasma incubated at 37°C, they found that the degradation rate of nanocomposite was more than 10 times (when measuring the lactic acid release) or 22 times (when measured weight change) higher than that of neat PLA. The degradation rate was dependent on the clay types and their concentration, i.e., the degree of degradation was higher in the better intercalated nanocomposite which was dependent on the type of clays. During the hydrolytic degradation of PLA and PLA nanocomposites, the formation of lactic acid oligomers obtained from the chain scission of PLA increases carboxylic acid end groups concentration and these carboxylic groups are known to catalyze the degradation reaction. Since the hydrolytic degradation of PLA is a self-catalyzed and self maintaining process, the hydrolysis of PLA can be affected by not only such structural factors as stereo structure, molar mass and their distribution, crystallinity, and purity, but also type of fillers. Similar results were observed with other bionanocomposite films including PHB (Maiti et al. 2007), soy protein-based nanocomposite (Sasmal et al. 2009), and nano-silica/starch/polyvinyl alcohol films (Tang et al. 2008).

Thin TiO_2 films exhibit excellent mechanical and chemical durability in the visible and near-infrared region and incorporation of TiO_2 into synthetic plastic matrix has shown to increase the biodegradability (Kubacka et al. 2007). Zhou et al. (Zhou et al. 2009) first incorporated TiO_2 into biopolymer (whey protein isolate) and they tested the effect of content of TiO_2 on mechanical and water vapor barrier properties without testing antimicrobial effect.

To date, the data regarding the issue of nanoparticles toxicity humans are still lacking, except for isolated case-reports on accidental high dose exposure in the workplace (Phillips et al. 2010). Effects seen in animals cannot be automatically translated to humans. On the other hand, data are not univocal. Some well-performed studies show no toxicity from the same nanoparticles showing adverse effects in other experiments (Travan et al. 2009, Hauck et al. 2010). This is due to the lack of standardization both in characterizing the nanoparticle and in the methods used to challenge the nanoparticles with biological systems. Some nanoparticles cytotoxic to human cells like carbon nanotubes can exhibit epidermal (Shvedova et al. 2003, Monteiro-Riviere et al. 2005) or pulmonary toxicity (Warheit et al. 2004, Li et al. 2011), and TiO_2 (Jani et al. 1994, Gurr et al. 2005) and Ag nanoparticles (Kim 2008) can enter blood circulation from the gastro-intestinal tract. The physico–chemical properties of the nanoparticles (such as size) or the physiological state of the organs are also responsible for these processes, the most affected organs seems to be the liver and the spleen due to the blood circulation (Silvestre et al. 2011).

In order to assess the risks associated with the presence nanoparticles in the human body or in the environment, many studies have been conducted, but significant research is still required to evaluate the potential toxicity as well as the environmental safety of the use of nanocomposite materials.

Designing the surface proprieties of the poly-L-lactic acid and polyhydroxybutyrate by conventional and non-conventional methods

Materials and methods

BSA (purity > 98%), silver nitrate (AgNO₃, 0,01 mol/l - 0,01 N volumetric standard solution) and sodium hydroxide (NaOH, 2 mol/l - 2 N volumetric standard solution) were purchased from Roth. These reagents were used as received. Deionized water was used throughout the entire work.

Poly-L-lactic acid (PLLA, density 1.25 g cm⁻³) and polyhydroxybutyrate (PHB, with 8% polyhydroxyvalerate, density 1.25 g cm⁻³) in the form of 50 μm thick foils (Goodfellow, Ltd., UK) were cut in square shapes of 6 x 6 mm size.

BSA containing silver nanoparticles solution

The solution of BSA containing silver nanoparticles (BSA/Ag Np) was obtained based on a procedure developed by Xie et al. for fluorescent gold nanoclusters (Xie et al. 2009), with some modifications. Briefly, 5 ml of 0.01 M AgNO₃ aqueous solution was added to 5 ml of 50 mg/ml BSA aqueous solution obtained from albumin fraction V, under vigorous stirring. Five minutes later, 1 ml of 1 M NaOH aqueous solution was added to the above solution to adjust the solution pH value to 12. Then the resulting solution was maintained in water-bath (37 C) for 6 h, under vigorous stirring.

Plasma treatments

The two steps experimental method assures the optimal content of silver at the polymer surface. The first step is the plasma treatment, performed in an EMITECH RF plasma device at different powers and time. Argon was used as the background gas at a pressure p = 10^{-3} m bar, the polymer being immersed into the plasma that fills the gas vessel. For the second step, the plasma-treated polymer was immersed for 2 days at room temperature in BSA/AgNp solution, protected from light. Then the samples were rinsed with deionized water and analyzed by the different characterization techniques.

The selected samples to be presented in this chapter are:

- neat PLLA (PLLA0) and PHB (PHB0) films,
- PLLA and PHB films treated in plasma for 4 min at 40 W RF power, immersed in chitosan/AgNp (PLLA1 and PHB1, respectively),
- PLLA and PHB films treated in plasma for 6 min at 40 W RF power, immersed in chitosan/AgNp (PLLA2 and PHB2, respectively).

Characterization methods

The spectra were recorded on a Bruker Vertex 70 FTIR spectrophotometer, equipped with a diamond ATR device (Golden Gate, Bruker). ATR-FTIR (Attenuated Total Reflection - Fourier Transform Infrared) analysis was performed in the range 600–4000 cm^{-1}, at a resolution of 2 cm^{-1} at incidence angle of 45°. For each spectrum, 128 scans were taken, with a baseline correction.

Scanning electron microscopy (SEM) was used to analyze the surface morphology and topography. The electron micrographs of PLLA samples were registered with a Quanta 200 microscope at an accelerating voltage of 15 kV.

Transmission electron microscopy (TEM) images of CS/Ag Np solution were obtained on a Hitachi High-Tech HT1700 microscope using an acceleration voltage of 120 kV.

The surface composition analysis was run by X-ray photoelectron spectroscopy (XPS) by using an PHI-5000 Versa Probe photoelectron spectrometer ULVACPHI, Inc., Japan with a hemispherical energy analyzer (0.85 eV binding energy resolutions for organic materials). A monochromatic Al K X-ray radiation was used as excitation source. The standard takeoff angle used for the analysis was 45°, producing a maximum analysis depth in the range of 3–5 nm. The spectra were recorded from at least three different locations on each sample, with a 1 x 1 mm^2 area of analysis. The low-resolution survey spectra were recorded in 0.5 eV steps with 117.4 eV analyzer pass energy and, in addition, the high-resolution carbon (1s) spectra were recorded in 0.1 eV steps with 58.7 eV analyzer pass energy.

Particle sizes of Ag nanoparticle solution were measured using a Zetasizer instrument (Model Zetasizer Nano ZS) using UV Grade cuvettes after treatment in an ultrasonic water bath (Model FB11012, Fisher-brand) for 30 min to break up any aggregates present. All measurements were performed in triplicate.

Results and discussions

The FT-IR spectroscopy was used to study the changes in the secondary structure of BSA upon interaction with the AgNPs. The characteristic absorption band of amide I (the most sensitive probe for monitoring changes in the protein secondary structures) is located at 1600–1700 cm^{-1}. Figure 1 shows the spectra of albumin in the absence of the AgNPs and of BSA in the presence of the AgNPs. The spectrum of albumin shows a strong band between 1600 and 1700 cm^{-1}, while the spectrum of the BSA/AgNP solution indicates changes in the intensity and position of this band, which can confirm the change in the secondary structure of BSA. As shown in Figure 1, the peak position of amide I (1652 cm^{-1}) for the albumin indicates that the secondary structure is an α-helix form. A shift to lower frequency (1625 cm^{-1}) after the incorporation of AgNp shows less compact structures and more random and open chains. Moreover, the broadening to the right of the center frequency indicates an increase in the β-sheet

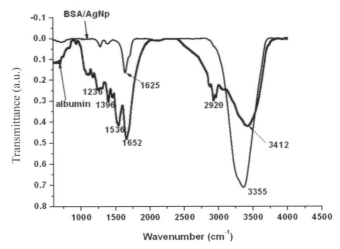

Figure 1. FTIR results for pristine albumin and BSA/AgNp solution.

content, random chains, or extended chains. From the above considerations, it can be concluded that the secondary structure of albumine changed due to the reaction with AgNPs (Binaymotlagh et al. 2016).

Figure 2 shows the IR spectra of the PLLA samples. The typical spectrum of PLLA, peak at 1746 cm^{-1}, is given by the stretching of C = O; the 1185 and 1077 cm^{-1} bands are attributable to C–O–C asymmetric and symmetric stretchings, respectively. The peak at 1128 cm^{-1} is ascribable to the C–H (of CH$_3$ groups) rocking mode, while the peak at 1038 cm^{-1} is caused by C–CH$_3$ stretching. After the treatment, we can observe an increase of the 1746 cm^{-1} band intensity and of band 1381 cm^{-1}. A shift with 2 cm^{-1} to lower wavenumbers of those bands can be observed. Those bands are assigned to the carboxylic groups and demonstrate an increase of those functional groups quantity after plasma treatments and the presence of silver ions at the polymer surface. It is known that the crystal modification of PLLA is easily formed from the melt, therefore, the increase in the intensity of the absorption bands characteristic for the crystal structure of PLLA is an indication that the process of PLLA crystallization is caused by heating of the polymer surface as a result of the interaction with high-energy plasma particles.

The FTIR spectra for PHB plasma treated samples and plasma followed by chemical treatment samples were compared with the spectrum obtained from commercial PHB in Figure 3. The most prominent marker band for the identification of PHB is the ester carbonyl band at 1720 cm^{-1}. The bands present at 1179 cm^{-1}, 1226 cm^{-1} and 1260 cm^{-1} are bands sensitive to crystallinity and are characteristic

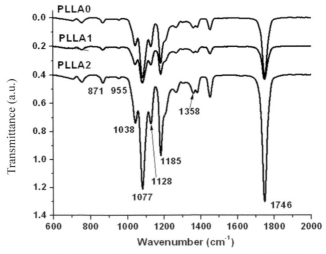

Figure 2. FTIR results for samples PLLA0, PLLA1 and PLLA2.

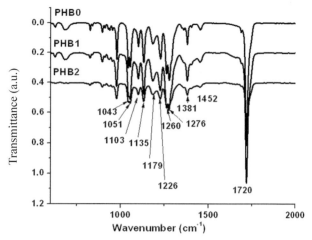

Figure 3. FTIR results for samples PHB0, PHB1 and PHB2.

of C-O-C. In contrast with the behavior of PLLA films, the peak intensities observed for commercial PHB were greater than those for treated PHB. A greater transmittance correlates to a higher degree of crystallinity; this observation could lead to the inclusion that the treated PHB have lower degree of crystallinity compared to that of the commercial PHB (Furukawa et al. 2005, Randriamahefa et al. 2003). The presence of silver nanoparticles is demonstrated by the shift and the change in intensity of the FTIR peaks.

The representative TEM images of AgNPs show that the particles are randomly distributed in the solution (Figure 4), have a spherical core-shell shape (Figure 4a), and a mean core size of about 60 nm of silver with a shell of 2.5 nm of BSA (Figure 4b), in good concordance with the particle distribution obtained by dynamic light scattering method (Figure 4a right up).

The morphological changes in PLLA and PHB surface after the two-step treatments are presented in Figure 5. In Figure 5a, pristine PLLA and PHB film has a smooth surface without any irregularities. The PLLA films surfaces, after the combined plasma-wet chemical treatment, have patterns of different size and shape due to the surface interactions with different reactive species formed in plasma (Figure 5b and Figure 5c, respectively) and due to the presence of AgNp and chitosan (Figure 5d and Figure 5e respectively).

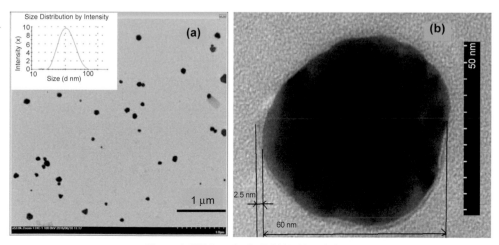

Figure 4. TEM results for BSA/AgNp solution.

The elemental and chemical composition of the samples, before (native films) and after BSA adsorption, was determined by XPS. Figure 6 shows typical XPS spectra recorded for the same samples as the one studied by FTIR and SEM in previous sections. From the survey spectrum (Figure 6a and 6b), it was found that both the native PLLA and PHB films were composed of carbon and oxygen, consistent with previous findings (Saulou et al. 2012). The intensity in the Ag 3d region was markedly lower in the samples with the high content of BSA (Figure 6c). The Ag 3d3/2 peak binding energy is 365 eV, while the Ag 3d5/2 peak binding energy is 370 eV (Figure 6c). As expected, N 1s was detected after BSA adsorption. The N 1s peak, centered at 397 eV as shown in Figure 6d, is symmetric as expected for the amine or amide groups of BSA (Dufrêne et al. 1999, Rouxhet et al. 1994).

In Table 1 are presented the results of XPS elemental analysis. The highest content of BSA (determined by the highest content of N and N/C ratio) is present in PLLA2 sample, while the highest content of AgNp is present in PHB2 sample.

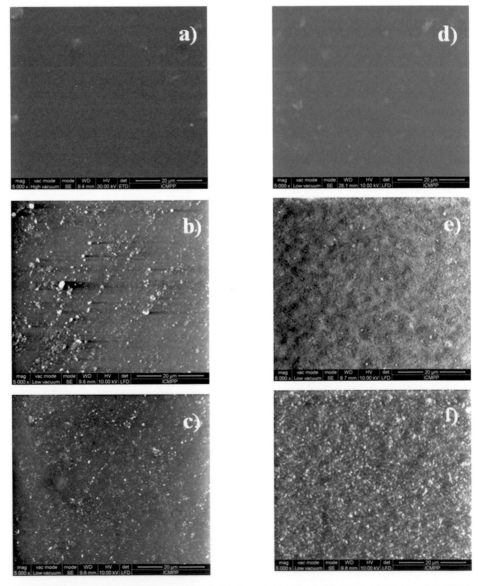

Figure 5. SEM results for samples: (a) PLLA0, (b) PLLA1, (c) PLLA2, (d) PHB0, (e) PHB1 and (f) PHB2.

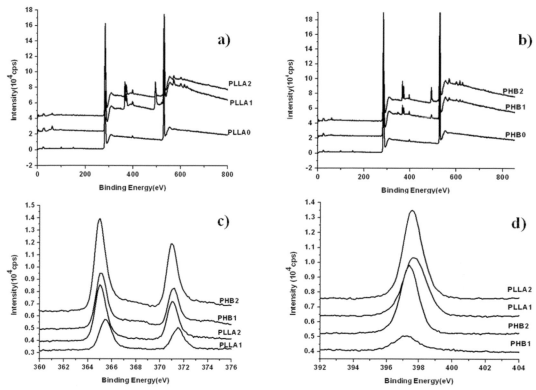

Figure 6. XPS results for all samples: (a) PLLA wide angle, (b) PHB wide angle, (c) Ag3d spectra for the treated samples, and (d) N1s spectra for the treated samples.

Table 1. Composition for all samples from XPS measurements.

Element (%)	CK	OK	NK	AgL	N/C
PLLA0	80.04	19.96	-	-	
PLLA1	70.79	12.87	15.47	00.87	0.21
PLLA2	71.58	10.81	16.52	01.09	0.23
PHB0	79.38	20.62	-	-	
PHB1	74.10	11.03	13.58	01.29	0.18
PHB2	71.47	11.47	14.93	02.13	0.20

Final thoughts

The biopolymer-based nanocomposites reviewed in this article are particularly attractive in the biomedical field due to their biocompatibility and adjustable biodegradation kinetics. Combined conventional and nonconventional materials processing methods have been adapted and incorporation of silver nanoparticles into polymer films was reported. Stable spherical AgNPs (about 60 nm diameter) were obtained in BSA solution. FTIR, SEM and XPS characterization methods demonstrated that silver nanoparticles were successfully adsorbed on PLLA and PHB films exposed to plasma treatments, by simply immersing the treated films in the solution containing silver nanoparticles. This is a time saving and eco-friendly process which minimizes the use of toxic chemicals.

Concerns for the effects of nano-sized particles on the human body and the environment are increasing recently. Generally, beneficial effects of nanocomposite materials are well recognized, but the potential (eco-)toxicological effects and impacts on human health of nanoparticles have so far received little attention. The high speed of penetration of nanocomposite-based consumer products bring about the need for a better understanding about the potential impacts that nanoparticles may have on biological systems. The high surface-to-volume ratio of nanomaterials makes them more reactive and potentially more toxic.

The incorporation of nanoparticles and immobilization of biological components on the surface to enhance antimicrobial functions, or cellular adhesion and proliferation are promising and currently under extensive research. In this direction, the new approach of biopolymer-based nanocomposite enables the scaffold surface to mimic complex local biological functions and may lead in the near future to *in vitro* and *in vivo* growth of tissues and organs.

References

Aflori, M., D. Serbezeanu, I.-D. Carja and G. Fortunato. 2015. Gold nanoparticles incorporated into electrospun polyimide fibers. Chem. Lett. 44: 1440–1442.

Applerot, G., N. Perkas, G. Amirian, O. Girshevitz and A. Dedanken. 2009. Coating of glass with ZnO via ultrasonic irradiation and a study of its antibacterial properties. Appl. Surf. Sci. 256(Suppl.): S3–8.

Armentano, I., M.A. Alvarez-Pérez, B. Carmona-Rodríguez, I. Gutiérrez-Ospina, J.M. Kenny, H. Armentano et al. 2010. Biodegradable polymer matrix nanocomposites for tissue engineering: a review. Polym. Degrad. Stabil. 95: 2126–2146.

Arvanitoyannis, I., E. Psomiadou, A. Nakayama, S. Aiba and N. Yamamoto. 1997. Edible films made from gelatin, soluble starch and polyols, part 3. Food Chem. 60: 593–604.

Averous, L. and N. Boquillon. 2004. Biocomposites based on plasticized starch: thermal and mechanical behaviors. Carbohydr. Polym. 56: 111–122.

Armentano, I., M.A. Álvarez-Pérez, B. Carmona-Rodríguez, I. Gutiérrez-Ospina, J.M. Kenny and H. Arzate. 2008. Analysis of the biomineralization process on SWNT-COOH and F-SWNT films. Mat. Sci. Eng. C. 28: 1522–1529.

Bakare, R., S. Hawthrone, C. Vails, A. Gugssa, A. Karim, J. Stubbs III et al. 2016. Antimicrobial and cell viability measurement of bovine serum albumin capped silver nanoparticles (Ag/BSA) loaded collagen immobilized poly (3-hydroxybutyrate-co-3-hydroxyvalerate) (PHBV). J. Colloid Interface Sci. 465: 140–148.

Bertan, L.C., P.S. Tanada-Palmu, A.C. Siani and C.R.F. Grosso. 2005. Effect of fatty acids and 'Brazilian elemi' on composite films based on gelatin. Food Hydrocolloids 19: 73–82.

Bi, L., L. Yang, G. Narsimhan, A.K. Bhunia and Y. Yao. 2011. Designing carbohydrate nanoparticles for prolonged efficacy of antimicrobial peptide. J. Con. Rel. 150: 150–156.

Binaymotlagh, R., H. Hadadzadeh, H. Farrokhpour, F.H. Haghighi, F. Abyar and S.Z. Mirahmadi-Zare. 2016. *In situ* generation of the gold nanoparticlesebovine serum albumin (AuNP/BSA) bioconjugated system using pulsed-laser ablation (PLA). Mat. Chem. Phys. 177: 360–370.

Bondeson, D. 2007. Biopolymer-based Nanocomposites: Processing and Properties. Ph.D. Thesis, Norwegian University of Science and Technology, Trondheim, Norway.

Bordes, P., E. Pollet and L. Averous. 2009. Nano-biocomposites: biodegradable polyester/nanoclay systems. Prog. Polym. Sci. 34: 125–155.

Bordianu, I.E., G. David, B. Simionescu, M. Aflori, C. Ursu, A. Coroaba et al. 2015. Functional silsesquioxane-based hierarchical assemblies for antibacterial/antifungal coatings. J. Mat. Chem. B. 3: 723–727.

Butkus, M.A., L. Edling and M.P. Labare. 2003. The efficacy of silver as a bactericidal agent: advantages, limitations and considerations for future use. J. Water Supply: Res. Tech. – AQUA 52: 407–416.

Carlson, C., S.M. Hussain, A.M. Schrand, L.K. Braydich-Stolle, K.L. Hess, R.L. Jones et al. 2008. Unique cellular interaction of silver nanoparticles: size-dependent generation of reactive oxygen species. J. Phys. Chem. B112: 13608–13619.

Chen, G.-Q. and Q. Wu. 2005. The application of polyhydroxyalkanoates as tissue engineering materials. Biomaterials 26: 6565–6578.

Chen, Q., S. Liang and G.A. Thouas. 2013. Elastomeric biomaterials for tissue engineering. Prog. Polym. Sci. 38: 584–671.

Dallas, P., V.K. Sharma and R. Zboril. 2011. Silver polymeric nanocomposites as advanced antimicrobial agents: classification, synthetic paths, applications, and perspectives. Adv. Colloid Interface Sci. 166: 119–135.

Dong, C., D. Song, J. Cairney, O.L. Maddan, G. He and Y. Deng. 2011. Antibacterial study of $Mg(OH)_2$ nanoplatelets. Mat. Res. Bull. 46: 576–582.

Du, W.L., S.S. Niu, Y.L. Xu, Z.R. Xu and C.L. Fan. 2009. Antibacterial activity of chitosan tripolyphosphate nanoparticles loaded with various metal ions. Carbohydr. Polym. 75: 385–389.

Dufrêne, Y.F., T.G. Marchal and P.G. Rouxhet. 1999. Probing the organization of adsorbed protein layers: complementarity of atomic force microscopy, X-ray photoelectron spectroscopy and radiolabeling. Appl. Surf. Sci. 144-145: 638–643.

Feng, Q.L., J. Wu, G.Q. Chen, F.G. Cui, T.N. Kim and J.O. Kim. 2000. A mechanistic study of the antibacterial effect of silver ions on *Escherichia coli* and *Staaphylococcus aureus*. J. Biomed. Mat. Res. 52: 662–668.

Flores, C.Y., A.G. Miñán, C.A. Grillo, R.C. Salvarezza, C. Vericat and P.L. Schilardi. 2013. Citrate-capped silver nanoparticles showing good bactericidal effect against both planktonic and sessile bacteria and a low cytotoxicity to osteoblastic cells. ACS Appl. Mater. Interfaces 5: 3149–3159.

Freire, P.L.L., T.M.C. Stanford, A.J.R. Albuquerque, F.C. Sampaio, H.M.M. Cavalcante, R.O. Macedo et al. 2015. Action of silver nanoparticles towards biological systems: cytotoxicity evaluation using hen's egg test and inhibition of *Streptococcus mutans* biofilm formation. Int. J. Antimicrob. Agents 45: 183–187.

Fujishima, A., T.N. Rao and D.A. Truk. 2000. Titanium dioxide photocatalysis. J. Photochem. Photobiol. C: Photochem. Rev. 1: 1–21.

Fukushima, H., L.T. Drzal, B.P. Rook and M.J. Rich. 2006. Thermal conductivity of exfoliated graphite nanocomposites. J. Therm. Anal. Calorim. 85: 235–238.

Furukawa, T., H. Sato, R. Murakami, J. Zhang, Y.-X. Duan, I.n. Shuckichi et al. 2005. Structure, dispersibility, and crystallinity of poly(hydroxybutyrate)/poly(l-lactic acid) blends studied by FT-IR microspectroscopy and differential scanning calorimetry. Macromolecules 38: 6445–6454.

Gay, S., S. Arostegui and J. Lemaitre. 2009. Preparation and characterization of dense nanohydroxyapatite/PLLA composites. Mat. Sci. Eng. C. 29: 172–177.

Goddard, J.M. and J.H. Hotchkiss. 2007. Polymer surface modification for the attachment of bioactive compounds. Prog. Polym. Sci. 32: 698–725.

Guilbert, S., B. Cuq and N. Gontard. 1997. Recent innovations in edible film and/or biodegradable packaging materials. Food Addit. Contam. 14: 741–51.

Gurr, J.R., A.S. Wang, C.H. Chen and K.Y. Jan. 2005. Ultrafine titanium dioxide particles in the absence of photoactivation can induce oxidative damage to human bronchial epithelial cells. Toxicology 213: 66–73.

Hackenberg, S., A. Scherzed, M. Kessler, S. Hummel, A. Technau, K. Froelich et al. 2011. Silver nanoparticles: evaluation of DNA damage, toxicity and functional impairment in human mesenchymal stem cells. Toxicol. Lett. 201: 27–33.

Harrison, B.S. and A. Atala. 2007. Review, carbon nanotube applications for tissue engineering. Biomaterials 28: 344–353.

Hauck, T.S., R.E. Anderson, H.C. Fischer, S. Newbigging and W.C.W. Chan. 2010. *In vivo* quantum-dot toxicity assessment. Small 6: 138–144.

Hong, Z.K., X.Y. Qiu, J.R. Sun, M.X. Deng, X.S. Chen and X.B. Jing. 2004. Grafting polymerization of l-lactide on the surface of hydroxyapatite nanocrystals. Polymer. 45: 6705–6713.

Ikada, Y. 1994. Surface modification of polymers for medical applications. Biomaterials 15: 725–736.

Jani, P., D. McCarthy and A.T. Florence. 1994. Titanium dioxide (rutile) particle uptake from the rat GI tract and translocation to systemic organs after oral administration. Int. J. Pharmac. 105: 157–168.

Kasálková, N.S., P. Slepičkaa, P. Sajdl and V. Švorčík. 2014. Surface changes of biopolymers PHB and PLLA induced by Ar+ plasma treatment and wet etching. Nuclear Instruments and Methods in Physics Research B 332: 63–67.

Kester, J.J. and O.R. Fennema. 1986. Edible films and coatings: a review. Food Technol. 40: 47–59.

Kim, G.G., J.A. Kang, J.H. Kim, S.J. Kim, N.H. Lee and S.J. Kim. 2006. Metallization of polymer through a novel surface modification applying a photocatalytic reaction. Surf. Coat. Tech. 201: 3761–3766.

Kim, J.S., K.E. Kuk, K.N. Yu, J.H. Kim, S.J. Park and H.J. Lee. 2007. Antimicrobial effects of silver nanoparticles. Nanomedicine: Nanotechnology. Biology and Medicine 3: 95–101.

Kim, Y.S., J.S. Kim, H.S. Cho, D.S. Rha, J.M. Kim et al. 2008. Twentyeight-day oral toxicity, genotoxicity, and gender-related tissue distribution of silver nanoparticles in Sprague-Dawley rats. Inhalation Toxicology 20: 575–583.

Knowles, J.C. 2003. A review article: phosphate glasses for biomedical applications. J. Mat. Chem. 13: 2395–2401.

Kubacka, A., C. Serrano, M. Ferrer, H. Lunsdorf, P. Bielecki et al. 2007. High-performance dualaction polymer-TiO$_2$ nanocomposite films via melting processing. Nano Letters 7: 2529–2534.

Krochta, J.M. and C.D. De-Mulder-Johnston. 1997. Edible and biodegradable polymer films: challenges and opportunities. Food Technol. 51: 61–74.

Lee, J.Y., J.L.R. Nagahata and S. Horiuchi. 2006. Effect of metal nanoparticles on thermal stabilization of polymer/metal nanocomposites prepared by a one-step dry process. Polymer 47: 7970–7979.

Li, J., X.L. Lu and Y.F. Zheng. 2008. Effect of surface modified hydroxyapatite on the tensile property improvement of HA/PLA composite. Appl. Surf. Sci. 255: 494–497.

Li, W., X. Li, P. Zhang and Y. Xing. 2011. Development of nano-ZnO coated food packaging film and its inhibitory effect on *Eschericia coli in vitro* and in actual tests. Adv. Mat. Res. 152-153; 489–492.

Lunt, J. 1998. Large-scale production, properties and commercial applications of polylactic acid polymers. Polym. Degrad. Stab. 59: 145–152.

Maiti, P., K. Yamada, M. Okamoto, K. Ueda and K. Okamoto. 2002. New polylactide/layered silicate nanocomposites: role of organoclays. Chem. Mater. 14: 4654–4661.

Maiti, P., C.A. Batt and E.P. Giannelis. 2007. New biodegradable polyhydroxybutyrate/layered silicate nanocomposites. Biomacromolecules 8: 3393–3400.

Mochalin, V.N., O. Shenderova, D. Ho and Y. Gogotsi. 2012. The properties and applications of nanodiamonds. Nature Nanotechnol. 7: 11–23.

Mohanty, A.K., L.T. Drzal and M. Misra. 2003. Nano reinforcements of bio-based polymers—the hope and the reality. Polym. Mater. Sci. Eng. 88: 60–61.

Monteiro-Riviere, N.A., R.J. Nemanich, A.O. Imman, Y.Y. Wang and J.E. Riviere. 2005. Multi-walled carbon nanotube interactions with human epidermal keratinocytes. Toxicol. Lett. 155: 377–384.

Murugaraj, P., D.E. Mainwaring, T. Jakubov, N.E. Mora-Huertas, N.A. Khelil and R. Siegele. 2006. Electron transport in semiconducting nanoparticle and nanocluster carbon–polymer composites. Solid State Commun. 137: 422–426.

Mwenifumbo, S., M.S. Shaffer and M.M. Stevens. 2007. Exploring cellular behavior with multi-walled carbon nanotube constructs. J. Mat. Chem. 17: 1894–1902.

Nair, G.K. and A. Dufresne. 2003. Crab shell chitin whisker reinforced natural rubber nanocomposites. 1. Processing and swelling behavior. Biomacromolecules 4: 657–665.

Nejati, E., H. Mirzadeh and M. Zandi. 2008. Synthesis and characterization of nanohydroxyapatite rods/poly(l-lactic acid) composite scaffolds for bone tissue engineering. Composites Part A Appl. Sci. Manufact. 39: 1589–96.

Nieddu, E., L. Mazzucco, P. Gentile, T. Benko, V. Balbo, R. Mandrile and G. Ciardelli. 2009. Preparation and biodegradation of clay composite of PLA. Reactive & Functional Polym. 69: 371–379.

Pandey, J.K., A.P. Kumar, M. Misra, A.K. Mohanty, L.T. Drzal and R.P. Singh. 2005. Recent advances in biodegradable nanocomposites. J. Nanosci. Nanotechnol. 5: 497–526.

Paul, M.A., C. Delcourt, M. Alexandre, Ph. Degée, F. Monteverde and Ph. Dubois. 2005. Polylactide/montmorillonite nanocomposites: study of the hydrolytic degradation. Polym. Degrad. Stabil. 87: 535–542.

Perego, G., G.D. Cella and C. Bastioli. 1996. Effect of molecular weight and crystallinity on poly(lactic acid) mechanical properties. J. Appl. Polym. Sci. 59: 37–43.

Phillips, J.I., F.Y. Green, J.C. Davies and J. Murray. 2010. Pulmonary and systemic toxicity following exposure to nickel nanoparticles. Am. J. Ind. Med. 53: 763–767.

Rai, M., A. Yadav and A. Gade. 2009. Silver nanoparticles as a new generation of antimicrobials. Biotechnology Advances 27: 76–83.

Ranby, B.G. 1952. The cellulose micelles. Tappi. 35: 53–58.

Randriamahefa, S., E. Renard, P. Guérin and V. Langlois. 2003. Fourier transform infrared spectroscopy for screening and quantifying production of PHAs by *Pseudomonas* grown on sodium octanoate. Biomacromolecules 4: 1092–1097.

Rimpelová, S., N. Slepicková Kasálková, P. Slepicka, H. Lemerová, V. Švorcík and T. Ruml. 2013. Plasma treated polyethylene grafted with adhesive molecules for enhanced adhesion and growth of fibroblasts. Mat. Sci. Eng. 33: 1116–1124.

Rouxhet, P.G., N. Mozes, P.B. Dengis, Y.F. Dufrêne, P.A. Gerin and M.J. Genet. 1994. Application of X-ray photoelectron spectroscopy to microorganisms. Colloids Surf. B: Biointerfaces 2: 347–369.

Sasmal, A., P.L. Nayak and S. Sasmal. 2009. Degradability studies of green nanocomposites derived from soy protein isolate (SPI)-furfural modified with organoclay. Polymer – Plastics Techn. Eng. 48: 905–909.

Saulou, C., B. Despax, P. Raynaud, S. Zanna, A. Seyeux, P. Marcus et al. 2012. Plasma-mediated nanosilver-organosilicon composite films deposited on stainless steel: synthesis, surface characterization, and evaluation of anti-adhesive and anti-microbial properties on the model yeast *Saccharomyces cerevisiae*. Plasma Process. Polym. 9: 324–338.

Schadler, L.S. 2003. Polymer-based and polymer-filled nanocomposites. *In*: P.M. Ajayan, L.S. Schadler and P.V. Braun (eds.). Nanocomposite Science and Technology. Wiley-VCH Verlag, Weinheim.

Silvestre, C., Duraccio and S. Cimmino. 2011. Food packaging based on polymer nanomaterials. Prog. Polym. Sci. 36: 1766–1782.

Sinha Ray, S., K. Yamada, M. Okamoto and K. Ueda. 2003. New polylactide layered silicate nanocomposites. 2. Concurrent improvements of materials properties, biodegradability and melt rheology. Polymer. 44: 857–866.

Shvedova, A., V. Castranova, E. Kisin, D. Schwegler-Berry, A. Murray, V. Gandelsman et al. 2003. Exposure to carbon nanotube material: assessment of nanotube cytotoxicity using human keratinocyte cells. J. Toxicol. Environm. Health, Part A. 66: 1909–1926.

Slepicka, P., S. Trostová, N.S. Kasálková, Z. Kolská, P. Malinsky, A. Macková et al. 2004. Silver nanoparticles as antimicrobial agent: a case study of *E. coli* as a model for Gram-negative bacteria. J. Colloid Interface Sci. 275: 177–182.

Švorcík. 2012. Nanostructuring of polymethylpentene by plasma and heat treatment for improved biocompatibility. Polym. Degrad. Stabil. 97: 1075–1082.

Supronowicz, P.R., P.M. Ajayan, K.R. Ullmann, B.P. Arulanandam, D.W. Metzger and R. Bizios. 2002. Novel current-conducting composite substrates for exposing osteoblasts alternating current stimulation. J. Biomed. Mat. Res. 59: 499–506.

Tan, L., X. Yu, P. Wan and K. Yang. 2013. Biodegradable materials for bone repairs: a review. J. Mater. Sci. Technol. 29: 503–513.

Tang, S., P. Zou, H. Xiong and H. Tang. 2008. Effect of nano-SiO₂ on the performance of starch/polyvinyl alcohol blend films. Carbohydr. Polym. 72: 521–526.

Tharanathan, R.N. 2003. Biodegradable films and composite coatings: past, present and future. Trends Food Sci. Technol. 14: 71–78.

Thompson, C.M., H.M. Herring, T.S. Gates and J.W. Connell. 2003. Preparation and characterization of metal oxide/polyimide nanocomposites. Compos. Sci. Technol. 63: 1591–1598.

Tian, H., Z. Tang, X. Zhuang, X. Chen and X. Jing. 2012. Biodegradable synthetic polymers: preparation, functionalization and biomedical application. Prog. Polym. Sci. 37: 237–280.

Tian, X., X. Zhang, W. Liu, J. Zheng, C. Ruan and P. Cui. 2006. Preparation and properties of poly(ethylene terephthalate)/silica nanocomposites. J. Macromol. Sci. Part B: Phys. 45: 507–513.

Travan, A., C. Pelillo, I. Donati, E. Marsich, M. Benincasa, T. Scarpa et al. 2009. Non-cytotoxic silver nanoparticle-polysaccharide nanocomposites with antimicrobial activity. Biomacromolecules 10: 1429–1435.

Vainionpä, S., P. Rokkanen and P. Törmälä. 1989. Surgical applications of biodegradable polymers in human tissues. Prog. Polym. Sci. 14: 679–716.

Vink, E.T.H., K.R. Rábago, D.A. Glassner and P.R. Gruber. 2003. Applications of life cycle assessment to NatureWorksTM polylactide (PLA) production. Polym. Degrad. Stab. 80: 403–419.

Warheit, D.B., B.R. Laurence, K.L. Reed, D.H. Roach, D.A.M. Reynolds and T.R. Webb. 2004. Comparative pulmonary assessment of single-wall carbon nanotubes in rats. Toxicol. Sci. 77: 117–125.

Watkins, J.J. and T.J. McCarthy. 1995. Polymer/metal nanocomposite synthesis in supercritical CO_2. Chem. Mater. 7: 1991–1994.

Webster, T.J., C. Ergun, R.H. Doremus, R.W. Siegel and R. Bizios. 2000. Enhanced functions of osteoblasts on nanophase ceramics. Biomaterials 21: 1803–1810.

Xie, J., Y. Zheng and J.Y. Ying. 2009. Protein-directed synthesis of highly fluorescent gold nanoclusters. J. Am. Chem. Soc. 131: 888–889.

Xie, S., W. Li, Z. Pan, B. Chang and L. Sun. 2000. Mechanical and physical properties on carbon nanotubes. J. Phys. Chem. Solids 61: 1153–1158.

Xu, X., Q. Yang, Y. Wang, H. Yu, X. Chen and X. Jing. 2006. Biodegradable electrospun poly(l-lactide) fibers containing antibacterial silver nanoparticles. Eur. Polym. J. 42: 2081–2087.

Yoksan, R. and S. Chirachanchai. 2010. Silver nanoparticle-loaded chitosan starch based films: fabrication and evaluation of tensile, barrier and antimicrobial properties. Mat. Sci. Eng. C. 30: 891–897.

Yoshida, S., K. Hagiwara, T. Hasebe and A. Hotta. 2013. Surface modification of polymers by plasma treatments for the enhancement of biocompatibility and controlled drug release. Surf. Coat. Tech. 233: 99–107.

Zhang, L., Y. Jiang, Y. Ding, N. Daskalakis, L. Jeuken et al. 2010. Mechanistic investigation into antibacterial behaviour of suspensions of ZnO nanoparticles against *E. coli*. J. Nanopart. Res. 12: 1625–1636.

Zhang, Q., V.N. Mochalin, I. Neitzel, K. Hazeli, J. Niu et al. 2012. Mechanical properties and biomineralization of multifunctional nanodiamond-PLLA composites for bone tissue engineering. Biomaterials 33: 5067–5075.

Zhou, Q. and M. Xanthos. 2008. Nanoclay and crystallinity effects on the hydrolytic degradation of polymers. Polym. Degrad. Stabil. 93: 1450–1459.

Zhou, J.J., S.Y. Wang and S. Gunasekaran. 2009. Preparation and characterization of whey protein film incorporated with TiO_2 nanoparticles. J. Food Sci. 74: N50–6.

Dental Nanocomposites

Eric Habib,[†] *Ruili Wang*[†] *and Julian Zhu**

Introduction

Throughout history, many different methods have been used for repairing dental caries. There is anecdotal evidence indicating that dental amalgam was used for such restorations as early as 618 CE in the Tang Dynasty (http://www.caringtreechildrensdentistry.com/the-history-of-dental-amalgams/), using alloys of tin and silver. However, only by the 16th or 19th century (depending on the source) did the more modern mercury-containing version of dental amalgams appear (http://www.toothbythelake.net/wellness-center/amalgam-fillings/a-brief-history-of-amalgams/), alongside other metals like tin, gold, and even lead and thorium. It was not until the 1960s, though, that dental resin composites emerged (Bowen 1963). Since then, they have seen steady advances in technology and clinical use for the restoration of dental caries and other defects.

These composites are made from a mix of polymerizable resin matrix, including monomers and photosensitizers, and surface modified inorganic filler particles (Klapdoh and Moszner 2005, Valente et al. 2013, Habib et al. 2016). The resin matrix is comprised of multimethacrylate monomer blends that, once polymerized, result in highly crosslinked and highly resilient polymer networks. Filler particles, on the other hand, vary a lot more, but are generally comprised of inorganic silicate particles of nanometer to micrometer size. In order to ensure complete integration and attachment to the resin matrix, these are usually modified with methacrylated silanes that allow covalent crosslinking between the filler and the resin upon polymerization. Most of the recent developments in filler technology have been within the domain of nanomaterials, whether simply in their size, or including more complex nanostructures as described further.

A large number of commercial and experimental formulations have been developed since the original, aiming at superior properties to fulfill the demanding and specific requirements for these materials. This chapter will principally review the applications of nanotechnology that have been explored as part of dental resin composite restoratives.

Department of Chemistry, University of Montreal, C.P. 6128, Succ. Centre-Ville, Montreal, QC, H3C 3J7, Canada.
  Emails: eric.christophe-habib@umontreal.ca; ruili.wang@umontreal.ca
* Corresponding author: julian.zhu@umontreal.ca
† Eric Habib and Ruili Wang are joint First Authors.

> ***Terminology for 'fillers':*** **The inorganic particles used in dental materials would normally be termed 'reinforcement' since they serve to strengthen the properties of these materials, rather than just reducing the production cost. In this field, they are termed 'fillers', and so the words will be used as it is in the context of dentistry.**

The requirements for these composite materials are very stringent, due, in part, to the high stresses experienced by the teeth, estimated at 3.9 to 17.9 MPa (Yap et al. 2004a). After etching and priming the existing tooth structure, dental resin-based composites are designed to be used by dentists as viscous pastes that can be inserted into cavities with ease to accommodate any surface shape. Once they are in place, a blue light is used to photoinitiate the polymerization to cross-link the resin component, resulting in the final hard material that can then be sanded and polished to its final shape. Consequently, these materials must be soft enough for easy handling before they are polymerized and as hard and resilient as possible after that. Furthermore, due to the constant exposure to water in the oral cavity, these materials must be hydrophobic enough to minimize water uptake, and show high levels of biocompatibility to be acceptable for use (ISO 7405 2008). Finally, dental composites are required to have high X-ray opacity so as to allow diagnostic examinations of the state of the material using dental X-ray radiography.

Clinically speaking, the current primary cause of failure for these materials is the development of secondary caries, which is thought to be primarily due to shrinkage that occurs upon the material's polymerization (Ferracane and Hilton 2016). Therefore, one of the current targets of new dental composite restoratives is the reduction of polymerization shrinkage, which can be achieved in two main ways: (1) producing a resin that contracts less upon polymerization, or (2) increasing the relative fraction of (non-shrinking) inorganic filler particles in the material.

The advances presented below (Scheme 1) all aim at increasing these desirable characteristics while decreasing the drawbacks.

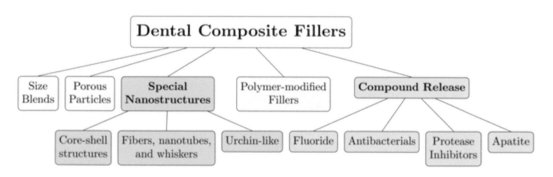

Scheme 1. Chapter summary.

Dental restorative matrix

The composition of the resin matrix of these materials has not changed much since their advent, being typically comprised of a mixture of different dimethacrylate monomers. The mix of bisphenol A glycerolate dimethacrylate (Bis-GMA) and triethylene glycol dimethacrylate (TEGDMA) which was used in the initial formulation has remained the basis for most modern dental composite formulations (Bowen 1962, Moszner and Salz 2001, Lavigueur and Zhu 2012), as shown in Figure 1. Nonetheless, a great many number of derivatives of these have been developed to alter the properties of the materials, both before and after polymerization (Wang et al. 2013a). Aside from the slightly newer diurethane dimethacrylate (UDMA) and derivatives thereof, there have been few radically different novel resin monomers.

One notable exception to that is the so-called silorane monomers (mix of oxiranes and siloxanes, shown in Figure 2) that were developed by 3M ESPE as a solution to the polymerization shrinkage

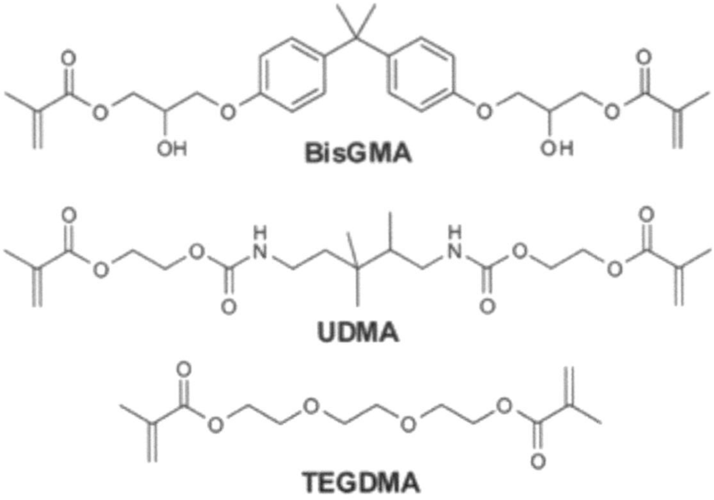

Figure 1. Structure of the most commonly used monomers in dental composite resins (Bowen 1962, Moszner and Salz 2001).

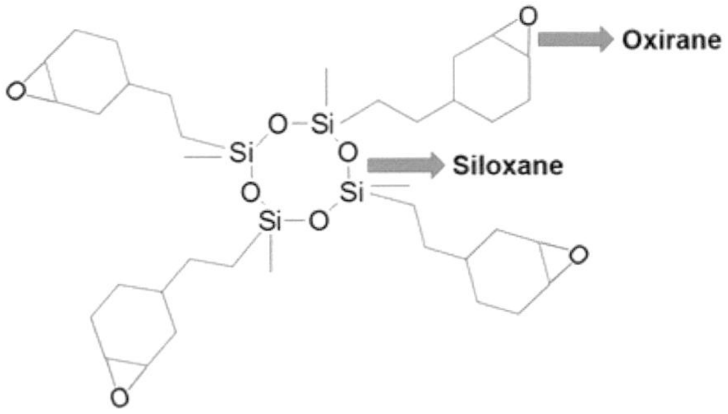

Figure 2. Silorane chemistry (Filtek Silorane Low Shrink Posterior Restorative).

problem (Weinmann et al. 2005). However, market adoption has been slow due to the incompatibility with other commercially available materials.

Aside from the monomers for general use, in order to accommodate higher depths of cure and reduce shrinkage stress upon polymerization, one of the strategies adopted to address this problem is the inclusion of a small fraction of 'addition-fragmentation' monomers. These monomers are designed to open and subsequently close after a short time interval, during which time the matrix of the material can relax, thereby alleviating some of the internal stresses that can later favor material fractures (Filtek Bulk Fill Posterior Restorative). Other companies may have also included similar materials but due to trade secrets, many of these remain proprietary and confidential.

Dental restorative fillers

Initial formulations of dental resin composites used ground vitreous quartz (silicon dioxide) fillers that were approximately 150 microns in diameter (Bowen 1963). This large size of fillers is now classed as 'macro' fillers. As is still the case with fillers used today, they can be produced either through by top-down approaches which consist of taking bulk matter with the desired composition, and milling and sieving it down to the desired size, like the ground quartz, or by bottom-up approaches, using sol-gel, pyrogenesis, or other solution-based methods to directly synthesize the desired particle characteristics.

Chemical composition

The composition of the fillers that are used today vary a lot, but silicates dominate the field. The most prevalent type is termed alkaline glass, which consists of alkali-doped silicon dioxide, such as barium or strontium. In addition to increasing the refractive index, which is beneficial in the case of Bis-GMA based resins, it also increases X-ray radiopacity, negating the need for additional radiopacifiers.

Other metal oxides are also used as filler particles, for the same reasons stated above, the most prominent of which are zirconium dioxide, titanium dioxide, and aluminium dioxide. In addition to metal oxides, certain groups have explored the use of biomimetic fillers such as hydroxyapatite (HAP), the same compound that makes up natural teeth and bones. However, they generally have a poorer performance than the more traditional metal oxide fillers.

Since the objective here is to explore the uses of nanostructured fillers, the readers are directed to previous reviews that more thoroughly cover the chemical composition of these materials (Habib et al. 2016).

Silane coupling agents

The polar nature of the M-OH bonds that cover the surface of all of the previously mentioned metal oxide fillers can cause discordance and agglomeration when integrated in the hydrophobic matrix (Darvell 2009). Studies have shown that when using fillers without surface modifications, fractures in the composite tend to propagate along these interfacial boundaries, reducing overall strength (Lin et al. 2000, Debnath et al. 2004).

Consequently, filler particles are modified with silane coupling agents that can covalently bind to both the surface of the particles and the resin matrix. The most commonly used silane is 3-(methacryloyloxy) propyltrimethoxysilane (MPS, Figure 3b). During the surface modification reaction, the siloxane groups hydrolyze, yielding silanol groups that can then condense to the metal oxide surface; the methacrylate group can then copolymerize with the resin monomers (Nihei 2016). Many other silane coupling agents have also been explored, exposing the effect of variations of hydrophobicity, length, and reactivity (Wilson et al. 2005).

Aside from enhancing the resin-filler bonding characteristics, the coupling agents also serve as a solvation layer around the filler particles. As previously mentioned, due to incompatibility between the resin and the filler surface, a given amount of resin will always be occupied around the particles as a solvation layer. Using a coupling agent with an optimal mix of bonding and solvating power produces the best properties in the resulting composites.

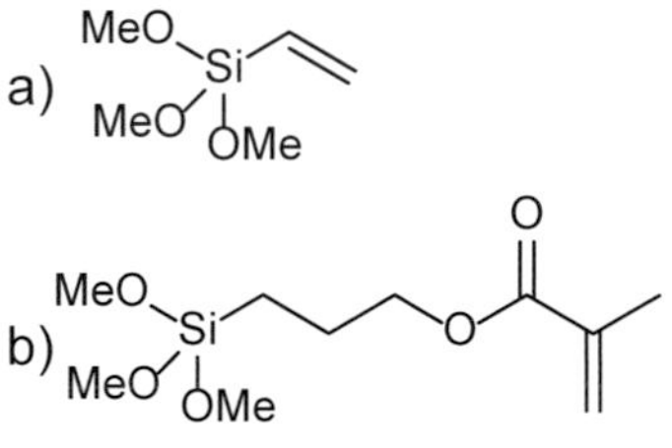

Figure 3. Initially-used silane coupling agent trimethoxyvinylsilane (a), and the most currently-used coupling agent, 3-methacryloylpropyl trimethoxysilane (b).

Size and loading

In terms of size, as technology has progressed, filler particles have become increasingly small, going from *macro fillers* (> 5 μm), through *micro fillers* (100 nm to 5 μm), to *nano fillers* (< 100 nm). Investigations have shown that the principal advantage conferred by smaller filler particles is better wear resistance, and superior polishability and sheen retention. However, these advantages come at the cost of lower maximum filler loading due to increasing filler surface area, such that mechanical properties are decreased and shrinkage is increased (Sakaguchi 2012).

In terms of maximum filler loading of the composites, close packing theory dictates that for identical spheres, the maximum filler loading is set at 74.1 vol%. This value considers fully ordered particle packing with regular voids in between (Wang et al. 2017). A large body of experimental data has shown that such values cannot be attained in a real system, even for non-spherical fillers (Song et al. 2008). Indeed, most of the filler that are used in commercial composites have irregular morphology due to their higher possible loading (up to 100%, theoretically, if perfectly interlocking), despite the evidence that spherical particles produce superior properties (Kim et al. 2002, Lu et al. 2006).

Aside from packing considerations, the main obstacle for high loading fractions is due to viscosity which increases with loading due to interactions between the resin and the filler surface as mentioned above (Darvell 2009).

Most dental composites include as much reinforcing filler as is possible, often around 55–60 vol%, and going up to 75 vol% (Sakaguchi 2012). In order to obtain such high filler loading and minimize void formation, the composites include a mix of many different particle sizes. Conversely, the maximum loading with nanoparticle-filled materials is more dependent on material viscosity than it does on filler packing. The interaction between the resin matrix and nanoparticle filler is much greater than with larger particles due to their much higher relative surface area, which macroscopically leads to dramatically higher paste viscosity.

To further address this unwanted viscosity increase, several solutions have been adopted to exploit the benefits of both size categories: (1) hybrid composites that include both large and very small filler particles; (2) nanoparticle agglomeration or sintering into so-called nanoclusters; and (3) composite pre-polymerization. These three methods will be discussed in greater detail below.

Hybrid fillers

The particle size determines the properties of final resin composites. Generally, larger particles (1–40 μm) present higher filler loading and therefore strength (Ferracane 2011, Kumar et al. 2012, Habib et al. 2017), whereas smaller particles (< 1 μm) allow superior wear resistance and polishability (Lutz and Phillips 1983). The most straightforward—and commonly used—method of increasing filler loading and obtaining the best properties from both large and small particles is to simply mix the two, resulting in *hybrid fillers*. These are composed of a mix of two or more different size classes described above, although sometimes they are different sizes from within a given class.

According to the close packing theory, the size of smaller particles can be determined with theoretical calculations of the maximum filler size for spherical particles. The voids in the packing differ in size and shape, tetrahedral ($d_t = 0.45R$) and octahedral ($d_g = 0.828R$), where R is the radius of identical spheres, but model shows that the maximum theoretical filler loading is at 74.1 vol%. Further increase in filler loading may be achieved slightly by the incorporation of smaller particles filling the voids (Wang et al. 2017).

Depending on the specific mix of sizes used, they can be classified into several families (Willems et al. 1992). Typically, micro-hybrid composites are those that will yield the best flexural strength and modulus, while nano-hybrids will yield the highest depth of cure, transparency, and polishability, which makes them more useful for anterior restorations (Ferracane 2011).

Porous fillers

Porous materials can be classified by the pore size, dimensions, and chemical composition; however, the most pertinent classification method for nano-porous materials is by size (Beretta 2009). There are three classes of pore that fall into this size regime: micropores (< 2 nm), mesopores (2–50 nm), and macropores (> 50 nm) (Sing et al. 1985). By definition, a nanostructured porous material is a system with a regular pattern of pore diameters in the range 1–100 nm (Loni et al. 2015), which possesses unique surface, structural, and bulk properties, and finds widespread applications such as drug delivery (Horcajada et al. 2010, Jarosz et al. 2016), catalysis (Titze et al. 2015, Debnath et al. 2016), and separation technologies (Humplik et al. 2011, Cai et al. 2012). The application of nano-porous fillers for use in dental resin composites is illustrated as follows.

Porous particles

Porous particles have found applications in dental resin composites, which was first proposed by Bowen and Reed (Bowen and Reed 1976a,b). Recently, porous structures were introduced to the surface of glass-ceramic fillers by chemical etching with hydrofluoric acid, as shown in Figure 4a,b (Zandinejad et al. 2006). The results demonstrated that more porous filler has a positive effect on flexural strength. However, the etching process gave poor control on pore size and uniformity. In addition, mesoporous silica possessing interconnected porous channels (~ 4 nm diameter) were synthesized and used as fillers for dental composites (Figure 4c) due to their potential for creating micromechanical filler/resin matrix bonding (Wei et al. 1998, Praveen et al. 2006, Samuel et al. 2009). The results showed that the composites reinforced with a combination of mesoporous and nonporous silica with the optimal weight ratio of 30:70 (filler loading: 70 wt%) exhibited higher filler loading and better mechanical properties than those filled

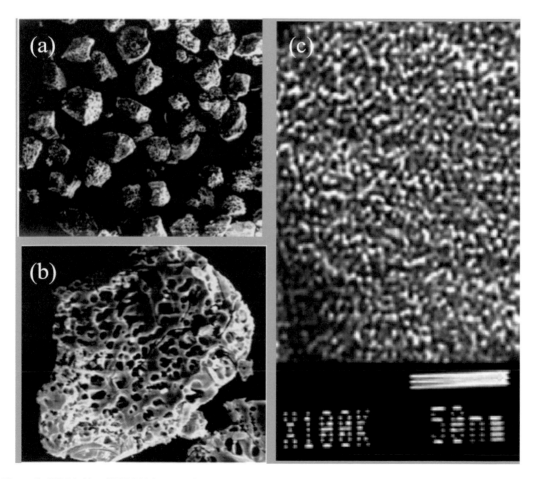

Figure 4. SEM (a,b) and TEM (c) images of (a,b) glass-ceramic porous fillers after HF acid etching, (b) shows the surface details at a greater magnification (adapted from Zandinejad et al. 2006), (b) mesoporous silica. The light features in figure represent interconnected channels with regular diameters of about 4 nm (adapted from Samuel et al. 2009, Copyright 2009, with permission from Elsevier).

with either of these fillers alone. However, flexural strength (68 ± 9 MPa) of this composite still needs to be increased, considering the minimum value of 80 MPa (ISO 4049-2009).

Nanoclusters

Although the use of hybrid fillers retains some of the advantages of small and large particles (Ferracane 2011), developing resin composites with excellent strength and aesthetics remains the objective for both manufacturers and researchers. With the development of nanotechnology, nanoclusters have been used as an alternative to conventional hybrid fillers. They were first introduced in the commercial Filtek Supreme (3M ESPE, USA) in 2003 (Mitra et al. 2003), and consist of nanoparticles such as SiO_2 and ZrO_2 that are bound together into tight agglomerates, and act as a single unit to achieve higher strength, while allowing some breakaway to retain some of the advantages of the small constituent particles. They have been prepared by self-assembly (Kraft et al. 2009), spray-drying (Lee et al. 2010), aerosol-assisted technology (Cho et al. 2007), sintering (Mitra et al. 2003, Curtis et al. 2009a,b), and covalent coupling (Wang et al. 2013b, 2015); however, only the last two methods are suitable for resin restorative applications, considering the scalability of the processes.

The manufacturers of Filtek Supreme obtained silica nanoclusters and silica/zirconia nanoclusters by the sintering process using a bottom-up approach. The corresponding nanocomposites showed better wear resistance and polish retention than the hybrids and microhybrids that were tested, while exhibiting equivalent or higher physical-mechanical properties (Mitra et al. 2003). Conversely, rather than using solution synthesis methodology, Atai et al. used amorphous fumed silica as primary particles (~ 12 nm) to obtain porous clusters through a sintering process at 1300°C (Atai et al. 2012). SEM images of the resulting clusters are shown in Figure 5.

Further methods for nanocluster synthesis were explored by Wang et al. who synthesized silica nanoclusters through a coupling reaction between amine and epoxy functionalized silica nanoparticles at room temperature (Figure 6). Using these nanoclusters blended with silica nanoparticles, bimodal filler resin composites were prepared (Wang et al. 2013b, 2015). Results showed that the properties of resin composites were affected by the filler ratios of bimodal silica nanostructures. Among all composites, the maximum filler loading of silica nanoclusters alone was 60 wt%, whereas that of silica nanoparticles was 70 wt%. When added at the weight ratio at 70% loading, 20% of which were nanoclusters, the obtained composite showed the improvement of 28, 38, and 19% in flexural strength, wear volume and polymerization shrinkage, respectively, compared to those of nanoclusters alone, which is most likely due to the reduced interparticle spacing and the increased filler packing density.

Polymer-modified fillers

Many of the methods that have been explored in the recent literature for particle modification consist of polymer reinforcements either on, around, or through traditional fillers, and can be separated into simple surface modification, linear or crosslinked polymer brush modifications, and through-particle modifications, as illustrated in Figure 7.

Pre-polymerized filler

To minimize the polymerization shrinkage, maximize filler loading, and overcome the viscosity problem, caused by the agglomeration of colloidal filler particles in traditional hybrid and microfilled composites while maintaining their advantageous properties, many manufacturers have modified their formulations to include prepolymerized fillers (also called organic fillers) along with uncured resin (Mitra et al. 2003, Blackham et al. 2009, Ferracane 2011). This method makes nanoparticles disperse in higher filler concentrations and polymerized into the resin matrix, allowing higher filler loading and excellent handling characteristics (Mitra et al. 2003), and lower shrinkage (Angeletakis et al. 2005, Kleverlaan and Feilzer 2005) of final resin composites. In addition, because both the resin matrix for prepolymers and the size and morphology used for prepolymerized fillers vary significantly, mechanical properties can be finely

Figure 5. SEM micrographs of nano silica sintered at different temperatures: (a) 1200°C, (b) 1300°C, (c) 1400°C. The insets show the surface details at a greater magnification (Reprinted from Atai et al. 2012, Copyright 2012, with permission from Elsevier).

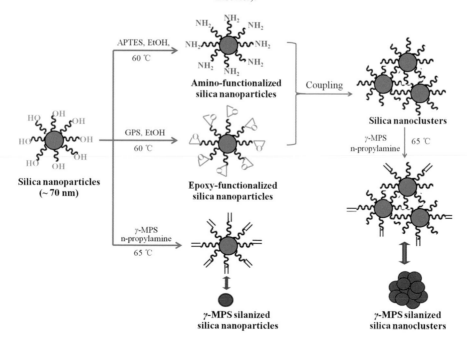

Figure 6. Representative scheme of the preparation of silanized silica nanoparticles and silica nanoclusters (Reprinted from Wang et al. 2013b, Copyright (2013), with permission from Elsevier).

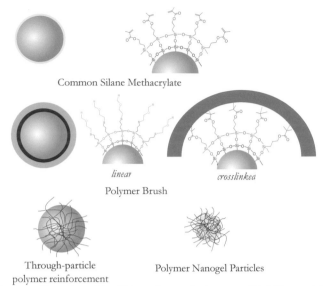

Common Silane Methacrylate

linear   *crosslinked*

Polymer Brush

Through-particle   Polymer Nanogel Particles
polymer reinforcement

Figure 7. Illustration of the different forms of polymer-modified filler particles.

tuned to the desire of manufactures (Kim et al. 2000, Leprince et al. 2010, Donova et al. 2016), which is useful for load bearing restorations. While this is more a case of trapping filler particles in polymer, it resembles some of the other methods described below.

Polymeric surface modification

An intermediate method between traditional surface modification and pre-polymerized filler particle method described above are polymer brush methods. These methods involve the synthesis of a polymer brush layer (sometimes crosslinked) that is either grafted from or to the surface of existing filler particles.

Several versions of such modifications have been explored in the literature. The Stansbury and Bowman groups developed flexible dendrimer-based surface modifications where they assembled and attached 16 monomer units to a core that was then grafted to the particle surfaces, fully using thiol-ene reactivity (Ye et al. 2012). Their modifications resulted in 93% surface coverage with 0.7–4.5 nm thickness. The resulting particles were incorporated into a thiol-ene-methacrylate resin and displayed a 30% reduction in shrinkage stress when compared to the normal filler particles, while retaining the same modulus.

Other work by Liu used hydroxyapatite whiskers fillers and modified their surface with BisGMA-TEGDMA shells to improve mechanical properties (Liu et al. 2013a). Shell characterization showed uneven surface coverage that varied with reaction time. When the materials were subjected to mechanical tests, their results showed that there were only small differences in the final properties of the materials with this modification, showing marginally higher flexural strength values with the shortest reaction time. Otherwise, these fillers were equal or inferior to the unmodified versions.

Though a few attempts have been made using these methods, many other similar modifications of this type remain to be tried.

Through-particle polymeric filler reinforcement

Given the interfacial issues that were observed with unmodified filler particles, and the perceived weakness at the resin-filler interface and the low toughness of fillers, some groups have investigated growing filler particles around polymer threads or networks, resulting in very strong filler-resin bonding (Moszner and Klapdohr 2004).

The most prominent examples of this type of filler have been termed *organically modified ceramics* or ORMOCERs (Haas and Wolter 1999), which are synthesized by the sol-gel process of organofunctional metal alkoxides that contain low molecular weight or oligomeric organic groups (Buestrich et al. 2001, Houbertz et al. 2003, Moszner et al. 2008). Such structures can be considered prepolymers and are therefore less susceptible to polymerization contraction due to a lower number of polymerization sites (Gayossoa et al. 2004, Silva et al. 2007). Initially developed at the Fraunhöfer silicate research institute, these materials have indeed shown improved surface wetting properties, and superior compressive strength when compared to the normal MPS-modified silica. Furthermore, when comparing commercial composites, those containing ORMOCERs showed higher microhardness and wear resistance (Manhart et al. 2000, Hilton et al. 2004, Hahnel et al. 2010), but lower polishability (Tagtekin et al. 2004, Yap et al. 2004a). Moszner et al. synthesized three different types of ORMOCERs from amine (O1) or amide (O2 and O3) dimethacrylate trialkoxysilanes (Figure 8, Moszner et al. 2008). The prepared restorative composites showed similar flexural strengths and improved flexural moduli compared to the ORMOCERs-based commercial filling composite. In addition, the ORMOCERs O1 to O3 exhibited lower cytotoxicity and lower water solubility than Bis-GMA or TEGDMA (Yoshi 1997).

Other work with such materials using BisGMA-based polymers consistently showed lower volumetric shrinkage as compared with other commercially available composites (Rosin et al. 2002, Yap and Soh 2004b).

Nanogel particles

Most of the reinforcing filler that are being evaluated are inorganic particles; however, one of the more recent innovative fillers was termed nanogels (Moraes et al. 2011). These are internally crosslinked and cyclised single or multi-chain polymeric particles, typically well below micrometer size. Their composition can be completely distinct from that of the monomer resin, and tunable depending on the desired properties. Their advantages are similar to those presented in the pre-polymerized filler section, but have less drawbacks due to the nanoscopic particle size.

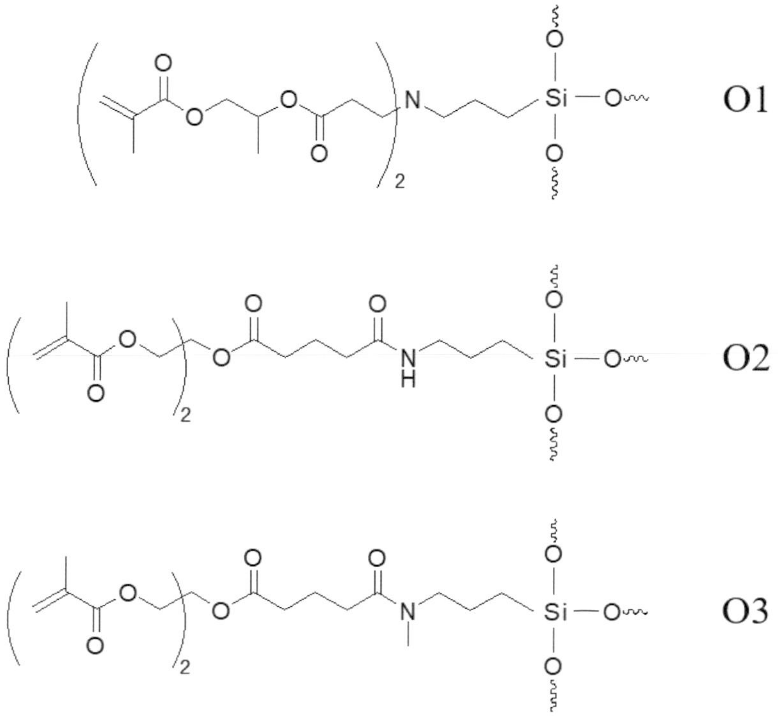

Figure 8. Structure of the used dimethacrylate ORMOCERs (Reprinted from Moszner et al. 2008, Copyright (2008), with permission from Elsevier).

Many tests were performed with these materials, and increasing concentrations of nanogel fillers led to lower polymerization shrinkage and stress, and similar flexural modulus, while only marginally decreasing the flexural strength. Therefore, while work on these materials has only recently begun, they show promise as next-generation additives for dental composite restoratives.

POSS

An interesting advance in nanotechnology that has emerged recently is the appearance of the family of polyhedral oligomeric silsesquioxanes (POSS). POSS is a well-defined 3D nanocage, consisting of an inner inorganic framework of silicon and oxygen atoms, and an outer shell of organic functional groups, as shown in Figure 9 (Ghanbari et al. 2011, Habib et al. 2016). Previous results showed that POSS-modified polymers could increase glass transition temperature, modulus, thermal stability, and oxygen permeability (Lichtenhan et al. 1995, Jang et al. 2001, Wheeler et al. 2006).

Fong et al. (Fong et al. 2005) used methacrylated POSS (POSS-MA) monomers in dental resins, in which POSS-MA was used to partially (or completely) replace Bis-GMA. It was found that 10 wt% POSS-MA substitution of Bis-GMA in the resin systems improved flexural strength of the final resin composite by 20%, which was also confirmed in the work of Wu et al. (Wu et al. 2010). Additionally,

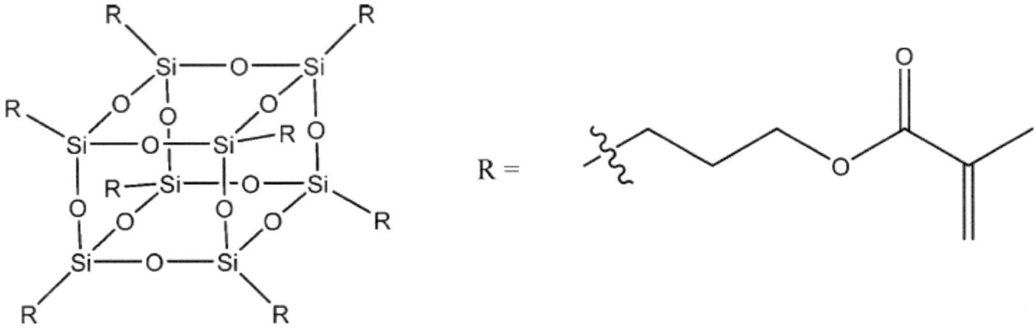

Figure 9. Structure of POSS-methacrylate.

another advantage of POSS-MA was demonstrated in terms of reducing polymerization shrinkage from 3.53 to 2.18% when 15 wt% POSS-MA was used. Therefore, taking advantage of the aforementioned properties and its rigidity, POSS can be used as a potential candidate in dental composites, but further work is still needed to investigate the uniform dispersion of POSS in resin matrix so as to better understand the physical-mechanical properties of composites.

Nanostructured inorganic fillers

Even though nanoparticles have been used quite extensively in hybrid fillers and nanoclusters, several examples appear in the literature for applications of particles with unique morphologies. While standard nano-fillers have been mostly spherical or low aggregation-number clusters, Figure 10 shows examples of nanoscale structures with a greater degree of organization and asymmetry.

Silicate shell metal oxides

Other than simple silica, many other metal oxides have been explored for use as fillers, but the difference in their surface chemistries or surface stabilities may result in inferior composites (Thorat et al. 2012). However, some groups have begun to look at silica shells around other metal oxide particles.

In a recent publication, Kaizer et al. (Kaizer et al. 2016) showed that by covering the surface of the oxides of aluminum, titanium, or zirconium with silica shells (Figure 10a), they mildly increased

Figure 10. SEM images of (a) Silicon carbide whisker reinforcements (adapted from Xu et al. 2003), and (b) urchin-like HAP (adapted from Liu et al. 2014b, Copyright 2014, with permission from Elsevier).

the mechanical strength and modulus of the resulting composites, particularly after aging in aqueous solutions. Though few other published works exist in academic literature, many patents are found on the topic of core-shell particles. Their higher opacity to X-rays is often required for dental restoratives in order to distinguish restorations from normal dental tissue. These patents have covered simple metal oxide particle mixing (Zhang et al. 2002, Temin 1982), to mixed oxide nanoparticles (Burtscher et al. 2008, Lambert and Bringley 2012), to core-shell type particles (Ohtsuka and Tanaka 2011).

Fibers, nanotubes and whiskers

Drawing inspiration from traditional composite reinforcements, fibers, tubes, and whiskers have been explored as reinforcement for dental restoratives. As in traditional composites, these fillers are added to the material in the hope of exploiting the reinforcement mechanisms that are often observed for fibers, namely fiber pull-out and crack bridging. However, due to the methodology used in installing these materials (see *Introduction* section), much of the anisotropy that is often the goal in these materials cannot be maintained. The result is that such additives are only advantageous in small percentages (< 10%), since higher percentages often lead to agglomeration and overall material weakening. Nonetheless, many different types of these high aspect ratio materials have been explored as follows.

Due to their long history and frequent use in more traditional composites, glass fibers have been the subject of some research as composite reinforcements. Their similarity in composition to silica and glass fillers also certainly contributed to their use in dental composites. Several sizes of glass fibers have been attempted; larger micron-size fiber reinforcement resulted in improved strength and toughness, but also high intrinsic roughness due to the fibers (Willems et al. 1992). Further research using smaller electrospun glass fibers with diameters of 0.5 μm and lengths of 50 μm showed improved strength, elasticity, and toughness. The best properties were obtained with a low percentage of fibers, combined with traditional filler particles (Gao et al. 2008). Finally, diverging from traditional dental composite filler methods, Ruddell (Ruddell et al. 2002) prepared glass fiber mats that were then impregnated with resin, but had poor mechanical properties due to incomplete resin integration, though special methodologies may circumvent this limitation (Abdulmajeed et al. 2011). Therefore, having known surface chemistries, small glass fibers, integrated in small fractions, allow for increased overall material performance.

Several polymer fibers have also been used as composite reinforcements for dental composites, most notable among which are poly(vinyl acetate) (Dodiuk-Kenig et al. 2008), nylon-6 (Tian et al. 2007), polyethylene (van Heumen et al. 2008), and polyaramid (Bae et al. 2001). Reaching a similar conclusion to those working with glass fibers, small weight fractions (1–5 wt %) yielded optimal properties, principally due to problems in fiber aggregation and void formation when larger amounts were used.

Beyond fibers, nanostructured materials with lower aspect ratios have also been attempted, often called whiskers. Several materials were used in this class of additives, made from silicon carbide and silicon nitride in Figure 9b (Xu et al. 1999b, 2003), hydroxyapatite (Liu et al. 2013a, 2014a, Zhang and Darvell 2012), zirconia/silica (Guo et al. 2012), as well as a few types of mineral nanocrystals (Atai et al. 2007, Mucci et al. 2011, Zandinejad et al. 2006). Their results were very similar to those obtained for fibers, such that small percentages of filler (< 10 wt %) were beneficial to the strength and toughness of the composites, but higher percentages caused agglomeration and voids, leading to inferior properties.

The final entry in this category is nanotubes, the best known example of which is carbon nanotubes. Researchers have explored the use of both single (Zhang et al. 2008) and multiwalled carbon nanotubes (Borges et al. 2015), as well as some of their derivatives. Despite showing increases of up to 23% in flexural strength, the color of these additives severely limits their use due to problems of depth of cure as well as more mundane aesthetic considerations. Halloysite nanotubes have also been attempted, and addition of 1–2.5 wt % of these yielded a slightly higher flexural strength, and greatly increased fracture toughness of the resulting composite (Chen et al. 2012).

All of the high aspect ratio fillers seem to contribute similarly when added to resins. Their contributions are significant, and merit further investigation. Ideally, further work will highlight the differences between these different materials to better understand their strengths and weaknesses.

Urchin-like hydroxyapatite

The most uniquely nanostructured material of the list are the so-called urchin-like hydroxyapatite particles (Figure 10c). Using a hydrothermal microwave-based synthesis, these consist of a central core with whisker-like spokes radiating out from it, reminiscent of its namesake. These could theoretically offer mixed advantages of fibers and of spherical particles by utilizing both the fiber and the particulate reinforcement mechanisms. Liu et al. integrated these particles into dental resin composite in the hope of obtaining unique material properties (Liu et al. 2014b), and indeed improvements were observed for flexural strength and modulus, compressive strength, and hardness when compared to particulate HAP fillers. However, analysis of fracture surfaces showed that, similar to the fiber-reinforced composites, the resin integration in these structures was incomplete, leading to void formation. Despite this lack of integration, the properties were very advantageous and further exploration of integration methods may indeed result in stronger materials.

Compound release fillers

Another emerging trend in dental restoratives is slow and limited release materials. The compounds that are released range from antibacterial agents to fluorides. These materials have been almost exclusively tested in terms of their compound release properties, but not of their long-term mechanical stability, which might be expected to deteriorate over time. Although the amount of compounds to be released is usually small in the materials, the resulting voids after release that may affect the mechanical strength of the materials are under stress. Such formulations often aimed at fluoride release, but the release compounds also included antibacterial agents, protease inhibitors, and HAP.

Fluoride

Due to the beneficial effects of fluorides in teeth, maintaining basal levels of fluorides is beneficial to general dental health. In turn, one of the targets has been the fluoride buffering materials. These materials would absorb fluorides from toothpastes or drinking water, and then release those ions progressively throughout the day. Though this behavior has been mainly exploited in ionomers due to their intrinsic ion capturing characteristics, resin have also used it via the use of water soluble salts and sparingly soluble salts (Wiegand et al. 2007). While the presence of fluoride is generally beneficial, there is a trade-off due to the hydrophilicity that the salts introduce that can, in turn, cause decreased mechanical properties. Fluoride release thus remains an 'extra feature' of dental composites rather than a requirement.

Antibacterial agents

The other important class of released materials is antibacterial compounds. These compounds are designed to reduce biofilm (plaque) formation on the exposed restoration surface. These composites have come in two types: antibacterial resin monomers, and antibacterial fillers.

Two types of antibacterial resins exist, those with antibacterial monomers that are polymerized along with the resin, and those that simply include free antibiotics. The resin monomers that have been explored for use as antibacterial compounds are usually quaternary ammonium salts that include methacrylate groups to allow polymerization. Other composites have included known antibacterial compounds such as chlorhexidine to add the antibacterial effect to the composites. Since these are part of the resin not nanostructured, the reader is referred elsewhere for more information on this type of resin (Farrugia and Camilleri 2015).

Antibacterial fillers, on the other hand, have been made via the integration of inorganic ions with known antibacterial activity. Silver ions have been the most used due to their well-known antibacterial activity. Liu et al. explored the use of silver nanocrystals (Ag NCs), both for their mechanical enhancement and antibacterial effects (Liu et al. 2013b). Their antibacterial assays indeed showed that as little as 50 ppm of silver nanocrystals reduced bacterial load by up to 58%, as shown in Figure 11, in addition to a small increase in mechanical performance. Similar results were obtained in other research using silver-supported materials (Yoshida et al. 2008). Other work has examined the antibacterial effect of zinc oxide tetrapods (ZOT), so-named for their unique three-dimensional structure. Research by Niu et al. explored their effect on bacterial growth, and found that integrating 5 wt% ZOT had an antibacterial effect without adversely affecting the mechanical properties of the composite (Niu et al. 2010).

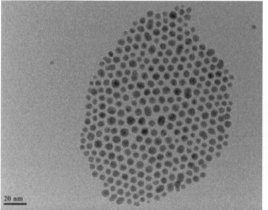 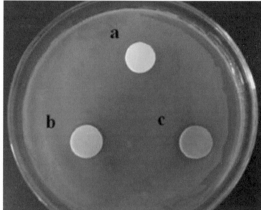

Figure 11. TEM image of the obtained Ag NCs (left), and results of agar disc-diffusion test using cured composites and *Escherichia coli* (right): the composite without Ag NCs (a); the composite with 50 ppm Ag NCs (b); the composite with 100 ppm Ag NCs (c) (Adapted from Liu et al. 2013b, Copyright (2013), with permission from Elsevier).

Protease inhibitors

This class of release materials is designed to release inhibitors of matrix metalloprotease (MMP) and cathepsin K inhibitors. These proteins are present in the natural tooth tissue, and when not properly inactivated, can also lead to composite debonding, due to their proteolytic activity on the hybrid layer, the collagen layer to which restorations are bound (Mazzoni et al. 2014). Studies have shown that including protease inhibitors in dental adhesives showed decreased debonding and higher hybrid layer stability (Breschi et al. 2008). These inhibitors are typically quaternary ammonium salts, and therefore, often also exhibit antibacterial activity. Though these have a useful effect, they are principally useful for adhesives and primers, or at the edges of the restorations where marginal gaps are most likely to occur.

Apatite

Due to the de/remineralization mechanism that occurs perpetually in teeth, several groups have tried to use calcium phosphate containing materials in order to bolster remineralization and hinder demineralization. The most obvious candidate being HAP itself, many groups have used it as a filler. Due to its solubility, surface treatments are not as effective as with silica fillers (Santos et al. 2001), yielding weaker mechanical properties, but showing effective mineralization and phosphate release (Xu et al. 2010). However, larger particles with lower surface area were less disadvantageous with regards to solubility and showed better mechanical performance (Zhang and Darvell 2012). As previous work has shown (Marovic et al. 2014), an optimal balance seems to exist between HAP and other fillers to maintain the mechanical properties while retaining the beneficial bioactivity of the materials.

Biosafety of dental composites

As the nanotechnology becomes increasingly prevalent, safety assessment of dental nanocomposites needs to include the potential toxicity of nanoparticles.

Kostoryz et al. evaluated the cytotoxicity of nanocomposite Z250 (3M-ESPE, Minnesota, US) and silica Aerosil fillers (Aerosil 200 and Aerosil OX-50 with the mean size of 10 and 40 nm, respectively) (Kostoryz et al. 2007). In their procedure, Z250 was polymerized and polished with a finishing bur, and the generated dust was collected for MTT tests. The results showed that nanocomposite particles were mildly cytotoxic to L929 mouse fibroblast cells. There is no significant difference between the cell viability of nanocomposite particles suspended in water (66%) and after washing with ethanol (82%). SEM images (Figure 12a) showed that the particles in Z250 adhered to the cells' surface while TEM analysis (Figure 12b) showed that the cellular uptake of these particles was restricted to the cytoplasm, indicating some cytotoxicity to cells. Additionally, IC50 of Aerosil OX-50 was 58 µg/ml while Aerosil 200 showed no cytotoxicity up to 1 mg/ml.

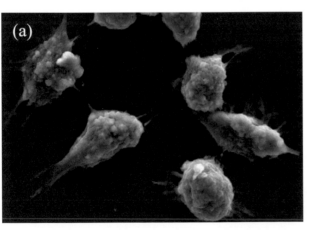
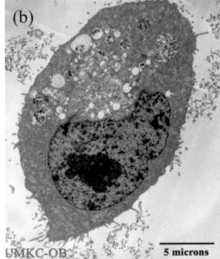

Figure 12. SEM image of the adherence of nanocomposite particles to surface of cells (a), and TEM image of L929 cell uptake of nanocomposite particles (b) (Kostoryz et al. 2007. © 2007 NSTI http://nsti.org. Reprinted and revised, with permission, from the Title of Conference Proceedings, pp. 647–650, 2007, Santa Clara, USA).

Conclusion and perspectives

Currently used dental resin composites provide an alternative to conventional mercury amalgams, but retain two important problems in their clinical application: secondary caries and restoration fractures, both thought to result from polymerization shrinkage and insufficient mechanical properties (Ferracane and Hilton 2016). Although recent research has largely addressed the first issue through the exploration of novel resin monomers systems such as silorane monomers and Bis-GMA derivatives, there have not been many systematic studies. Furthermore, due to the toxicity of bisphenol A (BPA), its incorporation in the BisGMA resins has become a concern for both health and environmental considerations. For this reason, methacrylate derivatives of natural products have been under development, to replace the BPA moiety in the composites (Hu et al. 2005, Gauthier et al. 2007, 2009).

In this context, with the development of nanotechnology, the use of nanostructured fillers, porous fillers, nanogel particles, and pre-polymerized fillers can contribute to the reduction of polymerization shrinkage and the improvement of mechanical properties. The properties of such composites can be further enhanced by the addition of reactive POSS monomers. In addition, small fractions of novel fillers with nanoscale structures including fibers, nanotubes and whiskers may further increase the mechanical properties.

All of these materials have contributed to the advancement of knowledge in the field, and towards optimization of the characteristics of nanostructured fillers. As research explores new technologies for producing and tuning nanostructures, these properties can be further explored and improved, bringing about new inorganic fillers that can meet the growing need for dental resin composites that are safe, strong, and long-lived.

References

Abdulmajeed, A.A., T.O. Narhi, P.K. Vallittu and L.V. Lassila. 2011. The effect of high fiber fraction on some mechanical properties of unidirectional glass fiber-reinforced composite. Dent. Mater. 27: 313–321.

Angeletakis, C., M.D.S. Nguyen and A.I. Kobashigawa. 2005. Prepolymerized filler in dental restorative composite. U.S. Patent # 6,890,968 B2.

Atai, M., E. Yassini, M. Amini and D.C. Watts. 2007. The effect of a leucite-containing ceramic filler on the abrasive wear of dental composites. Dent. Mater. 23: 1181–1187.

Atai, M., A. Pahlavan and N. Moin. 2012. Nano-porous thermally sintered nano silica as novel fillers for dental composites. Dent. Mater. 28: 133–145.

Bae, J.M., K.N. Kim, M. Hattori, H. Koji, Y. Masao, K. Eiji et al. 2001. The flexural properties of fiber-reinforced composite with light-polymerized polymer matrix. Int. J. Prosthodont. 14: 33–39.

Beretta, M. 2009. Nanostructured Mesoporous Materials Obtained by Template Synthesis and Controlled Shape Replica. Ph.D. Thesis, University of Milano-Bicocca, Milan, Italy.

Blackham, J.T., K.S. Vandewalle and W. Lien. 2009. Properties of hybrid resin composite systems containing prepolymerized filler particles. Oper. Dent. 34: 697–702.

Borges, A.L., E.A. Munchow, A.C. de, O. Souza, T. Yoshida, P.K. Vallittu et al. 2015. Effect of random/aligned nylon-6/MWCNT fibers on dental resin composite reinforcement. J. Mech. Behav. Biomed. Mater. 48: 134–144.

Bowen, R.L. 1962. Dental filling material comprising vinyl silane treated fused silica and a binder consisting of the reaction product of bis phenol and glycidyl acrylate. U.S. Patent # 3,066,112.

Bowen, R.L. 1963. Properties of a silica-reinforced polymer for dental restorations. J. Am. Dent. Assoc. 66: 57–64.

Bowen, R.L. and L.E. Reed. 1976a. Semi-porous reinforcing fillers for composite resins: I. Preparation of provisional glass formulations. J. Dent. Res. 55: 738–747.

Bowen, R.L and L.E. Reed. 1976b. Semi-porous reinforcing fillers for composite resins: II. Heat treatments and etching characteristics. J. Dent. Res. 55: 748–756.

Breschi, L., A. Mazzoni, A. Ruggeri, M. Cadenaro, R. Di Lenarda and E. De Stefano Dorigo. 2008. Dental adhesion review: aging and stability of the bonded interface. Dent. Mater. 24: 90–101.

Buestrich, R., F. Kahlenberg, M. Popall, P. Dannberg, R. Mueller-Fiedler and O. Roesch. 2001. ORMOCERs for optical interconnection technology. J. Sol-Gel Sci. Technol. 10: 181–186.

Burtscher, P., L. Mädler, N. Moszner, S.E. Pratsinis and V.M. Rheinberger. 2008. Dental composites based on X-ray-opaque mixed oxides prepared by flame spraying. U.S. Patent # 20050176843 A1.

Cai, J.J., Y.L. Xing and X.B. Zhao. 2012. Quantum sieving: feasibility and challenges for the separation of hydrogen isotopes in nanoporous materials. RSC Adv. 2: 8579–8586.

Chen, Q., Y. Zhao, W. Wu, T. Xu and H. Fong. 2012. Fabrication and evaluation of Bis-GMA/TEGDMA dental resins/composites containing halloysite nanotubes. Dent. Mater. 28: 1071–1079.

Cho, Y.S., G.R. Yi, Y.S. Chung, S.B. Park and S.M. Yang. 2007. Complex colloidal microclusters from aerosol droplets. Langmuir 23: 12079–12085.

Curtis, A.R., W.M. Palin, G.J.P. Fleming, A.C.C. Shortall and P.M. Marquis. 2009a. The mechanical properties of nanofilled resin-based composites: characterizing discrete filler particles and agglomerates using a micromanipulation technique. Dent. Mater. 25: 180–187.

Curtis, A.R., W.M. Palin, G.J.P. Fleming, A.C.C. Shortall and P.M. Marquis. 2009b. The mechanical properties of nanofilled resin-based composites: The impact of dry and wet cyclic pre-loading on bi-axial flexure strength. Dent. Mater. 25: 188–197.

Darvell, B.W. 2009. Materials Science for Dentistry. pp. 83–108. Rheology (Ninth Edition), CRC Press LLC. FL, USA.

Debnath, S., R. Ranade, S.L. Wunder, J. McCool, K. Boberick and G. Baran. 2004. Interface effects on mechanical properties of particle-reinforced composites. Dent. Mater. 20: 677–686.

Debnath, S., S.K. Saxena and V. Nagabhatla. 2016. Facile synthesis of crystalline nanoporous $Mg_3(PO_4)_2$ and its application to aerobic oxidation of alcohols. Catal. Commun. 84: 129–133.

Ding, T., K. Song, K. Clays and C.H. Tung. 2009. Fabrication of 3D photonic crystals of ellipsoids: convective self-assembly in magnetic field. Adv. Mater. 21: 1936–1940.

Dodiuk-Kenig, H., K. Lizenboim, S. Roth, B. Zalsman, W.A. McHale, M. Jaffe et al. 2008. Performance enhancement of dental composites using electrospun nanofibers. J. Nanomater. 2008: 1–6.

Donova, J.B., S. Garoushi, L.V.J. Lassila, F. Keulemans and P.K. Vallittu. 2016. Mechanical and structural characterization of discontinuous fiber-reinforced dental resin composite. J. Dent. 52: 70–78.

Farrugia, C. and J. Camilleri. 2015. Antimicrobial properties of conventional restorative filling materials and advances in antimicrobial properties of composite resins and glass ionomer cements—A literature review. Dent. Mater. 31: e89–e99.

Ferracane, J.L. 2011. Resin composite—state of the art. Dent. Mater. 27: 29–38.

Ferracane, J.L. and T.J. Hilton. 2016. Polymerization stress—is it clinically meaningful? Dent. Mater. 32: 1–10.

Filtek Bulk Fill Posterior Restorative. 3M ESPE. Technical product profile.

Filtek Silorane Low Shrink Posterior Restorative. 3M ESPE. Technical product profile.

Fong, H., S.H. Dickens and G.M. Flaim. 2005. Evaluation of dental restorative composites containing polyhedral oligomeric silsesquioxane methacrylate. Dent. Mater. 21: 520–529.

Gao, Yi, S. Sagi, L.F. Zhang, Y.L. Liao, D.M. Cowles, Y.Y. Sun et al. 2008. Electrospun nano-scaled glass fiber reinforcement of bis-GMA/TEGDMA dental composites. J. Appl. Polym. Sci. 110: 2063–2070.

Gauthier, M.A., P. Simard, Z. Zhang and X.X. Zhu. 2007. Bile acids as constituents for dental composites: *In vitro* cytotoxicity of (meth)acrylate and other ester derivatives of bile acids. J. Royal Soc. Interface 4: 1145–1150.

Gauthier, M.A., Z. Zhang and X.X. Zhu. 2009. New dental composites containing multi-methacrylate derivatives of bile acids: A comparative study with commercial monomers. ACS Appl. Mat. Interfaces 1: 824–832.

Gayossoa, C.A., F.B. Santana, J.G. Ibarra, G.S. Espinola and M.A.C. Marttinez. 2004. Calculation of contraction rates due to shrinkage in light-cured composites. Dent. Mater. 20: 228–235.

Ghanbari, H., B.G. Cousins and A.M. Seifalian. 2011. A nanocage for nanomedicine: polyhedral oligomeric silsesquioxane (POSS). Macromol. Rapid Commun. 32: 1032–1046.

Guo, G., Y. Fan, J.F. Zhang, J.L. Hagan and X. Xu. 2012. Novel dental composites reinforced with zirconia-silica ceramic nanofibers. Dent. Mater. 28: 360–368.

Haas, K.H. and H. Wolter. 1999. Synthesis, properties and applications of inorganic-organic copolymers (ORMOCERs). Curr. Opin. Solid State Mater. Sci. 4: 571–580.

Habib, E., R.L. Wang, Y.Z. Wang, M.F. Zhu and X.X. Zhu. 2016. Inorganic fillers for dental resin composites: present and future. ACS Biomater. Sci. Eng. 2: 1–11.

Habib, E., R.L. Wang and X.X. Zhu. 2017. Monodisperse silica-filled composite restoratives mechanical and optical properties. Dent. Mater. 33: 280–287.

Hahnel, S., A. Henrich, R. Bürgers, G. Handel and M. Rosentritt. 2010. Investigation of mechanical properties of modern dental composites after artificial aging for one year. Oper. Dent. 35: 412–419.

Hilton, T., D. Hilton, R. Randall and J.L. Ferracane. 2004. A clinical comparison of two cements for levels of post-operative sensitivity in a practice-based setting. Oper. Dent. 29: 241–248.

Horcajada, P., T. Chalati, C. Serre, B. Gillet, C. Sebrie, T. Baati et al. 2010. Porous metal-organic frameworks nanoscale carriers as a potential platform from drug delivery and imaging. Nature Mater. 9: 172–178.

Houbertz, R., L. Frohlich, M. Popall, U. Streppel, P. Dannberg, A. Brauer et al. 2003. Inorganic-organic hybrid polymers for information technology: from planar technology to 3D nanostructures. Adv. Eng. Mater. 5: 551–555.

Hu, X.Z., Z. Zhang, X. Zhang, Z.Y. Li and X.X. Zhu. 2005. Selective acylation of cholic acid derivatives with multiple methacrylate groups. Steroids 70: 531–537.

Humplik, T., J. Lee, S.C. O' Hern, B.A. Fellman, M.A. Baig, S.F. Hassan et al. 2011. Nanostructured materials for water desalination. Nanotechnology 22: 292001.

ISO 7405. 2008. Dentistry - Evaluation of biocompatibility of medical devices used in dentistry.

ISO 4049. 2009. Dentistry - In Polymer-based restorative materials.

Jang, J., J. Bae and D. Kang. 2001. Phase-separation prevention and performance improvement of poly (vinyl acetate)/TEOS hybrid using modified sol-gel process. J. Appl. Polym. Sci. 82: 2310–2318.

Jarosz, M., A. Pawlik, M. Szuwarzyński, M. Jaskula and G.D. Sulka. 2016. Nanoporous anodic titanium dioxide layers as potential drug delivery systems: Drug release kinetics and mechanism. Colloid. Surf. B 143: 447–454.

Kaizer, M.R., J.R. Almeida, A.P. Goncalves, Y. Zhang, S.S. Cava and R.R. Moraes. 2016. Silica coating of nonsilicate nanoparticles for resin-based composite materials. J. Dent. Res. 95: 1394–1400.

Kassaee, M.Z., A. Akhavan, N. Sheikhand and A. Sodagar. 2008. Antibacterial effects of a new dental acrylic resin containing silver nanoparticles. J. Appl. Polym. Sci. 110: 1699–1703.

Kim, K.H., Y.B. Kim and O. Okuno. 2000. Microfracture mechanisms of composite resins containing prepolymerized particle fillers. Dent. Mater. J. 19: 22–33.

Kim, K.H., J.L. Ong and O. Okuno. 2002. The effect of filler loading and morphology on the mechanical properties of contemporary composites. J. Prosthet. Dent. 87: 642–649.

Klapdohr, S. and N. Moszner. 2005. New inorganic components for dental filling composites. Monatsh. Chem. 136: 21–45.

Kleverlaan, C.J. and A.J. Feilzer. 2005. Polymerization shrinkage and contraction stress of dental resin composites. Dent. Mater. 21: 1150–1157.

Kostoryz, E.L., C.J. Utter, Y. Wang, V. Dusevich and P. Spencer. 2007. Cytotoxicity of dental nanocomposite particles. NSTI-Nanotech. USA 2: 647–650.

Kraft, D.J., W.S. Vlug, C.M. van Kats, A. van Blaaderen, A. Imhof and W.K. Kegel. 2009. Self-assembly of colloids with liquid protrusions. J. Am. Chem. Soc. 131: 1182–1186.

Kumar, N., N.A. Khoso, L. Sangi, F. Bhangar and F.A. Kalhoro. 2012. Dental resin-based composites: A transition from macrofilled to nanofilled. J. Pak. Dent. Assoc. 21: 39–44.

Lambert, P.M. and J.F. Bringley. 2012. Silica-alumina mixed oxide compositions. U.S. Patent # 20120004342 A1.

Lavigueur, C. and X.X. Zhu. 2012. Recent advances in the development of dental composite resins. RSC Adv. 2: 59–63.

Lee, S.Y., L. Gradon, S. Janeczko, F. Iskandar and K. Okuyama. 2010. Formation of highly ordered nanostructures by drying micrometer colloidal droplets. ACS Nano 4: 4717–4724.

Leprince, J., W.M. Palin, T. Mullier, J. Devaux, J. Vreven and G. Leloup. 2010. Investigating filler morphology and mechanical properties of new low-shrinkage resin composite types. J. Oral Rehabil. 37: 364–376.

Lichtenhan, J.D., Y.O. Otonari and M.J. Carr. 1995. Linear hybrid polymer building blocks: methacrylate-functionalized polyhedral oligomeric silsesquioxane monomers and polymers. Macromolecules 28: 8435–8437.

Lin, C.T., S.Y. Lee, E.S. Keh, D.R. Dong, H.M. Huang and Y.H. Shih. 2000. Influence of silanization and filler fraction on aged dental composites. J. Oral Rehabil. 27: 919–926.

Liu, F.W., R.L. Wang, Y.H. Cheng, X.Z. Jiang, Q.H. Zhang and M.F. Zhu. 2013a. Polymer grafted hydroxyapatite whisker as a filler for dental composite resin with enhanced physical and mechanical properties. Mater. Sci. Eng. C 33: 4994–5000.

Liu, F.W., R.L. Wang, Y.Y. Shi, X.Z. Jiang, B. Sun and M.F. Zhu. 2013b. Novel Ag nanocrystals based dental resin composites with enhanced mechanical and antibacterial properties. Prog. Nat. Sci. 23: 573–578.

Liu, F.W., X.Z. Jiang, Q.H. Zhang and M.F. Zhu. 2014a. Strong and bioactive dental resin composite containing poly(Bis-GMA) grafted hydroxyapatite whiskers and silica nanoparticles. Compos. Sci. Technol. 101: 86–93.

Liu, F.W., B. Sun, X.Z. Jiang, S.S. Aldeyab, Q.H. Zhang and M.F. Zhu. 2014b. Mechanical properties of dental resin/composite containing urchin-like hydroxyapatite. Dent. Mater. 30: 1358–1368.

Liu, F.W., S. Bao, Y. Jin, X.Z. Jiang and M.F. Zhu. 2014c. Novel bionic dental resin composite reinforced by hydroxyapatite whisker. Mater. Res. Innov. 18: 854–858.

Loni, A., T. Defforge, E. Caffull, G. Gautier and L.T. Canham. 2015. Porous silicon fabrication by anodisation: Progress towards the realisation of layers and powders with high surface area and micropore content. Micropor. Mesopor. Mater. 213: 188–191.

Lu, H., Y.K. Lee, M. Oguri and J.M. Powers. 2006. Properties of a dental resin composite with a spherical inorganic filler. Oper. Dent. 31: 734–740.

Lutz, F. and R.W. Phillips. 1983. A classification and evaluation of composite resin systems. J. Prosthet. Dent. 50: 480–488.

Manhart, J., K.H. Kunzelmann, H.Y. Chen and R. Hickel. 2000. Mechanical properties and wear behavior of light-cured packable composite resins. Dent. Mater. 16: 33–40.

Marovic, D., Z. Tarle, K.A. Hiller, R. Müller, M. Rosentritt, D. Skrtic et al. 2014. Reinforcement of experimental composite materials based on amorphous calcium phosphate with inert fillers. Dent. Mater. 30: 1052–1060.

Mazzoni, A., L. Tjaderhane, V. Checchi, R. Di Lenarda, T. Salo, F.R. Tay et al. 2014. Role of dentin MMPs in caries progression and bond stability. J. Dent. Res. 94: 241–251.

Mitra, S.B., D. Wu and B.N. Holmes. 2003. An application of nanotechnology in advanced dental materials. J. Am. Dent. Assoc. 134: 1382–1390.

Moraes, R.R., J.W. Garcia, M.D. Barros, S.H. Lewis, C.S. Pfeifer, J. Liu et al. 2011. Control of polymerization shrinkage and stress in nanogel-modified monomer and composite materials. Dent. Mater. 27: 509–519.

Moszner, N. and U. Salz. 2001. New developments of polymeric dental composites. Prog. Polym. Sci. 26: 535–576.

Moszner, N. and S. Klapdohr. 2004. Nanotechnology for dental composites. Int. J. Nanotechnology 1: 130–156.

Moszner, N., A. Gianasmidis, S. Klapdohr, U.K. Fischer and V. Rheinberger. 2008. Sol-gel materials 2. Light-curing dental composites based on ormocers of cross-linking alkoxysilane methacrylates and further nano-components. Dent. Mater. 24: 851–856.

Mucci, V., J. Pérez and C.I. Vallo. 2011. Preparation and characterization of light-cured methacrylate/montmorillonite nanocomposites. Polym. Int. 60: 247–254.

Nihei, T. 2016. Dental applications for silane coupling agents. J. Oral Sci. 58: 151–155.

Niu, L.N., M. Fang, K. Jiao, L.H. Tang, Y.H. Xiao, L.J. Shen et al. 2010. Tetrapod-like zinc oxide whisker enhancement of resin composite. J. Dent. Res. 89: 746–750.

Ohtsuka, K. and H. Tanaka. 2011. Dental filler. U.S. Patent # 7981513 B2.

Praveen, S., Z.F. Sun, J.G. Xu, A. Patel, Y. Wei. R. Ranade et al. 2006. Compression and aging properties of experimental dental composites containing mesoporous silica as fillers. Mol. Cryst. Liq. Cryst. 448: 223–231.

Rosin, M., A.D. Urban, C. Gärtner, O. Bernhardt, C. Splieth and G. Meyer. 2002. Polymerization shrinkage-strain and microleakage in dentin-bordered cavities of chemically and light-cured restorative materials. Dent. Mater. 18: 521–528.

Ruddell, D.E., M.M. Maloney and J.Y. Thompson. 2002. Effect of novel filler particles on the mechanical and wear properties of dental composites. Dent. Mater. 18: 72–80.

Samuel, S.P., S.X. Li, I. Mukherjee, Y. Guo, A.C. Patel, G. Baran et al. 2009. Mechanical properties of experimental dental composites containing a combination of mesoporous and nonporous spherical silica as fillers. Dent. Mater. 25: 296–301.

Santos, C., Z.B. Luklinska, R.L. Clarke and K.W.M. Davy. 2001. Hydroxyapatite as a filler for dental composite materials: mechanical properties and *in vitro* bioactivity of composites. J. Mater. Sci. Mater. Med. 12: 565–573.

Sideridou, I.D. and M.M. Karabela. 2009. Effect of the amount of 3-methacyloxypropyltrimethoxysilane coupling agent on physical properties of dental resin nanocomposites. Dent. Mater. 25: 1315–1324.

Sakaguchi, R.L. 2012. Restorative materials-composites and polymers. pp. 161–198. *In*: R.L. Sakaguchi and J.M. Powers (eds.). Craig's Restorative Dental Materials (Thirteenth Edition). Saint Louis: Mosby.

Silva, F.F., L.C. Mendes, M. Ferreira and M.R. Benzi. 2007. Degree of conversion versus the depth of polymerization of an organically modified ceramic dental restoration composite by Fourier transform infrared spectroscopy. J. Appl. Polym. Sci. 104: 325–330.

Sing, K.S.W., D.H. Everett, R.A.W. Haul, L. Moscou, R.A. Pierotti, J. Rouquérol et al. 1985. Reporting physisorption data for gas/solid systems with special reference to the determination of surface area and porosity. Pure Appl. Chem. 57: 603–619.

Song, C., P. Wang and H.A. Makse. 2008. A phase diagram for jammed matter. Nature 453: 629–632.

Tagtekin, D.A., F.C. Yanikoglu, F.O. Bozkurt, B. Kologlu and H. Sur. 2004. Selected characteristics of an Ormocer and a conventional hybrid resin composite. Dent. Mater. 20: 487–497.

Temin, S.C. 1982. Radio-opaque dental compositions. CA Patent # 1134527 A1.

Thorat, S.B., N. Patra, R. Ruffilli, A. Diaspro and M. Salerno. 2012. Preparation and characterization of a BisGMA-resin dental restorative composites with glass, silica and titania fillers. Dent. Mater. J. 31: 635–644.

Thorat, S.B., A. Diaspro and M. Salerno. 2014. *In vitro* investigation of coupling-agent-free dental restorative composite based on nano-porous alumina fillers. J. Dent. 42: 279–286.

Tian, M., Y. Gao, Y. Liu, Y. Liao, R. Xu, N.E. Hedin et al. 2007. Bis-GMA/TEGDMA dental composites reinforced with electrospun nylon 6 nanocomposite nanofibers containing highly aligned fibrillar silicate single crystals. Polymer (Guildf) 48: 2720–2728.

Titze, T., C. Chmelik, J. Kullmann, L. Prager, E. Miersemann, R. Gläser et al. 2015. Microimaging of transient concentration profiles of reactant and product molecules during catalytic conversion in nanoporous materials. Angew. Chem. Int. Ed. 54: 5060–5064.

Valente, L.L., S.L. Peralta, F.A. Ogliari, L.M. Cavalcante and R.R. Moraes. 2013. Comparative evaluation of dental resin composites based on micron- and submicron-sized monomodal glass filler particles. Dent. Mater. 29: 1182–1187.

van Heumen, C.C., C.M. Kreulen, E.M. Bronkhorst, E. Lesaffre and N.H. Creugers. 2008. Fiber-reinforced dental composites in beam testing. Dent. Mater. 24: 1435–1443.

Wang, R.L., M. Zhu, S. Bao, F.W. Liu, X.Z. Jiang and M.F. Zhu. 2013a. Synthesis of two Bis-GMA derivates with different size substituents as potential monomer to reduce the polymerization shrinkage of dental restorative composites. J. Mater. Sci. Res. 2: 12–22.

Wang, R.L., S. Bao, F.W. Liu, X.Z. Jiang, Q.H. Zhang, B. Sun et al. 2013b. Wear behavior of light-cured resin composites with bimodal silica nanostructures as fillers. Mater. Sci. Eng. C 33: 4759–4766.

Wang, R.L., M.L. Zhang, F.W. Liu, S. Bao, T.T. Wu, X.Z. Jiang et al. 2015. Investigation on the physical-mechanical properties of dental resin composites reinforced with novel bimodal silica nanostructures. Mater. Sci. Eng. C 50: 266–273.

Wang, R.L., E. Habib and X.X. Zhu. 2017. Application of close-packed structures in dental resin composites. Dent. Mater. 33: 288–293.

Wei, Y., D.L. Jin, T.Z. Ding, W.H. Shih, X.H. Liu, Stephen Z.D. Cheng et al. 1998. A non-surfactant templating route to mesoporous silica materials. Adv. Mater. 10: 313–316.

Weinmann, W., C. Thalacker and R. Guggenberger. 2005. Siloranes in dental composites. Dent. Mater. 21: 68–74.

Wheeler, P.A., B.X. Fu, J.D. Lichtenhan, W.T. Jia and L.J. Mathias. 2006. Incorporation of metallic POSS, POSS copolymers and new functionalized POSS compounds into commercial dental resins. J. Appl. Polym. Sci. 102: 2856–2862.

Wiegand, A., W. Buchalla and T. Attin. 2007. Review on fluoride-releasing restorative materials—fluoride release and uptake characteristics, antibacterial activity and influence on caries formation. Dent. Mater. 23: 343–362.

Willems, G., P. Lambrechts, M. Braem, J.P. Celis and G. Vanherle. 1992a. A classification of dental composites according to their morphological and mechanical characteristics. Dent. Mater. 8: 310–319.

Wilson, K.S., K. Zhang and J.M. Antonucci. 2005. Systematic variation of interfacial phase reactivity in dental nanocomposites. Biomaterials 26: 5095–5103.

Wu, X.R., Y. Sun, W.L. Xie, Y.J. Liu and X.Y. Song. 2010. Development of novel dental nanocomposites reinforced with polyhedral oligomeric silsesquioxane (POSS). Dent. Mater. 26: 456–462.

Xu, H.H.K. 1999a. Dental composite resins containing silica-fused ceramic single-crystalline whiskers with various filler levels. J. Dent. Res. 78: 1304–1311.

Xu, H.H.K., T.A. Martin, J.M. Antonucci and E. Eichmiller. 1999b. Ceramic whisker reinforcement of dental resin composites. J. Dent. Res. 78: 706–712.

Xu, H.H.K., J.B. Quinn, D.T. Smith, A.A. Giuseppetti and F.C. Eichmiller. 2003. Effects of different whiskers on the reinforcement of dental resin composites. Dent. Mater. 19: 359–367.

Xu, H.H.K., M.D. Weir, L. Sun, J.L. Moreau, S. Takagi, L.C. Chow et al. 2010. Strong nanocomposites with Ca, PO_4, and F release for caries inhibition. J. Dent. Res. 89: 19–28.

Yap, A.U.J., C.H. Tan and S.M. Chung. 2004a. Wear behavior of new composite restoratives. Oper. Dent. 29: 269–274.

Yap, A.U.J. and M.S. Soh. 2004b. Post-gel polymerization contraction of "Low Shrinkage" composite restoratives. Oper. Dent. 29: 182–187.

Ye, S., S. Azarnoush, I.R. Smith, N.B. Cramer, J.W. Stansbury and C.N. Bowman. 2012. Using hyperbranched oligomer functionalized glass fillers to reduce shrinkage stress. Dent. Mater. 28: 1004–1011.

Yoshi, E. 1997. Cytotoxic effects of acrylates and methacrylates: relationships of monomer structures and cytotoxicity. J. Biomed. Mater. Res. 37: 517–524.

Yoshida, K., M. Tanagawa and M. Atsuta. 1999. Characterization and inhibitory effect of antibacterial dental resin composites incorporating silver-supported materials. J. Biomed. Mater. Res. 47: 516–522.

Zandincjad, A.A., M. Atai and A. Pahlevan. 2006. The effect of ceramic and porous fillers on the mechanical properties of experimental dental composites. Dent. Mater. 22: 382–387.

Zhang, F., Y. Xia, L. Xu and N. Gu. 2008. Surface modification and microstructure of single-walled carbon nanotubes for dental resin-based composites. J. Biomed. Mater. Res., Part B 86: 90–97.

Zhang, H.Q. and B.W. Darvell. 2012. Mechanical properties of hydroxyapatite whisker-reinforced bis-GMA-based resin composites. Dent. Mater. 28: 824–830.

Zhang, X., B.U. Kolb, D.A. Hanggi, S.B. Mitra, P.D.N. Ario and R.P. Rusin. 2002. Radiopaque dental materials with nano-sized particles. U.S. Patent # WO 2001030305 A1.

6

Graphene-based Nanocomposites for Electrochemical Energy Storage

Chunwen Sun

Introduction

Electrochemical energy storage has played important roles in energy storage technologies (Armand et al. 2008, Goodenough et al. 2010, Sun et al. 2011). Since graphene was first isolated in 2004, researchers have developed various methods to synthesize it, which can be classified as either 'bottom-up' strategy, like epitaxial growth, chemical vapor deposition and chemical synthesis, or 'top-down' methods, including micromechanical cleavage (original scotch-tape method), liquid-phase exfoliation, un-zipping carbon nanotubes (CNTs) (Ji et al. 2016) and the reduction of graphene oxide (Ferrari et al. 2015). However, extending the excellent properties of individual graphene sheets to a macroscopic scale is still the most important issue for realizing its practical applications. Sheet assemble is a thermodynamic process in which the interactions between single sheets, such as van der Waals forces, π–π stacking, electrostatic interactions, hydrogen bonding and hydrophobic interactions, have decisive roles. Various unique macroscopic assemblies of graphene can be prepared that cover all "four" dimensions, in the form of quantum dots (0D), wires (1D), films (2D), monoliths (3D), and potentially 4D self-healing and/or self-folding structures (see Figure 1, EI-Kady et al. 2016). Graphene and graphene-based materials have attracted great attention owing to their unique properties of high mechanical flexibility, large surface area, chemical stability, superior electronic and thermal conductivities that render them great choices as alternative electrode materials for electrochemical energy storage systems (EI-Kady et al. 2016, Qu et al. 2014). The structure of graphene can be modified by various techniques to tune their properties and performance (Zhu et al. 2014). These methods include: (i) Direct assembly of graphene sheets into porous electrode materials; (ii) Deposition and coating of metal, metal oxides, or other electrochemically active materials on graphene sheets; (iii) Surface modification of graphene sheets with surfactant molecules; and (iv) Self-assembly of ordered nanocomposites (see Figure 2, Raccichini et al. 2015). However, the prevention of re-stacking is still a big challenge in the direct assembly of graphene, using a homogeneous dispersion that requires a proper balance between salvation and hydration forces of the solvent. The use of surfactants provides precise control of graphene dispersion, leading to the formation of new hybrid

Beijing Institute of Nanoenergy and Nanosystems, No. 30 Xueyuan Road, Haidian District, Beijing 100083, P. R. China,
School of Nanoscience and Technology, University of Chinese Academy of Sciences, Beijing 100049, P. R. China.
Email: sunchunwen@binn.cas.cn

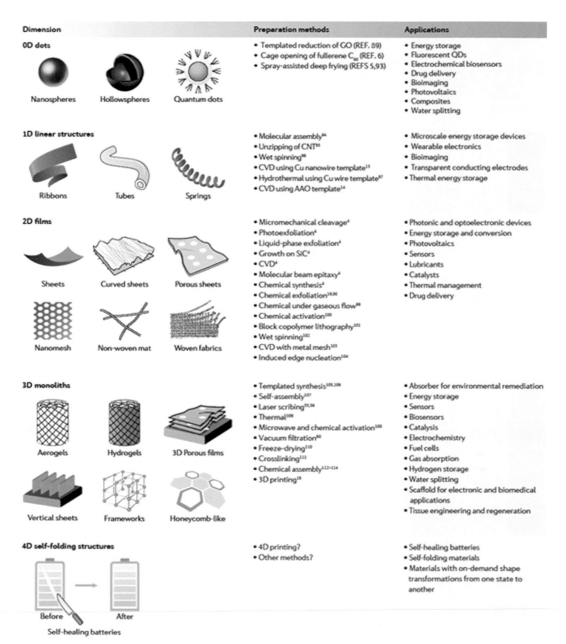

Dimension	Preparation methods	Applications
0D dots Nanospheres Hollowspheres Quantum dots	• Templated reduction of GO (REF. 89) • Cage opening of fullerene C₆₀ (REF. 6) • Spray-assisted deep frying (REFS 5,93)	• Energy storage • Fluorescent QDs • Electrochemical biosensors • Drug delivery • Bioimaging • Photovoltaics • Composites • Water splitting
1D linear structures Ribbons Tubes Springs	• Molecular assembly[94] • Unzipping of CNT[95] • Wet spinning[96] • CVD using Cu nanowire template[33] • Hydrothermal using Cu wire template[97] • CVD using AAO template[34]	• Microscale energy storage devices • Wearable electronics • Bioimaging • Transparent conducting electrodes • Thermal energy storage
2D films Sheets Curved sheets Porous sheets Nanomesh Non-woven mat Woven fabrics	• Micromechanical cleavage[4] • Photoexfoliation[4] • Liquid-phase exfoliation[4] • Growth on SiC[4] • CVD[4] • Molecular beam epitaxy[4] • Chemical synthesis[4] • Chemical exfoliation[18,96] • Chemical under gaseous flow[99] • Chemical activation[100] • Block copolymer lithography[101] • Wet spinning[102] • CVD with metal mesh[103] • Induced edge nucleation[104]	• Photonic and optoelectronic devices • Energy storage and conversion • Photovoltaics • Sensors • Lubricants • Catalysts • Thermal management • Drug delivery
3D monoliths Aerogels Hydrogels 3D Porous films Vertical sheets Frameworks Honeycomb-like	• Templated synthesis[105,106] • Self-assembly[107] • Laser scribing[55,56] • Thermal[109] • Microwave and chemical activation[100] • Vacuum filtration[60] • Freeze-drying[110] • Crosslinking[111] • Chemical assembly[112–114] • 3D printing[58]	• Absorber for environmental remediation • Energy storage • Sensors • Biosensors • Catalysis • Electrochemistry • Fuel cells • Gas absorption • Hydrogen storage • Water splitting • Scaffold for electronic and biomedical applications • Tissue engineering and regeneration
4D self-folding structures Before After Self-healing batteries	• 4D printing? • Other methods?	• Self-healing batteries • Self-folding materials • Materials with on-demand shape transformations from one state to another

Figure 1. Synthesis and assembly of graphene. Graphene sheets can be manipulated to form macroscopic structures that take advantage of the unique features of the individual graphene sheets to enable the construction of new graphene architectures that are useful for many different applications. 2D graphene sheets can be assembled into macroscopic structures of all dimensionalities: 0D dots, 1D linear structure, 2D films, 3D monoliths and potentially 4D self-folding structures. Methods for the preparation of these structures and how they affect the properties and applications of graphene are described. AAO, anodic aluminum oxide; CNT, carbon nanotube; CVD, chemical vapor deposition; GO, graphene oxide; QDs, quantum dots. (Reprinted from Ref. (EI-Kady et al. 2016) with permission from Nature Publishing Group.)

architectures through self-assembly. In the case of metal oxide deposition, it is important to control the surface defects and functional groups because they can serve as nucleation sites for crystal growth and dictate the deposition thickness. This chapter summarizes the recent progress in graphene-based composite materials for electrochemical energy storage devices, including lithium/sodium-ion batteries, lithium-sulfur batteries, lithium-air batteries and supercapacitors.

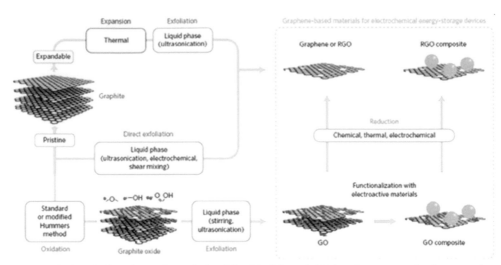

Figure 2. A typical synthetic pathways for producing GO, rGO, GO-, and rGO-based nanocomposites for use as electronic active materials for energy storage devices. (Reprinted from Ref. (Raccichini et al. 2015) with permission from Nature Publishing Group.)

Lithium/sodium ion batteries

Graphene is preferably used for the encapsulation of electrochemically active materials for energy storage and conversion owning to its extraordinary conductivity, large surface area, excellent flexibility and high chemical stability (Wu et al. 2010, Zhang et al. 2010). Graphene can be functionalized by various simple methods to produce localized highly reactive regions which result in impressive properties for respective applications. Silicon is considered as a most promising anode material for next generation LIBs, but its structural and interfacial stability issues still remain a big challenge. One way to address these problems is to build energetic silicon architectures supported with elastic and conductive materials that could be adaptable. Wang et al. (Wang et al. 2013) developed a novel self-supporting binder free silicon-based electrode through the encapsulation of silicon nanowires with dual adaptable apparels (overlapped graphene sheaths and reduced GO overcoats). The resulting architecture gives two advantages to the electrode. Firstly, sealed and adaptable coated graphene sheets avoid the direct contact of encapsulated silicon and electrolyte, enabling the structural and interfacial stabilization. Secondly, the flexible and conductive reduced GO controls the pulverization of the electrode and provides the conductive homogeneity to the composite. Thus, the composite electrodes exhibit excellent reversible specific capacity and rate capability with high capacity retention. It is generally accepted that the formation of defects in the graphitic planes of graphene by replacing the carbon atom with hetero-atoms increases its electrochemical performance (Reddy et al. 2010). For this purpose, Zhou et al. (Zhou et al. 2012) directly grew NiO nanosheets (NSs) on graphene via the oxygen bridges formed between graphene and NiO NSs. This combination is one step towards realizing thin-based anodes for LIBs. The resulting composite has a high reversible capacity and good cycle life due to the pinning effect of hydroxyl/epoxy groups on the Ni atoms of NiO NSs. Meanwhile, graphene provides electron transport paths that enhance the conductivity and stabilize the reversible storage process of lithium ions (Zhou et al. 2012, Cai et al. 2013).

Vanadium pentoxide is widely studied as a cathode for lithium-ion batteries. To address the low intrinsic electronic conductivity, slow lithium-ion diffusion and irreversible phase transition of vanadium pentoxide on deep discharge, graphene sheets were incorporated into vanadium pentoxide nanoribbons via a sol-gel process (Liu et al. 2015). V_2O_5 nanoribbons with a 5–10 nm diameter were anchored or sandwiched between the graphene sheets, indicated by the yellow dashed circle region (see Figure 3, Liu et al. 2015). In addition, a low-voltage, aberration-corrected high-resolution TEM image of the synthesized V_2O_5-G (see Figure 3b, Liu et al. 2015) clearly shows the V_2O_5 nanoribbons sandwiched between the graphene sheets.

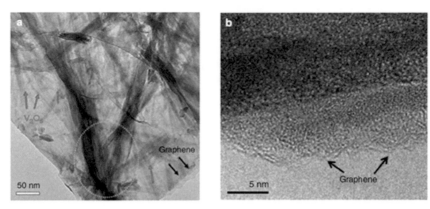

Figure 3. TEM characterization of the V$_2$O$_5$-G hybrids. (a) TEM of as-synthesized V$_2$O$_5$-G hybrid after calcinations. The black arrows point to the graphene sheets, the red arrows point to the V$_2$O$_5$ nanoribbons and the yellow dashed circle point to the area where the graphene and nanoribbons twist together. (b) HRTEM of as-synthesized V$_2$O$_5$-G hybrid after calcinations. The red dashed line points out the area where a V$_2$O$_5$ nanoribbon is located. (Reprinted from Ref. (Liu et al. 2015) with permission from Nature Publishing Group.)

The introduction of a little amount of graphene (that is, 2 wt.%) into the V$_2$O$_5$ has an extraordinary effect on its electrochemical performance (see Figure 4, Liu et al. 2015). In addition to the high specific capacity, this V$_2$O$_5$-G hybrid shows an excellent reversibility from the voltage profile, which indicates that most of the inserted lithium ions can be removed during the discharge/charge process. Even at a higher rate, 1C, such V$_2$O$_5$-G still delivered 315 mAh g^{-1} (corresponding to 768 Wh kg^{-1}), which is 2.23 times that of the pure V$_2$O$_5$, 137 mAh g^{-1} (corresponding to 299 Wh kg^{-1}).

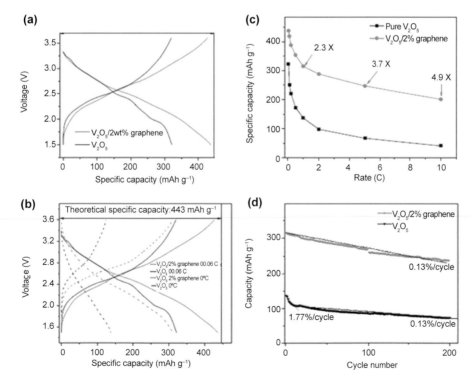

Figure 4. Electrochemical characterization. Charge/discharge curves of pure V$_2$O$_5$ and V$_2$O$_5$-2%wt.%G graphene cells at (a) 0.05 and (b) 0.05 and 1.0C; (c) Rate performance of pure V$_2$O$_5$ and V$_2$O$_5$-2%wt.%G graphene cells based on C-rate and (d) Cycle life of pure V$_2$O$_5$ and V$_2$O$_5$-G with 2% graphene cells at a 1C rate. (Reprinted from Ref. (Liu et al. 2015) with permission from Nature Publishing Group.)

Gwon et al. (Gwon et al. 2011) reported a new approach to prepare flexible energy devices based on graphene paper. Graphene paper is a functional material, which does not only act as a conducting agent, but also as a current collector. The unique combination of its outstanding properties such as high mechanical strength, large surface area, and superior electrical conductivity make graphene paper a promising base material for flexible energy storage devices. The battery with the V_2O_5/graphene paper was discharged (Li insertion) and charged (Li extraction) between 3.8 and 1.7 V at a constant density of $10\ \mu A\ cm^{-2}$. The corresponding capacities were normalized to the geometrical surface area of the electrode ($S = 0.495\ cm^2$). For comparison, the 2D electrode with V_2O_5 deposited on conventional current collector Al foil under the same depositions was also tested. The charge-discharge profile of both V_2O_5 electrodes showed apparent voltage plateaus, which represents the characteristic of an amorphous V_2O_5 cathode. It shows that the V_2O_5/graphene paper retains higher capacity at relatively faster rates, from 10 to 600 μA cm^{-2}. The nanostructured V_2O_5/graphene paper still provides 40% of its initial capacity even when the current density increases to 60 times the value, while a conventional V_2O_5/Al electrode retains below 10% of its initial capacity with the same experimental conditions. Compared with a conventional electrode, an increase in volumetric capacity at all charge-discharge rates when a nanostructured graphene hybrid electrode is used can be clearly observed. These results indicate that the kinetics of lithium insertion/extraction and electron conduction in V_2O_5/graphene paper is significantly improved. Moreover, the capacity retention of the V_2O_5/graphene paper is superior to that of a conventional electrode on Al foil. Moreover, the fading of the capacity is less than 0.1% per cycle. It is attributed to the excellent chemical/mechanical robustness of the nanostructure V_2O_5/graphene paper.

MoS_2, with a typical layered structure, has strong covalent bonds within the layers, which is very similar to graphite (Rapoport et al. 2003, Srivastava et al. 1993, Rapoport 1997). The spacing between neighboring layers is 0.615 nm, significantly larger than that of graphite (0.335 nm), and the weak van der Waals forces between the layers allows Li ions to diffuse without a significant increase in volume. Due to these properties, MoS_2 is more suitable for Li-ion insertion than graphite (Shidpour et al. 2010, Feng et al. 2009, Hwang et al. 2011, Ding et al. 2012, Du et al. 2010). However, the rapid capacity fading and poor high-rate properties of MoS_2 limit its practical applications. A widely used approach to overcome this problem is to design and optimize nanocomposites for good electrical conductivity and thermal conductivity. Yu et al. (Yu et al. 2013) prepared three-dimensional (3D) architectures of graphene/MoS_2 nanoflake arrays by a facile one-step hydrothermal method. The MoS_2 nanoflakes are grown vertically on both sides of graphene sheets (see Figure 5, Yu et al. 2013), allowing the 3D architectures to be more stable during charging and discharging (see Figure 6, Yu et al. 2013). Even at a high current density of $8000\ mA\ g^{-1}$, the discharge capacity of the 3D porous architectures is as much as that at $516\ mAh\ g^{-1}$, and the discharging or charging time is only 3.9 min. Thus, rapid charging and discharging of LIBs is realized by this novel 3D architecture.

Sodium-ion (Na^+) batteries have recently attracted much attention as an alternative to lithium-ion (Li^+) batteries for electric energy storage applications owing to the low cost and abundant sodium resources (Ma et al. 2014, Hou et al. 2016). Strategies to increase the charge carrier transport kinetics in NASICON-type

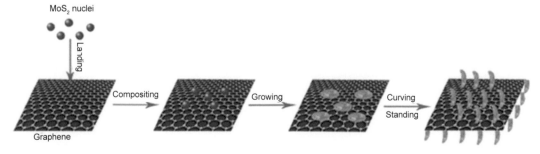

Figure 5. Illustration of the growth of MoS_2 nanoflake arrays on graphene. (Reprinted from Ref. (Yu et al. 2013) with permission from John Wiley and Sons.)

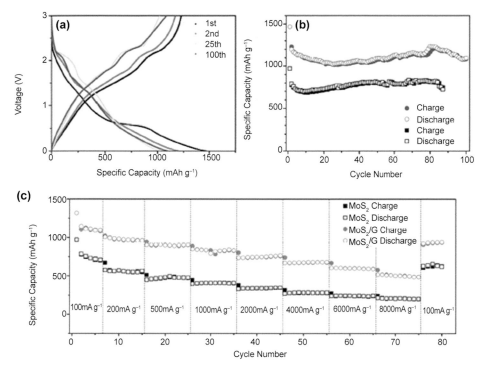

Figure 6. (a) The charge/discharge curves of the 3D architectures; (b) Comparison of the cycling performance of the 3D architectures (upper curves, circles) with MoS_2 flowers (bottom curves, squares); (c) Rate capability of the 3D architectures (circles) and MoS_2 flowers (squares) at different charge/discharge current densities. (Reprinted from Ref. (Yu et al. 2013) with permission from John Wiley and Sons.)

$Na_3V_2(PO_4)_3$ (NVP) are mainly focused on coating the NVP active materials with a conductive carbon layer (i.e., enhancing the surface conductivity and the electrical contact in the electrode), and downsizing the particles (i.e., reducing the electron and Na ion transport lengths). Rui et al. (Rui et al. 2015) reported the synthesis of a 3D hierarchical meso- and macroporous NVP-based hybrid cathode with connected Na ion/electron pathways for ultrafast charge and discharge NIBs, consisting of NVP nanocrystals coated with amorphous carbon and then wrapped by reduced graphene oxide nanosheets (abbreviated as NVP@C@rGO) using a freeze-drying-assisted method. The NVP@C@rGO nanocomposite cathode exhibited a long cycle life up to 10,000 cycles at a high rate of 100C. The discharge capacity remained around 55 mAh g^{-1} after 10,000 cycles with a capacity retention of 64% of its initial capacity.

The combined advantages of graphene network are beneficial to the electrochemical performance of the rGO-based composite. Fang et al. (Fang et al. 2016) proposed a promising structural design of $NaTi_2(PO_4)_3$ through a hierarchical graphene-embedding process by using a facile spray-drying method with post calcination. The foam-like residuals contain a quasi-spherical 3D grapheme network, which inherits the shape of the pristine NTP@rGO composite and indicates a homogenous distribution of graphene sheets in the secondary particles. The as-prepared NTP@rGO composite exhibits high reversible capacity of 130 mAh g^{-1} at 0.1C, ultrahigh rate capability (38 mAh g^{-1} at 200C), as well as long stability rate capability (77% capacity retention over 1000 cycles at 20C). A NTP@rGO//$Na_3V_2(PO_4)_3$/C full cell also shows high discharge capacity (128 mAh g^{-1}) at 0.1C based on anode mass, high rate performance (88 mAh g^{-1} at 50C), and long-term cycling life (80% capacity retention over 1000 cycles at 10C). The outstanding electrochemical performance of the NTP@rGO composite is attributed to the combined advantages of grapheme-coated nanosized particles with the presence of a 3D grapheme network, which significantly improves the ionic and electronic transport and accommodates structural variation during sodiation and desodiation.

Lithium-sulfur batteries

Lithium-sulfur (Li-S) batteries are being considered as the next-generation high-energy-storage system due to their high theoretical energy density of 2600 Wh kg^{-1} (Bruce et al. 2012, Manthiram et al. 2014, Yang et al. 2013). Sulfur, a petroleum by-product, is inexpensive and abundant with no or little environmental impact. However, the practical implementation of Li-S batteries has been impeded by the intrinsic electronically insulating properties of sulfur and its discharge products, rapid capacity fading/low Coulombic efficiency originating from polysulfide shuttle effect, and the large volume change occurring between sulfur and lithium sulfide (Li$_2$S), causing electrode structure damage as a result of different densities (Peng et al. 2014, Lv et al. 2014, Yin et al. 2013, Zhou et al. 2015a, Li et al. 2015, Zhao et al. 2015, Qie et al. 2015). Another growing concern for efficient Li-S batteries is the non-uniform deposition of lithium during cycling, leading to lithium dendrite formation, thus causing short-circuiting within the cells (Cheng et al. 2014, Zheng et al. 2014). Zhou et al. (Zhou et al. 2016) developed a self-supported, high capacity, long-life cathode material for Li-S batteries by coating Li$_2$S onto doped graphene aerogels via a simple liquid infiltration-evaporation coating method. The obtained cathodes can lower the initial charge voltage barrier and attain a high specific capacity, good rate capability, and excellent cycling stability. The improved performance can be attributed to the cross-linked, porous graphene network enabling fast electron/ion transfer, coated Li$_2$S on graphene with high utilization and a reduced energy barrier, and doped heteroatoms with a strong binding affinity toward Li$_2$S/lithium polysulfides with reduced polysulfide dissolution based on first-principles calculations.

Chen et al. prepared a multiwalled carbon nanotube/sulfur (MWCNT@S) composite embedded into the interlay galleries of graphene sheets (GS) with core-shell structure through a facile two-step assembly process (see Figure 7a, Chen et al. 2013). As a cathode material for Li/S batteries, the GS-MWCNT@S composite exhibits a high initial capacity of 1396 mAh/g at a current density of 0.2C (1C = 1672 mA/g), corresponding to 83% usage of the sulfur active material (see Figure 7b, Chen et al. 2013). The superior electrochemical performance of the GS-MWCNT@S composite is mainly attributed to the synergistic effects of GS and MWCNTs, which provide a 3D conductive network for electron transfer, open channels for ion diffusion, strong confinement of soluble polysulfides, and effective buffer for volume expansion of the S cathode during discharge.

Recently, graphene-sulfur (G-S) hybrid materials with sulfur nanocrystals anchored on interconnected fibrous graphene were reported by Cheng's group (see Figure 8, Zhou et al. 2013). With the aid of a

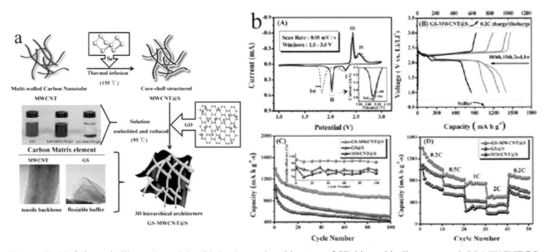

Figure 7. (a) Schematic illustration of the fabrication and architecture of 3D hierarchically structured GS-MWCNT@S composite. (b) Electrochemical performances: (A) Cyclic voltammetry of the GS-MWCNT@S composite cathode over a voltage range of 1.0–3.0 V at a scanning rate of 0.05 mV/s. (B) Discharge/charge voltage profiles of GS-MWCNT@S composite cathode at a current rate of 0.2C. (C) Cycling performance and (D) Rate capabilities of MWCNT@S, GS@S, and GS-MWCNT@S composite cathodes. (Reprinted from Ref. (Chen et al. 2013) with permission from American Chemical Society.)

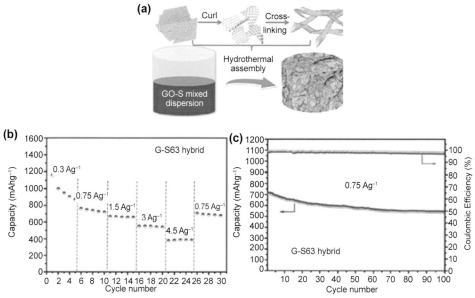

Figure 8. (a) Illustration of the formation process of the graphene-sulphur hybrid and schematic of fabrication of a self-supporting electrode. (b) Capacity at different current densities of the graphene with 63% sulfur (G-S63) cathode. (c) Cyclic performance and Coulombic efficiency of the G-S63 cathode at 0.75A g⁻¹ for 100 cycles after the high current density test. (Reprinted from Ref. (Zhou et al. 2013) with permission from American Chemical Society.)

freeze-drying process, the assembled structure could be further cut and pressed into pellets to be directly used as lithium-sulfur battery cathodes without using a metal current collector, binder, and additional conductive additive. The hybrid showed a high capacity and an excellent high-rate performance, which may be attributed to the porous network and sulfur nanocrystals that enable rapid ion and lithium-ion transport, the highly conductive electron pathway provided by interconnected fibrous graphene, and strong binding between polysulfides and oxygen-containing groups.

Lithium-air batteries

Metal-air batteries are another attractive energy-storage and conversion system owing to their high energy and power densities, safe chemistries, and economic viability (see Cheng et al. 2012). Considering the high conductivity, excellent mechanical properties as well as good anti-corrosion properties of graphene, highly active and stable graphene (G)-Co_3O_4 nanocomposite electrocatalysts were synthesized via a facile hydrothermal route and subsequent thermal treatment process by our group (see Sun et al. 2014). Low-magnification scanning electron microscopy (SEM) and transmission electron microscopy (TEM) images show that Co_3O_4 and Co nanoparticles are well dispersed on the surface of the graphene sheets. The annular dark-field (ADF) scanning transmission electron microscopy (STEM) images of the as-prepared graphene-Co_3O_4 prepared under Ar atmosphere display that the Co_3O_4 particles have a diameter of about 200 nm and consist of many smaller NPs, forming a porous structure (see Figure 9b and c, Sun et al. 2014). The oxygen reduction reaction (ORR) and oxygen evolution reaction (OER) properties of graphene (G)-Co_3O_4 nanocomposite as well as its applications as a bifunctional catalyst for Li-air batteries were systematically studied. The graphene-Co_3O_4 nanocomposite catalyst demonstrates an excellent catalytic activity toward ORR including a much more positive half-wave potential (–0.23 V) than the pristine graphene (–0.39 V) as well as higher cathodic currents. Importantly, this catalyst shows a better long-term durability than the commercial Pt/C catalyst in an alkaline solution. The preliminary results indicate that the graphene-Co_3O_4 nanocomposite is an efficient and stable bifunctional catalyst for a Li-air battery (see Figure 9e–g, Sun et al. 2014). It may be as an alternative to the high-cost commercial Pt/C catalyst for the ORR/OER in alkaline solutions.

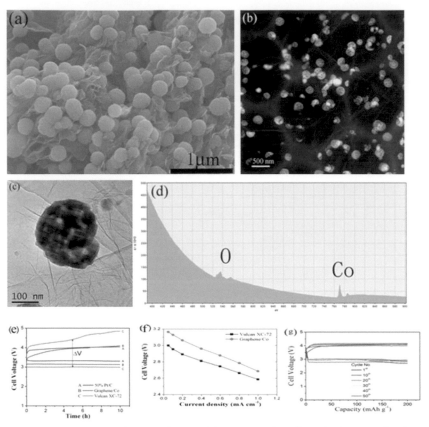

Figure 9. (a) SEM, (b) ADF-STEM, and (c) TEM images of the as-prepared graphene-Co₃O₄ prepared under Ar atmosphere, (d) The corresponding electron energy loss spectroscopy (EELS) analysis taken at the nanoparticle, (e) Comparison of the first charge and discharge curves of the prepared Li-air batteries with various catalysts at a current density of 80 mA g⁻¹, (f) comparison of discharge voltage as a function of current densities at different current densities between Vulcan XC-72 and the graphene-Co₃O₄ catalyst, (g) The discharge-charge curves of the prepared Li-air batteries with the graphene-Co₃O₄ catalyst at a current density of 160 mA g⁻¹ at different cycles. Catalyst loading was 1.25 mg. (Reprinted from Ref. (Sun et al. 2014) with permission from Royal Society of Chemistry.)

Xiao and co-works adopted colloidal microemulsion to fabricate hierarchically porous 3D graphene cathode with interconnected pore channels at both the micro and nanometer scales (Xiao et al. 2011). The SEM images indicate the presence of porous architectures (see Figure 10a and b, Xiao et al. 2011). It suggests that lattice defects on functionalized graphene play a critical role through hosting the small, nanometer-sized discharge products (Li₂O₂) (see Figure 10c and d, Xiao et al. 2011), which is also observed by DFT modeling (see Figure 10e and f, Xiao et al. 2011). The combination of hierarchical pore structure and graphene surface defects produces an exceptionally high capacity of ≈ 15000 mAh g⁻¹ at the first discharge cycle (see Figure 10g, Xiao et al. 2011). The challenge is to obtain a high density porous graphene material for an improved volumetric capacity and also to make the system rechargeable.

Supercapacitors

Electrochemical capacitors (ECs), also known as supercapacitors, have attracted much attention due to superior power density, fast charge/discharge rates and long cycle lifetime compared to other chemical energy storage devices. However, commercial supercapacitors suffer from low energy density. To meet the increasing energy demands for next-generation ECs, the energy density should be substantially increased without sacrificing the power density and cycle life (Simon et al. 2008). For this purpose, graphene and graphene-based nanocomposites have been proposed as attractive and alternative nanostructured

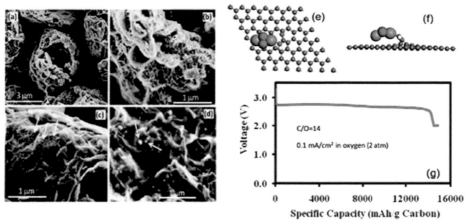

Figure 10. (a) SEM image of egg shell structured FGS with large pores, (b) SEM image of the nanopores in FGS, (c) Small Li_2O_2 nanoparticles produced on FGS with C/O = 14. (d) Large Li_2O_2 nanoparticles of FGS with C/O = 100. (e, f) Top and side views of optimized Li_2O_2 on the 5-8-5 defect graphene with COOH functional groups. (g) Discharge curve for Li-air battery using FGS (C/O = 14). (Reprinted from Ref. (Xiao et al. 2011) with permission from American Chemical Society.)

electrode materials for supercapacitors due to their excellent chemical stability, and superior electrical and thermal conductivities. The large open pores in graphene and graphene composites facilitate the fast transport of hydrate ions that can lead to high double-layer capacitance in different electrolytes. Ruoff and co-works reported a facile chemical activation method to produce high-quality activated microwave exfoliated graphene oxide (a-MEGO) with numerous micropores and extremely large surface area ($\sim$ 3100 m^2 g^{-1}) (Zhu et al. 2011). The specific capacitances achieved in the as-obtained porous graphene oxide were as high as 165, 166 and 166 F g^{-1} at current densities of 1.4, 2.8, and 5.7A g^{-1}, respectively. The large gravimetric capacitance originated from the sp^2 bonded carbon with a continuous 3D network of highly curved, atomic thick graphene walls that form 0.6–5 nm pores with the low oxygen and hydrogen content.

He et al. reported the use of freestanding, lightweight (0.75 mg/cm^2), ultrathin (< 200 μm), highly conductive (55 S/cm), and flexible three-dimensional (3D) graphene networks, loaded with MnO_2 by electrode position, as the electrodes of a flexible supercapacitor. The 3D graphene networks are an ideal supporter for active materials and permitted a large MnO_2 mass loading of 9.8 mg/cm^2 ($\sim$ 92.9% of the mass of the entire electrode), with a high area capacitance of 1.42 F/cm^2 at a scan rate of 2 mV/s. The assembled supercapacitor was light-weight (less than 10 mg), thin ($\sim$ 0.8 mm), and highly flexible (see Figure 11a, He et al. 2013). The flexible supercapacitor exhibits an energy density of 6.8 Wh/kg at a power density of 62 W/kg for a 1V window voltage (see Figure 11e, He et al. 2013). The energy density can keep $\sim$ 55% when the power density increases to 2500 W/kg. These values are superior to other symmetric systems reported previously.

Dubal et al. (Dubal et al. 2014) designed hybrid energy storage devices using hybrid electrodes based on rGO and pseudocapacitive $Ni(OH)_2$ and $Co(OH)_2$ deposited on the skeleton of 3D macroporous sponge support (see Figure 12, Dubal et al. 2014). The rGO ink in an aqueous solution was first coated onto the bare sponge substrate by a "dip and dry" method. Then nanostructured $Ni(OH)_2$ and $Co(OH)_2$ were deposited by the chemical both deposition (CBD) strategy (see Figure 12a, Dubal et al. 2014). The 3D macroporous conducting framework uniformly covered with aggregated of rGO sheet with high surface area is formed, which provides enough space for the further large and uniform deposition of $Ni(OH)_2$ and $Co(OH)_2$ onto the skeleton of the sponge (Dubal et al. 2014). The characteristic of hybrid electrode is the hybridization of pseudocapacitive component $Ni(OH)_2$ and $Co(OH)_2$ and nonfaradic component rGO in a single electrode so that energy can be stored through both mechanisms (see Figure 12b and c, Dubal et al. 2014). The as-assembled SP@rGO//SP@Co(OH)$_2$ hybrid devices with a cell voltage of 1.5 V delivered a high specific energy of 42.02 Wh kg^{-1} with a maximum specific power of 11 kW kg^{-1} while the SP@rGO//SP@rGO@Co cell exhibited 33.01 Wh kg^{-1} with a specific power of 8 kW

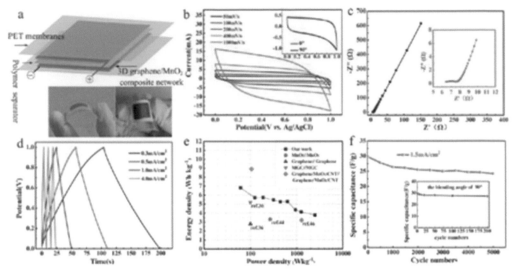

Figure 11. (a) Schematic of the structure of our flexible supercapacitors consisting of two symmetrical graphene/MnO$_2$ composite electrodes, a polymer separator, and two PET membranes. The two digital photographs show the flexible supercapacitors when bended. (b) CVs of the flexible supercapacitors at scan rates of 50, 100, 200 and 1000 mV/s. Inset shows the CVs of the flexible supercapacitors with bending angles of 0° and 90° at a fixed scan rate of 10 mW/s. (c) Nyquist plot of the flexible supercapacitor. (d) Galvanostatic charging/discharging curves of the flexible supercapacitor devices at different current densities. (e) Rogone plots of the flexible supercapacitor, compared with the values of similar symmetrical systems from ref. 36, 44, and 46. (f) Cycling performance of the flexible supercapacitors for charging and discharging at a current density of 1.5 mA/cm^2. Inset shows cycling performance of the flexible supercapacitors for bending cycles with a bending angle of 90°. (Reprinted from Ref. (He et al. 2013) with permission from American Chemical Society.)

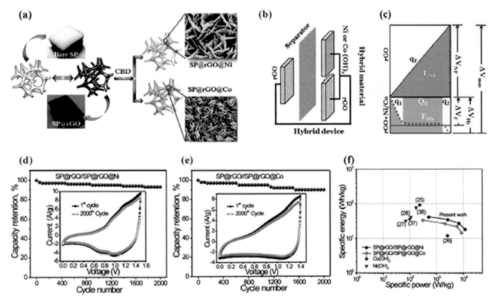

Figure 12. (a) Schematic representation of the fabrication of hybrid materials based on rGO and transition metal hydroxides (Ni(OH)$_2$ and Co(OH)$_2$) onto skeleton of 3D macroporous sponge. (b,c) Schematic representation of hybrid devices of hybrid electrode (supercapacitor (SP)@rGO@Ni(OH)$_2$ or SP@rGO@Co(OH)$_2$) and SP@rGO electrode with propylene carbonate paper as a separator along with charge-potential profile. Variation of capacity retention of (d) SP@rGO//SP@rGO@Ni(OH)$_2$, and (e) SP@rGO@Co(OH)$_2$ hybrid devices with number of cycles at 100 mV s^{-1} scan rate and the corresponding CV curves (inset) at first and 2000th cycles. (f) The power density versus energy density of SP@rGO//SP@rGO@Ni(OH)$_2$ and SP@rGO//SP@rGO@Co(OH)$_2$ hybrid devices in a Ragone plot. (Reprinted from Ref. (Dubal et al. 2014) with permission from Nature Publishing Group.)

kg⁻¹. Furthermore, these devices can retain an energy density > 10 Wh kg⁻¹ even at a high power density. The energy density reported in this work was significantly higher than those obtained for other carbon-based symmetric capacitors in aqueous electrolytes.

There is growing interest in thin, lightweight, and flexible energy storage devices to meet the special needs for next-generation, high-performance, flexible electronics. Li et al. (Li et al. 2012) reported a thin, lightweight, and flexible lithium ion battery made from graphene foam, a three-dimensional, flexible, and conductive interconnected network, as a current collector, loaded with $Li_4Ti_5O_{12}$ and $LiFePO_4$ for use as anode and cathode, respectively. No metal current collectors, conducting additives, or binders are used. The excellent electrical conductivity and pore structure of the hybrid electrodes enable rapid electron and ion transport. The $Li_4Ti_5O_{12}$ graphene foam electro shows a high rate discharge up to 200C, equivalent to a full discharge in 18s (see Figure 13, Li et al. 2012). They demonstrate a thin, lightweight, and flexible full lithium ion battery with a high-rate performance and energy density that can be repeatedly bent to a radius of 5 mm without structural failure and performance loss.

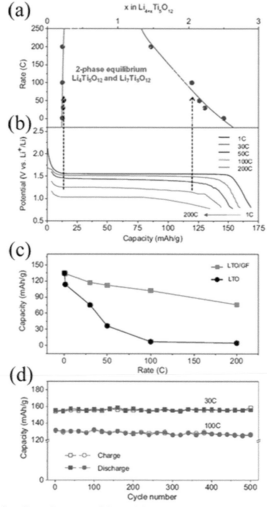

Figure 13. Discharge rate and cyclic performance of the LTO/GF electrode. (a) Two-phase equilibrium region of the LTO/GF with different charge/discharge rates; C/n denotes the rate at which a full charge or discharge takes n hours. (b) Discharge voltage curves of the LTO/GF with different charge/discharge rates. (c) Specific capacities of the LTO/GF and reference LTO at various charge/discharge rates within a flat plateau segment shown in B. (d) Capacities of the LTO/GF charged/discharged at constant 30C and 100C rates for 500 cycles. (Reprinted from Ref. (Li et al. 2012) with permission from the Copyright (2012) National Academy of Sciences, USA.)

Micro-supercapacitors are promising energy storage devices that can complement or even replace batteries in miniaturized portable electronics and microelectromechanical systems. Their main limitation, however, is the low volumetric energy density when compared with batteries. Yu et al. (Yu et al. 2014) reported a hierarchically structured carbon microfiber made of an interconnected network of aligned single-walled carbon nanotubes with interposed nitrogen-doped reduced graphene oxide sheets. A full micro-supercapacitor with a gel PVA/H_3PO_4 electrolyte, free of binder, current collector and separator, has a volumetric energy density of ~ 6.3 mWh cm^{-3} (a value comparable to that of 4V–500 µAh thin-film lithium batteries) while maintaining a power density more than two orders of magnitude higher than that of batteries, as well as a long cycle life. Fiber-3 (with the best capacitive performance measured in the three-electrode cell) was used as both active materials and current collectors to construct flexible micro-SCs (see Figure 14a, Yu et al. 2014). Typically, two parallel fibre-3 electrodes were mounted onto a flexible polyester (PET) substrate using PVA-H_3PO_4 electrolyte without binder, current collector, separator or any packaging material. The total volume of each micro-SC, including two fibres and the surrounding solid electrolytes, was estimated to be ~ 7.5 × 10^{-5} cm^3. Its galvanostatic charge/discharge curves (see Figure 14b, Yu et al. 2014) have a triangular shape with a coulombic efficiency of ~ 98%, indicating excellent reversibility and good charge propagation between the two fiber electrodes. The volumetric capacitance of the micro-SC ($C_{cell,V}$, normalized to the whole devices volume) is ~ 45.0 F cm^{-3} at ~ 26.7 mA cm^{-3} and ~ 25.1 F cm^{-3} at ~ 800 mA cm^{-3}, corresponding to area capacitances ($C_{cell,A}$) of ~ 116.3 mF cm^{-2} and ~ 64.6 mF cm^{-2}, respectively, outperforming all previously reported carbon-based micro-SCs with capacitances in the range of 1–18 F cm^{-3} or 0.5–86 mF cm^{-2} (see Figure 14c, Yu et al. 2014). The specific volumetric capacitances of a single-fiber electrode (C_{sp}) in a two-electrode cell was calculated using galvanostatic discharge curves to be ~ 300 F cm^{-3} at ~ 26.7 mA cm^{-3}. It shows that the micro-SC retains 93% of its initial capacitance after 10,000 charge/discharge cycles (see Figure 14d, Yu et al. 2014), demonstrating its impressive performance stability with a long cycle life, as is the case for other nitrogen-doped graphene materials (Jeong et al. 2011).

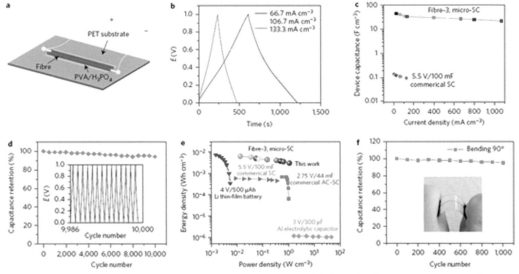

Figure 14. Electrochemical performances of all-solid-state micro-SCs. (a) Schematic of a micro-SC constructed using two fibre-3 electrodes on a polyester (PET) substrate, (b) Galvanostatic charge/discharge curves at various current densities, (c) Volumetric devices capacitance at different current densities. Data obtained from a commercial activated carbon supercapacitor are also shown for comparison, (d) Cycle life of the micro-SC. Inset: Galvanostatic charge/discharge curve after 10,000 cycles between 0 and 1V at 250 mA cm^{-3}. (e) Energy and power densities of the micro-SC compared with commercially available state-of-the-art energy storage systems. (f) Capacitance retention after 1000 cycles up to 90° bending angle. Inset: Photograph of a bent micro-SC. (Reprinted from Ref. (Yu et al. 2014) with permission from Nature Publishing Group.)

The volumetric power/energy density of the whole SC is a more meaningful parameter for evaluating the energy storage performance of the microdevices than the gravimetric power/energy density based on active electrode materials (Gogotsi et al. 2011). The Ragon plots (see Figure 14e, Yu et al. 2014) compare the volumetric performance of our micro-SCs to those of commercially available energy-storage devices. This micro-SC has a volumetric energy density ($E_{cell, V}$, normalized to the whole device volume) of ~ 6.3 mWh cm^{-3}, which is about ten times higher than the energy densities of commercially available supercapacitors (2.75 V/44 mF and 5.5 V/100 mF, < 1 mWh cm^{-3}) and even comparable to the 4V/500 μAh thin-film lithium battery (0.3–10 mWh cm^{-3}) (Pech et al. 2010). This energy density value is also higher than that of recently reported thin-film supercapacitors based on different two-dimensional materials, including laser-scribed graphene (~ 1.36 mWh cm^{-3} in ionic liquid) (El-Kady et al. 2012) and transition-metal carbides and carbonitrides (Ti$_3$C$_2$ MXenes, ~ 1 mWh cm^{-3} in K$_2$SO$_4$) (Lukatskaya et al. 2013), and is comparable to that of macroscale SCs based on graphene-derived three-dimensional porous carbon (Tao et al. 2013). The maximum volumetric power density ($P_{cell,V}$) for the micro-SCs is 1,085 mW cm^{-3}, comparable to the commercially available supercapacitors and more than two orders of magnitude higher than that of lithium thin-film batteries.

David et al. prepared layered free-standing papers composed of acid-exfoliated few-layer molybdenum disulfide (MoS$_2$) and reduced graphene oxide (rGO) flakes for use as a self-standing flexible electrode in sodium-ion batteries (see David et al. 2014). The free-standing papers were synthesized through vacuum filtration of homogeneous dispersions consisting of varying weight percent of acid-treated MoS$_2$ flakes in GO in DI water, followed by thermal reduction at elevated temperatures. The electrode showed good cycling stability with a charge capacity of approximately 230 mAh g^{-1} with respect to total weight of the electrode with Coulombic efficiency reaching approximately 99% (see Figure 15, David et al. 2014).

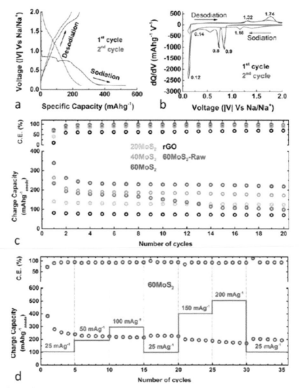

Figure 15. (a) Voltage profile of 60MoS$_2$ free-standing electrode along with its corresponding (b) differential capacity curves for the first two cycles, (c) Sodium charge capacity of various electrodes at a constant current density of 25 mA g^{-1}, (d) Sodium charge capacity and corresponding Coulombic efficiency of 60MoS$_2$ electrode cycled at varying current densities. (Reprinted from Ref. (David et al. 2014) with permission from American Chemical Society.)

Summary and outlook

The past decade has witnessed a great interest in electrochemical energy storage devices to meet the ever-increasing demand of modern society for portable electronic devices and electric vehicles. Graphene has shown an exceptional chemical structure, and outstanding electronic, optical, thermal and mechanical properties, which demonstrates potential and wide applications in various batteries (e.g., lithium/sodium ion batteries, lithium-sulfur batteries and lithium-air batteries) and supercapacitors. A brief summary on the current state-of-the-art progress and challenges of graphene-based composites for next-generation battery has been presented. The large number and variety of research efforts in this field are very encouraging and will undoubtedly prompt the progress of energy-storage devices.

Numerous promising results have demonstrated that graphene-based composite materials will play an increasingly active role in the next generation of energy storage devices due to its superior electrical conductivity, high surface area, and mechanical robustness. The enhanced properties of these composite materials are not only ascribed to the individual components, but are also attributable to the interactions between them. Although great progresses have been made in using graphene and graphene-based materials in electrochemical energy storage devices, to realize the commercial potential of graphene, it is necessary to develop reliable, cost-effective and facile processes for the industry-scale fabrication of graphene electrodes with good performance. Facile wet chemical reactions have also proven advantageous due to their high yield and simplicity. Moreover, great effort should be paid to effectively enlarge the specific surface area, porosity, reduce the O/C ratio and increase the conductivity of the electrode by heteroatom doping (like N, P, S, Cl, etc.).

Most of the present energy storage devices with various sizes and shapes are all rigid; thus, bending them may lead to cell damage and electrolyte leakage. With the 2D one-atom-thick structure of graphene, it can adapt to mechanical stress by deforming in the direction normal to its surface. The inherent mechanical flexibility together with exceptional electrical properties and large surface area makes graphene attractive for flexible energy-storage devices.

Acknowledgment

The author acknowledges the financial support of the National Science Foundation of China (Nos. 51372271, 51672029 and 51172275), National Key R & D Project from Minister of Science and Technology, China (2016YFA0202702) and the National Key Basic Research Program of China (No. 2012CB215402). This work was also supported by the *Thousands Talents Program* for the pioneer researcher and his innovation team in China.

References

Armand, M. and J.M. Tarascon. 2008. Building better batteries. Nature 451: 652–657.
Bruce, P.G., S.A. Freunberger, L.J. Hardwick and J.M. Tarascon. 2012. $Li-O_2$ and Li-S batteries with high energy storage. Nat. Mater. 11: 19–29.
Cai, D., D. Li, S. Wang, X. Zhu, W. Yang, S. Zhang et al. 2013. High rate capability of TiO_2/nitrogen-doped nanocomposite as anode material for lithium-ion batteries. J. Alloys Compd. 561: 54–58.
Chen, R.J., T. Zhao, J. Lu, F. Wu, L. Li, J. Chen et al. 2013. Graphene-based three-dimensional hierarchical sandwich-type architecture for high-performance Li/S batteries. Nano Lett. 13: 4642–4649.
Cheng, F.Y. and J. Chen. 2012. Metal-air batteries: from oxygen reduction electrochemistry to cathode catalysts. Chem. Soc. Rev. 41: 2172–2192.
Cheng, X.B., H.J. Peng, J.Q. Huang, F. Wei and Q. Zhang. 2014. Dendrite-free nanostructured anode: entrapment of lithium in a 3D fibrous matric for ultra-stable lithium-sulfur batteries. Small 10: 4257–4263.
David, L., R. Bhandavat and G. Singh. 2014. MoS_2/graphene composite paper for sodium-ion battery electrode. ACS Nano 8: 1759–1770.
Ding, S.J., D.Y. Zhang, J.S. Chen and X.W. Lou. 2012. Facile synthesis of hierachical MoS_2 microspheres composed of few-layered nanosheets and their lithium storage properties. Nanoscale 4: 95–98.
Du, G.D., Z.P. Guo, S.Q. Wang, R. Zeng, Z.X. Chen and H.K. Liu. 2010. Superior stability and high capacity of restacked molybdenum disulfide as anode material for lithium ion batteries. Chem. Commun. 46: 1106–1108.

Dubal, D.P., R. Holze and P. Gomez-Romero. 2014. Development of hybrid materials based on sponge supported reduced graphene oxide and transition metal hydroxides for hybrid energy storage devices. Sci. Rep. 4: 7349.

El-Kady, M.F., V. Strong, S. Dubin and R.B. Kaner. 2012. Laser scribing of high performance and flexible graphene-based electrochemical capacitors. Science 335: 1326–1330.

El-Kady, M.F., Y. Shao and R.B. Kaner. 2016. Graphene for batteries, supercapacitors and beyond. Nature Rev. 1: 16033.

Fang, Y., L. Xiao, J. Qian, Y. Cao, X. Ai, Y. Huang et al. 2016. 3D graphene decorated $NaTi_2(PO_4)_3$ microspheres as a superior high-rate and ultracycle-stable anode material for sodium ion batteries. Adv. Energy Mater. 6: 1502197.

Feng, C.Q., J. Ma, H. Li, R. Zeng, Z.P. Guo and H.K. Liu. 2009. Synthesis of molybdenum disulfide (MoS_2) for lithium ion battery applications. Mater. Res. Bull. 44: 1811–1815.

Ferrari, A.C., F. Bonaccorso, V. Fal'ko, K.S. Novoselov, S. Roche, P. Bøggild et al. 2015. Science and technology roadmap for graphene, related two-dimensional crystals, and hybrid systems. Nanoscale 7: 4598–4810.

Gogotsi, Y. and P. Simon. 2011. True performance metrics in electrochenmical energy storage. Science 334: 917–918.

Goodenough, J.B. and Y. Kim. 2010. Challenges for rechargeable Li batteries. Chem. Mater. 22: 587–603.

Gwon, H., H. Kim, K.U. Lee, D. Seo, Y.C. Park, Y. Lee et al. 2011. Flexible energy storage devices based on graphene paper. Energy Environ. Sci. 4: 1277–1283.

He, Y.M., W.J. Chen, X.D. Li, Z.X. Zhang, J.C. Fu, C.H. Zhao et al. 2013. Freestanding three-dimensional graphene/MnO_2 composite networks as ultralight and flexible supercapacitor electrodes. ACS Nano 7: 174–182.

Hou, H.D., B.H. Gan, Y.D. Gong, N. Chen and C.W. Sun. 2016. P2-type $Na_{0.66}Ni_{0.23}Mg_{0.1}Mn_{0.67}O_2$ as a high performance cathode for sodium-ion battery. Inorg. Chem. 55: 9033–9037.

Hwang, H., H. Kim and J. Cho. 2011. MoS_2 nanoplates consisting of disordered graphene-like layer for high rate lithium battery anode materials. Nano Lett. 11: 4826–4830.

Jeong, H.M., J.W. Lee, W.H. Shin, Y.J. Choi, H.J. Shin, J.K. Kang et al. 2011. Nitrogen-doped graphene for high-performance ultracapacitors and the importance of nitrogen-doped sites at basal planes. Nano Lett. 11: 2472–2477.

Ji, L., P. Meduri, V. Agubra, X. Xiao and M. Alcoutlabi. 2016. Graphene-based nanocomposites for energy storage. Adv. Energy Mater. 1502159.

Li, N., Z. Chen, W. Ren, F. Li and H.M. Cheng. 2012. Flexible graphene-based lithium ion batteries with ultrafast charge and discharge rates. PNAS 109: 17360–17365.

Li, Z., Y. Huang, L. Yuan, Z. Hao and Y. Huang. 2015. Status and prospects in sulfur-carbon composites as cathode materials for rechargeable lithium-sulfur batteries. Carbon 92: 41–63.

Liu, Q., Z. Li, Y. Liu, H. Zhang, Y. Ren, C. Sun et al. 2015. Graphene-modified nanostructured vanadium pentoxide hybrids with extraordinary electrochemical performance for Li-ion batteries. Nature Commun. 6: 6127.

Lukatskaya, M.R., O.L. Taberna, M. Naguib, P. Simon, M.W. Barsoum and Y. Gogotsi. 2013. Cation intercalation and high volumetric capacitance of two-dimensional titanium carbide. Science 341: 1502–1505.

Lv, W., Z. Li, G. Zhou, J.J. Shao, D. Kong, X. Zheng et al. 2014. Tailoring microstructure of graphene-based membrane by controlled removal of trapped water inspired by the phase diagram. Adv. Funct. Mater. 24: 3456–3463.

Ma, Z., Y. Wang, C.W. Sun, J.A. Alonso, M.T. Fernández-Díaz and L. Chen. 2014. Experimental visualization of the diffusion pathway of sodium ions in the $Na_3[Ti_2P_2O_{10}F]$ anode for sodium-ion battery. Sci. Reports 4: 7231.

Mahmood, N., C. Zhang, H. Yin and Y. Hou. 2014. Graphene-based nanocomposites for energy storage and conversion in lithium batteries, supercapacitors and fuel cells. J. Mater. Chem. A. 2: 15–32.

Manthiram, A., Y. Fu, S.H. Chung, Z. Fu and Y.S. Su. 2014. Rechargeable lithium-sulfur batteries. Chem. Rev. 114: 11751–11787.

Pech, D., M. Brunet, H. Durou, P.H. Huang, V. Mochalin, Y. Gogotsi et al. 2010. Ultrahigh-power micrometer-sized supercapacitors based on onion-like carbon. Nat. Nanotechnol. 5: 651–654.

Peng, H.J., J.Q. Huang, M.Q. Zhao, Q. Zhang, X.B. Cheng, X.Y. Liu et al. 2014. Nanoarchitectured graphene/CNT@porous carbon with extraordinary electrical conductivity and interconnected micro/mesopores for lithium-sulfur batteries. Adv. Func. Mater. 24: 2772–2781.

Qie, L. and A. Manthiram. 2015. A facile layer-by-layer approach for high-areal-capcity sulfur cathodes. Adv. Mater. 27: 1694–1700.

Qu, B., C. Ma, G. Ji, C. Xu, J. Xu, Y.S. Meng et al. 2014. Layered SnS_2-reduced graphene oxide composite—a high-capacity, high-rate, and long-cycle life sodium-ion battery anode material. Adv. Mater. 26: 3854–3859.

Raccichini, R., A. Varzi, S. Passerini and B. Scrosati. 2015. The role of graphene for electrochemical energy storage. Nature Mater. 14: 271–279.

Rapoport, L., Y. Bilik, Y. Feldman, M. Homyonfer, S.R. Cohen and R. Tenne. 1997. Hollow nanoparticles of WS_2 as potential solid-state lubricants. Nature 387: 791–793.

Rapoport, L., V. Leshchinsky, M. Lvovsky, I. Lapsker, Y. Volovik, Y. Feldman et al. 2003. Superior tribological properties of powder materials with solid lubricant nanoparticles. Wear 255: 794–800.

Reddy, A.L.M., A. Srivastava, S.R. Gowda, H. Gullapalli, M. Dubey and P.M. Ajayan. 2010. Synthesis of nitrogen-doped graphene films for lithium battery application. ACS Nano. 4: 6337–6342.

Rui, X.H., W.P. Sun, C. Wu, Y. Yu and Q.Y. Yan. 2015. An advanced sodium-ion battery composed of carbon coated $Na_3V_2(PO_4)_3$ in a porous grapheme network. Adv. Mater. 27: 6670–6676.

Shidpour, R. and M. Manteghian. 2010. A density functional study of strong local magnetism creation on MoS_2 nanoribbon by sulfur vacancy. Nanoscale 2: 1429–1435.

Simon, P. and Y. Gogotsi. 2008. Materials for electrochemical capacitors. Nat. Mater. 7: 845–854.

Srivastava, S.K. and B.N. Avasthi. 1993. Preparation and characterization of molybdenum-disulfide catalysts. J. Mater. Sci. 28: 5032–5035.

Sun, C.W., S. Rajasekhara, J.B. Goodenough and F. Zhou. 2011. Monodisperse porous $LiFePO_4$ microspheres for a high power Li-ion battery cathode. J. Am. Chem. Soc. 133: 2132–2135.

Sun, C.W., F. Li, C. Ma, Y. Wang, Y. Ren, W. Yang et al. 2014. Graphene-Co_3O_4 nanocomposite as an efficient bifunctional catalyst for lithium-air batteries. J. Mater. Chem. A 2: 7188–7196.

Tao, Y., X.Y. Xie, W. Lv, D.M. Tang, D.B. Kong, Z.H. Huang et al. 2013. Towards ultrahigh volumetric capacitance: graphene derived highly dense but porous carbons for supercapacitors. Sci. Rep. 3: 2975.

Wang, B., X. Li, X. Zhang, B. Luo, M. Jin, M. Liang et al. 2013. Adaptable silicon-carbon nanocables sandwiched between reduced graphene oxide sheets as lithium ion battery anodes. ACS Nano. 7: 1437–1445.

Wu, Z.S., W. Ren, L. Wen, L. Gao, J. Zhao, Z. Chen et al. 2010. Graphene anchored with Co_3O_4 nanoparticles as anode of lithium ion batteries with enhanced reversible capacity and cyclic performance. ACS Nano 4: 3187–3194.

Xiao, J., D. Mei, X. Li, W. Xu, D. Wang, G.L. Graff et al. 2011. Hierarchically porous graphene as a lithium-air battery electrode. Nano Lett. 11: 5071–5078.

Yang, Y., G. Zheng and Y. Cui. 2013. Nanostructured sulfur cathodes. Chem. Soc. Rev. 42: 3018–3032.

Yin, Y.X., S. Xin, Y. Guo and L.J. Wan. 2013. Lithium-sulfur batteries: electrochemistry, materials, and prospects. Angew. Chem. Int. Ed. 52: 13186–13200.

Yu, D., K. Goh, H. Wang, L. Wei, W. Jiang, Q. Zhang et al. 2014. Scalable synthesis of hierarchically structured carbon nanotube-graphene fibres for capacitive energy storage. Nat. Nanotechnol. 9: 555–562.

Yu, H., C. Ma, B. Ge, Y. Chen, Z. Xu, C. Zhu et al. 2013. Three-dimensional hierarchical architectures constructed by graphene/MoS_2 nanoflake arrays and their rapid charging/discharging properties as lithium-ion battery anodes. Chem.-A Euro. J. 19: 5815–5823.

Zhang, B., Q.B. Zheng, Z.D. Huang, S.W. Oh and J.K. Kim. 2011. SnO_2-graphene-carbon nanotube mixture for anode material with improved rate capacities. Carbon 49: 4524–4534.

Zheng, G., S.W. Lee, Z. Liang, H.W. Lee, K. Yan, H. Yao et al. 2014. Interconnected hollow carbon nanospheres for stable lithium metal anodes. Nat. Nanotechnol. 9: 618–623.

Zhou, G., D.W. Wang, L.C. Yin, N. Li, F. Li and H.M. Cheng. 2012. Oxygen bridges between NiO nanosheets and graphene for improvement of lithium storage. ACS Nano 6: 3214–3223.

Zhou, G., L.C. Yin, D.W. Wang, L. Li, S.F. Pei, I.R. Gentle et al. 2013. Fibrous hybrid of graphene and sulfur nanocrystals for high-performance lithium-sulfur batteries. ACS Nano 7: 5367–5375.

Zhou, G., L. Li, D.W. Wang, X.Y. Shan, S. Pei, F. Li et al. 2015. Li-S batteries: A flexible sulfur-graphene-polypropylene separator integrated electrode for advanced Li-S batteries. Adv. Mater. 27: 641–647.

Zhou, G., Y. Zhao and A. Manthiram. 2015. Dual-confined flexible sulfur cathodes encapsulated in nitrogen-doped double-shelled hollow carbon spheres and wrapped with graphene for Li-S batteries. Adv. Energy Mater. 5: 1402263.

Zhou, G., E. Paek, G.S. Hwang and A. Manthiram. 2016. High-performance lithium-sulfur batteries with a self-supported, 3D Li_2S-doped graphene aerogel cathodes. Adv. Energy Mater. 6: 1501355.

Zhu, J., D. Yang, Z. Yin, Q. Yan and H. Zhang. 2014. Graphene and graphene-based materials for energy storage applications. Small 10: 3480–3498.

Zhu, Y., S. Murali, M.D. Stoller, J.J. Ganesh, W. Cai, P.J. Ferreira et al. 2011. Carbon-based supercapacitors produced by activation of graphene. Science 332: 1537–1541.

7

Non-Conventional Techniques for Characterization of Nanohybrid Materials Based on Clays

Denis T. Araújo,[1] Breno F. Ferreira,[1] Tiago H. da Silva,[1]
Maisa A. Moreira,[1] Katia J. Ciuffi,[1] Eduardo J. Nassar,[1]
Vicente Rives,[2] Miguel A. Vicente,[2] Raquel Trujillano,[2] Antonio Gil,[3]
Sophia Korili[3] and Emerson H. de Faria[1],*

Introduction

Grafting of organic units onto an inorganic matrix leads to materials combining the functionalities of the organic moiety (i.e., light absorption, reactivity and flexibility) with the specific properties of the support (chemical and thermal stability), resulting in new materials with high functionalities and complementary properties, while new ones are different from those of the isolated compounds emerge. We aim to provide here a general overview of the topic of nanohybrid materials (clay-based hydrogels, nanostructured clay polymer composites, heterogeneous catalysts, photocatalysts and adsorbents) based on different clays (kaolinite, saponite, montmorillonite and laponite), characterized by using Small Angle X-Ray Scattering and X-Ray Photoelectron Spectroscopy, among other techniques.

Characterization of layered compounds by small angle X-ray scattering (SAXS)

Natural and synthetic clay minerals are extremely versatile materials with many industrial applications. The interaction of layered clay minerals with organic species has been the topic of studies in both

[1] Universidade de Franca, Grupo de Pesquisa em Materiais Lamelares Híbridos (GPMatLam), Av., Dr. Armando Salles Oliveira, Parque Universitário, 201, Franca, SP, Brazil, 14404-600.
Emails: denistalarico@gmail.com; brenofreitasferreira@gmail.com; eqhonorato@gmail.com; ma.06.moreira@gmail.com; katia.ciuffi@unifran.edu.br; eduardo.nassar@unifran.edu.br
[2] Universidade de Salamanca (GIR-QUESCAT), Departamento de Química Inorgánica, Salamanca, Spain, E-37008.
Emails: vrives@usal.es; mavicente@usal.es; rakel@usal.es
[3] Universidade Pública de Navarra, Departamento de Química Aplicada, Campus de Arrosadía, Pamplona, Spain, E-31006.
Emails: andoni@unavarra.es; sofia.korili@unavarra.es
* Corresponding author: emerson.faria@unifran.edu.br

the academic and industrial fields and has called for new characterization techniques. In this context, composites consisting of clay minerals and polymeric materials stand out as being a special niche for the use of layered clays. Investigating the interaction between the inorganic (clay) and organic (substituents) domains in these systems is necessary because this interaction may provide the final material with various target properties (Hernandez et al. 2007). In a similar way, polymeric hybrids, which correspond to a suspension that uses clays, represent an important class of complex fluids with application in the oil industry; these hybrids can also be used as dyes and as ceramic additives in the cosmetic and pharmaceutical industries. The structure of the material, a result of the interaction between the inorganic matrix and the ligand, determines the characteristics of the hybrid (Zhang et al. 2003). Therefore, the characterization techniques usually applied may not provide a sufficiently comprehensive knowledge of these composites.

The use of new techniques to investigate materials obtained from clays is essential. Small Angle X-ray Scattering (SAXS) is a technique that permits to confirm and to deep in the results obtained by other existing methodologies. It helps to determine particle dimension, interparticle spacing, and particle organization in solution or in suspension. Among the many physical properties of a clay mineral, interparticle distance or interparticle spacing of organized (or oriented) domains is a particularly important parameter to characterize the colloidal organization in clays in response to the concentration of clay-ligand particles, the type of counter ion, and the ionic strength of the medium (Shang and Rice 2005). SAXS has been increasingly used to characterize clay materials (O'Brien et al. 2010, Akkal et al. 2013, Hansen et al. 2012, Kaneko et al. 2007, Motokawa et al. 2014, Negrette et al. 2004, Sandí et al. 2002, Segad et al. 2012, Shang et al. 2001, Shang and Rice 2005).

Figure 1 illustrates the growing application of this technique to the study of clays. The number of publications that use "SAXS-Clay" or "Small Angle X-ray Scattering-Clay" as keywords has increased since 2000, albeit not constantly. The data indicate the interest of academia in this field and point to the need for further studies relating this technique to clay minerals. The sensitivity of SAXS might have contributed to its increased use. SAXS is a remarkably efficient methodology to analyze systems at the nanometric scale (Momani et al. 2016), and it is particularly useful in understanding how structure evolves during the exfoliation of nanocomposites.

According to Oliveira (2011), one strategy to understand the SAXS technique is to analyze individual particles, as depicted in Figure 2. In this image, the primary light beam (vector $\vec{K}$) reaches the particle at positions O and P, separated by vector $\vec{r}$. As the behavior of the scattering process is close to an elastic behavior, the scattering attributed to vector $\vec{K}$ has the same module as the scattering incident vector, and the properties of the scattering beam are given by the following equations:

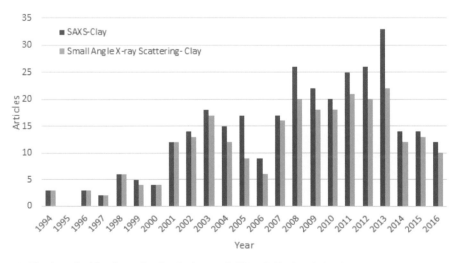

Figure 1. Number of publications related to the keywords "SAXS-Clay" and "Small Angle X-ray Scattering-Clay" (https://www.scopus.com/, consulted in September 2016).

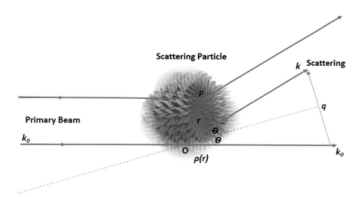

Figure 2. Schematic representation of the scattering process on a single particle (© 2011 Cristiano Luis Pinto Oliveira. Adapted from Oliveira, 2011; originally published under CC BY 3.0 license. Available from: http://www.intechopen.com/ books/current-trends-in-x-raycrystallography/investigating-macromolecular-complexes-in-solution-by-small-angle-x-ray-scattering).

$$\vec{q} = \vec{k} - \vec{k_0}, \qquad\qquad |\vec{k}| = |\vec{k_0}| \qquad\qquad q = 2k \sin \Theta$$

$$k = \frac{2\pi}{\lambda} \quad q = \frac{4\pi}{\lambda}\sin\theta$$

With this relationship, it is possible to define the reciprocal momentary transfer space (vector q). According to the author, this relationship could be submitted to several mathematical treatments, like Fourier transform. This possibility would provide a large amount of information and could broaden the use of this technique. Moreover, these mathematical operations could generate unidimensional (1D) and bidimensional (2D) representations because the mathematical treatment from light scattering could consider the plane only (Cartesian plane) or the other coordinates (spectral or polar coordinates), as well.

Figure 3 represents a simplified scheme of how SAXS works. However, the equipment has a series of possible conformations, and other characteristics may be added. During analysis, it is necessary to measure the intensity of the whole system (sample → matrix and particles) and to subtract the intensity of the standard system (blank) from this measurement. Normalization of the data to the absolute scale calls for dispersion standards. Oliveira (2011) reported the use of these two procedures. For a known protein, the author measured the blank in the same conditions as the sample (buffer, temperature, etc.); then, the dispersion obtained for this sample was used to normalize the other unknown data. In another procedure, water at 20°C was used as the primary standard, which is suitable if the value of the transversal section of the dispersion can be calculated from the fundamental macroscopic properties of water with high precision. In both cases, the data had to be normalized by using the values obtained from the standard under the same

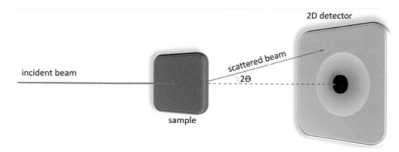

Figure 3. Simplified representation of the conformation of SAXS equipment (© 2011 Cristiano Luis Pinto Oliveira. Adapted from Oliveira, 2011; originally published under CC BY 3.0 license. Available from: http://www.intechopen.com/ books/current-trends-in-x-raycrystallography/investigating-macromolecular-complexes-in-solution-by-small-angle-x-ray-scattering).

experimental conditions used for the sample and multiplied by the theoretical intensity value (Oliveira 2011). In general, applications that use SAXS follow these principles.

This technique offers many possibilities for the analysis of clays. For polymeric aggregates dispersed in clay, SAXS can be applied to evaluate the advance of the exfoliation process in a nanocomposite consisting of a polymer functionalized/intercalated into the clay. Complete exfoliation may take place, and the layer-by-layer arrangement is lost. Momani et al. (2016) used SAXS to study the exfoliation of clays as a function of temperature. Their studies, conducted at 83°C, extended for different periods, as indicated in Figure 4. SAXS study, focused on the interlayer space of the clay, were used to determine the crystalline domains of clays and to assess the interaction between the polymer and the clay as well as the modifications promoted by the polymer in the clay. The results revealed that the clay-clay interaction weakened along time. Surprisingly, most of the increase in the rheological properties occurred when the system was intercalated, but not exfoliated. This result was unexpected, as the authors had designed this system to derive from the reinforcement provided by the network that crossed the solid and consisted of bonds between the polymer and the exfoliated clay layers (Momani et al. 2016). Thus, this unexpected SAXS finding improved the comprehension of the structure of the clay-polymer materials.

By using SAXS and other analytical techniques, Thuresson et al. (2016) obtained information on the structural conformation of Laponite tactoids. Clays can form lamellar structures, grouped as tactoids, whose formation has recently been shown to depend on the diameter of the platelets. On the basis of this information, combined with computational chemistry and SAXS analysis, the interaction between laponite platelets and the polymers was demonstrated. Computational chemistry also constitutes an important tool; it helps to understand several mechanisms, and may provide a better insight into lamellar materials. The structural peaks obtained by SAXS allowed Thuresson et al. to simulate the clay-polymer conformation, as illustrated in Figure 5 (Thuresson et al. 2016).

SAXS is normally used to measure scattering angles smaller than 5°, originating from variations in the electron density within the material as a function of the structure scale, which varies from 1 to 100 nm. Thus, this technique has also been applied, for instance, to investigate powder clays used in pharmaceutical applications (Marçal et al. 2015, Okada et al. 2015, Tian et al. 2014). On the other hand, clay-polymer materials can form aggregates or even undergo compaction; this behavior, as well as the granulometry of powder clays, has also been investigated. Laity and coworkers (2015) used SAXS to

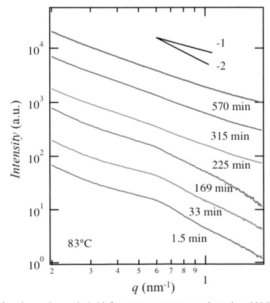

Figure 4. SAXS results for the clay-polymer hybrid for measurements conducted at 83°C (Reprinted from Polymer 93, Momani, B., M. Sen, M. Endoh, X. Wang, T. Koga and H.H. Winter. Temperature dependent intercalation and self-exfoliation of clay/polymer nanocomposite, 204–212, Copyright 2016, with permission from Elsevier).

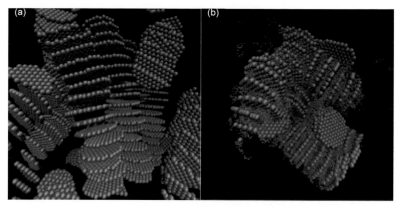

Figure 5. Structure of (a) Laponite and (b) Laponite-polymer obtained from SAXS data and computational chemistry (Reprinted from Journal of Colloid and Interface Science, 466, Flocculated Laponite-PEG/PEO dispersions with monovalent salt, a SAXS and simulation study, Thuresson, A., M. Segad, M. Turesson and M. Skepö, 330–342, Copyright 2016, with permission from Elsevier).

examine the behavior of smectite (gMaSm) at angles lower than 3° (Figure 6). 2D-SAXS analyses revealed that the non-compacted material (Figure 6a) was symmetric, and that there was no preferential scattering orientation in this morphology, as expected for clays with random granulometry. The other clays exhibited scattering along the vertical axis, which indicated compaction. Figure 6d evidenced a radial intensity for non-compacted gMaSm. Hence, SAXS can be used to describe the density of the material according to Figure 7 (Laity et al. 2015).

Results from Figure 7 evidenced the density of the material originating from the compaction process (Figure 7a). It was possible to assess the orientation adopted by the clay (b and c) and to estimate the geometry of the solid (e and f). Information obtained from SAXS also enabled the evaluation of the stress

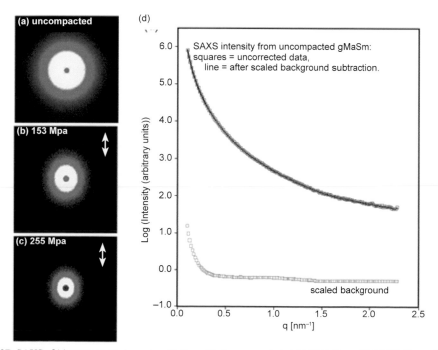

Figure 6. 2D-SAXS of (a) uncompacted powder gMaSm; gMaSm compacted at (b) 153 MPa and (c) 255 MPa; and 1D-SAXS comparing the blank to gMaSm (Reprinted from Applied Clay Science 108, Laity, P.R., K. Asare-Addo, F. Sweeney, E. Šupuk and B.R. Conway. Using small-angle X-ray scattering to investigate the compaction behavior of a granulated clay. 149–164, Copyright 2015, with permission from Elsevier).

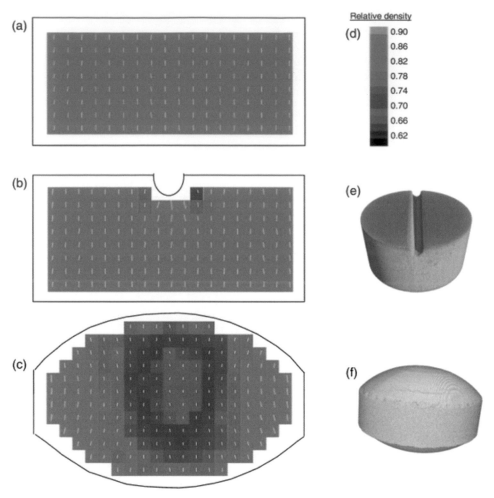

Figure 7. Color-coded maps showing variations in the relative density of the SAXS data analysis. (Reprinted from Applied Clay Science 108, Laity, P.R., K. Asare-Addo, F. Sweeney, E. Šupuk and B.R. Conway. Using small-angle X-ray scattering to investigate the compaction behavior of a granulated clay. 149–164, Copyright 2015, with permission from Elsevier.)

due to compaction of the material. Therefore, SAXS analysis provides relevant information on clays, hardly obtained by other techniques, especially when combined with other characterization tools (Laity et al. 2015).

Besides the aforementioned publications, other papers have reported the use of SAXS to characterize clay materials and their applications. Table 1 summarizes some of these studies.

Characterization of layered compounds by X-ray photoelectron spectroscopy (XPS)

X-Ray Photoelectron Spectroscopy (XPS), also known as Electron Spectroscopy for Chemical Analysis (ESCA), is a spectroscopic technique that can be used to characterize the surface and chemistry of inorganic solids, even though application of XPS to study clay materials is not common. Depending on the intended characterization, the XPS technique offers an array of possibilities and can become a powerful tool to provide additional information about materials, especially about extremely complex nanohybrids.

To understand the potential application of XPS in the characterization of inorganic solids, Figure 8 illustrates some of the most important aspects of XPS experiments.

Table 1. General view of the applications of SAXS to study clay systems.

Clay	Amplitude q (nm⁻¹)	Characteristics	Reference
Montmorillonite/Cloisite	0.1–3.8	The degree of clay dispersion in polymers was determined. The intercalation of polymers into clays was studied at the nanometric scale.	Silva et al. 2016
Organo-Montmorillonite	0.0–1.0	The functionalization of clays with polymers and the conformation of the clays along the process were studied.	Vargas et al. 2016
Laponite	0.001–0.01	By using the SAXS technique, the polymer/clay percentage (in mass) was estimated.	Takeno and Sato 2016
Cordierite	0.002–0.01	Based on zeolite B, SAXS was used to study the formation of cordierite by crystallization.	Sankar et al. 2016

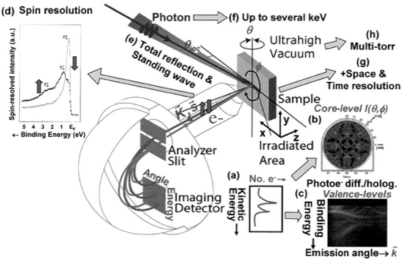

Figure 8. Illustration of a typical experimental configuration for X-ray photoelectron spectroscopy experiments, together with the various types of possible measurements, including (a) simple spectra or energy distribution curves, (b) core-level photoelectron diffraction, (c) valence-band mapping or binding energy vs. k plots, (d) spin-resolved spectra, (e) excitation with incident X-rays such that there is total reflection and/or a standing wave in the sample, (f) use of much higher photon energies than have been typical in the past, (g) use of space and/or time resolution, and (h) surrounding of the sample with high ambient sample pressures of several Torr (Reprinted from Journal of Electron Spectroscopy and Related Phenomena, 178–179, Fadley, C.S. X-ray photoelectron spectroscopy: Progress and perspectives. 2–32, Copyright 2010, with permission from Elsevier).

XPS experiments furnish data about the kinetic energies of the photoelectrons ejected after X-rays strike on the sample. Each chemical element has its own kinetic bond energy, so the photoelectrons ejected from different chemical elements have different energies (Figure 9). Measuring these energies by XPS can therefore help to identify the chemical elements existing in a sample and understand its chemical configuration (Fadley 2009, 2010, Kloprogge 2016).

The incident energy is transported by an X-ray photon (hv) and is absorbed by a certain target atom, which consequently takes the atom to the excited state (higher energy). Relaxation then takes place, and a photoelectron (atom ionization) from the more internal electronic layers of the atom is emitted (Watts and Wolstenholme 2003, Fadley 2009, 2010, Kloprogge 2016). Eq. 1 (Fadley 2009) gives the energy involved in this process.

$$hv = E_{binding}^{Fermi} + \varnothing_{spectrophotomer} + E_{kinetic} \qquad (1)$$

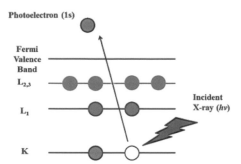

Figure 9. Schematic representation of the photoemission of a 1s electron (adapted from Geochimica et Cosmochimica Acta, 61, Graveling, G.J., K.V. Ragnarsdottir, G.C. Allen, J. Eastman, P.V. Brady, S.D. Balsley and D.R. Skuse. Controls on polyacrylamide adsorption to quartz, kaolinite, and feldspar. 3515–3523, Copyright 1997, with permission from Elsevier).

This electron projection energy (emission of the photoelectron) is the kinetic energy ($E_{kinetic}$), which depends on the incident energy ($h\nu$), the Fermi energy ($E_{binding}^{Fermi}$), and the work function of the spectrophotometer ($\varnothing_{spectrophotomer}$), a factor that corrects the electrostatic medium where the electron is generated and measured (Watts and Wolstenholme 2003, Fadley 2009, 2010).

Figure 10 depicts a typical XPS measurement equipment. Measurements are accomplished in vacuum to control the purity of the surface and to reduce the dispersion of electrons by the gas molecules. To provide a photon beam with certain characteristics, the device is equipped with an X-ray source that is focused on the surface of the sample. The photoelectrons emitted from the sample have a characteristic energy and are measured in an analyzer. The $E_{kinetic}$ at which the electrons are emitted follows Eq. 1, as described previously (Martínez 2012).

Usually, nanohybrids based on clay minerals are characterized by X-ray diffractometry (XRD), infrared (FTIR) and UV/Visible (UV/Vis) spectroscopies, thermal analyses, textural analyses (nitrogen adsorption), and scanning and transmission electron microscopies (SEM and TEM). These techniques aid in the structural elucidation of the material and help to understand how the organic components interact with the inorganic matrixes. Some authors (Da Silva et al. 2016, Watts and Wolstenholme 2003, Fadley 2009, 2010, Kloprogge 2016) have used information obtained by XPS to characterize materials and to gain better understanding of their chemical behavior and of the interaction between the organic and inorganic units in these materials.

Da Silva and coworkers reported new strategies to synthesize and immobilize phthalocyanines on kaolinite in a single step (Da Silva et al. 2016). Their work consisted of heating kaolinite (Ka) previously intercalated with dimethylsulfoxide with one of the precursors of the synthesis of metallophthalocyanines,

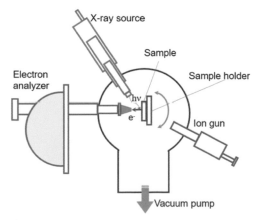

Figure 10. Schematic drawing of an experimental XPS system (©2012 Lidia Martínez, Elisa Román and Roman Nevshupa. Adapted from Martínez et al. 2012; originally published under CC BY 3.0 license. Available from: http://dx.doi.org/10.5772/48101).

namely 4-nitrophtalonitrile ($PCNO_2$), to obtain $KaPCNO_2$. The spectroscopy study in the infrared (IR) and ultraviolet/visible (UV/Vis) regions suggested the formation of the metallophthalocyanine and its immobilization on kaolinite—bands typical of metallophthalocyanine emerged in the spectra, but they were shifted as compared to the bands of free metallophthalocyanine. However, one of the starting materials ($PCNO_2$) showed IR and UV/Vis bands very close to those of the metallophthalocyanine because both compounds contain C-N chemical bonds. Hence, IR and UV/VIS spectroscopies were not enough to confirm the formation and immobilization of metallophthalocyanine on kaolinite. The authors then turned to XPS, which attested that the typical chemical bonds of metallophthalocyanines were actually formed, Figure 11.

Figure 11a shows the N 1s XPS spectra recorded for the precursor $PCNO_2$ and for the hybrid $KaPCNO_2$. For both materials, the signal relative to the energy of the component NO_2 was recorded at 405.5 eV. In the N 1s XPS spectrum of the precursor $PCNO_2$, the signal due to the energy of the N atom belonging to the nitrile group (C≡N) was recorded at 399.2 eV. In the N 1s XPS spectrum of the hybrid material, the energy decreased, and the signal was recorded at 398.7 eV. This decrease can be attributed to the formation of pyrrole bonds (C=N−C) (Da Silva 2016).

These authors also reported the Al 2p XPS spectra of $Ka-PCNO_2$, to understand how kaolinite influenced the formation of metallophthacyanine. The ^{27}Al NMR results had indicated that the octahedral aluminum layer of kaolinite dissolved and released Al(III) to the reaction medium, contributing to the cyclomerization of the metallophthalocyanine. XPS confirmed these findings: the energy signal for Al in kaolinite was recorded at 75.3 eV, a value typical of Al existing in an octahedral layer in kaolinite, while after the reaction, the energy component decreased to 74.2 eV, assigned to aluminum located in the center of the metallophthacyanine (Da Silva 2016).

Schampera and coworkers studied the properties of organophilic clays prepared from natural bentonite and alkylammonium cations such as hexadecylpyridinium, $HDPy^+$, or hexadecyltrimethylammonium, $HDTMA^+$ (Schampera et al. 2016). The materials were prepared by constantly mixing bentonite (with > 90% montmorillonite) suspended in deionized water with HDPy or HDTMA salts at different concentrations. This study was accomplished on the basis of the cation exchange capacity (CEC) of the ions in the clay. XPS analysis was used to monitor the changes in the external surface of the modified clays as a function of the concentration of the organic salts. Table 2 lists some of the XPS data, which helped to determine the percentage of chemical components in the samples.

In this table, when naming the samples, the numbers refer to the amount of organic cations ($HDpy^+$ or $HDTMA^+$) used in the treatments, given as percentage of the CEC of the original clay. The authors studied the amounts of N, C, and Cl in the final materials for the different percentages of organic matter, finding that the atomic percentage of these elements increased similarly for the solids up to 100% CEC (HDPy-32 and 93 and HDTMA-34 and 69). For samples prepared with amounts higher than 100% CEC,

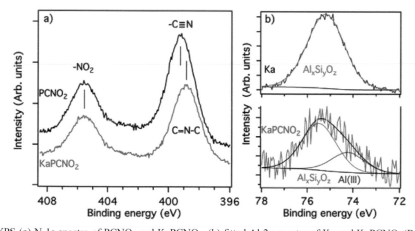

Figure 11. XPS (a) N 1s spectra of $PCNO_2$ and $KaPCNO_2$, (b) fitted Al 2p spectra of Ka and $KaPCNO_2$ (Reprinted from Dyes and Pigments, 134, da Silva, T.H., T.F.M. de Souza, A.O. Ribeiro, P.S. Calefi, K.J. Ciuffi, E.J. Nassar, E.F. Molina, P. Hammer and E.H. de Faria. New strategies for synthesis and immobilization of methalophtalocyanines onto kaolinite: Preparation, characterization and chemical stability evaluation. 41–50, Copyright 2016, with permission from Elsevier).

Table 2. Surface elemental composition of natural clay and organoclays as derived from XPS survey scans (in atom%) (Reprinted from Journal of Colloid and Interface Science, 478, Schampera, B., D. Tunega, R. Šolc, S.K. Woche, R. Mikutta, R. Wirth, S. Dultz and G. Guggenberger. External surface structure of organoclays analyzed by transmission electron microscopy and X-ray photoelectron spectroscopy in combination with molecular dynamics simulations. 188–200, Copyright 2016, with permission from Elsevier).

Element	Clay	HDPy-32	HDPy-93	HDPy-111	HDPy-180	HDTMA-34	HDTMA-69	HDTMA-93	HDTMA-112	HDTMA-150
Na 1s	0.97 ± 0.14	0.10 ± 0.10	0.02 ± 0.04	0.07 ± 0.05	0.03 ± 0.03	0.05 ± 0.07	0.05 ± 005	0.09 ± 0.08	0.06 ± 0.09	0.06 ± 0.06
Ca 2p	0.69 ± 0.09	0.47 ± 0.11	0.27 ± 0.05	0.21 ± 0.06	0.21 ± 0.08	0.35 ± 0.13	0.25 ± 0.11	0.20 ± 0.04	0.22 ± 0.20	0.05 ± 0.07
Fe 2p	0.75 ± 0.13	0.56 ± 0.11	0.40 ± 0.00	0.30 ± 0.19	0.28 ± 0.18	0.57 ± 0.11	0.62 ± 0.25	0.58 ± 0.15	0.47 ± 0.13	0.33 ± 0.21
Mg 2s	0.99 ± 0.15	0.72 ± 0.18	0.44 ± 0.01	0.48 ± 0.17	0.40 ± 0.07	0.69 ± 0.20	0.54 ± 0.15	0.58 ± 0.13	0.46 ± 0.09	0.56 ± 0.17
Al 2p	6.19 ± 0.96	5.24 ± 0.41	4.39 ± 0.42	3.51 ± 0.27	3.13 ± 0.25	5.45 ± 0.32	4.45 ± 0.54	4.18 ± 0.31	4.45 ± 0.27	4.15 ± 027
Si 2p	15.44 ± 2.43	13.12 ± 0.80	10.57 ± 0.39	8.58 ± 0.41	7.01 ± 0.28	12.56 ± 0.61	10.74 ± 1.39	9.84 ± 0.65	9.85 ± 0.25	9.52 ± 0.32
O 1s	57.71 ± 4.14	47.61 ± 1.47	39.03 ± 1.41	28.97 ± 1.41	23.33 ± 0.51	46.77 ± 1.22	39.18 ± 2.94	33.41 ± 1.46	34.12 ± 1.07	33.80 ± 0.91
N 1s	0.08 ± 0.09	0.46 ± 0.18	1.12 ± 0.23	1.87 ± 0.47	2.15 ± 0.17	0.92 ± 0.12	1.33 ± 0.36	2.24 ± 0.40	2.09 ± 0.13	1.80 ± 0.35
C 1s	17.14 ± 7.67	31.63 ± 2.62	43.65 ± 2.35	55.32 ± 1.67	62.48 ± 0.61	32.59 ± 2.02	42.74 ± 5.25	48.39 ± 2.70	47.53 ± 1.31	49.24 ± 1.40
Cl 2p	0.03 ± 0.04	0.08 ± 0.11	0.10 ± 0.05	0.69 ± 0.08	0.99 ± 0.09	0.04 ± 0.05	0.11 ± 0.11	0.49 ± 0.04	0.75 ± 0.10	0.49 ± 0.02
%	100.0	99.99	99.99	100.0	100.0	100.0	100.0	100.0	100.0	100.0

differences in the composition of the materials were observed. The materials organically modified with HDTMA had only slightly altered their elemental composition as a function of the percentage of CEC, while for the materials modified with HDPy, the elemental composition varied as a function of the amount of cations used. The graphical correlation of the percentage of C (organic cations) and O (silicate layer) in the surface composition of the materials is shown in Figure 12.

On the basis of XPS, the authors reported the thickness of the layers of the solids. For HDTMA-containing solids, the thickness of the layer on the surface of the samples prepared with high amounts of cations (HDTMA-93, HDTMA-112, and HDTMA-150) did not increase significantly because the atomic ratio between O and C remained constant. In the case of the samples containing HDPy, the O atomic percentage decreased and the C atomic percentage increased, which altered the thickness of the surface layer of the clay mineral significantly. Thus, different concentrations of the organic salts elicited distinct behavior and generated particles with different surface composition, which led to materials with heterogeneous surfaces.

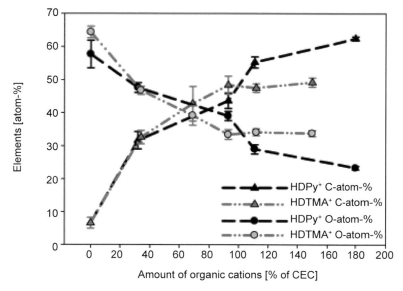

Figure 12. Element content of O and C at the surfaces of the natural clay and of both organoclays when using in the synthesis of different amounts of organic cations, as derived from XPS scans (in atom%) (Reprinted from Journal of Colloid and Interface Science, 478, Schampera, B., D. Tunega, R. Šolc, S.K. Woche, R. Mikutta, R. Wirth, S. Dultz and G. Guggenberger. External surface structure of organoclays analyzed by transmission electron microscopy and X-ray photoelectron spectroscopy in combination with molecular dynamics simulations. 188–200, Copyright 2016, with permission from Elsevier).

Characterization of layered compounds by extended X-ray absorption fine structure (EXAFS)

X-ray Diffraction (XRD) is the most commonly used technique to characterize materials in the areas of Materials Chemistry and Inorganic Chemistry. XRD provides reasonably precise structural information about the crystallinity of compounds. However, in the case of "little crystallites" or amorphous materials, like exfoliated or delaminated clays, XRD can only provide qualitative information about the crystalline structure.

Analysis of X-ray absorption fine structure, known as X-ray absorption spectroscopy (XAS), has emerged as a tool to study the structure of poorly crystalline and amorphous materials. XAS provides information not only about the crystallinity of the materials, but also about their symmetry, interatomic distances, coordination number of the atoms, oxidation states, and bond nature. De Broglie was the first to observe X-ray absorption in 1912; Fricke later observed the same phenomenon. Nevertheless, researchers only attempted to improve the technique in 1975, when synchrotron radiation sources started being developed to replace conventional X-ray tubes, which required long exposure times (Mastelaro 2004). In 1970, Sayers et al. (1970) had formulated the theory that involved the physical and mathematical principles of X-ray absorption and of the two distinct X-ray absorption techniques, namely XANES and EXAFS (Mazali 1998).

X-ray absorption spectroscopy

X-ray absorption spectroscopy (XAS) uses X-ray absorption to excite electrons located at levels closer to the nucleus (K or L shells) of the absorbing atom. The spectra measure the X-ray absorption coefficient, $\mu(E)$, as a function of the incident radiation energy ($E = h\nu$) (Ribeiro et al. 2003).

The X-ray absorption spectrum can be divided into three regions, Figure 11 (Mazali 1998):

Pre-edge region: This region is located between 2 and 10 eV below the absorption edge and refers to the electronic transitions from internal levels (1s, 2s, etc.) to partially occupied or unoccupied external levels with lower absorption energy than the bond energy. The measured absorption intensity can provide information about the oxidation state, bond nature, and symmetry site of the absorbing atom.

XANES region: The region spanning from the absorption edge up to 50 eV above the absorption edge is known as XANES (X-ray Absorption Near-Edge Structure). Its theoretical analysis involves interaction of multiple photoelectron scattering and transitions to unoccupied levels (Figure 14). In this case, the

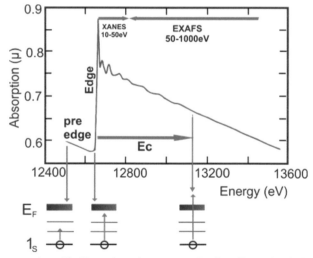

Figure 13. Schematic representation of the X-ray absorption spectrum (K edge of Se) and typical spectral transitions (©1998 Italo Odone Mazali. Adapted from Mazali 1998).

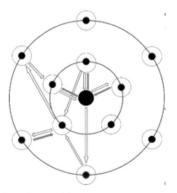

Figure 14. Schematic illustration of the multiple scattering process observed in the XANES region.

scattering centers are located outside the first coordination sphere, which makes them sensitive to changes in the geometrical arrangement of the closest neighboring atoms. This may provide a large amount of information about site symmetry and crystallinity.

EXAFS region: The region spanning from 50 to 1000 eV above the absorption edge is known as EXAFS (Extended X-Ray Absorption Fine Structure). This region presents oscillations originating mainly from simple diffusion processes (Figure 15). The oscillations are milder than those taking place in the XANES region. Only two atoms participate in the process, the absorbing atom and the backscattering atom of the first neighboring atoms. EXAFS analysis furnishes information about the interatomic distance (< 5Å) and the number and type of neighboring atoms located around the absorbing atom.

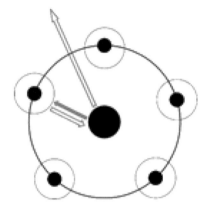

Figure 15. Schematic illustration of the simple scattering process observed in the EXAFS region.

Theoretical basis and EXAFS equation

The physical process involving oscillations in the EXAFS region originates from the interference between the wave emitted by the photoelectron of the absorbing atom and the wave backscattered by the neighboring atoms. During oscillation, peaks correspond to the wave backscattered in phase with the wave emitted by the absorbing atom, to generate a constructive interference. Valleys emerge when the two waves are not in phase, which causes a destructive interference. In the XANES region, photoelectrons propagate with a mean-free path that can generate a multiple diffusion effect. In the EXAFS region, the mean-free path of the photoelectrons is limited, and only the simple scattering effects on a limited number of neighboring atoms contribute to the signal amplitude significantly. If the absorbing atom does not have a close neighbor, the absorption edge of the spectrum will increase abruptly and then decrease smoothly after the edge. However, the absorption coefficient will be modulated when a close neighbor exists (Figure 16).

The $\chi(k)$ function is the sum of individual waves due to the different number of neighboring atoms or the different distances for the same type of neighbors (Ribeiro et al. 2003). Therefore, the general expression that is valid for all the K edge absorption and for systems with random spatial orientation is given by Eq. 2:

$$\chi(k) = -\Sigma \frac{Ni}{kR_i^2} |f_i(\pi, k)| \sin(2kR_i \mid \psi) \cdot e^{-2\sigma_i^2 \cdot k^2} \cdot e^{-\frac{2R_i}{\lambda i(k)}} \tag{2}$$

Equation 2. General expression valid for the K edge absorption and for systems with random spatial orientation.

The meaning of the parameters in this Equation is as follows:

k : vector of the photoelectron wave, given by $k = \sqrt{\frac{2 \cdot m}{\hbar^2} \cdot (E - E_0)}$,

M : mass of the electron,

$\hbar$: Planck constant,

E$_0$: kinetic energy of the electron, which is close to the absorption edge but does not necessarily coincide with it,

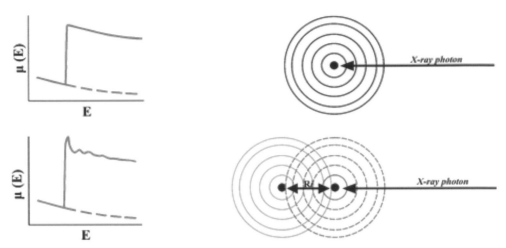

Figure 16. Absorption spectrum of an isolated atom (in blue) and of a diatomic molecule (in red) (©1998 Italo Odone Mazali. Adapted from Mazali 1998).

E   :   energy of the incident photoelectron,

N   :   number of neighboring atoms,

$f(\pi, \mathbf{k})$:   backscattering amplitude,

R   :   distance between the absorbing atom and the backscattering atom,

ψ   :   total delay during the backscattering process given by $\psi = 2.\delta(k) + \theta(k)$, where δ is the delay due to the absorbing atom, and θ is the delay due to the neighboring atom.

This equation includes two damping terms:

$-\dfrac{2R_i}{\lambda_i(k)}$: Factor that considers the limitation of the mean-free path of the photoelectron of the material (λ);

$-2\sigma_i^2 \cdot k^2$: Debye-Waller factor, where σ^2 is the mean quadratic variation in the relative position between the absorbing atom and the backscattering atom.

Multiple scattering in EXAFS spectral analysis

One of the major advances in EXAFS spectral analysis has been the development of the theory that deals with multiple diffusions in spectra (Rehr et al. 1991). This is an important advance with respect to considering only the simple scattering effects, even though scattering by other atoms may occur before the X-ray returns to the absorbing atom.

Such theory was proposed by Sayers (Sayers and Stern 1971, Sayers et al. 1972), who used Fourier transform to analyze the contributions from different atomic layers. Later, the FEFF algorithm was developed (Zabinsky et al. 1995), which enabled the acquisition of theoretical EXAFS detailing all the multiple scattering phenomena.

Criteria for data analysis

The EXAFS signal represents only a small part of the total absorption. This signal consists of the sum of waves generated by the different neighboring atoms and the absorbing atom. The amplitude function can be transferred from a theoretical or experimental model compound to the type of atom in a sample with unknown structure, and its phase function can also be transferred from a model compound to a pair of atoms in which one atom is the absorbing atom and the other atom is the scattering atom.

To analyze the data, it is necessary to extract the signal $\chi(k)$ from the total absorption coefficient by following the steps shown in Figure 17.

To obtain the atomic absorption coefficient (μ_0), the *background* must be subtracted and, after normalization, the absorption edge energy must be defined by using the value determined in the calculation of the wave vector k. Then, the signal $\chi(k)$ can be obtained.

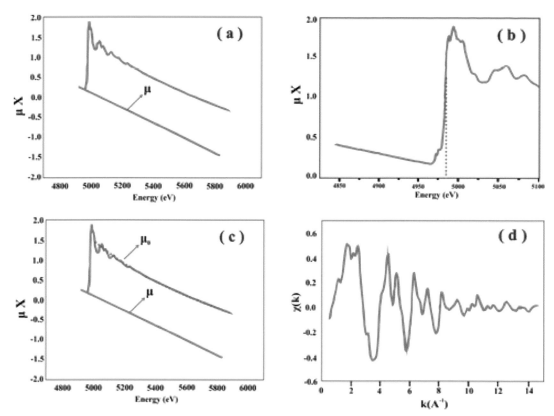

Figure 17. Steps to obtain the EXAFS oscillations from XAS. (a) *Background* subtraction (b) determination of absorption edge (*edge*), (c) determination of atomic absorption coefficient, and (d) EXAFS spectrum after normalization (©2004 Valmor Roberto Mastelaro. Adapted from Mastelaro 2004).

Fourier transform

In EXAFS, the Fourier transform (FT) is used to solve the scattering of each distance in a fine band with Fourier coefficient along the R axis, to result in a radial distribution function defined in the R space (central atom-absorbing atom distance). This includes the "window", W(k), known as the cutoff window that limits the spectral region, Eq. 3:

$$FT(R) = \frac{1}{\sqrt{2\pi}} \int_{k_{min}}^{k_{max}} W(k) k^n \chi(k) e^{i2kR} \, dk \tag{3}$$

FT can be obtained for different k weights (k^n multiplication), which facilitates distinction and highlights the region with higher intensity among the scattering atoms located around the absorbing atom. Because FT is a complex function, a real and an imaginary part are obtained. The real part (absolute) is determined by the number of neighboring atoms and by the degree of disorder, whereas the imaginary part helps to determine the distances in space-R.

EXAFS data collection

Data can be collected by transmission or fluorescence. The most common method is transmission, during which the incident beam on the sample and the beam transmitted by the detectors before and after the sample are monitored. This method is indicated for more concentrated samples because it can be used with fine films. In the case of diluted or extremely fine samples, fluorescence (detection of the fluorescence of the sample or of the secondary electrons generated during the absorption process) is used.

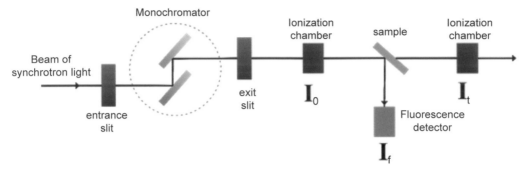

Figure 18. Schematic representation of the EXAFS line for measurement of signal transmission ($\ln(I_0/I_t)$) and for the fluorescence method (I_f/I_0).

Advantages and disadvantages of XAS

XAS studies the local order and aids structural characterization. Disagreement between the theory and the experimental results is small for the XAS technique as a whole, which allows determination of the interatomic distance with a precision of 0.01 Å and of the number of neighboring atoms with exactness close to 15% (Maeda 1987). As other techniques, XAS has advantages and disadvantages (Table 3), and its use will depend on the type of information pursued.

Table 3. Advantages and disadvantages of using the XAS technique.

Information	Advantages	Disadvantages
Crystallinity	XAS allows to study non-crystalline and liquid materials. XAS provides structural information that complements XRD data.	In the case of crystalline materials, the XRD technique is preferred and it is cheaper.
Selectivity	XAS displays high atomic selectivity because the bond energy in the more internal layers is typical of each atom. Each element has unique absorption edge.	Energy-tunable X-ray sources, like synchrotron light sources, are necessary for high selectivity to be achieved.
Structure	XAS provides information about the oxidation state and site symmetry of the absorbing atom. XAS provides highly precise structural information like the interatomic distance. Specific spin states are not necessary.	It is possible to obtain symmetry only on the first coordination sphere, depending on the degree of ordering. It is not possible the distinction of the scattering atoms that differ by two or more units of atomic number (Z).

Applying EXAFS to characterize clay-based nanohybrids

Considering that clay-based hybrid materials comprise an inorganic matrix that interacts with organic ligands, polymers, or organometallic complexes, among others, EXAFS can provide information about the crystallinity, oxidation states, interatomic distances between the central atom and ligands, and symmetry of clay minerals, for example. In other words, EXAFS is an attractive technique to characterize the nanostructure of this class of materials with a view to an array of applications.

Kremleva and coworkers investigated the interactions of uranyl(IV) with groups (especially aluminol and silicon oxides) on the surface of clay minerals such as kaolinite and pyrophyllite (Kremleva et al. 2008). DFT (*Density Functional Theory*) computational studies and EXAFS data of kaolinite and other clay minerals were compared in an attempt to understand the interatomic characteristics, mainly the U-O bond length. The authors aimed to apply the materials as adsorbents of actinides, particularly radioactive elements like uranium, in the context of environmental chemistry. The results allowed separation of the coordination spheres and distinction between the U-O bonds (shorter spaces) and interactions with water (longer spaces). The formal charge on the main oxygen centers on the surface was calculated by computational methods. All the models considered the coordination number of uranium as five, and only data for the more energetically favorable complexes (more stable) were selected besides the surface of the mineral. The formal charge and the U-O interatomic distances correlated almost linearly (Figure 19).

Comparison of these data with those obtained by EXAFS showed that a variety of adsorption complexes coexisted in the various forms of kaolinite and pyrophyllite. This resulted from hydrolysis, which generated OH– ligands in the first coordination sphere of uranyl. These complexes originated mainly in the Al_2O group on the basal surface and on the AlOH– in the interlayer space (Table 4). The EXAFS data agreed with the interpretation of the DFT results, thus confirming the usefulness of the latter technique.

Over the past years, some adsorbents that can potentially remove contaminants like toxic metals from solids and liquids have been extensively investigated. In this context, Vasconcelos and collaborators

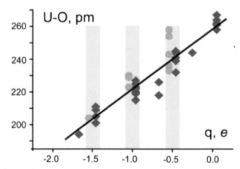

Figure 19. U-O bond length for the surface oxygen centers in uranyl(IV) adsorption complexes as a function of the formal charge of the oxygen centers on the surface of these complexes (Reprinted from Surface Science, 615, Kremleva, A., S. Krüger and N. Rösch. Assigning EXAFS results for uranyl adsorption on minerals via formal charges of bonding oxygen centers. 21–25, Copyright 2008, with permission from Elsevier).

Table 4. EXAFS data for the adsorption of uranyl onto clays: U-O, U-Al, and U-Si distances and coordination numbers (in parentheses). The experimental conditions are also presented (Reprinted from Surface Science, 615, Kremleva, A., S. Krüger and N. Rösch. Assigning EXAFS results for uranyl adsorption on minerals via formal charges of bonding oxygen centers. 21–25, Copyright 2008, with permission from Elsevier).

	Conditions	pH_{PZC}	U-O	U-Al/Si	U-O$_{eq}$ [a]
Silica	pH 3.5, CO_2	4.1	227 (2.2), 252 (3.2)		242 (5.4)
	pH 4.5	4.1	226 (2.5), 251 (2.0)		237 (4.5)
	pH 3.1	4.1	228 (2.1), 246 (3.0)		239 (5.1)
	pH 6.5	4.1	226 (3.5), 248 (1.7)	308 (1.0)	233 (5.2)
γ-Al_2O_3	pH 3.5, CO_2	7.2	237 (2.8), 253 (2.0)		244 (4.8)
	pH 6.5, CO_2	7.2	232 (2.6), 247 (3.1)		240 (5.7)
α-Al_2O_3 ($1\bar{1}02$)	pH 5	8.5	226 (2.0), 243 (3.0)		236 (5.0)
Kaolinite	pH 7.5, Ar	5.5	228 (2.1), 249 (2.9)	~ 330	240 (5.0)
Montmorillonite	pH 6.4, CO_2	2.5	230 (3.0), 248 (2.7)		239 (5.7)
	pH 7.0	2.5	232 (2.8), 248 (2.1)	342 (0.2)	239 (4.9)
	pH 6.6, N_2	2.5	229 (2.1), 247 (2.2)	331 (0.6)	238 (4.2)

[a] Coordination numbers and average equatorial U-O$_{eq}$ bonds estimated using experimental values from a two-shell fitting procedure (fourth column).

studied the surface structure of the Cd(II) complex formed after contact with kaolinite in aqueous medium to evaluate the transport and removal of contaminants (Vasconcelos et al. 2008). To this end, the authors varied the pH and the initial concentration of Cd(II) along a short period and conducted measurements in the transmission and in the fluorescence mode for standard compounds and clay samples, respectively.

A former study had revealed that an increase on the pH of the solution was directly related to an increased adsorption of Cd(II) onto kaolinite. Cd(II) tends to approach the oxygen atom in the hydration water (external sphere) or the complex that contains a mixture of water and oxygen atoms on the surface of the clay (Table 5 and Figure 20). The EXAFS signals cannot distinguish between Al and Si because they have similar backscattering amplitude; therefore, they were assigned as M.

At pH 7, the signal was strongly weighed for adsorbed Cd(II) and not for aqueous Cd(II). Therefore, the authors concluded that the cation adsorbed as a complex of external sphere, and that the hydration sphere did not interact with the surface. At pH 9, adsorption did not depend on the ionic strength, which suggested that Cd(II) adsorbed onto the internal sphere of kaolinite. The results achieved for the samples at pH 8 were not satisfactory. Therefore, pH influenced adsorption directly. Cd(II) adsorbed as a complex in the internal sphere of aluminol at pH 7 even though external sphere complexes interacted only electrostatically. At higher pH, deprotonation of the AlOH and SiOH occurred. At pH 9, Cd(II) adsorption increased.

In the case of intercalation, the reaction might be reversible, but reversibility may not be attractive for some applications. Hence, studies on exchange reactions in the matrix have been intensified in an attempt to make intercalation irreversible. Muñoz-Páez and coworkers submitted layered matrices intercalated with lanthanides to thermal and hydrothermal treatments to avoid reversibility in hybrid materials (Muñoz-Páez et al. 1995). Using EXAFS (Figure 21), it was proposed that lutetium intercalated as an aquocomplex between the montmorillonite layers at the initial treatment temperature, and that this structure was maintained at up to 300°C. At higher temperatures, oxide structures emerged around the lanthanide ion, a process that was complete at approximately 700°C. The reversibility of intercalated lutetium in lutetium-montmorillonite samples without thermal treatment, with thermal treatment in air, and with hydrothermal treatment was

Table 5. EXAFS results for Cd(II) perchlorate and for Cd(II)-kaolinite at different pH (Reprinted from Chemical Geology, 249, Vasconcelos, I.F., E.A. Haack, P.A. Maurice and B.A. Bunker. EXAFS analysis of cadmium(II) adsorption to kaolinite. 237–249, Copyright 2008, with permission from Elsevier).

Shell	N	σ^2 (10^{-3} Å^{-2})	R (Å)	ΔE_0 (eV)	χ_v^2	R
(a) Cd(II) perchlorate						
O	6.0 ± 0.2	8.7 ± 0.5	2.27 ± 0.01	-1.06 ± 0.29	698	0.021
H	$2 \times N_O$	17.6 ± 7.2	2.93 ± 0.04			
(b) Cd(II) + kaolinite − $[Cd_{aq}]_{in}$ 100 μM − pH 7						
O	5.9 ± 0.2	9.0 ± 0.9	2.27 ± 0.01	-1.96 ± 0.38	603	0.015
H	$2 \times N_O$	10.5 ± 6.2	2.91 ± 0.04			
(c) Cd(II) + kaolinite − $[Cd_{aq}]_{in}$ 100 μM − pH 8						
Model 1						
O	4.8 ± 0.3	9.0 ± 1.4	2.26 ± 0.01	-2.40 ± 0.67	3582	0.055
H	$2 \times N_O$	13.5 ± 13.0	2.90 ± 0.07			
Model 2						
O	5.1 ± 0.3	10.1 ± 1.3	2.27 ± 0.01	-1.74 ± 0.53	3500	0.053
M[a]	1 (fixed)	15.8 ± 12.5	3.32 ± 0.08			
(d) Cd(II) + kaolinite − $[Cd_{aq}]_{in}$ 100 μM − pH 9						
O	4.9 ± 0.3	10.1 ± 1.0	2.26 ± 0.01	-2.14 ± 0.54	1614	0.038
M[a]	1 (fixed)	12.3 ± 5.8	3.34 ± 0.04			

χ_v^2 and R restricted to the range R = 2.2 Å to R = 3.2 Å. [a] Stands for either Al or Si.

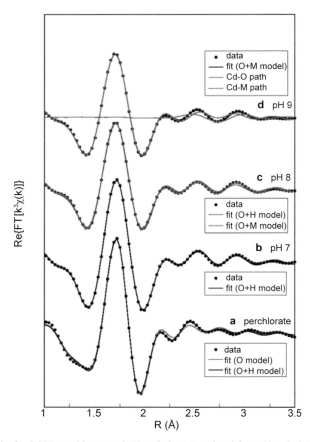

Figure 20. EXAFS results for Cd(II) perchlorate and pH variations (Reprinted from Chemical Geology, 249, Vasconcelos, I.F., E.A. Haack, P.A. Maurice and B.A. Bunker. EXAFS analysis of cadmium(II) adsorption to kaolinite. 237–249, Copyright 2008, with permission from Elsevier).

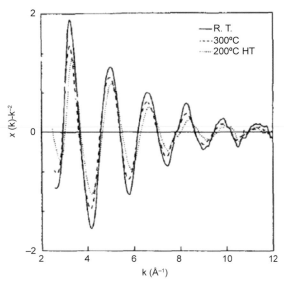

Figure 21. EXAFS (function k^2) of untreated lutetium-montmorillonite sample (R.T.) and of lutetium-montmorillonite sample submitted to thermal treatment in air (300°C) and hydrothermal treatment (200°C HT) (Reprinted from Nuclear Instruments and Methods in Physics Research Section B: Beam Interactions with Materials and Atoms, 97, Muñoz-Páez, A., M.D. Alba, R. Alvero, A.I. Becerro, M.A. Castro and J.M. Trillo. 142–144, Copyright 1995, with permission from Elsevier).

also analyzed, in an important study for future applications of these materials in the adsorption of nuclear residues.

The EXAFS data complemented the XRD data of the samples. The XRD profile of the sample submitted to hydrothermal treatment at 200°C did not contain any peak that could be attributed to phases containing lutetium ions. The EXAFS spectrum clearly showed a drastic change for the sample treated at 300°C—the signal did not shift, but its amplitude diminished. The authors proposed that the sample submitted to hydrothermal treatment was more complex and displayed additional backscattering than the other samples. To prove this effect, the authors applied Fourier transform ($\Delta k = 3\text{–}12$ Å^{-1}) to the results in Figure 21, to show that the main peak of the hydrothermally treated sample also decreased (Figure 22). This peak, which had a single backscattering and was shifted to shorter distances, gave a coordination number of 4.5. This value was considered small given that lutetium with coordination numbers ranging from 6 to 8 is more stable.

The conditions found in the lutetium-montmorillonite sample submitted to hydrothermal treatment were close to the conditions expected for nuclear residue deposits of other studied clay minerals like bentonite. Hydrothermally treated clay minerals such as bentonite, smectite, and kaolinite, among others, have been studied. Shirai and coworkers reported on many divalent cations, including cobalt, intercalated between the octahedral layers of smectite and submitted to hydrothermal treatment, using EXAFS to evaluate how pH affected the structure of the pores of smectite containing Co(II) (Shirai et al. 2001) (Figure 23).

Smectite calcined at temperatures higher than 1000°C had a large specific surface area and considerable pore volume. Variation in the pH could aid to control the pore structure. The Fourier transform of the EXAFS spectrum of the K edge of the Co(II)-smectite sample showed two peaks at the same distance. These peaks were assigned to Co-O (0.1 and 0.2 nm) and to Co-O-Co and Co-O-Si (0.2 and 0.3 nm). The peaks were more intense for the sample calcined at 800°C as compared to those for the sample calcined at 700°C. On the basis of these results, it was concluded that the size of the silicate fragments in the smectite samples was larger at higher temperature, and that the pores emerged between the layers.

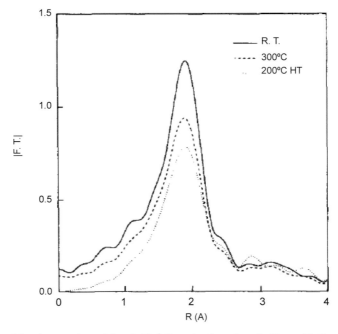

Figure 22. Application of Fourier transform ($\Delta k = 3\text{–}12$ Å^{-1}) to the data given in Figure 21 (Reprinted from Nuclear Instruments and Methods in Physics Research Section B: Beam Interactions with Materials and Atoms, 97, Muñoz-Páez, A., M.D. Alba, R. Alvero, A.I. Becerro, M.A. Castro and J.M. Trillo. 142–144, Copyright 1995, with permission from Elsevier).

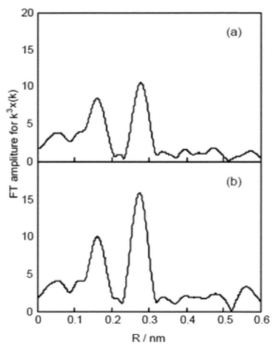

Figure 23. Fourier transform of the EXAFS oscillations in the K edge region of Co(II) for the Co(II)-smectite samples hydrothermally treated at (a) 700°C for 2 h and (b) 800°C for 4 h (from Shirai, M., K. Aoki, K. Torii and M. Arai. *In situ* EXAFS study on the formation of smectite-type clays containing cobalt cations in lattice. Journal of Synchrotron Radiation, 8: 743–745, 2001. Reproduced with permission of the International Union of Crystallography).

Final thoughts

Based on the examples presented above, the potential of non-conventional techniques such as SAXS, XAS, EXAFS and XPS to evaluate and to complete the information for a better knowledge of nanoclay hybrid materials and nanocomposites is fully demonstrated. Very significant and important information can be extracted from each of these characterization techniques, although only some possibilities were presented here. The use of these techniques provides outstanding information about nanostructure and usually the information is acquired very quickly compared to conventional techniques, with higher quality and better resolution. In addition, combining these techniques will improve the knowledge about the structure and specific properties of nanoclay materials. Of course, new developments are on course to increase the quality of the data and to help in understanding the information extracted from these non-conventional techniques to characterize nanoclays. Probably, one of their main drawbacks is the enormous cost of the equipments, in many cases impossible to be afforded by a single laboratory, thus leading to enhancing cooperation.

Acknowledgements

This work was carried out in the frame of a Spain-Brazil Interuniversity Cooperation Grant, financed by MEC (PHBP14/00003) and CAPES (317/15). Spanish authors thank additional financial support from Ministerio de Economía y Competitividad (MAT2013-47811-C2-R). Brazilian authors thank the Brazilian Synchrotron Light Laboratory (LNLS) for providing beamtime at the D01ASAXS2 beamline and for assistance with the X-ray scattering experiments (Projects: SAXS 1/16981 and SAXS 1/17845) financial support from FAPESP 2013/19523-3, CAPES and CNPq.

References

Akkal, R., N. Cohaut, M. Khodja, T. Ahmed-Zaid and F. Bergaya. 2013. Rheo-SAXS investigation of organoclay water in oil emulsions. Colloids Surfaces A Physicochem. Eng. Asp. 436: 751–762.

Da Silva, T.H., T.F.M. De Souza, A. Orzari, E. Ferreira, P. Sergio, K.J. Ciuffi et al. 2016. New strategies for synthesis and immobilization of methalophtalocyanines onto kaolinite: Preparation, characterization and chemical stability evaluation. Dyes Pigments 134: 41–50.

Fadley, C.S. 2009. X-ray photoelectron spectroscopy: From origins to future directions. Nucl. Instrum. Meth. A 601: 8–31.

Fadley, C.S. 2010. X-ray photoelectron spectroscopy: Progress and perspectives. J. Electron Spectros. Relat. Phenom. 178-179: 2–32.

Graveling, G.J., K.V. Ragnarsdottir, G.C. Allen, J. Eastman, P.V. Brady, S.D. Balsley et al. 1997. Controls on polyacrylamide adsorption to quartz, kaolinite, and feldspar. Geochim. Cosmochim. Ac. 61: 3515–3523.

Hansen, E.L., H. Hemmen, D.M. Fonseca, C. Coutant, K.D. Knudsen, T.S. Plivelic et al. 2012. Swelling transition of a clay induced by heating. Sci. Rep. 2: 1–2.

Hernandez, M., B. Sixou, J. Duchet and H. Sautereau. 2007. The effect of dispersion state on PMMA-epoxy-clay ternary blends: *In situ* study and final morphologies. Polymer 48: 4075–4086.

Kaneko, M.L.Q.A., Í.L. Torriani and I.V.P. Yoshida. 2007. Morphological evaluation of silicone/clay slurries by small-angle/wide-angle X-ray scattering. J. Braz. Chem. Soc. 18: 765–773.

Kloprogge, J.T. 2016. Characterisation of Halloysite by Spectroscopy. Elsevier.

Kremleva, A., S. Krüger and N. Rösch. 2008. Assigning EXAFS results for uranyl adsorption on minerals via formal charges of bonding oxygen centers. Surf. Sci. 615: 21–25.

Laity, P.R., K. Asare-Addo, F. Sweeney, E. Šupuk and B.R. Conway. 2015. Using small-angle X-ray scattering to investigate the compaction behaviour of a granulated clay. Appl. Clay Sci. 108: 149–164.

Maeda, H. 1987. Accurate bond length determination by EXAFS method. J. Phys. Soc. Jpn. 56: 2777–2787.

Marçal, L., E.H. De Faria, E.J. Nassar, R. Trujillano, N. Martín, M.A. Vicente et al. 2015. Organically modified saponites: SAXS study of swelling and application in caffeine removal. ACS Appl. Mater. Interfaces 7: 10853–10862.

Martínez, L., E. Román and R. Nevshupa. 2012. X-ray photoelectron spectroscopy for characterization of engineered elastomer surfaces. pp. 165–194. *In*: M.A. Farrukh (ed.). Advanced Aspects of Spectroscopy. InTech Open Science.

Mastelaro, V.R. 2004. A Espectroscopia de Absorção de Raios-X Aplicada ao Estudo da Estrutura Atômica de Materiais Inorgânicos. Ph.D. Thesis. Universidade de São Paulo, Brazil.

Mazali, I.O. 1998. EXAFS Como Técnica de Caracterização Estrutural de Materiais: Fundamentos Teóricos e Aplicações. Universidade Estadual de Campinas, Brazil.

Momani, B., M. Sen, M. Endoh, X. Wang, T. Koga and H.H. Winter. 2016. Temperature dependent intercalation and self-exfoliation of clay/polymer nanocomposite. Polymer 93: 204–212.

Motokawa, R., H. Endo, S. Yokoyama, S. Nishitsuji, T. Kobayashi, S. Suzuki et al. 2014. Collective structural changes in vermiculite clay suspensions induced by cesium ions. Sci. Rep. 4: 6585.

Muñoz-Páez, A., M.D. Alba, R. Alvero, A.I. Becerro, M.A. Castro and J.M. Trillo. 1995. EXAFS study of the interaction of lanthanide cations with layered clays upon hydrothermal treatments. Nucl. Instrum. Meth. B 97: 142–144.

Negrete, N.H., J. Letoffe, J. Putaux, L. David, E. Bourgeat-Lami, C.B. Lyon et al. 2004. Aqueous dispersions of silane-functionalized laponite clay platelets. A first step toward the elaboration of water-based polymer/clay nanocomposites. Langmuir 20: 1564–1571.

O'Brien, M.G., A.M. Beale and B.M. Weckhuysen. 2010. The role of synchrotron radiation in examining the self-assembly of crystalline nanoporous framework materials: from zeolites and aluminophosphates to metal organic hybrids. Chem. Soc. Rev. 39: 4767–4782.

Okada, T., J. Oguchi, K.I. Yamamoto, T. Shiono, M. Fujita and T. Iiyama. 2015. Organoclays in water cause expansion that facilitates caffeine adsorption. Langmuir 31: 180–187.

Oliveira, C.L.P. 2011. Investigating macromolecular complexes in solution by small angle X-Ray scattering. pp. 367–392. *In*: A. Chandrasekaran (ed.). Current Trends in X-Ray Crystallography. InTech Open Science.

Rehr, J.J., J.M. Leon, S.I. Zabinsky and R.C. Albers. 1991. Theoretical X-ray absorption fine structure standards. J. Am. Chem. Soc. 113: 5135–5140.

Ribeiro, E.S., S.P. Francisco, Y. Gushikem and J.E. Gonçalves. 2003. Princípio Básico de XAS e XPS. Universidade Estadual de Campinas, Brazil.

Sandí, G., R.E. Winans, S. Seifert and K.A. Carrado. 2002. *In situ* SAXS studies of the structural changes of sepiolite clay and sepiolite-carbon composites with temperature. Chem. Mater. 14: 739–742.

Sankar, G., A.J. Dent, B. Dobson and W. Bras. 2016. Influence of dopant metal ions on the formation of cordierite using combined SAXS/WAXS and EXAFS/WAXS techniques. J. Non-Cryst. Solids 451: 16–22.

Sayers, D.E., E.A. Stern and F.W. Lytle. 1970. Point scattering theory of X-ray absorption fine structure. Adv. X-Ray Anal. 13: 248–256.

Sayers, D.E. and E.A. Stern. 1971. New technique for investigating noncrystalline structures: Fourier analysis of the extended X-Ray—Absorption fine structure. Phys. Rev. Lett. 27: 1204–1207.

Sayers, D., F. Lytle and E. Stern. 1972. Structure determination of amorphous Ge, GeO_2 and GeSe by Fourier analysis of extended X-ray absorption fine structure (EXAFS). J. Non. Cryst. Solids 8-10: 401–407.

Schampera, B., D. Tunega, R. Šolc, S.K. Woche, R. Mikutta, R. Wirth et al. 2016. External surface structure of organoclays analyzed by transmission electron microscopy and X-ray photoelectron spectroscopy in combination with molecular dynamics simulations. J. Colloid Interf. Sci. 478: 188–200.

Segad, M., S. Hanski, U. Olsson, J. Ruokolainen, T. Åkesson and B. Jönsson. 2012. Microstructural and swelling properties of Ca and Na Montmorillonite: (*In situ*) observations with Cryo-TEM and SAXS. J. Phys. Chem. C 116: 7596–7601.

Shang, C. and J.A. Rice. 2001. Interpretation of small-angle x-ray scattering data from dilute montmorillonite suspensions using a modified Guinier approximation. Phys. Rev. E. Stat. Nonlin. Soft Matter Phys. 64: 21401.

Shang, C., J.A. Rice and J.-S. Lin. 2001. Thickness and surface characteristics of colloidal 2:1 aluminosilicates using an indirect Fourier transform of small-angle X-ray scattering data. Clays Clay Miner. 49: 277–285.

Shang, C. and J.A. Rice. 2005. Invalidity of deriving interparticle distance in clay-water systems using the experimental structure factor maximum obtained by small-angle scattering. J. Colloid Interf. Sci. 283: 94–101.

Shirai, M., K. Aoki, K. Torii and M. Arai. 2001. *In situ* EXAFS study on the formation of smectite-type clays containing cobalt cations in lattice. J. Synchrotron Radiat. 8: 743–745.

Silva, A.A., B.G. Soares and K. Dahmouche. 2016. Organoclay-epoxy nanocomposites modified with polyacrylates: The effect of the clay mineral dispersion method. Appl. Clay Sci. 124-125: 46–53.

Takeno, H. and C. Sato. 2016. Effects of molecular mass of polymer and composition on the compressive properties of hydrogels composed of Laponite and sodium polyacrylate. Appl. Clay Sci. 123: 141–147.

Thuresson, A., M. Segad, M. Turesson and M. Skepö. 2016. Flocculated Laponite-PEG/PEO dispersions with monovalent salt, a SAXS and simulation study. J. Colloid Interf. Sci. 466: 330–342.

Tian, X., B. Vestergaard, M. Thorolfsson, Z. Yang, H.B. Rasmussen and A.E. Langkilde. 2015. In-depth analysis of subclass-specific conformational preferences of IgG antibodies. IUCrJ 2: 9–18.

Vargas, M.A., R. Montiel and H. Vázquez. 2016. Effect of ammonium and aminosilane montmorillonites organo-clays on the curing kinetics of unsaturated polyester (UP) resin nanocomposites. Thermochim. Acta 630: 21–30.

Vasconcelos, I.F., E.A. Haack, P.A. Maurice and B.A. Bunker. 2008. EXAFS analysis of cadmium(II) adsorption to kaolinite. Chem. Geol. 249: 237–249.

Watts, J.F. and J. Wolstenholme. 2003. An Introduction to Surface Analysis by XPS and AES. John Wiley & Sons.

Zabinsky, S.I., J.J. Rehr, A. Ankudinov, R.C. Albers and M.J. Eller. 1995. Multiple-scattering calculations of X-ray-absorption spectra. Phys. Rev. B 52: 2995–3016.

Zhang, L-M., C. Jahns, B.S. Hsiao and B. Chu. 2003. Synchrotron SAXS/WAXD and rheological studies of clay suspensions in silicone fluid. J. Colloid Interf. Sci. 266: 339–345.

8

Nanostructured Composite Materials for CO_2 Activation

Anurag Kumar,[1,2] *Pawan Kumar*[1,2] *and Suman L. Jain*[1,]*

Introduction

Due to tremendous exploitation of natural fuel reserves to fulfil our energy demands, the concentration of green house gases has increased to manifold in the last few decades. A plethora of frequently released reports on climate change gives clear evidence of harmful impact on the environment due to anthropogenic emissions of greenhouse gases. The most easily accessible and widely exploited source of energy is hydrocarbon having high energy density (33 GJ/m^3 for gasoline) and is derived from fossil fuel (see Brand and Blok 2015, Johnson et al. 2007, Peura and Hyttinin 2011, Giesekam et al. 2014). Approximately 81% of energy comes from burning of fossil fuel while renewable source accounts for only 13% of the total energy produced. Among various green house gases, carbon dioxide (CO_2) has the most prominent effect on environment because it is a major component of emission (see Manne and Richels 2001, Panwar et al. 2011). The carbon dioxide (CO_2) concentration in the atmosphere has been rising steadily since the beginning of the industrial revolution. As a result, the current CO_2 level is the highest in at least the past 8,00,000 years (see Rehan and Nehdi 2005, Cempbell et al. 2008, Zachos et al. 2008, Wang et al. 2011). Due to this, the global earth temperature and sea level are continuously rising along with depletion of fossil fuel reserves. Various techniques for capturing and storing of CO_2 have been developed like in underground abandoned oil wells, storage under sea water, etc. (see Dincer 1999, Bachu 2008, Gouedard et al. 2009, Markewitz et al. 2012). But sudden spilling and acidification of sea water can deteriorate natural flora and fauna. Thus, conversion of CO_2 to high value chemicals is a promising option and in this regard, valuable products like polycarbonates, polyols esters, etc. have been synthesized from carbon dioxide (see Zhou et al. 2008, Aresta et al. 2014, Lanzafame et al. 2014, Olah et al. 2009). However, CO_2 is highly thermodynamically stable molecules and, therefore, its further conversion to value added chemical is energy intensive and requires heat and catalyst which add extra cost and makes process less viable for the realization of technology. Sunlight is an inexhaustible source of energy and can be used

[1] Chemical Sciences Division, CSIR-Indian Institute of Petroleum, Dehradun-248005, India.
[2] Academy of Scientific and Industrial Research (AcSIR), New Delhi-110001, India.
  Emails: anukmnbd@gmail.com; choudhary.2486pawan@yahoo.in
* Corresponding author: suman@iip.res.in

for the production of solar fuel (see Momirlan and Veziroglu 2005, Barber 2009). The mean solar light irradiance at normal incidence outside the atmosphere is 1360 W/m², and the total annual incidence of solar energy in India alone is about 107 kW, and for the southern region, the daily average is about 0.4 kW/m². If we are able to harvest 10% of sunlight falling on 0.3% of earth's land surface, it will be sufficient to meet our energy demand by 2050 (see Cook et al. 2010, Jacobson et al. 2013, Powell and Lenton 2012). Hydrogen has been considered an ideal fuel with high calorific value and water is the only by-product of its combustion. But due to high mass to volume ratio storage, transfer of hydrogen is difficult (see Felderhoff et al. 2007, Atabani et al. 2012). Till date, most of the industrial hydrogen is produced by methane reforming of natural gas (see Barelli et al. 2008). Solar water splitting for hydrogen production can be a sustainable route for the production of hydrogen by the cheapest, widely available source water (see Krol et al. 2008, Peharz et al. 2007). However, the problem of storage still exists. Many adsorbing agents like hydrides, MOFs, etc. have been used for the storage of hydrogen via temporary bond formation but desorption from the surface requires high temperature that is disadvantageous in the viewpoint of energy. Chemical storage in the lower hydrocarbons or oxygenated hydrocarbons can solve the problem of hydrogen storage (see Davda et al. 2005, Dresselhaus and Thomas 2001). Photocatalytic reduction of CO_2 using water as a source of hydrogen to hydrocarbons is the most promising approach for levelling the concentration of greenhouse CO_2 along with the storage of hydrogen in the form of fuel (see Du et al. 2009, Roy et al. 2010). In this process, water splitting generates electrons and protons which are utilized for the reduction of CO_2 to hydrocarbons. This process is similar to photosynthesis where higher products are produced from water and CO_2, so this process is called artificial photosynthesis.

Since the discovery of photocatalytic water splitting over TiO_2 by Fujishima and Honda in 1972, a great deal of research has been focused on semiconductor photocatalysis (see Fujishima and Honda 1972). However, for the first time, Halmann (1978) reported that CO_2 can also be reduced electrochemically over GaP electrode by using UV light to various lower hydrocarbons, i.e., formic acid, methanol, methane, formaldehyde, etc. (Halmann 1978). After that, Inoue et al. (1979) used various semiconductors like WO_3, TiO_2, ZnO, CdS, etc. for CO_2 reduction by using Xe lamp. Most of the identified products were C1 products (Inoue et al. 1979). These initial findings have opened path for using various semiconductor materials like ZnS, CdS, Cu_2O, ZnO, Fe_2O_3, WO_3, etc. for efficient reduction of CO_2 to valuable products (see Mao et al. 2013, Li et al. 2014, Navalon et al. 2013, Neațu et al. 2014). It has been observed that semiconductors mainly produce formic acid, CO and methane. Gaseous products are less desirable because of the problem associated with their storage. Among the all liquid products formed, methanol has been identified as the best liquid C1 oxygenates because of high calorific value, higher octane number and suitability to internal combustion engines (see Ganesh 2014).

Basic principle of CO_2 reduction

The reduction of CO_2 is highly unfavourable because of its linear geometry and closed shell structure which makes it highly stable (Heat of formation $\Delta H°$ gas = –393.5 KJ/mol). One electron reduction of CO_2 to $CO_2^{-\circ}$ radical is highly unfavorable because of generation of bent structure in which acquired additional electron imposes repulsive force on lone pair of electrons at oxygen atoms. This arrangement makes the one electron reduced species highly unstable which requires high reduction potential (–1.90 V vs. NHE) (see Zhang et al. 2004, Balcerski et al. 2007, Wang et al. 2009a). Although proton-assisted multiple-electron transfers (MET) is a much easier way to reduce CO_2 at lower reduction potential than one electron reduction (see Tahir and Amin 2013, Xiang et al. 2011, Liu et al. 2010a). It can be seen from Eqs. 1–7 that proton assisted reduction can be achieved at lower reduction potential. Further for getting hydrocarbons with higher C:H ratio, transfer of multiple electrons and protons is essential. The required electrons and protons can be obtained from water oxidation or water splitting (H_2O/O_2 (+0.82 V vs. NHE at pH 7).

$$CO_2 + 2H^+ + 2e^- \rightarrow CO + H_2O \qquad E^0 = -0.53 \text{ V} \ldots\ldots\ldots\ldots \tag{1}$$

$$CO_2 + 2H^+ + 2e^- \rightarrow HCOOH \qquad E^0 = -0.61 \text{ V} \ldots\ldots\ldots\ldots \tag{2}$$

$$CO_2 + 4H^+ + 4e^- \rightarrow HCHO + H_2O \qquad E_0 = -0.48 \text{ V} \ldots\ldots\ldots\ldots \tag{3}$$

$$CO_2 + 6H^+ + 6e^- \rightarrow CH_3OH + H_2O \qquad E^0 = -0.38 \text{ V} \ldots\ldots\ldots\ldots \qquad (4)$$

$$CO_2 + 8H^+ + 8e^- \rightarrow CH_4 + 2H_2O \qquad E^0 = -0.24 \text{ V} \ldots\ldots\ldots\ldots \qquad (5)$$

$$CO_2 + e^- \rightarrow CO_2^{-\circ} \qquad E^0 = -1.90 \text{ V} \ldots\ldots\ldots\ldots \qquad (6)$$

$$2H^+ + 2e^- \rightarrow H_2 \qquad E^0 = -0.41 \text{ V} \ldots\ldots\ldots\ldots \qquad (7)$$

The reduction potential of CO_2/CH_3OH is –0.38 V (at pH 7 vs. NHE), while for reducing proton to hydrogen, the value is 0.00 V at pH-0. However, –0.41 V over potential is needed for the production of hydrogen in aqueous solutions at pH 7 vs. NHE (H^+/H_2 (–0.41 V vs. NHE at pH-7). Because the values of reduction potential for proton reduction and CO_2 reduction are almost similar, so the photocatalyst that can reduce CO_2 can also reduce protons (see Izumi 2013, Sato et al. 2015). Therefore, hydrogen is always observed as a by-product which makes the process less efficient because protons compete for the electrons.

It is noteworthy to mention here that the reduction potential changes by changing pH value according to Nernst equation (Eq. 8).

$$E^0 \text{ (pH)} = E^0 \text{ (pH0)} - 0.06\text{pH} \ldots\ldots\ldots\ldots\ldots \qquad (8)$$

Semiconductor materials work as photocatalyst due to the presence of band gap. The band gap of semiconductors is determined by the energy difference between hybridized system of HOMO (highest occupied molecular orbital) and LUMO (lowest unoccupied molecular orbital) of the material (see Wehling et al. 2008, Ajayaghosh 2003). In semiconductors, after absorption of light of appropriate wavelength, electrons get excited from valance band to conduction band which creates positively charged vacancy in valance band (holes) (see Chestnoy et al. 1986). Electrons can reduce any species while holes are responsible for the oxidation (see Dukovic et al. 2004).

For the efficient water splitting or CO_2 reduction, the band gap of semiconductor should be higher than 1.23 V. In other words, the position of valance band should be more positive than +0.82 V vs. NHE at pH-7, so it can oxidize water while the position of conduction band should be more negative than –0.41 V vs. NHE at pH-7, so it can reduce CO_2 or protons (see Karamian and Sharifnia 2016). Very few semiconductors meet this requirement of suitable band position and most of the semiconductors have wide band gaps and absorb in the UV region. However, the solar spectrum consists of only about 4% of UV light (see Nagaveni et al. 2004, Kumar and Devi 2011, Hashimoto et al. 2005), but of about 45% visible light, so the basic need is to develop photocatalyst that can absorb in visible region. Further, charge recombination is a prominent phenomenon in the semiconductors which is also responsible for lowering the photo-efficiency (see Hochbaum and Yang 2010, Yu et al. 2010). Only less than 10% of produced electron and protons are available for the water splitting or CO_2 reduction process. The recombination may be of two types, i.e., the volume recombination and surface recombination (see Linsebigler et al. 1995). Volume recombination can be prevented by reducing the particles size of semiconductor so more electrons and holes can reach at the surface while surface recombination can be stopped by electron and hole capturing agents like doping with metals. Doping with various metals like Cu, Ag, Au, Pt, Ru etc. and non-metals like C, N, S, I, etc. can be a viable approach for reducing the fast electron hole recombination along with band gap modification (see Zhang et al. 2013, Kumar and Devi 2011). Doped metal has Fermi level below the conduction band of semiconductor so it can accept electrons efficiently. Due to this effect, the Fermi level of semiconductor shifts slightly upward and semiconductor becomes more reductive in nature. Back transfer of electron may reduce catalytic performance, so reducing the size of semiconductor material can slow down the process of back electron transfer. Some metal oxides like IrO_2, CoO_x can capture holes (see Zhong et al. 2011, Li et al. 2013a, Wang et al. 2015b). Non-metal dopants like B, C, N, S, etc. have 2p orbital with higher energy than O 2p orbital in TiO_2, so hybridization of these orbital generates a new hybridized orbital with higher energy and the valance band position shifts to higher energy (less positive). This resulted to the lowering of band-gap and therefore material can absorb in the visible region. Wu et al. (2011) reported that the band gap of N doped and N–B co-doped TiO_2 was 2.16 eV and 2.13 eV, respectively (Zhang et al. 2011a), which is much smaller than that of pure TiO_2 (3.18 eV for anatase) (see Serpone et al. 1995). Further, reduction of semiconductor in hydrogen at higher temperature creates vacancy due to abstraction of oxygen atom which creates colour centre and material can absorb in the visible region (see Varghese et al. 2009, Wang et al. 2010). Apart from electron and hole capturing agents, sacrificial donors can be added for getting higher yield of CO_2 reduction products. Tertiary amines like triethylamine, triethanolamine, BNAH and NaOH, Na_2SO_3, Na_2S, EDTA, etc. have been used as sacrificial agents.

Most of the semiconductors, due to the inappropriate position of their band edge position, can initiate either water splitting or CO_2 reduction only. Very few semiconductors like TiO_2 ZnS, ZnO, etc. meet the demand of suitable band edge position but their large band gap limits their application (see Pradhan and Sharma 2011, Choudhary et al. 2012). This limitation can be overcome by synthesizing heterojunctions by mixing of low band gap semiconductors with high band gap semiconductors in which one can reduce CO_2 and other can oxidize water. This addition creates a p-n heterojunction and photogenerated electrons and holes can move from low band gap semiconductor to high band gap semiconductor. Low band gap semiconductor can transfer electron to the high band gap semiconductor via two mechanisms in which the first mechanism is direct and the second is Z-scheme depending on band edge position of high band gap semiconductors (see Yang et al. 2012, Qu and Duan 2013, Wang et al. 2013a). In direct scheme, low band gap semiconductor, after absorption of visible light, transfer electrons and holes to high band gap semiconductor and indirectly produce charge separated species at high band gap semiconductor. While in Z scheme, a semiconductor which can only oxidize water gets excited by absorption of photon, so the generated electrons move to its conduction band while holes are used for oxidation of water (see Takanabe and Domen 2012). However, due to lower position of conduction band, they cannot reduce CO_2. So electrons in conduction band are transferred to valance band of another semiconductor which has

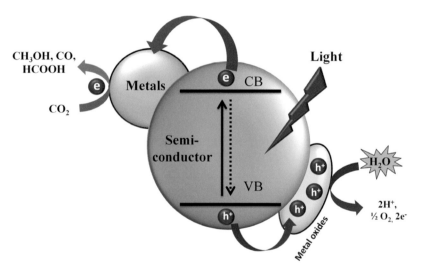

Figure 1. Mechanism of electrons and holes capture by metal and metal oxides doped semiconductor.

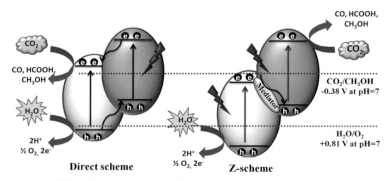

Figure 2. Schematics of CO_2 reduction and water oxidation over direct scheme and Z-scheme photocatalytic systems.

higher position of valance band (see Wang et al. 2015a). Generally, in electron mediator redox system like IO_3^-/I^-, Fe^{3+}/Fe^{2+}, $[Co(bpy)_3]^{3+/2+}$, etc. redox couples are required in Z scheme photocatalysis which play an important role in transfer of electrons from water oxidation photocatalyst to CO_2 reduction catalyst. For example, Z-scheme photosystem composed of $Pt/SrTiO_3$:Rh as hydrogen evolving catalyst and $BiVO_4$ as oxygen evolving catalyst was visible light responsive in the presence of Fe^{3+}/Fe^{2+} redox couple and can operate under 520 nm wavelength, which corresponds to the band gap of $SrTiO_3$:Rh and $BiVO_4$ (see Konta et al. 2004). In some cases, nanoparticles and conductive graphene sheets have been reported to work as electron mediator which facilitates better charge transfer (see Brownson et al. 2011, Yang et al. 2014). For example, in their work, He et al. (2015) used Ag_3PO_4/g-C_3N_4 composites for CO_2 reduction which works as Z-scheme photocatalyst and *in situ* generated Ag nanoparticles that work as electron facilitator (He et al. 2015). In this system, Ag_3PO_4 transfer electron to the valance band of carbon nitride via exited state and carbon nitride, after absorption of photon, transfer electrons to conduction band which are used for the CO_2 reduction while holes in valance band of $AgPO_4$ are used for water oxidation. The optimal Ag_3PO_4/g-C_3N_4 photocatalyst gave CO_2 reduction rate of 57.5 μmol h^{-1} gcat^{-1}, which was 6.1 and 10.4 times higher than g-C_3N_4 and P25, respectively. Another work by Iwase et al. (2011) used $BiVO_4$ and $Ru/SrTiO_3$:Rh Z scheme photocatalyst in which photoreduced graphene oxide (PRGO) work as solid electron mediator (Iwase et al. 2011).

Oxides based nanostructured composite for CO_2 reduction

Various semiconductors and their composites are extensively studied for the photocatalytic reduction of CO_2. Among them, titanium oxide (TiO_2) is the most widely used because of its abundance, nontoxic nature, wide band gap, non corrosive nature and higher activity (see Takayama et al. 2017, Qu and Duan 2013). TiO_2 is found in two main forms in which the first one, anatase, has tetragonal geometry where Ti^{+4} ions are surrounded by O^{-2} ions in octahedral manner and the second one is rutile form having slightly orthorhombic distortion (see Izumi 2013). Owing to this distortion, TiO_2 has different band gap in both anatase and rutile form. The band gap of anatase TiO_2 is 3.2 eV associated with 389 nm wavelength while the band gap of rutile form is 3.0 eV associated with 413 nm wavelength. In natural P25 TiO_2, both anatase and rutile forms exist in the ratio of 80:20 with particle size 25 nm and 56 $m^2 g^{-1}$ BET surface area. Rutile form is more stable at higher temperature and annealing at temperature above 400°C promotes the transformation of anatase to rutile form. Surprisingly, anatase form is more photoactive in comparison to rutile, which has low band gap because the charge recombination rate is 10 times slower in anatase form. It has been found that mixture of rutile and anatase form was more photoactive in comparison to individual component that was assumed due to transfer of electrons from CB of anatase form to CB of rutile form. However, in some literatures, it has been proposed that in anatase form trapping sites are present at 0.8 eV below the conduction band and electrons can flow from CB of rutile to anatase. Anatase TiO_2 has CB band edge position –0.58 V (vs. NHE at pH-7), so it can reduce CO_2 while VB position is +2.52 V (vs. NHE at pH-7) which facilitates water oxidation (see Kondratenko et al. 2013).

In TiO_2, some physical properties like adsorption, catalytic reactivity, selectivity, etc. were determined by surface atomic configuration and the degree of exposure of reactive crystal facets. Typically, during the synthesis of anatase TiO_2, higher energy facets {001} (0.90 J/m²) get diminished quickly and the crystals are dominated by the thermodynamically stable {101} facets with lower surface free energy (0.44 J/m²). Normal synthesis conditions produce truncated octahedral bipyramid (TOB) seed, exposing eight {101} facets and two {001} facets. However, high-energy {001} facets can be stabilized by the use of capping agents typically fluorides (see Liu et al. 2011, Yang et al. 2009, Yang et al. 2008, Liu et al. 2010d). The {001} facet anatase TiO_2 was found to be more active for the reduction of CO_2. So increasing the {001} facet should drive more CO_2 reduction products. But contradictory to a report by Yu et al. (2014a), the 58% {001} facet TiO_2 was found to be more active for CO_2 reduction in comparison to 83% {001} facet TiO_2. The highest rate of methane production obtained for HF 4.5 is 1.35 μmol g^{-1} h^{-1}. The possible reason of increased yield was explained on the basis of energy difference ("surface heterojunction") determined by DFT calculation. It has been found that CB and VB position of {101} facet TiO_2 is lower in energy than {001} facet TiO_2 so photogenerated electrons can move from the conduction band of {001} facet TiO$_2$ to {101} facet TiO$_2$ while hole moves in opposite direction (Yu et al. 2014a). Similarly, {111} facet TiO$_2$ has much higher energy (1.61 J/m^2) that arises from the under coordinated Ti atoms and O atoms that existed on the {111} surface which acts as active sites in the photoreaction (see Sun et al. 2011, Liu et al. 2012). Experimentally, it has been proved that {111} facet TiO$_2$ is the most photoactive among other TiO$_2$ samples exposed with majority {010}, {101}, and {001} facets (see Xu et al. 2013a). A less explored form of anatase TiO$_2$, the {100} facet having intermediate energy (0.44 J/m^2) is considered to be the best attractive active facet because of its superior surface atomic structure and electronic structure (see Li and Xu 2010, Pan et al. 2014, Li et al. 2012). Theoretical prediction reveals that the {100} facet should appear as the "belt"; however, in most of cases single crystals with exposed {100} facet, anatase TiO$_2$ was obtained as rods or cuboids. The cuboids structure has {001} and {101} facet exposed on the upper and bottom of the cuboids. Apart from this, the small surface area as the utmost important property also limits their application. So the promising approach to increase surface area and {100} facet is to reduce the thickness of the TiO$_2$ cuboids which resulted in the decrease in the percentages of the {101} and {001} facets. In a report, Xu et al. (2013b) has synthesized ultrathin anatase TiO$_2$ nanosheets with high percentage of the exposed {100} facet and surface area (57.1 m^2 g^{-1}). The photoactivity of {100} facet TiO$_2$ sheets T$_{Sheets}$ (362 µmol h^{-1}) was found to be about 3.5 times higher than T$_{Cuboids}$ (104 µmol h^{-1}) (Xu et al. 2013b).

Doping with various dopants further improve the photocatalytic character which also prevent the electron hole annihilation. For example, photochemically sensitive iodine-doped TiO$_2$ nanoparticles as synthesized by hydrothermal method were found to be active in visible light irradiation (see Zhang et al. 2011b). The photocatalytic activities of the I-TiO$_2$ powders were investigated for photocatalytic reduction of CO$_2$ with H$_2$O under visible light ($\lambda > 400$ nm) and also under UV–vis illumination. The photoreduction experiment showed that CO was a major photoreduction product (highest CO yield equivalent to 2.4 mol g^{-1} h^{-1}) which was much higher as compared in undoped TiO$_2$. In case of Cu/TiO$_2$ synthesized by improved sol-gel method, using CuCl$_2$ was found to be more active than Cu/TiO$_2$ synthesized by (CH$_3$COOH)$_2$Cu for reduction of CO$_2$ to methanol. TPR, XPS, and XAS measurements showed that Cu(I) served as an active site. Further, higher zeta potential at pH 7 was found to be responsible for increased activity. The yield of methanol was found to be 600 µmol g^{-1} cat after 30 hr under UVC (254 nm) irradiation (see Tseng et al. 2004). Other semiconductors like ZnO, Fe$_2$O$_3$, CdS, ZnS, Nb$_2$O$_5$, Ta$_2$O$_5$, and BiTaO$_4$ have been explored as photocatalysts. In particular, oxides with perovskite structure formed by TaO$_6$ or NbO$_6$ octahedra layers have shown photocatalytic activity in tantalates and niobates of structural formula NaMO$_3$ (M = Ta and Nb) for the stoichiometric decomposition of water. Photocatalytic activity better than TiO$_2$ has been observed on laminar oxides such as BaLi$_2$Ti$_6$O$_{14}$, MTaO$_3$ (M = Li, Na, K), and SrM$_2$O$_7$ (M = Nb, Ta) for the degradation of organic pollutants. Some mixed oxides such as In$_{1-x}$Ni$_x$TaO$_4$, CaIn$_2$O$_4$, InVO$_4$, BiVO$_4$, and Bi$_2$MoO$_6$ have shown better visible light driven catalytic activity. In recent years, oxynitride solid-solution photocatalysts, such as (Ga$_{1-x}$Zn$_x$)(N$_{1-x}$O$_x$) and (Zn$_{1+x}$Ge)(O$_x$N$_2$), are established to be most efficient photocatalysts that work under visible-light irradiation (see Jensen et al. 2008, Chouhan et al. 2013). Although ZnO is a wide band gap (3.37 eV) material that is less investigated than TiO$_2$ for photocatalytic application, but in recent years it has emerged as an interesting alternative due to its photocatalytic activity, long lived charged species, low cost and noncorrosive nature (see Das and Khushalani 2010, Georgekutty et al. 2008). For reducing band gap, nitrogen doped ZnO (N-ZnO) nanobundles with visible-light photocatalytic activity has been synthesized by thermal treatment of ZnOHF nanobundles under NH$_3$ (see Zong et al. 2013). X-ray photoelectron spectroscopy (XPS) analysis indicated that N was bound to Zn as nitride (Zn-N) and oxynitride (O-Zn-N). The band gap of ZnO was reduced from 2.20 to 1.95 eV as a consequence of nitrogen doping which raises the VB position upward. Water splitting reaction was tested to check the performance of photocatalyst under visible light irradiation ($\lambda > 420$ nm) which showed significant increase in the rate of hydrogen evolution. Wang et al. (2014) synthesized the films of N-ZnO by reactive magnetron sputtering (Wang et al. 2014). The undoped ZnO films exhibited n-type conduction, while the N-ZnO films showed p-type conduction. N was involved mostly in Zn-N bonds, substituting O atoms to form NO acceptors in the N-ZnO films which were also confirmed by XPS analysis. The ZnO:N film has high optical quality and displays a stronger near band edge (NBE) emission in the temperature-dependent photoluminescence spectrum.

Porous nature of these materials is beneficial for enhanced concentration of CO_2 adsorbed on catalyst's surface which enhances yield of CO_2 reduction products due to access of more active sites. Park et al. (2012) has synthesized highly porous gallium oxide (Ga_2O_3) with mesopores and macropores by using TTAB (Myristyltrimethylammonium bromide) and utilized for the reduction of CO_2. The calculated surface area was found to be 42.7 m^2 g^{-1} as determined with BET. The amount of methane in the absence of any dopant found after 10 hr was found to be 2.09 μmol g^{-1} or 156 ppm which was much higher than Ga_2O_3 prepared without any template (Park et al. 2012).

Mixed nanostructured materials for CO_2 reduction

Nanoscale integration of multiple functional components in which one having low band-gap and another having high band-gap is beneficial because it increases the lifetime of excited state indirectly by continuous supply of electrons in the conduction band and providing holes in the valence band (see Chang et al. 2014). Careful designed core shell structure of two or more different semiconductors has proven a better photocatalyst because of better interfacial contact which facilitates better charge separation. For example, Ni@NiO core shell structure-modified nitrogen-doped $InTaO_4$ ($Ni@NiO/InTaO_4$-N) was found to be better catalyst for reduction of CO_2 to methanol (see Tsai et al. 2011). After 2 hr, the yield of methanol was found to be 350 μmol g^{-1} cat. Despite that, the CB of NiO (E_{CB} = –0.96 V vs. NHE) is more negative than $InTaO_4$ (ECB = –0.8 V vs. NHE), so the electron transfer from $InTaO_4$ to NiO is thermodynamically not possible, but modification with Ni double layered structure assists the transfer of photogenerated electrons to CB of $InTaO_4$ (see Kato and Kudo 2003). Pan et al. (2007) reported that nickel oxide working as a co-catalyst deposited on the surface of $InTaO_4$ and 1 wt% loading was optimal concentration used in the photoreduction of CO_2 to CH_3OH which afforded 1.394 μmol g^{-1} h^{-1} of methanol in the first 20 hr (Pan and Chen 2007). Lee et al. (2010) synthesized $InNbO_4$ modified with NiO and Co_3O_4 by incipient-wetness impregnation method and used for the CO_2 activation (Liu et al. 2010b). The 0.5 wt% NiO loaded $InNbO_4$ gave methanol up to 1.577 μmol g^{-1} h^{-1} cat while 0.5 wt% Co_3O_4 loaded $InNbO_4$ gave 1.503 μmol g^{-1} h^{-1} of methanol (see Lee et al. 2012). Wang and co-workers (2010) synthesized CdSe quantum dot sensitized TiO_2 and the Pt were incorporated by the wet impregnation methods onto the TiO_2 for the photoreduction of CO_2 to CH_3OH, CH_4, H_2 and CO as the secondary product in visible light (Wang et al. 2010). Liu et al. (2010c) has shown that ternary metal oxide like zinc orthogermanate (Zn_2GeO_4) can reduce CO_2 into methane. The yield was further increased by addition of 1 wt% RuO_2 and 1 wt% Pt nanoparticles (Liu et al. 2010c). In a report by Feng et al. (2010), ultrafine Pt nanoparticles supported on hollow TiO_2 nanotubes were synthesized *in situ* by microwave-assisted solvothermal approach. For Pt nanoparticles loaded, nanotubes' photocatalytic methane production rate of 25 ± 4 ppm/(cm^2 h) was obtained from the CO_2/water vapour atmosphere (see Feng et al. 2011, Xie et al. 2015). In another study by Yan et al. (2012), $ZnAl_2O_4$-modified mesoporous ZnGaNO solid solution was synthesized by a two-step reaction template route (Yan et al. 2012). The first step involves $NaGa_{1-x}Al_xO_2$ solid solution, which was prepared by heating the mixed gel of $NaGaO_2$ and $NaAlO_2$; however in the second step, $Zn(CH_3COO)_2$ aqueous solution was introduced into the $NaGa_{1-x}Al_xO_2$ colloidal suspension. The ion exchange reaction produced $ZnAl_2O_4$ modified mesoporous ZnGaNO after nitridation step. The $ZnAl_2O_4$-modified mesoporous ZnGaNO loaded with 0.5 wt% Pt as the co-catalyst exhibited a CH_4 generation rate of 9.2 μmol g^{-1} h^{-1} after 1 hr of visible light illumination ($\lambda \geq 420$ nm). BET and TPD results showed that high surface area and higher adsorption of CO_2 on catalyst surface promoted higher yield. Hollow nanotubes or other hollow structures possess superior photocatalytic performance because of low recombination in bulk volume. However, their design and structure are difficult to maintain. In another study, CuO-$TiO_{2-x}N_x$ hollow nanocubes with exceptionally high photoactivity were synthesized for reduction of CO_2 to CH_4 under visible light irradiation (see In et al. 2012). The yield of methane was found to be 41.3 ppm g^{-1} h^{-1} which was 2.5 times higher than that of Degussa P25 TiO_2 (16.2 ppm g^{-1} h^{-1}).

Layered double hydroxides (LDHs) with a general formula $[M^{2+}_{1-x}M^{3+}_x(OH)_2]^{x+}(A_n)_{x/n}\cdot mH_2O)$ in as such form or after modification with different ions like M^{2+} = Mg^{2+}, Zn^{2+}, Ni^{2+}; M^{3+} = Al^{3+}, Ga^{3+}, In^{3+}, etc. were also investigated for the CO_2 reduction. For example Teramura et al. (2012) used Mg-In LDH as photocatalyst, which yields CO and O_2 (3.21 and 17.0 μmol g^{-1} h^{-1}, respectively) from 100 mg of Mg-

In LDH (Teramura et al. 2012). Ahmed and coworkers reported the use of various LDHs incorporated with different metal ions like Cu, Zn, Al, Ga, etc. for reduction of CO_2 at pressurized reactor under UV-Vis light irradiation. By using $[Zn_{1.5}Cu_{1.5}Ga(OH)_8]^+_2[Cu(OH)_4]^{2-} \cdot mH_2O$ (Zn-Cu-Ga) layered double hydroxides as photocatalyst, the yield of methanol was found to be 170 nmol g^{-1} cat h^{-1} (see Ahmed et al. 2011). Further addition of light absorbers can increase the yield of the products. In this context, Hong et al. (2014) synthesized LDHs modified with carbon nitride (Mg-Al-LDH/C$_3$N$_4$) for the photocatalytic reduction of CO_2. The nitrate ions were replaced with carbonates which facilitates CO_2 concentration over LDHs while carbon nitride generates electron hole pairs (Hong et al. 2014).

Sensitization of semiconductors with metal complexes

For enhancing the visible light absorption, other methods like sensitization with metal complexes can be used as light harvesting units (see Kumar et al. 2012). Homogeneous photocatalysts, including transition metal complexes such as ruthenium(II) polypyridine carbonyl complex, cobalt(II) trisbipyridine, and cobalt(III) macrocycles have been widely investigated for this reaction. Transition metal based molecular complexes are advantageous due to their high quantum efficiencies and high selectivity of products. However, transition metal complexes suffer from the drawback of non-recyclability and higher costs which limit their practical applications. In order to overcome these limitations, immobilization of transition metal complexes to a photoactive support is a promising approach as it provides facile recovery, recyclability as well as enhanced activity due to synergistic effect of both components. Metal complexes after absorption of visible light get excited via MLCT (metal to ligand charge transfer) transition and can transfer electrons to the conduction band of semiconductors which are used for the reduction of CO_2 or protons. For the efficient transfer of electrons from metal complex molecule to semiconductor, the position of LUMO of the metal complex should be higher in energy than the conduction band edge of the semiconductor. However, its rate depends on a number of parameters such as the orientation and distance of the light absorbing unit with respect to the surface or size and flexibility of the anchoring unit. The positively charged metal complex that is generated gets back to its initial state by extracting electrons from the valence band (or by transferring positive charge in valance band). So, the hole is created in the valance band of semiconductor which oxidizes water to derive necessary electrons and protons. Immobilization of metal complexes indirectly generates electron and hole pairs in the conduction band and valance band, respectively.

So far, a number of photosensitizers such as ruthenium polyazine, metal phthalocyanines, porphyrines, etc. have been used widely for the immobilization on photoactive semiconductor supports. For the efficient sensitization, the life time of excited state should be long enough to transfer electrons

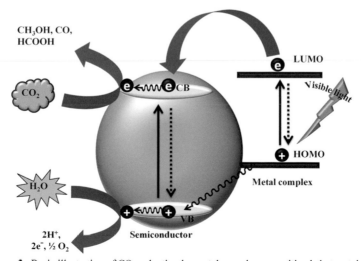

Figure 3. Basic illustration of CO₂ reduction by metal complexes sensitized photocatalyst.

in the conduction band of semiconductors. In this regard, nature of metal ions play a pivotal role for determining the life time of excited state; for example, tin phthalocyanine has longer life time of excited state than other metal phthalocyanines. So, there is sufficient time available for the transfer of electrons from the phthalocyanine molecule to the conduction band of semiconductor. Another major factor that influences electron transfer is the vicinity of metal complex with semiconductor. Attachment with linkers having long alkyl chain is less efficient in comparison to complexes attached with short chain linkers. Similarly, if metal complex is coordinated with semiconductor via a group having conjugated linker, the electron transfer from metal complex to conduction band is faster. But back electron transfer may be a detrimental process, so presence of sacrificial donor is of utmost importance which prevents the back transfer of electrons via donating electrons through redox quenching.

In general, semiconductor photocatalysts provide hydrocarbon products having lower C:H ratio (< 0.5) because of cessation of reaction on the surface of catalyst. So, in order to produce hydrocarbons with higher C:H ratio (> 0.5), multi-electron transfer is essential. The multi-electron transfer is possible by sensitization of semiconductors with molecular catalysts, which have the ability to generate more than one electron-hole pair simultaneously (see Mohamed and Bahnemann 2012).

Mainly, two approaches for the immobilization of metal complexes on the semiconductor supports have been utilized; the first one is non-covalent immobilization and the second one is covalent attachment. In a non-covalent approach, the metal complex can be immobilized to support material by ionic interaction, physical adsorption via van-der-Waals forces or through π–π interaction. Simple adsorption by van-der-Waals or π–π interaction of metal complex is less robust due to the weak interaction between support material and metal complex, so leaching of the catalyst from the solid support is an obvious drawback. Ionic interaction is somewhat stronger but ionization in the aqueous reaction medium limits its practical utility. Thus, covalent immobilization of metal complex to support is more attractive in the viewpoint of stability. For the covalent attachment of complex to the support, the surface of semiconductor should have lots of functionalities to accommodate enough concentration of metal complex. Furthermore, most of the semiconductor surface is decorated with –OH groups that form M-OH groups which are weaker than C-OH group; hence, multi-site attachment is more favourable to achieve highly stable photocatalysts. In this regard, the use of linkers provides benefits of strong and leaching proof multisite attachment of metal complex to the solid support matrix. The most commonly used linkers are 3-aminopropyl trimethoxysilane (APTMS), dopamine, glycine, etc. Furthermore, the stability of any immobilized complex on semiconductor is strongly dependent on the number of sites and nature of groups through which it is coordinated.

O. Ishitani and co-workers have done lot of work in the field of immobilization of homogeneous metal complexes of ruthenium $[Ru(dcbpy)(bpy)(CO)_2]^{2+}$, rhenium $[Re(dcbpy)(CO)_2]^{1+}$, etc. on various semiconductor supports like tantalum oxynitride, carbon nitride, etc. In a recent work, Sato et al. linked $[Ru-(dcbpy)_2(CO)_2]^{2+}$ complex, N-doped Ta_2O_5 (N-Ta_2O_5) as a p-type semiconductor for the photoreduction of CO_2 under visible light and obtained HCOOH selectively with a TON value of 89 (see Yamanaka et al. 2011). Here the p-type semiconductor is beneficial since it fastens the transport of electrons to the complex. Recently, Woolerton et al. (2010) developed ruthenium complex sensitized TiO_2 attached enzymatic system in which Ch-CODH I: carboxydothermus hydrogenoformans, expressed as carbon monoxide dehydrogenase, were able to transfer two electrons simultaneously to the CO_2 (Woolerton et al. 2010). This enzyme coupled catalytic system was used to reduce CO_2 to CO selectively. Sekizawa et al. (2013) reported Ru metal complex attached with semiconductor for the photocatalytic reduction of CO_2 to CH_3OH, HCOOH via Z-scheme mechanism (Sekizawa et al. 2013).

It has been well documented in the literature that phthalocyanines have a longer life time of excited state in comparison to ruthenium complex so they can effectively transfer electrons to semiconductors. In this regard, recently Kumar et al. (2015a) have synthesized tin phthalocyanine dichloride ($SnPcCl_2$) immobilized to mesoceria ($SnPc@CeO_2$) by considering the labile chlorine atoms as the linking sites (Kumar et al. 2015a). It has been observed that in the solution form, SnPc remains in agglomerated form and gives broad Q band from 650 nm to 750 nm while after attachment to meso-ceria support, it converted to monomeric form and gave sharp Q band which is a clear indication of attachment of SnPc to mesoceria. The synthesized photocatalyst was used for the visible light assisted photoreduction of CO_2

to methanol along with minor amount of CO and H_2 using triethylamine as sacrificial donor. By using SnPc@CeO$_2$, the yield of methanol, CO and H_2 was found to be 2342 μmol g^{-1} cat, 840 mmol g^{-1} cat and 13.5 mmol g^{-1} cat after 24 hr of vis-irradiation. However, it has been observed that after a certain period, the rate of methanol formation started to deplete that was due to consumption of CO_2 in the reaction mixture. In order to confirm that the saturation point is reached due to consumption of CO_2 and not due to the deactivation of the photocatalyst, the reaction mixture was re-purged with CO_2 and again irradiated under visible light. The yield of methanol again increased which confirmed that the photocatalyst was not deactivated during irradiation.

The higher photocatalytic performance was explained on the basis of better electron injection from SnPc to conduction band of mesoceria. After absorption of visible light, SnPc get excited via MLCT transition and electrons get transferred from HOMO to LUMO. This excited state transfers electrons to the conduction band of CeO$_2$. Triethylamine works as sacrificial donor and get degraded to its degradation products like acetaldehyde and diethyl amine, etc.

In recent years, magnetically separable nanocomposite materials have gained considerable attention because of the easy recovery of the catalyst after reaction by using external magnet. In this regard,

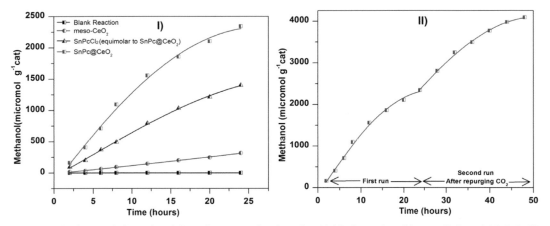

Figure 4. (I) Photocatalytic methanol formation versus time by using (a) blank reaction, (b) meso-CeO$_2$ and (c) SnPcCl$_2$ equimolar amount as in SnPc@CeO$_2$ and (d) SnPc@CeO$_2$. (II) Methanol yield after repurging CO_2. Reprinted with permission from ref. Kumar, P., A. Kumar, C. Joshi, R. Singh, S. Saran and S.L. Jain. 2015. Heterostructured nanocomposite tin phthalocyanine@mesoporous ceria (SnPc@CeO$_2$) for photoreduction of CO_2 in visible light. RSC Adv. 5: 42414–42421. Copyright@Royal Society of Chemistry.

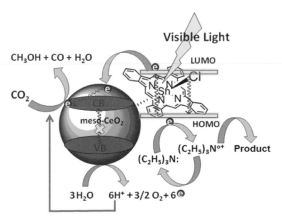

Figure 5. Plausible mechanism of CO_2 reduction over SnPc@CeO$_2$ catalyst. Reprinted with permission from ref. Kumar, P., A. Kumar, C. Joshi, R. Singh, S. Saran and S.L. Jain. 2015. Heterostructured nanocomposite tin phthalocyanine@mesoporous ceria (SnPc@CeO$_2$) for photoreduction of CO_2 in visible light. RSC Adv. 5: 42414–42421. Copyright@Royal Society of Chemistry.

Kumar et al. (2015b) developed a core shell structured bimetallic Ru complex and cobalt phthalocyanine attached $TiO_2@SiO_2@Fe_3O_4$ microspheres (Ru-CoPc@$TiO_2@SiO_2@Fe_3O_4$) by step wise coating of SiO_2, TiO_2, CoPc and Ru complex on Fe_3O_4 nanoparticles (Kumar et al. 2015b). The synthesized photocatalyst was found to be much more photoactive in comparison to any component and gave 2570 µmol g^{-1} cat after 48 hr by using water/triethylamine mixture under visible light irradiation. The photocatalyst was magnetically separable and could be reused several times without loss of photoactivity and no leaching of active metallic components such as Ru and Co was observed due to covalent attachment. The particular arrangement was chosen because Ru complex has short life time in comparison to CoPc, so it can transfer

Figure 6. FESEM images of (a) Fe_3O_4, (b) $SiO_2@Fe_3O_4$, (c) $TiO_2@SiO_2@Fe_3O_4$, (d) Ru-CoPc@$TiO_2@SiO_2@Fe_3O_4$. Reprinted with permission from ref. Kumar, P., R.K. Chauhan, B. Sain and S.L. Jain. 2015. Photo-induced reduction of CO_2 using a magnetically separable Ru-CoPc@$TiO_2@SiO_2@Fe_3O_4$ catalyst under visible light irradiation. Dalton Trans. 44: 4546–4553. Copyright@Royal Society of Chemistry.

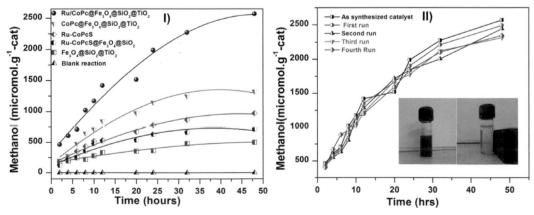

Figure 7. (I) Methanol yield using (a) Ru/CoPc@$TiO_2@SiO_2@Fe_3O_4$, (b) CoPc@$TiO_2@SiO_2@Fe_3O_4$, (c) Ru-CoPcS, (d) Ru-CoPcS@$SiO_2@Fe_3O_4$, (e) $TiO_2@SiO_2@Fe_3O_4$ and (f) blank run. (II) Recycling of the catalyst. Inset: Magnetic separation of the catalyst using an external magnetic. Reprinted with permission from ref. Kumar, P., R.K. Chauhan, B. Sain and S.L. Jain. 2015. Photo-induced reduction of CO_2 using a magnetically separable Ru-CoPc@$TiO_2@SiO_2@Fe_3O_4$ catalyst under visible light irradiation. Dalton Trans. 44: 4546–4553. Copyright@Royal Society of Chemistry.

electrons to CoPc faster which has higher life time, so it can efficiently transfer electrons to conduction band of TiO_2. SiO_2 was coated in between TiO_2 and Fe_3O_4 to provide plenty of –OH functionalities for the better coating of TiO_2. Further, it prevents recombination of electrons and protons at Fe_3O_4 core.

In some instances of post grafting approach of immobilization of metal complexes to semiconductors, auto degradation of metal complexes in the presence of light is observed. Thus, to overcome such limitations, *in situ* synthesis of metal complexes grafted semiconductor is a fascinating approach to prevent the loss of metal complex from catalyst support as well as from self degradation. In this regard, Zhao et al. (2009) have synthesized CoPc grafted TiO_2 (CoPc/TiO_2) and elaborate on the reduction of CO_2 to methanol and formic acid. The CoPc/TiO_2 photocatalyst containing 0.7% of CoPc afforded 9.38 mole g cat^{-1} h^{-1} of methanol and 148.81 mole g cat^{-1} h^{-1} of formic acid (Zhao et al. 2009). Kumar and co-workers (2015c) synthesized *in situ* Ru(bpy)$_3$Cl$_2$ complex grafted TiO_2 (*in situ* Ru(bpy)$_3$/TiO_2) by precipitation of $TiCl_4$ in triethanolamine containing solution followed by dissolution in water and then re-precipitation with ammonia solution (Kumar et al. 2015c). TEM analysis showed fringes of TiO_2 crystal lattice with 0.35 nm (101 plane) interplaner distance that confirmed anatase form (Figure 8-I). Further, wide scan XPS spectra in Ti2p region illustrated characteristics of Ti2p$_{1/2}$ and Ti2p$_{3/2}$ peaks at 464.23 and 458.39 eV binding energy (Figure 8-II). The synthesized *in situ* Ru(bpy)$_3$/TiO_2 photocatalyst was found to be an efficient CO_2 photoreduction catalyst in comparison to TiO_2 synthesized by following same procedure without metal complex. Methanol was identified as the selective reduction product of CO_2 and the obtained formation rate (R$_{MeOH}$) of methanol was found to be 78.1 µmol g^{-1} h^{-1} after 24 hr under visible light irradiation in the presence of triethylamine as sacrificial donor. The apparent quantum yield (AQY) at 450 nm was calculated to be 0.26 mol Einstein^{-1}. Better electron injection to the conduction band of TiO_2 was assumed to be responsible for enhanced photocatalytic performance.

Graphene based materials for solar fuel

A single layer of graphite called graphene has attracted wide attention of scientific community since its discovery by Novoselov et al. (2004). In gaphene, a 2D network of sp² hybridized carbon atoms make long range order conjugation which provides exceptional properties like high electron conductivity, thermal conductivity, mechanical strength, optical properties and very high surface area (see Rao et al. 2009, Chen et al. 2012, Hu et al. 2010). Due to these specific properties, graphene has been utilized in

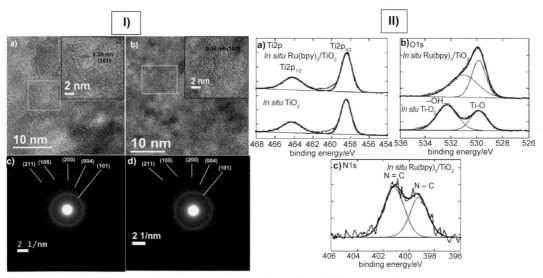

Figure 8. (I) HR-TEM images of (a) *in situ* TiO₂, (b) Ru(bpy)₃/TiO₂, SAED patterns of (c) *in situ* TiO₂ and (d) *in situ* Ru(bpy)₃/TiO₂. (II) Wide scan XPS spectra of *in situ* TiO₂ and Ru(bpy)₃/TiO₂, (a) Ti 2p, (b) O 1s and (c) N 1s region. Reprinted with permission from ref. Kumar, P., C. Joshi, N. Labhsetwar, R. Boukherroub and S.L. Jain. 2015. A novel Ru/TiO₂ hybrid nanocomposite catalyzed photoreduction of CO₂ to methanol under visible light. Nanoscale 7: 15258–15267. Copyright@Royal Society of Chemistry.

various applications like electronic devices, solar energy conversion, super capacitors, sensors, catalysis and particularly for the preparation of composite materials (see Ji et al. 2012, Zhuang et al. 2014, Bai and Shen 2012, Hou et al. 2012, Sahoo et al. 2012). Graphene sheets, due to high mobility of electron on its surface, have been used for the synthesis of nano-hybrid with various semiconductors like TiO_2, ZnO, and Cu_2O nanostructure semiconductors (see Zhang et al. 2012, Cowan and Durrant 2013, Li et al. 2013b, Low et al. 2015).

Ong et al. (2014) used exposed 001 facet anatase doped with nitrogen-doped TiO_2 (N–TiO_2-001) for hybridization with graphene sheets (N–TiO_2-001/GR). The photocatalyst was synthesized by hydrothermal method in one-step using NH_4F as the structure determining agents. In previous studies, it has been showed that 001 facet TiO_2 is more active toward CO_2 reduction. The photocatalytic results showed that N–TiO_2-001/GR nanostructure catalyst gave CH_4 selectively with a yield of 3.70 $\mu mol\ g^{-1}$ cat that was 11-fold higher than the TiO_2-001. The enhanced activity was assumed due to three facts: (i) 001 facet TiO_2, (ii) nitrogen doping which reduce band gap, (iii) blending with graphene sheets which provide higher mobility to electrons (Ong et al. 2014).

Harsh oxidation of graphite with strong oxidizing agents like $KMnO_4$ and H_2SO_4 or H_3PO_4 and H_2SO_4 gave graphene oxide, an oxidized form of graphene (see Hummers and Offeman 1958). During the harsh oxidation of graphene, some of the sp^2 carbons get converted into sp^3 hybridized carbons. Thus, enormous sp^2 and sp^3 hybridized domains get evolved on the surface of graphene sheets. The domains having sp^3 carbons due to the presence of tightly held electrons work as valance band while sp^2 domain, due to the presence of mobile electrons, works as conduction band. So oxidation of graphene oxide convert conductive graphene sheets to semiconductor sheets (see Loh et al. 2010, Bonaccorso et al. 2010, Eda et al. 2009). The charge separation phenomenon can take place on the surface of graphene oxide which can be used for the reduction of CO_2 and oxidation of water. In a study, Hsu et al. (2013) has showed that graphene oxide synthesized by $KMnO_4$ and H_2SO_4 using H_3PO_4 as mild oxidizing agent has low band gap 2.9 eV. The synthesized graphene oxide was used for the reduction of CO_2 to methanol under visible light irradiation of xenon lamp as a source of light. The rate of methanol formation was found to be 0.172 $\mu mol\ g^{-1}$ cat h^{-1} after 4 hr of visible irradiation that was six-fold higher than pure TiO_2 (see Hsu et al. 2013). The position of conduction band was negative enough (–0.79 V vs. NHE) to facilitate CO_2 reduction to methanol while the position of the valence band (+2.91 V vs. NHE) was positive enough to perform water splitting.

Yeh et al. (2010) have explored the use of graphene oxide for the reduction of protons to hydrogen under visible light irradiation (Yeh et al. 2010). As graphene oxide works as semiconductor, so the phenomenon of charge recombination still exists and so the electron or hole capturing agents can improve the photocatalytic performance. So, addition of platinum nanoparticles as electron capturing agents and methanol as hole scavenger improve the yield of hydrogen from 280 μmol after 6 hr for GO to 17000 μmol after 6 hr for Pt/GO having methanol as hole scavenger. In another study by Shown et al. (2014), copper nanoparticles (4–5 nm) modified GO (10 wt% Cu-NPs) has been found to increase the rate of methanol production from 0.172 $\mu mol\ g^{-1}$ cat h^{-1} after 4 hr to 2.94 $\mu mol\ g^{-1}\ h^{-1}$ after 2 hr with acetaldehyde as byproduct (3.88 $\mu mol\ g^{-1}\ h^{-1}$) (Shown et al. 2014). The increased yield was assumed due to capturing of electrons which lower the electron hole pair recombination.

Nanostructured composite of graphene oxide with various semiconductor materials exhibits higher photocatalytic performance because of transfer of electrons from semiconductors to GO which can move apart on the surface. High specific surface area provides better adsorption of CO_2 and H_2O molecule. Further modified semiconductors have narrow band gap and can induce charge production at higher wave length. In this regard, Tan et al. (2015a) synthesized a composite of oxygen rich TiO_2 and graphene oxide (GO-OTiO_2) for CO_2 reduction. The oxygen rich TiO_2 was prepared by precipitation of titanium butoxide in the presence of hydrogen peroxide (Tan et al. 2015a). Due to the narrowing of band gap and creation of defects, the TiO_2 was visible light active with a band gap of 2.95 eV. The optimal loading of GO which promotes better CO_2 reduction was calculated to be 5 wt%. After 6 hr of visible irradiation, the yield of CH_4 over GO-TiO_2 nanocomposite was found to be 1.718 $\mu mol\ g^{-1}$ cat which was 14 times higher than commercial grade P25 TiO_2.

Due to rich surface chemistry, graphene oxide also serves as semiconductor support for attachment of various semiconductors and metal complexes. Graphene oxide has absorption range in 230 to 300 nm, so it can absorb light of lower wave length. Hybridization with metal complex further improves the visible light absorption profile along with the significant enhancement in CO$_2$ conversion efficiency. Kumar et al. (2015d) have immobilized a ruthenium heteroleptic complex to GO by using chloroacetic acid as linker (Kumar et al. 2015d). The developed photocatalyst showed increased yield of methanol in comparison to graphene oxide alone. The methanol formation rate and quantum yield at 510 nm for GO-Ru catalyst was found to be 85.4 µmol g^{-1} cat h^{-1} after 24 hr and (ϕ_{MeOH}) of 0.09, respectively, while for GO alone this value was 20.0 µmol g^{-1} cat h^{-1} and 0.044, respectively. The increased conversion efficiency was assumed due to continuous pumping of electrons from excited ruthenium complex to conduction band of GO. In a similar approach, CoPc was also used as anchoring photosensitizer unit because of its wide absorption pattern in visible region and longer life time of excited state (see Kumar et al. 2014a). Ruthenium trinuclear complex, which works as antenna and collects photogenerated electrons, was also immobilized to GO for photocatalytic CO$_2$ activation to methanol (see Kumar et al. 2014b). It can transfer multiple electrons so it creates possibility for getting higher hydrocarbons. The synthesized photocatalyst was used for the photocatalytic reduction of CO$_2$ to methanol using triethylamine as a reductive quencher by using 20 W LED as visible light source. The yield of methanol was found to be 3977.57 ± 5.60 µmol g^{-1} cat after 48 hr.

Reduced graphene oxide as obtained by reduction of graphene oxide is also well documented for the photocatalytic applications. In this regard, Tan and co-workers (2013) synthesized reduced graphene oxide TiO$_2$ (rGO-TiO$_2$) nano-hybrid by solvothermal method and obtained CH$_4$ product in 0.135 µmol g$_{cat}^{-1}$ h^{-1} yield after 4 hr of reaction (Tan et al. 2013). Composite of Cu$_2$O semiconductor with reduced graphene oxide (Cu$_2$O/RGO) posseses higher photocatalytic performance for reduction of CO$_2$. Addition of 0.5% RGO was found to be of the optimal amount and the Cu$_2$O/0.5% RGO composites produce CO at an average of 50 ppm g^{-1} h^{-1} after 20 h, with an apparent quantum yield of 0.34% at 400 nm. This yield was approximately 50 times higher than bare Cu$_2$O which clearly explains the role of graphene oxide as charge separator (see An et al. 2014). During the reduction step of graphene oxide, oxygen carrying moieties are removed which create defects in graphene sheets and diminish charge mobility over the sheets. In a study, Liang et al. showed that the solvent exfoliated graphene (SEG) synthesized by ultrasonication in *N,N*-dimethyl formamide (DMF) has less defect in comparison to rGO synthesized with chemical reduction methods, which improves electrons' mobility on the sheets. A composite of SEG with TiO$_2$ (P-25) SEG (0.27%)-TiO$_2$ synthesized by using ethyl cellulose as a stabilizing and film forming polymer can reduce CO$_2$ to methane with a formation rate of 8.3 µmol h^{-1} m^{-2} under UV light irradiation that was 4.5 times higher than bare TiO$_2$ (see Liang et al. 2011). Yu et al. (2014) reported that reduced graphene (RGO)–CdS nanorod composite prepared by a one-step microwave-hydrothermal method in an ethanolamine–water solution exhibited a higher activity for the photocatalytic reduction of CO$_2$ to CH$_4$, without any co-catalyst (Yu et al. 2014b).

Noble metals like Pt, Pd, Ag and Au nanoparticles decorated on reduced graphene oxide/TiO$_2$ (GT) were also tested for photoreduction of CO$_2$ to explore the better charge capturing performance of noble metals in composite. Among the noble metals used, the Pt-doped GT nanocomposites were found to be the most active and afforded CH$_4$ yield up to 1.70 µmol g^{-1} cat after 6 hr of light irradiation, that was 2.6 and 13.2 folds higher in comparison to GT and commercial P25, respectively (see Tan et al. 2015b).

In a recent report by Gusain et al. (2016), reduced graphene oxide (rGO)–copper oxide (rGO-CuO) nanocomposites were prepared by grafting of CuO nanorods on the rGO sheets (Gusain et al. 2016). Surface modification of CuO naorods by APTMS (3-aminopropyl-trimethoxysilane) led to accumulation of positive charge which interacts with negatively charged graphene oxide sheets. These GO modified CuO nanorods after hydrothermal treatment produce rGO-CuO nanocomposite. The NaOH to copper salt ratio determines the width of CuO nanorods in composite. By using different NaOH:(CH$_3$COOH)$_2$Cu ratio, the thickness of CuO nanorods in rGO-CuO nanocomposite was found to be 3–6, 5–9, 9–11 and 10–15 nm for rGO-CuO$_{14}$, rGO-CuO$_{18}$, rGO-CuO$_{116}$ and rGO-CuO$_{124}$ nanocomposites, respectively (Figure 9-I). High resolution XPS spectra in Cu2p region gave an intense peak at ~933.5 eV with associated satellite peaks for Cu^{+2} state confirming the presence of CuO in composite (Figure 9-II). Methanol was obtained selectively in CO_2 photoreduction experiment and rGO-CuO$_{116}$ composite gave the highest yield of methanol with a formation rate (R_{MeOH}) of 51.1 µmol g^{-1} h^{-1} after 24 h. Recycling experiments showed that the synthesized photocatalyst was stable and recyclable for several runs without compromising the performance of catalyst (Figure 9-III). The increased performance was explained due to higher mobility of electrons on rGO sheets, which prevents electron and hole recombination (Figure 10).

Making composite of graphene not only improves charge separation on its surface but also, in some cases it has been observed that graphene can raise the position of conduction band of semiconductor. A

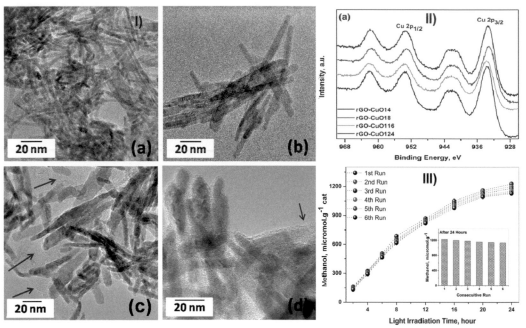

Figure 9. (I) HRTEM images of (a) rGO–CuO14, (b) rGO–CuO18, (c) rGO–CuO116 and (d) rGO–CuO124 nanocomposites. (II) High resolution XPS in Cu 2p region of rGO–CuO14, rGO–CuO18, rGO–CuO116 and rGO–CuO124 nanocomposites. (III) Recycling results by using rGO–CuO116 photocatalyst for the methanol yield as a function of irradiation time. Reprinted with permission from ref. Gusain, R., P. Kumar, O.P. Sharma, S.L. Jain and O.P. Khatri. 2016. Reduced graphene oxide–CuO nanocomposites for photocatalytic conversion of CO_2 into methanol under visible light irradiation. Appl. Catal. B 181: 352–362. Copyright@Elsevier.

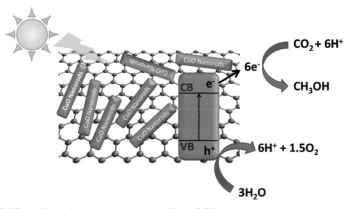

Figure 10. Schematic illustration of photocatalytic conversion of CO_2 to methanol, by using rGO–CuO nanocomposites under visible light irradiation. Reprinted with permission from ref. Gusain, R., P. Kumar, O.P. Sharma, S.L. Jain and O.P. Khatri. 2016. Reduced graphene oxide–CuO nanocomposites for photocatalytic conversion of CO_2 into methanol under visible light irradiation. Appl. Catal. B 181: 352–362. Copyright@Elsevier.

recent report by Wang et al. 2013b showed that blending of graphene with WO_3 raises the position of conduction band which makes it more reductive for the efficient reduction of CO_2. The CB band edge value of WO_3 was found to be -0.10 V vs. NHE at pH-7 which was low for the reduction of CO_2, while for WO_3/rGO composite the conduction band edge position was found to be -0.47 V (obtained by XPS VB spectra) which was negative enough to promote CO_2 reduction to methane. After 8 hours of visible irradiation, the yield of CH_4 was found to be 0.89 µmol g^{-1} cat.

Doping of heteroatoms such as N, S, B, P, etc. to graphene oxide is a well known approach to enhance its photocatalytic activity (see Lin et al. 2012). Nitrogen doping removes carbon atoms from the surface of graphene and three types of nitrogen pyridinic, quarternary nitrogen, and pyrrolic were evolved depending upon their position on the sheets. The contribution of electron pair by pyridinic nitrogen in π conjugated network of graphene transforms sheets in electron rich sheets and Fermi level shift above the Dirac point which distorts the symmetry of graphene sub lattice and creates a band gap. The band gap of N-doped graphene is highly dependent on the nitrogen content and a band gap value up to 5 eV can be reached. Nitrogen doping can be achieved by either *in situ* synthesis in the presence of nitrogen containing substrate or via post grafting approach using gaseous nitrogen source. Nitrogen doping disturbs the charge distribution on neighbouring atoms and negative charge gets accumulated on the surface of graphene which creates "activation region" on the graphene sheets. Due to the presence of these activation regions, N-doped graphene can be used for various reactions and attachment of metal particles/semiconductors/metal complexes, etc. The nanostructured composite of N-doped graphene with various semiconductor materials has been proven to show superior photocatalytic performance. Reduction of graphene oxide in the presence of phosphorous source like phosphoric acid, P-containing ionic liquids, etc. produce P-doped graphene. Phosphorous atoms due to loosely held electrons on the sheets make sheets more electron rich in character and afford higher band gap in comparison to N-doped graphene which can be used for the proton or CO_2 reduction. Sanchez et al. synthesized P-doped graphene (P)G-4 by pyrolysis of $H_2PO_4^-$ modified alginate at 900°C under inert atmosphere and were able to tune band gap up to 2.85 eV. Further, phosphorous content was calculated with the help of XPS which reveal 12.73% phosphorous. The catalyst was able to produce hydrogen from water in the presence triethanolamine as sacrificial donor. Furthermore, immobilization of metal complexes to N-doped graphene for photoreduction of CO_2 to methanol is also documented in the literature. In this regard, Kumar et al. (2015e) reported copper complex [Cu(bpy)$_2$(H$_2$O)$_2$]Cl$_2$.2H$_2$O supported on N-doped graphene (GrN$_{700}$-CuC) as efficient photocatalyst for photoreduction of CO_2 into CH_3OH with a conversion rate of 0.77 µmol g^{-1} h^{-1} with the associated quantum yield of 5.8 x 10^{-4} after 24 hr under visible light irradiation. The nitrogen content of N-doped graphene was found to be 6.01% which was enough to create a band gap required for the reduction of CO_2 (Kumar et al. 2015e).

A nanocomposite of boron doped graphene and P25 TiO$_2$ (P25/B-GR) was used for the reduction of CO_2 to methane (see Xing et al. 2014). Boric acid (H$_3$BO$_3$) was used as a source of boron for the synthesis of B-doped graphene. It has been well documented in literature that exposure of ZZ-edges on nanoscaled graphene give rise to semi-metallic property. B-doping in graphene has many exposed ZZ-edges and thus, the photo-generated electrons of B-GR make its Fermi level (E'$_f$-B-GR nanosheets) higher than the conduction band (EC-GR sheets) of GR sheets (from E$_f$-B-GR nanosheets to E'$_f$-B-GR nanosheets). The Fermi level of B-doped graphene lies in between the conduction band of TiO$_2$ and reduction potentials of CO_2/CH$_4$, so electrons can be transferred from CB of TiO$_2$ to B-doped nanosheets. The as prepared composite after 120 min solar light irradiation showed the highest photo generation of CH$_4$ (> 2.50 µmol g^{-1}).

Photocatalysis by metal clusters based nanostructured composite

Metal clusters, a class of Polyoxometalates (POMs), are molecular metal oxide aggregates and are typically formed by oligo-condensation reactions of small oxometalate precursors, often in the presence of templating anions (see Kibsgaard et al. 2014). Octahedral metal clusters of Mo and Re can be good photocatalyst for various applications including CO_2 reduction and hydrogen evolution. These are an attractive alternative because of the presence of multi-metallic centres that can transfer multiple electrons to reduce CO_2 to give higher hydrocarbons. Metal clusters are fully oxidized with d_0 configuration and show LMCT (ligand to metal charge transfer) transition in the range of 200–500 nm due to O→M transition. Recently, work group of Hill et al. (2012) reported cobalt based Polyoxometalates (POMs, clusters) as photocatalysts for the solar water splitting. Further, $[Ru(bpy)_3]^{2+/3+}$ redox photosensitizer modified cobalt-POM-cluster $[CoII_4(H_2O)_2(PW_9O_{34})_2]^{10-}$ was found to be good photocatalyst for water oxidation (Zhu et al. 2012). The photosensitizer unit acts as electron shuttle that transfers electrons to water-oxidising Co-POM cluster. A work by Kumar et al. (2014c) used octahedral molybdenum $[Mo_6Br_{14}]^{2-}$ having Cs^+ and TBA^+ (tetrabutylammonium) counter ions as CO_2 reduction catalyst with higher quantum efficiency (Kumar et al. 2014c). However, due to their homogeneous nature, they are difficult to recover and recycle. Thus, to overcome these drawbacks metal clusters have been immobilized to photoactive semiconductor supports. Fabre et al. (Fabre et al. 2014, Cordier et al. 2010) immobilized octahedral cluster on n-type and p-type Si (111) surface modified with pyridine terminated alkyl layer by taking advantage of labile nature of apical halogen atoms (Fabre et al. 2009, Cordier et al. 2010). Graphene oxide, due to presence of plenty oxygenated groups, has been identified as suitable supporting material. Kumar et al. (2015f) have immobilized octahedral Mo clusters on GO by replacement of apical bromine atoms by oxygen carrying functionalities present on the surface of graphene oxide (Kumar et al. 2015f). Methanol was observed as the sole CO_2 reduction product by using $GO-Cs_2Mo_6Br^i_8Br^a_x$ (i = icosahedral, a = apical) and $GO-(TBA)_2Mo_6Br^i_8Br^a_x$ as photocatalysts and the formation rate was found to be 68.5 and 53.9 μmol g^{-1} h^{-1}, respectively, after 24 hr irradiation under visible light without using any sacrificial donor. The obtained quantum yield and turn over number by using $GO-Cs_2Mo_6Br^i_8Br^a_x$ was calculated to be 0.015 and 19.0 respectively, while for $GO-(TBA)_2Mo_6Br^i_8Br^a_x$ these values were 0.011 and 10.38, respectively. The plausible mechanism defines that metal clusters after absorption of visible light get excited and transfer electrons and positive charge to CB and VB of GO, respectively, so indirectly charge pairs get evolved on GO that was used for CO_2 reduction and water oxidation.

In a study, Kumar et al. (2015g) synthesized octahedral hexacyano rhenium $K_4[Re_6S_8(CN)_6]$ cluster complexes which were grafted onto photoactive $Cu(OH)_2$ cluster modified TiO_2 {$Cu(OH)_2/TiO_2$} support (Kumar et al. 2015g). In a previous report, $Cu(OH)_2$ cluster modified TiO_2 has been identified as good hydrogen evolving catalyst (see Yu and Ran 2011), so it was used as an active support for the immobilization of Re clusters. Re cluster molecules were immobilized covalently by Re-CN-M bridges bond formation. The methanol yield after 24 hr irradiation was found to be 149 μmol/0.1 g cat for Re-cluster@$Cu(OH)_2/TiO_2$ photocatalyst that is much higher than 35 μmol/0.1 g cat for $Cu(OH)_2/TiO_2$ and 75 μmol/0.1 g cat for equimolar rhenium cluster in the presence of triethanolamine (TEOA) as a sacrificial donor. The quantum yields (ϕ_{MeOH}) of Re-cluster@$Cu(OH)_2/TiO_2$ and $Cu(OH)_2/TiO_2$ were

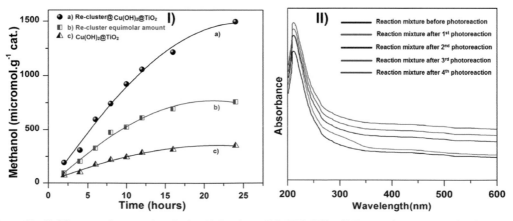

Figure 11. (I) CO_2 conversion to methanol using (a) Re-cluster@$Cu(OH)_2/TiO_2$, (b) Re complex at same equimolar amount and (c) $Cu(OH)_2/TiO_2$. (II) UV–vis spectra of reaction mixtures of recycling experiments. Reprinted with permission from ref. Kumar, P., N.G. Naumov, R. Boukherroub and S.L. Jain. 2015. Octahedral rhenium $K_4[Re_6S_8(CN)_6]$ and $Cu(OH)_2$ cluster modified TiO_2 for the photoreduction of CO_2 under visible light irradiation. Appl. Catal. A 499: 32–38. Copyright@Elsevier.

found to be 0.018 and 0.004 mol Einstein^{-1}, respectively (Figure 11-I). The catalyst was recyclable and no loss in activity was observed. Further, after every recycling, the UV-Vis spectra of solution was measured which showed no leaching of cluster from the surface of modified $Cu(OH)_2/TiO_2$ semiconductor support (Figure 11-II).

The $Cu(OH)_2$ clusters modified TiO_2 has band gap 3.11 eV which was somewhat lower than TiO_2. The reduction potential of $Cu(OH)_2/Cu$ is -0.634 V (vs. NHE at pH-7), which was negative enough for the reduction of CO_2 to methanol {CO_2/CH_3OH (-0.38 V vs. NHE at pH-7)} and H^+/H_2(-0.41 V vs. NHE at pH-7). So electrons can be transferred from TiO_2 CB to $Cu(OH)_2$ and, subsequently, to CO_2 or proton.

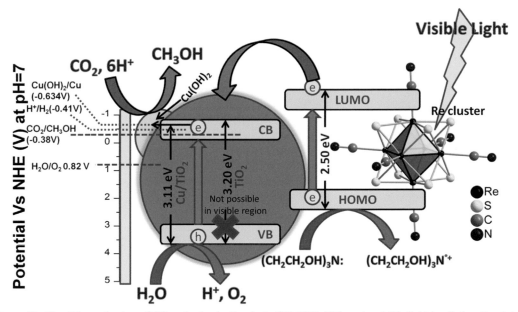

Figure 12. Plausible mechanism of CO_2 reduction by Re-cluster@$Cu(OH)_2/TiO_2$ under visible light irradiation. Reprinted with permission from ref. Kumar, P., N.G. Naumov, R. Boukherroub and S.L. Jain. 2015. Octahedral rhenium $K_4[Re_6S_8(CN)_6]$ and $Cu(OH)_2$ cluster modified TiO_2 for the photoreduction of CO_2 under visible light irradiation. Appl. Catal. A 499: 32–38. Copyright@Elsevier.

However, the value of reduction potential of CO_2 is more negative than reduction potential of proton, so CO_2 reduction win over proton reduction and methanol was observed as a major reaction product (Figure 12).

Graphitic carbon nitride as photocatalyst

Semiconductor polymeric materials have gained considerable importance in recent years due to their flexibility for structure modification and tuning of band gap by degree of polymerization. Among various polymeric materials known, graphitic carbon nitride is important due to its low band gap (2.7 eV) and suitable position of conduction band (-1.1 eV) and valence band ($+1.6$ eV) (see Ong et al. 2016). Carbon nitride is a 2D polymer consisting of interconnected tri-s-triazines units via tertiary amines. Carbon nitride is mainly synthesized with dicyanamide, melamine, etc. For example, Wang et al. (2009b) synthesized carbon nitride by heating melamine in air and used it for photocatalytic water splitting (Wang et al. 2009b). However, the precursor melamine is expensive, difficult to obtain and highly explosive, so urea can be used as a cost effective and non-toxic alternative for preparation of carbon nitride. Dong and coworkers (2012) synthesized a layered g-C_3N_4 with a high surface area by directly heat treating the urea in air, which exhibited much higher visible light photoactivity (Dong et al. 2012). The band gap can be further reduced via doping with heteroatoms like P, S and I, etc. (see Zhang et al. 2014). Recently, Zhang et al. (2010) synthesized P-doped g-C$_3$N$_4$ by using 1-butyl-3-methylimidazolium hexafluorophosphate (BmimPF$_6$) as a source of phosphorous and obtained an improved photocurrent response (Zhang et al. 2010). Hong et al. (2012) prepared S-doped mesoporous carbon nitride (mpgCNS) by *in situ* method and observed higher hydrogen evolution rate in comparison to mpgCN (Hong et al. 2012).

As carbon nitride possesses graphene like sheets for better mobility to electrons and small band gap, therefore, it has been combined with other semiconductor materials like BiVO$_4$, TaON, BiPO$_4$, TiO$_2$, WO$_3$, NaNbO$_3$, Cu$_2$O, Bi$_2$WO$_6$, to obtain nano-hybrids for various applications including hydrogen evolution and CO$_2$ reduction. Porosity is the main feature which determines the concentration of reactant on the photocatalyst surface and therefore the photocatalytic performance (see Xu et al. 2013c, Portehault et al. 2010, Yan et al. 2010, Pan et al. 2012, Ge et al. 2011).

Kailasam et al. (2011) synthesized highly mesoporous carbon nitride by sol gel method using silica as template. Ordered structured mesoporous carbon nitride was prepared by facile sol-gel route using precursor TEOS as source of silica and cyanamide as the source for carbon nitride. The synthesized porous carbon nitride was used for solar hydrogen evolution experiment and it has been found that sample CN-6 (1:6, TEOS:CN) gave the highest yield of hydrogen (40.5 H$_2$/mL) after 24 h using K$_2$PtCl$_6$ as catalyst and TEOA as sacrificial donor (Kailasam et al. 2011).

In another report, Kailasam et al. (2015) synthesized WO$_3$/carbon nitride (W-TEOS-CN) composite by simple mixing of precursor cyanamide, TEOS, followed by addition of WO$_3$ (Kailasam et al. 2015). The synthesized composite was found to be very active for hydrogen evolution and the rate of hydrogen evolution could be reached up to 326 μmol H$_2$ h^{-1}, which was 2.4 times of simple nanoporous carbon nitride. The enhanced activity was assumed due to the efficient charge separation via Z-scheme and this was confirmed by photoluminescence (PL) spectroscopic analysis.

Shi and co-workers (2014) synthesized visible-light-responsive g-C$_3$N$_4$/NaNbO$_3$ nanowires by fabrication of polymeric g-C$_3$N$_4$ on NaNbO$_3$ nanowires. The calculated value of optical band gap for NaNbO$_3$ and g-C$_3$N$_4$ was found to be 3.4 and 2.7 eV, respectively, so the addition of g-C$_3$N$_4$ generates heterojunction which facilitates electron transfer from carbon nitride to NaNbO$_3$. During the first 4 hr of irradiation, the rate of CH$_4$ generation over Pt-g-C$_3$N$_4$/NaNbO$_3$ was as high as 6.4 μmol h^{-1} g^{-1} in contrast to that of Pt-g-C$_3$N$_4$ (0.8 μmol h^{-1} g^{-1}) (Shi et al. 2014).

As can be seen in the existing literature, most of semiconductor photocatalysts use expensive noble metals like Pt, Pd, Ru, Ir, etc. as charge capturing species. These noble metals can activate substrate molecules by bond formation which further facilitates CO$_2$ reduction/protons reduction. However, the addition of noble metal adds extra cost which makes the process less viable from economical viewpoints. So, efforts are directed for designing photocatalyst which can work without noble metal dopants. The core shell LaPO$_4$/g-C$_3$N$_4$ nanowires synthesized via an *in situ* hydrothermal growth of LaPO$_4$ nanorods in tubular g-C$_3$N$_4$ was investigated for photocatalytic conversion of CO$_2$ to CO in the absence of any noble metal (see Li et al. 2017). The irradiation of 30 mg nanocomposite yielded 0.433 μmol of CO from CO$_2$ within 1 hr irradiation. Photoluminescence spectra (PL spectra) of LaPO$_4$ showed the lowest PL intensity due to the small amount of charge generation. The g-C$_3$N$_4$ showed the highest PL intensity because of charge generation under visible light. However, core shell structure formation by on LaPO$_4$ quenched PL signal which was probably due to charge transfer from g-C$_3$N$_4$ to LaPO$_4$ (Li et al. 2017).

The photo-efficiency of carbon nitride material can be further increased by sensitization with metal complexes. Recently, Takanabe et al. (2010) have immobilized magnesium phthalocyanine on carbon nitride by π–π stacking interaction and observed enhanced photocatalytic performance for water splitting (Takanabe et al. 2010). Ruthenium complex (trans(Cl)-[Ru(bpyX$_2$)(CO)$_2$Cl$_2$]) immobilized on carbon nitride via phosphate linkers has been found a potent catalyst for photoreduction of CO$_2$ to CO under visible light (see Kuriki et al. 2015). Graphitic carbon nitride also works as CO$_2$ capturing agent so it can capture and reduce CO$_2$ simultaneously. Lin et al. (2014) developed a cobalt bipyridine redox catalyst modified carbon nitride, Co(bpy)$_3^{+2}$/C$_3$N$_4$, system, having cobalt oxide and hole capturing agent for photocatalytic reduction of CO$_2$ to CO under visible light irradiation (Lin et al. 2014). In a work by Kuriki et al. (2016), they have synthesized ruthenium binuclear complex supported on graphitic carbon nitride via phosphate moiety (RuRu'/C$_3$N$_4$) for the photoreduction of CO$_2$ to HCOOH in visible light region (Kuriki et al. 2016). Further, modification of carbon nitride with silver nanoparticles increases the yield of formic acid (TON > 33,000 with respect to the amount of RuRu'/Ag/C_3N_4). Kumar and co-workers (2016) have synthesized iron bipyridyl complex grafted nanoporous carbon nitride (Fe(bpy)$_3$/npg-C_3N_4) which was used for visible light, mediating the amines oxidative coupling (Kumar et al. 2016).

Other carbonaceous materials for CO₂ activation

Apart from graphene, carbon nitride and other carbonaceous materials like fullerene, carbon nanotube (CNT), carbon quantum dots (CDs), etc. have also been explored for the CO_2 activation applications. For example, Wu et al. (2015) reported nitrogen doped carbon nanotube which due to the presence of pyridinic nitrogen generated electron rich defects that work as active sites for the reduction of CO_2 (Wu et al. 2015). Liang et al. (2012) synthesized a composite of single walled carbon nanotube–titania nanosheet (SWCNT–TiNS) with low carbon defect densities for investigating role of carbon nanomaterial dimensionality on photocatalytic response. However, the photoactivity of SWCNT–TiNS composite for methane production was found to be less than solvent exfoliated graphene-titania nanosheet SEG–TiNS but this activity was much higher than bare titania nanosheet (Liang et al. 2012).

Carbon nanodots, due to the quantum confinement effect, have been emerged as new photocatalytic materials. C Dots have been used for various application, fluorescence emission, dye degradation, energy conversion and storage, etc. (see Fernando et al. 2015). Carbon dots, due to the presence of graphene like oxidized sheets, can work as semiconductor and the optical properties are strongly dependent on their size. Further, like graphene, their band gap can be reduced by doping with nitrogen. A work by Sahu et al. (2014) has synthesized carbon quantum dots decorated with Au nanoparticles for solar light harvesting. Various small hydrocarbon molecules were observed with higher selectivity for formic acid. Interestingly, the increase in CO_2 pressure enhances the yield of products (formic acid and acetic acid were 1.2 and 0.06 mmol h^{-1} g^{-1}), at 700 psi and 405–720 nm wavelength for 4 hr due to better absorption of CO_2 at elevated pressure (Sahu et al. 2014). In a similar study, Choi et al. (2013) prepared Ag decorated carbon dots by decomposition of cyclodextrin under UV light in the presence of AgNO$_3$ (Choi et al. 2013).

Very recently, Yadav et al. (2016) reported 6-amino-2-(9,10-dioxo-6-(2-(perylen-3-yl)-4,5-di-p-tolyl-4,5-dihydro-1H-imidazol-1-yl)-9,10-dihydroanthracen-2-yl)-1Hbenzo[de]isoquinoline-1,3(2H)-dione (ANP) functionalized graphene quantum dots coupled with formate dehydrogenase (FDH) enzyme as potent photocatalyst for selective reduction of CO_2 to formic acid (Yadav et al. 2016).

Conclusion

Various nanostructured composites including semiconductors and carbon materials hybrids have been explored thoroughly for photocatalytic activation of water, carbon dioxide and other molecules. Photocatalytic conversion of CO_2 to solar fuel by means of artificial photosynthesis is a challenging aspect which has the potential to solve both the issues related to global warming as well as depleting energy resources. Thus, the field of development of photocatalytic systems for CO_2 reduction using solar light is likely to be a topic of tremendous importance in the prospects of current situation. CO_2 reduction over unmodified semiconductor material afforded poor yield of photoreduced products due to the large band gap which makes this approach far from realization. However, band gap engineering with doping, mixing with low band gap semiconductor, sensitization with metal complexes or hybridization with various nanomaterials can produce nanohybrid photocatalysts with tuned band edge positions that can sustain CO_2 reduction and water oxidation process. The efficient separation of photogenerated charge is the key for achieving higher conversion efficiency, so electron and hole capturing agents play a pivotal role, and facilitate charge separation. Carbon materials such as graphene oxide, reduced graphene oxide, and carbon nitride have shown to provide enhanced quantum yield due to higher mobility of charge carrier which prevents recombination. Doping with heteroatoms and composite formation with different band gap materials showed higher photocatalytic performance. Among the various known approaches, hybridization of semiconductors with metal complexes seems to be the most promising one to get higher hydrocarbon products in enough concentration to reduce impact of CO_2 concentration on environment.

References

Ahmed, N., Y. Shibata, T. Taniguchi and Y. Izumi. 2011. Photocatalytic conversion of carbon dioxide into methanol using zinc–copper–M (III) (M = aluminum, gallium) layered double hydroxides. J. Catal. 279: 123–135.

Ajayaghosh, A. 2003. Donor–acceptor type low band gap polymers: polysquaraines and related systems. Chem. Soc. Rev. 32: 181–191.

An, X., K. Li and J. Tang. 2014. Cu_2O/reduced graphene oxide composites for the photocatalytic conversion of CO_2. ChemSusChem. 7: 1086–1093.

Aresta, M., A. Dibenedetto and A. Angelini. 2014. Catalysis for the valorization of exhaust carbon: from CO_2 to chemicals, materials, and fuels. Technological use of CO_2. Chem. Rev. 14: 1709–1742.

Atabani, A.E., A.S. Silitonga, I.A. Badruddin, T.M.I. Mahlia, H.H. Masjuki and S. Mekhilef. 2012. A comprehensive review on biodiesel as an alternative energy resource and its characteristics. Renew. Sustain. Energy Rev. 16: 2070–2093.

Bachu, S. 2008. CO_2 storage in geological media: role, means, status and barriers to deployment. Pro. Energy Combus. Sci. 34: 254–273.

Bai, S. and X. Shen. 2012. Graphene–inorganic nanocomposites. RSC Adv. 2: 64–98.

Balcerski, W., S.Y. Ryu and M.R. Hoffmann. 2007. Visible-light photoactivity of nitrogen-doped TiO_2: photo-oxidation of HCO_2H to CO_2 and H_2O. J. Phys. Chem. C 111: 15357–15362.

Barber, J. 2009. Photosynthetic energy conversion: natural and artificial. Chem. Soc. Rev. 38: 185–196.

Barelli, L., G. Bidini, F. Gallorini and S. Servili. 2008. Hydrogen production through sorption-enhanced steam methane reforming and membrane technology: a review. Energy 33: 554–570.

Bonaccorso, F., Z. Sun, T. Hasan and A.C. Ferrari. 2010. Graphene photonics and optoelectronics. Nat. Photonics 4: 611–622.

Brand, B. and K. Blok. 2015. Renewable energy perspectives for the North African electricity systems: a comparative analysis of model-based scenario studies. Energy Strat. Rev. 6: 1–11.

Brownson, D.A.C., D.K. Kampouris and C.E. Banks. 2011. An overview of graphene in energy production and storage applications. J. Power Sources 196: 4873–4885.

Campbell, J.E., D.B. Lobell, R.C. Genova and C.B. Field. 2008. The global potential of bioenergy on abandoned agriculture lands. Environ. Sci. Tech. 42: 5791–5794.

Chang, Y.H., W. Zhang, Y. Zhu, Y. Han, J. Pu, J.K. Chang et al. 2014. Monolayer $MoSe_2$ grown by chemical vapor deposition for fast photodetection. ACS Nano 8: 8582–8590.

Chen, D., H. Feng and J. Li. 2012. Graphene oxide: preparation, functionalization, and electrochemical applications. Chem. Rev. 112: 6027–6053.

Chestnoy, N., T.D. Harris, R. Hull and L.E. Brus. 1986. Luminescence and photophysics of cadmium sulfide semiconductor clusters: the nature of the emitting electronic state. J. Phys. Chem. 90: 3393–3399.

Choi, H., S.-J. Ko, Y. Choi, P. Joo, T. Kim, B.R. Lee et al. 2013. Versatile surface plasmon resonance of carbon-dot-supported silver nanoparticles in polymer optoelectronic devices. Nat. Photonics 7: 732–738.

Choudhary, S., S. Upadhyay, P. Kumar, N. Singh, V.R. Satsangi, R. Shrivastav et al. 2012. Nanostructured bilayered thin films in photoelectrochemical water splitting—A review. Int. J. Hydro. Energy 37: 18713–18730.

Chouhan, N., R.S. Liu and S.F. Hu. 2013. Cd-ZnGeON solid solution: the effect of local electronic environment on the photocatalytic water cleavage ability. J. Mater. Chem. A 1: 7422–7432.

Cook, T.R., D.K. Dogutan, S.Y. Reece, Y. Surendranath, T.S. Teets and D.G. Nocera. 2010. Solar energy supply and storage for the legacy and nonlegacy worlds. Chem. Rev. 110: 6474–6502.

Cordier, S., B. Fabre, Y. Molard, A.-B.F.-Djomkam, N. Tournerie, A. Ledneva et al. 2010. Covalent anchoring of Re6Sei8 cluster cores monolayers on modified n-and p-type Si (111) surfaces: Effect of coverage on electronic properties. J. Phys. Chem. C 114: 18622–18633.

Cowan, A.J. and J.R. Durrant. 2013. Long-lived charge separated states in nanostructured semiconductor photoelectrodes for the production of solar fuels. Chem. Soc. Rev. 42: 2281–2293.

Das, J. and D. Khushalani. 2010. Nonhydrolytic route for synthesis of ZnO and its use as a recyclable photocatalyst. J. Phys. Chem. C. 114: 2544–2550.

Davda, R.R., J.W. Shabaker, G.W. Huber, R.D. Cortright and J.A. Dumesic. 2005. A review of catalytic issues and process conditions for renewable hydrogen and alkanes by aqueous-phase reforming of oxygenated hydrocarbons over supported metal catalysts. Appl. Catal. B 56: 171–186.

Dincer, I. 1999. Environmental impacts of energy. Energy Policy 27: 845–854.

Dong, F., Y. Sun, L. Wu, M. Fu and Z. Wu. 2012. Facile transformation of low cost thiourea into nitrogen-rich graphitic carbon nitride nanocatalyst with high visible light photocatalytic performance. Catal. Sci. Technol. 2: 1332–1335.

Dresselhaus, M.S. and I.L. Thomas. 2001. Alternative energy technologies. Nature 414: 332–337.

Du, P., J. Schneider, G. Luo, W.W. Brennessel and R. Eisenberg. 2009. Visible light-driven hydrogen production from aqueous protons catalyzed by molecular cobaloxime catalysts. Inorg. Chem. 48: 4952–4962.

Dukovic, G., B.E. White, Z. Zhou, F. Wang, S. Jockusch, M.L. Steigerwald et al. 2004. Perovskites as catalysts in the reforming of hydrocarbons: a review. J. Am. Chem. Soc. 126: 15269–15275.

Eda, G., C. Mattevi, H. Yamaguchi, H. Kim and M. Chhowalla. 2009. Insulator to semimetal transition in graphene oxide. J. Phys. Chem. C 113: 15768–15771.

Fabre, B., S. Cordier, Y. Molard, C. Perrin, S.A.-Girard and C. Godet. 2009. Electrochemical and charge transport behavior of molybdenum-based metallic cluster layers immobilized on modified n-and p-type Si (111) surfaces. J. Phys. Chem. C 113: 17437–17446.

Felderhoff, M., C. Weidenthaler, R.V. Helmolt and U. Eberle. 2007. Hydrogen storage: the remaining scientific and technological challenges. Phys. Chem. Chem. Phys. 9: 2643–2653.

Feng, X., J.D. Sloppy, T.J. LaTempa, M. Paulose, S. Komarneni, N. Bao et al. 2011. Synthesis and deposition of ultrafine Pt nanoparticles within high aspect ratio TiO_2 nanotube arrays: application to the photocatalytic reduction of carbon. J. Mater. Chem. 21: 13429–13433.

Fernando, K.A.S., S. Sahu, Y. Liu, W.K. Lewis, E.A. Guliants, A. Jafariyan et al. 2015. Carbon quantum dots and applications in photocatalytic energy conversion. ACS Appl. Mater. Interfaces 7: 8363–8376.

Fujishima, A. and K. Honda. 1971. Electrochemical evidence for the mechanism of the primary stage of photosynthesis. Nature 44: 1148–1150.

Fujishima, A. and K. Honda. 1972. Electrochemical photolysis of water at a semiconductor electrode. Nature 238: 37–38.

Ganesh, I. 2014. Conversion of carbon dioxide into methanol—a potential liquid fuel: Fundamental challenges and opportunities (a review). Renew. Sustainable Energy Rev. 31: 221–257.

Ge, L., C.C. Han and J. Liu. 2011. Novel visible light-induced gC_3N_4/Bi_2WO_6 composite photocatalysts for efficient degradation of methyl orange. Appl. Catal. B 108-109: 100–107.

Georgekutty, R., M.K. Seery and S.C. Pillai. 2008. A highly efficient Ag-ZnO photocatalyst: synthesis, properties, and mechanism. J. Phys. Chem. C 112: 13563–13570.

Giesekam, J., J. Barrett, P. Taylor and A. Owen. 2014. The greenhouse gas emissions and mitigation options for materials used in UK construction. Energy Build. 78: 202–214.

Gouedard, V.B., G. Rimmele, O. Porcherie, N. Quisel and J. Desroches. 2009. A solution against well cement degradation under CO_2 geological storage environment. Int. J. Greenhouse Gas Control. 3: 206–216.

Gusain, R., P. Kumar, O.P. Sharma, S.L. Jain and O.P. Khatri. 2016. Reduced graphene oxide–CuO nanocomposites for photocatalytic conversion of CO_2 into methanol under visible light irradiation. Appl. Catal. B 181: 352–362.

Halmann, M. 1978. Photoelectrochemical reduction of aqueous carbon dioxide on p-type gallium phosphide in liquid junction solar cells. Nature 275: 115–116.

Hashimoto, K., H. Irie and A. Fujishima. 2005. TiO_2 photocatalysis: a historical overview and future prospects. Jpn. J. Appl. Phys. 44: 8269–8285.

He, Y., L. Zhang, B. Teng and M. Fan. 2015. New application of Z-scheme Ag_3PO_4/g-C_3N_4 composite in converting CO_2 to fuel. Environ. Sci. Technol. 49: 649–656.

Hochbaum, A.I. and P. Yang. 2010. Semiconductor nanowires for energy conversion. Chem. Rev. 110: 527–546.

Hong, J., X. Xia, Y. Wang and R. Xu. 2012. Mesoporous carbon nitride with *in situ* sulfur doping for enhanced photocatalytic hydrogen evolution from water under visible light. J. Mater. Chem. 22: 15006–15012.

Hong, J., W. Zhang, Y. Wang, T. Zhou and R. Xu. 2014. Photocatalytic reduction of carbon dioxide over self-assembled carbon nitride and layered double hydroxide: the role of carbon dioxide enrichment. ChemCatChem. 6: 2315–2321.

Hou, J., Z. Wang, W. Kan, S. Jiao, H. Zhu and R.V. Kumar. 2012. Efficient visible-light-driven photocatalytic hydrogen production using CdS@TaON core–shell composites coupled with graphene oxide nanosheets. J. Mater. Chem. 22: 7291–7299.

Hsu, H.C., I. Shown, H.Y. Wei, Y.C. Chang, H.Y. Du, Y.G. Lin et al. 2013. Graphene oxide as a promising photocatalyst for CO_2 to methanol conversion. Nanoscale 5: 262–268.

Hu, Y.H., H. Wang and B. Hu. 2010. Thinnest two-dimensional nanomaterial—graphene for solar energy. ChemSusChem. 3: 782–796.

Hummers, W.S. and R.E. Offeman. 1958. Preparation of graphitic oxide. J. Am. Chem. Soc. 80: 1339.

In, S., D.D. Vaughn and R.E. Schaak. 2012. Photocatalytic conversion of CO_2 in water over layered double hydroxides. Angew. Chem. 124: 3981–3984.

Inoue, T., A. Fujishima, S. Konishi and K. Honda. 1979. Photoelectrocatalytic reduction of carbon dioxide in aqueous suspensions of semiconductor powders. Nature 277: 637–638.

Iwase, A., Y.H. Ng, Y. Ishiguro, A. Kudo and R. Amal. 2011. Reduced graphene oxide as a solid-state electron mediator in Z-scheme photocatalytic water splitting under visible light. J. Am. Chem. Soc. 133: 11054–11057.

Izumi, Y. 2013. Recent advances in the photocatalytic conversion of carbon dioxide to fuels with water and/or hydrogen using solar energy and beyond. Coord. Chem. Rev. 257: 171–186.

Jacobson, M.Z., R.W. Howarth, M.A. Delucchi, S.R. Scobie, J.M. Barth, M.J. Dvorak et al. 2013. Examining the feasibility of converting New York State's all-purpose energy infrastructure to one using wind, water, and sunlight. Energy Policy 57: 585–601.

Jensen, L.L., J.T. Muckerman and M.D. Newton. 2008. First-principles studies of the structural and electronic properties of the $(Ga_{1-x}Zn_x)(N_{1-x}O_x)$ solid solution photocatalyst. J. Phys. Chem. C 112: 3439–3446.

Ji, Z., X. Shen, G. Zhu, H. Zhou and A. Yuan. 2012. Reduced graphene oxide/nickel nanocomposites: facile synthesis, magnetic and catalytic properties. J. Mater. Chem. 22: 3471–3477.

Johnson, J.M.F., A.J. Franzluebbers, S.L. Weyers and D.C. Reicosky. 2007. Agricultural opportunities to mitigate greenhouse gas emissions. Environ. Pollution 150: 107–124.

Kailasam, K., J.D. Epping, A. Thomas, S. Losse and H. Junge. 2011. Mesoporous carbon nitride–silica composites by a combined sol–gel/thermal condensation approach and their application as photocatalysts. Energy Environ. Sci. 4: 4668–4674.

Kailasam, K., A. Fischer, G. Zhang, J. Zhang, M. Schwarze, M. Schrçder et al. 2015. Mesoporous carbon nitride-tungsten oxide composites for enhanced photocatalytic hydrogen evolution. ChemSusChem. 8: 1404–1410.

Karamian, E. and S. Sharifnia. 2016. On the general mechanism of photocatalytic reduction of CO_2. Journal of CO_2 Utilization 16: 194–203.

Kato, H. and A. Kudo. 2003. Photocatalytic water splitting into H_2 and O_2 over various tantalate photocatalysts. Catal. Today 78: 561–569.

Kibsgaard, J., T.F. Jaramillo and F. Besenbacher. 2014. Building an appropriate active-site motif into a hydrogen-evolution catalyst with thiomolybdate $[Mo_3S_{13}]^{2-}$ clusters. Nat. Chem. 6: 248–253.

Kondratenko, E.V., G. Mul, J. Baltrusaitis, G.O. Larrazabal and J.P. Ramırez. 2013. Status and perspectives of CO_2 conversion into fuels and chemicals by catalytic, photocatalytic and electrocatalytic processes. Energy Environ. Sci. 6: 3112–3135.

Konta, R., T. Ishii, H. Kato and A. Kudo. 2004. Photocatalytic activities of noble metal ion doped $SrTiO_3$ under visible light irradiation. J. Phys. Chem. B 108: 8992.

Krol, R., Y. Liang and J. Schoonman. 2008. Solar hydrogen production with nanostructured metal oxides. J. Matter. Chem. 18: 2311–2320.

Kumar, B., M. Llorente, J. Froehlich, T. Dang, A. Sathrum and C.P. Kubiak. 2012. Photochemical and photoelectrochemical reduction of CO_2. Annu. Rev. Phys. Chem. 63: 541–69.

Kumar, P., A. Kumar, B. Sreedhar, B. Sain, S.S. Ray and S.L. Jain. 2014a. Cobalt phthalocyanine immobilized on graphene oxide: An efficient visible-active catalyst for the photoreduction of carbon dioxide. Chem. Eur. J. 20: 6154–61611.

Kumar, P., B. Sain and S.L. Jain. 2014b. Photocatalytic reduction of carbon dioxide to methanol using a ruthenium trinuclear polyazine complex immobilized on graphene oxide under visible light. J. Mater. Chem. A 2: 11246–11253.

Kumar, P., S. Kumar, S. Cordier, S. Paofai, R. Boubherroub and S.L. Jain. 2014c. Photoreduction of CO_2 to methanol with hexanuclear molybdenum $[Mo_6Br_{14}]^{2-}$ cluster units under visible light irradiation. RSC Adv. 4: 10420–10423.

Kumar, P., A. Kumar, C. Joshi, R. Singh, S. Saran and S.L. Jain. 2015a. Heterostructured nanocomposite tin phthalocyanine@ mesoporous ceria ($SnPc@CeO_2$) for photoreduction of CO_2 in visible light. RSC Adv. 5: 42414–42421.

Kumar, P., R.K. Chauhan, B. Sain and S.L. Jain. 2015b. Photo-induced reduction of CO_2 using a magnetically separable Ru-$CoPc@TiO_2@SiO_2@Fe_3O_4$ catalyst under visible light irradiation. Dalton Trans. 44: 4546–4553.

Kumar, P., C. Joshi, N. Labhsetwar, R. Boukherroub and S.L. Jain. 2015c. A novel Ru/TiO_2 hybrid nanocomposite catalyzed photoreduction of CO_2 to methanol under visible light. Nanoscale 7: 15258–15267.

Kumar, P., A. Bansiwal, N. Labhsetwar and S.L. Jain. 2015d. Visible light assisted photocatalytic reduction of CO_2 using a graphene oxide supported heteroleptic ruthenium complex. Green Chem. 17: 1605–1609.

Kumar, P., H.P. Mungse, O.P. Khatri and S.L. Jain. 2015e. Nitrogen-doped graphene-supported copper complex: a novel photocatalyst for CO_2 reduction under visible light irradiation. RSC Adv. 5: 54929–54935.

Kumar, P., H.P. Mungse, S. Cordier, R. Boukherroub, O.P. Khatri and S.L. Jain. 2015f. Hexamolybdenum clusters supported on graphene oxide: Visible-light induced photocatalytic reduction of carbon dioxide into methanol. Carbon 94: 91–100.

Kumar, P., N.G. Naumov, R. Boukherroub and S.L. Jain. 2015g. Octahedral rhenium $K_4[Re_6S_8(CN)_6]$ and $Cu(OH)_2$ cluster modified TiO_2 for the photoreduction of CO_2 under visible light irradiation. Appl. Catal. A 499: 32–38.

Kumar, P., A. Kumar, C. Joshi, S. Ponnada, A.K. Pathak, A. Ali et al. 2016. A $[Fe(bpy)_3]^{2+}$ grafted graphitic carbon nitride hybrid for visible light assisted oxidative coupling of benzylamines under mild reaction conditions. Green Chem. 18: 2514–2521.

Kumar, S.G. and L.G. Devi. 2011. Review on modified TiO_2 photocatalysis under UV/visible light: selected results and related mechanisms on interfacial charge carrier transfer dynamics. J. Phys. Chem. A 115: 13211–13241.

Kuriki, R., K. Sekizawa, O. Ishitani and K. Maeda. 2015. Visible-light-driven CO_2 reduction with carbon nitride: enhancing the activity of ruthenium catalysts. Angew. Chem. Int. Ed. 54: 2406–2409.

Kuriki, R., H. Matsunaga, T. Nakashima, K. Wada, A. Yamakata, O. Ishitani et al. 2016. Nature-inspired, highly durable CO_2 reduction system consisting of a binuclear ruthenium (II) complex and an organic semiconductor using visible light. J. Am. Chem. Soc. 138: 5159–5170.

Lanzafame, P., G. Centi and S. Perathoner. 2014. Catalysis for biomass and CO_2 use through solar energy: opening new scenarios for a sustainable and low-carbon chemical production. Chem. Soc. Rev. 43: 7562–7580.

Lee, D.-S., H.-J. Chen and Y.-W. Chen. 2012. Photocatalytic reduction of carbon dioxide with water using $InNbO_4$ catalyst with NiO and Co_3O_4 cocatalysts. J. Physics and Chemistry of Solids 73: 661–669.

Li, J. and D. Xu. 2010. Tetragonal faceted-nanorods of anatase TiO_2 single crystals with a large percentage of active {100} facets. Chem. Commun. 46: 2301–2303.

Li, J., K. Cao, Q. Li and D. Xu. 2012. Tetragonal faceted-nanorods of anatase TiO_2 with a large percentage of active {100} facets and their hierarchical structure. CrystEngComm. 14: 83–85.

Li, K., X. An, K.H. Park and M. Khrasheh. 2014. A critical review of CO_2 photoconversion: catalysts and reactors. Catal. Today 224: 3–12.

Li, M., L. Zhang, X. Fan, M. Wu, M. Wang, R. Cheng et al. 2017. Core-shell $LaPO_4/gC_3N_4$ nanowires for highly active and selective CO_2 reduction. Appl. Catal. B 201: 629–635.

Li, R., Z. Chen, W. Zhao, F. Zhang, K. Maeda, B. Huang et al. 2013a. Sulfurization-assisted cobalt deposition on $Sm_2Ti_2S_2O_5$ photocatalyst for water oxidation under visible light irradiation. J. Phys. Chem. C 117: 376–382.

Li, X., Q. Wang, Y. Zhao, W. Wu, J. Chen and H. Meng. 2013b. Green synthesis and photo-catalytic performances for ZnO-reduced graphene oxide nanocomposites. J. Colloid Interface Sci. 411: 69–75.

Liang, Y.T., B.K. Vijayan, K.A. Gray and M.C. Hersam. 2011. Minimizing graphene defects enhances titania nanocomposite-based photocatalytic reduction of CO_2 for improved solar fuel production. Nano Lett. 11: 2865–2870.

Liang, Y.T., B.K. Vijayan, O. Lyandres, K.A. Gray and M.C. Hersam. 2012. Effect of dimensionality on the photocatalytic behavior of carbon–titania nanosheet composites: charge transfer at nanomaterial interfaces. J. Phys. Chem. Lett. 3: 1760–1765.

Lin, Z., G. Waller, Y. Liu, M. Liu and C.P. Wong. 2012. Facile synthesis of nitrogen-doped graphene via pyrolysis of graphene oxide and urea, and its electrocatalytic activity toward the oxygen-reduction reaction. Adv. Energy Mater. 2: 884–888.

Lin, J., Z. Pan and X. Wang. 2014. Photochemical reduction of CO_2 by graphitic carbon nitride polymers. ACS Sustainable Chem. Eng. 2: 353–358.

Linsebigler, A.L., G. Lu and J.T. Yates, Jr. 1995. Photocatalysis on TiO_2 surfaces: principles, mechanisms, and selected results. Chem. Rev. 95: 735–758.

Liu, G., C. Sun, H.G. Yang, S.C. Smith, L. Wang, G.Q. Lu et al. 2010a. Nanosized anatase TiO_2 single crystals for enhanced photocatalytic activity. Chem. Commun. 46: 755–757.

Liu, J., Z. Zhao, C. Xu, A. Duan and G. Jiang. 2010b. Simultaneous removal of soot and NOx over the $(La_{1.7}Rb_{0.3}CuO_4)x/nmCeO_2$ nanocomposite catalysts. Ind. Eng. Chem. Res. 49: 3112–3119.

Liu, Q., Y. Zhou, J. Kou, X. Chen, Z. Tian, J. Gao et al. 2010c. High-yield synthesis of ultralong and ultrathin Zn_2GeO_4 nanoribbons toward improved photocatalytic reduction of CO_2 into renewable hydrocarbon fuel. J. Am. Chem. Soc. 132: 14385–14387.

Liu, S., J. Yu and M. Jaroniec. 2010d. Tunable photocatalytic selectivity of hollow TiO_2 microspheres composed of anatase polyhedra with exposed {001} facets. J. Am. Chem. Soc. 132: 11914–11916.

Liu, S., J. Yu and M. Jaroniec. 2011. Anatase TiO_2 with dominant high-energy {001} facets: synthesis, properties, and applications. Chem. Mater. 23: 4085–4093.

Liu, X., H. Zhang, X. Yao, T. An, P. Liu, Y. Wang et al. 2012. Visible light active pure rutile TiO_2 photoanodes with 100% exposed pyramid-shaped (111) surfaces. Nano Res. 5: 762–769.

Loh, K.P., Q. Bao, G. Eda and M. Chhowalla. 2010. Graphene oxide as a chemically tunable platform for optical applications. Nature Chem. 2: 1015–1024.

Low, J., J. Yu and W. Ho. 2015. Graphene-based photocatalysts for CO_2 reduction to solar fuel. J. Phys. Chem. Lett. 6: 4244–4251.

Manne, A.S. and R.G. Richels. 2001. An alternative approach to establishing trade-offs among greenhouse gases. Nature 401: 675–677.

Mao, J., K. Li and T. Peng. 2013. Recent advances in the photocatalytic CO_2 reduction over semiconductors. Catal. Sci. Technol. 3: 2481–2498.

Markewitz, P., W. Kuckshinrichs, W. Leitner, J. Linssen, P. Zapp, R. Bongartz et al. 2012. Worldwide innovations in the development of carbon capture technologies and the utilization of CO_2. Energy Environ. Sci. 5: 7281–7305.

Mohamed, H.H. and D.W. Bahnemann. 2012. The role of electron transfer in photocatalysis: Fact and fictions. App. Cat. B 128: 91–104.

Momirlan, M. and T. Veziroglu. 2005. The properties of hydrogen as fuel tomorrow in sustainable energy system for a cleaner planet. Int. J. Hyrdo. Energy 30: 795–802.

Nagaveni, K., M.S. Hegde, N. Ravishankar, G.N. Subbanna and G. Madras. 2004. Synthesis and structure of nanocrystalline TiO_2 with lower band gap showing high photocatalytic activity. Langmuir 20: 2900–2907.

Navalon, S., A. Dhakshinamoorthy, M. Alvaro and H. Garcia. 2013. Photocatalytic CO_2 reduction using non-titanium metal oxides and sulfides. ChemSusChem. 6: 562–577.

Neaţu, S., J.A. M.-Agullo and H. Garcia. 2014. Solar light photocatalytic CO_2 reduction: general considerations and selected bench-mark photocatalysts. Int. J. Mol. Sci. 15: 5246–5262.

Novoselov, K.S., A.K. Geim, S.V. Morozov, D. Jiang, Y. Zhang, S.V. Dubonos et al. 2004. Electric field effect in atomically thin carbon films. Science 306: 666–669.

Olah, G.A., A. Goeppert and G.K.S. Prakash. 2009. Chemical recycling of carbon dioxide to methanol and dimethyl ether: From greenhouse gas to renewable, environmentally carbon neutral fuels and synthetic hydrocarbons. J. Org. Chem. 74: 487–498.

Ong, W.-J., L.-L. Tan, S.-P. Chail, S.-T. Yong and A.R. Mohamed. 2014. Self-assembly of nitrogen-doped TiO_2 with exposed {001} facets on a graphene scaffold as photo-active hybrid nanostructures for reduction of carbon dioxide to methane. Nano Res. 7: 1528–1547.

Ong, W.-J., L.-L. Tan, Y.H. Ng, S.-T. Yong and S.-P. Chai. 2016. Graphitic carbon nitride (g-C_3N_4)-based photocatalysts for artificial photosynthesis and environmental remediation: are we a step closer to achieving sustainability? Chem. Rev. 116: 7159–7329.

Pan, C.S., J. Xu, Y.J. Wang, D. Li and Y.F. Zhu. 2012. Dramatic activity of C_3N_4/$BiPO_4$ photocatalyst with core/shell structure formed by self-assembly. Adv. Funct. Mater. 22: 1518–1524.

Pan, F., K. Wu, H. Li, G. Xu, and W. Chen. 2014. Synthesis of {100} facet dominant anatase TiO_2 nanobelts and the origin of facet-dependent photoreactivity. Chem. Eur. J. 20: 15095–15101.

Pan, P.-W. and Y.-W. Chen. 2007. Photocatalytic reduction of carbon dioxide on NiO/InTaO$_4$ under visible light irradiation. Catal. Commun. 8: 1546–1549.

Panwar, N.L., S.C. Kaushik and S. Kothari. 2011. Role of renewable energy sources in environmental protection: a review. Renew. Sustain. Energy Rev. 15: 1513–1524.

Park, H., J.H. Choi, K.M. Choi, D.K. Lee and J.K. Kang. 2012. Highly porous gallium oxide with a high CO$_2$ affinity for the photocatalytic conversion of carbon dioxide into methane. J. Mater. Chem. 22: 5304.

Peharz, G., F. Dimorth and U. Wittstadt. 2007. Solar hydrogen production by water splitting with a conversion efficiency of 18%. Int. J. Hydro. Energy 32: 3248–3252.

Peura, P. and T. Hyttinin. 2011. The potential and economics of bioenergy in Finland. J. Cleaner Pro. 19: 927–945.

Portehault, D., C. Giordano, C. Gervais, I. Senkovska, S. Kaskel, C. Sanchez et al. 2010. High-surface-area nanoporous boron carbon nitrides for hydrogen storage. Adv. Funct. Mater. 20: 1827–1833.

Powell, T.W.R. and T.M. Lenton. 2012. Future carbon dioxide removal via biomass energy constrained by agricultural efficiency and dietary trends. Energy Environ. Sci. 5: 8116–8133.

Pradhan, N. and D.D. Sharma. 2011. Advances in light-emitting doped semiconductor nanocrystals. J. Phys. Chem. Lett. 2: 2818–2826.

Qu, Y. and X. Duan. 2013. Progress, challenge and perspective of heterogeneous photocatalysts. Chem. Soc. Rev. 42: 2568–2580.

Rao, C.N.R., K. Biswas, K.S. Subrahmanyama and A. Govindaraj. 2009. Graphene, the new nanocarbon. J. Mater. Chem. 19: 2457–2469.

Rehan, R. and M. Nehdi. 2005. Carbon dioxide emissions and climate change: policy implications for the cement industry. Environ. Sci. Policy 8: 105–114.

Roy, S.C., O.K. Varghese, M. Paulose and C.A. Grimes. 2010. Toward solar fuels: photocatalytic conversion of carbon dioxide to hydrocarbons. ACS Nano 4: 1259–1278.

Sahoo, N.G., Y. Pan, L. Li and S.H. Chan. 2012. Graphene-based materials for energy conversion. Adv. Mater. 24: 4203–4210.

Sahu, S., Y. Liu, P. Wang, C.E. Bunker, K.A.S. Fernando, W.K. Lewis et al. 2014. Visible-light photoconversion of carbon dioxide into organic acids in an aqueous solution of carbon dots. Langmuir 30: 8631–8636.

Sato, S., T. Arai and T. Morikawa. 2015. Toward solar-driven photocatalytic CO$_2$ reduction using water as an electron donor. Inorg. Chem. 54: 5105–5113.

Sekizawa, K., K. Maeda, K. Domen, K. Koike and Osamu Ishitani. 2013. Artificial Z-scheme constructed with a supramolecular metal complex and semiconductor for the photocatalytic reduction of CO$_2$. J. Am. Chem. Soc. 135: 4596–4599.

Serpone, N., D. Lawless and R. Khairutdinov. 1995. Size effects on the photophysical properties of colloidal anatase TiO$_2$ particles: size quantization versus direct transitions in this indirect semiconductor? J. Phys. Chem. 99: 16646–16654.

Shi, H., G. Chen, C. Zhang and Z. Zou. 2014. Polymeric g-C$_3$N$_4$ coupled with NaNbO$_3$ nanowires toward enhanced photocatalytic reduction of CO$_2$ into renewable fuel. ACS Catal. 4: 3637–3643.

Shown, I., H.C. Hsu, Y.C. Chang, C.H. Lin, P.K. Roy, A. Ganguly et al. 2014. Highly efficient visible light photocatalytic reduction of CO$_2$ to hydrocarbon fuels by Cu-nanoparticle decorated graphene oxide. Nano Lett. 14: 6097–6103.

Sun, L., Y. Qin, Q. Cao, B. Hu, Z. Huang, L. Ye et al. 2011. Novel photocatalytic antibacterial activity of TiO$_2$ microspheres exposing 100% reactive {111} facets. Chem. Commun. 47: 12628–12630.

Tahir, M. and N.S. Amin. 2013. Advances in visible light responsive titanium oxide-based photocatalysts for CO$_2$ conversion to hydrocarbon fuels. Energy Convers. Manage 76: 194–214.

Takanabe, K., K. Kamata, X. Wang, M. Antonietti, J. Kubota and K. Domen. 2010. Photocatalytic hydrogen evolution on dye-sensitized mesoporous carbon nitride photocatalyst with magnesium phthalocyanine. Phys. Chem. Chem. Phys. 12: 13020–13025.

Takanabe, K. and K. Domen. 2012. Preparation of inorganic photocatalytic materials for overall water splitting. ChemCatChem. 4: 1–14.

Takayama, T., K. Sato, T. Fujimura, Y. Kojima, A. Iwase and A. Kudo. 2017. Photocatalytic CO$_2$ reduction using water as an electron donor by a powdered Z-scheme system consisting of metal sulfide and an RGO–TiO$_2$ composite. Faraday Discuss 198: 397–407.

Tan, L.-L., W.-J. Ong, S.-P. Chai and A.R. Mohamed. 2013. Reduced graphene oxide-TiO$_2$ nanocomposite as a promising visible-light-active photocatalyst for the conversion of carbon dioxide. Nanoscale Res. Lett. 8: 465.

Tan, L.-L., W.-J. Ong, S.-P. Chai, B.T. Goh and A.R. Mohamed. 2015a. Visible-light-active oxygen-rich TiO$_2$ decorated 2D graphene oxide with enhanced photocatalytic activity toward carbon dioxide reduction. Appl. Catal. B 179: 160–170.

Tan, L.-L., W.-J. Ong, S.-P. Chai and A.R. Mohamed. 2015b. Noble metal modified reduced graphene oxide/TiO$_2$ ternary nanostructures for efficient visible-light-driven photoreduction of carbon dioxide into methane. Appl. Catal. B 166–167: 251–259.

Teramura, K., S. Iguchi, Y. Mizuno, T. Shishido and T. Tanaka. 2012. Photocatalytic conversion of CO$_2$ in water over layered double hydroxides. Angew Chem. Int. Ed. 51: 8001–8018.

Tsai, C.W., H.M. Chen, R.S. Liu, K. Asakura and T.S. Chan. 2011. Ni@NiO core–shell structure-modified nitrogen-doped InTaO$_4$ for solar-driven highly efficient CO$_2$ reduction to methanol. J. Phys. Chem. C 115: 10180–10186.

Tseng, I.-H., J.C.S. Wu and H.-Y. Chou. 2004. Effects of sol-gel procedures on the photocatalysis of Cu/TiO$_2$ in CO$_2$ photoreduction. J. Catal. 221: 432–440.

Varghese, O.K., M. Paulose, T.J. Latempa and C.A. Grimes. 2009. High-rate solar photocatalytic conversion of CO$_2$ and water vapor to hydrocarbon fuels. Nano Lett. 9: 731–737.

Wang, C., R.L. Thompson, J. Baltrus and C. Matranga. 2010. Visible light photoreduction of CO_2 using CdSe/Pt/TiO₂ heterostructured catalysts. J. Phys. Chem. Lett. 1: 48–53.

Wang, J., D.N. Tafen, J.P. Lewis, Z. Hong, A. Manivannan, M. Zhi et al. 2009a. Origin of photocatalytic activity of nitrogen-doped TiO₂ nanobelts. J. Am. Chem. Soc. 131: 12290–12297.

Wang, J.C., L. Zhang, W.-X. Fang, J. Ren, Y.-Y. Li, H.-C. Yao et al. 2015a. Enhanced photoreduction CO_2 activity over direct Z-scheme α-Fe₂O₃/Cu₂O heterostructures under visible light irradiation. ACS Appl. Mater. Interfaces 7: 8631–8639.

Wang P.-Q., Y. Bai, P.-Y. Luo and J.-Y. Liu. 2013b. Graphene–WO3 nanobelt composite: Elevated conduction band toward photocatalytic reduction of CO_2 into hydrocarbon fuels. Catal. Commun. 38: 82–85.

Wang, W., S. Wang, X. Ma and J. Gong. 2011. Recent advances in catalytic hydrogenation of carbon dioxide. Chem. Soc. Rev. 40: 3703–3727.

Wang, X., K. Maeda, X. Chen, K. Takanabe, K. Domen, Y. Hou et al. 2009b. Polymer semiconductors for artificial photosynthesis: hydrogen evolution by mesoporous graphitic carbon nitride with visible light. J. Am. Chem. Soc. 131: 1680–1681.

Wang, Y., Q. Wang, X. Zhan, F. Wang, M. safdar and J. He. 2013a. Visible light driven type II heterostructures and their enhanced photocatalysis properties: a review. Nanoscle 5: 8326–8339.

Wang, Z.W., Y. Yue and Y. Cao. 2014. Preparation and properties of nitrogen doped p-type zinc oxide films by reactive magnetron sputtering. Vacuum 101: 313–316.

Wang, Z., G. Liu, C. Ding, Z. Chen, F. Zhang, J. Shi et al. 2015b. Synergetic effect of conjugated Ni(OH)₂/IrO₂ cocatalyst on titanium-doped hematite photoanode for solar water splitting. J. Phys. Chem. C 119: 19607–19612.

Wehling, T.O., K.S. Novoselov, S.V. Morozov, E.E. Vdovin, M.I. Katsnelson, A.K. Giem et al. 2008. Molecular doping of grapheme. Nano. Lett. 8: 173–177.

Woolerton, T.W., S. Sheard, E. Reisner, E. Pirece, S.W. Ragsdale and F.A. Armstrong. 2010. Efficient and clean photoreduction of CO_2 to CO by enzyme-modified TiO₂ nanoparticles using visible light. J. Am. Chem. Soc. 132: 2132–2133.

Wu, J., R.M. Yadav, M. Liu, P.P. Sharma, C.S. Tiwary, L. Ma et al. 2015. Achieving highly efficient, selective, and stable CO_2 reduction on nitrogen-doped carbon nanotubes. ACS Nano 9: 5364–5371.

Xiang, Q., J. Yu, W. Wang and M. Jaroniec. 2011. Nitrogen self-doped nanosized TiO₂ sheets with exposed {001} facets for enhanced visible-light photocatalytic activity. Chem. Commun. 47: 6906–6908.

Xie, S., Y. Wang, Q. Zhang, W. Deng and Y. Wang. 2015. SrNb₂O₆ nanoplates as efficient photocatalysts for the preferential reduction of CO_2 in the presence of H₂O. Chem. Commun. 51: 3430–3433.

Xing, M., F. Shen, B. Qiu and J. Zhang. 2014. Highly-dispersed boron doped graphene nanosheets loaded with TiO₂ nanoparticles for enhancing CO_2 photoreduction. Sci. Rep. 4: 6341.

Xu, H., P. Reunchan, S. Ouyang, H. Tong, N. Umezawa, T. Kako et al. 2013a. Anatase TiO₂ single crystals exposed with high-reactive {111} facets toward efficient H₂ evolution. Chem. Mater. 25: 405–411.

Xu, H., S. Ouyang, P. Li, T. Kako and J. Ye. 2013b. High-active anatase TiO₂ nanosheets exposed with 95% {100} facets toward efficient H₂ evolution and CO_2 photoreduction. ACS Appl. Mater. Interfaces 5: 1348–1354.

Xu, J., Y. Wang and Y. Zhu. 2013c. Nanoporous graphitic carbon nitride with enhanced photocatalytic performance. Langmuir 29: 10566–10572.

Yadav, D., R.K. Yadav, A. Kumar, N.-J. Park and J.-O. Baeg. 2016. Functionalized graphene quantum dots as efficient visible-light photocatalysts for selective solar fuel production from CO_2. Chemcatchem. 8: 3389–3393.

Yamanaka, K.I., S. Sato, M. Iwaki, T. kajino and T. Morikawa. 2011. Photoinduced electron transfer from nitrogen-doped tantalum oxide to adsorbed ruthenium complex. J. Phys. Chem. C 115: 18348–18353.

Yan, S.C., S.B. Lv, Z.S. Li and Z.G. Zou. 2010. Organic–inorganic composite photocatalyst of gC₃N₄ and TaON with improved visible light photocatalytic activities. Dalton Trans. 39: 1488–1491.

Yan, S., H. Yu, N. Wang, Z. Li and Z. Zou. 2012. Efficient conversion of CO_2 and H₂O into hydrocarbon fuel over ZnAl₂O₄-modified mesoporous ZnGaNO under visible light irradiation. Chem. Commun. 48: 1048–1050.

Yang, H.G., C.H. Sun, S.Z. Qiao, J. Zou, G. Liu, S.C. Smith et al. 2008. Anatase TiO₂ single crystals with a large percentage of reactive facets. Nature 453: 638–641.

Yang, H.G., G. Liu, S.Z. Qiao, C.H. Sun, Y.G. Jin, S.C. Smith et al. 2009. Solvothermal synthesis and photoreactivity of anatase TiO₂ nanosheets with dominant {001} facets. J. Am. Chem. Soc. 131: 4078–4083.

Yang, M.Q., N. Zhang, M. Pagliaro and Y.J. Xu. 2014. Artificial photosynthesis over graphene–semiconductor composites. Are we getting better? Chem. Soc. Rev. 43: 8240–8254.

Yang, S., D. Prendergast and J.B. Neaton. 2012. Tuning semiconductor band edge energies for solar photocatalysis via surface ligand passivation. Nano Lett. 12: 383–388.

Yeh, T.-F., J.-M. Syu, C. Cheng, T.-H. Chang and H. Teng. 2010. Graphite oxide as a photocatalyst for hydrogen production from water. Adv. Funct. Mater. 20: 2255–2262.

Yu, J. and J. Ran. 2011. Facile preparation and enhanced photocatalytic H₂-production activity of Cu(OH)₂ cluster modified TiO₂. Energy Environ. Sci. 4: 1364–1371.

Yu, J., J. Low, W. Xiao, P. Zhou and M. Jaroniec. 2014a. Enhanced photocatalytic CO_2-reduction activity of anatase TiO₂ by coexposed {001} and {101} facets. J. Am. Chem. Soc. 136: 8839–8842.

Yu, J., J. Jin, B. Cheng and M. Jaroniec. 2014b. A noble metal-free reduced graphene oxide–CdS nanorod composite for the enhanced visible-light photocatalytic reduction of CO_2 to solar fuel. J. Mater. Chem. A. 2: 3407–3416.

Yu, Q., Y. Wang, Z. Yi, N. Zu, J. Zhang, M. Zhang et al. 2010. High-efficiency dye-sensitized solar cells: the influence of lithium ions on exciton dissociation, charge recombination, and surface states. ACS Nano 4: 6032–6038.

Zachos, J.C., G.R. Dickens and R.E. Zeebe. 2008. An early Cenozoic perspective on greenhouse warming and carbon-cycle dynamics. Nature 451: 279–283.

Zhang, G., M. Zhang, X. Ye, X. Qiu, S. Lin and X. Wang. 2014. Iodine modified carbon nitride semiconductors as visible light photocatalysts for hydrogen evolution. Adv. Mater. 26: 805–809.

Zhang, N., Y. Zhang and Y.-J. Xu. 2012. Recent progress on graphene-based photocatalysts: current status and future perspectives. Nanoscale 4: 5792–5813.

Zhang, Q., J. Wang, S. Yin, T. Sato and F. Saito. 2004. Synthesis of a visible-light active $TiO_{2-x}S_x$ photocatalyst by means of mechanochemical doping. J. Am. Ceram. Soc. 126: 1161–1163.

Zhang, Q., Y. Li, E.A. Ackerman, M.G. Josifovska and H. Li. 2011a. Visible light responsive iodine-doped TiO_2 for photocatalytic reduction of CO_2 to fuels. Appl. Catal. A 400: 195–202.

Zhang, Q., Y. Li, E.A. Ackerman, M.G. Josifovska and H. Li. 2011b. Visible light responsive iodine-doped TiO_2 for photocatalytic reduction of CO_2 to fuels. Appl. Catal. A 400: 195–202.

Zhang, Y., T. Mori, J. Ye and M. Antonietti. 2010. Phosphorus-doped carbon nitride solid: enhanced electrical conductivity and photocurrent generation. J. Am. Chem. Soc. 132: 6294–6295.

Zhang, Z., Z. Wang, S.-W. Cao and C. Xue. 2013. Au/Pt nanoparticle-decorated TiO_2 nanofibers with plasmon-enhanced photocatalytic activities for solar-to-fuel conversion. J. Phys. Chem. C 117: 25939–25947.

Zhao, Z., J. Fan, M. Xie and Z. Wang. 2009. Photo-catalytic reduction of carbon dioxide with *in situ* synthesized $CoPc/TiO_2$ under visible light irradiation. J. Clean. Prod. 17: 1025–1029.

Zhong, D.K., S. Choi and D.R. Gamelin. 2011. Near-complete suppression of surface recombination in solar photoelectrolysis by "Co-Pi" catalyst-modified W: $BiVO_4$. J. Am. Chem. Soc. 133: 18370–18377.

Zhou, C.H., J.N. Beltramini, Y.X. Fan and G.Q. Lu. 2008. Chemoselective catalytic conversion of glycerol as a biorenewable source to valuable commodity chemicals. Chem. Soc. Rev. 37: 527–549.

Zhu, G., Y.V. Geletii, P. Kogerler, H. Schilder, J. Song, S. Lense et al. 2012. Water oxidation catalyzed by a new tetracobalt-substituted polyoxometalate complex: $[\{Co_4(\mu OH)(H_2O)_3\}(Si_2W_{19}O_{70})]^{11-}$. Dalton Trans. 41: 2084–2090.

Zhuang, S., X. Xu, B. Feng, J. Hu, Y. Pang, G. Zhou et al. 2014. Photogenerated carriers transfer in dye–graphene–SnO_2 composites for highly efficient visible-light photocatalysis. ACS Appl. Mater. Interfaces 6: 613−621.

Zong, X., C. Sun, H. Yu, Z.G. Chen, Z. Xing, D. Ye et al. 2013. Activation of photocatalytic water oxidation on N-doped ZnO bundle-like nanoparticles under visible light. J. Phys. Chem. C. 117: 4937–4942.

9

Polymer/Filler Composites for Optical Diffuse Reflectors

Yue Shao and *Frank Shi**

Introduction

In the last three decades, scientific and technological emphasis has been placed on the development of polymer-filler nanocomposites. The polymer-filler composites are materials consisting of nanometer-sized and submicron-sized inorganic fillers uniformly dispersed in the polymer matrices. With advanced nanotechnology, inorganic nanostructured materials have been developed and fabricated with important cooperative physical phenomena such as size-dependent band-gap, ferromagnetism, electron and phonon transport. However, these nanostructured materials are usually very expensive in manufacturing, and are often difficult to be shaped and further processed (Althues et al. 2007). The issues in using inorganic nanostructured materials can be solved by employing a polymer matrix to embed a relatively small content of inorganic nanofillers. In general, polymers have the advantages of low cost, ease of fabrication, high throughput, potential good UV, thermal and environmental stability. They are also known to allow easy processing and can be fabricated in the form of coatings and thin films by various techniques such as dip-coating, spin-coating, film-casting, and printing (Aegerter and Mennig 2004). The incorporation of inorganic nanofillers into polymer matrices leads to combination and improvement in both the properties of inorganic nanofillers and polymer, and can provide high-performance advanced materials that find applications in many industrial fields.

Frequently employed inorganic nanofillers include metals and metal alloys (e.g., Au, Ag, Cu, Ge, Pt, Fe, CoPt), semiconductors (e.g., PbS, CdS, CdSe, CdTe, ZnO), other oxides (e.g., TiO_2, SiO_2, ferric oxide) (Caseri 2008), and carbon-based materials (e.g., carbon nanotube (CNT), graphite, carbon nanofiber) (Agarwal et al. 2008, Grossiord et al. 2008). The choice of polymer matrix is also manifold depending on the applications that can be generally divided into industrial plastics (e.g., nylon 6, nylon MXD6, polyimide, polypropylene (PP)), conducting polymers (e.g., polypyrrole, polyaniline (PANI)), and transparent polymers (e.g., silicone, polymethyl methacrylate (PMMA), polystyrene (PS)) (Li et al. 2010). The nanocomposites obtained by the integration of these types of materials can lead to improvements in several areas, such as optical, thermal, mechanical, electrical, magnetic, rheological,

Optoelectronics Packaging & Materials Lab, University of California, Irvine.
Department of Chemical Engineering and Materials Science, 916 Engineering Tower, Irvine, CA, USA, 92697.
    Email: shaoy@uci.edu
* Corresponding author: fgshi@uci.edu

Table 1. Properties and application of selected inorganic fillers (Kango et al. 2013).

Fillers	Properties	Applications
TiO_2	Optical, electronic, spectral, structural, mechanical and anticorrosion properties	Photocatalysts, solar cells, cosmetics, waste water treatment and antimicrobial application
ZnO	Optical, thermal conductivity, electrical, sensing, transport, magnetic and electronic properties	Electronic and optoelectronic device applications, cosmetics, medical filling materials, antimicrobial and anticancerous applications
Al_2O_3	Optical, transport, mechanical and fracture properties	Waste water treatment, antimicrobial applications, in catalysis and absorption processes and drug delivery
SiO_2	Physicochemical, optical, luminescent, thermal and mechanical properties	Drug delivery, tissue engineering, biosensing
Ag	Optical, structural, thermal, electrical and catalytic properties	Coatings, antibacterial application
Au	Optical and photothermal, thermal, electrical, magnetic properties	Antibacterial application

and fire retardancy properties. Popular applications of polymer-filler composites include optical applications, thermal applications, magnetic applications, mechanical applications, electrical application and biomedical applications.

There are mainly three different methods to fabricate polymer nanocomposites. In the first method, nanofillers are prepared separately and they are subsequently blended with a desired polymer, either in solution or in the melt (Agrawal et al. 2010). The second method is *in situ* polymerization which involves dispersing the inorganic nanofillers directly in the monomer solution prior to a polymerization process (Satraphan et al. 2009). The third method involves bulk polymerization of a suitable monomer in the presence of fillers. It is known that nanofillers have a strong tendency to undergo agglomeration followed by insufficient dispersal in the polymer matrix. In order to minimize the degradation of optical and mechanical properties of composites caused by agglomeration of fillers, fillers must be integrated in a way leading to isolated, well-dispersed primary nanofillers inside the matrix. There is a need for selecting processing techniques that are effective on the nanoscale, yet are applicable to macroscopic processing.

Currently, polymer-filler nanocomposites are widely employed in the optoelectronics industry and are playing important roles in packaging materials. The packaging materials are critical for optoelectronic device packaging in terms of optical and thermal performance. Continuous innovation and improvement of packaging materials have been introduced to improve the overall efficiency of LEDs, but there are still some critical issues which need to be addressed for their application in general lighting and LCD backlight displays. In this chapter, recent fundamental and technological advances in our Optoelectronics Packaging & Materials lab and other labs in developing a new class of nanocomposites for optical diffuse reflectors used in advanced general LED lighting, LCD display backlighting, solar cells and other energy related applications will be reviewed, and challenges and directions for further developments will be outlined.

Optical application and properties of nanocomposites

Optical application: reflectors

The market of energy related devices, including light-emitting diodes (LEDs) and solar cells, has grown rapidly in the past few decades because of the global trends toward more environmental friendly technology. The packaging design and packaging materials play very important roles in determining optical performance of LEDs and solar cells. Packaging materials and technologies, including the addition of reflective layers, novel chip surface optical structures for light extraction and new encapsulation materials with enhanced thermal and radiation stability, has experienced continuous development to meet the requirements of novel optoelectronic devices. Optical reflectors are key components to the development of high performance lighting modules, display backlighting modules, as well as solar cells (Kim et al. 2010, Starry 2010). The reflectors offer advanced diffusion, distribution and guiding of light and thus have a significant impact on the brightness, uniformity, color and stability of LEDs lighting and backlight displays (Kaminsky and Bourdelais 2003). The absorption and trapping of light within the optoelectronics package is one of the major cause for the light loss and should be minimized by placing a optical reflector between the substrate and the LED active layers. Light emitted from the active region towards the substrate will then be redirected and has a higher chance to escape from the optoelectronic devices through the top surface. High light reflectivity is desired for optical reflector to reduce the loss of exiting light for achieving high light extraction efficiency with low energy consumption. Besides high reflectivity, long operational lifetime, simple manufacturing process and low cost are also important requirements for development of optical reflector for LEDs and solar cells applications (Käläntär et al. 2001, Pan et al. 2007).

LCD backlight units and general light are two main markets for LEDs application. For backlights units used in portable devices such as laptops and cell phones, the reflective materials are required to be used as thin films, for example, less than 250 microns, in order to minimize the thickness of the whole display illumination systems (O'Brien 2010). For direct view backlight, the reflective materials are required to maintain high visible light reflectance, and good optical and mechanical performance under the operation temperature in the range of 50 to 70°C. There are still many rooms for improvement of reflector materials (Anderson et al. 2007).

Metallic mirrors, conventional dielectric and diffusive dielectric reflectors are high performance optical reflectors, widely used in industrial optoelectronic devices (Lin et al. 2014). The simplified structures of above-mentioned reflectors are shown in Figure 1. Metallic mirrors, including silver and aluminum, are the oldest type of optical reflectors. They are characterized by a broad spectral reflectivity band and a weak angular dependence of the reflectivity, but also have the serious shortcoming of poor long-term stability (Kumei et al. 2012). Although metallic mirrors are simple and viable reflectors, the reflection losses are quite high, which is about 5% loss of one reflection event for metal-semiconductor reflectors. Conventional dielectric mirrors, including distributed Brag reflectors (DBR), cannot provide a random light reflecting due to their well-defined geometries. The reflectivity of DBRs can be tailored by increasing or decreasing the number of reflector pairs so long as these reflector pairs are fully transparent. However, unlike the broad band with high reflectivity exhibited by metallic mirrors, DBRs display only a narrow band of high reflectivity denoted as the stop band. The limited reflection bandwidth of DBR is a major issue for its applicability for solar cells. All-dielectric omnidirectional mirrors can totally reflect electromagnetic waves at all angles with nearly free of loss at optical frequencies. However, the applicability of the omnidirectional mirrors to LEDs is limited due to the insulating electrical characteristics of the constituent materials. Also, the multilayer structures of DBR and omnidirectional mirrors lead to a high production complexity. Diffusive dielectric composites including polymer-filler reflectors have been demonstrated to be effective reflectors in LED and solar cell technology (Kim et al. 2005).Whereas specular reflection can only reflect light at an angle equal to that of the incident light, diffuse reflection is able to reflect light at multiple angles leading to an even and uniform distribution of light from the surface. Accordingly, diffuse reflectors are more preferred by LCD and LED backlight

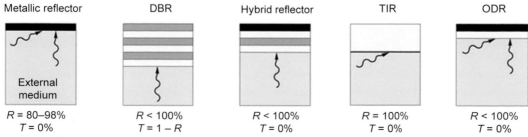

Figure 1. Different types of reflectors including metallic reflector, distributed Bragg reflector (DBR), hybrid reflector, total internal reflector (TIR), and omni-directional reflector (ODR). Also given are angles of incidence for high reflectivity and typical reflectances and transmittances (Schubert 2006).

displays in terms of improving uniformity and brightness of light distribution, eliminating hotspots for better overall optical aesthetics, and reducing the energy consumption. Among all types of reflectors described above, polymer-based diffuse reflectors have the advantages of high reflectance over a broad spectral range, low cost, ease of fabrication and high throughput.

Efforts have been made to develop highly reflective polymer-based composites for many years because relatively small improvements in the percent reflectivity can have a substantial effect on the amount of energy needed to produce the required lighting. Here we introduce a brief review of commercial products of diffuse reflective materials. Currently, many high quality commercial diffuse reflectors are very expensive and only available in standard shapes, such as white inorganic compounds in the form of pressed cake or ceramic tile, microporous materials made from sintered polytetrafluoroethylene (PTFE) (Masuda et al. 2013), microporous materials with thermally induced phase separation technology (Kretman and Kayter 2002), and microvoided materials filled with particles and controlled air voids (Ing and Hou 2011), etc. For examples, SPECTRALON® by Labsphere, Inc., North Sutton, N.H., USA (Springsteen 1988), is a well known diffuse reflector with excellent diffuse reflectivity. In this reflector, lightly packed granules of PTFE having a void volume of about 30 to 50% was sintered into a relatively hard cohesive block so as to maintain such void volume. Exceptionally high diffuse visible light reflectance characteristics, a photopic reflectance over the visible wavelengths of light better than 99%, can be achieved with this material (Springsteen 1988). However, SPECTRALON reflector is not generally available in very thin films of less than 250 μm, such as those needed for the laptop LCD market, and furthermore at these thickness levels, adequate reflection performance is not obtained. Gore™ DRP®, produced by W.L. Gore & Associates, Inc., DE, USA, is an expanded PTFE reflective material with a microporous structure defined by polymeric nodes interconnected by fibrils. Despite high flexibility and excellent diffuse reflectance properties of this material, its drawbacks are higher cost and reduced reflectivity over the blue range of the visible spectrum at a thickness desirable for many optical display applications (Kaytor et al. 1999). Such drawback requires display manufacturers to modify the display in order to transmit more light in that region in the direction of the viewer, which undesirably consumes more energy. Nonwoven polyethylene fabric diffuse light reflector with a thickness from 150 μm to 250 μm was reported to have an average reflectance varying from 77% to 85%, depending on the thickness, over the wavelength range of 380 to 720 nm. Commercial diffuse reflectors used in optical display applications are "White PET", which are filled with microvoided poly(ethylene terephthalate) (PET) films available in different thickness with reflectivity varying with thickness. White PET films around 190 μm thick find utility in notebook, personal computer (PC), LCDs and desktop PC LCDs. These films typically have an average reflectance in the visible light wavelengths of about 95%. A 190 μm thick White PET reflector is sold by Toray Industries, Inc. of Chiba, Japan, commercially available as "E60L" (Miyakawa et al. 1997). However, E60L suffers from poor resistance to UV radiation and requires a UV coating which raises the cost of the reflector as well as causes a reduction in the reflectivity at wavelengths in the blue region of the visible spectrum (i.e., wavelengths less than about 400 nm).

Other commercial diffuse reflectors realize high diffuse reflectivity by controlling the surface shape, such as reflectors with concave-convex surface formed by deposition and patterning the insulating film (Mizobata et al. 2000), reflectors with surface pattern formed by contacting with a chill roller, and reflectors with complex and random three-dimensional structures created by exposure to radiation (Fleming et al. 2001), etc. However, all these commercial products require complicated manufacturing methods which are time consuming and cost prohibitive.

High diffusive and inexpensive polymer-filler reflectors are needed for visible light management applications that will allow for the production of more affordable and energy efficient optical displays. Common polymer matrices used for LED lead frame reflectors are polyphthalamide (PPA), polycyclohexylenedimethylene terephthalate (PCT) and epoxy. PPA shows high reflectance (> 95%) in visible wavelength spectrum, but it is mainly applicable for LEDs with power lower than 0.5 watts due to its poor resistance of heat and radiation. Epoxy mounting compound (EMC) has lower initial reflectance than PPA, but it shows better resistance to heat and radiation. EMC is gradually replacing PPA for mid power LEDs. Silicone molding compound (SMC) is the latest technology in packaging material for high power LEDs. Table 2 shows the high initial reflectance, good chemical stability and

Table 2. Comparison between silicone resin and other polymers (Source: Shin-Etsu).

	Thermoplastic		Thermosetting	
	PPA	**PCT**	**Epoxy**	**Silicone**
Advantages	• High whiteness > 95% • Good moldability • Low material cost	• Good resistance to heat and light damage	• Excellent resistance to heat and light damage • Initial whiteness ~ 94.8%	• Excellent resistance to heat and light damage • Excellent reliability • Good adhesion with silicone • High whiteness ~ 97%
Disadvantages	• Poor resistance to heat and light damage	• Low whiteness ~ 93% • Poor moldability	• High material cost (~ $170/kg) • Transfer molding	• High material cost (~ $500/kg) • Transfer molding
LED powers	Under < 0.5 W	0.4–0.7 W	0.4–1 W	> 1 W

superior thermal and UV resistance of SMC. Other advantages of SMC, including durability, flexibility, good resistance to discoloration over time and excellent adhesion with silicone encapsulants, makes silicone-based reflectors more cost-effective and reliable for applications in higher power devices and short wavelength radiation. There is a great interest in the development of cost-effective and high performance silicone-filler nanocomposite reflectors in order to decrease the price-performance ratio for optoelectronic devices. Leading LED manufactures have been developing LED package technology, thermosetting resin composition for lead frame/reflector cup used in LEDs and white LEDs for display backlight applications (Shimizu et al. 2011, 2012, Ichikawa et al. 2013).

Optical properties

A good understanding of the theories for optical applications is a prerequisite for designing and optimizing polymer-filler nanocomposite reflectors. It is known that the interaction of light with a material includes scattering (reflection and absorption) and transmission of the incident light. Figure 2 presents a simplified schematic diagram of the interaction of an incident light with polymer-filler composites. The polymer-filler composite reflector is treated as spherical fillers of certain radius homogeneously dispersed within the weak-absorbing polymer matrix. When a beam of light strikes on such a material, it is partially scattered from the surface and partially transmitted into the material. The relative fraction of specular reflection and diffuse reflection determines the visual appearance of a bulk material. Usually, the surface of a material looks glossy if specular reflected light is predominated, while it will be matt if the surface-scattered light is the major component. The ceramic fillers (TiO_2, SiO_2, Al_2O_3, ZnO, etc.) are strong scattering and non-absorbing in the wavelength of visible light, and the composite reflectors have a largely diffuse appearance. The key factors affecting the reflectance of composites are particle size of fillers, difference of refractive indices between fillers and matrices, and thickness of reflectors (Berger et al. 2007).

Modeling technique is necessary to evaluate polymer-filler composites in the role of diffuse reflectors in LEDs and solar cells and to optimize their reflection properties. Different modeling approaches have already been used to investigate the diffuse composite reflectors. Light scattering of such composites

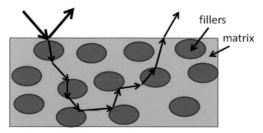

Figure 2. Schematic of light scattering in polymer-filler composite reflectors (Shao et al. 2015).

can be described precisely by Monte Carlo simulation and radiation transfer theory. Monte Carlo simulation is a numerical method with high accuracy but is very costly and time-consuming (Maheu et al. 1984). The exact radiation transfer theory based on radiation transfer equation (RTE) requires heavy calculations. Its rigorous solutions are often very difficult to solve without approximations. N-flux model such as Kebulka-Munk model is approximated analytical method where the number of N can be selected depending on the required accuracy. Such approximated model provides explicit formulas which are easily utilizable to make a number of predictions for diffuse polymer-filler reflectors (Vargas and Niklasson 1997). For industrial application, a simple analytical model is preferred due to its high computational efficiency in analysis of diffuse reflectors and a deeper understanding of physics behind the high reflectance of polymer-filler composite reflectors.

Optimization of filler size and refractive indices ratio between filler and matrix

Size of inorganic fillers is the key factor affecting optical properties of composite reflectors. The ratio of particle size to light wavelength determines the light scattering efficiency at the filler/matrix interface (Tran et al. 2009). With the optimized particle size, a polymer-based composite reflector can obtain the theoretical maximum reflectance with the lowest filler volume fraction and reflector thickness.

Kubelka-Munk model of reflectance is an almost universally used model for analyzing the optical properties of polymer-filler composite reflectors (Murphy 2006). This model works well for optically thick materials with high light reflectivity and low transparency. The basic assumptions including material have a constant finite thickness, illumination is diffuse and homogenous, and no reflection occurs at the surface of materials. In this two-flux model, it is assumed that light consists of two isotropic diffuse fluxes travelling both upwards (I_d) and downwards (J_d) through the material. The differential equations for the intensities of two diffuse fluxes are:

$$\frac{dI_d}{dx} = -(S+K)I_d + SJ_d \tag{1}$$

$$\frac{dJ_d}{dx} = (S+K)J_d + SI_d \tag{2}$$

One solution of Kubelka-Munk theory is given:

$$Sx = \frac{1}{\left(\dfrac{1}{R_\infty} - R_\infty\right)} \ln\left(\frac{1 - RR_\infty}{1 - \dfrac{R}{R_\infty}}\right) \tag{3}$$

where x is the thickness of sample, R is the reflectance of sample and R_∞ is the reflectivity of sample materials at infinite thickness. The scattering coefficient per unit thickness (S) is a heuristic parameter and can be defined as an expression of scattering efficiency (Moersch and Christensen 1995). The scattering behavior of incident light on particles depends on the size parameter (X), which is defined as $X = \pi d/\lambda$. In the present model, the scattering efficiency was calculated from anomalous diffraction theory (ADT) for particles with large size parameters ($X \geq 1$), and Rayleigh approximation for small size parameters ($X < 1$). ADT is a very approximate but computationally fast technique to calculate extinction, scattering, and absorption efficiencies for many typical size distributions.

For large particles, we calculated the extinction efficiency by:

$$Q_{ext} = 2 - \frac{2\lambda}{(n-1)\pi d}\sin\left[\frac{2(n-1)\pi d}{\lambda}\right] + \frac{4\lambda^2}{\left[2(n-1)\pi d\right]^2}\left\{1 - \cos\left[\frac{2(n-1)\pi d}{\lambda}\right]\right\} \tag{4}$$

For small particles in the Rayleigh regime, the scattering efficiency for a single particle is given by:

$$Q_{sca} = \frac{8\pi^4}{3}\left(\frac{d^4}{\lambda^4}\right)\left(\frac{n^2-1}{n^2+2}\right)(n_m)^4 \qquad (5)$$

where Q_{sca} is the scattering efficiency, which is defined as the ratio of the scattering cross section and geometrical cross section πr^2, λ is the light wavelength, d is the diameter of spherical particle, n is the ratio of refractive indices of fillers and matrices, and n_m is the refractive index of matrix. Based on our assumption, the extinction efficiency (Q_{ext}) here is equal to the scattering efficiency considering there is no absorption. For spherical particles, the scattering coefficient (S) is usually evaluated as the particle volume fraction times the volumetric scattering cross section of the particle. Here we take into consideration the effects of interfaces between fillers and matrix material and accordingly modified the expression of scattering coefficient as follows:

$$S = \frac{3f}{4V_p}C_{sca}(1-g)Y = \frac{9f}{8d}Q_{sca}(1-g)Y \qquad (6)$$

where f is the volume fraction of particles in composites, C_{sca} is the scattering cross section of spheres as a function of diameter d, V_p is the volume of a single particle, and g is the asymmetry parameter which can be calculated with Mie theory (Auger et al. 2003). Y is the correction factor determined by experimental measurements.

By entering Eqs. (4)–(6) into Kubelka-Munk general Eq. (3), a simple model for analyzing the reflectance of polymer-based diffuse reflectors is developed (Shao et al. 2015). For large filler particles, the formula is:

$$R = \frac{1-e^{[\frac{9f}{8d}\{2-\frac{2\lambda}{(n-1)\pi d}\sin\frac{2(n-1)\pi d}{\lambda}+\frac{4\lambda^2}{[2(n-1)\pi d]^2}[1-\cos\frac{2(n-1)\pi d}{\lambda}]\}x\left(\frac{1}{R_\infty}-R_\infty\right)](1-g)Y}}{R_\infty - e^{[\frac{9f}{8d}\{2-\frac{2\lambda}{(n-1)\pi d}\sin\frac{2(n-1)\pi d}{\lambda}+\frac{4\lambda^2}{[2(n-1)\pi d]^2}[1-\cos\frac{2(n-1)\pi d}{\lambda}]\}x\left(\frac{1}{R_\infty}-R_\infty\right)](1-g)Y}} \qquad (7)$$

And for small filler particles, the formula is:

$$R = \frac{1-e^{\{\frac{9f}{8d}[\frac{8\pi^3}{3}\left(\frac{d^4}{\lambda^4}\right)\left(\frac{n^2-1}{n^2+2}\right)(n_m)^4]x\left(\frac{1}{R_\infty}-R_\infty\right)](1-g)Y}}{R_\infty - e^{\{\frac{9f}{8d}[\frac{8\pi^3}{3}\left(\frac{d^4}{\lambda^4}\right)\left(\frac{n^2-1}{n^2+2}\right)(n_m)^4]x\left(\frac{1}{R_\infty}-R_\infty\right)](1-g)Y}} \qquad (8)$$

Figure 3 shows the comparison between the model's predictions and experimental data for reflectance of Al_2O_3-silicone and TiO_2-silicone composite reflectors as a function of filler size at a wavelength of 450 nm. The plots clearly present that the effect of filler size on light reflectance is non-monotonic. The light reflectance of composite reflector first increases as the particle size increases from nano-size to submicron-size and then decreases as the filler size increases to micron sizes. The critical size for fillers to effectively reflect incident radiation at a 450 nm wavelength is 0.8 microns for Al_2O_3 pigment, and 0.2 microns for TiO_2 pigment. In the case of nano-size fillers, a high amount of visible light is transmitted by composites, especially if the filler size is less than one tenth of wavelength. The reason for the maximum reflectance of composites with submicron-size fillers is because of its high light scattering coefficient. It is also known from Mie theory that when the size of particle is comparable to the wavelength of incident light, the magnitude of scattering efficiency reaches its maximum. This light reflectance decreases with increasing particle size due to the decrease in effective scattering cross sections. The total surface area for larger particles is lower because at the same volume fraction, the number of smaller particles is much more than that of larger particles. It is demonstrated that those modeling results give us a good indication of what particle sizes of fillers are most effective than others for specific fillers at a particular wavelength.

The relationship between critical filler and the refractive index ratio between filler and matrix for different inorganic fillers at a wavelength of 450 nm is shown in Figures 4 and 5. The results show that the optimized filler size is negatively related to refractive index ratio. The smaller the filler size, the larger the refractive index ratio between filler and matrix is required to obtain the maximum theoretical reflectance of

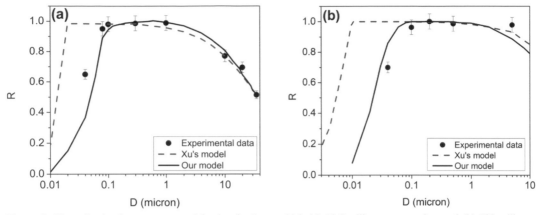

Figure 3. Normalized reflectance vs. particle size for 3 mm thick (a) Al_2O_3-silicone composites and (b) TiO_2-silicone composites at wavelength of 450 nm with 0.1 filler volume fraction. For better comparison, reflectance is normalized by setting the highest value as 1. The circles indicate the measured reflectance and the lines show the calculated reflectance based on model (Shao et al. 2015).

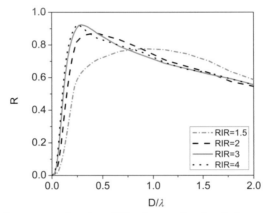

Figure 4. The prediction of reflectance as a function of filler size-wavelength ratio for composite reflectors with different filler-matrix refractive index ratio (RIR = n_{filler}/n_{matrix}) at wavelength of 450 nm (Shao et al. 2015).

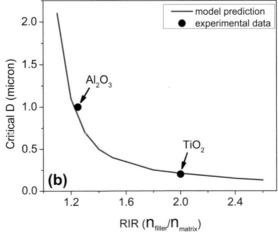

Figure 5. The prediction of critical filler size as a function of RIR (a) for polymer-based reflectors at different wavelength and (b) for silicone-based reflectors at wavelength of 450 nm. The symbols indicate the measured optimized filler size with different fillers (left: Al_2O_3, right: TiO_2) and the line shows the calculated optimized filler size based on model.

polymer-filler composite reflectors. Proper selection of inorganic nano-sized and submicron-sized fillers is crucial for advanced nanocomposite reflectors. If nano-sized fillers smaller than the wavelength of incident light are homogeneously dispersed in the composite, the scattering of incident light is considered to be small. However, significant scattering will still arise from even small particles individually dispersed in polymers due to a large refractive index mismatch between fillers and surrounding polymer matrix. Maximization of refractive index mismatching among the material components can be an effective method to increase light scattering and improve the reflectance. This essentially means that the incoming

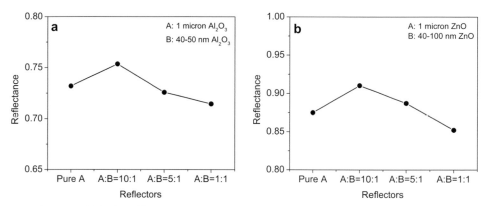

Figure 6. Reflectance of (a) Al$_2$O$_3$-silicone and (b) ZnO-silicone nano/microcomposites at wavelength of 450 nm with 0.1 filler volume fraction.

light can easily distinguish the fillers from the matrix and get scattered at polymer/filler interfaces while traveling through the material.

Effect of nanofillers

During the past two decades, inorganic nano particles have gained increasing attention as transparent nanofillers for polymer-filler composites (Althues et al. 2007, Hakimelahi et al. 2010). However, it has been demonstrated that addition of nanofillers into microcomposites can effectively enhance the light reflectance of composite reflectors. Adding a small amount (less than 10%) of Al$_2$O$_3$ and ZnO nanofillers increases the reflectance by 2% and 4% for Al$_2$O$_3$ and ZnO composite reflectors. With increasing content of the Al$_2$O$_3$ and ZnO nanofillers, their contribution to improvement in light reflectance is diminished. If excessive (larger than 50%) nanofillers were added, due to the agglomeration of nanofillers, the light reflectance of nano/microcomposite reflectors become lower than pure microcomposite reflectors. The results provide an important insight for the selection of both filler sizes and refractive index ratio for producing high performance polymer-filler composite reflectors.

Optimization of thickness

The thickness is another critical factor affecting many optical properties of nanocomposites reflectors. For backlights used in portable devices, the reflectors are required to have high levels of reflection of visible light from very thin materials. It is desirable to have large amount of diffuse reflectance across the reflection film, to compensate for uneven brightness across a backlit display (Starry 2010, Kaminsky and Bourdelais 2003). Therefore, much effort is required to develop thin reflective films or coatings with high diffuse reflection. A model study on critical thickness for maximum reflectance is essential for developing cost effective high performance reflectors with optimized thickness.

Considering there is radiation leak out through the back side of reflectors and back reflected radiation (R') from the substrate, the solution of diffuse reflection (R) using Kubelka-Muck model is:

$$Sx = \frac{1}{\frac{1}{R_\infty} - R_\infty} \ln \left| \frac{R' - R_\infty}{R' - \frac{1}{R_\infty}} \times \frac{1 - RR_\infty}{1 - \frac{R}{R_\infty}} \right| \tag{9}$$

Applying Eqs. (4)–(6) gives the expression of reflectance for large fillers as (Shao 2016):

$$R(x) = \frac{\left(\frac{R' - \frac{1}{R_\infty}}{R' - R_\infty} \right) R_\infty e^{\left[\frac{9f}{8d} \{ 2 - \frac{2\lambda}{(n-1)\pi d} \sin \frac{2(n-1)\pi d}{\lambda} + \frac{4\lambda^2}{[2(n-1)\pi d]^2} [1 - \cos \frac{2(n-1)\pi d}{\lambda}] \} x \left(\frac{1}{R_\infty} - R_\infty \right)](1-g)Y} - \frac{1}{R_\infty}}{\left(\frac{R' - \frac{1}{R_\infty}}{R' - R_\infty} \right) e^{\left[\frac{9f}{8d} \{ 2 - \frac{2\lambda}{(n-1)\pi d} \sin \frac{2(n-1)\pi d}{\lambda} + \frac{4\lambda^2}{[2(n-1)\pi d]^2} [1 - \cos \frac{2(n-1)\pi d}{\lambda}] \} x \left(\frac{1}{R_\infty} - R_\infty \right)](1-g)Y} - 1} \tag{10}$$

And the expression for small fillers as (Shao and Shi 2016):

$$R(x) = \frac{\left(\frac{R' - \frac{1}{R_\infty}}{R' - R_\infty} \right) R_\infty e^{\{ \frac{9f}{8d} [\frac{8\pi^3}{3} \left(\frac{d^4}{\lambda^4} \right) \left(\frac{n^2-1}{n^2+2} \right) (n_m)^4] x \left(\frac{1}{R_\infty} - R_\infty \right)](1-g)Y} - \frac{1}{R_\infty}}{\left(\frac{R' - \frac{1}{R_\infty}}{R' - R_\infty} \right) e^{\{ \frac{9f}{8d} [\frac{8\pi^3}{3} \left(\frac{d^4}{\lambda^4} \right) \left(\frac{n^2-1}{n^2+2} \right) (n_m)^4] x \left(\frac{1}{R_\infty} - R_\infty \right)](1-g)Y} - 1} \tag{11}$$

It is shown that a non-linear relationship between increase in reflectance and thickness of reflector. As can be seen from Figure 7a, the reflective coating sample on Cu substrate requires a thickness of about 200 µm to reach its saturation reflectance, while for the sample on Al substrate, the corresponding thickness is only about 100 µm. A thinner layer of reflective coating is needed for a substrate with higher reflectance in order to obtain the same effect.

The theoretical and experimental reflectance as a function of reflector thickness for BN-silicone and ZnO-silicone reflectors are shown in Figure 8. The theoretical results are found to be consistent with experimental data, verifying the optimization trends calculated by the modeling. For both types of

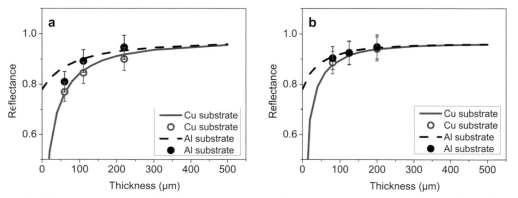

Figure 7. Reflectance change of (a) BN-silicone coatings and (b) ZnO-silicone coatings on Cu substrate (R' = 0.07) and Al substrate (R' = 0.78) at wavelength of 450 nm. The filler volume fraction is 0.1. The symbols indicate the measured reflectance with different thickness (x) and the lines show the theoretical reflectance based on model (Shao and Shi 2016).

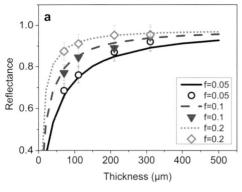

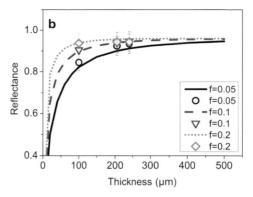

Figure 8. Reflectance vs. reflector thickness for (a) BN-silicone coatings and (b) ZnO-silicone coatings on Cu substrate (R' = 0.07) at wavelength of 450 nm. The symbols indicate the measured reflectance with different filler volume fraction (f) and the lines show the theoretical reflectance based on model (Shao and Shi 2016).

reflectors, the reflectance increases rapidly from very small thickness to a critical thickness, and then reaches stabilization as the reflector gets thicker. The scattering interface between fillers and matrix within the composite reflectors increases as total thickness increases, resulting in a boost of reflectance for reflectors with a thickness smaller than the critical value. As the thickness further increases, the absorption of light by silicone matrix causes the saturation of reflectance, and therefore it is unnecessary to make a reflector thicker.

In general, the reflectance (R) increases as back reflection (R'), thickness (x), refractive index ratio between matrices (n), reflectivity of reflectors (R_∞) and filler volume fraction (f) increase. Therefore, it is desired to have high values of all these parameters in order to obtain a high reflectance of polymer-filler reflectors.

Table 3. Properties of materials used for composite reflectors in Figures 3–7.

	Silicone resin	Al_2O_3 powders	TiO_2 powders	ZnO powders	BN powders	SiO_2 powders
Refractive index ($\lambda = 450$ nm)	1.41	1.78	2.81	2.11	1.81	1.47
Density (g/cm³)	1.03	3.95	4.23	5.61	2.1	2.65

Thermal application and properties of nanocomposites

Thermal application: cooling coatings

With the trend of optoelectronic devices such as LEDs, LCD display backlight and solar cells towards thinner and higher power, thermal management under the condition of higher heat flux density has become the main challenge for packaging materials (Taguchi and Sawada 2013, Mozzochette and Amaya 2005). In optoelectronic devices, a portion of electric energy is converted into waste heat during the operation. The conversion efficiency from electric power to light in LEDs is only 30 percent, while the remaining 70 percent is transferred to heat (Ahn et al. 2015). Heat generated in high power LEDs is conducted from the chip to the printed circuit board (PCB) and then dissipated to surrounding air. As the waste heat accumulates, the high temperature of printed circuit board and the electronic elements result in a shorter lifetime, lower luminance flux or even failure of the entire electronic device. Thus, it is necessary to eliminate the heat by means of heat radiation, cooling and others.

It is possible to effectively minimize heat by enhancing the thermal transfer of heat energy generated in the electronic element components. There are three methods to effectively remove heat from the surface of an object: heat conduction, heat convection and heat radiation. Air natural convection, forced air, cold

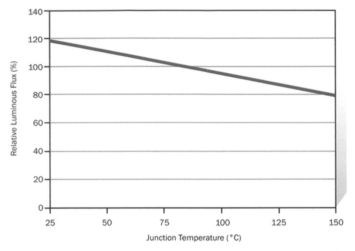

Figure 9. Typical white LED flux vs. junction temperature (Source: Vektrex).

air, forced and indirect phase transition cooling, and evaporative cooling are the common methods for system cooling of electronic devices. Heat dissipation via forced heat convection, such as installation of a mechanical fan, is not suitable for those devices and equipments where space is limited and power supply is not available. Heat dissipation via heat conduction, such as the copper-based heat is not suitable for LED lamps and luminaries because of the limited system weight and size. These traditional cooling methods have the drawbacks such as low heat radiation efficiency, complex structure, large volume and high cost. Therefore, radiation cooling becomes attractive for LEDs application since no additional space or power supply is required. Thermal control coating materials with high cooling effect are critical in many applications such as energy conversion and a wide variety of future micro- and nano-technologies (Suryawanshi et al. 2010, Mauer et al. 2012). Passive radiation cooling refers to a process in which a subject will emit radiation heat energy absorbed through normal convection and conduction process. Heat dissipation through radiation has its advantage in being passive, and no power supply is required and is thus ideal for space sensitive devices. For a natural convection finned heat sink, it is reported that typically around 25 percent of the total heat can be dissipated via radiation. Also, the passive cooling shows the possibility for water-free and electricity-free approaches to cooling building and vehicles at all hours of the day to decrease the carbon emissions caused by air conditioning (Raman et al. 2014).

In real engineering applications, the cooling ability of a coating layer is realized by heat dissipation due to radiation, convection and conduction. Thus, the combined cooling effect of heat radiation, conduction and convection is what really matters in engineering applications of a passive device coating material. Recently, the polymer-filler composites have been investigated for passive cooling of electronic systems, cars and buildings (Yu et al. 2009, Suryawanshi and Lin 2009, Eyassu et al. 2014). The roof cooling coatings proposed in those studies can effectively reduce the surface temperature of objects if the heat source is solar radiation. However, their effectiveness in cooling the electronic devices is very questionable. Conventional heat radiating coating materials generally comprise fillers like carbon black or graphite which have high emissivity. Unfortunately, such radiating coating material is inferior in design and may often cause a problem in designing products of electronic devices and equipments due to black colored carbon black or graphite it contained. Radiation cooling coatings materials colored other than black have been also studied. Such coating materials include a far infrared radiation coating material containing metal oxide ceramics such as Al_2O_3, SiO_2 and TiO_2 (see JP H10-279845A), a heat radiating body having a coating film containing sodium silicate, potassium silicate and metal oxides such as silicon oxide and aluminum oxide (see JP 2003-309383A), a heat radiating material containing a powder of a metal compound selected from a hydrotalcite-series compound, zirconium silicate and zirconium carbide, and a resin ingredient (see JP 2006-124597A), as well as a heat releasing coating material containing SnO_2-Sb_2O_5-based semiconductor particles in a coating material vehicle (see JP 2004-043612A).

LED packages are conventionally mounted on a PCB in the form of array by surface mount technology and then attached on a heat sink, to manufacture a LED array unit used in a LCD backlight. Such manufacturing process of LED units is complicated and expensive. In addition, the increased overall thickness of LED units is an obstacle to the thinning of LCD backlights. The employment of a layer of dual-functional polymer-filler composite coating on the PCB can possibly address the above-mentioned problem. The high reflectance of dual-functional composite coating can increase the surface reflection of PCB to improve the brightness and uniformity. The passive cooling ability of such composite coating can also help dissipate the heat from LEDs via radiation to increase the light output and lifetime of LEDs. The potential to eliminate heat sink from the LED units packaging can largely simplify the manufacturing process, save the manufacturing cost and reduce the overall thickness of LED units and, ultimately, backlight units.

However, in view of the significance, both technologically and scientifically, it is surprising to find that there is no report on polymer-filler composite coatings with both optical and thermal properties. For the first time, our lab developed a dual-functional optically reflective and thermally cooling composites, and carried out a systematic experimental study to investigate dependence of the light reflectance and apparent cooling effect on filler type, filler size and filler volume fraction of polymer-filler composite coatings. The new materials we developed can be used as diffuse reflector to provide better overall optical aesthetics for displays. Meanwhile, it can be used as passive cooling coatings in advanced general LED lighting, LCD display backlighting, solar cells and other energy related applications.

Thermal properties

Figure 10 presents the temperature of LED emitter against time. The filler used in composite coating is ZnO powders. The plot shows that the LED emitter has lower temperature rise if the PCB was coated with composite cooling coatings. Figure 11 shows the relationship between filler volume fraction in silicone composites, and the temperature reduction of a LED unit consist of LED emitter and PCB coated with silicone composite coatings filled with ZnO, TiO_2 and BN fillers. The temperature drop of LED is defined as the difference of LED emitter temperature between LED units, with and without coating on PCBs. As indicated in plots a, b and c, by increasing the filler volume fraction, the temperature of LED unit experience a gradual decrease. For LED unit with ZnO and TiO_2 filled silicone composite coatings, the reduction in temperature is slow as the filler volume fraction increases from 0.025 to 0.1, then, the reduction becomes rapid as the filler volume fraction further increases from 0.1 to 0.3. The addition of fillers into silicone resin coatings improves heat dissipation by reducing the temperature of LED by 9.3°C, 9.2°C, and 11.4°C for ZnO, TiO_2 and BN fillers, respectively.

Two contributions need to be taken into account during the increase of filler volume fraction: heat radiation and heat conduction. (1) Heat radiation: the emissivity is 0.15 for Al substrate, 0.65 for pure

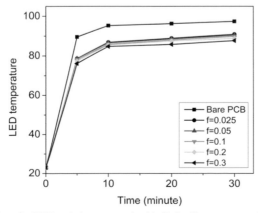

Figure 10. Temperature vs. time for LED unit incorporated with ZnO-silicone composite coatings with different filler volume fraction.

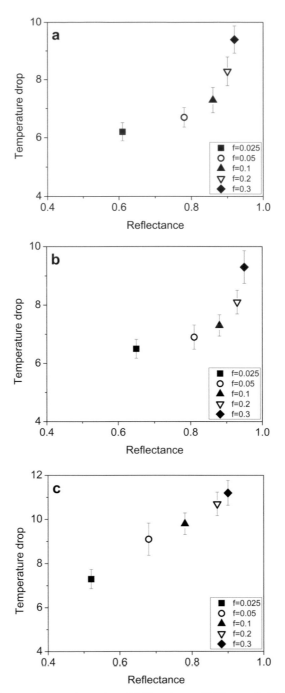

Figure 11. Temperature reduction of LED unit consists of LED emitter and PCB coated with (a) ZnO, (b) TiO_2 and (c) BN filled silicone composites vs. light reflectance of silicone composite coatings. The filler volume fractions are 0.025, 0.05, 0.1, 0.2 and 0.3. The coating thickness is 50 ± 5 microns. The power of LED is 10 Watts.

silicone resin and 0.90–0.95 for silicone composites filled with inorganic fillers. The radiation emissivity of composite coating gradually increases as the filler volume fraction increases. The high emissivity of the silicone composites effectively enhanced the contribution of radiation to heat transfer. (2) Heat conduction: the thermal conductivity of silicone resin, ZnO filler, and TiO_2 filler and BN filler is 0.18 W/m·K, 37 W/m·K, 11.8 W/m·K and 30–600 W/m·K, respectively. Since the thermal conductivity of pure silicone resin is much lower than those of inorganic fillers, the pure silicone coating acts as a thermal barrier with high heat resistance, preventing the heat generated in the system being conducted to the coating surface quickly and efficiently. The dispersed filler particles act as spots of lower thermal resistance microscopically. With an increase in filler volume fraction, heat tends to dissipate along the low-thermal-resistance channels consisting of clusters of micro-particle fillers. Therefore, an increase in the filler volume fraction is essential to improve the apparent cooling effect of silicone composite coatings.

The variation in light reflectance with the filler volume fraction for silicone composite coatings filled with ZnO, TiO_2 and BN fillers, is also shown in Figure 10. With a given thickness of 50 micrometers, the light reflectance first rapidly increases as the volume fraction increases from 0.025 to a critical value around 0.2, and then increases slowly as the volume fraction continues to increase. The plots indicate that the critical volume fractions for ZnO, TiO_2 and BN fillers to achieve a 90% reflectance at 450 nm wavelength are 0.2, 0.2 and 0.3, respectively. The common commercial PCB white coating has a reflectance of 70–80% at visible light wavelength. Our silicone composite coating, with a filler volume fraction higher than 0.3, can provide a higher reflectance to reduce the light loss, increase light output, improve the brightness, uniformity, color and stability. Meanwhile, the high volume fraction of filler offers silicone composite coating a good passive cooling ability to reduce the heat load to LEDs and LCD backlight units, increase the energy efficiency and extend the operation lifetime. Furthermore, the utilization of dual-functional silicone composite coatings as alternative reflectors and heat dissipation layers can largely simplify the manufacturing process, lowering the manufacturing cost and reducing the overall thickness of LED units and, ultimately, backlight units.

References

Aegerter, M.A. and M. Mennig. 2004. Sol–Gel Technologies for Glass Producers and Users. Kluwer Academic Publishers, Boston/Dordrecht/New York/London.

Agarwal, S., M.M.K. Khan and R.K. Gupta. 2008. Thermal conductivity of polymer nanocomposites made with carbon nanofibers. Polymer Eng. Sci. 48: 2474–2481.

Agrawal, M., S. Gupta, N.E. Zafeiropoulos, U. Oertel, R. Hassler and M. Stamm. 2010. Nano-level mixing of ZnO into poly(methyl methacrylate). Macromol. Chem. Phys. 211: 1925–1932.

Ahn, B.L., J.W. Park, S. Yoo, J. Kim, S.B. Leigh and C.Y. Jang. 2015. Saving in cooling energy with a thermal management system for LED lighting in office buildings. Energies 8: 6658–6671.

Althues, H., J. Henle and S. Kaskel. 2007. Functional inorganic nanofillers for transparent polymers. Chem. Soc. Rev. 36: 1454–1465.

Anderson, J., C. Schardt, J. Yang, B. Koehler, B. Ostlie, P. Watson et al. 2007. New back reflector and front film for improved efficiency of direct-lit LED backlights for LCD TV. SID Symposium Digest of Technical Papers 38: 1236–1239.

Auger, J.C., R.G. Barrera and B. Stout. 2003. Scattering efficiency of clusters composed by aggregated spheres. J. Quant. Spectrosc. Radiat. Transf. 79-80: 521–531.

Berger, O., D. Inns and A.G. Aberle. 2007. Commercial white paint as back surface reflector for thin-film solar cells. Sol. Energy Mater. Sol. Cells 91: 1215–1221.

Caseri, W. 2008. Inorganic nanoparticles as optically effective additives for polymers. Chem. Eng. Comm. 196: 549–572.

Eyassu, T., T.J. Hsiao, K. Henderson, T. Kim and C.T. Lin. 2014. Molecular cooling fan: factors for optimization of heat dissipation devices and applications. Ind. Eng. Chem. Res. 53: 19550–19558.

Fleming, P.R., A.J. Ouderkirk and E.J. Borchers. 2001. Method and apparatus for step and repeat exposures. U.S. Patent # 6285001.

Grossiord, N., J. Loos, L. van Laake, M. Maugey, C. Zakri, C.E. Koning et al. 2008. High-conductivity polymer nanocomposites obtained by tailoring the characteristics of carbon nanotube fillers. Adv. Func. Mater. 18: 3226–3234.

Hakimelahi, H.R., L. Hu, B.B. Rupp and M.R. Coleman. 2010. Synthesis and characterization of transparent alumina reinforced polycarbonate nanocomposite. Polymer 51: 2494–2502.

Ichikawa, H., M. Hayashi, S. Sasaoka and T. Miki. 2013. Light emitting device, resin package, resin-molded body, and methods for manufacturing light emitting device, resin package and resin-molded body. U.S. Patent # 8530250.

Ing, W. and W. Hou. 2011. Lighting device, display, and method for manufacturing the same. U.S. Patent # 20110006316.

Käläntär, K., S. Matsumoto and T. Onishi. 2001. Functional light-guide plate characterized by optical micro-deflector and micro-reflector for LCD backlight. IEICE Trans. Electron. E84-C: 1637–1646.

Kaminsky, C.J. and R.P. Bourdelais. 2003. Light reflector with variable diffuse light reflection. U.S. Patent # 0214718.

Kango, S., S. Kalia, A. Celli, J. Njuguna, Y. Habibi and R. Kumar. 2013. Surface modification of inorganic nanoparticles for development of organic–inorganic nanocomposites—a review. Prog. Polym. Sci. 38: 1232–1261.

Kaytor, S.R., K.A. Epstein, J.C. Harvey, S.J. Pojar, N.T. Strand, C.P. Waller et al. 1999. Diffuse reflective articles. U.S. Patent # 5976686.

Kim, B., J. Kim, W. Ohm and S. Kang. 2010. Eliminating hotspots in a multi-chip LED array direct backlight system with optimal patterned reflectors for uniform illuminance and minimal system thickness. Opt. Express 18: 8595–8603.

Kim, J.K., H. Luo, E.F. Schubert, J. Cho, C. Sone and Y. Park. 2005. Strongly enhanced phosphor efficiency in GaInN white light-emitting diodes using remote phosphor configuration and diffuse reflector cup. Japan. J. Appl. Phys. 44: 649–651.

Kretman, W. and S. Kayter. 2002. Diffuse reflective articles. U.S. Patent # 6497946.

Kumei, M., T. Sakai and M. Sato. 2012. Light emitting device with a porous alumina reflector made of aggregation of alumina particles. U.S. Patent # 8106413.

Li, S., M.M. Lin, M.S. Toprak, D.K. Kim and M. Muhammed. 2010. Nanocomposites of polymer and inorganic nanoparticles for optical and magnetic applications. Nano Rev. 1: 5214.

Lin, A., Y.K. Zhong, S.M. Fu, C.W. Tseng and S.L. Yan. 2014. Aperiodic and randomized dielectric mirrors: alternatives to metallic back reflectors for solar cells. Opt. Express 22: A880–A894.

Maheu, B., J.N. Letoulouzan and G. Gouesbet. 1984. Four-flux models to solve the scattering transfer equation in terms of Lorenz-Mie parameters. Appl. Opt. 23: 3353–3362.

Masuda, J., M. Ohkura, S. Tanaka, R. Morita and H. Fukushima. 2013. Microporous polypropylene film and process for producing the same. U.S. Patent # 8491991.

Mauer, M., P. Kalenda, M. Honner and P. Vacikova. 2012. Composite fillers and their influence on emissivity. J. Phys. Chem. Solids 73: 1550–1555.

Miyakawa, K., K. Tsunashima and S. Aoki. 1997. Polyester film reflector for a surface light source. U.S. Patent # 5672409.

Mizobata, E., H. Ikeno and H. Kanoh. 2000. Reflective LCD having a particular scattering means. U.S. Patent # 6018379.

Moersch, J.E. and P.R. Christensen. 1995. Thermal emission from particulate surfaces: a comparison of scattering models with measured spectra. J. Geophys. Res. 100: 7465–7477.

Mozzochette, J. and E. Amaya. 2005. Light emitting diode arrays with improved light extraction. U.S. Patent # 0225222.

Murphy, A.B. 2006. Modified Kubelka–Munk model for calculation of the reflectance of coatings with optically-rough surfaces. J. Phys. D: Appl. Phys. 39: 3571–3581.

O'Brien, W.G. 2010. Diffuse reflector, diffuse reflective article, optical display, and method for producing a diffuse reflector. U.S. Patent # 20100014164.

Pan, C., C. Su, H. Cheng and C. Pan. 2007. Backlight module and brightness enhancement film thereof. U.S. Patent # 7290919.

Raman, A.P., M.A. Anoma, L. Zhu, E. Rephaeli and S. Fan. 2014. Passive radiative cooling below ambient air temperature under direct sunlight. Nature 515: 540–544.

Satraphan, P., A. Intasiri, V. Tangpasuthadol and S. Kiatkamjornwong. 2009. Effects of methyl methacrylate grafting and *in situ* silica particle formation on the morphology and mechanical properties of natural rubber composite films. Polym. Adv. Technol. 20: 473–486.

Schubert, E.F. 2006. Light-Emitting Diodes. 2nd Ed., New York, NY, USA: Cambridge Univ. Press, 163.

Shao, Y. and F.G. Shi. 2016. Exploring the critical thickness for maximum reflectance of optical reflectors based on polymer-filler composites. Opt. Mater. Express 6: 1106–1113.

Shao, Y., Y.C. Shih, G. Kim and F.G. Shi. 2015. Study of optimal filler size for high performance polymer-filler composite optical reflectors. Opt. Mater. Express 5: 423–429.

Shimizu, Y., K. Sakano, Y. Noguchi and T. Moriguchi. 2011a. Light emitting device and display. U.S. Patent # 7915631.

Shimizu, Y., K. Sakano, Y. Noguchi and T. Moriguchi. 2011b. Liquid crystal display and back light having a light emitting diode. U.S. Patent # 7901959.

Shimizu, Y., K. Sakano, Y. Noguchi and T. Moriguchi. 2012. Light emitting device and display. U.S. Patent # 8309375.

Springsteen, A.W. 1988. Laser cavity material. U.S. Patent # 4912720.

Starry, A.B. 2010. Diffuse reflective article. U.S. Patent # 7,660,040.

Suryawanshi, C.N. and C.T. Lin. 2009. Radiative cooling: lattice quantization and surface emissivity in thin coatings. ACS Appl. Mater. Interface 1: 1334–1338.

Suryawanshi, C.N., T. Kim and C.T. Lin. 2010. An instrument for evaluation of performance of heat dissipative coatings. Rev. Sci. Instrum. 81: 035105.

Taguchi, Y. and J. Sawada. 2013. White heat-curable silicone resin composition and optoelectronic part case. U.S. Patent # 8173053.

Tran, N.T., J.P. You and F.G. Shi. 2009. Effect of phosphor particle size on luminous efficacy of phosphor-converted white LED. J. Lightwave Technol. 27: 5145–5150.

Vargas, W.E. and G.A. Niklasson. 1997. Applicability conditions of the Kubelka-Munk theory. Appl. Opt. 36: 5580–5586.

Yu, H., G. Xu, X. Shen, X. Yan, C. Shao and C. Hu. 2009. Effect of size, shape and floatage of Cu particles on the low infrared emissivity coatings. Prog. Org. Coat. 66: 161–166.

10

Synthesis and Applications of LDH-Based Nanocomposites

Alireza Khataee* and Samira Arefi-Oskoui

Introduction

Layered double hydroxides (LDHs), also known as hydrotalicite-like compounds, are an important class of two-dimensional anionic clay, which can be expressed by general formula of $[M^{2+}_{1-x}M^{3+}_x(OH)_2]^{x+}[A^{n-}_{x/n}.mH_2O]^{x-}$ (Fan et al. 2014, Xu et al. 2011). LDHs are composed of positively charged brucite-like layers in which divalent metal cations (M^{2+}: Mg^{2+}, Co^{2+}, Fe^{2+}, Ni^{2+}, Cu^{2+}, or Zn^{2+}) and trivalent metal cations (M^{3+}: Al^{3+}, Mn^{3+}, Fe^{3+}, Cr^{3+}, In^{3+} or Ga^{3+}) are coordinated octahedrally by hydroxyl groups (Fan et al. 2014, Kura et al. 2014). The positive charge of the layers, induced by partial replacement of divalent cations with trivalent cationic, is counter-balanced with A^{n-} anions (e.g., SO_4^{2-}, CO_3^{2-}, NO_3^-, or Cl^-) located in interlayer region together with water molecules (Cao et al. 2016, Fan et al. 2014). The value of x in this formula is equal to molar ratio of $M^{3+}/M^{2+}+M^{3+}$ (Fan et al. 2014). The hydrogen bond and electrostatic interaction between layers and interlayer anions and water molecules results in three dimensional structure of LDHs (see Figure 1). These weak interactions between layers induce excellent expanding properties to LDHs. In addition, the flexibility of LDHs in the nature of metal cations, molar ratio of divalent and trivalent cations and type of interlayer anions result in preparing compounds with different physical and chemical properties (Fan et al. 2014). Moreover, LDHs can be synthesized using simple and cost effective routes, e.g., co-precipitation which facilitates their scaling up for industrial production (Cavani et al. 1991). Owning to the presence of positive charged layers and numerous hydroxyl groups (–OH) on the surface of layers, LDHs can interact with other nanomaterials and polymers for generating LDH-based nanocomposites. In recent years, combining the LDHs with other functional components have provided new approach for design and fabrication of efficient materials with great potential in different fields, e.g., environment, healthcare, energy storage and energy conversion (Kura et al. 2014, Lan et al. 2014, Long et al. 2016). In LDH-based nanocomposites, there is a notable synergistic effect between LDHs and other materials, e.g., metal nanoparticles, metal oxides, carbon based nanomaterials and polymers, which give them unique properties and make them attractive.

Research Laboratory of Advanced Water and Wastewater Treatment Processes, Department of Applied Chemistry, Faculty of Chemistry, University of Tabriz, 51666-16471 Tabriz, Iran.
 Email: S.arefi_oskoui@yahoo.com
* Corresponding author: a_khataee@tabrizu.ac.ir

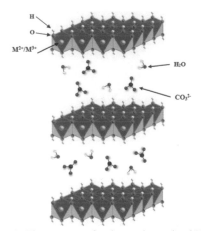

Figure 1. The structure of carbonate-intercalated LDH.

In this book chapter, we comprehensively summarize the recent progress and advances in design, preparation and applications of LDH-based nanocomposites. In first section,the combination of LDH with different materials, e.g., metal nanoparticles, metal oxides, polyoxometalates (POM), carbon based materials and polymers, have been discussed. In the second section, the methods which can be used for fabrication of LDH-based nanocomposites have been introduced. In the final section, the applications of LDH-based nanocomposites in the different fields, e.g., environment, healthcare, chemical reactions and energy storage and conversion, have been discussed. The different types of the LDH-based nanocomposites and their applications have been summarized in Figure 2.

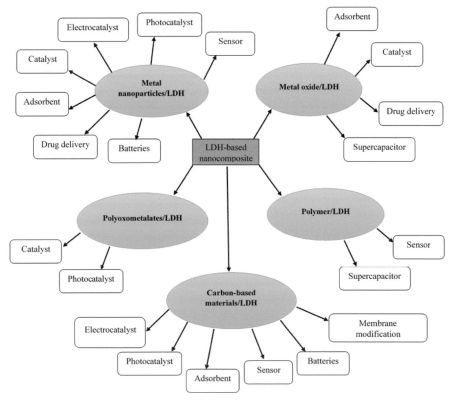

Figure 2. LDH-based nanocomposites and their applications.

LDH-based nanocomposites

Metal nanoparticles/LDH nanocomposites

LDHs, owning to their excellent properties such as high adsorption capacity, exfoliation capability, tunable acidity-basicity surface, environment-friendly property, high stability and cost effectivity, have been widely applied as support for immobilization of different metallic nanoparticles (Feng et al. 2015). Immobilization of metallic nanoparticles, e.g., Au, Pt, Pd and … on the LDHs results in generating nanocomposites with notable properties and reactivity. The interaction between metal nanoparticles and LDH and the synergy of metal-LDH give excellent properties to the generated nanocomposites (Fan et al. 2014). Metal nanoparticles/LDH nanocomposites have been widely used as catalyst in different chemical industrial reactions, e.g., catalytic oxidation and hydrogenation, electrocatalysis and photocatalysis (Feng et al. 2015, Taei et al. 2016b, Wang et al. 2012). In addition, there are reports indicating that these nanocomposites have potential to be applied as adsorbent (Jin et al. 2012). These types of nanocomposites are mainly prepared using *in situ* chemical reduction method.

Metal oxide/LDH nanocomposites

Combining different metal oxides with LDH compounds results in formation of nanocomposites with various applications in the fields of catalysis, electrocatalysis, photocatalysis, drug delivery, superconductor and adsorption (Asif et al. 2017, Bo et al. 2015, Huang et al. 2013, Lv et al. 2015, Ning et al. 2014, Pan et al. 2011). Semiconductors are one of the main groups of the metal oxides. Combining them with LDH compounds results in improving the photocatalytic activity performance through efficient separation of photo-induced electron and holes (Carja et al. 2010, Hadnadjev-Kostic et al. 2014, Zhao et al. 2015a). Moreover, combining the metal oxides with magnetic property, e.g., Fe_3O_4 with LDH, results in formation of magnetic metal oxide/LDH nanocomposite with enhanced separation and re-dispersion performance in aqueous solutions (Chen et al. 2012, Shan et al. 2014, Shou et al. 2015). These nanocomposites can be used in different fields by high reusability performance. The metal oxide/LDH nanocomposites can be prepared using different methods including self-assembly, *in situ* crystallization of LDH and reconstitution.

Polyoxometalates/LDH nanocomposites

Polyoxometalates (POMs) are an important family of anionic metal oxides of group 5 and 6 (Omwoma et al. 2014). Condensation of metal oxides polyhedral (MOx, M = V^V, Nb^V, Mo^V, Mo^{VI}, W^{VI}, etc., and x = 4–7) with each other results in forming POMs (Omwoma et al. 2014). Although the atoms of oxygen are the common ligand coordinated with the metal atoms in most of the POMs, other atoms or groups, e.g., alkoxy, nitrosyl, bromine and sulphor, can be substituted in some clusters of POMs. When the POM consists of only metal and oxygen atoms, the cluster is known as isopolymetalate. However, when there is an additional atom besides metal and oxygen, the cluster is called heteropolymetalate (Omwoma et al. 2014).

The intercalation of POM anions into the LDHs results in generation of POM/LDH nanocomposites with unique properties. The stability and the recyclability of the POMs can be significantly improved by embedding POMs into the gallery of LDH and forming POM/LDH nanocomposite. As illustrated in Figure 3, both hydrogen bonding and electrostatic interactions can be seen between the LDH's layers and intercalated POM anions.

The research review demonstrates that POM/LDH nanocomposites have been widely used as heterogeneous catalysts in different catalytic reactions. The high catalytic activity of these nanocomposites can be attributed to the large gallery space, homogeneous dispersion of active species and unique chemical environment in the nanocomposite (Fan et al. 2014, Omwoma et al. 2014). Moreover, these nanocomposites present excellent photocatalytic activity and adsorption properties (Bi et al. 2011, Guo et al. 2001, Mohapatra et al. 2014, Omwoma et al. 2014).

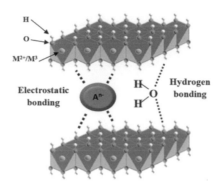

Figure 3. The interaction between the LDH's layers and intercalated POM anions.

The POM/LDH nanocomposites can be fabricated using different methods including co-precipitation, ion exchange, reconstitution, electrochemical reduction, delamination technology and *in situ* crystallization of LDH in the presence of POM.

Carbon based materials/LDH nanocomposites

Hybridization of different LDHs with carbon-based materials such as graphene, carbon nanotube, carbon quantum dots and carbon fibers due to their interesting physicochemical properties including high specific surface area, good electronic conductivity and facile surface modification capability has attracted the interest of many researches (Cao et al. 2016, Guo et al. 2015, Tang et al. 2014, Zhao et al. 2013a, Zhu et al. 2015). These nanocomposites have been widely used in different catalytic reactions, in which the carbon-based materials have enhanced heat and mass transfer during the reaction and facilitate the electron transfer through the composite, which finally results in enhanced performance of nanocomposite in catalytic reactions (Cao et al. 2016, Fan et al. 2014). In the field of energy storage, introducing carbon based materials into the LDH improves the power density and the life cycles of supercapacitors (Long et al. 2016). Moreover, these nanocomposites have attracted great attention in the field of sensors, batteries, adsorbent electrocatalysis and photocatalysis (Gunjakar et al. 2013, Hou et al. 2016, Wang et al. 2015, Yang et al. 2013, Zhang et al. 2014). Various synthesis methods can be applied for preparation of different metal based/LDH nanocomposites. Depending on the nature of the carbon based materials and architecture of the nanocomposite, different methods including self-assembly, *in situ* crystallization of the LDH and reconstitution can be mainly applied.

Polymer/LDH nanocomposites

In recent years, polymer/LDH nanocomposites have found different applications in various fields including drug delivery, catalysis, supercapacitors, sensors, etc. (Kura et al. 2014, Li et al. 2016b, Nam et al. 2016, Shao et al. 2015a). There are two main methods for fabrication of polymer/LDH nanocomposites including adsorption of pre-synthesized polymer and *in situ* polymerization (Costa et al. 2007, Gu et al. 2015). A typical adsorption synthesis method is shown in Figure 4. As can be seen, the negatively charged polymer chains can be adsorbed on the sheets of the LDH via electrostatic interaction, resulting in fabrication of polymer/LDH nanocomposite.

In situ polymerization process is a common method for fabrication of polymer/LDH nanocomposite. This process is usually performed in aqueous solution, so it is a solution-based method. The primary step in this method is the intercalation of monomers into the interlayer space of LDH. As can be seen in Figure 5, various methods can be used for this purpose including ion exchange (path 1 and 4 of Figure 5), reconstitution of LDH in the presence of monomers (path 2 of Figure 5) and crystallization of LDH in the presence of monomers (path 3 of Figure 5). The subsequent step is the polymerization of intercalated monomers, resulting in fabrication polymer in the interlayer gallery of LDH, resulting in production of polymer/LDH nanocomposite.

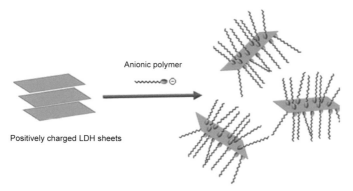

Figure 4. Schematic pathways of electrostatic adsorption of negatively charged polymer chains on the LDH sheets to synthesize polymer/LDH nanocomposites.

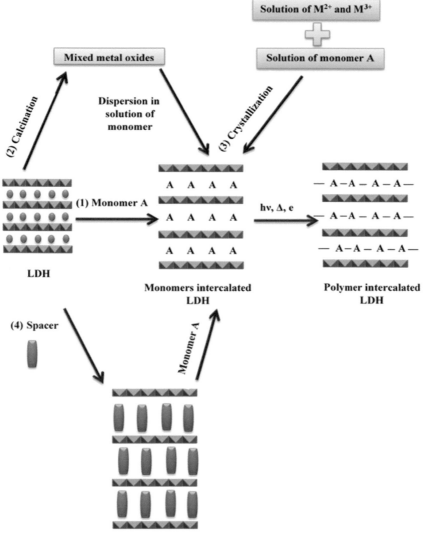

Figure 5. Schematic pathways of *in situ* polymerization within LDH layers to synthesize polymer/LDH nanocomposites. (Reprinted with permission from Costa, F.R., M. Saphiannikova, U. Wagenknecht and G. Heinrich. Springer. 101 (2007). Copyright@. Springer. License Number: 3978941080967).

Synthesis of LDH-based nanocomposites

In this section, convenient and efficient methods for the fabrication and preparation of different types of the LDH-based nanocomposites have been summarized.

Ion exchange

In this method, LDH is dispersed in the solution containing the anions, which we want to intercalate in the structure of LDH. As can be seen in Figure 6, the anions which initially exist in the interlayer space of LDH are replaced with the objective anions. By ion exchange method, various organic anions can be intercalated into the interlayer gallery of LDH, leading to the construction of a variety of interesting nanocomposites with host-guest intercalation structure, in which LDH acts as host material. The ion exchange process is efficiently affected by the nature of the anions which were initially present in the structure of the LDH. The anions which have weak interaction with the layers of the host LDH can be easily replaced with other anions to form nanocomposite with desirable properties. Based on the researches (Omwoma et al. 2014), the order of the anions which can be easily replaced through ion exchange process is $NO_3^- > Cl^- > SO_4^{2-} > CO_3^{2-}$. The research review shows that this method is widely used for fabrication of POM/LDH nanocomposites (Liu et al. 2009, Wei et al. 2008, Zhao et al. 2012, Zhao et al. 2011).

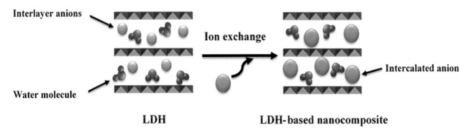

Figure 6. Schematic pathways of ion exchange to prepare LDH-based nanocomposite.

Self-assembly

For preparation of LDH-based nanocomposites through self-assembly method, in the first step, the used precursors should be exfoliated. LDHs can be delaminated into the positively charged two-dimensional nanosheets, which can be spontaneously assembled with various negatively charged species, e.g., polymers, metal complexes, carbon-based materials and metal nanoparticles, resulting in formation of nanocomposites with well-defined architecture. It should be noted that because of the strong electrostatic force between the positively charged LDH sheets and negatively charged interlayer anions, the exfoliation of the LDH is difficult. To overcome this problem, some molecules, e.g., formamide and short chain alcohols, can be introduced into the interlayer gallery to enhance the distance of the two neighbored layers, and consequently facilitate the exfoliation process (Adachi-Pagano et al. 2000, Hibino 2004). A typical self-assembly method, representing the preparation of layer by-layer ordered TiO_2/LDH nanocomposite is shown in Figure 7 (Gunjakar et al. 2011).

Reconstitution pathway

As can be seen in Figure 8, this method contains two main steps, including calcination and rehydration. In the first step, calcination of LDH at intermediate temperature (450–600°C) results in producing well dispersed mixed metal oxide (MMO). In the next step, rehydration of obtained MMO via the dispersion of MMO in the solution containing the objective anionic species results in reformation of LDH with anions incorporated in the interlayer space of LDH. The reconstitution of calcined LDH to the original

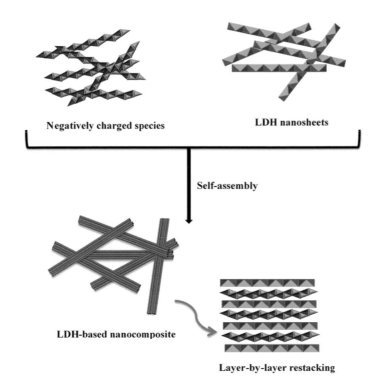

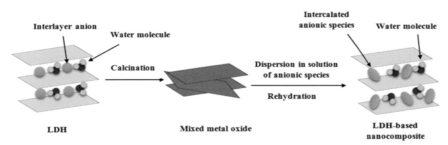

Figure 7. Schematic pathway of self-assembly to synthesise layer-by-layer LDH-based nanocomposite.

Figure 8. Schematic pathway of reconstitution to synthesis of LDH-based nanocomposites.

LDH structure in the presence of aqueous solution is known as memory effect. It should be noted that the memory effect is only feasible for LDHs calcined below 600°C (Omwoma et al. 2014). This method can be used for fabrication of POM/LDH and metal oxide/LDH nanocomposite.

Electrochemical deposition (Electrodeposition)

Electrochemical deposition is a fast and facile method, in which sulfate and nitrate solutions containing appropriate divalent and trivalent cations are used. In this method, sulfate and nitrate ions are reduced through electrochemical reactions as shown in Eqs. 1 and 2, respectively:

$$SO_4^{2-} + H_2O + 2e \rightarrow SO_3^{2-} + 2OH^- \qquad (1)$$

$$NO_3^- + H_2O + 2e \rightarrow NO_2^- + 2OH^- \qquad (2)$$

The produced hydroxide anions result in local pH enhancement in the working electrode, which consequently induce precipitation of LDH. The crystallinity and the purity of the prepared LDH material is significantly controlled by the nature of the metal cations, ratio of the divalent and trivalent cations, deposition potential and pH of the plating solution. This method can be applied for deposition of LDH on the other conductive materials in order to fabricate LDH-based nanocomposites.

In situ chemical reduction

This method is mainly used for preparation of metal nanoparticles/LDH nanocomposites. Typically in this method, the LDH nanosheets are dispersed in a solution containing metal cations (e.g., Ag^+, Pd^{2+} and …) or ionic compounds containing the metal element (e.g., $AuCl_4^-$, $PdCl_4^{2-}$ and …). Subsequently, by adding reductant agent, e.g., $NaBH_4$ and N_2H_4, the nanoparticles of the metal are immobilized on the nanosheets of LDH, resulting in fabrication of nanocomposite with appropriate properties. Schematic representation of this method for fabrication of metal nanoparticle/LDH nanocomposite is shown in Figure 9.

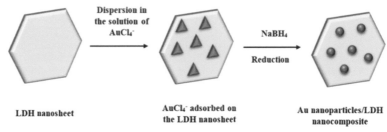

Figure 9. Schematic pathway of *in situ* reduction to synthesis of Au nanoparticles/LDH nanocomposites.

In situ crystallization of LDH

This method is one of the most commonly used methods for the fabrication of LDH-based nanocomposites. In this method, the divalent and trivalent cations which are necessary for synthesis of chosen LDH are crystalized *in situ* on the appropriate support, e.g., graphene, carbon nanotube and metal oxides using co-precipitation process, hydrothermal process or microwave irradiation in order to form LDH-based nanocomposites. By using microwave irradiation, the reaction time for nanocomposite fabrication is significantly reduced compared with conventional method (co-precipitation and hydrothermal processes). Usually, a rigid nanoplatelet array and well oriented structure is obtained by *in situ* crystallization method.

Applications of LDH-based nanocomposites

Catalysis

There are various reports indicating catalytic activity of metal nanoparticles/LDH nanocomposites (see Table 1). Pt, Au and Pd are the common metallic materials which have been immobilized on the different LDHs and have been used as catalysts in different chemical reactions. Li et al. (Li et al. 2014a) successfully immobilized the ultrafine Au nanoclusters with the average size of 1.5 ± 0.6 nm on the M-Al LDH (M = Mg, Ni or Co). The catalytic activity of prepared nanocomposite was tested for aerobic oxidation of alcohols. Au nanoclusters/Ni-Al LDH exhibited the highest activity, attributed to the strong Au-support synergy effect. Miao et al. (Miao et al. 2015) prepared Au/NiAL-LDH/grapheme nanocomposite for aerobic selective oxidation of benzyl alcohol. In this research, NiAl-LDH/grapheme was used as support for Au nanoparticles. The detailed characterization indicated that simultaneous presence of defect sites and oxygenic functional groups in support resulted in generation of small particle size of Au. In the recent years, in order to improve the recyclability and stability of LDH-based nanocomposites, novel core-shell structure has attracted extensive attention. In this structure, a magnetic core, e.g., Fe_3O_4, is covered with an LDH shell. Mi and his co-workers (Mi et al. 2011) loaded Au nanoparticles on the Fe_3O_4@ MgAl-LDH to generate three-phase Fe_3O_4@MgAl-LDH/Au nanocomposite for catalytic oxidation of 1-phenylethanol (see Figure 10).

Figure 10. The synthesis pathway of Fe_3O_4@MgAl-LDH@Au nanocomposite. (Reprinted with permission from Mi, F., X. Chen, Y. Ma, S. Yin, F. Yuan and H. Zhang. Chem. Commun. 47: 12804 (2011). Copyright@RSC. License Number: 3978911455195.)

POM/LDH nanocomposites have been widely used as catalyst in different catalytic reactions. Carriazo et al. (Carriazo et al. 2007) intercalated $H_2W_{12}O_4^{6-}$ and $W_4Nb_2O_{19}^{4-}$ between the layers of Mg-Al LDH and Zn-Al LDH through ion exchange. The resulted POM/LDH nanocomposites were used as catalyst for epoxidation of cyclooctene in the presence of oxidants (H_2O_2 or t-BuOOH). The $H_2W_{12}O_4^{6-}$/Zn-Al LDH nanocomposite showed the best results for epoxide yield (17% at 24 h). Maciuca et al. (Maciuca et al. 2008a) investigated the catalytic oxidation of thioethers and thiophene derivatives using H_2O_2 as oxidant and V-, Mo- and W-containing layered double hydroxides as catalyst. For preparing catalysts, metal-oxoanions, including WO_4^{2-}, $W_7O_{24}^{6-}$, $V_2O_7^{4-}$, $V_{10}O_{28}^{6-}$, MnO_4^{2-}, $Mo_7O_{24}^{6-}$, intercalated in the interlayer gallery of Mg-Al LDH using direct ion exchange. The results demonstrated that LDHs containing W-based POMs are more active and stable compared with LDHs containing V-based and Mo-based POMs. In addition, Maciuca et al. (Maciuca et al. 2008b) in another research investigated the catalytic activities of mentioned V-, Mo- and W-containing layered double hydroxides as catalyst for mild oxidation of tetrahydrothiophene to sulfolane. WO_4^{2-}/LDH nanocomposite showed the best catalytic activity and stability for this oxidation process.

Das and Parida (Das et al. 2007) intercalated molybdophosphoric acid ($H_3PMo_{12}O_{40}$) and tungstophosphoric acid ($H_3PW_{12}O_{40}$) as heteropoly acids in ZnAl LDH through *in situ* crystallization of LDH in the presence of these POMs. These nanocomposites were used as catalyst for esterification of acetic acid using n-butanol under autogenously condition. $H_3PMo_{12}O_{40}$-ZnAl LDH represented highest conversion (84.15%) with 100% selectivity for n-butyl acetate. Adwani et al. (Adwani et al. 2015) prepared a novel inorganic-bio nanocomposite of chitosan and MgAl-LDH through *in situ* crystallization of LDH in the presence of chitosan. The resulted nanocomposite exhibited efficient catalytic activity for selective synthesis of jasminaldehyde via condensation of benzaldehyde and 1-heptanal under solvent free condition. The results demonstrated that the initial rate of synthesis reaction in the presence of as-prepared nanocomposite is ~ 20% higher than the rates of reactions performed in the presence of individual chitosan and LDH.

Nam et al. (Nam et al. 2016) prepared polydopamine/LDH nanocomposite as a novel polymer/LDH nanocomposite via interlayer polymerization under N_2 atmosphere in a basic buffer solution. The XRD analysis confirmed the intercalation of polydopamine chains into the interlayer space of LDH. The catalytic activity of the prepared nanocomposite was evaluated by reducing the p-nitraphenol in the presence of $NaBH_4$. Polydopamine protected the surface of the nanocomposite against the contamination.

Electrocatalysis

LDH-based nanocomposites have been widely used as electrocatalyst in different electrocatalytic reactions (see Table 2). Zhao et al. (Zhao et al. 2013b) fabricated well designed Pd/CoAl-LDH nanocomposite with a mesoporous structure through *in situ* redox reaction between CoAl-LDH as support and K_2PdCl_4 as metal salt precursor. The TEM image of nanocomposite confirmed the uniform dispersion of Pd nanoparticles on the surface of LDH nanosheets. The resultant nanocomposite displayed excellent catalytic activity for electrooxidation of ethanol.

Nowadays, splitting of water through direct oxidation of water molecules is of great importance owning to provide renewable and clean fuel for energy devices (Cheng et al. 2012). In the field of renewable energy conversion technologies, designing novel cost-effective electrocatalysts for simultaneous generation of O_2 and H_2 is of great importance. However, low reaction rate and high kinetic

Table 1. Catalytic activity of LDH-based nanocomposites.

Nanocomposite	Type	Preparation method	Reaction	Ref.
Au/MgAl-LDH	Metal nanoparticle/ LDH	*In situ* chemical reduction	Synthesis of lactones from diols	(Mitsudome et al. 2009c)
Au/MgAl-LDH	Metal nanoparticle/ LDH	*In situ* chemical reduction	Oxidation of 1-phenylethanol	(Mitsudome et al. 2009a)
Au/MgAl-LDH	Metal nanoparticle/ LDH	*In situ* chemical reduction	Oxidation of benzyl alcohol	(Wang et al. 2011b)
Flower-likeAu/NiAl-LDH	Metal nanoparticle/ LDH	*In situ* chemical reduction	Oxidation of benzyl alcohol	(Du et al. 2015)
Au/MgAlCr-LDH	Metal nanoparticle/ LDH	*In situ* chemical reduction	Oxidation of alcohols	(Liu et al. 2014)
Au/MgAl-LDH	Metal nanoparticle/ LDH	*In situ* chemical reduction	Low temperature CO oxidation	(Li et al. 2012)
Au/MgAl-LDH	Metal nanoparticle/ LDH	*In situ* chemical reduction	Oxidation of glycerol	(Takagaki et al. 2011)
Au/MgAl-LDH	Metal nanoparticle/ LDH	*In situ* chemical reduction	Oxidative tandem synthesis of methyl esters and imines from primary alcohols	(Liu et al. 2012)
Au/MgAl-LDH	Metal nanoparticle/ LDH	*In situ* chemical reduction	Selective oxidation of silanes to silanols in water	(Mitsudome et al. 2009b)
Au nanoclusters/NiAl-LDH Au nanoclusters/NiMn-LDH	Metal nanoparticle/ LDH	*In situ* chemical reduction	Oxidation of 1-phenylethanol to acetophenone	(Wang et al. 2016c)
Au–Pd alloy/MgAl-LDH	Alloy nanoparticles/ LDH	*In situ* chemical reduction	Aerobic oxidative dehydrogenation of cyclohexanols and cyclohexanones to phenols	(Jin et al. 2016)
Ag/ZnAl-LDH	Metal nanoparticle/ LDH	*In situ* chemical reduction	Oxidation of alcohol	(Islam et al. 2015)
Pd/MgAl-LDH	Metal nanoparticle/ LDH	*In situ* chemical reduction	Oxidation of benzyl alcohol	(Chen et al. 2013)
Pd/MgAl-LDH	Metal nanoparticle/ LDH	*In situ* chemical reduction	Hydrogenations of ketones	(Tao et al. 2010)
Pd/MgAl-LDH	Metal nanoparticle/ LDH	Precipitation– reduction method	Hydrogenation of acetylene	(Feng et al. 2012)
Pd/MgAl-LDH	Metal nanoparticle/ LDH	Ion exchange-*in situ* chemical reduction	Suzuki cross-coupling reaction	(Zhang et al. 2013a)
Pd/amine functionalized ZnAl-LDH	Metal nanoparticle/ LDH	*In situ* chemical reduction	Suzuki coupling reaction	(Singha et al. 2011)
Pd/MgAl-LDH	Metal nanoparticle/ LDH	*In situ* chemical reduction	Synthesis of a-alkylated nitriles with carbonyl compounds (aldol reaction/hydrogenation)	(Motokura et al. 2005)
Pd/MgAl-LDH	Metal nanoparticle/ LDH	Ion exchange followed by *in situ* chemical reduction	Heck-type, Suzuki-type, Sonogashira-type, and Stille-type coupling reactions of chloroarenes	(Choudary et al. 2002)
Pd/MgAl-LDH/carbon	Metal nanoparticle/ LDH/carbon based material	*In situ* chemical reduction	Selective hydrogenation of citral	(Han et al. 2015)

Table 1 contd. ...

...Table 1 contd.

Nanocomposite	Type	Preparation method	Reaction	Ref.
Pd/dodecylsulfate embedded MgAl-LDH	Metal nanoparticle/ LDH	Ion exchange followed by *in situ* chemical reduction	Suzuki coupling reactions	(Shiyong et al. 2010)
Pd/MgAl-LDH/Al$_2$O$_3$	Metal nanoparticle/ LDH/metal oxide	*In situ* precipitation-reduction	Selective hydrogenation of acetylene	(Ma et al. 2011)
Nanowire Pd/MgAl-LDH	Metal nanoparticle/ LDH	*In situ* crystallization of LDH and *in situ* chemical reduction	Hydrogenation of acetylene	(He et al. 2012)
Pd/MgAl-LDH Pt/MgAl-LDH	Metal nanoparticle/ LDH	*In situ* chemical reduction	Aldol reaction/ hydrogenation	(Motokura et al. 2005)
Pt/MgAl-LDH	Metal nanoparticle/ LDH	*In situ* chemical reduction	Hydrogenation of cinnamaldehyde	(Xiang et al. 2013)
Pt/MgAl-LDH	Metal nanoparticle/ LDH	*In situ* chemical reduction	Oxidation of glycerol	(Tsuji et al. 2011)
Co/MgAl-LDH	Metal nanoparticle/ LDH	Wetness impregnation method	Hydrogenation of CO	(Tsai et al. 2011)
MnO$_2$/silylated MgAl-LDH	Metal oxide/LDH nanocomposites	Reconstitution pathway	Aerobic oxidation of ethylbenzene	(Lv et al. 2015)
γ-Al$_2$O$_3$/NiMgAl-LDH	Metal oxide/LDH nanocomposites	*In situ* crystallization of LDH on the Al$_2$O$_3$ using hydrothermal method	Carbon dioxide reforming of methane	(Xu et al. 2015)
[WZnMn$_2$ (ZnW$_9$O$_{34}$)$_2$]$^{12-}$/ MgAl-LDH [WZn$_3$ (ZnW$_9$O$_{34}$)$_2$]$^{12-}$/MgAl-LDH [WCo$_3$ (CoW$_9$O$_{34}$)$_2$]$^{12-}$/MgAl-LDH [PW$_{11}$O$_{39}$]$^{7-}$/MgAl-LDH	POM/LDH	Ion exchange	Epoxidation of allylic alcohols	(Liu et al. 2009)
K$_{11}$WFe$_3$ (H$_2$O)$_2$O[ZnW$_9$O$_{34}$]$_2$·44H$_2$O/ ZnAl-LDH K$_{11}$WMn$_3$ (H$_2$O)$_2$O[ZnW$_9$O$_{34}$]$_2$·27H$_2$O/ Zn-Al LDH	POM/LDH	Ion exchange	Oximation of aromatic aldehydes	(Zhao et al. 2011)
[WZn$_3$ (H$_2$O)$_2$ (ZnW$_9$O$_{34}$)$_2$]$^{12-}$/ MgAl-LDH	POM/LDH	Ion exchange	Oximation of aldehydes	(Zhao et al. 2012)
[CoW$_{12}$O$_{40}$]$^{5-}$/MgAl-LDH	POM/LDH	Ion exchange	Oxidation of benzaldehyde	(Wei et al. 2008)
MoO$_4$$^{2-}$/MgAl-LDH	POM/LDH	Ion exchange	Generation of singlet molecular oxygen from H$_2$O$_2$	(Van Laar et al. 2001)
[PFeW$_{11}$O39]$^{4-}$/ZnAl-LDH	POM/LDH	Ion exchange	Oxidation of alcohols	(Hasannia et al. 2015)
[WZn$_3$ (ZnW$_9$O$_{34}$)2]$^{12-}$/ MgAlLDH	POM/LDH	Ion exchange	Heterogeneous epoxidation of olefins	(Liu et al. 2008)
WO$_4$$^{2-}$/MgAl-LDH	POM/LDH	Ion exchange	N-oxidation of tertiary amines	(Choudary et al. 2001)
Carbon dot/NiAl-LDH	Carbon based material/ LDH	Self-assembly	Oxidation of 3,3',5,5'-tetramethylbenzidine (TMB) in the presence of H$_2$O$_2$	(Guo et al. 2015)

Table 2. Electrocatalytic activity of LDH-based nanocomposites.

Nanocomposite	Type	Preparation method	Reaction	Ref.
Au/MgAl-LDH	Metal nanoparticle/ LDH	*In situ* chemical reduction	Electrooxidation of glucose	(Cho et al. 2016)
Au/MgAl-LDH	Metal nanoparticle/ LDH	*In situ* chemical reduction	Electrooxidation of methanol in alkaline solution	(Wang et al. 2010c)
Au/CuMgFe-LDH	Metal nanoparticle/ LDH	Au nanoparticles through electrodeposition and LDH through physical coating were deposited on electrode	Electrooxidation of ethanol	(Taei et al. 2016a)
Pd/MgAl-LDH	Metal nanoparticle/ LDH	*In situ* chemical reduction	Electrooxidation of methanol in alkaline solution	(Jia et al. 2015)
Carbon Quantum Dot/ NiFe-LDH	Carbon based material/LDH	*In situ* crystallization of LDH in the presence of Carbon Quantum Dot	O_2 generation	(Tang et al. 2014)
Reduced graphene oxide/NiMn-LDH	Carbon based material/LDH	Self-assembly	O_2 generation	(Ma et al. 2016)
Reduced graphene oxide/ZnCo-LDH	Carbon based material/LDH	*In situ* crystallization of LDH using co-precipitation in the presence of reduced graphene oxide	O_2 generation	(Tang et al. 2014)
Graphene oxide/CoAl-LDH	Carbon based material/LDH	*In situ* crystallization of LDH using co-precipitation in the presence of graphene oxide	Electroreduction of oxygen	(Wang et al. 2016d)
Carbon nanotube/NiFe-LDH	Carbon based material/LDH	*In situ* crystallization of LDH using hydrothermal in the presence of carbon nanotube	O_2 generation	(Gong et al. 2013)
Mildly oxidized graphene/single-walled carbon nanotube/NiFe-LDH	Carbon based material/LDH	*In situ* crystallization of LDH using co-precipitation in the presence of mildly oxidized graphene and single-walled carbon nanotube	O_2 generation	(Zhu et al. 2015)
$[Mo^{VI}O_2 (O_2CC (S) Ph_2]^{2-}$/ZnAl-LDH	Metal complex/LDH	Ion exchange	Air oxidation of thiols	(Cervilla et al. 1994)
CuO/MnAl-LDH	Metal oxide/LDH	*In situ* crystallization of LDH using co-precipitation in the presence of CuO	Reduction of H_2O_2	(Asif et al. 2017)

of overpotential limit the development of these devices. Therefore, development of novel electrocatalysts for efficient water splitting has got great attention in recent years. Shao et al. (Shao et al. 2014b) synthesized ZnO@CoNi-LDH nanocomposite with core-shell structure through facile electrosynthesis method. The obtained nanocomposite displayed excellent photoelectrochemical behavior for water splitting, which can be attributed to the excellent photocatalytic activity of LDH shells and effective separation of generated electrons and holes. In addition, the hierarchical structure of this nanocomposite facilitates the ion diffusion and convenient charge transfer on the interface of electrode and electrolyte, where the oxidation of water molecules occurs. Carbon-based materials, e.g., graphene, carbon nanotubes and carbon nanoparticles, have great potential for incorporating with LDHs in order to form efficient catalysts for electrochemical reaction owning to their high specific surface area, mechanical strength and electronic conductivity. Hou et al. (Hou et al. 2016) developed a novel three-dimensional hierarchical exfoliated graphene/CoSe/NiFe-LDH nanocomposite for electrocatalytic water splitting. For this purpose, CoSe nanosheets were grown vertically on the exfoliated graphene foil, and then, NiFe-LDH nanosheets were deposited on the obtained hybrid using hydrothermal treatment. The results showed that 20 mA cm^{-2} current density at 1.71 V was obtained by applying as-prepared nanocomposite as both cathode and anode for water splitting, which is comparable with the performance of Pt/C and Ir/C catalysts. Wang et al. (Wang et al. 2010a) prepared carbon nanotube/NiAl-LDH via dispersion of positively charged LDH nanocrystallites on the surface of negatively charged carbon nanotubes using electrostatic interaction. The experimental tests indicated that the electrode, improved by carbon nanotube/LDH nanocomposite, exhibited excellent electrocatalytic activity for electrooxidation of glucose.

Photocatalysis

O_2 generation

Gunjakar et al. (Gunjakar et al. 2011) synthesized layer by layer nanocomposite using self-assembly between layered TiO$_2$ and two-dimensional ZnCr-LDH nanosheets with opposite charges. The generated heterolayered nanocomposite represented high activity for O$_2$ generation under visible light irradiation with a rate of ~ 1.18 mmol h^{-1} g^{-1}. Highly porous structure and visible-light-harvesting ability were observed for resultant nanocomposite, which were ascribed to electronic coupling between the TiO$_2$ and ZnCr-LDH nanosheets. In addition, the results demonstrated that the chemical stability of the ZnCr-LDH improved owing to the protection effect of layered titanate. Recently, combining the graphene with LDHs to generate composites with high photocatalytic activity has received great attention owning to its visible light-responsive activity and good electron transfer capability. Gunjakar et al. (Gunjakar et al. 2013) prepared ZnCr/RGO nanocomposite with highly efficient photocatalytic activity for visible light-induced O$_2$ generation. In this research, positively charged ZnCr-LDH were immobilized on the negatively charged two-dimensional graphene nanosheets through self-assembly method, resulting in producing a highly porous stacked structure. The obtained results demonstrated that the ZnCr-LDH/RGO nanocomposite represented significantly high photocatalytic activity for O$_2$ generation under visible light irradiation with a rate of ~ 1.2 mmol h^{-1} g^{-1}, which was far greater than that of pristine ZnCr-LDH (~ 0.67 mmol h^{-1} g^{-1}).

H_2 generation

Hong and co-workers (Hong et al. 2011) prepared a novel visible-light responsive nanocomposite for H$_2$ evolution by immobilization of photosensitizer (rose-bengal, RB) and Pt as photocatalyst on the well-dispersed MgAl-LDH. In this system, the self-quenching is suppressed by immobilization of dye photosensitizer, efficient electron transfer occurred due to the close arrangement between catalyst nanoparticles and the molecules of photosensitizer and the catalyst nanoparticles are well-dispersed on the support. The aforementioned advantages enhance the photocatalytic activity of such a system. Zhang et al. (Zhang et al. 2015) prepared CdS/ZnCr-LDH nanoparticles via incorporating the CdS nanoparticles into interlayer space of ZnCr-LDH nanosheets using an exfoliation-restacking method. The obtained nanocomposite exhibited significantly high photocatalytic activity for evolution of H$_2$ under visible and UV-visible irradiation, which was attributed to the efficient separation of photogenerated electrons and holes which resulted due to strong electronic coupling between CdS and ZnCr-LDH.

Pollutant elimination

Chen et al. (Chen et al. 2014) intercalated Pt nanoparticles into gallery of ZnTi-LDH using ion exchange and photochemical reduction method in order to prepare an efficient photocatalyst for degradation of rhodamine B under sunlight irradiation. Pt/ZnTi-LDh nanocomposite represented 17-fold higher degradation efficiency compared with pure ZnTi-LDH which can be ascribed to the high specific area, wide photoresponding range, negative shift in the flat-band potential and efficient charge separation owing to high migration of photo-induced electrons.

POMs/LDH nanocomposites show photocatalytic activity. For example, MoO_4^{2-}/Zn-Y LDH and WO_4^{2-}/Zn-Y LDH were prepared using ion exchange method by Mohapatra et al. (Mohapatra et al. 2012), and their photocatalytic activity was tested for degradation of Rhodamine 6G under visible-light irradiation. The metal-to-metal charge transition from the Zn-O-Y oxo-bridged linkages resulted in absorption in the visible region. In addition, the reactivity enhancement of investigated POMs intercalated Zn-Y LDH demonstrated that the interlayer space is the reaction site. Table 3 summarized the photocatalytic activity of different LDH-based nanocomposites.

Table 3. Photocatalytic activity of LDH-based nanocomposites.

Nanocomposite	Type	Preparation method	Reaction	Irradiation source	Ref.
TiO_2/ZnAl-LDH	Metal oxide/ LDH	Reconstruction of the zinc-containing clay in a $TiOSO_4$ aqueous	Photodegradation of phenol	UV	(Carja et al. 2010)
TiO_2/MgAl-LDH	Metal oxide/ LDH	*In situ* crystallization of LDH in the presence of TiO_2	Photodegradation of dimethyl phthalate	UV	(Huang et al. 2013)
TiO_2/MgAl-LDH	Metal oxide/ LDH	Reconstitution pathway	Photodegradation of methyl orange	UV	(Seftel et al. 2013)
TiO_2/CuMgAl-LDH	Metal oxide/ LDH	Selective reconstitution pathway	Photodegradation of methylene blue	Both UV and visible	(Lu et al. 2012)
Fe_3O_4/ZnCr-LDH	Metal oxide/ LDH	*In situ* crystallization of LDH using hydrothermal method in the presence of Fe_3O_4	Photodegradation of methylene blue	UV	(Chen et al. 2012)
Ag/AgBr/CoNi-LDH	Metal nanoparticles/ metal halide/ LDH	Ion exchange and *in situ* reduction	Photodegradation of phenol, methyl orange and rhodamine B	Visible	(Fan et al. 2013)
Ag/AgCl/ZnAl-LDH	Metal nanoparticles/ metal halide/ LDH	Ion exchange and *in situ* reduction	Photodegradation of Gram-negative and Gram-positive bacteria and yeast	Visible	(Nocchetti et al. 2013)
α-Ag_2WO_4/ZnCr-LDH	Metal tungstate/ LDH	Ion exchange followed by precipitation	Photodegradation of rhodamine B	Visible	(Zhu et al. 2013)
$[W_7O_{24}]^{6-}$/MgAl-LDH	POM/LDH	Ion exchange	Photodegradation of hexachlorocyclohexane	UV	(Guo et al. 2001)
$[SiW_{11}O_{39}Ni(H_2O)]^{6-}$/ZnCr-LDH	POM/LDH	Ion exchange	Photodegradation of hexachlorocyclohexane	UV	(Wang et al. 2010b)
$[SiW_{11}O_{39}]^{8-}$/ZnAl-LDH	POM/LDH	Ion exchange	Photodegradation of hexachlorocyclohexane	UV	(Guo et al. 2001)
$[P_2W_{17}O_{61}Mn(H_2O)]^{8-}$/ZnAl-LDH	POM/LDH	Ion exchange	Photodegradation of hexachlorocyclohexane	UV	(Guo et al. 2001)
$V_{10}O_{28}^{6-}$/ZnBi-LDH	POM/LDH	Ion exchange	Photodegradation of malachite green, rhodamine 6G and 4-chloro-2-nitrophenol	Solar light	(Mohapatra et al. 2014)
Carbon nanotubes/ ZnAl-LDH	Carbon based material/LDH	*In situ* crystallization of LDH using co-precipitation method in the presence of CNT	Photodegradation of methyl orange	UV	(Wang et al. 2010b)

Table 3 contd....

...*Table 3 contd.*

Nanocomposite	Type	Preparation method	Reaction	Irradiation source	Ref.
Graphene/ZnCr-LDH	Carbon based material/LDH	Self-assembly	O_2 generation	Visible	(Gunjakar et al. 2013)
Graphene/ZnCr-LDH	Carbon based material/LDH	*In situ* crystallization of LDH using co-precipitation method in the presence of graphene	Photodegradation of rhodamine B	Visible	(Lan et al. 2014)
Carboxyl grapheme/ZnAl-LDH	Carbon based material/LDH	Self-assembly	Photodegradation of methylene blue and orange G	Visible	(Huang et al. 2014)
Reduced graphene oxide/NiTi-LDH	Carbon based material/LDH	*In situ* crystallization of LDH using co-precipitation method in the presence reduced graphene oxide	O_2 generation	Visible	(Li et al. 2013)

Adsorbent

Adsorption is one of the efficient and cost effective methods for water and wastewater treatment, which has been widely used for dye and heavy metal removal from water effluents (Abas et al. 2013, Gupta 2009). In adsorption processes, the molecules of the pollutant are concentrated at a solid surface from their gaseous or liquid surroundings (Abas et al. 2013). In recent years, LDH-based nanocomposites owing to their remarkable advantageous including large surface area, high sorption, good thermal stability and high regeneration efficiency have been widely used for adsorbing toxic materials from aquatic solutions (see Table 4). Moreover, the adsorption of proteins on the surface of solids has received great attention in recent years due to its application in different fields, e.g., biomedicine, interfacial engineering and material science. Au/MgAl-LDH nanocomposite, prepared via *in situ* reduction method, was applied for adsorption of different proteins including hemoglobin, bovin serum albumin and lysozyme by Jin et al. (Jin et al. 2012). The results demonstrated that this nanocomposite displays a significantly high capacity for adsorption of hemoglobin and the adsorption process can be modeled pseudo second order kinetic model. Bi et al. (Bi et al. 2011) fabricated $[PW_{10}Mo_2O_{40}]^{5-}$ using ion exchange pathway and applied it as POM/LDH nanocomposite for elimination of methylene blue, a cationic dye, from wastewater through adsorption. The efficiency of dye removal using this nanocomposite was found to be 75%. Zhang et al. (Zhang et al. 2014) fabricated carbon dot/MgAl-LDH as an environmental friendly nanocomposite via self-assembly method, and its adsorption performance was evaluated for removal of methyl blue as anionic dye. The as-prepared nanocomposite displayed high capacity for adsorption of methylene blue (185 mg/g), and the adsorption behavior was successfully fitted with pseudo–second-order kinetic model and Langmuir isotherm. Guo et al. (Guo et al. 2011) prepared a novel inorganic-bio nanocomposite in which MgAl-LDH nanosheets were immobilized on the fibers of eggshell membrane via *in situ* crystallization method. The obtained nanocomposite was used as adsorbent for removal of Cr (VI). The adsorption isotherm was well fitted with the Langmuir isotherm model.

Drug delivery

In recent years, LDH-based nanocomposites have received great attention in the field of drug delivery owning to their biocompatibility and low toxicity (Kura et al. 2014). Tian et al. (Tian et al. 2016) developed a promising LDH-based nanocomposite for anti-tumor drug delivery and synergistic therapy. The synthesis method is summarized in Figure 11. As can be seen, methotrexate (MTX) and poly (diallyldimethylammonium chloride) (PDDA) were alternatively deposited on the Au nanoparticles through the layer by layer technology, resulting in formation of Au@PDDA-MTX hybrid with core-shell structure. In the next step, the obtained hybrid was conjugated on the surface of MTX/LDH hybrid via the

Table 4. LDH-based nanocomposites used as adsorbent material.

Nanocomposite	Type	Preparation method	Pollutant	Ref.
TiO_2/ZnAl-LDH	Metal oxide/LDH	Simultaneous *in situ* crystallization of LDH and TiO_2	Methylene blue	(Shao et al. 2014a)
MnO_2/MgAl-LDH	Metal oxide/LDH	Simultaneous *in situ* crystallization of LDH and TiO_2 using hydrothermal method	Pb (II)	(Bo et al. 2015)
Fe_3O_4/ZnCr-LDH	Metal oxide/LDH	*In situ* crystallization of LDH using hydrothermal method in the presence of Fe_3O_4	Methyl orange	(Chen et al. 2012)
Fe_3O_4/MgAl-LDH	Metal oxide/LDH	Self-assembly	Congo red	(Chen et al. 2011)
Fe_3O_4/MgAl-LDH	Metal oxide/LDH	*In situ* crystallization of LDH using co-precipitation method in the presence of Fe_3O_4	Phosphate	(Koilraj et al. 2016)
Fe_3O_4/MgAl-LDH	Metal oxide/LDH	*In situ* crystallization of LDH using co-precipitation method in the presence of Fe_3O_4	Reactive red, congo red and acid red 1	(Shan et al. 2014)
Fe_3O_4@MgAl-LDH	Metal oxide/LDH	*In situ* crystallization of LDH using co-precipitation method in the presence of Fe_3O_4	Acid yellow 219	(Yan et al. 2015)
Biochar/NiFe-LDH	Carbon based material/LDH	*In situ* crystallization of LDH using co-precipitation method in the presence of biochar	Arsenic	(Wang et al. 2016b)
Fe_3O_4/graphene/ MgAl-LDH	Metal oxide/carbon based material/LDH	*In situ* crystallization of LDH using hydrothermal method in the presence of Fe_3O_4/graphene composite	Arsenate	(Wu et al. 2011)
Ag/AgBr/CoNi-LDH	Metal nanoparticles/ metal halide/LDH	Ion exchange and *in situ* reduction	Methyl orange	(Fan et al. 2013)

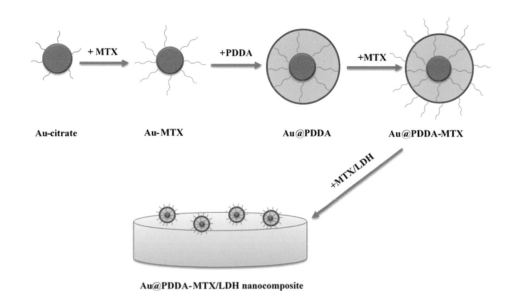

Au@PDDA-MTX/LDH nanocomposite

Figure 11. The synthesis pathway of Au@PDDA-MTX/LDH nanocomposite. (Reprinted with permission from Tian, D.Y., W.Y. Wang, S.P. Li, X.D. Li and Z.L. Sha. Int. J. Pharm. 505: 96 (2016). Copyright@Elsevier. License Number: 3978911045886.)

electrostatic attraction to generate Au@PDDA-MTX/LDH nanocomposite. The obtained nanocomposite represented a high capacity for loading of drug, interesting pH-responsive release profile and excellent colloidal stability.

In another research, Gd-doped MgAl-LDH/Au nanocomposite was developed by Wang and co-workers (Wang et al. 2013) as an efficient drug carrier and a diagnostic agent. The produced nanocomposite not only had a high capacity for loading of doxorubicine (DOX) as a non-anionic anti-cancer drug, but also showed an interesting pH-responsive release profile. In addition, the obtained nanocomposite represented better *in vitro* CT and T1-weighted MR imaging capabilities compared with the commercial MRI and CT contrast. Zhang et al. (Zhang et al. 2009) prepared a novel magnetic nanocomposite with a core-shell structure for developing a magnetically controlled drug release system. For this purpose, diclofenac as a non-steroid anti-inflammatory drug was intercalated into the interlayer space of the MgAl-LDH, which was coated as shell on the core of ferrite particles. The TEM analysis showed that the magnetic nanocomposites were successfully synthesized in core-shell structure with diameter of 90–150 nm. The dimension of the diclofenac-LDH nanocrystallites in magnetic nanocomposites were smaller compared with that of pure diclofenac-LDH, attributed to the heterogeneous nuclcation in the prescnce of magnetic core. The results demonstrated that in the presence of external magnetic field, the drug release rate of magnetic nanocomposite decreased due to the aggregation of the magnetic nanocomposite particles (see Figure 12). The kinetic data demonstrated that the release rate of the diclofenac from the magnetic nanocomposite is efficiently controlled by diffusion of particles, particle size and the aggregation extent of nanocomposite particles.

In another research, Pan et al. (Pan et al. 2011) prepared Fe_3O_4@doxifluridine (DFUR)-MgAl-LDH nanocomposite through co-precipitation-calcination-reconstruction pathway. Doxifluridine is an anticancer agent. The obtained nanocomposite with well-defined core-shell structure presented good magnetic property and good magnetically controlled drug release and drug delivery. Du et al. (Du et al. 2007) reported the preparation of MgAl-LDH encapsulated in vesicles as a novel nanocomposite with high potential for drug delivery and gene therapy. The results demonstrated that positively charged LDH nanoparticles can induce the spontaneous generation of vesicles in a mixture of dodecyl betaine, zwitterionic surfactant and sodium bis(2-ethylhexyl)sulfosuccinate. LDH nanoparticles are capsulated in generated vesicles, resulting in fabrication of a novel nanocomposite.

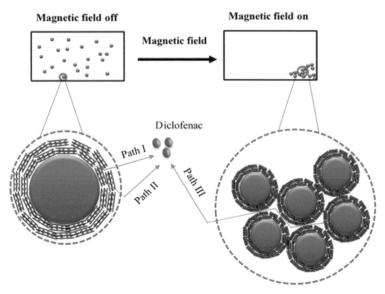

Figure 12. A schematic drawing of diclofenac release from the core-shell structured nanocomposite in the presence and absence of external magnetic field. (Reprinted with permission from Zhang, H., D. Pan, K. Zou, J. He and X. Duan. J. Mater. Chem. 19: 3069 (2009). Copyright@ RSC. License Number: 3978780861888.)

Supercapacitor

Supercapacitors, which are known as electrochemical capacitors, have received great attention as power device for energy storage because of fast charging and discharging properties, long cycle life, high power density, high power delivery and low maintenance cost (Long et al. 2016, Shao et al. 2015a). Based on the mechanism of storage, the electrochemical supercapacitors can be classified in two main groups, including electrical double layer capacitors (EDLCs) and pseudocapacitors. In pseudocapacitors, reversible faradic redox reactions occur on the surface of electro-active material including metal hydroxides, transition metal oxides and polymers for charge storage (Fang et al. 2010, Shao et al. 2015a, Yang et al. 2008). Pseudocapacitors represent 10–100-fold higher specific capacitance compared with EDLCs (Shao et al. 2015a). LDHs are considered as a promising candidate for supercapacitors because of their low cost and high theoretical specific capacity (Shao et al. 2015a). However, LDHs have some disadvantages which limit their application in supercapacitors including low cycle life and low power density (Long et al. 2016, Shao et al. 2015a). To overcome these limitations, the composites of LDHs, including carbon based materials/LDH and polymer/LDH nanocomposites, have been developed (see Table 5). Carbon based materials/LDH nanocomposites contain graphene/LDH, carbon nanoparticle/LDH and carbon nanotube/ LDH. Graphene, owing to its high surface area and high charge mobility, has been widely used as support for LDHs to generate an effective nanocomposite for producing efficient supercapacitors. Dong et al. (Dong et al. 2011) prepared graphene/CoAl-LDH multilayer film using self-assembly method, and the potential of prepared nanocomposite was investigated to be used as electrode material in flexible supercapacitor devices. The resultant nanocomposite film represented a high area capacitance of 70 F/ m^2 and specific capacitance of 800 F/g under the scan rate of 5 mV/s. The improved specific capacitance of the prepared film can be ascribed to the efficient electron transport which resulted due to the face-to-face contact of GO sheets and CoAl-LDH nanosheets. In addition, the results demonstrated that by partial reduction of GO at 200°C in H_2 atmosphere, the area capacitance and specific capacitance of prepared film were significantly improved up to 90 F/m^2 and 1204 F/g, respectively. Liu et al. (Liu et al. 2015) synthesized NiAl-LDH@carbon nanoparticles composite using a facile and cost effective method, involving the growth of LDH nanosheets through a hydrothermal reaction following the coating of carbon nanoparticles via a vacuum filtration process. The obtained nanocomposite was used to prepare an asymmetric supercapacitor with excellent capacitive performance containing high energy density, high specific capacitance and good cycling capability. The observed capacitive performance enhancement can be attributed to the electrical conductivity improvement, resulted by coating of carbon nanoparticles. Bai and his co-workers (Bai et al. 2016) fabricated carbon nanotube/NiAl-LDH using a simple one-step homogeneous precipitation method. The obtained nanocomposite represented an excellent specific capacitance of 694 F/g at the current density of 1 A/g, which still retained 87% by increasing the current density from 1 A/g to 10 A/g. In addition, the results showed that after 3000 cycles at 20 A/g, at least 92% of nanocomposite capacitance was retained, demonstrating excellent cycling durability. Li et al. (Li et al. 2015d) reported the fabrication of an asymmetric supercapacitor with super-high energy density based on NiCo-LDH@carbon nanotube nanocomposite with three-dimensional core-shell structure. The prepared LDH@carbon nanotube/nickel foam electrode exhibited an excellent capacitance of 2046 F/g at current density of 1 A/g.

In addition to mentioning LDH-based nanocomposites applied in supercapacitors, Wang et al. (Wang et al. 2016e) developed a novel three-dimensional porous MXene/LDH nanocomposite for high performance supercapacitors. Mxene are a novel group of two-dimensional transition metal carbides or carbonitrides, including V_2C, Nb_2C, Ti_2C, Ti_3C and Ti_3CN. These compounds have received great attention in recent years due to their chemical stability, ultrahigh electrical conductivity and high specific capacitance. Ti_3C/Nial-LDH nanocomposite was fabricated through *in situ* crystallization of LDH using co-precipitation method in the presence of Ti_3C. The optimum MXene/LDH nanocomposite represented an excellent high specific capacitance of 1061 F/g at the current density of 1 A/g. In addition to carbon based material/LDH nanocomposites, the nanocomposites constructed from LDH and conductive polymer are of great importance in the field of supercapacitors because of providing excellent charge transport in the electrochemical process. In this regard, Zhao et al. (Zhao et al. 2015b) developed a novel

Table 5. LDH-based nanocomposites used as supercapacitor material.

Nanocomposite	Type	Preparation method	Specific capacitance	Ref.
Co_3O_4@NiAl-LDH	Metal oxide/LDH	*In situ* crystallization of LDH using hydrothermal method in the presence of Co_3O_4@AlOOH	1772 F/g at current density of 2 A/g	(Ning et al. 2014)
Graphene/NiAl-LDH	Carbon based materials/LDH	*In situ* crystallization of LDH using hydrothermal method in the presence of graphene	781.5 F/g at scan rate of 5 mV/s	(Gao et al. 2011)
Activated graphene/ NiAl-LDH	Carbon based materials/LDH	*In situ* crystallization of LDH using hydrothermal method in the presence of activated graphene	1730.2 F/g at current density of 0.1 A g^{-1}	(Yulian et al. 2013)
Reduced graphene oxide/CoAl-DH	Carbon based materials/LDH	Simultaneous *in situ* crystallization of LDH and reduction of graphene oxide using hydrothermal method	~ 87 F/g at scan rate of 5 mV/s	(Zhang et al. 2013b)
Layered assembly graphene oxide/ CoAl-LDH	Carbon based materials/LDH	Self-assembly	1031 F/g at current density of 1 A/g	(Wang et al. 2011a)
Graphene/CoAl-DH	Carbon based materials/LDH	*In situ* crystallization of LDH using microwave irradiation in the presence of graphene	772 F/g at current density of 1 A/g	(Fang et al. 2012)
Reduced graphene oxide nanocup/NiAl-LDH	Carbon based materials/LDH	*In situ* crystallization of LDH using hydrothermal method in the presence of reduced graphene oxide nanocup	2712.7 F/g at the current density of 1 A/g	(Lin et al. 2013)
Reduced graphene oxide/CoAl-LDH	Carbon based materials/LDH	Exfoliation-restacking	1296 F/g at a current density of 1 A/g	(Huang et al. 2015)
Reduced graphene oxide/NiCo-LDH	Carbon based materials/LDH	*In situ* crystallization of LDH using hydrothermal method in the presence of reduced graphene oxide	1911.1 F/g at the current density of 2.0 A/g	(Cai et al. 2015)
Reduced graphene oxide/NiAl-LDH	Carbon based materials/LDH	*In situ* crystallization of LDH using microwave irradiation in the presence of reduced graphene oxide	1630 F/g at current density of 1 A/g	(Li et al. 2014b)
Reduced graphene oxide /Flower-like NiCoMn-LDH	Carbon based materials/LDH	Simultaneous *in situ* crystallization of LDH and reduction of graphene oxide using co-precipitation method	912 F/g at current density of 1 A/g	(Li et al. 2015b)
Reduced graphene oxide/flower-like CoMn-LDH	Carbon based materials/LDH	Simultaneous *in situ* crystallization of LDH and reduction of graphene oxide using co-precipitation method	1635 F/g at current density of 1 A/g	(Li et al. 2016a)
Reduced graphene oxide/flower-like NiMn-LDH	Carbon based materials/LDH	Simultaneous *in situ* crystallization of LDH and reduction of graphene oxide using co-precipitation method	84.26 F/g at current density of 1 A/g	(Li et al. 2016a)
Reduced graphene oxide/CoAl-LDH	Carbon based materials/LDH	Self-assembly	450 F/g at current density of 5 A/g	(Ma et al. 2014)
Reduced graphene oxide/CoNi-LDH	Carbon based materials/LDH	Self-assembly	450 F/g at current density of 5 A/g	(Ma et al. 2014)
Sandwich-type reduced graphene oxide/NiAl-LDH	Carbon based materials/LDH	*In situ* crystallization of LDH using co-precipitation method in the presence of reduced graphene	1329 F/g at a current density of 3.57 A/g	(Xu et al. 2014)
Carbon fiber/CoMn-LDH	Carbon based materials/LDH	*In situ* crystallization of LDH using co-precipitation method in the presence of carbon fiber	1079 F/g at current density of 2.1 A/g	(Zhao et al. 2013a)
Carbon nanotube/ NiAl-LDH	Carbon based materials/LDH	*In situ* crystallization of LDH using co-precipitation method in the presence of carbon nanotube	1500 F/g at current density of 1 A/g	(Li et al. 2015c)

Table 5 contd....

...Table 5 contd.

Nanocomposite	Type	Preparation method	Specific capacitance	Ref.
Carbon nanotube/ CoAl-LDH	Carbon based materials/LDH	Self-assembly	884 F/g at current density of 0.55 A/g	(Yu et al. 2014)
Carbon nanotube/ NiMn-LDH	Carbon based materials/LDH	*In situ* crystallization of LDH using co-precipitation method in the presence of carbon nanotube	2960 F/g at current density of 1.5 A/g	(Zhao et al. 2014)
Carbon nanoparticle/ NiAl-LDH	Carbon based materials/LDH	*In situ* crystallization of LDH using hydrothermal method in the presence of carbon nanoparticles	1355 F/g at 5 mV/s	(Liu et al. 2015)
Polypyrrole@CoNi-LDH	Polymer/LDH	Electrosynthesis	1137 F/g at current density of 1 A/g	(Shao et al. 2015a)

nanocomposite via self-assembly of CoNi-LDH monolayers and the poly(3,4-ethyleneioxythiophene): poly(styrene sulfonate) as conductive polymer. In addition, a significantly increased energy density of 46.1 Wh/kg at 11.9 kW/kg resulted, by using fabricated nanocomposite in supercapacitor device. CoAl-LDH@poly(3,4-ethylenedioxythiophene), denoted as PEDOT, nanocomposite with core-shell structure was grown on the flexible foil of the Ni by Han et al. (Han et al. 2013) to develop a high-performance pseudocapacitor. The maximum specific capacitance of 649 F/g was observed for LDH@PEDOT nanocomposite by cyclic voltammetry with scan rate of 2 mV/s and 672 F/g by galvanostatic discharge with current density of 1 A/g. The improved pseudocapacitor behavior of fabricated nanocomposite can be attributed to the synergistic effect of LDH and conductive polymer. Indeed, LDH nanosheets provide capacity for energy storage and the conductive polymer and porous structure facilitate the transportation of electron and mass in the redox reaction.

Batteries

The nickel-Zinc rechargeable batteries have attracted great attention in the field of energy storage due to their high open-circuit voltage, high specific energy and excellent specific power (Lee et al. 2011, Yang et al. 2014). However, the poor cycling characteristic of these batteries, which is due to the solubility of ZnO anode in alkaline electrolyte, limits their widespread applications. To overcome this limitation, new anode material is needed to be developed. The nanocomposites, constructed of LDH and conductive materials, e.g., carbon based materials and metal nanoparticles, seem to be promising candidate for this purpose. Yang et al. (Yang et al. 2014) deposited silver nanoparticles on the nanosheets of ZnAl-LDH using *in situ* chemical reduction method in order to fabricate Ag/ZnAl-LDH nanocomposite. The electrochemical performance of obtained nanocomposite as an anode in Ni-Zn batteries was investigated. The optimum nanocomposite displayed excellent high discharge capacity and cycling stability, which can be ascribed to the synergistic effect between high conductivity of silver nanoparticles and large specific surface area of LDH nanosheets. Yang et al. (Yang et al. 2013) fabricated carbon nanotube/ZnAl-LDH nanocomposite via facile self-assembly approach, and applied the resultant nanocomposite as a high-performance anode in Ni-Zn secondary batteries for the first time. The results of electrochemical studies obviously indicated that the fabricated nanocomposite used as anode exhibited good electrochemical cycle stability, high discharge capacity, low charge plateau voltage and high discharge plateau voltage. The observed improvement can be attributed to the great stability of nanocomposite in alkaline electrolyte and formation of conductive network due to the replacement of carbon nanotubes on the surface of LDH nanosheets. Indeed, the generated conductive network can provide an efficient charge transfer between the electrode and Zn centers, leading to an improvement in electrochemical performance. In addition, the LDH-based nanocomposites can be used as cathode in Ni-based batteries in order to improve these kinds of batteries. Chen et al. (Chen et al. 2016) dispersed carbon black nanoparticles (CB) on the sheets of NiFe-LDH in order to fabricate CB/LDH nanocomposite, applied as cathode material. The obtained nanocomposite represented not only high specific capacity, but also excellent cycling stability under high-rate charge/ discharge conditions. Hu et al. (Hu et al. 2015) fabricated graphene/NiAl-LDH nanocomposite through self-assembly of delaminated LDH and the exfoliated graphene. The resultant nanocomposite was used as alkaline battery cathode and showed excellent rate performance.

Sensors

Today, development of cost-effective and reliable sensors for measuring and controlling the target molecules is of great importance in scientific and engineering communities. In recent years, LDH-based nanocomposites have been the subject of researches in the field of fabrication of sensors (see Table 6). Wu et al. (Wu et al. 2015) deposited Au nanoparticle/NiAL-LDH nanocomposite film on the glass carbon electrode using electrodeposition method. The nanoparticles of the Au enhanced the conductivity of the film, and the synergistic effect between the Au nanoparticles and LDHs improved the electrooxidation of L-cysteine, exhibiting high current density and low oxidation peak potential. The fabricated electrode was applied to sense l-systeine, exhibiting good selectivity and sensitivity.

In recent years, electrochemical biosensors, which are based on the direct transfer of electron from redox enzymes, have found many applications in different fields, e.g., food analysis, pharmaceutical monitoring and clinical diagnosis owning to their excellent sensitivity, selectivity and conveniences (Gholivand et al. 2014). However, there are some problems limiting their application, e.g., realizing the direct electrochemistry of these systems is difficult because the redox centers of these systems are placed deeply in their three-dimensional structure. In addition, most of the enzymes are denatured during the immobilization onto the electrode. To overcome these problems, appropriate material should be used for design and fabrication of electrochemical biosensors. LDH-based nanocomposites seem to be a promising candidate for this purpose. Wang et al. (Wang et al. 2015) used carbon nanodot/CoFe-LDH nanocomposite as support for immobilization of horseradish peroxidase (HRP). The obtained HRP/carbon nanodot/LDH bio-nanocomposite was immobilized on the glass carbon electrode in order to investigate the electrochemical behavior of the resultant electrode and its performance as hydrogen peroxide biosensor. The rate constant of electron transfer was found to be 8.46 s^{-1}, indicating fast electron transfer. In addition, the results demonstrated that the activity of HRP enzyme was retained by immobilization on carbon nanodot/LDH nanocomposite, and a reversible redox behavior was observed for obtained system. The produced H_2O_2 biosensor displayed a linear range of 0.1–23.1 ($R^2 = 0.9942$),

Table 6. LDH-based nanocomposites used as active phase of sensors.

Nanocomposite	Type	Preparation method	Detected compounds	Ref.
Hemoglobin/DNA/NiAl-LDH	Biomaterial/LDH	Self-assembly	H_2O_2 and NO_2^-	(Liu et al. 2013)
1-amino-8-naphthol-3,6-disulfonic acid sodium/ZnAl-LDH	Aromatic compound/LDH	Ion exchange	Hg^{2+}	(Sun et al. 2011)
Graphene/ZnAl-LDH	Carbon based materials/LDH	Casting the mixture of graphene and ZnAl-LDH on the electrode	Dopamine	(Wang et al. 2011c)
Graphene/NiAl-LDH	Carbon based materials/LDH	*In situ* crystallization of LDH using co-precipitation method in the presence of graphene	Chlorpyrifos	(Qiao et al. 2015)
Graphene/NiAl-LDH	Carbon based materials/LDH	*In situ* crystallization of LDH using co-precipitation method in the presence of graphene	Dopamine	(Li et al. 2011)
Carbon dot/NiAl-LDH	Carbon based materials/LDH	Self-assembly	Acetylcholine	(Wang et al. 2016a)
Polyaniline/MgAl-LDH and Polyaniline/NiAl-LDH	Polymer/LDH	*In situ* crystallization of LDH using hydrothermal method in the presence of polyaniline	Humidity	(Li et al. 2016b)

and the detection limit was calculated to be 0.04 μmol/L (S/N = 3). Some natural anionic biopolymers including xanthan gum, alginic acid, l-carrageenan, k-carrageenan and pectin were intercalated into gallery of ZnAl-LDH using *in situ* crystallization of LDH in the presence of mentioned biopolymers by Darder et al. (Darder et al. 2005). The prepared nanocomposites were used as active phase of sensor for calcium ions recognition through direct potentiometry. The results demonstrated that the sensors based on l-carrageenan/LDH and alginate/LDH nanocomposites exhibited the best responses. Li et al. (Li et al. 2015a) reported the luminescence sensor application of graphene/ZnAl-LDH nanocomposite, which was prepared using *in situ* crystallization of LDH through co-precipitation method. In this sensor, $Ru(Phe)_3Cl$ is realized from the graphene/ZnAl-LDH nanocomposite in the presence of certain amount of DNA. The realized molecules of the $Ru(Phe)_3Cl$ enter into reaction with the DNA, resulting in luminescence recovery of the sensor (see Figure 13).

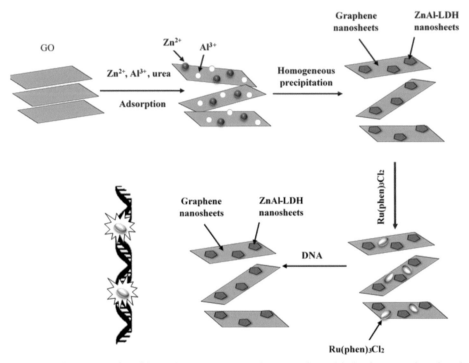

Figure 13. Schematic representation of the RGO-LDH nanocomposite preparation and the luminescence detection of DNA using the obtained nanocomposite. (Reprinted with permission from Li, H., J. Wen, R. Yu, J. Meng, Cong Wang, Chaoxia Wang and S. Sun. RSC Adv. 5: 9341 (2015). Copyright@ RSC. License Number: 3978770393033.)

Membrane modification

Although the application of LDH-based nanocomposites as nanofiller for performance improvement of membranes is still in preliminary stage, they absolutely are a promising candidate for this purpose owing to their high hydrophilicity, low toxicity and cost effectiveness. Lu et al. (Lu et al. 2016) prepared graphene oxide/MgAl-LDH nanocomposite using *in situ* crystallization of LDH method in the presence of graphene oxide. The prepared nanocomposite was used as nanofiller for improving polysolfonic thin film nanocomposite forward osmosis membrane. The results indicated that the hydrophilicity, porosity, surface pore diameter and chemical strength were enhanced by incorporating the fabricated graphene oxide/MgAl-LDH nanocomposite into the matrix of the membrane. In addition, the water flux of thin film forward osmosis membrane increased in the presence of prepared nanocomposite which can be attributed to the changes in the structure and hydrophilicity of fabricated membrane.

Acknowledgment

The authors thank the University of Tabriz (Iran) for all the support provided. We also acknowledge the support of Iran Science Elites Federation.

References

Abas, S.N.A., M.H.S. Ismail, M.L. Kamal and S. Izhar. 2013. Adsorption process of heavy metals by low cost adsorbent: A review. World Appl. Sci. J. 28: 1518–1530.

Adachi-Pagano, M., C. Forano and J.-P. Besse. 2000. Delamination of layered double hydroxides by use of surfactants. Chem. Commun. 1: 91–92.

Adwani, J.H., H.K. Noor-ul and R.S. Shukla. 2015. An elegant synthesis of chitosan grafted hydrotalcite nano-bio composite material and its effective catalysis for solvent-free synthesis of jasminaldehyde. RSC Adv. 5: 94562–94570.

Asif, M., W. Haitao, D. Shuang, A. Aziz, G. Zhang, F. Xiao et al. 2017. Metal oxide intercalated layered double hydroxide nanosphere: With enhanced electrocatalyic activity towards H_2O_2 for biological applications. Sens. Actuators B Chem. 239: 243–252.

Bai, C., S. Sun, Y. Xu, R. Yu and H. Li. 2016. Facile one-step synthesis of nanocomposite based on carbon nanotubes and Nickel-Aluminum layered double hydroxides with high cycling stability for supercapacitors. J. Colloid Interface Sci. 480: 57–62.

Bi, B., L. Xu, B. Xu and X. Liu. 2011. Heteropoly blue-intercalated layered double hydroxides for cationic dye removal from aqueous media. Appl. Clay Sci. 54: 242–247.

Bo, L., Q. Li, Y. Wang, L. Gao, X. Hu and J. Yang. 2015. One-pot hydrothermal synthesis of thrust spherical Mg–Al layered double hydroxides/MnO_2 and adsorption for Pb (II) from aqueous solutions. J. Environ. Chem. Eng. 3: 1468–1475.

Cai, X., X. Shen, L. Ma, Z. Ji, C. Xu and A. Yuan. 2015. Solvothermal synthesis of NiCo-layered double hydroxide nanosheets decorated on RGO sheets for high performance supercapacitor. Chem. Eng. J. 268: 251–259.

Cao, Y., G. Li and X. Li. 2016. Graphene/layered double hydroxide nanocomposite: Properties, synthesis, and applications. Chem. Eng. J. 292: 207–223.

Carja, G., A. Nakajima, S. Dranca, C. Dranca and K. Okada. 2010. TiO_2/ZnLDH as a self-assembled nanocomposite with photoresponsive properties. J. Phys. Chem. C 114: 14722–14728.

Carriazo, D., S. Lima, C. Martín, M. Pillinger, A. Valente and V. Rives. 2007. Metatungstate and tungstoniobate-containing LDHs: Preparation, characterisation and activity in epoxidation of cyclooctene. J. Phys. Chem. Solids 68: 1872–1880.

Cavani, F., F. Trifirò and A. Vaccari. 1991. Hydrotalcite-type anionic clays: Preparation, properties and applications. Catal. Today 11: 173–301.

Cervilla, A., A. Corma, V. Fornes, E. Llopis, P. Palanca, F. Rey et al. 1994. Intercalation of $[Mo^{VI}O_2 (O_2CC (S) Ph_2)_2]_2$-in a Zn (II)-Al (III) layered double hydroxide host: a strategy for the heterogeneous catalysis of the air oxidation of thiols. J. Am. Chem. Soc. 116: 1595–1596.

Chen, C., P. Gunawan and R. Xu. 2011. Self-assembled Fe_3O_4-layered double hydroxide colloidal nanohybrids with excellent performance for treatment of organic dyes in water. J. Mater. Chem. 21: 1218–1225.

Chen, D., Y. Li, J. Zhang, J.-Z. Zhou, Y. Guo and H. Liu. 2012. Magnetic Fe_3O_4/ZnCr-layered double hydroxide composite with enhanced adsorption and photocatalytic activity. Chem. Eng. J. 185: 120–126.

Chen, G., S. Qian, X. Tu, X. Wei, J. Zou, L. Leng et al. 2014. Enhancement photocatalytic degradation of rhodamine B on nanoPt intercalated Zn–Ti layered double hydroxides. Appl. Surf. Sci. 293: 345–351.

Chen, S., M. Mao, X. Liu, S. Hong, Z. Lu, S. Sang et al. 2016. A high-rate cathode material hybridized by in-site grown Ni–Fe layered double hydroxides and carbon black nanoparticles. J. Mater. Chem. A 4: 4877–4881.

Chen, T., F. Zhang and Y. Zhu. 2013. Pd nanoparticles on layered double hydroxide as efficient catalysts for solvent-free oxidation of benzyl alcohol using molecular oxygen: Effect of support basic properties. Catal. lett. 143: 206–218.

Cheng, F. and J. Chen. 2012. Lithium-air batteries: Something from nothing. Nat. Chem. 4: 962–963.

Cho, H.R., J.H. Lee, S.H. Cho and H.G. Ji. 2016. Gold nanoparticles on layered double hydroxide nanosheets and its electrocatalysis for glucose oxidation. Bull. Korean Chem. Soc. 37: 401–403.

Choudary, B., B. Bharathi, C.V. Reddy, M.L. Kantam and K. Raghavan. 2001. The first example of catalytic N-oxidation of tertiary amines by tungstate-exchanged Mg–Al layered double hydroxide in water: a green protocol II CT communication No: 4804. Chem. Commun. 18: 1736–1737.

Choudary, B.M., S. Madhi, N.S. Chowdari, M.L. Kantam and B. Sreedhar. 2002. Layered double hydroxide supported nanopalladium catalyst for Heck-, Suzuki-, Sonogashira-, and Stille-type coupling reactions of chloroarenes. J. Am. Chem. Soc. 124: 14127–14136.

Costa, F.R., M. Saphiannikova, U. Wagenknecht and G. Heinrich. 2007. Layered double hydroxide based polymer nanocomposites. Wax Crystal Control Nanocomposites Stimuli-Responsive Polymers, Springer pp. 101–168.

Darder, M., M. López-Blanco, P. Aranda, F. Leroux and E. Ruiz-Hitzky. 2005. Bio-nanocomposites based on layered double hydroxides. Chem. Mater. 17: 1969–1977.

Das, J. and K. Parida. 2007. Heteropoly acid intercalated Zn/Al HTlc as efficient catalyst for esterification of acetic acid using n-butanol. J. Mol. Catal. A: Chem. 264: 248–254.

Dong, X., L. Wang, D. Wang, C. Li and J. Jin. 2011. Layer-by-layer engineered Co–Al hydroxide nanosheets/graphene multilayer films as flexible electrode for supercapacitor. Langmuir 28: 293–298.

Du, N., W.-G. Hou and S.-E. Song. 2007. A novel composite: layered double hydroxides encapsulated in vesicles. J. Phys. Chem. B 111: 13909–13913.

Du, Y., Q. Jin, J. Feng, N. Zhang, Y. He and D. Li. 2015. Flower-like Au/Ni–Al hydrotalcite with hierarchical pore structure as a multifunctional catalyst for catalytic oxidation of alcohol. Catal. Sci. Tech. 5: 3216–3225.

Fan, G., F. Li, D.G. Evans and X. Duan. 2014. Catalytic applications of layered double hydroxides: recent advances and perspectives. Chem. Soc. Rev. 43: 7040–7066.

Fan, H., J. Zhu, J. Sun, S. Zhang and S. Ai. 2013. Ag/AgBr/Co–Ni–NO$_3$ layered double hydroxide nanocomposites with highly adsorptive and photocatalytic properties. Chem. Eur. J. 19: 2523–2530.

Fang, J., M. Li, Q. Li, W. Zhang, Q. Shou, F. Liu et al. 2012. Microwave-assisted synthesis of CoAl-layered double hydroxide/graphene oxide composite and its application in supercapacitors. Electrochim. Acta 85: 248–255.

Fang, Y., J. Liu, D.J. Yu, J.P. Wicksted, K. Kalkan, C.O. Topal et al. 2010. Self-supported supercapacitor membranes: Polypyrrole-coated carbon nanotube networks enabled by pulsed electrodeposition. J. Power Sources 195: 674–679.

Feng, J., Y. He, Y. Liu, Y. Du and D. Li. 2015. Supported catalysts based on layered double hydroxides for catalytic oxidation and hydrogenation: general functionality and promising application prospects. Chem. Soc. Rev. 44: 5291–5319.

Feng, J., X. Ma, Y. He, D.G. Evans and D. Li. 2012. Synthesis of hydrotalcite-supported shape-controlled Pd nanoparticles by a precipitation–reduction method. Appl. Catal. A 413: 10–20.

Gao, Z., J. Wang, Z. Li, W. Yang, B. Wang, M. Hou et al. 2011. Graphene nanosheet/Ni^{2+}/Al^{3+} layered double-hydroxide composite as a novel electrode for a supercapacitor. Chem. Mater. 23: 3509–3516.

Gholivand, M.B. and M. Khodadadian. 2014. Amperometric cholesterol biosensor based on the direct electrochemistry of cholesterol oxidase and catalase on a graphene/ionic liquid-modified glassy carbon electrode. Biosens. Bioelectron. 53: 472–478.

Gong, M., Y. Li, H. Wang, Y. Liang, J.Z. Wu, J. Zhou et al. 2013. An advanced Ni–Fe layered double hydroxide electrocatalyst for water oxidation. J. Am. Chem. Soc. 135: 8452–8455.

Gu, Z., J.J. Atherton and Z.P. Xu. 2015. Hierarchical layered double hydroxide nanocomposites: structure, synthesis and applications. Chem. Commun. 51: 3024–3036.

Gunjakar, J.L., T.W. Kim, H.N. Kim, I.Y. Kim and S.-J. Hwang. 2011. Mesoporous layer-by-layer ordered nanohybrids of layered double hydroxide and layered metal oxide: highly active visible light photocatalysts with improved chemical stability. J. Am. Chem. Soc. 133: 14998–15007.

Gunjakar, J.L., I.Y. Kim, J.M. Lee, N.-S. Lee and S.-J. Hwang. 2013. Self-assembly of layered double hydroxide 2D nanoplates with graphene nanosheets: an effective way to improve the photocatalytic activity of 2D nanostructured materials for visible light-induced O$_2$ generation. Energy Environ. Sci. 6: 1008–1017.

Guo, X., F. Zhang, Q. Peng, S. Xu, X. Lei, D.G. Evans et al. 2011. Layered double hydroxide/eggshell membrane: An inorganic biocomposite membrane as an efficient adsorbent for Cr (VI) removal. Chem. Eng. J. 166: 81–87.

Guo, Y., D. Li, C. Hu, Y. Wang and E. Wang. 2001. Layered double hydroxides pillared by tungsten polyoxometalates: synthesis and photocatalytic activity. Int. J. Inorg. Mater. 3: 347–355.

Guo, Y., X. Liu, X. Wang, A. Iqbal, C. Yang, W. Liu et al. 2015. Carbon dot/NiAl-layered double hydroxide hybrid material: facile synthesis, intrinsic peroxidase-like catalytic activity and its application. RSC Adv. 5: 95495–95503.

Gupta, V. 2009. Application of low-cost adsorbents for dye removal—a review. J. Environ. Manage. 90: 2313–2342.

Hadnadjev-Kostic, M., T. Vulic and R. Marinkovic-Neducin. 2014. Solar light induced rhodamine B degradation assisted by TiO$_2$–Zn–Al LDH based photocatalysts. Adv. Powder Technol. 25: 1624–1633.

Han, J., Y. Dou, J. Zhao, M. Wei, D.G. Evans and X. Duan. 2013. Flexible CoAl LDH@ PEDOT core/shell nanoplatelet array for high-performance energy storage. Small 9: 98–106.

Han, R., C. Nan, L. Yang, G. Fan and F. Li. 2015. Direct synthesis of hybrid layered double hydroxide-carbon composites supported Pd nanocatalysts efficient in selective hydrogenation of citral. RSC Adv. 5: 33199–33207.

Hasannia, S. and B. Yadollahi. 2015. Zn–Al LDH nanostructures pillared by Fe substituted Keggin type polyoxometalate: Synthesis, characterization and catalytic effect in green oxidation of alcohols. Polyhedron 99: 260–265.

He, Y.-F., J.-T. Feng, Y.-Y. Du and D.-Q. Li. 2012. Controllable synthesis and acetylene hydrogenation performance of supported Pd nanowire and cuboctahedron catalysts. ACS Catal. 2: 1703–1710.

Hibino, T. 2004. Delamination of layered double hydroxides containing amino acids. Chem. Mater. 16: 5482–5488.

Hong, J., Y. Wang, J. Pan, Z. Zhong and R. Xu. 2011. Self-assembled dye–layered double hydroxide–Pt nanoparticles: a novel H$_2$ evolution system with remarkably enhanced stability. Nanoscale 3: 4655–4661.

Hou, Y., M.R. Lohe, J. Zhang, S. Liu, X. Zhuang and X. Feng. 2016. Vertically oriented cobalt selenide/NiFe layered-double-hydroxide nanosheets supported on exfoliated graphene foil: an efficient 3D electrode for overall water splitting. Energy Environ. Sci. 9: 478–483.

Hu, J., G. Lei, Z. Lu, K. Liu, S. Sang and H. Liu. 2015. Alternating assembly of Ni–Al layered double hydroxide and graphene for high-rate alkaline battery cathode. Chem. Commun. 51: 9983–9986.

Huang, Z., P. Wu, Y. Lu, X. Wang, N. Zhu and Z. Dang. 2013. Enhancement of photocatalytic degradation of dimethyl phthalate with nano-TiO$_2$ immobilized onto hydrophobic layered double hydroxides: a mechanism study. J. Hazard. Mater. 246: 70–78.

Huang, Z., P. Wu, B. Gong, Y. Fang and N. Zhu. 2014. Fabrication and photocatalytic properties of a visible-light responsive nanohybrid based on self-assembly of carboxyl graphene and ZnAl layered double hydroxides. J. Mater. Chem. A 2: 5534–5540.

Huang, Z., S. Wang, J. Wang, Y. Yu, J. Wen and R. Li. 2015. Exfoliation-restacking synthesis of coal-layered double hydroxide nanosheets/reduced graphene oxide composite for high performance supercapacitors. Electrochim. Acta 152: 117–125.

Islam, D., D. Borah and H. Acharya. 2015. Controlled synthesis of monodisperse silver nanoparticles supported layered double hydroxide catalyst. RSC Adv. 5: 13239–13245.

Jia, Z., Y. Wang and T. Qi. 2015. Pd nanoparticles supported on Mg–Al–CO$_3$ layered double hydroxide as an effective catalyst for methanol electro-oxidation. RSC Adv. 5: 62142–62148.

Jin, L., D. He, Z. Li and M. Wei. 2012. Protein adsorption on gold nanoparticles supported by a layered double hydroxide. Mater. Lett. 77: 67–70.

Jin, X., K. Taniguchi, K. Yamaguchi and N. Mizuno. 2016. Au–Pd alloy nanoparticles supported on layered double hydroxide for heterogeneously catalyzed aerobic oxidative dehydrogenation of cyclohexanols and cyclohexanones to phenols. Chem. Sci. 7: 5371–5383.

Koilraj, P. and K. Sasaki. 2016. Fe$_3$O$_4$/MgAl-NO$_3$ layered double hydroxide as a magnetically separable sorbent for the remediation of aqueous phosphate. J. Environ. Chem. Eng. 4: 984–991.

Kura, A.U., M.Z. Hussein, S. Fakurazi and P. Arulselvan. 2014. Layered double hydroxide nanocomposite for drug delivery systems: bio-distribution, toxicity and drug activity enhancement. Chem. Cent. J. 8: 47–55.

Lan, M., G. Fan, L. Yang and F. Li. 2014. Significantly enhanced visible-light-induced photocatalytic performance of hybrid Zn–Cr layered double hydroxide/graphene nanocomposite and the mechanism study. Ind. Eng. Chem. Res. 53: 12943–12952.

Lee, S.-H., C.-W. Yi and K. Kim. 2011. Characteristics and electrochemical performance of the TiO$_2$-coated ZnO anode for Ni–Zn secondary batteries. J. Phys. Chem. C 115: 2572–2577.

Li, B., Y. Zhao, S. Zhang, W. Gao and M. Wei. 2013. Visible-light-responsive photocatalysts toward water oxidation based on NiTi-layered double hydroxide/reduced graphene oxide composite materials. ACS Appl. Mater. Interfaces 5: 10233–10239.

Li, H., J. Wen, R. Yu, J. Meng, C. Wang, C. Wang et al. 2015a. Facile synthesis of a nanocomposite based on graphene and ZnAl layered double hydroxides as a portable shelf of a luminescent sensor for DNA detection. RSC Adv. 5: 9341–9347.

Li, L., Q. Chen, Q. Zhang, J. Shi, Y. Li, W. Zhao et al. 2012. Layered double hydroxide confined Au colloids: High performance catalysts for low temperature CO oxidation. Catalysis Communications 26: 15–18.

Li, L., L. Dou and H. Zhang. 2014a. Layered double hydroxide supported gold nanoclusters by glutathione-capped Au nanoclusters precursor method for highly efficient aerobic oxidation of alcohols. Nanoscale 6: 3753–3763.

Li, M., J.E. Zhu, L. Zhang, X. Chen, H. Zhang, F. Zhang et al. 2011. Facile synthesis of NiAl-layered double hydroxide/graphene hybrid with enhanced electrochemical properties for detection of dopamine. Nanoscale 3: 4240–4246.

Li, M., J. Cheng, J. Fang, Y. Yang, F. Liu and X. Zhang. 2014b. NiAl-layered double hydroxide/reduced graphene oxide composite: microwave-assisted synthesis and supercapacitive properties. Electrochim. Acta 134: 309–318.

Li, M., J. Cheng, F. Liu and X. Zhang. 2015b. 3D-architectured nickel–cobalt–manganese layered double hydroxide/reduced graphene oxide composite for high-performance supercapacitor. Chem. Phys. Lett. 640: 5–10.

Li, M., F. Liu, J. Cheng, J. Ying and X. Zhang. 2015c. Enhanced performance of nickel–aluminum layered double hydroxide nanosheets/carbon nanotubes composite for supercapacitor and asymmetric capacitor. J. Alloys Compd. 635: 225–232.

Li, M., J. Cheng, J. Wang, F. Liu and X. Zhang. 2016a. The growth of nickel-manganese and cobalt-manganese layered double hydroxides on reduced graphene oxide for supercapacitor. Electrochim. Acta 206: 108–115.

Li, X.-Z., S.-R. Liu and Y. Guo. 2016b. Polyaniline-intercalated layered double hydroxides: synthesis and properties for humidity sensing. RSC Adv. 6: 63099–63106.

Li, X., J. Shen, W. Sun, X. Hong, R. Wang, X. Zhao et al. 2015d. A super-high energy density asymmetric supercapacitor based on 3D core-shell structured NiCo-layered double hydroxide@carbon nanotube and activated polyaniline-derived carbon electrodes with commercial level mass loading. J. Mater. Chem. A 3: 13244–13253.

Lin, Y., L. Ruiyi, L. Zaijun, L. Junkang, F. Yinjun, W. Guangli et al. 2013. Three-dimensional activated reduced graphene oxide nanocup/nickel aluminum layered double hydroxides composite with super high electrochemical and capacitance performances. Electrochim. Acta 95: 146–154.

Liu, L.-M., L.-P. Jiang, F. Liu, G.-Y. Lu, E. Abdel-Halim and J.-J. Zhu. 2013. Hemoglobin/DNA/layered double hydroxide composites for biosensing applications. Anal. Methods 5: 3565–3571.

Liu, M., S. He, Y.-E. Miao, Y. Huang, H. Lu, L. Zhang et al. 2015. Eco-friendly synthesis of hierarchical ginkgo-derived carbon nanoparticles/NiAl-layered double hydroxide hybrid electrodes toward high-performance supercapacitors. RSC Adv. 5: 55109–55118.

Liu, P., H. Wang, Z. Feng, P. Ying and C. Li. 2008. Direct immobilization of self-assembled polyoxometalate catalyst in layered double hydroxide for heterogeneous epoxidation of olefins. J. Catal. 256: 345–348.

Liu, P., C. Wang and C. Li. 2009. Epoxidation of allylic alcohols on self-assembled polyoxometalates hosted in layered double hydroxides with aqueous H$_2$O$_2$ as oxidant. J. Catal. 262: 159–168.

Liu, P., C. Li and E.J. Hensen. 2012. Efficient tandem synthesis of methyl esters and imines by using versatile hydrotalcite-supported gold nanoparticles. Chem. Eur. J. 18: 12122–12129.

Liu, P., V. Degirmenci and E.J. Hensen. 2014. Unraveling the synergy between gold nanoparticles and chromium-hydrotalcites in aerobic oxidation of alcohols. J. Catal. 313: 80–91.

Long, X., Z. Wang, S. Xiao, Y. An and S. Yang. 2016. Transition metal based layered double hydroxides tailored for energy conversion and storage. Mater. Today 19: 213–226.

Lu, P., S. Liang, T. Zhou, X. Mei, Y. Zhang, C. Zhang et al. 2016. Layered double hydroxide/graphene oxide hybrid incorporated polysulfone substrate for thin–film nanocomposite forward osmosis membranes. RSC Adv. 6: 56599–56609.

Lu, R., X. Xu, J. Chang, Y. Zhu, S. Xu and F. Zhang. 2012. Improvement of photocatalytic activity of TiO_2 nanoparticles on selectively reconstructed layered double hydroxide. Appl. Catal. B 111: 389–396.

Lv, W., L. Yang, B. Fan, Y. Zhao, Y. Chen, N. Lu et al. 2015. Silylated MgAl LDHs intercalated with MnO_2 nanowires: highly efficient catalysts for the solvent-free aerobic oxidation of ethylbenzene. Chem. Eng. J. 263: 309–316.

Ma, R., X. Liu, J. Liang, Y. Bando and T. Sasaki. 2014. Molecular-scale heteroassembly of redoxable hydroxide nanosheets and conductive graphene into superlattice composites for high-performance supercapacitors. Adv. Mater. 26: 4173–4178.

Ma, W., R. Ma, J. Wu, P. Sun, X. Liu, K. Zhou et al. 2016. Development of efficient electrocatalysts via molecular hybridization of NiMn layered double hydroxide nanosheets and graphene. Nanoscale 8: 10425–10432.

Ma, X.-Y., Y.-Y. Chai, D.G. Evans, D.-Q. Li and J.-T. Feng. 2011. Preparation and selective acetylene hydrogenation catalytic properties of supported Pd catalyst by the *in situ* precipitation−Reduction method. J. Phys. Chem. C 115: 8693–8701.

Maciuca, A.-L., C.-E. Ciocan, E. Dumitriu, F. Fajula and V. Hulea. 2008a. V-, Mo- and W-containing layered double hydroxides as effective catalysts for mild oxidation of thioethers and thiophenes with H_2O_2. Catal. Today 138: 33–37.

Maciuca, A.-L., E. Dumitriu, F. Fajula and V. Hulea. 2008b. Mild oxidation of tetrahydrothiophene to sulfolane over V-, Mo- and W-containing layered double hydroxides. Appl. Catal. A 338: 1–8.

Mi, F., X. Chen, Y. Ma, S. Yin, F. Yuan and H. Zhang. 2011. Facile synthesis of hierarchical core–shell Fe_3O_4@ MgAl–LDH@ Au as magnetically recyclable catalysts for catalytic oxidation of alcohols. Chem. Commun. 47: 12804–12806.

Miao, M., J. Feng, Q. Jin, Y. He, Y. Liu, Y. Du et al. 2015. Hybrid Ni–Al layered double hydroxide/graphene composite supported gold nanoparticles for aerobic selective oxidation of benzyl alcohol. RSC Adv. 5: 36066–36074.

Mitsudome, T., A. Noujima, T. Mizugaki, K. Jitsukawa and K. Kaneda. 2009a. Efficient aerobic oxidation of alcohols using a hydrotalcite-supported gold nanoparticle catalyst. Adv. Synth. Catal. 351: 1890–1896.

Mitsudome, T., A. Noujima, T. Mizugaki, K. Jitsukawa and K. Kaneda. 2009b. Supported gold nanoparticle catalyst for the selective oxidation of silanes to silanols in water. Chem. Commun. 0: 5302–5304.

Mitsudome, T., A. Noujima, T. Mizugaki, K. Jitsukawa and K. Kaneda. 2009c. Supported gold nanoparticles as a reusable catalyst for synthesis of lactones from diols using molecular oxygen as an oxidant under mild conditions. Green Chem. 11: 793–797.

Mohapatra, L., K. Parida and M. Satpathy. 2012. Molybdate/Tungstate intercalated oxo-bridged Zn/Y LDH for solar light induced photodegradation of organic pollutants. J. Phys. Chem. C 116: 13063–13070.

Mohapatra, L. and K. Parida. 2014. Dramatic activities of vanadate intercalated bismuth doped LDH for solar light photocatalysis. Phys. Chem. Chem. Phys. 16: 16985–16996.

Motokura, K., N. Fujita, K. Mori, T. Mizugaki, K. Ebitani and K. Kaneda. 2005. One-pot synthesis of α-alkylated nitriles with carbonyl compounds through consecutive aldol reaction/hydrogenation using a hydrotalcite-supported palladium nanoparticle as a multifunctional heterogeneous catalyst. Tetrahedron Lett. 46: 5507–5510.

Nam, H.J., E.B. Park and D.-Y. Jung. 2016. Bioinspired polydopamine-layered double hydroxide nanocomposites: controlled synthesis and multifunctional performance. RSC Adv. 6: 24952–24958.

Ning, F., M. Shao, C. Zhang, S. Xu, M. Wei and X. Duan. 2014. Co_3O_4@ layered double hydroxide core/shell hierarchical nanowire arrays for enhanced supercapacitance performance. Nano Energy 7: 134–142.

Nocchetti, M., A. Donnadio, V. Ambrogi, P. Andreani, M. Bastianini, D. Pietrella et al. 2013. Ag/AgCl nanoparticle decorated layered double hydroxides: synthesis, characterization and antimicrobial properties. J. Mater. Chem. B 1: 2383–2393.

Omwoma, S., W. Chen, R. Tsunashima and Y.-F. Song. 2014. Recent advances on polyoxometalates intercalated layered double hydroxides: from synthetic approaches to functional material applications. Coord. Chem. Rev. 258: 58–71.

Pan, D., H. Zhang, T. Fan, J. Chen and X. Duan. 2011. Nearly monodispersed core–shell structural Fe_3O_4@ DFUR–LDH submicro particles for magnetically controlled drug delivery and release. Chem. Commun. 47: 908–910.

Qiao, L., Y. Guo, X. Sun, Y. Jiao and X. Wang. 2015. Electrochemical immunosensor with NiAl-layered double hydroxide/graphene nanocomposites and hollow gold nanospheres double-assisted signal amplification. Bioprocess. Biosyst. Eng. 38: 1455–1468.

Sailaxmi, G. and K. Lalitha. 2015. Impact of a stress management program on stress perception of nurses working with psychiatric patients. Asian J. Psychiatr. 14: 42–45.

Seftel, E., M. Mertens and P. Cool. 2013. The influence of the Ti^{4+} location on the formation of self-assembled nanocomposite systems based on TiO_2 and Mg/Al-LDHs with photocatalytic properties. Appl. Catal., B 134: 274–285.

Shan, R.-R., L.-G. Yan, K. Yang, S.-J. Yu, Y.-F. Hao, H.-Q. Yu et al. 2014. Magnetic Fe_3O_4/MgAl-LDH composite for effective removal of three red dyes from aqueous solution. Chem. Eng. J. 252: 38–46.

Shao, L., Y. Yao, S. Quan, H. Wei, R. Wang and Z. Guo. 2014a. One-pot *in situ* synthesized TiO_2/layered double hydroxides (LDHs) composites toward environmental remediation. Mater. Lett. 114: 111–114.

Shao, M., F. Ning, M. Wei, D.G. Evans and X. Duan. 2014b. Hierarchical nanowire arrays based on ZnO core−layered double hydroxide shell for largely enhanced photoelectrochemical water splitting. Adv. Funct. Mater. 24: 580–586.

Shao, M., Z. Li, R. Zhang, F. Ning, M. Wei, D.G. Evans and X. Duan. 2015a. Hierarchical conducting polymer@clay core–shell arrays for flexible all-solid-state supercapacitor devices. Small 11: 3530–3538.

Shao, M., R. Zhang, Z. Li, M. Wei, D.G. Evans and X. Duan. 2015b. Layered double hydroxides toward electrochemical energy storage and conversion: design, synthesis and applications. Chem. Commun. 51: 15880–15893.

Shiyong, L., Z. Qizhong, J. Zhengneng, H. Jiang and X. Jiang. 2010. Dodecylsulfate anion embedded layered double hydroxide supported nanopalladium catalyst for the Suzuki reaction. Chin. J. Catal. 31: 557–561.

Shou, J., C. Jiang, F. Wang, M. Qiu and Q. Xu. 2015. Fabrication of Fe_3O_4/MgAl-layered double hydroxide magnetic composites for the effective decontamination of Co (II) from synthetic wastewater. J. Mol. Liq. 207: 216–223.

Singha, S., M. Sahoo and K. Parida. 2011. Highly active Pd nanoparticles dispersed on amine functionalized layered double hydroxide for Suzuki coupling reaction. Dalton Trans. 40: 7130–7132.

Sun, Z., L. Jin, S. Zhang, W. Shi, M. Pu, M. Wei et al. 2011. An optical sensor based on H-acid/layered double hydroxide composite film for the selective detection of mercury ion. Anal. Chim. Acta 702: 95–101.

Taei, M., E. Havakeshian, F. Abedi and M. Movahedi. 2016a. The effect of Cu-Mg-Fe layered double hydroxide on the electrocatalytic activity of gold nanoparticles towards ethanol electrooxidation. Int. J. hydrogen Energy 41: 13575–13582.

Taei, M., E. Havakeshian, H. Salavati and F. Abedi. 2016b. Electrocatalytic oxidation of ethanol on a glassy carbon electrode modified with a gold nanoparticle-coated hydrolyzed CaFe–Cl layered double hydroxide in alkaline medium. RSC Adv. 6: 27293–27300.

Takagaki, A., A. Tsuji, S. Nishimura and K. Ebitani. 2011. Genesis of catalytically active gold nanoparticles supported on hydrotalcite for base-free selective oxidation of glycerol in water with molecular oxygen. Chem. Lett. 40: 150–152.

Tang, D., Y. Han, W. Ji, S. Qiao, X. Zhou, R. Liu et al. 2014. A high-performance reduced graphene oxide/ZnCo layered double hydroxide electrocatalyst for efficient water oxidation. Dalton Trans. 43: 15119–15125.

Tang, D., J. Liu, X. Wu, R. Liu, X. Han, Y. Han et al. 2014. Carbon quantum dot/NiFe layered double-hydroxide composite as a highly efficient electrocatalyst for water oxidation. ACS Appl. mater. Interfaces 6: 7918–7925.

Tao, R., Y. Xie, G. An, K. Ding, H. Zhang, Z. Sun et al. 2010. Arginine-mediated synthesis of highly efficient catalysts for transfer hydrogenations of ketones. J. Colloid Interface Sci. 351: 501–506.

Tian, D.-Y., W.-Y. Wang, S.-P. Li, X.-D. Li and Z.-L. Sha. 2016. A novel platform designed by Au core/inorganic shell structure conjugated onto MTX/LDH for chemo-photothermal therapy. Int. J. Pharm. 505: 96–106.

Tsai, Y.-T., X. Mo, A. Campos, J.G. Goodwin and J.J. Spivey. 2011. Hydrotalcite supported Co catalysts for CO hydrogenation. Appl. Catal., A 396: 91–100.

Tsuji, A., K.T.V. Rao, S. Nishimura, A. Takagaki and K. Ebitani. 2011. Selective oxidation of glycerol by using a hydrotalcite-supported platinum catalyst under atmospheric oxygen pressure in water. ChemSusChem. 4: 542–548.

Van Laar, F., D.E. De Vos, F. Pierard, A. Kirsch De Mesmaeker, L. Fiermans and P. Jacobs. 2001. Generation of singlet molecular oxygen from H_2O_2 with molybdate-exchanged layered double hydroxides: effects of catalyst composition and reaction conditions. J. Catal. 197: 139–150.

Wang, H., X. Xiang and F. Li. 2010a. Facile synthesis and novel electrocatalytic performance of nanostructured Ni–Al layered double hydroxide/carbon nanotube composites. J. Mater. Chem. 20: 3944–3952.

Wang, H., X. Xiang and F. Li. 2010b. Hybrid ZnAl-LDH/CNTs nanocomposites: Noncovalent assembly and enhanced photodegradation performance. AIChE J. 56: 768–778.

Wang, L., D. Wang, X.Y. Dong, Z.J. Zhang, X.F. Pei, X.J. Chen et al. 2011a. Layered assembly of graphene oxide and Co–Al layered double hydroxide nanosheets as electrode materials for supercapacitors. Chem. Commun. 47: 3556–3558.

Wang, L., J. Zhang, X. Meng, D. Zheng and F.-S. Xiao. 2011b. Superior catalytic properties in aerobic oxidation of alcohols over Au nanoparticles supported on layered double hydroxide. Catal. Today 175: 404–410.

Wang, L., H. Xing, S. Zhang, Q. Ren, L. Pan, K. Zhang et al. 2013. A Gd-doped Mg-Al-LDH/Au nanocomposite for CT/MR bimodal imagings and simultaneous drug delivery. Biomaterials 34: 3390–3401.

Wang, L., X. Chen, C. Liu and W. Yang. 2016a. Non-enzymatic acetylcholine electrochemical biosensor based on flower-like NiAl layered double hydroxides decorated with carbon dots. Sens. Actuators B Chem. 233: 199–205.

Wang, S., B. Gao, Y. Li, A.R. Zimmerman and X. Cao. 2016b. Sorption of arsenic onto Ni/Fe layered double hydroxide (LDH)-biochar composites. RSC Adv. 6: 17792–17799.

Wang, S., S. Yin, G. Chen, L. Li and H. Zhang. 2016c. Nearly atomic precise gold nanoclusters on nickel-based layered double hydroxides for extraordinarily efficient aerobic oxidation of alcohols. Catal. Sci. Tech. 6: 4090–4104.

Wang, Y., D. Zhang, M. Tang, S. Xu and M. Li. 2010c. Electrocatalysis of gold nanoparticles/layered double hydroxides nanocomposites toward methanol electro-oxidation in alkaline medium. Electrochim. Acta 55: 4045–4049.

Wang, Y., W. Peng, L. Liu, M. Tang, F. Gao and M. Li. 2011c. Enhanced conductivity of a glassy carbon electrode modified with a graphene-doped film of layered double hydroxides for selectively sensing of dopamine. Microchimica Acta 174: 41–46.

Wang, Y., H. Ji, W. Peng, L. Liu, F. Gao and M. Li. 2012. Gold nanoparticle-coated Ni/Al layered double hydroxides on glassy carbon electrode for enhanced methanol electro-oxidation. Int. J. hydrogen Energy 37: 9324–9329.

Wang, Y., Z. Wang, Y. Rui and M. Li. 2015. Horseradish peroxidase immobilization on carbon nanodots/CoFe layered double hydroxides: direct electrochemistry and hydrogen peroxide sensing. Biosens. Bioelectron. 64: 57–62.

Wang, Y., Z. Wang, X. Liu and M. Li. 2016d. Synergistic effect between strongly coupled CoAl layered double hydroxides and graphene for the electrocatalytic reduction of oxygen. Electrochim. Acta 192: 196–204.

Wang, Y., H. Dou, J. Wang, B. Ding, Y. Xu, Z. Chang et al. 2016e. Three-dimensional porous MXene/layered double hydroxide composite for high performance supercapacitors. J. Power Sources 327: 221–228.

Wei, X., Y. Fu, L. Xu, F. Li, B. Bi and X. Liu. 2008. Tungstocobaltate-pillared layered double hydroxides: Preparation, characterization, magnetic and catalytic properties. J. Solid State Chem. 181: 1292–1297.

Wu, L., J. Li and H.M. Zhang. 2015. One step fabrication of Au nanoparticles-Ni-Al layered double hydroxide composite film for the determination of L-cysteine. Electroanalysis 27: 1195–1201.

Wu, X.-L., L. Wang, C.-L. Chen, A.-W. Xu and X.-K. Wang. 2011. Water-dispersible magnetite-graphene-LDH composites for efficient arsenate removal. J. Mater. Chem. 21: 17353–17359.

Xiang, X., W. He, L. Xie and F. Li. 2013. A mild solution chemistry method to synthesize hydrotalcite-supported platinum nanocrystals for selective hydrogenation of cinnamaldehyde in neat water. Catal. Sci. Tech. 3: 2819–2827.

Xu, J., S. Gai, F. He, N. Niu, P. Gao, Y. Chen et al. 2014. A sandwich-type three-dimensional layered double hydroxide nanosheet array/graphene composite: fabrication and high supercapacitor performance. J. Mater. Chem. A 2: 1022–1031.

Xu, Z., N. Wang, W. Chu, J. Deng and S. Luo. 2015. *In situ* controllable assembly of layered-double-hydroxide-based nickel nanocatalysts for carbon dioxide reforming of methane. Catal. Sci. Tech. 5: 1588–1597.

Xu, Z.P., J. Zhang, M.O. Adebajo, H. Zhang and C. Zhou. 2011. Catalytic applications of layered double hydroxides and derivatives. Appl. Clay Sci. 53: 139–150.

Yan, Q., Z. Zhang, Y. Zhang, A. Umar, Z. Guo, D. O'Hare et al. 2015. Hierarchical Fe_3O_4 core–shell layered double hydroxide composites as magnetic adsorbents for anionic dye removal from wastewater. Eur. J. Inorg. Chem. 2015: 4182–4191.

Yang, B., Z. Yang, R. Wang and T. Wang. 2013. Layered double hydroxide/carbon nanotubes composite as a high performance anode material for Ni–Zn secondary batteries. Electrochim. Acta 111: 581–587.

Yang, B., Z. Yang, R. Wang and Z. Feng. 2014. Silver nanoparticle deposited layered double hydroxide nanosheets as a novel and high-performing anode material for enhanced Ni-Zn secondary batteries. J. Mater. Chem. A 2: 785–791.

Yang, G.-W., C.-L. Xu and H.-L. Li. 2008. Electrodeposited nickel hydroxide on nickel foam with ultrahigh capacitance. Chem. Commun. 48: 6537–6539.

Yu, L., N. Shi, Q. Liu, J. Wang, B. Yang, B. Wang et al. 2014. Facile synthesis of exfoliated Co–Al LDH–carbon nanotube composites with high performance as supercapacitor electrodes. Phys. Chem. Chem. Phys. 16: 17936–17942.

Yulian, N., L. Ruiyi, L. Zaijun, F. Yinjun and L. Junkang. 2013. High-performance supercapacitors materials prepared via *in situ* growth of NiAl-layered double hydroxide nanoflakes on well-activated graphene nanosheets. Electrochim. Acta 94: 360–366.

Zhang, G., B. Lin, W. Yang, S. Jiang, Q. Yao, Y. Chen et al. 2015. Highly efficient photocatalytic hydrogen generation by incorporating CdS into ZnCr-layered double hydroxide interlayer. RSC Adv. 5: 5823–5829.

Zhang, H., D. Pan, K. Zou, J. He and X. Duan. 2009. A novel core-shell structured magnetic organic-inorganic nanohybrid involving drug-intercalated layered double hydroxides coated on a magnesium ferrite core for magnetically controlled drug release. J. Mater. Chem. 19: 3069–3077.

Zhang, M., Q. Yao, C. Lu, Z. Li and W. Wang. 2014. Layered double hydroxide–carbon dot composite: High-performance adsorbent for removal of anionic organic dye. ACS Appl. Mater. Interfaces 6: 20225–20233.

Zhang, Q., J. Xu, D. Yan, S. Li, J. Lu, X. Cao et al. 2013a. The *in situ* shape-controlled synthesis and structure–activity relationship of Pd nanocrystal catalysts supported on layered double hydroxide. Catal. Sci. Tech. 3: 2016–2024.

Zhang, W., C. Ma, J. Fang, J. Cheng, X. Zhang, S. Dong et al. 2013b. Asymmetric electrochemical capacitors with high energy and power density based on graphene/CoAl-LDH and activated carbon electrodes. RSC Adv. 3: 2483–2490.

Zhao, C., L. Liu, G. Rao, H. Zhao, L. Wang, J. Xu et al. 2015a. Synthesis of novel MgAl layered double oxide grafted TiO_2 cuboids and their photocatalytic activity on CO_2 reduction with water vapor. Catal. Sci. Tech. 5: 3288–3295.

Zhao, J., J. Chen, S. Xu, M. Shao, D. Yan, M. Wei et al. 2013a. CoMn-layered double hydroxide nanowalls supported on carbon fibers for high-performance flexible energy storage devices. J. Mater. Chem. A 1: 8836–8843.

Zhao, J., M. Shao, D. Yan, S. Zhang, Z. Lu, Z. Li et al. 2013b. A hierarchical heterostructure based on Pd nanoparticles/layered double hydroxide nanowalls for enhanced ethanol electrooxidation. J. Mater. Chem. A 1: 5840–5846.

Zhao, J., J. Chen, S. Xu, M. Shao, Q. Zhang, F. Wei et al. 2014. Hierarchical NiMn layered double hydroxide/carbon nanotubes architecture with superb energy density for flexible supercapacitors. Adv. Funct. Mater. 24: 2938–2946.

Zhao, J., S. Xu, K. Tschulik, R.G. Compton, M. Wei, D. O'Hare et al. 2015b. Molecular-scale hybridization of clay monolayers and conducting polymer for thin-film supercapacitors. Adv. Funct. Mater. 25: 2745–2753.

Zhao, S., J. Xu, M. Wei and Y.-F. Song. 2011. Synergistic catalysis by polyoxometalate-intercalated layered double hydroxides: oximation of aromatic aldehydes with large enhancement of selectivity. Green Chem. 13: 384–389.

Zhao, S., L. Liu and Y.-F. Song. 2012. Highly selective oximation of aldehydes by reusable heterogeneous sandwich-type polyoxometalate catalyst. Dalton Trans. 41: 9855–9858.

Zhu, J., H. Fan, J. Sun and S. Ai. 2013. Anion-exchange precipitation synthesis of α-Ag$_2$WO$_4$/Zn–Cr layered double hydroxides composite with enhanced visible-light-driven photocatalytic activity. Sep. Purif. Technol. 120: 134–140.

Zhu, X., C. Tang, H.-F. Wang, Q. Zhang, C. Yang and F. Wei. 2015. Dual-sized NiFe layered double hydroxides *in situ* grown on oxygen-decorated self-dispersal nanocarbon as enhanced water oxidation catalysts. J. Mater. Chem. A 3: 24540–24546.

11

Magnetic Anisotropy of Nanocomposites Made of Magnetic Nanoparticles Dispersed in Solid Matrices

Costica Caizer

||||||||||||||||||||||||||||||||||||||||||||||||||||||||||||||||||||||||||||||||||||||||||||||||||||||||||||||||||||||||||||||||||||||||||||||||||||||||||||||||||||||||||||

Introduction

The magnetic anisotropy is a very important observable of the magnetic materials (Kneller 1962, Chikazumi 1964, Vonsovski 1971, Rado and Suhl 1973, Chikazumi and Graham 1997, Nogues and Schuller 1999, Skumryev et al. 2003, Jamet et al. 2004, Berkowitz and Kodama 2006, Restrepo et al. 2006, Mazo-Zuluaga et al. 2008, Cullity and Graham 2009, Peddis et al. 2009) and, in particular, of the nanomaterials (nanocomposites, nanopowders, nanoparticles, nanofluids, thin films, multilayers, nanostructures, etc.) or biomaterials, which should be considered in nanotechnology and theoretical studies, along with other specific observables, such as the saturation magnetization, magnetic permeability, coercive field, magnetic behavior in an external magnetic field or with temperature, etc. (Rinkevich et al. 2016). The magnetic anisotropy determines a certain magnetic behavior of the magnetic material and nanomaterial in external magnetic field, from hard magnetic to soft magnetic and even superparamagnetic for certain magnetic nanocomposites, determining multiple practical applications of them (Victora and Shen 2005, Garcia-Sancheza et al. 2006, Tartaj 2009, Antoniak et al. 2011, Raj et al. 2013, Ridi et al. 2014, Chen et al. 2016, Wang et al. 2016).

In this chapter, I will refer to the *magnetic anisotropy* of *magnetic nanocomposites* consisting of *nanoparticles* with *magnetic ordering*, having atomic *magnetic* moments *aligned* under the exchange or superexchange interaction, namely *ferro-* and *ferrimagnetic*. These also represent the most practical interest because of their high magnetization in the external field, surpassing, for example, the one for paramagnets by a high difference, or the one for antiferromagnets. The ordering of the atomic magnetic moments ($\vec{\mu}$) in the nanoparticles' crystalline ferromagnetic network under the action of the exchange interaction (Heisenberg-Dirac model) (Heisenberg 1926, Dirac 1926, Heisenberg 1928) is in the same

West University of Timisoara, Department of Physics, Bv. V. Parvan no. 4, 300223 - Timisoara, Romania, RO.
Email: costica.caizer@e-uvt.ro; ccaizer@physics.uvt.ro

direction (Figure 1a) in a magnetic domain (Weiss), and in the opposite directions (Figure 1b) in case of ferrimagnetic ordering under the superexchange interaction (Anderson 1950, Van Vleck 1951).

When it comes to *paramagnetic*, there is *no magnetic ordering, magnetic moments of atoms (or ions) being orientated in all directions* (Figure 2a). For the antiferromagnetic state, *the magnetic moments are oriented in opposite directions*, similar to the ferromagnetic state, but compensating each other (Figure 2b).

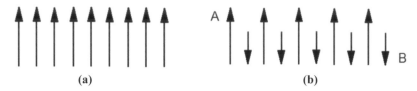

(a)   (b)

Figure 1. Schematic representation of the relative orientation of magnetic moments in a magnetic domain from a (a) ferromagnetic and (b) ferrimagnetic nanocrystal.

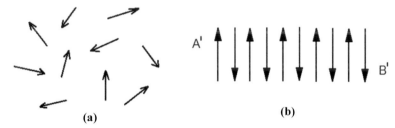

(a)   (b)

Figure 2. Schematic representation of the relative orientation of magnetic moments in a (a) paramagnetic and (b) antiferromagnetic material.

The magnetization ($\vec{M}$), calculated as the resultant vector of the magnetic moments (i) in the unit volume (V),

$$\vec{M} = \sum_i \vec{\mu}_i / V, \tag{1}$$

is considerably higher in the case of nanoparticles with ferromagnetic alignment (Figure 1a) ($M = \sum_i \mu_i / V$), being equal to the spontaneous magnetization ($\vec{M}_s$) of the material (which, in this particular case, equals with the saturation magnetization ($\vec{M}_{sat}$) as well). For the ferrimagnetic nanoparticles, the magnetization is much lower, which is the result of the value (vectorial sum) of the magnetizations of the two magnetic subllatices, A and B (Figure 1b),

$$\vec{M} = \vec{M}_A + \vec{M}_B, \tag{2}$$

according to the Néel model of ferrimagnetism (Néel 1948), and $M = M_A - M_B$ (in module), respectively. As an example, for the ferromagnetic iron nanoparticles, the saturation magnetization is $M_{sat,iron} = 1714$ kA/m (Cullity and Graham 2009), and for the ferrimagnetic magnetite nanoparticles the saturation magnetization is $M_{sat,magnetite} = 477.5$ kA/m (Smit and Wijin 1961). In the other two cases shown in Figure 2 the resultant magnetizations are null due to the compensation of magnetic moments.

A special case is that of the *superparamagnetic* "ordering" of the magnetic moments of small single-domain nanoparticle, with an order of magnitude of 10 nm or less, under the influence of an external magnetic field (Bean and Livingston 1959). In the absence of an external magnetic field, the spontaneous magnetization of the magnetic nanoparticles is no longer stable, but fluctuates along a direction (Figure 3a), under the action of the thermal activation (for example, at the room temperature) (Néel 1949).

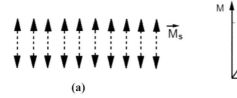

(a)   **(b)**

Figure 3. Schematic representation of the magnetic moments orientation in (a) a superparamagnetic nanostructure and (b) the magnetization (*M*) in an external magnetic field (*H*) of a nanocomposite consisting of superparamagnetic nanoparticles dispersed in a solid matrix.

In the case of a nanocomposite consisting of superparamagnetic nanoparticles dispersed in a solid matrix, each nanoparticle having a magnetic moment (resultant)

$$\overrightarrow{m}_{NP} = V_{NP}\overrightarrow{M}_s, \tag{3}$$

where V_{NP} is the nanoparticle volume, the magnetization of the nanocomposite is equal to zero in the absence of an external magnetic field, and increases according to the Langevin function (Jacobs and Bean 1963) when magnetic field is applied, and in the absence of interactions (Figure 3b) (similar to a system of paramagnetic atoms). In this case the saturation magnetization of nanocomposite is

$$M_{sat,NC} = nm_{NP}, \tag{4}$$

where n is the concentration of the nanoparticles.

Thus, depending on the type of magnetic ordering of the nanoparticles into the nanocomposite, namely ferromagnetic (Figure 1a), ferrimagnetic (Figure 1b) or superparamagnetic (Figure 3), the magnetization at saturation will be specific on the nanoparticles' material ($M_{sat,ferro/ferri}$). However, having in view the volume packing fraction (*Φ*) of the nanoparticles in composite, the saturation magnetization of nanocomposite ($M_{sat,NC}$) will always be less than the corresponding nanoparticle material,

$$M_{sat,NC} < M_{sat,ferro/ferri}, \tag{5}$$

because the volume occupied by the entire magnetic material (nanoparticles) from the composite ($V_{m,NP}$) is always smaller than the volume of the nanocomposite (V_{NC}).

The magnetic anisotropy originates in the existence of the spins (magnetic moments) from the crystalline network and the coupling between them and crystalline network, through the spin-orbit and orbit-network interactions under the crystalline field action, leading ultimately to *a preferred direction* of the spontaneous magnetization within the crystal, corresponding to a position of minimum energy with respect to the crystallographic axes. This causes a *magnetic anisotropy*. The echillibrum position of the spontaneous magnetization is in most cases very close to a direction in which the magnetization is the easiest to achieve in an external magnetic field, called *the easy magnetization axis* (e.m.a.). For expressing the anisotropy energy, a component of the crystal's free energy will be introduced, taking into account the symmetry conditions of the crystal, which will depend on the orientation of the spontaneous magnetization vector relative to the crystallographic axes of the crystal (Akulov 1928, Becker and Döring 1939). The anisotropy energy in this case, known as the *magnetocrystalline anisotropy* energy, will depend both on the crystal's symmetry, i.e., cubic, hexagonal, etc., and the nature of the material (Kneller 1962, Herpin 1968, Vonsovski 1971). The magnetocrystalline anisotropy depends also on the temperature, that usually being maximum at 0 K (where the entropy is minimal) and minimum at the Curie temperature, where the spontaneous magnetization does not exist.

An important component of the magnetic anisotropy is also the *surface anisotropy*, firstly introduced by Néel (Néel 1953, 1954). This occurs due to different symmetry in which the magnetic moments are found on the surface of a crystal relative to those inside. In bulk ferro- and ferrimagnetic materials this component is not too significant, and can be often neglected in relation to the magnetocrystalline component, but in the case of nanoparticles it can sometimes become dominant, exceeding in value by significant values the magnetocrystalline energy (Caizer 2004a). Therefore, this component of the magnetic anisotropy must be evaluated every time when using magnetic nanocomposites made out of ferro- or ferrimagnetic nanoparticles.

Moreover, when the nanocomposites contain nanoparticles having different shapes (physical shapes), regardless of where they are dispersed, it must also take into account the nanoparticles' shape, which could cause a *shape anisotropy* (Kneller 1962). The shape anisotropy always occurs when the magnetic nanocomposite are composed of ferro- or ferromagnetic nanoparticles which are not spherical. The value of this anisotropy depends very much on the shape of the nanoparticles contained and their orientation in the external magnetic field, rather than the material's composition (Antonel et al. 2011).

In the process of obtaining magnetic nanocomposites, *induced magnetic anisotropy* may arise, such as the anisotropy due to *stress* (Vonsovski 1971), or by magnetic annealing processes, irradiation, etc., when the spontaneous magnetization vector will have another position of equilibrium relative to the crystallographic axes. Also, a *uniaxial magnetic anisotropy* (Skumryev et al. 2003) could be obtained, although the crystalline symmetry is cubic. Such anisotropy appears in the case of very small nanoparticles, such as those of magnetite, with cubic symmetry and uniaxial anisotropy. A more special type of magnetic anisotropy is the *unidirectional magnetic anisotropy* (Rosencwaig et al. 1971), which appears in nanostructures with different magnetic ordering, due to exchange coupling between them (Nogues et al. 2005, Berkowitz and Kodama 2006, Peddis et al. 2009).

In the following sections, I will present and discuss the magnetic anisotropy of magnetic nanocomposites made of ferro- and ferrimagnetic nanoparticles, large (with magnetic domains structures) or small (single-domain), and those with a superparamagnetic behavior, dispersed in solid matrices. First, I present the important physical aspects regarding magnetic anisotropy, magneto-crystalline anisotropy, surface anisotropy, shape anisotropy and the induced anisotropy by stress and the exchange coupling. Further on, I will present and discuss the temperature dependence of the magnetocrystalline anisotropy constants, as well as the ferromagnetic resonance (FMR) method, currently used in research laboratories for accurate experimental determination of the magnetic anisotropy constants. Nevertheless, I will also present and discuss some important results concerning the issue of magnetic anisotropy in nanocomposites. Finally, I show the importance of knowing the magnetic anisotropy of the nanocomposites made of nanoparticles dispersed in solid matrices, in terms of advanced applications in nano- and biotechnology.

Physical aspects regarding magnetic anisotropy

The magnetic anisotropy of the nanocomposites composed of magnetic nanoparticles, ferro- or ferrimagnetic, dispersed in various solid matrices, crystalline or amorphous, mainly result from the nanoparticles' magnetic anisotropy itself, already existing in these composite. When it comes to the small-size magnetic nanoparticles (< 20–30 nm), their structure is generally monocrystalline (single-crystal), a fact confirmed by numerous experimental results (Figure 4b) (Thakur et al. 2009), they being crystallized in one of the well-known seven basic crystallographic systems: cubic, tetragonal, hexagonal, etc., depending on the type of material (the chemical composition). Figure 4 shows (a) the transmission electron microscopy (TEM) and (b) high resolution TEM (HR-TEM) images of magnetic nanocomposite made of $Co_{50}Ni_{50}$ alloy nanoparticles, monocrystalline (nanocrystals), dispersed in an amorphous SiO_2 matrix.

Thus, the magnetic anisotropy of nanocomposites should be seen in tight correlation with the magnetic nanoparticles contained, namely, in most cases, the ferro- or ferrimagnetic single-crystals (nanocrystals), which will be approached in this chapter. However, since the nanocomposites are composed of many magnetic nanoparticles (N) (as in Figure 5a), with a particular concentration (n) in composite, having the magnetic volume $V_{m,NP}$ ($V_{m,NP} = V_{NP,1} + V_{NP,2} + ... + V_{NP,N}$) relative to the volume of the entire nanocomposite V_{NC}, it is important to define *the volume magnetic packing fraction measure* of the nanocomposites,

$$\Phi = V_{m,NP}/V_{NC} . \text{ (in \%)} \tag{6}$$

For this measure, the condition $\Phi < 1$ is always fulfilled.

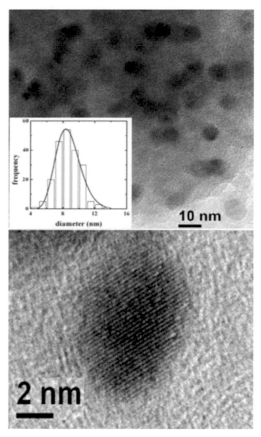

Figure 4. TEM images for $\Phi = 10\%$. The corresponding histograms of the particle size are shown in the inset. The continuous curves indicate the log-normal fits. High resolution TEM image (Reprinted from (Thakur et al. 2009), © AIP 2009, with the permission of AIP Publishing).

When studying or determining the magnetic anisotropy of these nanocomposites, one should consider, in addition to the magnetic packing fraction Φ, the dispersion of nanoparticles and, most importantly, the orientation of the crystallographic axes/crystalline planes of the nanocrystals relative to the external magnetic field applied. Thus, there will be a spatial distribution of the crystallographic axes of the nanocrystals in the composite and, at the same time, a distribution of the spontaneous magnetization vector (magnetic moments) of nanoparticles (Figure 5a), and, implicitly, of the magnetic anisotropy axes (see next Section) at a macroscopic level in the nanocomposite. Consequently, in the case of such nanocomposites, evaluating the magnetic anisotropy can become an extremely complex issue, and it should be treated specifically, on a case by case basis, taking into account the real distribution of magnetic anisotropy axis for the nanoparticle systems with different crystal symmetries. But there may be cases when the problem is simplified such as the case when the magnetic nanoparticles (nanocrystals) would have one of the crystallographic axes oriented in the same directions with the spontaneous magnetization (the magnetic moments) (Figure 5b). This would be the case of nanocomposites formed out of magnetic nanoparticles *aligned* or textured.

Nanocomposites of this kind can be obtained through various physical methods or after applying some very intense magnetic fields (of saturation) during their preparation, or by applying some subsequent thermomagnetic treatments on the nanocomposites. Thus, one can obtain a nanocomposite where the magnetic anisotropy can be studied starting from the magnetic anisotropy of a nanoparticle/nanocrystallite component (having in mind, in particular cases, identical nanoparticles, lack of interactions, etc.). It is therefore very important to know first the existing magnetic anisotropies at the level of a magnetic nanocrystal/magnetic nanoparticle, a problem which will be addressed in the following paragraph.

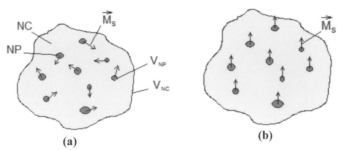

Figure 5. Composite formed out of magnetic nanoparticles with spontaneous magnetization (a) oriented in all directions and (b) aligned.

Below, I first present the magnetic anisotropy approached on the basis of the phenomenological and macroscopic models, which have in view the symmetry properties of monocrystals and then, by the end of the chapter, I will also present the theoretical approach of the magnetic anisotropy using the Heisenberg model, considering the magnetic crystal as a spin system (magnetic moments), a model used in more recent theoretical studies with good results.

The main forms of magnetic anisotropy found in composites made of magnetic nanoparticles are determined by the magnetic anisotropy of the composing nanoparticles, which are: the magnetocrystalline anisotropy, surface anisotropy, shape anisotropy. In addition, during the process for obtaining magnetic nanoparticles and composites, other magnetic anisotropies can be induced, such as the anisotropy caused by the external elastic tensions (stress), unidirectional anisotropy or uniaxial anisotropy in the case of small and very small nanoparticles (< 15 nm). Besides all these, sometimes other induced anisotropy can arise during the nanocomposites preparation, such as in the case of magnetic annealing, irradiation, growth in flux of the nanostructures, pulverizing preparation, or as a result of the existence of residual internal stress, etc.

In the following sections, I will present and discuss in more detail the main magnetic anisotropy, and those induced as a result of external stress, the coupling between the spins of nanostructures with different magnetic ordering (ferromagnetic – paramagnetic (FM – PM), ferromagnetic – antiferromagnetic (FM – AFM)), which are the most common in composites made of magnetic nanoparticles dispersed in solid matrices.

Magnetocrystalline anisotropy

Magnetizing a single crystal/nanocrystal ferromagnetic (or ferrimagnetic), which may even be a nanoparticle or a composite containing a magnetic nanoparticle, with an external magnetic field applied by different directions in relation to the main crystallographic axes, it obtains a different magnetization. Thus, it was found that there are preferential directions of spontaneous magnetization in a crystal, compared to the crystallographic axes, corresponding to a minimum energy (equilibrium energy). This shows that there is a magnetic anisotropy, called in this case *magnetocrystalline anisotropy*. The direction (axis) of magnetization based on which the magnetization is the easiest to have is called easy magnetization axis (e.m.a.). The direction (axis) of magnetization based on which the magnetization makes the most difficult, is hard magnetization axis (h.m.a.). In this respect, the classic experimental results obtained as a result of the magnetization of the ferromagnetic iron, nickel and cobalt single crystals (Honda and Kaya 1926, Kaya 1928a,b) are highly suggestive.

In Figure 6, the first magnetization curves of the iron single crystal are shown (in the cgsE units system) (Honda and Kaya 1926), which crystallizes in the face centered cubic system (fcc). The magnetization was done by applying an external magnetic field following the main directions in the crystal fcc: [100] (the direction of cube edge), [110] (the direction of cube face diagonal), [111] (the direction of cube space diagonal).

The results show that the magnetization of the iron single crystal is the easiest when done following the axis [100], when saturation is reached quickly after applying an external magnetic field rather weak,

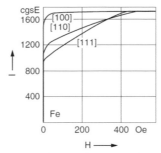

Figure 6. Magnetization curves of Fe-single crystal depending on the fundamental directions (Honda and Kaya 1926) (Reprinted from (Kneller 1962), © Springer-Verlag 1962; With permission of Springer).

and the hardest when done following the axis [111], when to reach the magnetic saturation a high magnetic field is needed. Any other direction of magnetization, such as the cube face diagonal [110], the magnetization is always harder than in the case of [100] axis, and easier than in the case of [111] axis, yielding different intermediate magnetization curves between the two extreme cases (e.g., case [110] in Figure 6).

If magnetizing the nickel single crystal with an external magnetic field is applied following the same crystallographic directions (Figure 7) (Kaya 1928a), the situation is reversed: the magnetization is the easiest when the magnetic field is applied after the cube spatial diagonal direction [111] and the hardest when the magnetic field is applied after the cube edge direction [100]. Compared with iron, the nickel crystallizes in volume centered cubic system (vcc) and has a saturation magnetization less sensitive (Ni: 500 Gs (Kaya 1928a); Fe: ~ 1700 Gs (Honda and Kaya 1926) (in cgsE units)). Thus, in the case of nickel, the axis [111] is the axis of easy magnetization and the axis [100] is the axis of hard magnetization. Like in the iron case, the magnetization after any other crystallographic direction, such as the cube face diagonal [110] (Figure 7), will be harder than if [111] and easier than if [100], the first magnetization curve occupying an intermediate position between the two extremes of magnetization, easy and hard magnetization.

For cobalt, the result is completely different (Figure 8) (Kaya 1928b) compared to both iron and nickel, the cobalt crystallizing in the hexagonal system and not cubic. In this case, the magnetization is most easily obtained when the external magnetic field is applied following the direction of the main axis of the hexagonal volume [0001] and the hardest when the magnetic field is applied following any other direction perpendicular on the direction [0001], such as, for example, the hexagonal face edge direction [10$\bar{1}$0]. In the case of cobalt with hexagonal symmetry, the [0001] direction is the easy magnetization axis, and the [10$\bar{1}$0] direction is the hard magnetization axis.

For a quantitative expression of the magnetocrystalline anisotropy energy using the properties of crystal symmetry, a component of the crystal's free energy, $F_k(\alpha_1, \alpha_2, \alpha_3)$, will be added to reflect this, where $\alpha_1, \alpha_2, \alpha_3$ are direction cosines of the spontaneous magnetization vector $(\vec{M}_s)$ relative to the main crystallographic axes (Akulov 1928). In order to rotate the spontaneous magnetization vector from the

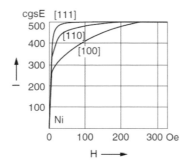

Figure 7. Magnetization curves of Ni-single crystal depending on the fundamental directions (Kaya 1928b) (Reprinted from (Kneller 1962), © Springer-Verlag 1962; With permission of Springer).

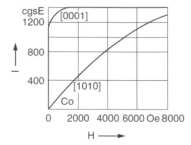

Figure 8. Magnetization curves of Co-single crystal depending on the fundamental directions (Kaya 1928a) (Reprinted from (Kneller 1962), © Springer-Verlag 1962; With permission of Springer).

direction of easy magnetization (e.m.a.) in another direction (any direction), fixed by the direction cosines $\alpha_1, \alpha_2, \alpha_3$ in relation to the main (fundamental) crystallographic axes of the crystal, there is need of energy. This energy is called *magnetocrystalline anisotropy energy*. This energy can be expressed in two cases.

I. The case of the *cubic symmetry* (Figure 6 and Figure 7 (Fe, Ni)) (Kneller 1962, Caizer 2004b).

The free energy component of the crystal with cubic symmetry for a spontaneous magnetization vector $\vec{M}_s$, fixed by the direction cosines $\alpha_1, \alpha_2, \alpha_3$ in relation to the main crystallographic axes 0x, 0y and 0z (Figure 9), can be written as a series of power of the direction cosines (Akulov 1928, Aubert 1968).

$$F_{k.c}(\alpha_1, \alpha_2, \alpha_3) = K_0 + K_1(\alpha_1^2 \alpha_2^2 + \alpha_1^2 \alpha_3^2 + \alpha_2^2 \alpha_3^2) + K_2\alpha_1^2 \alpha_2^2 \alpha_3^2$$
$$+ K_3(\alpha_1^4 \alpha_2^4 + \alpha_1^4 \alpha_3^4 + \alpha_2^4 \alpha_3^4) + \ldots,$$

(7)

where K_0, K_1, K_2, K_3 ... are magnetocrystalline constants (power densities, expressed in J/m^3).

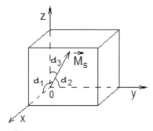

Figure 9. The spontaneous magnetization $\vec{M}_s$ fixed by the direction cosines $\alpha_1, \alpha_2, \alpha_3$ in a system with cubic symmetry.

When the spontaneous magnetization vector $\vec{M}_s$ is on the easy magnetization axis (position of equilibrium), the free energy component will be

$$F_{k.c}(1,0,0) = K_0,$$

(8)

because the cosines for this direction are $\alpha_1 = 1$, $\alpha_2 = \alpha_3 = 0$. This energy is a *constant*, which shows (physically) that in the case when magnetization follows the direction of easy axis, a small mechanical work different than zero will be performed.

The energy required for the rotation of the spontaneous magnetization vector from the direction of easy magnetization to the direction fixed by the cosines $\alpha_1, \alpha_2, \alpha_3$, will be

$$\mathcal{E}_{k.c}(\alpha_1, \alpha_2, \alpha_3) = F_k(\alpha_1, \alpha_2, \alpha_3) - F_k(1,0,0),$$

(9)

respectively,

$$\mathcal{E}_{k.c}(\alpha_1, \alpha_2, \alpha_3) = K_1(\alpha_1^2 \alpha_2^2 + \alpha_1^2 \alpha_3^2 + \alpha_2^2 \alpha_3^2) + K_2\alpha_1^2 \alpha_2^2 \alpha_3^2$$
$$+ K_3(\alpha_1^4 \alpha_2^4 + \alpha_1^4 \alpha_3^4 + \alpha_2^4 \alpha_3^4) + \ldots .$$

(10)

The Eq. (10) is the *magnetocrystalline anisotropy energy* for the *cubic symmetry*. This is actually a density energy, expressed in J/m³, as the constants K_1, K_2, K_3.

In practical cases, it was found that the values of the anisotropy constants higher than second order (e.g., K_3 respectively) are much lower than the values of the other two (first (K_1) and second order (K_2)), and, thus, it should be neglected. In these conditions, the expression of anisotropy energy used most often in data analysis, or in determining the constants of magnetocrystalline anisotropy, K_1 and K_2, remains in a power series development up to the second order

$$\mathcal{E}_{k.c}\,(\alpha_1, \alpha_2, \alpha_3) = K_1\,(\alpha_1^2\,\alpha_2^2 + \alpha_1^2\,\alpha_3^2 + \alpha_2^2\,\alpha_3^2) + K_2 \alpha_1^2\,\alpha_2^2\,\alpha_3^2. \tag{11}$$

The experimental results showed, in many cases, that even K_2 is significantly lower than K_1 when determining the energy of magnetocrystalline anisotropy, thus it can just use only the first term in Eq. (11), respectively

$$\mathcal{E}_{k.c}\,(\alpha_1, \alpha_2, \alpha_3) \cong K_1\,(\alpha_1^2\,\alpha_2^2 + \alpha_1^2\,\alpha_3^2 + \alpha_2^2\,\alpha_3^2). \tag{12}$$

For example, for Fe it was found that $K_1 = 4.8 \times 10^4$ Jm⁻³ (Graham Jr. 1960) and $K_2 = 7.1 \times 10^3$ Jm⁻³ (Sato and Chandrasekhar 1957) (at temperature of 293 K), where it can easily be seen that the K_2 constant is an order of magnitude smaller than K_1, and under these conditions, the second term from the equation (11) can be neglected without affecting the result.

Having in view Eqs. (9) and (11), and that is minimal (the lowest energy) for the easy magnetization axis [1,0,0] (Figure 6), it is concluded that constants K_1 and K_2 for Fe must be positive, or at least the highest constant, i.e., K_1, must be positive. Indeed, experimental data confirm this result for Fe, where both magnetocrystalline anisotropy constants (K_1, K_2) are positive.

An opposite situation was found in the case of Ni, where constant K_1 is negative ($K_1 = -0.49 \times 10^3$ Jm⁻³ (Rodbell 1965)). For nickel, the $F_k(1,0,0)$ component of the free energy is the maximum, corresponding to the axis of hard magnetization [1,0,0] (Figure 7), and the difference given by the Eq. (9) must be negative, which implies a negative magnetocrystalline anisotropy energy (indeed confirmed by the negative value of K_1 ($|k_1| > K_2$)).

II. The case of *hexagonal symmetry* (Figure 8 (Co)).

The hexagonal symmetry is typical for the single crystal of cobalt. For the same conditions of symmetry, it was also found (Carr 1960) that for the (uniaxial) hexagonal symmetry, the expression for the free energy of the crystal also depends on the single angle (φ) which the spontaneous magnetization $\vec{M}_s$ makes with the main symmetry axis [0001] (Figure 10):

$$F_{k.u}\,(\varphi) = K_0 + K_1\,\sin^2\varphi + K_2\,\sin^4\varphi + K_3\,\sin^6\varphi + \dots. \tag{13}$$

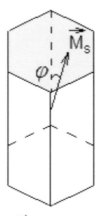

Figure 10. The spontaneous magnetization $\vec{M}$ fixed by angle φ in the system with hexagonal symmetry.

For easy magnetization axis, where $\varphi = 0$, it obtains

$$F_{k,u}(0) = K_0. \tag{14}$$

In this case, the *magnetocrystalline anisotropy energy for hexagonal symmetry* will result out of the following

$$\mathcal{E}_{k,u}(\varphi) = F_{k,u}(\varphi) - F_{k,u}(0), \tag{15}$$

and taking into account Eqs. (13) and (14), the energy will be:

$$\mathcal{E}_{k,u}(\varphi) = K_1 \sin^2 \varphi + K_2 \sin^4 \varphi + K_3 \sin^6 \varphi + \dots . \tag{16}$$

In the second order approximation of the power series, as discussed in the case of cubic anisotropy, the formula (16) may be reduced to

$$\mathcal{E}_{k,u}(\varphi) = K_1 \sin^2 \varphi + K_2 \sin^4 \varphi. \tag{17}$$

In this case as well, experimental data often shows that the condition $K_2 \ll K_1$ is fulfilled ($K_{1,Co} = 4.3 \times 10^5$ Jm^{-3} and $K_2 = 12 \times 10^4$ Jm^{-3} (Bozorth 1954)), a result that leads to the simple formula

$$\mathcal{E}_{k,u}(\varphi) \approx K_u \sin^2 \varphi, \tag{18}$$

where the anisotropy constant K_1 was marked with K_u in the case of *uniaxial symmetry*. The formula (18) is most often used for determining the uniaxial magnetocrystalline anisotropy energy, as well as to check the experimental data.

Surface anisotropy

The magnetic surface anisotropy is due to the different symmetry in which the surface spins are found (the magnetic moments) compared with those inside of the crystal. The surface anisotropy was studied by Néel (Néel 1954), who showed that for the surface type (100) or (111) in a cubic symmetry (Figure 11), *the surface anisotropy energy* can be determined by the following formula

$$\mathcal{E}_s(\beta) = K_s \cos^2 \beta, \tag{19}$$

where β is the angle between the spontaneous magnetization $\vec{M}_s$ and the external normal vector ($\vec{n}$) to the surface.

In formula (19), K_s is *the surface anisotropy constant* and is expressed in Jm^{-2}, generally showing values of the order 10^{-5} (Gazeau et al. 1998). The expression of surface anisotropy constant in Jm^{-2} unit is not always convenient when one wants to compare it with the magnetocrystalline anisotropy, where K_1 constant is expressed in Jm^{-3} (being a volume density). For a direct comparison, K_s can

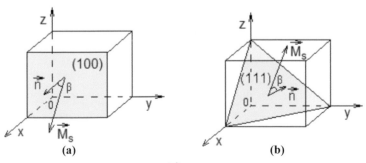

Figure 11. The orientation of spontaneous magnetization $\vec{M}_s$ relative to normal $\vec{n}$ on the surface (a) (100) and (b) (111) for the monocrystal with cubic symmetry.

also be expressed in Jm^{-3}. An example is the expression of K_S constant in Jm^{-3} expressed through the diameter (D) for spherical nanocrystals (situation that can be approximated in practical cases where magnetic nanoparticles and nanocomposites containing these nanoparticles are used), considering the mathematical expression between the surface and volume of the sphere: $K'_s = (6/D)K_s$ (in Jm^{-3}) (Bødker et al. 1994). A calculation shows that for small nanocrystals, usually under 15–20 nm, the surface area has a very important contribution, the surface magnetic anisotropy even exceeding the magnetocrystalline one in the case of soft magnetic materials. Therefore, this component of the magnetic anisotropy cannot be neglected in composites containing magnetic nanoparticles, especially when the nanoparticles are < 15 nm and are soft magnetic. This is proven by numerous experimental results, which showed that the contribution of spins on the surface often becomes even more dominant compared to those inside nanocrystals (Caizer 2004a).

Shape anisotropy

The magnetization of a single crystal depends on its shape, in addition to those discussed in previous sections. In case of general forms of the crystal/sample, which can be approximated to that of an *ellipsoid* (Figure 12), where a, b and c are the semi-axis of the ellipsoid on the three orthogonal directions that satisfy the condition a > b > c, the *energy due to the shape of the sample* using the crystal symmetry conditions was found as follows (Kneller 1962, Vonsovski 1971)

$$F_{sh}(\alpha_a, \alpha_b, \alpha_c) = (1/2)\,\mu_0 M_s^2(N_a\,\alpha_a^2 + N_b\,\alpha_b^2 + N_c\,\alpha_c^2), \tag{20}$$

where M_s is the spontaneous magnetization, α_a, α_b, α_c are the direction cosines of vector $\vec{M}_s$ relative to the ellipsoid axes a, b and c (Figure 12), and N_a, N_b and N_c are the demagnetizing coefficients along a, b and c axes.

The demagnetizing coefficients satisfy the equation

$$N_a + N_b + N_c = 1. \tag{21}$$

Depending on the concrete cases to consider, the formula (20) can have simplified forms.

i) In the case when the shape of the sample/crystal can be approximated with one of a *rotation ellipsoid* in which a > b = c, which leads to $N_b = N_c$, the formulas (21) and (20) for energy due to shape and the demagnetizing factors become

$$F_{sh}(\alpha_a, \alpha_b) = (1/2)\mu_0 M_s^2(N_a\,\alpha_a^2 + N_b\,\alpha_b^2) \tag{22}$$

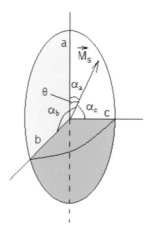

Figure 12. The crystal approximated by an ellipsoid (general case).

and

$$N_a + 2N_b = 1. \tag{23}$$

According to Figure 12, the F_{sh} energy can be expressed also according to the θ angle made by the spontaneous magnetization vector $\overrightarrow{M}_s$ with the c axis of the ellipsoid. In this case, it obtains

$$F_{sh}(\theta) = (1/2) \mu_0 M_s^2 (N_a \cos^2 \theta + N_b \sin^2 \theta). \tag{24}$$

Usually, this equation may be written as

$$F_{sh}(\theta) = (1/2)\mu_0 M_s^2 (N_b - N_a) \sin^2 \theta + const., \tag{25}$$

where *const.* is $(1/2)\mu_0 M_s^2 N_a$. The term dependent on the angle θ of this equation is the *shape anisotropy* component. The maximum value is obtained for $\pi/2$, when $\overrightarrow{M}_s$ is perpendicular on the axis *a*. The minimum energy is obtained for angle $\theta = 0$, when $\overrightarrow{M}_s$ is on the axis a, and will be

$$F_{sh}(0) = const. \tag{26}$$

By conveniently choosing the energy benchmark, the *const.* can be eliminated from the equation (25), or, as it saw earlier, considering as benchmark the energy (minimum one) along the axis *a* when the energy is $F_{sh}(0)$. Thus, the *shape anisotropy* will be

$$\mathcal{E}_{sh}(\theta) = F_{sh}(\theta) - F_{sh}(0) \tag{27}$$

and

$$\mathcal{E}_{sh}(\theta) = (1/2) \mu_0 M_s^2 (N_b - N_a) \sin^2 \theta, \tag{28}$$

respectively.

The *shape anisotropy constant* in this case (by analogy with formula (18)) is

$$K_{sh} = (1/2) \mu_0 M_s^2 (N_b - N_a). \tag{29}$$

The anisotropy constant K_{sh} may have $K_{sh} \gtrless 0$ values, depending on how conditions are met $N_b \gtrless N_a$ for demagnetizing coefficients.

ii) In the case of a *very elongated ellipsoid*, when a >> b = c, the inequality $N_a \ll N_b$ is fulfilled. Thus, according to Eq. (23) is obtained formula: $N_b \cong 1/2$. This leads to

$$K_{sh}^* \cong \pi M_s^2 \times 10^{-7}. \tag{30}$$

If the sample is made of iron, which has $M_s = 1714$ kA/m (Cullity and Graham 2009), a very high value is obtained for the shape anisotropy constant, $K_{sh}^* = 922.5 \times 10^3$ Jm^{-3}, which exceeds by an important value, the magnetocrystalline one ($K_{1,Fe} = 48 \times 10^3$ Jm^{-3}). In these conditions, the shape anisotropy becomes dominant, and the magnetocrystalline anisotropy will be insignificant.

Having in view this result, in practice one can obtain magnetic composite for permanent magnets using ferromagnetic nanoparticles elongated and embedded in various solid matrixes, able to lead to a high magnetic anisotropy for the nanocomposite (due to the shape of the nanoparticle), an anisotropy which is required in the case of permanent magnets for obtaining a very large hysteresis loop and remanence as close as possible to the magnetic saturation.

iii) In case the form of the crystal/sample is reduced to that of a *sphere*, and a = b = c, the demagnetizing factors will be

$$N_a = N_b = N_c = 1/3, \tag{31}$$

and based on the formula (29), in this case, $K_{sh} = 0$, and the shape anisotropy energy is zero. In other words, we cannot speak of shape anisotropy when the nanocrystallites have a spherical shape.

Given that, in general, in the process of obtaining magnetic nanocomposites consisting of magnetic nanoparticles, only rarely the nanoparticles/nanocrystals components' shape can be spherical. Thus, in most cases, there will be also a shape anisotropy and its value must be determined on a case by case basis. Usually, when the deviation from the spherical sample is not too much, this energy can be neglected compared to the magnetocrystalline, or surface one. For example, this is the case for the Ni-Zn ferrite nanoparticles of ~ 10 nm in the SiO_2 amorphous matrix (Caizer 2008). However, when the nanoparticles/ nanocrystals components of the composites have an elongated shape, this energy becomes significant and must be considered. Depending on the type of the material, such as for the case of soft magnetic materials, this form of magnetic anisotropy can become dominant in many cases, therefore it must be evaluated for each nanocomposite.

The approximation of rotation ellipsoid can be used successfully in the theoretical calculations, having in view Brown and Morrish's studies (Brown Jr. and Morrish 1957), which showed that the single domain nanoparticles (small magnetic particles) with various forms behave as ellipsoids.

Stress anisotropy

For the composites formed out of nanoparticles embedded in solid matrices, there are often situations when during the obtaining process due to the matrix, elastic tensions (*stress*) appear. Under the action of these stress, the spontaneous magnetization vector of a magnetic domain from a single crystal/ nanocrystal/nanoparticle, it can change its direction relative to the crystallographic axes, depending on the tension strength, thereby causing a magnetic *anisotropy due to stress*. The stress within the magnetic nanomaterial can also occur during the magnetization process, due to the effect of *magnetostriction* (Becker and Döring 1939, Cullity and Graham 2009). Without taking into account the latter, considering its reduced contribution compared to that caused by the stress, which in many cases could have the same magnitude with the magnetocrystalline anisotropy, or even more, I will briefly present below the stress anisotropy.

The appearance of a magnetic anisotropy under the action of tension is due to spin–orbit coupling; after applying a tension, the ions' orbital orientation will change, which leads, by coupling, to a change in spin direction (of the spontaneous magnetization relative to crystallographic axes). The *stress anisotropy energy* can be expressed in terms of a cubic crystal symmetry, under a *uniform tension* σ, by the following equation (Vonsovski 1971, Cullity and Graham 2009)

$$
\begin{aligned}
\mathcal{E}_\sigma = &-(3/2)\lambda_{100}\ \sigma(\alpha_1^2\ \gamma_1^2 + \alpha_2^2\ \gamma_2^2 + \alpha_3^2\ \gamma_3^2) \\
&-(3/2)\lambda_{111}\ \sigma(2\alpha_1\alpha_2\gamma_1\gamma_2 + 2\alpha_1\alpha_3\gamma_1\gamma_3 + 2\alpha_2\alpha_3\gamma_2\gamma_3),
\end{aligned}
\tag{32}
$$

where α_1, α_2, α_3 are the direction cosines for the spontaneous magnetization relative to the crystallographic axes, γ_1, γ_2, γ_3 are the direction cosines for the action direction of elastic tension (σ) relative to the crystallographic axes (Figure 13), and λ_{100}, λ_{111} are magnetostriction constants relative to the crystallographic directions [100] and [111].

When the external tensions are high compared to the magnetocrystalline anisotropy ($\sigma\lambda > K_1$), the anisotropy determined by the tensions becomes important. In this case, the magnetization of the sample under the influence of an external magnetic field can be changed significantly. There may be times when $\sigma\lambda \gg K_1$, the magnetic anisotropy in this case being determined only by the stress and the magnetocrystalline anisotropy becoming negligible. This event is observed in the magnetic nanocomposites of γ-Fe_2O_3 in SiO_2 silica matrix (Caizer and Hrianca 2003); therefore, this component of the magnetic anisotropy must also be considered when analyzing the entire magnetic anisotropy.

Moreover, when the magnetostriction is isotropic (or weak), as in the polycrystalline nanostructures (such as in some nanocrystals/large nanoparticles) or even amorphous,

$$
\lambda_{100} \approx \lambda_{111} \approx \lambda_s,
\tag{33}
$$

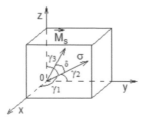

Figure 13. The elastic tension σ fixed by the direction cosines γ_1, γ_2, γ_3 in a system with cubic symmetry.

λ_s being the saturation magnetostriction constant of the material, the Eq. (32) becomes

$$\mathcal{E}_\sigma = -(3/2)\lambda_s\, \sigma(\alpha_1^2\, \gamma_1^2 + \alpha_2^2\, \gamma_2^2 + \alpha_3^2\, \gamma_3^2)$$
$$-(3/2)\lambda_s\, \sigma(2\alpha_1\alpha_2\, \gamma_1\gamma_2 + 2\alpha_1\alpha_3\, \gamma_1\gamma_3 + 2\alpha_2\alpha_3\, \gamma_2\gamma_3) \tag{34}$$

and

$$\mathcal{E}_\sigma = -(3/2)\lambda_s\, \sigma(\alpha_1\gamma_1 + \alpha_2\gamma_2 + \alpha_3\gamma_3)^2, \tag{35}$$

respectively.

Given the relation

$$\cos \delta = \alpha_1\gamma_1 + \alpha_2\gamma_2 + \alpha_3\gamma_3, \tag{36}$$

from Eq. (35) is obtained the formula

$$\mathcal{E}_\sigma = -(3/2)\lambda_s\, \sigma\cos^2 \delta, \tag{37}$$

where δ is the angle between the spontaneous magnetization vector $\vec{M}_s$ and the tension σ (Figure 13). The formula obtained (37) is often used in practice in the quantitative assessment of stress anisotropy energy. It is observed that in the case of a material with $\lambda_s < 0$ subject to a compressive stress ($\sigma < 0$), the same effect is obtained as for $\lambda_s > 0$ when the material is subject to a tensile stress.

For example, in the case of nickel, which has $\lambda_s < 0$, a stress of tensile or compressive produces totally different effect on the magnetization curve (Figure 14) (Cullity and Graham 2009): a compressive stress increases the magnetization, while the same stress, but one of tensile, leads to a considerable decrease of the magnetization.

The importance of the stress anisotropy results from the following example for nickel. The easy magnetization axis for nickel is [111] (e.m.a.) and $\lambda_s < 0$. When the magnetization is made following the [111] direction, the stress energy will be written as

$$\mathcal{E}_\sigma = -(3/2)\lambda_{111}\, \sigma\cos^2 \varphi. \tag{38}$$

For this energy to be of the magnetocrystalline magnitude, the following condition must be fulfilled

$$(3/2)|\lambda_{111}|\sigma \sim K_1. \tag{39}$$

Having known values for $K_1 = -5 \times 10^3$ Jm^{-3} and $\lambda_{111} = -20 \times 10^{-6}$, the resulting tension is $\sigma = 17$ kg/mm^2 (suggestively expressed in kg/mm^2). This is a reasonable value for the tension. The value of this tension can be overcome practically quite easily, which will lead to a stress anisotropy energy higher than the magnetocrystalline one and, for a difference of one order of magnitude, the stress anisotropy will be dominant.

Hence, it becomes clear that this anisotropy must be taken into account when it comes to nanocomposites containing magnetic nanoparticles/nanocrystals embedded in solid matrix, where, during preparation, mechanical tensions (stress) (tensile or compressive) on the nanocrystals components may often occur. The results of the experiments showed that the elastic tensions can induce very large

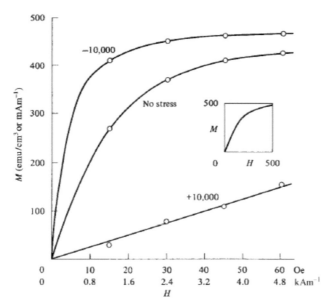

Figure 14. Effect of applied tensile (+) and compressive (−) stress on the magnetization curve of polycrystalline nickel; 10,000 lb/in² ≅ 69 Mpa (Republished with permission of John Wiley & Sons, Inc., from (Cullity and Graham 2009), © 2009 of the Institute of Electrical and Electronics Engineers, Inc.; permission conveyed through Copyright Clearance Center, Inc.).

magnetic anisotropy, with one to two orders of magnitude larger than the magnetocrystalline (Caizer 2016). For example, in the case of γ-Fe$_2$O$_3$ nanoparticles, Vassiliou et al. (Vassiliou et al. 1993) obtains the anisotropy constant value of *4.4 × 10⁵ J/m³* for 8.3 nm diameter nanoparticles embedded in polymer matrix, and Coey et al. (Coey and Khalafalla 1972) obtains the value of *1.2 × 10⁵ J/m³* for nanoparticles of 6.5 nm. These values are about two orders of magnitude higher than the magnetocrystalline anisotropy constant of *bulk* ferrite of γ-Fe$_2$O$_3$, which is K_V = *4.6 × 10³ J/m³* (Mørup 1983).

Unidirectional anisotropy

Another important result in terms of a more special type of magnetic anisotropy, which occurs only under certain conditions, is shown. In core-shell magnetic nanostructures, where the core is ferro- or ferrimagnetic and the shell is paramagnetic or antiferromagnetic (Meiklejohn and Bean 1956, Skumryev et al. 2003), a *unidirectional magnetic anisotropy* appears. For example, in Ref. (Skumryev et al. 2003), the nanostructure of Co nanoparticles embedded in a paramagnetic matrix (C or AlO$_3$) or in an antiferromagnetic matrix (CoO) is presented. I must specify that this form of anisotropy is different from the uniaxial magnetocrystalline anisotropy due to the crystalline symmetry, and should not be confused with the latter.

A ferromagnetic–antiferromagnetic (FM–AFM) core-shell nanostructure is shown in Figure 15 (Nogues et al. 2005).

The explanation for the emergence of this new type of magnetic anisotropy, *unidirectional*, was given on the basis of the *coupling between the spins* of the core and the surface, coupling that causes a shift of the spontaneous magnetization towards a *unique direction* in relation to the crystallographic axes. In this regard, the case of the fine particles of ferromagnetic cobalt coated with an antiferromagnetic cobalt oxide, a structure which shows a unidirectional anisotropy due to coupling by the exchange interaction between magnetic ions in the surrounding area of the two ferro- and antiferromagnetic nanostructures is known (Meiklejohn and Bean 1956, Meiklejohn and Bean 1957, Meiklejohn 1962). The interaction result leads to the reorientation of spins (the spontaneous magnetization) on a single direction, generating the uniaxial anisotropy.

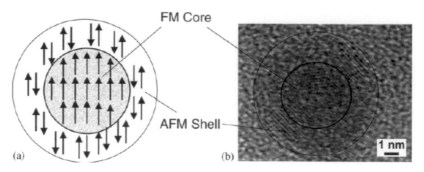

Figure 15. (a) Schematic drawing of a core-shell structure and (b) transmission electron microscopy (TEM) image of an oxidized Co particle (Reprinted from (Nogues et al. 2005), © 2005, with permission from Elsevier).

An interesting case is when the anisotropy of AFM structure is high, and as a result, there is a shift of the hysteresis loop along the magnetic field direction (Figure 16) (Nogues and Schuller 1999). "The intuitive spin configuration, for a FM–AFM couple, is shown schematically in Figure 16 for different stages of a hysteresis loop (Nogues and Schuller 1999). After the field cooling process, the spins in both the FM and the AFM lie parallel to each other at the interface (Figure 16a). When the magnetic field is reversed, the spins in the FM start to rotate. However, if the AFM anisotropy, K_{AFM}, is large enough, as is often the case, the spins in the AFM will remain fixed. Consequently, due to the interface coupling, they will exert a microscopic torque to the spins in the FM, trying to keep them in their original position (Figure 16b). Thus, the magnetic field required to completely reverse the magnetization in the FM will be higher than if the FM was not coupled to an AFM, i.e., an extra magnetic field will be required to overcome the microscopic torque exerted by the spins in the AFM. As a result, the coercive field in the negative field branch increases (Figure 16c). Conversely, when the magnetic field is reversed back to positive values, the rotation of spins in the FM will be easier than in an uncoupled FM, since the interaction with the spins in the AFM will now favor the magnetization reversal, i.e., the AFM will exert a microscopic torque in the same direction as the applied magnetic field (Figure 16d). Therefore, the coercive field in the positive field branch will be reduced. The net effect will be a shift of the hysteresis loop along the magnetic field

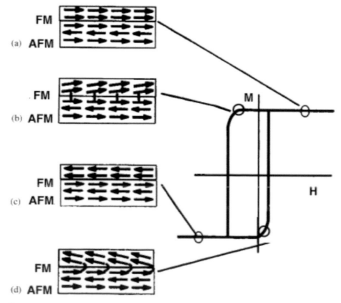

Figure 16. Schematic diagram of the spin configurations of an FM-AFM couple at the different stages of a shifted hysteresis loop for a system with large K_{AFM} (Nogues and Schuller 1999) (Reprinted from (Nogues et al. 2005), © 2005, with permission from Elsevier).

axis, H_E. Thus, the spins in the FM have only one stable configuration (i.e., unidirectional anisotropy)." (Nogues et al. 2005) (Physics Reports, Copyright 2005, with permission from Elsevier).

A review of the unidirectional (exchange) anisotropy can be found in Ref. (Berkowitz and Kodama 2006) for different FM-AFM nanostructures. A similar behavior was found in the case of $CoFe_2O_4/NiO$ ferrimagnetic/antiferromagnetic nanocomposites (Peddis et al. 2009).

Magnetic anisotropy in the Heisenberg model

Having in view the above forms of magnetic anisotropy, the total magnetic anisotropy energy ($\mathcal{E}_{ma}$) can be written as a sum of its components,

$$\mathcal{E}_{ma} = \mathcal{E}_k + \mathcal{E}_s + \mathcal{E}_{sh} + \mathcal{E}_\sigma + \mathcal{E}_{ua} + \mathcal{E}_i. \tag{40}$$

To the total magnetic anisotropy energy, sometimes a component of the induced magnetic anisotropy ($\mathcal{E}_i$) can be added (Eq. 40), which may appear within the material during the process of preparing the nanocomposites.

Considering the crystal as a system of spins ($\vec{S}_i$) (magnetic moments) and taking into consideration the possible magnetic interactions between them, and by using quantum mechanics, a theoretical study of magnetic anisotropy can be found in recent works (Jamet et al. 2004, Restrepo et al. 2006, Mazo-Zuluaga et al. 2008). Theoretically, the Heisenberg-Dirac hamiltonian ($\mathcal{H}$) is used to express the energy of magnetic anisotropy ($\mathcal{H}_{ma}$), with its main components being: *the energy of magnetocrystalline anisotropy* ($\mathcal{H}_k$), *the energy of surface anisotropy* ($\mathcal{H}_s$), *the energy of shape anisotropy* ($\mathcal{H}_{sh}$), *the energy of stress anisotropy* ($\mathcal{H}_\sigma$),

$$\mathcal{H}_{ma} = \mathcal{H}_k + \mathcal{H}_s + \mathcal{H}_{sh} + \mathcal{H}_\sigma \tag{41}$$

and, in some cases, the *induced anisotropy energy* ($\mathcal{H}_i$), such as the *unidirectional anisotropy energy* ($\mathcal{H}_{ua}$), or the *magnetostriction energy*, etc. For practical considerations, one can use the hamiltonian, when one or more components of the hamiltonian from the Eq. (41) can be neglected. In Ref. (Mazo-Zuluaga et al. 2008) a case is presented, when the contribution at hamitonian of the core spins ($\mathcal{H}_k$) and of those on the surface ($\mathcal{H}_s$), in a symmetrical cubic structure (with a spherical shape) and in the absence of external tensions, is considered. Thus, the hamiltonian due to the magnetic anisotropy in this case shows as follows:

$$\mathcal{H}_{ma} = \mathcal{H}_k + \mathcal{H}_s, \tag{42}$$

and

$$\mathcal{H}_{ma} = -K_k \sum_i (S_{x,i}^2 S_{y,i}^2 + S_{y,i}^2 S_{z,i}^2 + S_{x,i}^2 S_{z,i}^2) - K_s \sum_i (\vec{S}_i \cdot \vec{n}_i)^2, \tag{43}$$

respectively (Restrepo et al. 2006, Mazo-Zuluaga et al. 2008), where $\vec{n}$ is the unit vector normal on the surface. In Eq. (43), S represents the spin magnetic moment (i,j) with its components in a three-dimensional system (0xyz), and K_k and K_s are the constants of magnetocrystalline anisotropy and surface anisotropy, respectively. This case illustrates the contribution of surface magnetic anisotropy compared with magnetocrystalline anisotropy. Depending on the K_s/K_k ratio, the magnetic properties of nanosystems change considerably.

For example, in Figure 17a is shown the variation of spontaneous magnetization with temperature, and magnetic susceptibility (using Monte Carlo computer simulation), when the surface anisotropy becomes comparable with magnetocrystalline anisotropy ($K_s/K_k = 1$) in Fe_3O_4 ferrimagnetic nanoparticles (with ferromagnetic sublattices A (tetrahedral) and B (octahedral) (see Figure 1b)), having 5 nm diameter. In this case, the Curie temperature ($T_{c,NP} = 801$ K) is significantly lower than the one corresponding to bulk ferrite of Fe_3O_4 ($T_c = 860$ K). Moreover, with the decreasing diameter of the nanoparticles from 5 nm to 2.5 nm, the Curie temperature (ferrimagnetic – paramagnetic transition temperature) decreases further (Figure 17b), when the contribution of the surface anisotropy is greater than magnetocrystalline anisotropy ($K_s > K_k$).

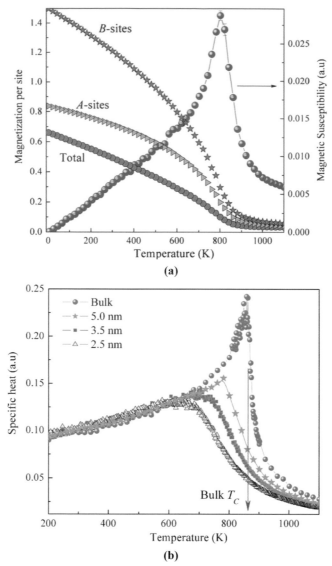

Figure 17. (a) (Color online) Magnetization per magnetic site as a function of temperature for the nanoparticle of 5 nm in diameter. Contributions from tetrahedral (*A*) and octahedral (*B*) sites are explicitly shown. The peak position of the magnetic susceptibility reveals a Curie temperature (~ 801 K) smaller than that of the bulk (~ 860 K). This feature agrees with the temperature at which the specific heat also exhibits a peak (Figure b) (Reprinted from (Mazo-Zuluaga et al. 2008), © AIP 2008, with the permission of AIP Publishing); (b) (Color online) Temperature dependence of the specific heat. A lambda-type behavior is observed, consistent with a thermally driven ferrimagnetic to paramagnetic phase transition. Positions of the peaks from both the susceptibility (Figure a) and the specific heat, give an average estimate of the Curie temperature at around 796 ± 5 K for the 5 nm nanoparticle. Smaller particles yield smaller critical temperatures. Error bars are smaller than symbol size (Reprinted from (Mazo-Zuluaga et al. 2008), © AIP 2008, with the permission of AIP Publishing).

In the theoretical approach, good results are often obtained which explain the magnetic anisotropy in many cases. However, in analyzing the experimental data, due to the complexity of a mathematical approach, the formulas arising out of the phenomenological anisotropy (considering the properties of crystal symmetry), presented above, are most often used.

The variation with temperature of anisotropy constants

The magnetic anisotropy depends on the temperature, meaning that it is maximum at 0 K and becomes zero (disappears) at the Curie temperature (T_c) for ferromagnetic materials, and the Néel temperature (T_N) for the ferrimagnetic materials, when the magnetic moments are no longer aligned under the action of exchange or superexchange interaction, but oriented in all directions (when the spin system becomes isotropic). The study of magnetic anisotropy with temperature has to measure the variation of anisotropy constants with temperature. The results show a wide variety of variations of the magnetic anisotropy constants with temperature (Akulov 1936, Tatsumoto et al. 1965, Furey 1967, Aubert 1968, Tokunaga 1974, Clarck 1978, Ono and Yamaha 1979), depending on the type of the material, crystalline symmetry, magnetic ordering (ferro- or ferromagnetic), type of magnetic anisotropy, etc., variations which can be both positive (sometimes which increases and then decreases) and negative, or even change sign at the zero crossing (such as nickel or some complex ferrimagnetic structures). The limited extent of this chapter does not allow for a more deep approach of this extremely complex problem, thus I will only make a brief reference to the variation with temperature of the constants of magnetocrystalline anisotropy considering some phenomenological models, which broadly indicates a variation close to the experimental one.

A temperature dependence of the magnetocrystalline anisotropy constants in ferromagnetic crystals, partially confirmed by experimental work, is in the study by Akulov (1936)

$$K_n(T) = K_n(0)\left[\frac{M_s(T)}{M_s(0)}\right]^{n(2n+1)} \tag{44}$$

where n is the order of the anisotropy constant. This law shows an universal dependence (independent of crystal symmetry) of the magnetocrystalline anisotropy constants K_1 and K_2 type

$$K_1(T) \sim M_s(T)^3 \quad \text{and} \quad K_2(T) \sim M_s(T)^{10}, \tag{45}$$

such as in the case of uniaxial symmetry for Co at low temperatures (Ono and Yamaha 1979).

The result indicates a faster variation of the K_1 constant with temperature (to the 3rd power) than the variation of the spontaneous magnetization with temperature (to the 2nd or 3/2 power), depending on the material (Bloch 1930, Eschenfelder 1962)), and faster for the K_2 constant (to the 10th power) (Figure 18).

In crystals with localized magnetic moments, such as the ferrimagnetic, having in view the symmetry of crystal, the following law seems to be more appropriate (Zener 1954, Van Vleck 1959) for the anisotropy constant K_1.

$$K_n(T) = K_n(0)\left[\frac{M_s(T)}{M_s(0)}\right]^{\frac{n(n+1)}{2}}. \tag{46}$$

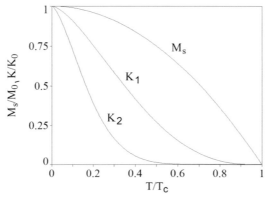

Figure 18. Reduced magnetisation (M_s/M_0, where M_0 is the spontaneous magnetisation at 0 K) and reduced anisotropy constant (K/K_0, where K_0 is the magnetocrystalline constant at 0 K and K is K_1 or K_2) as a function of reduced temperature (T/T_c).

A more general law of variation for $K_n(T)$, which does not depend on the type of symmetry of crystal, is found in Ref. (Vonsovski 1971)

$$K_n(T) = K_n(0)\left[1 - n(2n+1)\frac{\Delta M_s(T)}{M_s(0)}\right] \qquad (47)$$

where $\Delta M_s(T) = M_s(0) - M_s(T)$. The Eq. (47) is reduced approximately to Eq. (44) for $\Delta M_s(T) \ll M_s(0)$, or in the range of low temperatures.

In Figure 19 is shown the variation with temperature (T < 200 K) of the magnetic anisotropy constants obtained in the case of a nanocomposite consisting of nanoparticles of Fe_3O_4 (magnetite) embedded in a solid matrix of kerosene (kerosene frozen) (Hrianca et al. 2002), and in Figure 20 is shown the variation with temperature of the effective magnetic anisotropy for nanocomposites made of γ-Fe_2O_3 nanoparticles dispersed in SiO_2 solid matrix (Caizer and Hrianca 2003).

Due to the existence of surface anisotropy (K_S) (Figure 19d) and stress anisotropy (K_σ) (Figure 20) in nanocomposites, the effective magnetic anisotropy constants (K_{eff}) have the unusual variation with temperature. This behavior is generally a feature of nanostructures that contain ferrimagnetic nanoparticles.

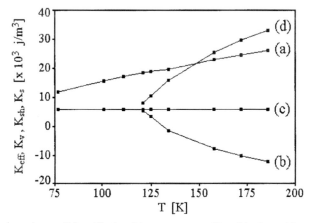

Figure 19. Temperature dependence of the effective (a), magnetocrystalline (b), shape (c), and surface (d) anisotropy constants (Reprinted from (Hrianca et al. 2002), with the permission of AIP Publishing).

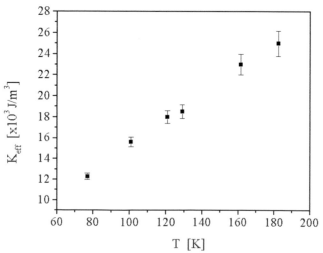

Figure 20. Effective anisotropy constant (K_{eff}) as a function of temperature (Reprinted from (Caizer and Hrianca 2003), © EDP Sciences, Societa Italiana di Fisica, Springer-Verlag 2003, with permission of Springer).

Magnetic anisotropies in nanocomposites containing magnetic nanoparticles dispersed in solid matrices

Some aspects regarding magnetic anisotropy of the nanocomposites

Knowing the magnetic anisotropy of the nanoparticle components of a composite, one can assess, by various techniques, the magnetic anisotropy of the composite. The assessment method will depend on the specific case of nanocomposite, as I showed in the previous section. In many cases, the representation of the magnetic anisotropy through the anisotropy field and its variation with different factors is more suggestive. The anisotropy field is a vector quantity, while the energy is a scalar quantity, and therefore the anisotropy field provides more complete information on anisotropy. Thus, given an $\overrightarrow{H}_a$ anisotropy field, acting on a certain direction in the crystal, different from the crystallographic ones, the magnetic anisotropy energy can be expressed by the formula (Chikazumi and Graham 1997, Cullity and Graham 2009)

$$\epsilon_{ma} = -\mu_0 (\overrightarrow{M}_s \cdot \overrightarrow{H}_{ma}), \tag{48}$$

and

$$\epsilon_{ma} = -\mu_0 H_{ma} M_s \cos\alpha, \tag{49}$$

respectively, where α is the angle between the spontaneous magnetization vector $\overrightarrow{M}_s$ and the direction followed by the magnetic anisotropy field $\overrightarrow{H}_{ma}$. For example, it is a well known case of the expression of the magnetocrystalline anisotropy field for a uniaxial symmetry, $\overrightarrow{H}_{k,u}$, using the uniaxial magnetocrystalline anisotropy energy $\epsilon_{k,u}$ (Eq. 11) when, usually, the spontaneous magnetization vector $\overrightarrow{M}_s$ is very close to the direction of easy magnetization (Cullity and Graham 2009)

$$H_{k,u} = \frac{2K_u}{\mu_0 M_s} \tag{50}$$

Next, it analyzes two cases of nanocomposite consisting of nanoparticles dispersed in solid matrices: *(i)* composite with nanoparticles having *aligned* uniaxial anisotropy axis, and *(ii)* composite with nanoparticles having *randomly* uniaxial anisotropy axis.

i) *The magnetic anisotropy in the case of composite with nanoparticles that have aligned axes of uniaxial magnetic anisotropy*: This is a case commonly found in composites when the nanoparticles are able to rotate under the action of a uniform external magnetic field, such as the case of magnetic nanofluids, or solid composites when during their preparation, prior orientation of the nanoparticles' magnetic moments in the field was obtained, similar to those in Figure 5b.

ii) *The magnetic anisotropy in the case of composite with nanoparticles having randomly uniaxial magnetic anisotropy axes:* In this case, since the uniaxial anisotropy axes of the nanoparticles are randomly similar to those in Figure 5a, it can use the theory described in Ref. (Caizer 2008) for a similar nanostructure.

In both cases, I consider the nanoparticles having uniaxial anisotropy a common situation in the case of nanoparticles, and in the absence of interactions between them, also homogeneously dispersed within the solid matrix with a relatively small packaging fraction Φ (generally below 10%) (Figure 21a and b).

In both cases, I consider the average diameter $<D>$ of nanoparticles resulting from a relatively narrow distribution. In this regard, the experimental results (Caizer et al. 2003) showed that in actual cases, there is often a lognormal distribution of the size of nanoparticle in the composite (Figure 21c).

In cases different than (*i*) and (*ii*), the problem is more complex and it should be treated as such, on a case by case basis, using a theory suitable for the specific conditions and the real structure of nanocomposites. For evaluating the magnetic anisotropy, the uniaxial anisotropy field or the uniaxial anisotropy constant of nanocomposite, I will use the ferromagnetic resonance (FMR), the accurate method currently used in research (see next section).

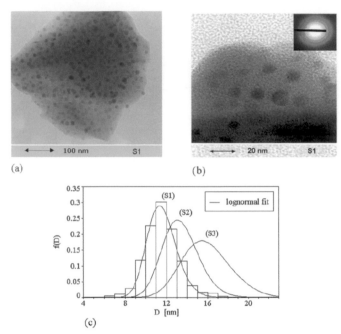

Figure 21. (a), (b) TEM images at two different scales; (c) Distribution of the nanoparticles' diameters (Reprinted from (Caizer et al. 2003), © 2003, with permission from Elsevier Science).

Determining accuracy of the uniaxial anisotropy constant using ferromagnetic resonance

In this section, I present the FMR in order to determining the accuracy of the uniaxial anisotropy constant of nanocomposites (Caizer 2008).[1]

Theory basis: In the case of the nanocomposite made up of noninteracting magnetic nanoparticles dispersed in a solid silica matrix (Figure 22), in the (effective) magnetic field $\vec{H}_e$, the (spontaneous) magnetization $\vec{M}$ of a single domain nanoparticle has a precession movement around it with the Larmor frequency (ω_L). This movement is described by the Landau-Lifshitz equation (Landau and Lifshitz 1960). In the absence of the damping, the precession movement is described by the equation

$$d\vec{M}/dt = \mu_0\gamma\,(\vec{M} \times \vec{H}_e),\tag{51}$$

where μ_0 is the magnetic permeability of vacuum, γ is the gyromagnetic ratio and $\vec{H}_e = \vec{H}_0 + \vec{H}_a + \vec{H}_d$.

Here it has to be considered that, besides the (external) static field $\vec{H}_0$, the magnetization of the nanoparticle is also aligned to the action of an anisotropy field $\vec{H}_a$ and a demagnetizing field $\vec{H}_d$ (due to the shape of the nanoparticle). If a low amplitude microwave field with the frequency $\omega_\sim$ is applied, having its' magnetic field component perpendicular to the applied field, the amplitude of the precession movement in the field perpendicular to $\vec{H}_e$ increases as a result of the energy absorption from the microwave field. The absorption reaches its maximum value at resonance (ferromagnetic resonance (FMR)) when the following condition is met $\omega_\sim = \omega_L = \omega_0$, where ω_0 is the frequency at resonance. If we approximate the nanoparticle to a ellipsoid, the main axes of which coincide with the axes of the tri-orthogonal coordinate system (Ox, Oy, Oz), and if $\vec{H}_0 = (0, 0, H_0)$, $\vec{H}_a = (H_{a,x}, H_{a,y}, 0)$,

$\vec{H}_d = (H_{d,x}, H_{d,y}, H_{d,z})$, then, after solving Eq. (1), it obtains the resonant frequency condition

$$\omega_0 = \mu_0\gamma\left[H_0 + H_{a,x} - \left(N_{d,z} - N_{d,x}\right)M\right]^{1/2} \cdot \left[H_0 + H_{a,y} - \left(N_{d,z} - N_{d,y}\right)M\right]^{1/2},\tag{52}$$

---

[1] With kind permission from Elsevier: Reprinted from (Caizer 2008), © 2007.

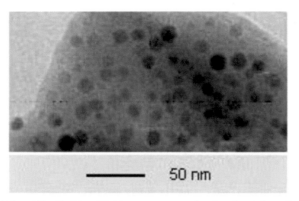

Figure 22. TEM image of $(Zn_{0.15}Ni_{0.85}Fe_2O_4)_{0.15}/(SiO_2)_{0.85}$ nanocomposite (Reprinted from (Caizer 2008), © 2007, with permission from Elsevier).

where $N_{d,x}$, $N_{d,y}$, $N_{d,z}$ are the demagnetizing factors along the directions Ox, Oy and Oz, respectively, and $H_{a,x}$, $H_{a,y}$ are the components of the anisotropy field along the directions Ox and Oy.

In the case of our sample, the nanoparticle considered as having a uniaxial anisotropy due to its spherical shape (Figure 1a), I can consider $N_{d,x} \cong N_{d,y} \cong N_{d,z}$.

Case (i): When the *uniaxial anisotropy field* $(\vec{H}_{au})$ *acts along the easy magnetization axis that is aligned along the direction of Oz* $(\vec{H}_{au} = (0, 0, H_{au,z}))$ *(direction in which the field* $\vec{H}_0$ *is applied)*, resonance is obtained when the external field is

$$H_{0r}(0°) = (\omega_0/\mu_0\gamma) - H_{au,z}. \tag{53}$$

However, if the anisotropy field is aligned perpendicularly to the direction of the external static field $(\vec{H}_{au} = (0, H_{au,y}, 0)$ or $\vec{H}_{au} = (H_{au,x}, 0, 0))$, the resonance will occur in the field

$$H_{0r}(90°) = (\omega_0/\mu_0\gamma) + H_{au,y}/2 \text{ (or } H_{au,x}). \tag{54}$$

By using the formula (53), one can accurately determine the uniaxial anisotropy field $H_{au,z}$, knowing the ω_0 pulsation (frequency $\nu_0 = \omega_0/2\pi$) at resonance (the observable that is known, and is indicated with precision by the microwave generator of FMR installation), and the value of the external uniform magnetic field H_{0r} where resonance is obtained (value precisely indicated by the magnetic field generator of FMR).

The uniaxial anisotropy constant, in this case, is determined using the Eqs. (53) and (50), resulting in

$$K_{u,i} = (M/2)[2\pi\nu_0/\gamma - B_{0r}]. \tag{55}$$

In this equation, and further on, I note the spontaneous magnetization (M_s) simply by M, and B_{0r} is the magnetic resonance induction, $B_{0r} = \mu_0 H_{0r}$, because, practically, its value is the one indicated at resonance (in T units). Experimentally, knowing the value of ν_0 and B_{0r} one can determine the value of the uniaxial magnetic anisotropy constant in this case, the other values in the formula (M and γ) being known.

Case (ii): For a *random angle θ between the easy magnetization axis of the nanoparticle and the static field* (Figure 23), in the absence of thermal fluctuations upon the Larmor precession, the following expression for the resonance field can be written (de Biasi and Devezas 1978, Raikher and Stepanov 1994):

$$H_{0r}(\theta) = (\omega_0/\mu_0\gamma) - H_{a,u}P_2(\cos\theta), \tag{56}$$

where $P_2(\cos\theta)$ represents the Legendre polynomials $P_n(x) = (1/2^n n!)(d^n/dx^n)(x^2-1)^n$ of the second order ($n = 2$ in the case of uniaxial symmetry and $x = \cos\theta$). This equation, which specifies the resonance field as a function of the angular distribution of the anisotropy axes, is in agreement with the empirical equation $H_{0r}(\theta) = H_{0r}(0°) + [H_{0r}(90°) - H_{0r}(0°)]\sin^2(\theta)$ which was determined by Gazeau et al. (Gazeau et al. 1998).

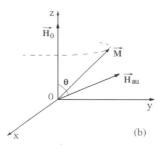

(b)

Figure 23. Spatial configuration of the magnetization ($\vec{M}$), the static magnetic field ($\vec{H}_0$) and the anisotropy magnetic field ($\vec{H}_{a,u}$) in the case of uniaxial anisotropy (Reprinted from (Caizer 2008), © 2007, with permission from Elsevier).

If I take into consideration the *angular distribution of the anisotropy axes for the entire system of randomly oriented single-domain nanoparticles* and then average the orientation of the easy magnetization axes of all the nanoparticles in relation to the anisotropy axis of a single particle, I can write

$$H_{0r} = \left(\omega_0 / \mu_0\gamma\right) - H_{a,u}\left\langle P_2\left(\cos\theta\right)\right\rangle \tag{57}$$

and

$$H_{0r} = \left(\omega_0 / \mu_0\gamma\right) - H_{a,u}\int_0^\pi \left[\left(3/2\right)\cos^2\theta - 1/2\right]d\theta . \tag{58}$$

From Eq. (58), after integration, I obtain

$$H_{0r} = \left(\omega_0 / \mu_0\gamma\right) - \pi K_u / \left(2\mu_0 M\right), \tag{59}$$

where the anisotropy field was assumed $H_{a,u} = 2K_u/(\mu_0 M)$ and K_u is the uniaxial anisotropy constant. Equation (59) can be used to determine, experimentally, the K_u anisotropy constant of the nanoparticle system with the anisotropy axes oriented randomly. Thus, considering the quantification of the orientation of the magnetic moments, I then obtain the following from Eq. (59)

$$K_{u,ii} = \left(2M / \pi\right)\left[2\pi\nu_0 / \gamma_1 - B_{0r}\right], \tag{60}$$

where $\gamma_1 = g\gamma$, g is the spectroscopic splitting factor, $\gamma = (1/2)(e/m_0) = 8.791 \times 10^{10}$ T^{-1}s^{-1} (e – electron charge, m_0 – electron mass), $\nu_0 = \omega_0/2\pi$ is the resonance frequency (of the microwave field) and $B_{0r} = \mu_0 H_{0r}$ is the value of the magnetic induction at resonance (this is changed externally by the experimenter).

Results and discussions: The FMR spectrum of the nanocomposite (the derivative of the absorption curve in relation to the field) is shown in Figure 24a. The shape of the spectrum tends to be asymmetrical, inherent to the randomly oriented system of uniaxially anisotropic nanoparticles (Morrison and Karayianis 1958, Schlomann and Zeender 1958). This suggests the presence of three resonance lines, with different linewidths and resonance fields. In order to distinguish the overlapping of the resonance lines, I have calculated the second derivative of the absorption curve. The corresponding curve is shown in Figure 24b.

The presence of the three lines can be seen very clearly here, namely: two intense and *very broad* lines ((l1) and (l2)), close to each other, obtained at higher resonance fields (due to their broadness, they cannot be seen separately) and a third line (l3), that is very weak and narrow, compared to the other two, which is obtained at a resonance field of ~ 0.12 T, appreciably lower than in the case of the intense lines. Furthermore, it can be seen that, for all the lines, the value of the magnetic induction at resonance is lower than the one obtained for free electrons

$$B_{0r,e} = \omega_{0r,e} / \gamma_1 = 0.3236 \text{ T}, \tag{61}$$

where $g = 2$.

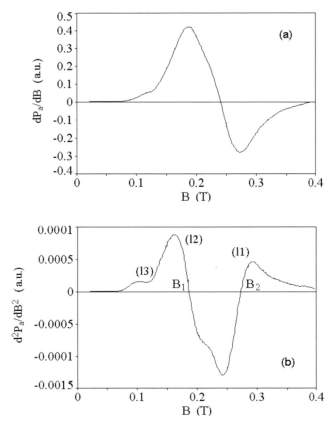

Figure 24. FMR spectra of the nanocomposite (Zn0.15Ni0.85Fe2O4)0.15/(SiO2)0.85 at room temperature (microwave frequency v_0 = 9.060 GHz); (a) first derivative and (b) second derivative of the absorption curve at resonance (Reprinted from (Caizer 2008), © 2007, with permission from Elsevier).

However, since the nanocomposite displays a distribution of the nanoparticles' diameters (Figure 21 and Figure 22), line (l1) can be attributed to *small* nanoparticles where there can be thermal fluctuations of the magnetization, and line (l2) is attributed to *large* particles where the magnetic moments are blocked (the magnetization is stable). The increase of nanoparticles volume leads to the increase of the barrier energy for the magnetic moments as a consequence of the anisotropy energy increase. The result will be the shift through smaller values of the resonance field, in the case of those nanoparticles. The result obtained is in agreement with the theoretical and experimental studies conducted by Gazeau et al. (Gazeau et al. 1999), who have shown that the value of the resonance field decreases and the width of the line increases when the diameter of the nanoparticles (γ-Fe$_2$O$_3$) increases from 4.8 nm to 10 nm.

The resonance line (l3) present at low fields can be attributed to the magnetic ions complex that has a disordered structure (spin canting) at the *surface* of the nanoparticles, in agreement with Ref. (Cannas et al. 1998). Thus, the FMR spectrum can constitute the proof for the existence of the layer on the surface of the nanoparticles in the SiO$_2$ matrix.

Using Eq. (60), where $g \sim 2$ (for nanoparticle systems (Dormann et al. 1997)), v_0 = 9.060 GHz and $M = 338 \times 10^3$ A/m (Smit and Wijin 1961, Valenzuela 1994) I have determined the value of the effective anisotropy constant

$$K_{u,ii} = 2.05 \times 10^4 \text{ Jm}^{-3}. \tag{62}$$

The magnetic induction at resonance, B_{0r} = 0.2285 T, was assumed to be the mean value between the values of the magnetic induction B_1 and B_2 (values that correspond to the intersection points of the second derivative with the abscissa) (Figure 24b). I must also specify that in order to determine the anisotropy constant, I have started from Eq. (57), neglecting the fluctuations on the Larmor precession (Gazeau 1999).

The value of the anisotropy constant that I have determined ($2.05 \times 10^4 \, Jm^{-3}$) was much higher than the magnetocrystalline anisotropy constant ($K_v = 5.5 \times 10^3 \, Jm^{-3}$) (Van Groenou et al. 1967). A value of the magnetic anisotropy constant ($1.8 \times 10^4 \, Jm^{-3}$), close to the one I have determined, was also obtained by Hendricksen et al. (Hendriksen et al. 1994). I have obtained similar high values of the anisotropy constant (in terms of order of magnitude) for γ–Fe_2O_3 nanoparticles dispersed in an amorphous silica matrix. This shows that the increased anisotropy does not depend on the nature of the material, but on other factors that are connected to the existence of the silica matrix (or that of the surfactant) around the nanoparticles and to the low diameter of the particles. These issues will be discussed below.

Since $K_{u,ii} \gg K_v$, it is concluded that in the case of $Zn_{0.15}Ni_{0.85}Fe_2O_4$ ferrite nanoparticles in a SiO_2 matrix, we must also consider forms of anisotropy other than the magnetocrystalline anisotropy, such as the anisotropy resulting from the *shape* of the nanoparticles. Moreover, the surface spins have a different structure from that in the core of the nanoparticle, which means that there can be a *surface magnetic anisotropy*.

Considering all the factors described above, the anisotropy energy can be written as

$$W_{aB} = W_v + W_{sh} + W_s , \qquad (63)$$

where W_v is the magnetocrystalline anisotropy energy, W_{sh} is the anisotropy energy determined by the shape of nanoparticle and W_s is the surface anisotropy energy.

However, if we analyze the electron microscopy image of the sample (Figure 22), we notice that the shape of the nanoparticles is spherical, so I can assume

$$W_{sh} \cong 0, \qquad (64)$$

because $K_{sh} \cong 0$ (in agreement with Ref. (Kneller 1962)), where K_{sh} is the anisotropy constant determined by the shape of the nanoparticle.

Considering the resonance field of ~ 0.12 T, which correspond to the line 13 from the FMR spectrum, an estimation of the anisotropy constant can be made; the result is the value of

$$K_s = 4.38 \times 10^4 \, Jm^{-3}. \qquad (65)$$

Transforming this result in Jm^{-2}, accordingly to the Ref. (Papusoi Jr. 1999), $K_s = K_s' 6 / \langle D_m \rangle$, the surface anisotropy constant $K_s' = 6.5 \times 10^{-5} \, Jm^{-3}$, is obtained. This result demonstrates that the most significant contribution for the total magnetic anisotropy belongs to the surface anisotropy. This is a result of the peculiar state of the magnetic moments on the surface of the nanoparticles compared to those in the core because the symmetry of the connections to their immediate neighbours is different, or a result of the ambiance of the spins on the surface. Certainly, the causes can be much more complex, such as the existence of interactions between the electron distribution and the crystalline electrostatic field, the spin-orbit coupling, etc.

In conclusion, in agreement with the facts described above and since $K_v \ll K_s$, from Eq. (63), I obtained $W_{aB} \approx K_s \langle V_m \rangle$. Expressing the anisotropy energy as an effective anisotropy constant multiplied by volume ($W_{aB} \approx K_{eff} \langle V_m \rangle$) results in

$$K_{eff} \approx K_s , \qquad (66)$$

which means that the high value of the (effective) anisotropy constant that I have found ($4.38 \times 10^4 \, Jm^{-3}$) is caused by the surface magnetic anisotropy.

However, the anisotropy constant value previously determined ($2.05 \times 10^4 \, Jm^{-3}$), using the anisotropy field B_{0r}, shows that there is another anisotropy component, besides the magnetocrystalline one ($5.5 \times 10^3 \, Jm^{-3}$), which contributes to the increase of the total anisotropy. Because the nanoparticles are embedded in the solid silica matrix, this component may be due to the *stress* made by *the matrix over the nanoparticles*; this stress may exist after the process of nanocomposites preparation.

Finally, taking into account these two components, the total (effective) anisotropy energy may be written as a formula

$$W_{aB,t} \approx W_s + W_\sigma , \tag{67}$$

where W_σ is the *anisotropy energy* due to *the stress*.

The relation must be considered referring to the observation that the surface anisotropy has the most significant contribution for the total magnetic anisotropy of the nanoparticles, the constant K_s being approximately three times bigger than K_σ.

The influence of magnetic anisotropy of nanocomposites consisting of magnetic nanoparticles, on potential applications in nano- and biotechnology

The magnetic anisotropy of magnetic nanocomposites made of magnetic nanoparticles dispersed in various matrices solid has a great influence on the properties and magnetic behavior. Consequently, knowing them is very important for different applications or potential applications of nanocomposites in engineering, nanotechnology or bionanotechnology. Depending on the strength of the magnetic anisotropy, the magnetic nanocomposites can cover a very wide range, from *soft* magnetic to *hard* magnetic. Thus, it could be used from applications in the electromagnetic waves field (RF) (Raj et al. 2013) to the permanent magnets (Antoniak et al. 2011) or nano-size magnetic memories (Garcia-Sancheza et al. 2006, Victora and Shen 2005) with fast magnetic switching, or in the biomedical field (Trindade and da Silva 2011, Summers 2013, Peng et al. 2015, Iordanskii et al. 2016, Martín-Saavedra et al. 2010), as nanomaterials for medical imaging contrast agents (Peng et al. 2015), targeted drugs (Iordanskii et al. 2016), magnetic hyperthermia (Martín-Saavedra et al. 2010) or even in the field of food (Manoochehri et al. 2015).

An important observable of magnetic nanocomposites made of small magnetic single domain nanoparticles, with volume V_{NP}, is the energy barrier (W_b) for the magnetic moments of nanoparticles $(m_{N,P})$, determined by magnetic anisotropy,

$$W_b = \mathcal{E}_{ma} V_{NP}(\text{in J}), \tag{68}$$

where $\mathcal{E}_{ma}$ is the density of magnetic anisotropy energy. The magnetic anisotropy may be caused by one or more of anisotropy components listed in previous sections (*magnetocrystalline anisotropy* $(\mathcal{E}_k)$, *surface anisotropy* $(\mathcal{E}_s)$, *shape anisotropy* $(\mathcal{E}_{sh})$, *stress anisotropy* $(\mathcal{E}_\sigma)$, *unidirectional anisotropy* $(\mathcal{E}_{ua})$, and *induced anisotropy* $(\mathcal{E}_i)$), or anisotropy of all components simultaneously, depending on the specific case of magnetic nanocomposite.

For example, in the case of nanoparticles with *uniaxial anisotropy*, where the anisotropy constant is K_u and the uniaxial anisotropy energy is given by Eq. (18) $(\mathcal{E}_{k,u}(\varphi) = K_u \sin^2 \varphi)$, the nanoparticles energy (W) has the variation in Figure 25a, depending on the angle φ between the direction of the magnetic moment of the nanoparticle $(\vec{m}_{NP})$ and the easy magnetization axis (e.m.a.) (Figure 25b).

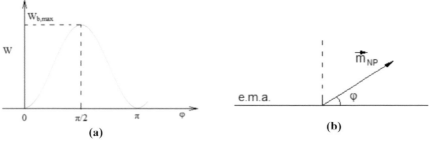

Figure 25. (a) The variation of energy with angle and (b) the orientation of the magnetic moment of nanoparticles relative to the e.m.a.

In this case, a *maximum* energy shows,

$$W_{max} = K_u V_{NP},$$

(69)

for $\varphi = \pi/2$ (for the direction perpendicular on e.m.a.), and a *minimum* (0 in this case),

$$W_{min} = 0,$$

(70)

for $\varphi = 0, \pi$.

In this case, depending on the value of the energy barrier,

$$W_{b,u} = W_{max} - W_{min} = k_u V_{NP}$$

(71)

or the uniaxial magnetic anisotropy energy compared with the thermal energy,

$$W_T = k_B T,$$

(72)

$W_{b,u} >/>> W_T$, $W_{b,u} \sim W_T$, $W_{b,u} </<< W_T$, where k_B is the Boltzmann's constant and T is temperature, the magnetic moments of the nanoparticles in composite (in the absence of an external magnetic field) can be blocked ($W_{b,u} >/>> W_T$) or not, can relax ($W_{b,u} \sim/< W_T$) or fluctuate along the easy magnetization axis ($W_{b,u} << W_T$) (Néel 1949). For magnetic relaxation, the relaxation time of magnetic moments (Néel) (Néel 1949) exponentially depends on the energy barrier value, determined by the magnetic anisotropy of nanoparticles in composite,

$$\tau = \tau_0 exp\left(\frac{W_{b,u}}{k_B T}\right)$$

(73)

where τ_0 is a time constant which usually has a value of 10^{-9} s (Back et al. 1998). This example shows the influence of strong magnetic anisotropy on the magnetic relaxation process.

When nanocomposites are *magnetized in an external magnetic field*, there may be magnetic behaviors totally different thereof, depending on the value of the energy barrier caused by the magnetic anisotropy (Caizer et al. 2003, Caizer 2005, Caizer and Tura 2006) and the volume of nanoparticles, the uniaxial anisotropy axes orientation of magnetic nanoparticles (aligned in one direction or oriented in all directions (Figure 5)) relative to the direction of magnetic field action. Thus, there may be magnetic behaviors (Figure 26): from a superparamagnetic behavior (Jacobs and Bean 1963), without hysteresis

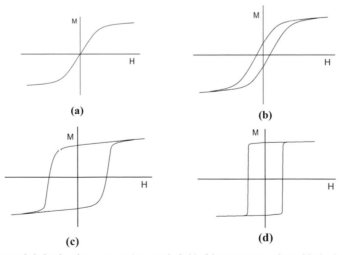

Figure 26. Various magnetic behaviors in an external magnetic field of the nanocomposites with single domain nanoparticles, depending on the anisotropy energy value compared to the thermal one and different orientations of the easy magnetization axes of nanoparticles relative to the field: (a) superparamagnetic, (b) with narrow hysteresis loop, (c) nanomagnet type or (d) magnetic nano-size memory.

loop (case corresponding to the magnetic relaxation) (a), to a behavior with a narrow hysteresis (b) or, under certain conditions, with a very large hysteresis loop (with high remanence and coercivity (hard magnet)) (c), or rectangular, with remanence close to the magnetic saturation and small coercivity (switching nano-size memory) (d).

In Figure 27 are shown the experimental curves of the magnetization of nanocomposites formed of $Ni_{1-x}Zn_xFe_2O_4$ ($x = 0.15$) and $CoFe_2O_4$ nanoparticles, with a mean diameter of ~12 nm, dispersed in silica amorphous matrix (Caizer et al. 2003, Caizer and Tura 2006), magnetized in the same external field conditions, where different magnetic anisotropy leads to different magnetic behaviors of the nanocomposites.

Having in view all these magnetic behaviors of the nanocomposites and their feature magnetic observable values (saturation magnetization, remanence magnetization, coercive field, magnetic susceptibility, etc.), and the pursued practical purpose, one can find multiple potential applications of the magnetic nanocomposites made of nanoparticles embedded in different crystalline or amorphous solid matrices.

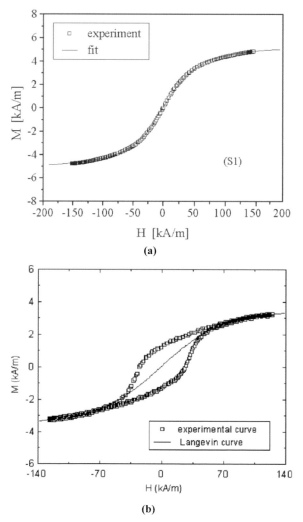

Figure 27. (a) M versus H for sample S1 (S1: nanocomposite with $Zn_{0.15}Ni_{0.65}Fe_2O_4$ nanoparticles in amorphous SiO_2 matrix) (Reprinted from (Caizer et al. 2003), © 2003, with permission from Elsevier Science); (b) Hysteresis loop of an assembly of Co ferrite nanoparticles embedded in amorphous SiO_2 particles, recorded at room temperature under a 50 Hz frequency alternating magnetic field (Reprinted from (Caizer and Tura 2006), © 2005, with permission from Elsevier).

Conclusions

The nanocomposites made of magnetic nanoparticles, ferro- or ferrimagnetic, is a highly topical field when it comes to nanomaterials, due to the multiple and potential applications in nano- and biotechnology. In this regard, the study of the physical (mechanical, electrical, magnetic, optical, etc.), and chemical properties is of great scientific interest. An important component of the magnetic properties of the *magnetic nanocomposites* is the *magnetic anisotropy*, due to their magnetic nanoparticles. Knowledge and accurate experimental determination of the magnetic anisotropy, with its components, on a case by case basis (magnetocrystalline anisotropy, surface anisotropy, shape anisotropy, stress anisotropy, unidirectional anisotropy, uniaxial anisotropy, or other induced anisotropies in designing the nanocomposites) is of major interest in order to use their full potential in nanotechnology, or in future applications. In this chapter, the physical aspects of magnetic anisotropy were presented, with reference to its components that may exist in the case of nanocomposites made of nanoparticles, exemplifying where possible with results, previous research results or experiments. Knowing the magnetic anisotropy of nanocomposites becomes vital because the magnetic anisotropy determines a certain magnetic behavior of the nanocomposites when they are magnetized in an external field, which can be from a *superparamagnetic* one to a *soft* and *hard* magnetic. In this respect, it is very important to accurately determine the magnetic anisotropy constants. Thus, the ferromagnetic resonance method was presented in the particular case of the $(Zn_{0.15}Ni_{1-0.85}Fe_2O_4)_{0.15}$-$(SiO_2)_{0.85}$, nanocomposite, having ferrimagnetic nanoparticles of Ni-Zn ferrite embedded in the amorphous silica matrix.

References

Akulov, N.S. 1928. Uber die magnetostriktion der eisenkristalle. Z. Phys. 52: 389–405.

Akulov, N.S. 1936. The quantum theory of the temperature dependence of the magnetization curve. Z. Phys. 100: 197–204.

Anderson, P.W. 1950. Generalizations of the Weiss molecular field theory of antiferromagnetism. Phys. Rev. 79: 705–710.

Antonel, P.S., G. Jorge, O.E. Perez, A. Butera, A.G. Leyva and R.M. Negri. 2011. Magnetic and elastic properties of $CoFe_2O_4$-polydimethylsiloxane magnetically oriented elastomer nanocomposites. J. Appl. Phys. 110: 043920-1–043920-8.

Antoniak, C., M.E. Gruner, M. Spasova, A.V. Trunova, F.M. Römer, A. Warland et al. 2011. A guideline for atomistic design and understanding of ultrahard nanomagnets. Nature Communications 2: 528.

Aubert, G. 1968. Torque measurements of the anisotropy of energy and magnetization of nickel. J. Appl. Phys. 39: 504.

Back, C.H., D. Weller, J. Heidmann, D. Mauri, D. Guarisco, E.L. Garwin et al. 1998. Magnetization reversal in ultrashort magnetic field pulses. Phys. Rev. Lett. 81: 3251–3254.

Bean, C.P. and L.D. Livingston. 1959. Superparamagnetism. J. Appl. Phys. 30: S120–S129.

Becker, R. and W. Döring. 1939. Ferromagnetismus. Springer-Verlag, Berlin.

Berkowitz, A.E. and R.H. Kodama. 2006. Exchange anisotropy. pp. 115−152. *In*: D.L. Mills and J.A.C. Bland (eds.). Contemporary Concepts of Condensed Matter Science, Volume 1. Nanomagnetism: Ultrathin Films, Multilayers and Nanostructures. Elesevier Science.

Bloch, F. 1930. Zur Theorie des Ferromagnetismus. Z. Phys. 61: 206–2019.

Bødker, F., S. Mørup and S. Linderoth. 1994. Surface effects in metallic iron nanoparticles. Phys. Rev. Lett. 72: 282–285.

Bozorth, R.M. 1954. Magnetostriction and crystal anisotropy of single crystals of hexagonal cobalt. Phys. Rev. 96: 311–316.

Brown Jr., W.F. and A.H. Morrish. 1957. Effect of a cavity on a single-domain magnetic particle. Phys. Rev. 105: 1198–1201.

Caizer, C. and I. Hrianca. 2003. Dynamic magnetization of γ-Fe_2O_3 nanoparticles isolated in an SiO_2 amorphous matrix. Eur. Phys. J. B 31: 391–400.

Caizer, C., M. Popovici and C. Savii. 2003. Spherical $(Zn_\delta Ni_{1-\delta}Fe_2O_4)_\gamma$ nanoparticles in an amorphous $(SiO_2)_{1-\gamma}$ matrix, prepared with the sol-gel method. Acta Mater. 51: 3607–3616.

Caizer, C. 2004a. Systems of Dispersed Ferrimagnetic Nanoparticles: Magnetic Behavior (in romanian). West University, Timisoara.

Caizer, C. 2004b. Magnetic Nanofluids (in romanian). Eurobit, Timisoara.

Caizer, C. 2005. The effect of external magnetic field on the thermal relaxation of magnetization. J. Phys.: Condens. Mater. 17: 2019–2034.

Caizer, C. and V. Tura. 2006. Magnetic relaxation/stability of Co ferrite nanoparticles embedded in amorphous silica particles. J. Magn. Magn. Mater. 301: 513–520.

Caizer, C. 2008. Magnetic properties of the novel nanocomposite $(Zn_{0.15}Ni_{0.85}Fe_2O_4)_{0.15}/(SiO_2)_{0.85}$ at room temperature. J. Magn. Magn. Mater. 320: 1056–1062.

Caizer, C. 2016. Nanoparticles size effect on some magnetic properties. pp. 475–519. *In*: M. Aliofkhazraei (ed.). Handbook of Nanoparticles. Springer.

Cannas, C., D. Gatteschi, A. Musinu, G. Piccaluga and C. Sangregorio. 1998. Structural and magnetic properties of Fe_2O_3 nanoparticles dispersed over a silica matrix. J. Phys. Chem. B 102: 7721–7726.

Carr, W.J. 1960. Hanbook of Physics XVIII: 274.

Chen, L., C.H. Zhou, S. Fiore, D.S. Tong, H. Zhang, C.S. Li et al. 2016. Functional magnetic nanoparticle/clay mineral nanocomposites: preparation, magnetism and versatile applications. Applied Clay Science 127–128: 143–163.

Chikazumi, S. 1964. Physics of Magnetism, 2nd Edition. John Wiley, New York.

Chikazumi, S. and C.D. Graham. 1997. Physics of Ferromagnetism, 2nd Edition. Oxford University Press, Oxford.

Clarck, A.E. 1978. Magnetostrictive rare earth–Fe_2 compounds. *In*: E.P. Wohlfarth (ed.). Handbook of Ferromagnetic Materials. North Holland Publishing Comp.

Coey, J.M.D. and D. Khalafalla. 1972. Superparamagnetic γ-Fe_2O_3. Phys. Stat. Solidi (a) 11: 229–241.

Cullity, B.D. and C.D. Graham. 2009. Introduction to Magnetic Materials. Second Edition, John Wiley & Sons Inc., NJ.

de Biasi, R.S. and T.C. Devezas. 1978. Anisotropy field of small magnetic particles as measured by resonance. J. Appl. Phys. 49: 2466–2469.

Dirac, P.M.A. 1926. On the theory of quantum mechanics. Proc. Roy. Soc. A 112: 661–677.

Dormann, J.L., D. Fiorani and E. Tronc. 1997. Magnetic relaxation in fine-particle systems. pp. 283–494. *In*: I. Prigogine and S.A. Rice (eds.). Advances in Chemical Physics. Volume 98. John Wiley & Sons.

Eschenfelder, A.H. 1962. Magnetic properties I. *In*: Landolt-Börnstein. Springer-Verlag, Berlin.

Furcy, W.N. 1967. Thesis Harvard University, Cambridge (1967). Bull. Am. Phys. Soc. 12: 311.

Garcia-Sancheza, F., O. Chubykalo-Fesenko, O.N. Mryasov and R.W. Chantrell. 2006. Multiscale models of hard-soft composite media. J. Magn. Magn. Mater. 303: 282–286.

Gazeau, F., J.C. Bacri, F. Gendron, R. Perzynski, Yu.L. Raikher, V.I. Stepanov et al. 1998. Magnetic resonance of ferrite nanoparticles: evidence of surface effects. J. Magn. Magn. Mater. 186: 175–187.

Gazeau, F., V. Shilov, J.C. Bacri, E. Dubois, F. Gendron, R. Perzynski et al. 1999. Magnetic resonance of nanoparticles in a ferrofluid: evidence of thermofluctuational effects. J. Magn. Magn. Mater. 202: 535–546.

Graham Jr., C.D. 1960. Temperature dependence of anisotropy and saturation magnetization in iron and iron-silicon alloys. J. Appl. Phys. 31: 150S.

Heisenberg, W. 1926. Mehrkörperproblem und resonanz in der quantenmechanik. Z. Phys. 38: 441–426.

Heisenberg, W. 1928. On the theorie of ferromagnetism (translated in English by D.H. Delphenich). Z. Phys. 49: 619–636.

Hendriksen, P.V., F. Bödker, S. Linderoth, S. Wells and S. Mörup. 1994. Ultrafine maghemite particles. I. Studies of induced magnetic texture. J. Phys.: Condens. Matter 6: 3081–3090.

Herpin, A. 1968. Théorie du Magnétisme. Univ. France, Paris.

Honda, K. and S. Kaya. 1926. On the magnetization of single crystals of iron. Sci. Rep. Tohoku Univ. 15(1926): 721–753.

Hrianca, I., C. Caizer and Z. Schlett. 2002. Dynamic magnetic behavior of Fe_3O_4 colloidal nanoparticles. J. Appl. Phys. 92: 2125–2132.

Iordanskii, A.L., A.V. Bychkova, K.Z. Gumargalieva and A.A. Berlin. 2016. Magnetoanisotropic biodegradable nanocomposites for controlled drug realease. pp. 171–196. *In*: A.M. Grumezescu (ed.). Nanobiomaterials in Drug Delivery: Applications of Nanobiomaterials. Volume 9. Elsevier.

Jacobs, I.S. and C.P. Bean. 1963. Fine particles, thin films and exchange anisotropy. pp. 271–350. *In*: G.T. Rado and H. Suhl (eds.). Magnetism. Volume 3. Academic Press, New York.

Jamet, M., W. Wernsdorfer, Ch. Thirion, V. Dupuis, P. Mélinon, A. Pérez et al. 2004. Magnetic anisotropy in single clusters. Phys. Rev. B 69: 024401.

Kaya, S. 1928a. On the magnetization of single crystals of cobalt. Sci. Rep. Tohoku Univ. 17: 639–663.

Kaya, S. 1928b. On the magnetization of single crystals of nickel. Sci. Rep. Tohoku Univ. 17: 1157–1177.

Kneller, E. 1962. Ferromagnetismus. Springer-Verlag, Berlin.

Landau, L.D. and E.M. Lifshitz. 1960. Electrodynamics of Continuous Media. Pergamon Press, Oxford.

Manoochehri, M., A.A. Asgharinezhad and N. Shekari. 2015. Synthesis, characterisation and analytical application of $Fe_3O_4@SiO_2$@polyaminoquinoline magnetic nanocomposite for the extraction and pre-concentration of Cd(II) and Pb(II) in food samples. Food Additives & Contaminants: Part A. 32: 737–747.

Martín-Saavedra, F.M., E. Ruíz-Hernández, A. Boré, D. Arcos, M. Vallet-Regí and N. Vilaboa. 2010. Magnetic mesoporous silica spheres for hyperthermia therapy. Acta Biomaterialia 6: 4522–4531.

Morrison, C.A. and N. Karayianis. 1958. Ferromagnetic resonance in uniaxial polycrystalline materials. J. Appl. Phys. 29: 339–340.

Mazo-Zuluaga, J., J. Restrepo and J. Mejía-López. 2008. Effect of surface anisotropy on the magnetic properties of magnetite nanoparticles: A Heisenberg-Monte Carlo study. J. Appl. Phys. 103: 113906-113906-8.

Meiklejohn, W.H. and C.P. Bean. 1956. New magnetic anisotropy. Phys. Rev. 102: 1413–1414.

Meiklejohn, W.H. and C.P. Bean. 1957. New magnetic anisotropy. Phys. Rev. 105: 904–913.

Meiklejohn, W.H. 1962. Exchange anisotropy—a review. J. Appl. Phys. 33: 1328–1335.

Mørup, S. 1983. Magnetic hyperfine splitting in mössbauer spectra of microcrystals. J. Magn. Magn. Mater. 37: 39–50.

Néel, L. 1948. Magnetic properties of ferrites: ferrimagnetism and antiferromagnetism (transleted). Ann. Phys. 3: 137–98.

Néel, L. 1949. Théorie du traînage magnétique des ferromagnétiques en grains fins avec application aux terres cuites. Ann. Geophys. 5: 99–136.

Neel, L. 1953. Les structures d'orientation. Compt. Rend. 237: 1613–1616

Neel, L. 1954. Anisotropie magnétique superficielle et sur structures d'orientation. J. Phys. Rad. 15: 225–239.

Nogues, J. and I.K. Schuller. 1999. Exchange bias. J. Magn. Magn. Mater. 192: 203–232.

Nogues, J., J. Sort, V. Langlais, V. Skumryev, S. Surinach, J.S. Munoz et al. 2005. Exchange bias in nanostructures. Phys. Rep. 422: 65–117.

Ono, F. and O. Yamaha. 1979. Temperature dependence of the uniaxial magneto-crystalline anisotropy energy of Co. J. Phys. Soc. Jpn. 46: 462–467.

Papusoi, Jr., C. 1999. The particle interaction effects in the field-cooled and zero-field-cooled magnetization processes. J. Magn. Magn. Mater. 195(1999): 708–732.

Peddis, D., S. Laureti, M.V. Mansilla, E. Agostinelli, G. Varvaro, C. Cannas et al. 2009. Exchange Bias in $CoFe_2O_4$/NiO nanocomposites. Superlattices and Microstructures 46: 125–219.

Peng, E., F. Wang and J.M. Xue. 2015. Nanostructured magnetic nanocomposites as MRI contrast agents. J. Mater. Chem B. 3: 2241–2276.

Rado, G.T. and H. Suhl. 1973. Magnetism. Academic Press, New York – London.

Raikher, Yu.L. and V.I. Stepanov. 1994. Ferromagnetic resonance in a suspension of single-domain particles. Phys. Rev. B 50: 6250–6259.

Rodbell, D.S. 1965. Magnetic resonance of high-quality ferromagnetic metal single crystals. Physics 1: 279–305.

Raj, P.M., H. Sharma, S. Samtani, D. Mishra, V. Nair and R. Tummala. 2013. Magnetic losses in metal nanoparticle-insulator nanocomposites. J. Mater. Sci.: Mater. Electron. 24: 3448–3455.

Restrepo, J., Y. Labaye and J.M. Greneche. 2006. Surface anisotropy in maghemite nanoparticles. Physica B: Condensed Matter 384: 221–223.

Ridi, F., M. Bonini and P. Baglioni. 2014. Magneto-responsive nanocomposites: preparation and integration of magnetic nanoparticles into films, capsules, and gels. Adv. Colloid Interface Sci. 207: 3–13.

Rinkevich, A.B., A.V. Korolev, M.I. Samoylovich, S.M. Klescheva and D.V. Perov. 2016. Magnetic properties of nanocomposites based on opal matrices with embedded ferrite-spinel nanoparticles. J. Magn. Magn. Mater. 399: 216–220.

Rosencwaig, A., W.J. Tabor, F.B. Hagedorn and L.C. Van Uitert. 1971. Noncubic magnetic anisotropies in flux-grown rare-earth iron garnets. Phys. Rev. Lett. 26: 775.

Sato, H and B.S. Chandrasekhar. 1957. Determination of the magnetic anisotropy constant K_2 of cubic ferromagnetic substances. J. Phys. Chem. Solids 1: 228–233.

Schlomann, E. and J.R. Zeender. 1958. Ferromagnetic resonance in polycrystalline nickel ferrite aluminate. J. Appl. Phys. 29: 341–343.

Skumryev, V., S. Stoyanov, Y. Zhang, G. Hadjipanayis, D. Givord and J. Nogués. 2003. Beating the superparamagnetic limit with exchange bias. Nature 423: 850–853.

Smit, J. and H.P.J. Wijin. 1961. Les ferites, Bibl. Tehn. Philips, Paris.

Summers, H. 2013. Nanomedicine. *In*: R.E. Palmer (ed.). Frontiers of Nanoscience, Volume 5. Elsevier.

Tartaj, P. 2009. Superparamagnetic composites: Magnetism with no memory. Eur. J. Inorg. Chem. 2009: 333–343.

Tatsumoto, E., T. Okamoto, N. Iwata and Y. Kadena. 1965. Temperature dependence of the magnetocrystalline anisotropy constants K_1 and K_2 of nickel. J. Phys. Soc. Jpn. 20: 1541–1542.

Thakur, M., M. Patra, S. Majumdar and S. Giri. 2009. Coexistence of superparamagnetic and superspin glass behaviors in $Co_{50}Ni_{50}$ nanoparticles embedded in the amorphous SiO_2 host. J. Appl. Phys. 105: 073905.

Tokunaga, T. 1974. Magnetocrystalline anisotropy of nickel and nickel-copper alloys. J. Sci. Hiroshima Univ. Ser. A 38: 215–237.

Trindade, T. and A.L.D. da Silva. 2011. Nanocomposite Particles for Bio-Applications: Materials and Bio-Interfaces. CRC Press, Taylor & Francis Group.

Valenzuela, R. 1994. Magnetic Ceramics. Cambridge Univ. Press, Cambridge.

Van Groenou, A.B., J.A. Schulkes and D.A. Annis. 1967. Magnetic anisotropy of some nickel zinc ferrite crystals. J. Appl. Phys. 38: 1133–1134.

Van Vleck, J.H. 1951. Recent developments in the theory of antiferromagnetism. J. Phys. Rad. 12: 262–274.

Vassiliou, J.K., V. Mehrotra, M.W. Russell and E.P. Giannelis. 1993. Magnetic and optical properties of γ-Fe_2O_3 nanocrystals. J. Appl. Phys. 73: 5109–5116.

Van Vleck, J.H. 1959. Some recent progress in the theory of magnetism for non-migratory models. J. Phys. Radium 20: 124–135.

Victora, R.H. and X. Shen. 2005. Composite media for perpendicular magnetic recording. IEEE Trans. Magn. 41: 537–542.

Vonsovski, S.V. 1971. Magnetism. Naika, Moscow.

Wang, G., Y. Ma, Z. Wei and M. Qi. 2016. Development of multifunctional cobalt ferrite/graphene oxide nanocomposites for magnetic resonance imaging and controlled drug delivery. Chem. Eng. J. 289: 150–160.

Zener, C. 1954. Classical theory of the temperature dependence of magnetic anisotropy energy. Phys. Rev. 96: 1335–7.

12

Biopolymer Layered Silicate Nanocomposites: Effect on Cloisite 10 A on Thermal, Mechanical and Optical Properties

Okan Akin, Hale Oguzlu, Onur Ozcalik and *Funda Tihminlioglu**

Introduction

This chapter is focused on the preparation and properties of biobased polymer nanocomposites composed of layered silicates as nanofiller. This review highlights the effect of nanofiller on mechanical, thermal and optical properties of the biodegradable polymers namely, poly(hydroxy-butyrate) (PHB), poly(hydroxyl-butyrate-co-hydroxy-valerate) (PHBHV), polylactic acid (PLA), Corn zein (CZ) and nanoclay (commercial organo-modified montmorillonite (OMMT-Cloisite 10A) systems.

In recent years, biopolymers have attracted growing interest due to waste problems and limited petroleum resources. Therefore, researchers from both academia and industry have focused on the development of renewable and environmental friendly biopolymeric materials as an alternative to petroleum based plastics. However, biopolymers are not cost effective and they suffer from insufficient properties in comparison to conventional petroleum-based plastics. Therefore, they have limited applications so far to date. Nano reinforcements of bio-based polymers have strong promise in production of environmental friendly nanocomposites with enhanced properties for various applications, especially in packaging and biomedical areas (Hule and Pochan 2007, Rhim et al. 2013).

Biodegradable polymer nanocomposites have attracted more attention since addition of only a few percent of nanoparticles to polymer matrix resulted in the tremendous enhancement in properties such as mechanical, thermal, barrier by lowering the cost variance compared to the traditional composites and biopolymers. In order to improve barrier properties of biodegradable polymer, layered silicates among the other nanofillers are preferred due to the high aspect ratio that has the potential to improve the barrier resistances of polymers as diffusion path in the polymer matrix are lengthened by the incorporation of layered silicates into the polymer (Cornwelle 2009). The key point to achieve enhanced properties in

İzmir Institute of Technology, Department of Chemical Engineering, Gulbahce Campus, Urla, İzmir, Turkey, 35430.
* Corresponding author: fundatihminlioglu@iyte.edu.tr

polymer nanocomposites is to properly disperse nanofillers in polymer matrix since nanofillers-polymer matrix interactions in nanocomposites are highly dependent on the dispersion levels and interface of polymer-filler matrix.

In most of the polymer nanocomposite studies, montmorillonite is the most commonly used layered silicate. Its dispersion in nanoscale dimensions (1–100 nm) in polymer matrix enhances the properties due to strong interaction in atomic level. Therefore, not only barrier properties but other properties such as mechanical and thermal properties are improved as well. Since the nanocomposite structure depends on the polymer-layered silicate compatibility and on the processing conditions, the enhancement in properties is highly dependent on the dispersion of layered silicates in polymer matrix. The dispersion of layered silicates is mainly defined as intercalated or exfoliated structures. Exfoliated structure is the desired one where interfacial interaction is achieved in nanometer scale. Structure of layered silicates in polymer matrix depends on the nature of the polymer (molecular weight, polarity, etc.) and layered silicate (organomodification, etc.). Besides, preparation method of polymer nanocomposites is also a key parameter that affects dispersion level of layered silicates in polymer matrix. There are various methods that have been studied to prepare polymer layered silicate nanocomposite films in literature. *In situ* polymerization (Nguyen and Baird 2006), melt extrusion (Maiti et al. 2007), solvent casting are the common (Gunaratne and Shanks 2005) methods in preparation of layered silicate nanocomposite polymer films (LSNP) (Botana et al. 2010). The first part of this chapter involves the preparation and characterization methods of biodegradable polymer-layered silicate nanocomposites in detail. In the second part, mechanical, thermal and optical properties of biodegradable polymer systems based on namely, poly(hydroxy-butyrate) (PHB), poly(hydroxyl-butyrate-co-hydroxy-valerate) (PHBHV), polylactic acid (PLA), Corn zein (CZ)–Cloisite 10A, layered silicate systems are reported in detail.

Layered silicates (montmorillonite) and nanocomposite structure

Layered silicates (LS) consist of layers made up two tetrahedral coordinated silicon atoms fused to an edge-shared octahedral sheet of either aluminum or magnesium hydroxide (Ray and Bousmina 2005a). The thickness of the layers of silicates is about 1 nm and its lateral dimensions vary from 50 nm to several microns depending on the type of the LS. In nature, layered silicates are formed by the volcanic eruptions and they exist as stacks of the layers that lead to regular van der Waals gaps between the layers. These gaps are called as interlayer galleries. Between these galleries, exchangeable cations such as Na^+ and K^+ are located. As the pristine form of layered silicates' surface composed of negative charges, montmorillonites are only miscible with hydrophilic polymers such as poly(vinyl alcohol), and poly(ethylene oxide) (Ray and Bousmina 2005b). In order to improve the miscibility of hydrophilic silicate layers with hydrophobic polymers, the surface of the layers is modified by using exchanged cations such as ammonium to organophilic that enables intercalation of layered silicates within hydrophobic polymers. The modification of LS surfaces with organic modifiers is called as organomodification. The modification takes place by ion exchange reactions with cationic surfactant, including primary, secondary, tertiary and quaternary alkylammonium or alkyphosponium cations (Alexandre and Dubois 2000). Organomodification does not only improve compatibility of polymer-LS interface, but also leads to higher interlayer spacing (d-space) between interlayers for the better penetration of polymer chains within LS stacks. The change in the basal plane spacing of stacks via surface modification was studied by Xi and coworkers (Xi et al. 2004). The properties of commercially available organomodified montmorillonites are listed in Table 1.

Incorporation of layered silicates into polymer matrix can result in several structures due to the dispersion level of layers by the penetration of polymer chains into the galleries. The dispersion of layered silicates is categorized in three types; *phase separated, intercalated* and *exfoliated structure*. The importance of dispersion level of layers and its effect on polymer properties has been investigated widely in different kinds of polymer-filler systems. Strong interaction between stacked layers of silicates does not allow the polymer chains to penetrate into galleries that end up phase separated structure of filler. Moreover, phase separation is generally observed for the composites of unmodified or inadequate modified silicates with hydrophobic polymers (Ray and Okamoto 2003, Choudalakis and Gotsis 2009). In phase separation, polymer chains cannot penetrate the interlayer galleries that resulted in weak bond

Table 1. Properties of montmorillonite.

Clay type	Organic modifier	Cation exchange capacity (m_{eq}/100 g)	d-spacing (A)
Cloisite Na	-		11.7
Cloisite 10A	2MBHT	125	19.2
Cloisite 15A	2M2HT	125	31.5
Cloisite 20A	2M2HT	95	24.2
Cloisite 30B	MT2EtOH	90	18.5
Cloisite 93A	M2HT	90	23.6

2MBHT: dimethyl, benzyl, hydrogenated tallow, quaternary ammonium
2M2HT: dimethyl, dehydrogenated tallow, quaternary ammonium
2M2HT: dimethyl, dehydrogenated tallow, quaternary ammonium
MT2EtOH: methyl, tallow, bis-2-hydroxyethyl, quaternary ammonium

between polymer and filler surface and hence, the resulting structure have similar properties to that of microcomposites. In the case of intercalated structure, better penetration of polymer chains due to the stronger interactions leads to dispersion of layered silicates in the form of stacks. However, intercalated structure is not enough to degrade the original conformation of nanoclays and the complete dispersion of layers in nanoscale cannot be achieved due to the stacks of layers. The intended and ideal structure in nanocomposite is the exfoliated structure in which silicate layers dispersed randomly and disordered within the polymer matrix. The dispersion of layers highly depends on the interfacial forces that become dominant in surface modifications (e.g., organically modified montmorillonite). Strong interaction between polymer chain and the modified silicate surface resulted in high penetration into the galleries that end-up well dispersed layers within the polymer matrix. Exfoliation maximizes the interaction of polymer and filler that leads to enhanced properties in the nanocomposites. The first reported study is from Toyota research group of Nylon-6-Montmorillonite (MMT) nanocomposite, wherein addition of a very small amount of layered silicate resulted in improved thermal and mechanical properties. Numerous studies have shown that incorporation of layered silicates in small quantities enhanced the properties of many polymers (Sanchez-Garcia et al. 2008, Erceg et al. 2009, Cretois et al. 2014).

Preparation methods of polymer layered silicate (PLS) nanocomposites

As mentioned before, nanocomposite properties highly depend on the structure of fillers within the polymer matrix. The targeted structure for layered silicate is exfoliation of layers. In order to achieve the targeted structure, different kinds of methods have been used to prepare PLS nanocomposites. Most common methods used in PLC preparations are as follows:

- Solution intercalation polymer/prepolymer solutions
- Melt Intercalation
- *In situ* intercalative polymerization

Solution intercalation

Solution intercalation method is the basic and the easiest way to prepare nanocomposites that, firstly, swells the layered silicates in the solvent phase. By this way, solvent molecules have the ability to penetrate into layered silicate stacks and increase the d-spacing (Ray and Okamoto 2003). Interfacial interaction of solvent and silicate layers can weaken the forces between stack of the layers; hence, layers can disperse easily in an adequate solvent before polymer addition. Once silicate layers swelled within the solvent, individually dissolved polymer mixture in the same solvent is added. The swollen layers of silicates lead to increase in basal spacing that contributes to absorption of polymer chains into galleries, and when solvent is evaporated, the sheet reassembles, sandwiching the polymer to form an ordered multilayered structure. Compatibility of solvents and nanoclays affects modified layered silicate swelling and d-spacing within organic solvents depending on the type of organomodification agent used (Burgentzle et al. 2004). This method has been widely used in literature to prepare layered silicate polymer nanocomposites (Zulfiqar et al. 2008, Hwang et al. 2012, Jaafar et al. 2012).

Melt intercalation method

As it is the most commonly used method in industry, melt intercalation has been widely investigated in preparation of polymer nanocomposites. Melt processing of nanocomposites is generally held in twin-screw extruders, either in co-or counter rotating configurations. In the extrusion process, applied heat and shear force can help to alter the structure of silicate layers. The penetration of polymer chains into galleries is supplied by the Shear force and heat during mixing of polymer and filler. A mixture of polymer and layered silicates are annealed above the softening point of polymer to achieve a uniform dispersion of filler. Since this method avoids the usage of organic solvent compared to *in situ* and solution intercalation method, it is environmental friendly. The melt intercalation method allows using polymers which were previously not suitable for *in situ* polymerization or the solution intercalated method (Okamoto 2005). However, in the case of surface modified clays, surface modification agents can degrade during melt compounding of the polymer. Degradation of surface modification could have a serious impact on the composite structure that affects the properties of prepared composites (Mittal 2010).

In situ intercalative polymerization

In this method, layered silicate is swollen within the liquid monomer or a monomer solution. Polymerization reaction takes place in the galleries of layers that leads to formation of polymer chains between layers. Polymerization can be initiated either by heat or radiation, diffusion of a suitable initiator, or by an organic initiator or catalyst fixed through cation exchange inside the interlayer before the monomer swelling step. *In situ* intercalative polymerization is advantageous since polymers are synthesized together with nanofillers at the same time. Another advantage of this technique is to facilitate the lower molecular weight of monomer or oligomer solution for enhanced dispersion efficiency of layered silicates. Some examples of intercalative polymerization method include the polymerization of a polyamide thermoplastic from its monomers, the cross-linking of a thermosetting epoxy and the vulcanization of rubber (Ray and Okamoto 2003).

Enhancement of properties in biopolymer layered silicate nanocomposites

Addition of only a few percent of layered silicates into a polymer matrix has shown tremendous enhancement in biopolymer properties by lowering the cost variance compared with other approaches and traditional composites (Xu et al. 2001, Thellen et al. 2005, Oguzlu and Tihminlioglu 2010). Biopolymer nanocomposites have good/excellent barrier properties against water vapor and gases. Studies in literature have shown that enhancement in barrier properties are strongly dependent on the clay type, structure of the nanocomposites and aspect ratio of clay platelets. Effect of layered silicates on water vapor, oxygen and carbon dioxide transmission rates through polylactide (PLA) nanocomposites films with the natural and organomodified form of clays has been investigated widely (Matusik et al. 2011, Issaadi et al. 2015). Zenkiewicz and coworkers reported that transmission rates of water vapor, oxygen and carbon dioxide of PLA-organomodified clay nanocomposites were reduced by the ratio of 43%, 39% and 82%, respectively. Besides, the reduction ratio of the natural clay (unmodified) incorporated nanocomposite was improved as 25%, 4% and 76% in water vapor, oxygen and carbon dioxide transmission rates, respectively. Increment in clay content resulted in continuous decrease in transmission rates (Zenkiewicz and Richert 2008). Similar results on the barrier properties of bionanocomposites have been reported (Oguzlu and Tihminlioglu 2010, Sanchez et al. 2008, Rhim et al. 2013, Ozcalik and Tihminlioglu 2013).

In order to enhance the mechanical properties of polymers such as Young's modulus, yield stress, ultimate stress and strain, fillers in nanosize dimensions are introduced to polymer matrix. Many research studies have reported that high aspect ratio of fillers supplies a strong interaction between filler and polymer. The stiffness of polymer nanocomposites generally increases with the nanoparticles volume fraction, as long as sufficient dispersion and degree of exfoliation of these particles are ensured. Many studies exist in the literature related to the mechanical improvements by incorporation of layered silicates into polymer matrix such as chitosan (Oguzlu and Tihminlioglu 2010), polyhydroxybutyrate (Sanchez-Garcia et al. 2008), and polyethyleneterephthalate (Bandyopadhyay et al. 2008).

Oguzlu and Tihminlioglu prepared chitosan-OMMT nanocomposite films where Cloisite 10A was used as an organically modified montmorillonite (OMMT). Besides the improvement in barrier and thermal properties, they reported an enhancement in mechanical properties with a low level of Cloisite 10A loading. Even small amount of clay loading (< 10 wt%) resulted in an improvement in tensile strength and strain at break of chitosan nanocomposites by about 80% and 50%, respectively, compared to pristine chitosan. This improvement is attributed to the structure of layered silicates in polymer matrix as exfoliated/intercalated as supported by X-Ray diffraction analysis (Oguzlu and Tihminlioglu 2010).

Thermal stability of polymers is of fundamental importance in processing since it defines the application areas. Thermal stability lies under the mechanism of thermal decomposition. The polymer decomposition process is a complex procedure that consists of numerous chemical reactions leading to the formation of gaseous and solid products. Therefore, thermal stability of polymers is highly important to application areas such as synthesis of fire-safe polymeric materials, development of new technologies for efficient energy management, and the thermal recycling of waste plastics. Numerous studies have been reported so as to improve the thermal properties of polymers, and incorporation of nanofillers into polymer matrix (Krishnamachari et al. 2009, Kim and Choi 2016, Kim et al. 2016). The thermal stability of polymeric materials is usually studied by thermogravimetric analysis (TGA). The weight loss due to the formation of volatile products during the degradation is evaluated as a function of temperature. Incorporation of layered silicates into polymer matrix improves the thermal stability in a way that layered silicates act as an insulator and barrier the volatile compounds generated during thermal decomposition reaction. When layered silicates are incorporated in polymer matrix, volatile compounds produced during degradation follows more a tortuous path than pristine polymers. Therefore, the barrier affect may slow down a rate of mass loss by retarding the escape of volatile products of thermal degradation outside the degrading material. Moreover, layer silicates hinder the diffusion of oxygen into polymer matrix in thermo-oxidative degradation. Thus, the barrier effect of nanolayers towards oxygen permeation results in quantitative changes of volatile compounds that evolved during oxidative degradation (Okamoto 2005, Mittal 2010).

The well dispersed and layered structure of clay in the polymer matrix is thought to be an effective barrier to the permeation of oxygen and combustion gas, which improves thermal stability (Choi et al. 2003). One of the mechanisms for thermal stability improvement is the restriction of the movement of polymer chains that changes the kinetic and mechanism of reactions during thermal degradation. Recent studies have confirmed that a change in the molecular dynamics of polymer chains is due to the filler-polymer interactions that results in an increase in the glass transition temperature (T_g). Wang and coworkers reported that incorporation of OMMT into PHBHV resulted in a higher glass transition temperature with increasing amount of OMMT. In addition, OMMT acted as a nucleating agent in the PHBHV matrix, which increases the nucleation and the over-all crystallization rate of nanocomposites (Wang et al. 2005). Improvement in thermal properties is believed to be due to the structure of silicate layers. Layered silicate acts a heat barrier; therefore, exfoliated structure improves the thermal stability better as it is in the case of mechanical and barrier properties of polymer nanocomposites.

Color changes of polymer composites due to the incorporation of fillers are also very important for their possible use as packaging films. In this aspect, many research studies have investigated the optical properties of prepared composites (Sanchez et al. 2003, Oguzlu and Tihminlioglu 2010, Hamadanian et al. 2014). The color differences due to the filler composition and type have been evaluated via Hunter method. The total color differences (ΔE) of the coated films were calculated using the Hunter parameters, L, a, and b, of nanocomposites. The parameters, "a and b", are the measure of redness-greenness and yellowness, respectively. If the color change is greater than 3 ($\Delta E > 3$, it can recognized by human naked eye), the changes in the color can be recognized by human naked eye (Oguzlu and Tihminlioglu 2010).

Polyhydroxyalkonates (PHAs) layered silicate nanocomposite films

PHAs are produced by different kind of microorganisms as energy storage material and classified as biodegradable and biocompatible thermoplastic polymer. Polyhydroxybutyrate (PHB) and its copolymers polyhydroxyvalerate (PHBV) are the most commonly used PHAs, which are promising polymers in especially food packaging applications due to its properties being similar to conventional synthetic plastics, e.g., polypropylene as a substitute to synthetic ones and recently in biomedical areas. However, PHAs suffer from brittleness and poor processing temperature range that limits its application areas (Reddy et al. 2003). In order to overcome these, various approaches have been used in literature such as by blending with other biodegradable plastics (Nguyen et al. 2010, Zhang and Thomas 2011) and oil based polymers with the aim of improving its mechanical properties, unfortunately with only limited success up until now (Garcia-Quesada et al. 2012). Therefore, to overcome the problems, nanocomposite technology could be used for improving the PHA's properties.

Botona and coworkers examined the effect of modification of montmorillonite (Cloisite 30B-M) on thermal, mechanical and morphological properties of PHB nanocomposites. Unfortunately, the layered silicates were not individually well dispersed, but formed intercalated structure. Consequently, no significant improvement in tensile strength was obtained with respect to pristine PHB. Only slight increase in Young's modulus was obtained with the clay content for the modified MMT incorporated samples (Botana et al. 2010). Moreover, the importance of organomodification was also revealed by testing oxygen transmission rate using various types of modifiers in PHB polymer (Lagaron et al. 2008).

Choi and coworkers prepared PHBV polymer nanocomposites in the presence of Cloisite 30B clay. They reported that even at low clay content (3 wt% Cloisite 30B), tensile strength was greatly improved, about 55% compared to pristine PHBHV. This improvement is attributed to the strong hydrogen bonding between PHBHV copolymer and Cloisite 30B, indicating the importance of the organomodification of silicate layer (Choi et al. 2003, Ray and Bousmina 2005a, Pavlidou and Papaspyrides 2008).

As mentioned above, modification of clay surface plays a key role in dispersion and interfacial interaction of polymer and filler. Fully exfoliation state or significant improvement in properties were not reported or demonstrated up to date (Bordes et al. 2009). For this purpose, Cloisite 10 A as a different modified clay was used in PHB and PHBHV polymers in this study. Nanocomposite films were prepared by solution intercalation method and effect of Cloisite 10A on mechanical, thermal and optical properties was investigated. It was found that the pristine PHB had a tensile strength of 11.6 MPa and strain at break of 0.43%. Small amount of clay addition (1 wt%) increased the tensile strength of PHB nanocomposites at about 156.1% (Figure 1). Moreover, incorporation of 1 wt% Cloisite 10A into PHB was enhanced Young's Modules by 54% compared to pristine PHB. An increase of about 214.8% in tensile strength and of strain at break 69% was achieved for 2 wt% clay loaded PHB (nanocomposite films due to the good interaction between Cloisite 10 A and PHB confirmed by the XRD and STEM analysis (Akin 2012). In addition, the copolymer PHBHV nanocomposite films exhibited same trend in which Young's modulus increased by addition of Cloisite 10A, with PHB nanocomposites samples. Moreover, small amount of Cloisite 10A addition (PHBHV-1 wt%) to pristine PHBHV improved Young's modulus at about 100% which is close to the improvement obtained in PHB nanocomposites. At higher concentration of Cloisite 10 A (5–7 wt%), mechanical properties diminished as seen Figure 1 due to the poor dispersion and agglomeration of clay stacks within the polymer matrix.

Thermal properties of PHB and PHBHV nanocomposites were investigated via thermogravimetric analysis (TGA). In this manner, effect of Cloisite 10A addition on thermal decomposition of PHB and PHBHV nanocomposites was examined. Thermal stability of PHB and PHBHV nanocomposites was discussed by considering thermal degradation properties:

- Onset temperature of thermal degradation (T_{onset})
- End temperature of thermal degradation (T_{end})
- The yield of charred residue (char%)
- 20% weight loss temperature ($T_{0.2}$)
- 50% weight loss temperature ($T_{0.5}$)

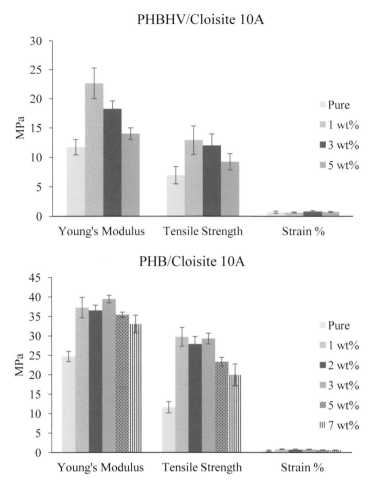

Figure 1. Young modulus, tensile strength and strain of break for PHB/Cloisite 10A and PHBHV/Cloisite 10A nanocomposite films.

The thermogravimetric weight loss curves of PHB/Cloisite 10A nanocomposites were tabulated in Table 2. As it is seen, the addition of 1 wt% of Cloisite 10A in polymer matrix resulted in significant improvement in thermal stability by increasing the thermal degradation onset temperature (T_{onset}) from

Table 2. Thermal degradation properties of PHB and PHBHV nanocomposites.

Sample	T_{onset} (°C)	T_{end} (°C)	Char %	$T_{0.2}$ (°C) 20% mass loss	$T_{0.5}$ (°C) 50% mass loss
Cloisite 10A	240.6	807.8	55.3	263.2	-
PHB-N	234.7	274.5	2.7	251.0	260.4
PHB-1	240.1	283.6	0.8	263.5	267.1
PHB-2	245.3	278.7	1.8	271.6	268.1
PHB-3	237.3	278.8	1.1	259.0	266.8
PHB-5	238.5	285.7	2.2	259.9	267.8
PHB-7	236.8	284.5	1.3	258.0	268.0
PHBHV-N	218.9	267.0	6.6	240.9	249.9
PHBHV-1	250.6	284.4	5.5	270.7	275.8
PHBHV-3	248.6	283.7	4.0	265.2	271.4
PHBHV-5	244.7	291.2	2.3	263.0	270.3

235°C to 240°C for PHB, and from 219°C to 251°C for PHBV polymer. However, the improvement was not as significant as in the case of higher clay loaded samples, but even at higher clay loadings, T_{onset} temperature was higher when compared to pristine PHB. This could be due to the restriction of thermomechanical motion of polymer chain by layered silicates.

Thermal degradation properties (T_{onset}, T_{end}, char%, $T_{0.2}$, and $T_{0.5}$) of Cloisite 10A, PHB and PHBHV nanocomposites were tabulated in Table 2. As seen in Table 2, onset decomposition temperature, 20% and 55% weight loss temperatures of Cloisite 10A were observed at 240.63°C, 263.7°C and 807°C, respectively. This showed that almost 50% of Cloisite 10A contains organic modifier.

The onset decomposition temperature of pristine PHB film increased by about 11°C by addition of 2 wt% Cloisite 10A. In addition, 50% weight loss temperature increased to 268.1°C from 260.4 C. In contrast, at higher Cloisite 10A loading, the onset and end decomposition temperature decreased to the value of pristine PHB. The improvement in thermal stability parameters of PHB films achieved at low content of Cloisite 10A can be attributed to dispersion level of layered silicates in polymer matrix decreasing the diffusion of volatile decomposition products. Moreover, the same trend was also observed for PHBHV nanocomposite film samples; in addition, 1 wt% Cloisite 10A resulted in significant increment in onset decomposition temperature, about 32°C compared to pristine PHBHV (Akin 2012). Moreover, at higher Cloisite 10A loaded samples, thermal decomposition parameters were found to be even higher than pristine PHBHV samples. It can be concluded that exfoliated structure of layered silicates enhanced thermal decomposition properties of PHB and PHBHV nanocomposites. Enhancement in thermal stability by incorporation of layered silicates to polymer matrix was also reported in several studies (Sinha Ray and Okamoto 2003, Maiti et al. 2007). When percent of char formation of nanocomposites is taken into account, Cloisite 10A lost its 50% of weight at the end of 1000°C. However, char formation was higher in PHB nanocomposite compared to PHB sample. This could be explained by catalytic effect of clay particles in polymer decomposition process. The products formed during the decomposition of the polymer can be changed due to catalytic effect of clay particles. Metals present in the clay and in acidic sites (lewis acid sides) inherently present on the clay surface, or formed as a result of the decomposition of alkylammonium salts, showed catalytic activity towards degradation reaction. Therefore, more volatile compounds can be created during decomposition process (Mittal 2010).

Application of PHB and PHBHV nanocomposites for food packaging applications has gained attention due to promising features. Hence, in food packaging application the color of the films is important and the change in color with the addition of the filler into polymer matrix should be considered. For this purpose, the color change in PHB and PHBHV Cloisite 10A nanocomposites were evaluated via Hunter method. The total color differences (ΔE) of the coated films were calculated using the parameters L, a and b of nanocomposites. The evaluated color differences in nanocomposites (PHB and PHBHV) were tabulated in Figure 2. The lightness parameter, L, did not change significantly by addition of clay into PHB and PHBHV matrix. The parameters "a and b", which are the measure of redness-greenness, yellowness, respectively, also did not change significantly compared to pristine PHB and PHBHV samples. When total color difference is greater than 3 (ΔE > 3), the changes in the color can be recognized by naked

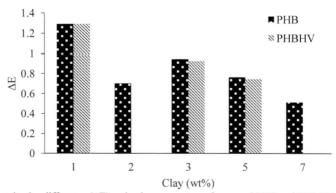

Figure 2. The total color difference (ΔE) and color parameters of prepared PHB and PHBHV nanocomposites.

human eye (Oguzlu and Tihminlioglu 2010). Since total color differences ΔE were much lower than 3 for all nanocomposites samples, it can be said that incorporation of Cloisite 10A to pristine PHB and PHBHV matrix did not alter color properties significantly.

Poly(lactic acid) (PLA) layered silicate nanocomposite films

Poly(lactic acid) is an aliphatic biodegradable polyester which has gained tremendous attention due to its availability on the market and its lower price as an alternative to conventional oil based polymers. The physical properties of PLA resemble the poly(styrene) and poly(ethylene terephthalate) synthetic polymers. However, in order to make PLA competitive for widespread applications, some of the properties such as thermal stability, flexural mechanical and gas barrier properties need to be improved. For this purpose, several approaches such as blending (Mohapatra et al. 2016), copolymerization (Bishai et al. 2013), and addition of nanofillers (Pantani et al. 2013) are generally applied.

Krikorian and Pochan attempted to prepare PLA nanocomposites by solvent interaction in the presence of Cloisite 30 B. Exfoliated structure with randomly distributed clay platelets were obtained due to the good interactions between the clay and PLA. Therefore, they reported improvement in mechanical properties such as the modulus increased by 61% (Krikorian and Pochan 2003). On the other hand, Mohapatra and coworkers investigated the effect of incorporation of commercial Cloisite 93A (C93A) and Cloisite 30B (C30B) modified clays on the mechanical, morphological and thermal properties of PLA nanocomposite by using melt mixing method (Mohapatra et al. 2014). It appeared that mechanical properties of PLA/layered silicates varied as a function of clay loading type and concentration. Incorporation of organoclay from 1 to 3 wt% resulted in increase in the tensile strength and modulus. Beyond 3 wt% of nanoclay loading, there was a drastic drop in the tensile strength which indicated an agglomeration of the clay particles. However, with increasing nanoclay loading from 1 to 5 wt%, the elastic modulus of the nanocomposites increased from 1,690 MPa to 2,480 and 2,362 MPa in case of PLA/C30B and PLA/C93A nanocomposites, respectively. The optimum tensile strength and modulus was reported at 3% of loading due to a better interfacial interaction which resulted in exfoliated structure of clays (Mohapatra et al. 2014).

Oguzlu prepared PLA based nanocomposites via solution intercalation method using two different modified clays: Cloisite 10A and Cloisite 93A (Oguzlu 2011). Effects of clay loading and modification on mechanical, thermal and optical properties were studied and given in this chapter. Mechanical results showed a small increment in tensile strength for only 1 wt% clay addition; however, significant improvements in Young modulus of the nanocomposite films were obtained compared to neat PLA (Figure 3). It appeared that the enhancement in Young's modulus was limited up to certain clay loadings; 7 wt% for Cloisite 10A and 5 wt% for Cloisite 93A. Beyond the critical clay loading, the addition of silicates resulted in the phase separated structure that weakens the mechanical properties of the nanocomposites above the critical content of filler addition (Figure 3). This behavior was ascribed to the resistance exerted by the clay itself and to the orientation and aspect ratio of the intercalated silicate layers (Mittal 2010). Additionally, the stretching resistance of the oriented backbone of the polymer chain in the gallery was also reported to contribute to modulus enhancements (Mittal 2010). The elongation at break of neat PLA clearly decreased with the incorporation of nanoclays, and also elasticity decrease became more expressed by the increase in the nanoclay loading as observed generally in many polymer–clay nanocomposite systems (Oguzlu 2011).

Thermal properties of PLA/Cloisite 10A nanocomposites were also investigated by TGA analysis. TGA results of T_{onset}, T_{end}, char percentage, temperature of 10% and 50% of weight loss were tabulated in Table 3. Onset decomposition temperatures of organomodifiers were generally observed at 200°C in the literature (Mittal 2010). In accordance with literature, decomposition of the organoclays started at around 200–250°C and the percent weight loss of the Cloisite 10A was observed as 45%. Addition of Cloisite 10A into PLA improved onset and final degradation temperatures of neat PLA films as seen in Table 3. Most promising improvement was reported for the 7 wt% of Cloisite 10A loading in which onset degradation increased by approximately 10°C. Similar enhancement was also observed at T_{end}, and temperatures of 10% and 50% of weight losses (Oguzlu 2011). Further addition of Cloisite 10 A did not enhance

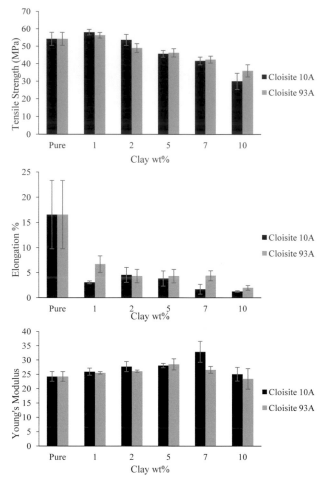

Figure 3. Mechanical properties of PLA/Cloisite 10A and PLA/Cloisite 93A, tensile strength, elongation at break and Young's modulus of PLA/layered silicate nanocomposites.

Table 3. Thermal properties of PLA/Cloisite 10A nanocomposite films.

Sample	T_{onset} (°C)	T_{end} (°C)	Char%	$T_{0.1}$ (°C)	$T_{0.5}$ (°C)
PLA	308.1	368.5	0.91	325.5	348.0
PLA-10A-1	312.8	473.9	7.25	330.6	353.3
PLA-10A-2	310.2	375.9	0.97	334.3	352.5
PLA-10A-5	313.0	376.8	0.60	321.9	349.1
PLA-10A-7	318.1	375.0	4.65	332.6	353.5
PLA-10A-10	316.7	379.7	2.00	334.7	356.1

the thermal properties as agreed with the mechanical results due to the dispersion level of clays. The enhancement in thermal properties could be due to the exfoliated structure of silicate layers that hinders the movement of polymer chains and hence, higher activation energy for the degradation processes is needed leading to higher degradation temperature. The clay acts as a heat barrier, which improves the thermal stability in the PLA layered silicate nanocomposites as well as helps in char formation. Similar improvements in polymer/layered silicate nanocomposite have been reported in many studies (Alexandre and Dubois 2000, Ray and Bousmina 2005a).

Change in the color of the prepared nanocomposites is as important as other properties such as barrier and mechanical properties for possible usage in food packaging applications. Therefore, the change in the color by incorporation of Cloisite 10A into PLA nanocomposites was investigated and parameters of Hunter method and total color difference (ΔE) were tabulated in Figure 4. As seen in Figure 4, an increase in clay content resulted in very slight change of ΔE calculated as high as 1.1 that was lower than 3. Therefore, the change in the color of nanocomposites due to Cloisite 10A loading is not significant as detected by naked human eyes. Hence, PLA/Cloisite 10A nanocomposites indicate good dispersion of filler within polymer matrix and it has promising features to be used as biobased food packaging films.

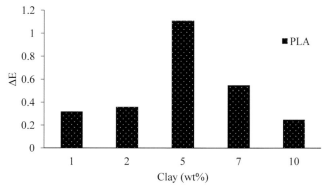

Figure 4. Optical parameters of PLA/Cloisite 10A nanocomposites.

Corn zein (CZ) layered silicate nanocomposite films

Corn zein (CZ) is a type of biopolymer that is obtained by extraction of the corn wet-milling industry by products named corn gluten meal. Corn zein has thermoplastic feature that makes it a promising candidate as an alternative to petroleum based polymers. Corn zein is a hydrophobic polymer and it has good mechanical properties and excellent barrier properties to gases, especially carbon dioxide and oxygen. However, its sensitivity to humidity affects the properties such as mechanical and barrier to gases, as well as its processability. Moreover, corn zein is brittle, especially at low water activity. In order to overcome the brittleness and processability of CZ, many studies have been reported in literature, using different plasticizers in CZ processing (Lawton 2004, Ghanbarzadeh et al. 2006), blending with starch (Chanvrier et al. 2005), and coating on polypropylene films (Tihminlioglu et al. 2011). A few studies are reported in the literature regarding corn zein nanocomposites (Luecha et al. 2010, Ozcalik and Tihminlioglu 2013). First study was reported by Luecha and coworkers (Luecha et al. 2010). Corn zein-OMMT (The Nanomer® I.34 TCN) nanocomposites were prepared by using solution intercalation and blown film extrusion methods. Improved elongation at break was reported for extrusion blown samples, while the samples prepared by solution intercalation increased the brittleness with the incorporation of modified clay. Tensile strength and elastic modulus of the prepared films improved up to critical clay loadings. Water vapor permeability of nanocomposites showed improvements in small OMMT concentrations whereas WVP of samples with OMMT content higher than 3 wt% were very close to that of pristine corn zein. Additionally, it was found that corn zein MMT films prepared by blown extrusion were more sensitive to MMT loadings than the samples prepared by solvent casting due to the well-distributed MMT platelets within zein polymer matrix by blown extrusion method. So, the nanoclay platelets provided thermal barrier throughout the zein matrix and eventually delayed the thermal degradation of zein films. Ozcalik and Tihminlioglu investigated the effect of organomodified montmorillonite (OMMT; Cloisite 10 A) addition to Corn zein coatings as a biobased gas barrier layer on poly(propylene) films. Incorporation of OMMT by solution intercalation into zein matrix significantly improved oxygen and water vapor barrier of coated PP films. Corn zein/OMMT nanocomposite coating of PP films has reduced the oxygen permeability nearly four times, with 3 wt OMMT incorporation and water vapor permeability reduced by 30% percentage with 5 wt.% OMMT content (Ozcalik and Tihminlioglu 2013).

Incorporation of layered silicates in CZ on PP films improved gas barrier (Ozcalik and Tihminlioglu 2013) as well as mechanical properties tremendously. The most significant effect of layered silicate addition on mechanical properties of corn zein coated poly(propylene) films were reported in the elastic modulus (Figure 5) (Ozcalik 2010). As shown in Figure 5, elastic modulus increased continuously with the clay incorporation up to 5 wt% loading from 8.21 x 10^2 MPa to 11.59 x 10^2 MPa with a 41% increase. Nanoclay content above 5 wt% resulted in decrease in the reinforcement effect due to the changes in the intermolecular forces that affect the structure of layered silicates in corn zein polymer, as well as confirmed by the results of barrier properties of the films (Ozcalik and Tihminlioglu 2013). Improvement in the tensile yield strength of the nanocomposite coated PP films was also shown in Figure 5 with Cloisite 10A incorporation that enhanced the mechanical durability of the PP films. While the tensile strength of corn zein coated films was 25 MPa, corn zein nanocomposite coated films increased up to 31 MPa. However, % strain values decreased continuously with the clay incorporation as reported in many polymer-clay systems.

To conclude, layered silicate structure within the polymer matrix is the key point in enhancement of the mechanical properties of polymer films. Degree of dispersion level as exfoliated or intercalated layered silicates was found to be responsible for property enhancement that depends on the critical loading ratios of clay in polymer matrix (Matusik et al. 2011, Ozcalik and Tihminlioglu 2013).

Corn zein polymer has yellowish color; therefore, color of the prepared nanocomposite films are as important as the other properties, especially if it is intended to be used in food packaging applications. Thus, the color change in Corn zein nanocomposite (CZ/OMMT) coating on PP films was investigated and found that CZ coating layers have slight effect on total color change. The most significant effect of total color change was reported by the yellowness index b. The applied polymer layers onto the substrate in laminated or coated structures change the color depending on the characteristics of applied layer (Weller et al. 1998, Lee et al. 2008). Corn zein coatings were also reported to change the color of PP

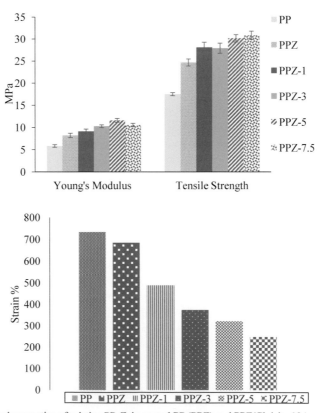

Figure 5. Mechanical properties of pristine PP, Zein coated PP (PPZ) and PPZ/Cloisite 10A nanocomposite films.

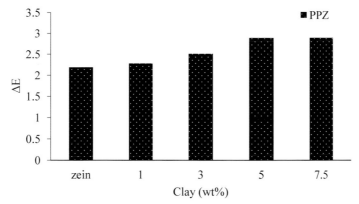

Figure 6. Hunter parameters of corn zein coated PP/Cloisite 10A nanocomposite films.

films to more yellow in different extends depending on the plasticizer type (Lee et al. 2008). Corn zein nanocomposite coatings prepared by using Cloisite 10A at different loadings exerted only slight changes to total color difference values (Figure 6). As the total difference value was at maximum 2.89, it can be said that the color change is not significant that can be realized by human eye.

Conclusion

Polymer nanocomposites have gained a great interest from both academia and industry due to their promising versatile properties. Among different kinds of fillers, layered silicates have been attracting attention because they exhibit dramatic enhancement in properties at even very low filler amounts. This chapter presents the mechanical, thermal and optical properties of Cloisite 10 A incorporated biodegradable polymers, PHB-PHBHV, PLA, and Corn Zein by considering structural importance of the layered silicates within the polymer matrix. It was found that level of dispersion mostly depends on the interfacial interaction between polymer and silicate layers that defines the compatibility of filler with polymer chains. It has been reported and shown that the higher reinforcing effect is generally limited to small % clay addition, generally less than 5 wt%. In the case of good dispersion and compatibility of clay, the barrier, mechanical and thermal properties are enhanced.

In summary, biopolymer/layered silicate nanocomposites offer promising features for versatile application areas such as biopackaging, biomedical devices, functional materials, etc. Therefore, the further developments in polymer science and nanotechnologies will offer new application areas for biopolymer/nanocomposites and will be competitive to commercial synthetic polymer based composites.

Acknowledgements

The authors acknowledge Scientific and Research Council of Turkey (Tubitak project, #108M335) for support of the project for PLA and PHB nanocomposite film studies.

References

Akin, O. 2012. Effect of Organo-Modified Clay Addition on Properties of Polyhydroxy Butyrate Homo and Copolymers Nanocomposite Films. Master of Science, İzmir Institute of Technology.

Alexandre, M. and P. Dubois. 2000. Polymer-layered silicate nanocomposites: preparation, properties and uses of a new class of materials. Materials Science & Engineering R-Reports 28(1-2): 1–63.

Bandyopadhyay, J., S.S. Ray and M. Bousmina. 2008. Nonisothermal crystallization kinetics of poly(ethylene terephthalate) nanocomposites. Journal of Nanoscience and Nanotechnology 8(4): 1812–1822.

Bishai, M., S. De, B. Adhikari and R. Banerjee. 2013. Copolymerization of lactic acid for cost-effective PLA synthesis and studies on its improved characteristics. Food Science and Biotechnology 22(1): 73–77.

Bordes, P., E. Pollet and L. Averous. 2009. Nanobiocomposites: Biodegradable polyester/nanoclay systems. Progress in Polymer Science 34: 125–155.

Botana, A., M. Mollo, P. Eisenberg and R.M. Torres Sanchez. 2010. Effect of modified montmorillonite on biodegradable PHB nanocomposites. Applied Clay Science 47(3-4): 263–270.

Burgentzle, D., J. Duchet, J.F. Gerard, A. Jupin and B. Fillon. 2004. Solvent-based nanocomposite coatings I. Dispersion of organophilic montmorillonite in organic solvents. Journal of Colloid and Interface Science 278(1): 26–39.

Chanvrier, H., P. Colonna, G. Della Valle and D. Lourdin. 2005. Structure and mechanical behaviour of corn flour and starch-zein based materials in the glassy state. Carbohydrate Polymers 59(1): 109–119.

Choi, W.M., T.W. Kim, o. O. Park, Y.K. Chang and J.W. Lee. 2003. Preparation and Characterization of Poly(hydroxybutyrateco-hydroxyvalerate)-Organoclay Nanocomposites.

Choudalakis, G. and A.D. Gotsis. 2009. Permeability of polymer/clay nanocomposites: a review. European Polymer Journal 45(4): 967–984.

Cornwelle, A.J. 2009. Barrier Properties of Polymer Nanocomposites. V. Mital. New York, Nova Science Publisher.

Cretois, R., N. Follain, E. Dargent, J. Soulestin, S. Bourbigot, S. Marais et al. 2014. Microstructure and barrier properties of PHBV/organoclays bionanocomposites. Journal of Membrane Science 467: 56–66.

Erceg, M., T. Kovacic and I. Klaric. 2009. Poly(3-hydroxybutyrate) nanocomposites: Isothermal degradation and kinetic analysis. Thermochimica Acta 485(1-2): 26–32.

Garcia-Quesada, J.C., I. Pelaez, O. Akin and I. Kocabas. 2012. Processability of PVC plastisols containing a polyhydroxybutyrate-polyhydroxyvalerate copolymer. Journal of Vinyl and Additive Technology 18(1): 9–16.

Ghanbarzadeh, B., A. Oromiehie, M. Musavi, K. Rezayi, E. Razmi and J. Milani. 2006. Investigation of water vapour permeability hydrophobicity and morphology of zein films plasticized by polyols. Iranian Polymer Journal 15(9): 691–700.

Gunaratne, L.M.W.K. and R.A. Shanks. 2005. Melting and thermal history of poly(hydroxybutyrate-co-hydroxyvalerate) using step-scan DSC. Thermochimica Acta 430(1-2): 183–190.

Hamadanian, M., M. Amani and V. Jabbari. 2014. Improving thermal and optical properties of biodegradable poly(ethyl vinyl ether-co-maleic anhydride) (PEVE/MA) copolymer by encapsulation of TiO_2 nanoparticles via *in situ* radical polymerization polymer-plastics technology and engineering 53(12): 1283–1289.

Hule, R.A. and D.J. Pochan. 2007. Polymer nanocomposites for biomedical applications. Mrs Bulletin 32(4): 354–358.

Hwang, J.J., S.M. Huang, H.J. Liu, H.C. Chu, L.H. Lin and C.S. Chung. 2012. Crystallization kinetics of poly(L-lactic acid)/montmorillonite nanocomposites under isothermal crystallization condition. Journal of Applied Polymer Science 124(3): 2216–2226.

Issaadi, K., A. Habi, Y. Grohens and I. Pillin. 2015. Effect of the montmorillonite intercalant and anhydride maleic grafting on polylactic acid structure and properties. Applied Clay Science 107: 62–69.

Jaafar, J., A.F. Ismail and T. Matsuura. 2012. Effect of dispersion state of Cloisite15A (R) on the performance of SPEEK/Cloisite15A nanocomposite membrane for DMFC application. Journal of Applied Polymer Science 124(2): 969–977.

Kim, S.W. and H.M. Choi. 2016. Morphology, thermal, mechanical, and barrier properties of graphene oxide/poly(lactic acid) nanocomposite films Korean. Journal of Chemical Engineering 33(1): 330–336.

Kim, Y.H., S.H. Kwon, H.J. Choi, K. Choi, N. Kao, S.N. Bhattacharya and R.K. Gupta. 2016. Thermal, mechanical, and rheological characterization of polylactic acid/halloysite nanotube nanocomposites. Journal of Macromolecular Science Part B-Physics 55(7): 680–692.

Krikorian, V. and D.J. Pochan. 2003. Polylactic acid/layered silicate nanocomposites: fabrication, characterization and properties. Chem. Mater. 15: 4317–4324.

Krishnamachari, P., J. Zhang, J.Z. Lou, J.Z. Yan and L. Uitenham. 2009. Biodegradable poly(lactic acid)/clay nanocomposites by melt intercalation: A study of morphological, thermal, and mechanical properties. International Journal of Polymer Analysis and Characterization 14(4): 336–350.

Lagaron, J.M., M.D. Sanchez-Garcia and E. Gimenez. 2008. Morphology and barrier properties of nanobiocomposites of poly(3-hydroxybutyrate) and layered silicates. Journal of Applied Polymer Science 108(5): 2787–2801.

Lawton, J.W. 2004. Plasticizers for zein: Their effect on tensile properties and water absorption of zein films. Cereal Chemistry 81(1): 1–5.

Lee, J.W., S.M. Son and S.I. Hong. 2008. Characterization of protein-coated polypropylene films as a novel composite structure for active food packaging application. Journal of Food Engineering 86(4): 484–493.

Luecha, J., N. Sozer and J.L. Kokini. 2010. Synthesis and properties of corn zein/montmorillonite nanocomposite films. Journal of Materials Science 45(13): 3529–3537.

Maiti, P., C.A. Batt and E.P. Giannelis. 2007. New biodegradable polyhydroxybutyrate/layered silicate nanocomposites. Biomacromolecules 8(11): 3393–3400.

Matusik, J., E. Stodolak and K. Bahranowski. 2011. Synthesis of polylactide/clay composites using structurally different kaolinites and kaolinite nanotubes. Applied Clay Science 51(1-2): 102–109.

Mittal, V. 2010. Optimization of Polymer Nanocomposite Properties. WILEY-VCH Verlag GmbH & Co.

Mohapatra, A.K., S. Mohanty and S.K. Nayak. 2014. Dynamic mechanical and thermal properties of polylactide-layered silicate nanocomposites. Journal of Thermoplastic Composite Materials 27(5): 699–716.

Mohapatra, A.K., S. Mohanty and S.K. Nayak. 2016. Properties and characterization of biodegradable poly(lactic acid) (PLA)/poly(ethylene glycol) (PEG) and PLA/PEG/organoclay: a study of crystallization kinetics, rheology, and compostability. Journal of Thermoplastic Composite Materials 29(4): 443–463.

Nguyen, Q.T. and D.G. Baird. 2006. Preparation of polymer–clay nanocomposites and their properties. Advances in Polymer Technology 25(4): 270–285.

Nguyen, T.P., A. Guinault and C. Sollogoub. 2010. Miscibility and morphology of poly(lactic acid)/poly(B-Hydroxybutyrate) blends international conference on advances in materials and processing technologies. Pts One and Two 1315: 173–178.

Oguzlu, H. and F. Tihminlioglu. 2010. Preparation and barrier properties of chitosan-layered silicate nanocomposite films. Polychar-18 World Forum on Advanced Materials 298: 91–98.

Oguzlu, H. 2011. Water Vapor and Gas Barrier Properties of Biodegradable Polymer Nanocomposite Films. Master of Science, İzmir Institute of Technology.

Okamoto, M. 2005. Biodegradable polymer layerd silicate nanocomposites: A review. American Scientific Publishers. 1.

Ozcalik, O. 2010. Preparation and Characterization of Corn Zein Nanocomposite Coated Polypropylene Films for Food Packaging Application Master of Science. İzmir Institute of Technology.

Ozcalik, O. and F. Tihminlioglu. 2013. Barrier properties of corn zein nanocomposite coated polypropylene films for food packaging applications. Journal of Food Engineering 114(4): 505–513.

Pantani, R., G. Gorrasi, G. Vigliotta, M. Murariu and P. Dubois. 2013. PLA-ZnO nanocomposite films: Water vapor barrier properties and specific end-use characteristics. European Polymer Journal 49(11): 3471–3482.

Pavlidou, S. and C.D. Papaspyrides. 2008. A review on polymer-layered silicate nanocomposites. Progress in Polymer Science 33(12): 1119–1198.

Ray, S.S. and M. Okamoto. 2003. Polymer/layered silicate nanocomposites: a review from preparation to processing. Progress in Polymer Science 28(11): 1539–1641.

Ray, S.S. and M. Bousmina. 2005a. Biodegradable polymers and their layered silicate nano composites: In greening the 21st century materials world. Progress in Materials Science 50(8): 962–1079.

Ray, S.S. and M. Bousmina. 2005b. Effect of organic modification on the compatibilization efficiency of clay in an immiscible polymer blend. Macromolecular Rapid Communications 26(20): 1639–1646.

Reddy, C.S.K., R. Ghai, Rashmi and V.C. Kalia. 2003. Polyhydroxyalkanoates: an overview. Bioresource Technology (87): 137–146.

Rhim, J.W., H.M. Park and C.S. Ha. 2013. Bio-nanocomposites for food packaging applications. Progress in Polymer Science 38(10-11): 1629–1652.

Sanchez-Garcia, M.D., E. Gimenez and J.M. Lagaron. 2008. Morphology and barrier properties of nanobiocomposites of poly(3-hydroxybutyrate) and layered silicates. Journal of Applied Polymer Science 108(5): 2787–2801.

Sanchez, C., B. Lebeau, F. Chaput and J.P. Boilot. 2003. Optical propertics of functional hybrid organic-inorganic nanocomposites. Advanced Materials 15(23): 1969–1994.

Sinha Ray, S. and M. Okamoto. 2003. Polymer/layered silicate nanocomposites: a review from preparation to processing. Progress in Polymer Science 28(11): 1539–1641.

Thellen, C., C. Orroth, D. Froio, D. Ziegler, J. Lucciarini, R. Farrell et al. 2005. Influence of montmorillonite layered silicate on plasticized poly(L-lactide) blown films. Polymer 46(25): 11716–11727.

Tihminlioglu, F., I.D. Atik and B. Ozen. 2011. Effect of corn-zein coating on the mechanical properties of polypropylene packaging films. Journal of Applied Polymer Science 119(1): 235–241.

Wang, S., C. Song, G. Chen, T. Guo, J. Liu, B. Zhang et al. 2005. Characteristics and biodegradation properties of poly(3-hydroxybutyrate--3-hydroxyvalerate)/organophilic montmorillonite (PHBV/OMMT) nanocomposite. Polymer Degradation and Stability 87(1): 69–76.

Weller, C.L., A. Gennadios and R.A. Saraiva. 1998. Edible bilayer films from Zein and grain sorghum wax or carnauba wax. Food Science and Technology-Lebensmittel-Wissenschaft & Technologie 31(3): 279–285.

Xi, Y.F., Z. Ding, H.P. He and R.L. Frost. 2004. Structure of organoclays—an X-ray diffraction and thermogravimetric analysis study. Journal of Colloid and Interface Science 277(1): 116–120.

Xu, W., M. Ge and P. He. 2001. Nonisothermal crystallization kinetics of polyoxymethylene/montmorillonite nanocomposite. Journal of Applied Polymer Science 82(9): 2281–2289.

Zenkiewicz, M. and J. Richert. 2008. Permeability of polylactide nanocomposite films for water vapour, oxygen and carbon dioxide. Polymer Testing 27(7): 835–840.

Zhang, M. and N.L. Thomas. 2011. Blending polylactic acid with polyhydroxybutyrate: The effect on thermal, mechanical, and biodegradation properties. Advances in Polymer Technology 30(2): 67–79.

Zulfiqar, S., Z. Ahmad and M.I. Sarwar. 2008. Preparation and properties of aramid/layered silicate nanocomposites by solution intercalation technique. Polymers for Advanced Technologies 19(12): 1720–1728.

Section II

CNT and Graphene Nanocomposites

13

Rubber—CNT Nanocomposites

M. Balasubramanian[1],* and P. Jawahar[2]

Introduction

Rubber components are being used for a wide range of applications starting from bottle caps to tires for automotives and aircrafts. The use of nanofillers in the rubber matrix has significantly improved the mechanical properties. Hence, the nanocomposites have gained momentum in the last decade. The high specific surface area of nanofillers enhanced the interface interaction with the polymer matrix, which increases the mechanical properties. The quality of filler interface, aspect ratio and dispersion state governs the efficient transfer of load from matrix to reinforcement. The mechanical properties of elastomeric composites also depend on filler interactions and filler-matrix interactions. All the factors should ideally meet to acquire the required high performance. In certain cases, the aggregation and networking of fillers also play an important role in enhancing the mechanical properties (Oberdisse et al. 2005). Carbon black is widely used as nanofillers in rubber. Other nanoparticles can also be added as second reinforcement along with carbon black. Researchers have developed hybrid nanocomposites with clay (Joly et al. 2002, Arroyo et al. 2003) and silica (Pal and De 1982, Murakami et al. 2003). Short fibers are also used as reinforcements for rubber (Silva et al. 2006). Irrespective of the outcome of these studies, nanoparticles like clay, silica and fibres are considered as potential reinforcements. A new class of carbonaceous material called carbon nanotube (CNT) seems to be a promising reinforcement in nano-level. CNTs possess high aspect ratio, high surface area and high mechanical properties. The incorporation of CNT in rubber matrix has improved the mechanical, thermal, electrical properties of elastomers. The enhancement occurs with the addition of CNTs in small quantities.

It has been reported that the incorporation of CNTs in natural rubber has improved the strength, modulus and hardness significantly (Razi et al. 2006, Takeuchi et al. 2015, George et al. 2015, Bhattacharyyaa et al. 2008, Sui et al. 2008). The incorporation of carbon nanotube has also improved the electrical conductivity significantly and the threshold value of CNT in elastomeric matrix is less than 1 phr (Tsuchiya et al. 2011), whereas for the carbon black, it varies from 10–25 phr (Klüppel 2003). The incorporation of CNT in rubber has brought a novel rubber nanocomposite, potentially suitable

---

[1] Department of Metallurgical & Materials Engineering, Indian Institute of Technology Madras, Chennai, Tamil Nadu, India, 600036.
[2] Department of Mechanical Engineering, Velammal Engineering College, Chennai, Tamil Nadu, India, 600066.
  Email: dr.p.jawahar@gmail.com
* Corresponding author: mbala@iitm.ac.in

for a wide range of engineering applications. In this chapter, some insights have been given on CNTs, functionalization of CNTs, the methods of processing of CNT-rubber nanocomposites, and physical, thermal and mechanical properties of CNT-rubber nanocomposites.

Carbon nanotube (CNT)

The CNTs possess superior electrical, thermal and mechanical properties. They are used as mechanical reinforcements in composites. They have applications in sensors, energy storage device, electronic devices, etc. (Endo et al. 2008). A CNT is one dimensional allotrope of carbon. It can also be called as a graphene cylinder, since it is similar to graphene sheet rolled into a cylinder with respect to an axis.

Classification of carbon nanotubes

Carbon nanotubes can be simply classified as follows:

1) Single-walled carbon nanotubes (SWCNT)
2) Multi-walled carbon nanotubes (MWCNT) and
3) Defective carbon nanotubes.

A brief description about these CNTs is given below.

Single-walled CNT

It was first synthesized in 1993 (Iijima and Ichihasi 1993). SWCNTs are prepared in larger volume by arc discharge method using graphite electrodes. A bimetallic Fe-Ni or Fe-Co alloy powder is used as a catalyst and it is filled in anode (Seraphin and Zhou 1994). The structure of SWCNT is very simple and it is nothing but a cylinder made of a graphene sheet.

Multi-walled CNT

It was found by researchers even before the processing of SWCNT. It consists of multiple layers of graphene and they are wound cylindrically in three forms.

1. Concentric cylinders : The structure consists of hollow cylinders of graphene stacked concentrically in increasing radius.
2. Spiral structure   : The graphene sheets are wound around the axis in a spiral manner.
3. Mixed structure   : Both the concentric and spiral structures of the graphene sheets are present (Amelinckx et al. 1995, Xu et al. 2006).

Defective CNT

The insertion of a pentagon or a heptagon ring in a hexagonal network bends the tube. The insertion of heptagon–pentagon pair in graphite sheets in appropriate combination will lead to the formation of defective CNT, as listed below.

1. Coiled carbon nanotubes
2. Y-junction carbon nanotubes.

Carbon atoms are arranged in regular hexagonal pattern in graphene sheets for single walled CNT and multiwalled CNT, whereas the arrangements of carbon atom in graphene sheets are not regular in defective CNT.

The presence of pentagon and heptagon rings has created the negative and positive curvature surfaces in simple bent sections (Fonseca et al. 1996). However, the Y junctions contain heptagon rings only. The positive curvature surface developed by the presence of heptagon rings in combination with hexagonal network leads to formation of Y junction (Gan et al. 2001). The heptagon-pentagon pairs are systematically arranged in coiled CNT (Dunlap 1992). For enhancing the mechanical and physical properties of rubber based systems, much focus is given only to SWCNT and MWCNT.

Synthesis of CNT

CNTs are generally produced by the following three methods (Sharma et al. 2015).

1. Chemical vapor deposition
2. Arc discharge
3. Laser ablation

Chemical vapor deposition

It is widely used to produce CNT on a larger scale. The materials required arc carbon containing gaseous materials, and catalyst or catalyst coated substrate. Methane, acetylene and carbon monoxide are widely used as gaseous source materials. Transition metals like iron, cobalt, nickel and vanadium are used as catalyst. The parameters such as the type of catalyst, gas flow rate and temperature control the quality and quantity of CNTs formed (Mukhopadhyay et al. 1999, Li et al. 2002). The type of catalyst influences the formation of either SWCNT or MWCNT (Cassell et al. 1999, Satishkumar et al. 1998). The CNT growth is also influenced by the particle size of the catalyst. The carbon vapor is formed by the dissociation of carbonaceous gases at high temperature. This carbon is then deposited on the catalyst coated substrate in a controlled manner, leading to the growth of CNT.

Arc discharge method

CNT can also be effectively produced by the arc discharge method. In arc discharge method, a reaction chamber filled with inert gas, such as argon is used. The reaction chamber also contains two graphite electrodes. DC arc discharge assists the vaporization of carbon from the graphite anode and the vaporized carbon gets deposited on the chamber wall. Different types of nanotubes can be fabricated by using transition metals, such as nickel, iron, cobalt, etc. as catalyst. For SWCNT formation, the metal in powder form is kept in the anode by drilling a hole. It evaporates along with graphite. SWCNT nucleates and grows on metal particles of different size. MWCNTs are formed in arc discharge method when a catalyst is not used (Suzuki et al. 2006).

The quality and quantity of CNT produced depends on the following factors.

1. Metal carbon mixture
2. Inert gas used
3. Arc current and
4. Gas pressure.

Laser ablation process

In this process, laser assists the vaporization of graphite under inert atmosphere (Journet and Bernier 1998). In laser ablation process, carbon is vaporized from the graphite disc surface with the assistance of laser beam under intense argon flow. Graphite disc is kept in the middle of long quartz tube, which is kept in furnace maintained at 1200°C in an inert atmosphere. Various carbon nanostructures are formed from vaporized carbon, which are collected from water cooled copper collector (Sharma et al. 2015). High yield of SWCNTs can be obtained by this technique (Thess et al. 1996).

In general, temperature, gas atmosphere, pressure and laser properties control the CNT properties. CNTs produced by Laser ablation technique possess good yield strength and purity level is comparatively better than other techniques (Baddour and Briens 2005). This process is not cost effective and large scale preparation is difficult.

Characterization of CNT

X-ray diffraction (XRD) studies

X-ray diffraction pattern consists of the intensity of the diffraction peak against the diffracted angle of the X-ray beam. The XRD helps in analyzing the following details.

1. Graphite layer arrangement
2. Degree of alignment of CNT
3. Diameter of CNT
4. Crystallinity
5. Defects and impurities

The intensity of peaks will assist in finding the degree of alignment (Cao et al. 2001).

Transmission electron microscope (TEM)

The TEM can help to determine the diameter of the CNT. It also helps in finding out the type of CNT formed, like SWCNT, MWCNT and Defective CNT. The TEM microstructure reveals the graphitic layer of MWCNT (Figure 1). The MWCNT is hollow and tubular in nature. The MWCNT shown in figure possesses high aspect ratio. It exhibits the strong tendency to agglomerate. It is difficult to disperse agglomerated CNTs in elastomeric (or) polymeric matrices.

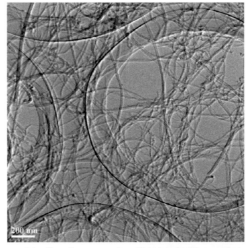

Figure 1. TEM picture of pristine multiwall carbon nanotubes. Reprinted with permission from, Lu, L., Y. Zhai, Y. Zhang, C. Ong and S. Guo. Appl. Surf. Sci. 255: 2162 (2008). Copyrights@Elsevier.

Raman spectroscopy

It is highly advantageous for microscopic analysis. It relies on Raman scattering of monochromatic light, like laser in the visible, near infrared or near ultraviolet range (Wikipedia). The laser beam interacts with molecular vibrations, phonons (or), other excitations in the system. Hence, the energy of the laser photons will be shifted up or down. This shift in energy level gives the information about the vibrational modes in the system. The Raman spectra of the pristine and treated CNT are shown in Figure 2. Two peaks are noted for MWCNT corresponding to D band and G band. The tangential vibration of the carbon atom is

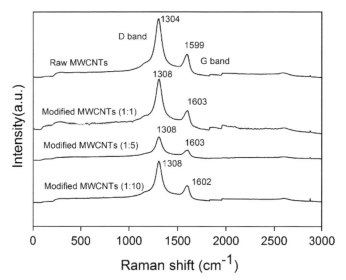

Figure 2. Raman spectra of the pristine MWCNT and modified MWCNT. Reprinted with permission from, Subramaniam, K., A. Das and G. Heinrich. Compos. Sci. Technol. 71: 1441 (2011). Copyrights@Elsevier.

referred to the G-band. It is referred to the peak at 1600 cm⁻¹. The peak at 1300 cm⁻¹ may be attributed to the presence of defects or disorderliness in the graphitic structure. It is referred to as D-band. It was also reported that the radial breathing mode was not present for MWCNT at 100–200 cm⁻¹. MWCNTs possess arrangement of tubes from large diameter to smaller diameter. The radial breathing mode (RBM) signal is too weak for large diameter tubes (Dresselhaus et al. 2005). Modified MWCNT shows slight shift in peak for both bands. The shift confirms the physical modification of CNT without alteration in chemical structure.

Mechanical properties of CNT

The mechanical properties of CNTs are highly significant. It was predicted that the elastic modulus of CNT is 1 TPa and shear modulus is 0.5 TPa, approximately (Lu 1997). The tensile strength is roughly estimated as 0.15 TPa (Demczyk et al. 2002). The single wall CNTs and multiwall CNTs possess higher stiffness and flexibility than any reinforcement. The interatomic distance or binding energy controls the mechanical properties. The high cohesive energy of C-C bonds assists in mechanical strength improvement. Alteration in lattice structure changes the modulus. The CNTs have ample avenues of being used as reinforcement to increase the mechanical properties. However, the dispersion of CNTs at nano-level in a polymer, especially in an elastomeric matrix is a real challenge.

It is necessary to give surface treatment to the CNTs for better dispersion and bonding. The various surface treatment methods of CNT are discussed in the following sections of this chapter.

Processing of CNT-rubber nanocomposites

The processing of rubber compounds involves the mixing/dispersion of fillers and other allied chemicals like antioxidants, accelerators, catalysts and vulcanizing agents. Researchers have found a convincing way of dispersing micron size fillers in both natural and synthetic rubbers. However, researchers are finding it difficult to disperse nanotubes in elastomeric matrix. Since the surface energy of nanotubes is very high, it leads to agglomeration of nano tubes in any processing method. Hence, chemical modifications are being made on nanotubes by surface treatment to enhance the dispersion of nanotubes in elastomeric matrix (George et al. 2015, Bhattacharyyaa et al. 2008, Mohamed et al. 2015).

The processing method also plays a key role in dispersing the pristine nanoparticles and surface treated nanoparticles. The commonly used methods of processing rubber nanocomposites are:

1. Melt compounding
2. Solvent dispersion method, and
3. Latex compounding

In this section, the functionalization of carbon nanotubes and the formation of nanocomposites by various processing methods have been discussed.

Functionalization of carbon nanotube

In order to enhance the dispersion of MWCNT in natural rubber, acid modification was carried out (George et al. 2015). The procedure followed by them is given below. One gram of MWCNT is dispersed in 200 ml of acid mixture made with sulfuric acid and nitric acid in the ratio of 3:1. It is mixed well using ultrasonicator for 1 hr and then refluxed at 100°C for 30 min. This mixture is then cooled, and diluted and washed with water. Finally, it is dried under vacuum. The modified CNT has polar carboxyl and hydroxyl groups (Figure 3). The hydrophilic group on CNT assists to attain electrostatic stability required for colloidal dispersion. During curing, it assists the formation of uniformly segregated network of CNT along the grain boundaries of latex beads (Figure 4).

In a similar manner, carboxyl group was incorporated in the CNT structure by nitric acid treatment for a period of 38 hr (Bhattacharyyaa et al. 2008). In another study, the CNT has been treated in 2 stages (Sui et al. 2008). In the first stage, CNT is treated with blended acid solution (3:1 volume ratio of sulfuric acid and nitric acid) for 30 min, washed and dried. In the second stage, the dried acid treated CNT is blended with HRH (hydrated silica, resorcinol and hexamethylenetetramine) in ratio of 25:3 and ball milled for 30 min. It introduces > C = O carbonyl functional group on surface of CNT due to oxidation. These carbonyl groups can produce physical interaction with NR.

The stable colloidal system of nanotube can be obtained by using ionic or non-ionic surfactant (Wang 2009). Junkong et al. (2015) used non-ionic surfactant for enhancing the dispersion of MWCNT in NR matrix. In that process, a predispersed MWCNT using non-ionic surfactant and ethyl alcohol is prepared for further processing of rubber nanocomposites using solution blending process. Subramaniam et al. (2011) modified the MWCNT using ionic surfactant, trifluromethylsulfonyl imide, for application

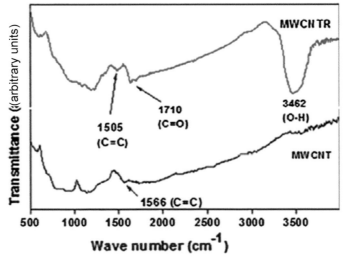

Figure 3. FTIR spectra of MWCNT (Untreated) and MWCNTR (Treated). Reprinted with permission from, George, N., C.S. Julie Chandra, A. Mathiazhagan and R. Joseph. Compos. Sci. Technol. 116: 33 (2015). Copyrights@Elsevier.

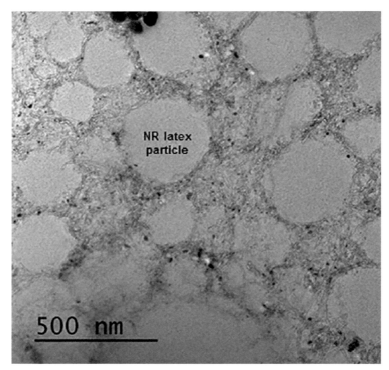

Figure 4. TEM images of NR–MWCNTR composite film. Reprinted with permission from, George, N., C.S. Julie Chandra, A. Mathiazhagan and R. Joseph. Compos. Sci. Technol. 116: 33 (2015). Copyrights@Elsevier.

in synthetic rubber (Chloroprene rubber). The CNTs are added into surfactant solution with thorough mixing until a black paste is achieved. This black paste is used for further dispersion in rubber to produce composites.

For latex compounding, the CNTs are treated in general with surfactants like, sodium dodecylsulfate (SDS) and sodium dodecylbenzenesulfonate (SDBS). The structure of the surfactants plays a key role in determining the ability of the nanotube to disperse and retain in the colloidal state. Chain length along with difference in molecular structure can significantly influence the dispersion. The presence of phenyl ring in the structure can be more effective for the dispersion of CNT than the conventional linear surfactant systems (Mohamed et al. 2015, Hu et al. 2009). The SDBS surfactant with phenyl ring is better than the SDS (Bystrzejewski et al. 2010). A study by Mohammed et al. (2015) confirms that the presence of phenyl group in surfactant increases the dispersion level of CNT in the following order (mono < di < tri chain). In general, the dispersion can be enhanced significantly using aromatic surfactants rather than aliphatic surfactants. The phenyl moieties can effectively diffuse in the space available in nanotube bundles and stay on the surface of nanotube by physical adsorption. The sulfonate groups are exposed to the latex molecules and assist in effective dispersion of CNT by disentangling them from the bundles of CNT.

Melt compounding

It is generally done using mechanical mixers such as two roll mill, brabender and kneader. The natural rubber/synthetic rubber is processed in a kneader or two roll mill with a roll spacing of 1 to 3 mm. During this process, the molecular weight of the natural rubber decreases. More free radicals will evolve out of polymer chain. The relative speed of the rollers lies in the ratio of 1:1.1 to 1:1.45. The activator like, zinc oxide or stearic acid is added to the rubber matrix, which enhances the free radical formation. Carbon nanotube is then added slowly into the masticated natural rubber. The mixing is carried out for 5 to 15 min. This process facilitates the dispersion of agglomerated CNTs by de-agglomerating them. The de-agglomeration of CNT in the natural rubber again depends on the processing time, processing speed and the type of surface treatment on carbon nanotube. Cooling water is continuously circulated through the rollers to maintain the temperature of roller at room temperature.

Other ingredients like, antioxidants, anti-degradants, pre-vulcanization inhibitors, post-vulcanization stabilizers are added and masticated further for 5 min. During this processing, oil is used as an aid for smooth dispersion and mixing. The vulcanizing agent, usually sulfur, is added in desired quantity and masticated further. Finally, the gap between the rollers is increased to the maximum. The masticated rubber compound is hot pressed in a metallic die at a temperature of 150 to 160°C for a period of 5 to 15 min to get desired rubber components.

During this process, vulcanization occurs and the component of desired strength forms. This process has been used by various researchers (Takeuchi et al. 2015, Lu et al. 2008, Subramaniam et al. 2011, Verge et al. 2010). A typical formulation of CNT reinforced rubber nanocomposite is given in Table 1. The safe mixing process for CNT-rubber nanocomposites through melt blending seems to be using the kneader because the contamination due to atmosphere during the dispersion of CNT can be avoided.

Table 1. Typical composition of CNT-NR rubber compound.

Material	Quantity in phr
Natural rubber	100
Activators : Stearic Acid	2
Activators : Zinc Oxide	4
CNT	0–5
Antioxidant	2
Sulfur	2.5

Solvent compounding

In this process, the carbon nanotubes are dispersed in a solvent usually tetrahydrofuran (THF) or toluene (Razi et al. 2006, Sui et al. 2008, Tsuchiya et al. 2011, Hu et al. 2012). The dispersion is assisted by ultrasonication or mechanical stirring. During this process, the entangled CNT is disentangled. The synthetic polymer like SBR is allowed to swell in THF for 24 hr (Tsuchiya et al. 2011). The CNT suspension and elastomeric solution are again mixed for a period of 30 min. Ultrasonication is generally used for mixing. Special mixing process like rotation-revolution mixing may also be used (Tsuchiya et al. 2011). The other essential ingredients like activators, aging inhibitors, accelerators and sulfur will be simultaneously mixed during this process. This homogenous mixture of CNT-rubber is dried under vacuum at elevated temperature to remove the solvent. The dried mixture is further cured at 150–160°C for a period of 5–20 min to get the desired nanocomposite samples.

In a similar manner, natural rubber CNT nanocomposites can also be prepared, which involves 5 stages.

In stage 1: nanotubes are dispersed in a solvent like toluene/THF.

In stage 2: NR is dissolved in a solvent in desired weight fraction.

In stage 3: both are mixed and dispersed ultrasonically to get a homogenous CNT-NR mixture. This mixture is dried in vacuum to remove the solvent.

In stage 4: the dried mixture of CNT–NR is then taken to two roll mill or any other mechanical mixture for further mastication. During this process, activators, antioxidants and vulcanization agents are added to get the final rubber compound ready for final process.

In stage 5: the rubber compound is then hot pressed at 150 to 160°C to get the components of required dimensions.

Latex compounding

The carbon nanotube handling should be extremely cautious, since it is carcinogenic in nature and affects human health. CNTs may be released to atmosphere during mastication of rubber in open type two roll mill. Hence, the processing route seems to be hazardous.

Similarly, in solvent processing method, the chemicals being used as solvent are THF, toluene, etc. These solvents are highly polluting and they are used in large quantities for dispersing rubber and nanotubes. Further, it is necessary to remove the solvent to get the composites out of it. Because of these detriments, processing of rubber nanocomposites through latex route seems to be a promising one. Here the dispersion of nanotubes in natural rubber is carried out in latex stage itself, where the viscosity of NR is very low. Several researchers have prepared CNT rubber nanocomposites using this technique (George et al. 2015, Bhattacharyyaa et al. 2008, Mohamed et al. 2015). They were able to control the dispersion and continuity of network as well.

In this process, the modified carbon nanotubes (treated one) are dispersed in ionized water (George et al. 2015). In certain cases, the CNTs in desired quantity are dispersed in surfactant solution (Mohamed et al. 2015). The aqueous dispersion is mixed mechanically or by ultrasonicator for a minimum period of 1 to 3 hr. This mixture is added to the compounded latex and mixed well using sonicator for a period of 1 hr. These samples are allowed to mature and then cast into films. These films are dried at room temperature and cured at 100°C for 1 hr (George et al. 2015).

The treated CNT may also be mixed with pristine latex along with coagulating agents. It is allowed to mature overnight, cold pressed and dried in oven at 60°C for a period of 24 hr. These dried sheets can be further taken for mastication in mechanical mill, where the required quantities of activators, vulcanizing agents, antioxidants and processing aid are added, compounded and then hot pressed at higher temperatures to get the required components or samples.

Physical properties of CNT—rubber nanocomposites

Microstructural studies

The dispersion level of CNT in elastomeric matrix can be analyzed by microstructural studies. Transmission electron microscope (TEM) is widely used for such analysis, since the dispersion is at the nano-level (Ma et al. 2010). Several researchers have used the transmission electron microscope to analyze the dispersion of CNTs in natural rubber (Razi et al. 2006, George et al. 2015, Bhattacharyyaa et al. 2008, Mohamed et al. 2015) and synthetic rubber (Tsuchiya et al. 2011, Lu et al. 2008, Subramaniam et al. 2011, Verge et al. 2010).

For TEM analysis, the samples of sizes in the order of 60 nm are sliced by using a diamond blade in a microtome. The electrons are allowed to pass through the sample in TEM, which facilitates the observation of the dispersion state of nanoparticle in rubber matrix. The degree of dispersion of CNTs in natural rubber matrix is shown in Figure 5. The spacing between CNTs is very high, when the quantity of CNTs is 1 wt.%. The dispersion of CNTs is homogeneous and there is low interaction between CNTs. However, at higher quantities of CNTs, the spacing between CNTs becomes very low.

The effect of surfactants can be understood from the dispersion of CNTs in rubber matrix (Mohamed et al. 2015). The length, size and aspect ratio of the dispersed CNTs can also be analyzed from the captured images using an image analyzer.

X-ray photoelectron spectroscopy (XPS)

The presence of defects and functional group on the surface of CNT can be effectively analyzed with the help of XPS. XPS study clearly depicts the presence of various functional groups on the treated CNT. The presence of peak at 284.1 eV corresponds to the graphitic structure of CNT (Ago et al. 1999). Even the defects on the nanotube surface can be identified by the presence of specific peak. Defect in nanotube is identified by the presence of peak at 285.1 eV (George et al. 2015). Carbon atom bonding with different

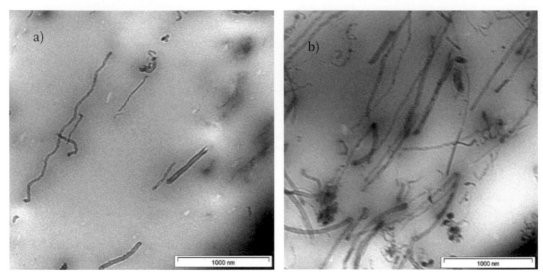

Figure 5. TEM images of CNTs in SMR (a) 1 wt.% of CNTs (b) 10 wt.% of CNTs. Reprinted with permission from, Razi, A.F., M.A. Atieh, N. Girun, T.G. Chuah, M. El-Sadig and D.R.A. Biak. Compos. Struct. 75: 496 (2006). Copyrights@ Elsevier.

oxygen containing moieties can be represented by the presence of peaks. For instance, the presence of C = O is identified by the peak at 287 eV and O – C = O is identified by the peak corresponding to 288.6 eV (George et al. 2015).

Curing characteristics

The study of curing mechanism and curing kinetics is essential to optimise the curing process (Chen et al. 2004, Ding and Leonov 1996). The curing process should be optimum to get better properties from CNT-rubber nanocomposites. The important parameters are:

a) Curing time
b) Curing agent, and
c) Curing temperature

The presence of CNT in natural rubber will alter the curing parameters. Researchers have analyzed the curing process in CNT-natural rubber (Sui et al. 2008) and CNT-synthetic rubber nanocomposites (Peddini et al. 2014). The typical curing of natural rubber with sulfur as vulcanizing agent has three regions.

Region 1: Scorch time. It corresponds to processing time. The changes in processing conditions can be effectively done during this period. Non-rotor rheometer is used to study the curing behavior of natural rubber. The curing curves of neat NR and CNT-NR nanocomposites are shown in Figure 6, which depicts the effect of processing temperature on curing time. As curing temperature increases, curing time decreases for both NR and CNT-NR nanocomposites. The vulcanizing kinetic parameters can be obtained from vulcanizing curve (Sui et al. 2008).

The degree of conversion (α) is represented as

$$\alpha = (M_t - M_o)/(M_\alpha - M_o)$$

where, M_t = Torque value corresponding to time t

M_o = Torque value at initial curing

M_α = Torque value after the curing reaction

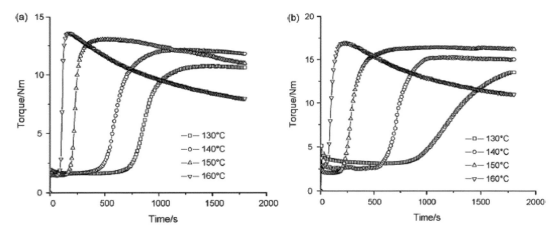

Figure 6. Curing characteristics curves of (a) neat NR and (b) CNT/NR nanocomposites. Reprinted with permission from, Sui, G., W.H. Zhong, X.P. Yang and Y.H. Yu. Mater. Sci. Engg. A. 485: 524 (2008). Copyrights@Elsevier.

The crosslink density is directly related to the stiffness of the rubber. The conversion rate decreases with the incorporation of CNT into the NR matrix. Reaction rate and conversion ratio increase with rise in vulcanizing temperature. The activation energy required to initiate the curing process also increases with the incorporation of CNT in rubber.

The CNT may interact or absorb the accelerator. The surface area of CNT is high and its interaction with the macromolecules of rubber will restrict the motion of the molecular chains of NR. It leads to decrease in reactivity. Hence, more energy is required to cure CNT-rubber nanocomposites. Considering the various factors of curing and safe processing time, the optimum curing temperature is generally 150°C for NR and CNT composites.

The presence of functional groups on the treated CNT will establish more physical interactions with the rubber matrix, which can further restrict the movement of rubber molecules and enhance the stiffness. Accordingly, the treated CNT increases the curing time. Sufficient trial experiments should be carried out to fix the optimum curing time for CNT-rubber nanocomposites, when different surface treatments are given for CNT. The type of matrix (NR/Synthetic rubber) will also alter the curing behavior and the optimum curing time will be different.

Region 2: Curing reaction time: Crosslinked network formation occurs during this time and the stiffness of rubber is increased.

Region 3: Maturing stage: Crosslinking is completed and in some cases, over curing occurs. Additional slow crosslinking occurs during this stage.

Thermogravimetric analysis

The sample is taken in a crucible and heated from RT to 800°C under inert atmosphere. The weight loss in the sample is corroborated with the decomposition of surfactant and other chemical reagents, which indicates the dispersion level and cross linking efficiency of CNT-rubber nanocomposites. Verge et al. (2010) studied the dispersed state of CNTs in synthetic rubber matrix. The supernatant of CNT-rubber was heated in thermogravimetric analyzer under inert atmosphere. The residue found after thermogravimetric analysis (TGA) is directly related to the presence of CNT content in supernatant. The degree of dispersion is also related to processing time and speed (Figure 7). Heat developed during mixing will induce free radical formation on rubber matrix due to thermo-oxidation and thermo-degradation. This will lead to the reaction with the surface of CNT and thus NBR chain gets grafted on the surface of the CNT. The vulcanization additive does not slow down the grafting process onto the CNT surface. However, it favors the reaction of free radicals with CNT to a certain extent. The vulcanization agent reacts with rubber molecules. The rubber molecules get grafted onto the CNTs. These grafted CNTs also take part in the pre-

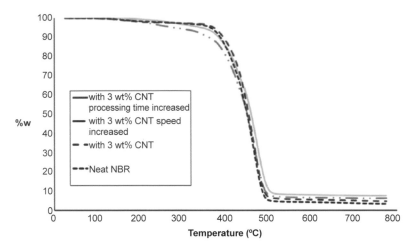

Figure 7. TGA curves of different melt processed NBR/CNT composites (before vulcanization step). Reprinted with permission from, Verge, P., S. Peeterbroeck, L. Bonnaud and P. Dubois. Compos. Sci. Technol. 70: 1453 (2010). Copyrights@ Elsevier.

vulcanization of rubber molecules. These reactions can be analyzed with the help of thermogravimetric analysis. The residue after TGA is around 5.5 wt.% more for rubber with vulcanizing additives than rubber without vulcanizing additives (Figure 8). The increase in weight is due to the presence of vulcanizing agents and polymer grafting on CNT surface. The presence of surface grafted CNT is higher in the system with vulcanizing agent than to the system without vulcanizing agent. The following relation will assist in calculating the quantity of polymer adsorbed on the filler surface (Verge et al. 2010).

$$Q_{ads} = (m_{dried} - m_{CNT})/m_{CNT}$$

where,

Q_{ads} = quantity of polymer adsorbed
m_{CNT} = initial mass of CNT
m_{dried} = mass of dried centrifuged residue

Hence, the TGA effectively helps to find the presence of surfactant and other molecules grafted on CNT.

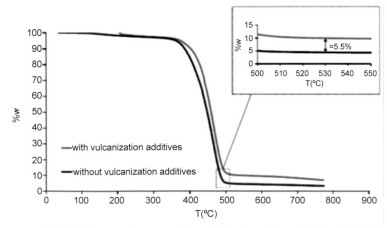

Figure 8. TGA curves of NBR/CNT composites prepared by melt blending with or without vulcanization additives (before vulcanization step). Reprinted with permission from, Verge, P., S. Peeterbroeck, L. Bonnaud and P. Dubois. Compos. Sci. Technol. 70: 1453 (2010). Copyrights@Elsevier.

Electrical properties of CNT-rubber nanocomposites

The electrical performance of rubber matrix has been improved by the incorporation of CNT into the elastomeric matrix. The electrical conductivity depends the following factors.

1. Dispersion level of CNT
2. Concentration of CNT in rubber matrix
3. Aspect ratio of CNT

The electrical conductivity increases with the increase in CNT content. Moreover, the electrical conductivity of CNT-rubber nanocomposites at lower filler concentration depends on frequency. The composites with higher CNT concentration exhibit high electrical conductivity, but it is not influenced by frequency, indicating the constant dielectric loss (Bhattacharyyaa et al. 2008, Tsuchiya et al. 2011). The carbon nanotube forms a continuous network, which results in decrease in percolation threshold value. With the increase in CNT content, the composites almost behave like ohmic conductor. Similar trend was noticed in natural rubber system as well (George et al. 2015). The electrical conductivity and dielectric permittivity increase near the percolation threshold (Dang et al. 2003, Pötschke et al. 2004). If the aspect ratio is maintained during mixing, the percolation threshold can be limited to a minimum value. Studies also confirm that a new technique called rotation-revolution mixing can assist effective dispersion of CNTs in the synthetic styrene-butadiene rubber. There is significant increase in electrical conductivity at lower filler concentration (0.67 phr) compared to conventional Banbury mixing technique, because of no reduction in aspect ratio (Figure 9).

The latex compounding route also assists the formation of segregated network along the grain boundary of latex molecules. It helps to increase the conductivity and results in low percolation value of 0.086 phr (George et al. 2015). The dielectric constant and dielectric loss are high for CNT-NR composites at low frequency and decrease exponentially with the increase in frequency. The dipole moment created/unit value is higher at low frequency resulting in high dielectric loss and dielectric constant.

The dielectric polarity induced at high frequency will reduce both dielectric constant and dielectric loss. The conduction is facilitated by conduction current at high frequency and by displacement current at low frequencies. The nature of dispersion of CNT also affects the conductivity of CNT-rubber nanocomposites. Aligned dispersion and segregated network near interfacial region of polymer molecule give better conductivity than random dispersion. Moreover, it gives a lower percolation threshold value than other non-segregated CNT-rubber composites (Tsuchiya et al. 2011, Das et al. 2012).

In certain cases, the electrical conductivity drops once the sample is stretched beyond 500% strain. The CNT dispersed in the elastomeric chain may get broken during stretching and lose its interfacial contact resulting in network disturbances, which can lead to a decrease in electrical conductivity (Bhattacharyyaa et al. 2008). For synthetic NBR rubber, the presence of acrylonitrile group assists the

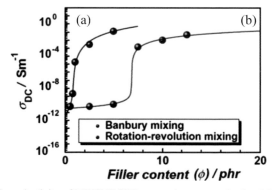

Figure 9. Dependence of DC conductivity of MWCNT-SBR composites prepared using (a) rotation–revolution mixer and (b) a conventional Banbury mixer. Reprinted with permission from, Tsuchiya, K., A. Sakai, T. Nagaoka, K. Uchida, T. Furukawa and H. Yajima. Compos. Sci. Technol. 71: 1098 (2011). Copyrights@Elsevier.

grafting of CNT. The dispersion of CNT is good in NBR containing higher acrylonitrile content. With the increase in acrylonitrile content, the electrical conductivity increases and it is corroborated well with the disaggregated CNT network (Verge et al. 2010).

Mechanical properties of CNT-rubber nanocomposites

Tensile properties

The CNT, being a nano-filler with the highest strength and modulus, can be an effective reinforcement to enhance the mechanical properties of rubber. The interfacial interaction between CNT and rubber assists the transfer of load from matrix to CNT. The tensile strength of rubber generally increases by the incorporation of nanofillers (Zhao et al. 2006). The same behavior was observed by many researchers when natural rubber or synthetic rubber is reinforced with carbon nanotube. The tensile strength of CNT-rubber nanocomposites increases with the increase in CNT concentration. However, the strain at failure decreases correspondingly (Razi et al. 2006, Takeuchi et al. 2015, George et al. 2015, Bhattacharyyaa et al. 2008). The improvement in tensile strength and modulus may depend on the following factors.

1. Homogenous dispersion of CNT in rubber matrix
2. Orientation of CNT in the rubber matrix
3. Diameter and aspect ratio of CNT in the dispersed state
4. Surface treatment of CNT
5. Concentration of CNT in rubber matrix
6. Structure of CNT in rubber matrix

The diffusion of elastomeric molecules into the space between nanotubes results in the disentanglement of CNT aggregates and enhances the interfacial contact between CNT and elastomeric molecules. This can cause the improvement in mechanical properties. The stress-strain behavior of CNT-rubber nanocomposite is shown in Figure 10.

In certain cases, especially at the higher concentration of nanotubes, CNT to CNT contact dominates the CNT to polymer interface and results in agglomerated CNT structure. Hence the enhancement in mechanical properties may not be significant.

Maintaining the high aspect ratio of CNT in elastomeric matrix also significantly enhances the mechanical properties. Conventional mixing processes generally break the CNTs. The aspect ratio of the CNT in CNT-synthetic rubber nanocomposites processed by rotation-revolution mixing is maintained even after the mixing process (Tsuchiya et al. 2011).

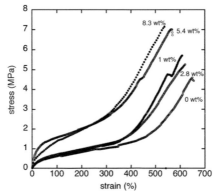

Figure 10. Stress-strain behavior of CNT-NR nanocomposites. Reprinted with permission from, Bhattacharyya, S., C. Sinturel, O. Bahloul, M.L. Saboungi, S. Thomas and J.P. Salvetat. Carbon. 46: 1037 (2008). Copyrights@Elsevier.

Likewise, the size of the CNT also significantly impacts the mechanical properties (Endo et al. 2008). With respect to latex compounding technique, the distributed rigid network of CNTs carries the applied load and results in better mechanical properties (George et al. 2015). The functionalization of CNTs by the introduction of OH and carbonyl groups also assists the dispersion of CNT in elastomers and enhances the mechanical properties.

A similar trend was also noticed with respect to the modulus of CNT-rubber nanocomposites. The modulus increases by several folds with the increase in CNT content (Razi et al. 2006). The modulus increases almost by 8 times for the CNT content of 10 wt.% (Figure 11). Similar trend is observed in

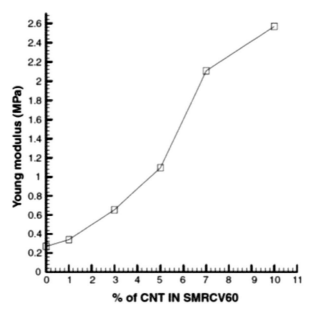

Figure 11. Elastic modulus of CNT–NR nanocomposites. Reprinted with permission from, Razi, A.F., M.A. Atieh, N. Girun, T.G. Chuah, M. El-Sadig and D.R.A. Biak. Compos. Struct. 75: 496 (2015). Copyrights@Elsevier.

CNT-natural rubber nanocomposites prepared through latex compounding (George et al. 2015). It has been reported that the distributed network of CNT and the strong interaction of CNT-rubber matrix are responsible for this enhancement.

The improvement in mechanical properties is evident from the fracture surface (Figure 12). Neat NR fracture surface remains smooth, whereas the fracture surface of 0.5 phr CNT-nanocomposites has rough facets. Better bonding between NR and the CNT network leads to the improvement in mechanical properties. The decrease in mechanical properties at higher CNT concentration may be attributed to the formation of aggregated network of CNT. Porosity could be the other factor that affects the mechanical properties (Jin et al. 2007).

Tear strength

The tear strength increases with the increase in CNT content. However, there is a slight decrease in tear strength at higher concentration (Table 2). Strong interaction between the NR and MWCNT is essential

Table 2. Tear strength of CNT–NR nanocomposites (George et al. 2015).

CNT concentration (phr)	0	0.05	0.1	0.3	0.5	1.0
Tear strength (MPa)	34.8	35.5	42.5	43.4	55.4	30.3

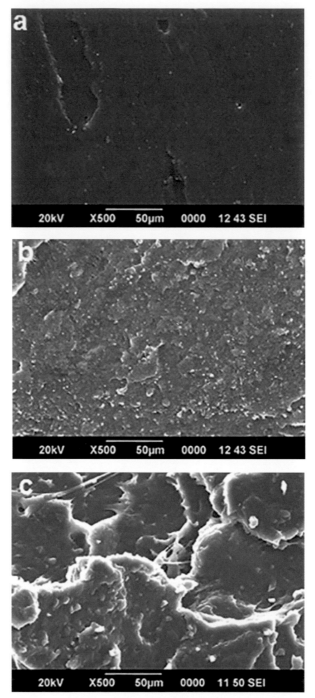

Figure 12. Fractographs from the tensile tested specimens of (a) NR, (b) NR–MWCNTR 0.1 and (c) NR–MWCNTR 0.5. Reprinted with permission from, George, N., C.S. Julie Chandra, A. Mathiazhagan and R. Joseph. Compos. Sci. Technol. 116: 33 (2015). Copyrights@Elsevier.

for the improvement in tear strength. At higher concentration, the aggregated networks of CNT make interpenetration of elastomer molecules more difficult, which leads to poor mechanical properties.

Hardness

It is a measure of the tendency of materials to resist indentation or abrasion. The hardness of elastomers in general is very low. However, it is being used in high abrasive applications, such as automotive tires, aircraft tires, conveyor belts, chappals, etc. after the incorporation of fillers. The wear resistance and hardness are directly related to each other in most of the cases. The hardness of the elastomers can be enhanced by the addition of CNT. The hardness value of natural rubber increases to a maximum for higher nanotube concentration from the value of 42 JIS A. The finer the CNT, the higher will be the enhancement in hardness (Takeuchi et al. 2015). The hardness of CNT-NR nanocomposites increases with CNT content not only because of the reinforcement effect, but also due to increase in crosslink density (Junkong et al. 2015).

Dynamic mechanical analysis

The dynamic mechanical properties of CNT-rubber nanocomposites are generally studied by using dynamic mechanical analyzer. The storage modulus measures the recoverable strain energy in a deformed specimen. The incorporation of CNT in the natural rubber has improved the storage modulus significantly (Takeuchi et al. 2015, Bhattacharyyaa et al. 2008, Sui et al. 2008). The high specific surface area of CNT and its modulus assist in stiffness improvement of CNT-rubber nanocomposites, which further enhances the storage modulus of CNT-rubber composites. The high temperature storage modulus significantly increases with the incorporation of CNT, whereas there is only a slight improvement at low temperature (Figure 13). The semicrystalline nature of natural rubber at lower temperature leads to high storage modulus (Takeuchi et al. 2015, Sui et al. 2008). The storage modulus of composites increases with increase in CNT content, but starts decreasing at higher CNT concentration with increasing strain amplitude. At lower concentration, strain amplitude has less impact on storage modulus (Bhattacharyyaa et al. 2008).

For synthetic rubbers like NBR (Lu et al. 2008), the storage modulus is high below T_g. The incorporation of CNT has increased the storage modulus significantly, but the influence is less above T_g. The improvement below T_g is due to the hydrodynamic reinforcement effect of CNT with the rubber matrix. For chloroprene rubber (CR) also, there is no significant change in storage modulus above T_g (Subramaniam et al. 2011).

Loss factor (tan δ) refers to energy dissipated as heat under an oscillating force. The effect of temperature on loss factor and also the influence of CNT incorporation are depicted in Figure 14. Loss factor curve helps to find the glass transition temperature of the natural rubber system. The temperature corresponding to the peak of the loss factor curve is glass transition temperature. The carbon nanotube has the tendency of arresting the mobility of rubber molecules, which facilitates the rise in glass transition temperature in most cases. The peak in the loss factor curve shifts slightly towards higher temperature by

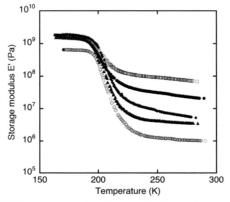

Figure 13. Storage modulus of CNT-NR nanocomposites. Reprinted with permission from, Bhattacharyya, S., C. Sinturel, O. Bahloul, M.L. Saboungi, S. Thomas and J.P. Salvetat. Carbon 46: 1037 (2008). Copyrights@Elsevier.

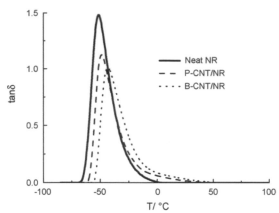

Figure 14. Loss factor of CNT-NR nanocomposites. Reprinted with permission from, Sui, G., W.H. Zhong, X.P. Yang and Y.H. Yu. Mater. Sci. Engg. A. 485: 524 (2008). Copyrights@Elsevier.

the incorporation of the CNT. The strong interfacial interaction between matrix and CNT is responsible for this rise. However, the CNT incorporation has brought a decreasing trend in T_g for latex based composite processing (Bhattacharyyaa et al. 2008).

The T_g value evaluated for CR does not change significantly with the addition of CNT (Subramaniam et al. 2011). Glass transition temperature decreases in certain polymer composite system and increases or remains the same in certain nanocomposite systems (Bokobza 2007, Das et al. 2008). It depends on the interfacial interaction between elastomers and fillers. The increase in tan δ peak with rise in MWCNT content is due to increase in interfacial interactions between CNT and rubber matrix. Increase in T_g also confirms the restriction of segmental motion of the polymeric chain (Peddini et al. 2014). Again, it also depends on the degree of dispersion of CNT in the rubber matrix, processing route of nanocomposites, type of CNT used, crosslinking efficiency, type of surfactant used and vulcanization process.

Application of CNT rubber nanocomposites

The CNT-rubber nanocomposites possess better mechanical properties and electrical conductivity. The high tensile strength of nanocomposites makes them suitable for pressure resistant applications. Florine rubber containing MWCNT is being used as O-ring in oil exploration (Endo et al. 2008). The incorporation of CNT with dispersion at nano-level has improved the electrical conductivity by decreasing the percolation threshold value, which makes them suitable for sensor and semiconductor applications. The incorporation of CNT in NR/SBR system also provides an opportunity of introducing high performance aircraft and automotive tires into the market.

References

Ago, H., T. Kugler, F. Cacialli, W. Salaneck, M. Shaffer, A. Windle et al. 1999. Work functions and surface functional groups of multiwall carbon nanotubes. J. Phys. Chem. B. 103: 8116–8121.

Amelinckx, S., D. Bernaerts, X.B. Zhang, G. Van Tendeloo and J. Van Landuyt. 1995. A structure model and growth mechanism for multishell carbon nanotubes. Science 267(5202): 1334–1338.

Arroyo, M., M.A. Lopez-Manchado and B. Herrero. 2003. Organomontmorillonite as substitute of carbon black in natural rubber compounds. Polymer. 44: 2447–2453.

Baddour, C. and C. Briens. 2005. Carbon nanotube synthesis: a review. Int. J. Chem. React. Eng. 3: DOI: https://doi.org/10.2202/1542-6580.1279.

Bhattacharyya, S., C. Sinturel, O. Bahloul, M.L. Saboungi, S. Thomas and J.P. Salvetat. 2008. Improving reinforcement of natural rubber by networking of activated carbon nanotubes. Carbon 46: 1037–1045.

Bokobza, L. 2007. Multiwall carbon nanotube elastomeric composites. Polymer. 48: 4907–4920.

Bystrzejewski, M., A. Huczko, H. Lange, T. Gemming, B. Buchner and M.H. Rummeli. 2010. Non-covalent functionalization of carbon nanotubes with surfactants and polymers. J. Colloid Interf. Sci. 345: 138–142.

Cao, A., C. Xu, J. Liang, D. Wu and B. Wei. 2001. X-ray diffraction characterization on the alignment degree of carbon nanotubes. Chem. Phys. Lett. 344: 13–17.

Cassell, A.M., J.A. Raymakers, J. Kong and H. Dai. 1999. Large scale CVD synthesis of single-walled carbon nanotubes. J. Phys. Chem. B. 103: 6484–6492.

Chen, W.Y., Y.Z. Wang, S.W. Kuo, C.F. Huang, P.H. Tung and F.C. Chang. 2004. Thermal and dielectric properties and curing kinetics of nanomaterials formed from POSS-epoxy and meta-phenylenediamine. Polymer 45: 6897–6908.

Dang, Z.M., L.Z. Fan, Y. Shen and C.W. Nan. 2003. Dielectric behavior of novel three-phase MWNTs/BaTiO$_3$/PVDF composites. Mater. Sci. Eng. B. 103: 140–144.

Das, A., K.W. Stöckelhuber, R. Jurk, M. Saphiannikova, J. Fritzsche, H. Lorenz et al. 2008. Modified and unmodified multiwalled carbon nanotubes in high performance solution styrene butadiene and butadiene rubber blends. Polymer. 49: 5276–5283.

Das, A., G.R. Kasaliwal, R. Jurk, R. Boldt, D. Fischer, K.W. Stöckelhuber et al. 2012. Rubber composites based on graphene nanoplatelets, expanded graphite, carbon nanotubes and their combination: a comparative study. Compos. Sci. Technol. 72: 1961–1967.

Demczyk, B.G., Y.M. Wang, J. Cumings, M. Hetman, W. Han, A. Zettl et al. 2002. Direct mechanical measurement of the tensile strength and elastic modulus of multiwalled carbon nanotubes. Mater. Sci. Eng. A. 334: 173–178.

Ding, R. and A.I. Leonov. 1996. A kinetic model for sulfur accelerated vulcanization of a natural rubber compound. J. Appl. Polym. Sci. 61: 455–463.

Dresselhaus, M.S., G. Dresselhaus, R. Saito and A. Jorio. 2005. Raman spectroscopy of carbon nanotubes. Phys. Rep. 409: 47–99.

Dunlap, B.L. 1992. Connecting carbon tubules. Phys. Rev. B: Condens. Matter. 46: 1933–1936.

Endo, M., M. Strano and P. Ajayan. 2008. Potential applications of carbon nanotubes. Carbon Nanotubes 111: 13–61.

Fonseca, A., K. Hernadi, J.B. Nagy, P. Lambin and A.A. Lucas. 1996. Growth mechanism of coiled carbon nanotubes. Synthetic Met. 77: 235–242.

Gan, B., J. Ahn, Q. Zhang, S.F. Rusli, S.F. Yoon, J. Yu et al. 2001. Y-junction carbon nanotubes grown by *in situ* evaporated copper catalyst. Chem. Phys. Lett. 333: 23–28.

George, N., C.S. Julie Chandra, A. Mathiazhagan and R. Joseph. 2015. High performance natural rubber composites with conductive segregated network of multiwalled carbon nanotubes. Compos. Sci. Technol. 116: 33–40.

Hu, C.Y., Y.J. Xu, S.W. Duo, R.F. Zhang and M.S. Li. 2009. Non-covalent functionalization of carbon nanotubes with surfactants and polymers. J. Chin. Chem. Soc. 56: 234–239.

Hu, H., L. Zhao, J. Liu, Y. Liu, J. Cheng, J. Luo et al. 2012. Enhanced dispersion of carbon nanotube in silicone rubber assisted by graphene. Polymer. 53: 3378–3385.

Iijima, S. and T. Ichihasi. 1993. Single-shell carbon nanotubes of 1-nm diameter. Nature 363(6430): 603–605.

Jin, S.H., Y.B. Park and K.H. Yoon. 2007. Rheological and mechanical properties of surface modified multi-walled carbon nanotube-filled PET composite. Compos. Sci. Technol. 67: 3434–3441.

Joly, S., G. Garnaud, R. Ollitrault, L. Bokobza and J.E. Mark. 2002. Organically modified layered silicates as reinforcing fillers for natural rubber. Chem. Mater. 14: 4202–4208.

Journet, C. and P. Bernier. 1998. Production of carbon nanotubes. Appl. Phys. A. 67: 1–9.

Junkong, P., P. Kueseng, S. Wirasate, C. Huynh and N. Rattanasom. 2015. Cut growth and abrasion behaviour, and morphology of natural rubber filled with MWCNT and MWCNT/carbon black. Polym. Test. 41: 172–183.

Klüppel, M. 2003. The role of disorder in filler reinforcement of elastomers on various length scales. Adv. Polym. Sci. 164: 1–86.

Li, W.Z., J.G. Wen and Z.F. Ren. 2002. Effect of temperature on growth and structure of carbon nanotubes by chemical vapor deposition. Appl. Phys. A. 74: 397–402.

Lu, J.P. 1997. Elastic properties of carbon nanotubes and nanoropes. Phys. Rev. Lett. 79: 1297–1300.

Lu, L., Y. Zhai, Y. Zhang, C. Ong and S. Guo. 2008. Reinforcement of hydrogenated carboxylated nitrile–butadiene rubber by multi-walled carbon nanotubes. Appl. Surf. Sci. 255: 2162–2166.

Ma, P.C., N.A. Siddiqui, G. Marom and J.K. Kim. 2010. Dispersion and functionalization of carbon nanotubes for polymer-based nanocomposites: a review. Compos. A. 41: 1345–1367.

Mohamed, A., A.K. Anas, S.A. Bakar, T. Ardyani, W.M.W. Zin, S. Ibrahim et al. 2015. Enhanced dispersion of multiwall carbon nanotubes in natural rubber latex nanocomposites by surfactants bearing phenyl groups. J. Colloid Interface Sci. 455: 179–187.

Mukhopadhyay, K., A. Koshio, T. Sugai, N. Tanaka, H. Shinohara, Z. Konya et al. 1999. Bulk production of quasi-aligned carbon nanotube bundles by the catalytic chemical vapour deposition (CCVD) method. Chem. Phy. Lett. 303: 117–124.

Murakami, K., S. Iio, Y. Ikeda, H. Ito, M. Tosaka and S. Kohjiya. 2003. Effect of silane-coupling agent on natural rubber filled with silica generated *in situ*. J. Mater. Sci. 38: 1447–1455.

Oberdisse, J., A. El Harrak, G. Carrot, J. Jestin and F. Boue. 2005. Structure and rheological properties of soft–hard nanocomposites: influence of aggregation and interfacial modification. Polymer 46: 6695–6705.

Pal, P.K. and S.K. De. 1982. Effect of reinforcing silica on vulcanization, network structure, and technical properties of natural rubber. Rubber Chem. Technol. 55: 1370–1388.

Peddini, S.K., C.P. Bosnyak, N.M. Henderson, C.J. Ellison and D.R. Paul. 2014. Nanocomposites from styrene-butadiene rubber (SBR) and multiwall carbon nanotubes (MWCNT) part 1: Morphology and rheology. Polymer. 55: 258–270.

Pötschke, P., M. Abdel-Goad, I. Alig, S. Dudkin and D. Lellinger. 2004. Rheological and dielectrical characterization of melt mixed polycarbonate-multiwalled carbon nanotube composites. Polymer. 45: 8863–8870.

Razi, A.F., M.A. Atieh, N. Girun, T.G. Chuah, M. El-Sadig and D.R.A. Biak. 2006. Effect of multi-wall carbon nanotubes on the mechanical properties of natural rubber. Compos. Struct. 75: 496–500.

Satishkumar, B.C., A. Govindaraj, R. Sen and C.N.R. Rao. 1998. Single-walled nanotubes by the pyrolysis of acetylene-organometallic mixtures. Chem. Phys. Let. 293: 47–52.

Seraphin, S. and D. Zhou. 1994. Single-walled carbon nanotubes produced at high yield by mixed catalysts. Appl. Phy. Lett. 64: 2087–2089.

Sharma, R., P. Benjwal and K.K. Kar. 2015. Carbon nanotubes: synthesis, properties and applications. pp. 89–138. *In*: S. Mohanty, S.K. Nayak, B.S. Kaith and S. Kalia (eds.). Polymer Nano-Composites Based on Inorganic and Organic Nanomaterials. Scrivener Publishing, Beverly, USA.

Silva, V.P., M.C. Goncalves and I.V.P. Yoshida. 2006. Biogenic silica short fibers as alternative reinforcing fillers of silicone rubbers. J. Appl. Polym. Sci. 101: 290–299.

Subramaniam, K., A. Das and G. Heinrich. 2011. Development of conducting polychloroprene rubber using imidazolium based ionic liquid modified multi-walled carbon nanotubes. Compos. Sci. Technol. 71: 1441–1449.

Sui, G., W.H. Zhong, X.P. Yang and Y.H. Yu. 2008. Curing kinetics and mechanical behavior of natural rubber reinforced with pretreated carbon nanotubes. Mater. Sci. Engg., A. 485: 524–531.

Suzuki, T., Y. Guo, S. Inoue, X. Zhao, M. Ohkohchi and Y. Ando. 2006. Multiwalled carbon nanotubes mass-produced by DC arc discharge in He–H$_2$ gas mixture. J. Nanoparticle Res. 8: 279–285.

Takeuchi, K., T. Noguchi, H. Ueki, K. Niihara, T. Sugiura, S. Inukai et al. 2015. Improvement in characteristics of natural rubber nanocomposite by surface modification of multi-walled carbon nanotubes. J. Phys. Chem. Solids 80: 84–90.

Thess, A., R. Lee, P. Nikolaev, H. Dai, P. Petit, J. Robert et al. 1996. Crystalline ropes of metallic carbon nanotubes. Science 273(5274): 483–487.

Tsuchiya, K., A. Sakai, T. Nagaoka, K. Uchida, T. Furukawa and H. Yajima. 2011. High electrical performance of carbon nanotubes/rubber composites with low percolation threshold prepared with a rotation–revolution mixing technique. Compos. Sci. Technol. 71: 1098–1104.

Verge, P., S. Peeterbroeck, L. Bonnaud and P. Dubois. 2010. Investigation on the dispersion of carbon nanotubes in nitrile butadiene rubber: Role of polymer-to-filler grafting reaction. Compos. Sci. Technol. 70: 1453–1459.

Wang, H. 2009. Dispersing carbon nanotubes using surfactants. Curr. Opin. Colloid Interf. Sci. 14: 364–371.

Xu, Z., X. Bai, Z.L. Wang and E. Wang. 2006. Multiwall carbon nanotubes made of monochirality graphite shells. J. Am. Ceram. Soc. 128: 1052–1053.

Zhao, Q., R. Tannenbaum and K.I. Jacob. 2006. Carbon nanotubes as Raman sensors of vulcanization in natural rubber. Carbon. 44: 1740–1745.

14

Nanocomposite Fibers with Carbon Nanotubes, Silver, and Polyaniline

Nuray Ucar* and *Nuray Kizildag*

Introduction

Functional textile fibers attract great interest because of their wide range of potential applications (Jianming et al. 2005) such as smart/intelligent garments (Carfagna and Persico 2006), tissue engineering (Cullen et al. 2008), medical textiles (Yeo and Jeong 2003), sensor applications (Devaux et al. 2011, Ferreira et al. 2013), etc. The use of nanotechnology in fiber spinning enables the production of fibers with improved performance and/or unique functions (Carfagna and Persico 2006). The addition of different types of nanoparticles such as metals, metal oxides, clays, carbon nanotubes, and other additives such as conductive polymers into fiber structure are reported to result in improved mechanical properties and desirable functionalities such as antistatic properties, antibacterial properties, flame retardancy, etc., which offer great potential in various fields. Carbon nanotubes, silver nanoparticles, and conductive polymers are frequently used in the production of the composite filaments.

Carbon nanotubes are widely added to fiber structure to improve mainly the mechanical properties and electrical properties, besides thermal properties, thermal stability, etc. Commodity fibers can replace super strong fibers, which are produced through much more complicated and expensive production techniques, if a significant improvement in strength can be obtained with the addition of CNTs (Moore et al. 2004). In addition, it is reported in literature that incorporation of CNTs to the fiber structure provides conductive properties (Li et al. 2004, Li et al. 2006). The conductivity of carbon nanotubes may be as high as 10^6 S/m (Min et al. 2009). Since they also have high aspect ratio (above 10^3), it is more feasible to obtain conductive networks with carbon nanotubes than with nanoparticles (Li et al. 2006).

Silver has been a popular material since ancient times due to its superior antibacterial properties against a wide variety of microorganisms. Silver inactivates microorganisms by combining with their cellular proteins (Gawish et al. 2012, Sondi and Salopek-Sondi 2004, Prabhu and Poulose 2012). The use of silver nanoparticles is relatively new but higher performance is expected from them due to their increased surface area (Dastjerdi et al. 2010). Although silver has the highest conductivity among metals, it is usually used in composite materials to impart antibacterial properties. However, there are some

Istanbul Technical University, Department of Textile Engineering, Inonu Street 69, Istanbul, Turkey, 34437.
  Email: kizildagn@itu.edu.tr
* Corresponding author: ucarnu@itu.edu.tr

studies reporting the conductive properties of composite materials, which are produced with the addition of silver nanoparticles (Ucar et al. 2015, Demirsoy et al. 2016).

Polyaniline (PANI) is one of the most widely studied and most promising conducting polymers due to its chemical stability and conductive properties (Cruz-Estrada and Folkes 2000). Its rigid structure and the difficulties in its processability lead to the use of PANI with other polymers. When used as a component in fiber structure, PANI macromolecules are expected to orient along the fiber axis and result in the formation of the conductive channels (Zhang et al. 2001). There are some recent studies performed with the combined use of the nanoparticles and polyaniline in fiber structure in order to form conductive paths between conductive nanoparticles and to obtain improved conductivity (Eren et al. 2016, Soroudi and Skrifvars 2010).

The concentration of the additives, their dispersion in the fiber structure and interaction with the host polymer affect the final properties of the nanocomposite fibers/filaments significantly. The structure of CNTs, their curvatures, chirality, and functional groups are important for CNTs as well. For PANI, the doping level is also an important parameter (Jianming et al. 2005).

This chapter provides a review of the studies about composite fibers produced by using carbon nanotubes, silver nanoparticles, and polyaniline and shows that it is possible to produce functional fibers/filaments with the incorporation of the additives into fiber/filament structure.

Fibers/filament production techniques

Fibers/filaments are produced mainly by two methods such as melt spinning and solution spinning.

In melt spinning (Figure 1), the polymer melt is extruded through a spinneret into air, cooled, and drawn into the fibers. It is used to produce fibers from several polymers such as polyethylene (PE), polypropylene (PP), polyamide (PA), polyester (PET), etc. (Zhang 2014).

Solution spinning technique can be classified into several methods such as wet spinning, dry spinning, or dry-wet spinning. Wet spinning (Figure 2) involves the extrusion of a polymer solution through the spinneret into a coagulation bath where the polymer solution precipitates as a porous fiber (Arbab et al. 2011). Wet spinning is often used to produce filaments from several polymers such as polyacrylonitrile (PAN), polyvinyl chloride (PVC), polyvinyl alcohol (PVA), etc. (Zhang 2014).

Dry spinning and dry-jet wet spinning are developed with the modification of wet spinning. In both of the methods, polymer solution is prepared and extruded through the spinneret. In dry spinning (Figure 3), the polymer solution is extruded through spinneret into hot air and the fiber in solid form is obtained as the solvent evaporates, whereas in dry-jet wet spinning (Figure 4), the polymer solution is extruded through the spinneret into air just before it enters the coagulation bath. Dry-jet wet spinning is also called gel spinning. Gel spinning enables the production of high performance fibers from polymers that has flexible polymer chains such as ultra-high molecular weight polyethylene (UHMWPE) (Wang et al. 2005).

Another method which is used in research studies is called co-flow wet spinning. It is a modified coagulation-based wet spinning method, which consists of a syringe pump, a needle, glass pipe and a

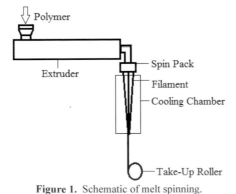

Figure 1. Schematic of melt spinning.

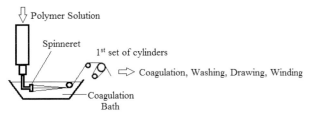

Figure 2. Schematic of wet spinning.

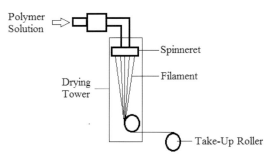

Figure 3. Schematic of dry spinning.

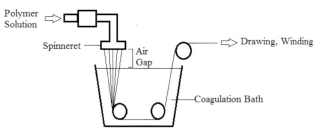

Figure 4. Schematic of dry-jet wet spinning.

rotating stage. The syringe pump is used to feed the solution through the needle into coagulant pipe in which the coagulant solution flows coaxially. Then the fibers coming out of the coagulant pipe are collected in the rotating water bath, washed, and dried. It has been used to produce composite fibers of PVA with high content of CNTs (Jee et al. 2012a,b).

Composite fibers/filaments with carbon nanotubes

Since their discovery in 1991 by Iijima (Iijima 1991), CNTs have been used as additives in a wide variety of polymers in order to develop advanced composite materials with improved mechanical, electrical, and thermal properties. The use of CNTs in filament reinforcement is of particular interest since the stresses applied during fiber spinning contributes to the alignment of the carbon nanotubes along the fiber axis which makes it possible to better utilize the excellent properties of carbon nanotubes (Moore et al. 2004, McIntosh et al. 2006, Poetschke et al. 2005, Meng and Hu 2008). Melt spinning and wet spinning, which are the two most common filament production methods, are also used to produce composite filaments with CNTs. CNTs are used as either in pristine form or after functionalization in both of the methods. Functionalization of CNTs enables the formation of interfacial bonding between polymer and CNTs, which is necessary to achieve load transfer from polymer to the CNTs (Wang et al. 2005). For dispersion of CNTs in the filament structure, melt compounding (Choi et al. 2009), *in situ* polymerization (Meng and Hu 2008), and ultrasonication in melt (Sulong et al. 2011) are used to disperse CNTs in the polymer before melt spinning, while ultrasonication in the solvent is the most preferred technique to disperse

CNTs in the solvent before wet spinning (Meng and Hu 2008, Mikołajczyk et al. 2010, Zhou et al. 2011, Mercader et al. 2012).

The studies about composite filaments with CNTs seem to be focused on meltspun commodity filaments such as PA, PE, PP, and PET. There are also some studies about PAN/CNT, PVA/CNT composite filaments, which are produced by wet spinning method. Some other polymers such as polycarbonate (PC), polymethylmetacrylate (PMMA), polyurethane (PU), shape memory polyurethane (SMPU), and ultrahigh molecular weight polyethylene (UHMWPE) are used in the composite filament production with CNTs as well.

A list of the studies about polymer/CNT composite fibers is presented in Table 1.

Sandler et al. (2004) produced composite filaments of PA 12 with a range of multi-walled carbon nanotubes (MWNTs) and carbon nanofibers in order to compare the dispersion quality and mechanical properties. The arc-grown nanotubes, aligned catalytically-grown nanotubes, and entangled catalytically-grown nanotubes were used for reinforcement. The polymer pellets were compounded with CNTs in a twin-screw microextruder and then forced through a capillary rheometer with a 1 mm-die in order to obtain composite PA 12 filaments with CNTs. Nanocomposites with a range of CNT fractions up to 10 wt% were produced. While all the catalytically-grown materials resulted in high quality of dispersion and were aligned, substrate-grown carbon nanotubes resulted in the greatest improvement in stiffness. The use of entangled MWNTs led to the most pronounced increase in yield stress. The arc-grown MWNTs exhibited the highest temperature stability reflecting the higher crystallinity of the composite material. The carbon nanotubes were found to act as nucleation sites under slow cooling conditions. No significant variations were observed in polymer morphology. Intrinsic crystalline quality as well as the straightness of the embedded nanotubes were found to be the significant factors affecting the reinforcement ability. Gao et al. (2005) polymerized PA 6 in the presence of single wall carbon nanotubes (SWNTs) and extruded through a spinneret to form meltspun composite filaments. The SWNT loading were changed as 0.1, 0.2, 0.5, 1, and 1.5 wt%. The PA6 chains were found to be grafted to the SWNTs. While the graft polymerization led to the uniform dispersion of the SWNTs, the presence of the graft copolymer strengthened the interfacial interaction between the nanotubes and PA6. The tensile strength, the Young's modulus, and thermal stability of the SWNT-reinforced composite PA 6 fibers were significantly improved. The tensile strength reached 92.7 MPa with an increase of 126.6% at the SWNT content of 0.2%. Rangari et al. (2008) used pristine carbon nanotubes (p-CNTs) and fluorinated carbon nanotubes (f-CNTs) to reinforce PA 6 filaments. PA 6 polymer powder and CNTs (0.5 and 1 wt%) were mixed as the first step, and then fed to a single-screw extruder for fiber extrusion process. The extruded fibers were stretched to their maxima, stabilized using a godet setup, and then characterized for CNT dispersion, morphology, mechanical and thermal properties. The alignment of CNTs and interfacial bonding to PA 6 polymer matrix was confirmed by SEM and Raman spectroscopy. The p-CNTs were found to be better aligned than f-CNTs. The surface of the fibers became rough with CNT addition and some f-CNTs were observed to be sticking out of the polymer. Increase was observed in glass transition temperature (T_g) with CNT addition which was attributed to the decrease in free volume that restricted the molecular motion. Slight increase was observed in crystallinity values. Tensile strength increased by about 231% with the addition of 0.5 wt% f-CNTs and 1.0 wt% p-CNTs. It was found that the degree of tensile strength was not only affected by the percentage loadings of the CNTs but also by the chemical reactivity of CNTs. While tensile strength increased with the increase in p-CNT content, it decreased with the increase in f-CNT content. Perrot et al. (2009) produced composite filaments of PA 12 with MWNTs and investigated the influence of several spinning factors, including spinning speed, extrusion rate, and draw ratio on the structure and properties of the filaments. The composite fibers exhibited a uniform texture and diameter. SEM images confirmed the alignment of the nanotubes along the fiber axis. While drawing, either during or after spinning, affected the polymer chain alignment and fiber mechanical properties, the spinning speed barely affected the structure and mechanical properties of the fibers. Both CNT and polymer alignment were improved with the increase in the draw ratio. While Young's modulus increased, elongation decreased with the addition of 7 wt% CNTs. As the draw ratio increased, the tensile strength and Young's modulus increased while elongation decreased for both pure and composite PA 12 filaments. The spinning speed had to be reduced with CNT addition. Meng et al. (2010) produced PA 6,6

Table 1. List of the studies about composite fibers/filaments with carbon nanotubes.

CNT type & CNT amount	Matrix material	Method of production	Focus of research	Reference
SWNT 0.5, 1, 1.5, 2 wt%	PP	melt spinning	mechanical properties	Kearns and Shambaugh 2002
arc-grown (1.25, 2.5, 10 wt%) aligned catalytically-grown (1.25, 2.5, 5 wt%) entangled catalytically-grown (1.25, 2.5, 5, 10 wt%) MWNTs	PA 12	melt spinning	dispersion morphology mechanical properties crystallinity thermal properties	Sandler et al. 2004
SWNT 0.5, 1 and 2 wt%	PP	melt spinning	mechanical properties	Moore et al. 2004
CNTs 0.5, 1, 2, 3 wt%	PP	melt spinning	antistatic ability	Li et al. 2004
Purified and functionalized CNTs 0.25, 1, 2, and 3 wt%	UHMWPE	gel spinning	dispersion mechanical properties thermal properties	Wang et al. 2005
SWNT MWNT DWNT Carbon nanofibers 5 wt%	PAN	wet spinning	morphology mechanical properties thermal shrinkage	Chae et al. 2005
MWNT	PP	melt spinning	CNT alignment electrical conductivity mechanical properties	Poetschke et al. 2005
CNTs 0.5, 2, 4, 8, and 10 wt%	PET	melt spinning	microstructure rheological behavior crystallization process electrical conductivity	Li et al. 2006
Pristine and fluorinated SWNTs 2.5, 5, 7.5, 10 wt%	PP	melt spinning	dispersion adhesion mechanical properties	McIntosh et al. 2006
MWNTs 0.5 and 1 wt%	PP	melt spinning	crystallization mechanical properties thermal properties	Jose et al. 2007
In situ functionalized SWNTs 2.5, 5, 7.5, 10 wt%	PP	melt spinning	dispersion adhesion mechanical properties	McIntosh et al. 2007
Pristine CNTs Fluorinated CNTs 0.5 and 1 wt%	PA 6	melt spinning	dispersion morphology structural properties thermal properties mechanical properties	Rangari et al. 2008
acid treated+functionalized MWNTs 0.3, 0.5, 1, 1.5 wt%	PET	melt spinning	dispersion morphology mechanical properties thermal properties electrical properties	Mun et al. 2008
Pristine, acid treated and alkyl chain grafted MWNTs 0.01, 0.02, 0.04, 0.06, 0.08, 1 wt%	PET	melt spinning	dispersion morphology structural properties thermal properties mechanical properties	Shen et al. 2008
MWNT 1, 3, 5, and 7 wt%	SMPU	melt spinning	CNT dispersion CNT alignment shape memory effect	Meng and Hu 2008

Table 1 contd. ...

...Table 1 contd.

CNT type & CNT amount	Matrix material	Method of production	Focus of research	Reference
MWNTs 3 and 7 wt%	PA 12	melt spinning	CNT dispersion CNT alignment morphology mechanical properties	Perrot et al. 2009
maleic anhydride polypropylene functionalized MWNTs 0.1, 0.5, and 1 wt%	PP	melt spinning	dispersion mechanical properties thermal properties	Choi et al. 2009
MWNTS 1.5, 3.7, and 7.5 wt%	PP	melt spinning	morphology electrical conductivity thermal properties	Skrifvars and Soroudi 2009
UV-treated MWNTs 0.5 wt%	PE	melt spinning	CNT alignment crystallization mechanical properties	Sulong et al. 2011
MWNTs 0.1, 0.5, 1, 2 and 5 wt%	PA 6,6	melt spinning	CNT dispersion morphology crystal size thermal properties mechanical properties	Meng et al. 2010
Chemically modified MWNTs SWNTs Short MWNTs Long MWNTs 1 wt%	PAN	wet spinning	rheological properties deformability	Mikołajczyk et al. 2010
MWNTs 2 wt%	LDPE	melt spinning	crystal size crystallinity mechanical properties thermal properties	Khan et al. 2011
MWNT 1, 2, 3, 4, and 5 wt%	PP	melt spinning	dispersion mechanical properties	Soitong and Pumchusak 2011
MWNTs 3 and 5 wt%	PAN	wet spinning	structural properties dispersion mechanical properties	Zhou et al. 2011
CNT 4 wt%	PLA	melt spinning	mechanical properties electrical conductivity water vapor sensitivity	Ferreira et al. 2011
SWNT (17 wt%) and MWNT (12 wt%)	PVA	wet spinning	mechanical properties electrical conductivity	Mercader et al. 2012
MWNT 60 wt%	PVA	co-flowing wet spinning	structural properties mechanical properties electrical conductivity	Jee et al. 2012a,b
MWNT 5, 10, and 20 wt%	PVA	wet spinning	structural properties mechanical properties thermal properties	Lai et al. 2015

filaments with oxidized MWNTs. They *in situ* polymerized PA 6,6 in the presence of MWNTs and then melt-spun into composite filaments. The CNT content was varied as 0.1, 0.5, 1, 2 and 5 wt%. The PA 6,6 compound with 5 wt% CNTs could not be melt-spun due to the lower viscosity of its melt. The CNTs were well dispersed in PA 6,6 matrix. Some CNTs were observed to be wrapped by the polymer. Investigation of the fractured surfaces showed broken CNTs, which implied the strong interaction between CNTs and the polymer matrix. While the incorporation of CNTs between 0.1 and 2 wt% did not have a significant effect on the thermal decomposition temperature, decrease was observed with 5 wt% CNT addition. The grain sizes of crystals increased as the CNT content increased. The storage modulus, the tensile strength and modulus of the fibers were significantly improved whereas the strain of the fibers decreased with the increase in CNT content. The highest tensile strength was observed at CNT content of 0.5 wt%.

Sulong et al. (2011) prepared composite filaments of PE with MWNTs, which were functionalized by ultraviolet (UV) ozone treatment. MWCNTs were sonicated with a horn-type ultra sonicator and then exposed to ultraviolet radiation for 60 min at radiation intensity of 38 mW/cm^2. MWNT content was selected as 0.5 wt%. The polymer powder was mixed with CNTs by stirring manually, melted, ultra sonicated by a horn-type ultrasonicator followed by solidification and finally cut into composite pellets. These pellets were extruded into fiber form using an in-house built melt spinning machine. Optimum settings for three melt spinning process parameters (spinning temperature, spinning distance, and the number of spinning revolutions) were investigated by Taguchi method with the objective of enhancing the mechanical strength. Spinning temperature was changed as 105, 110, 115°C; the spinning distance as 5, 10, 15 cm; and the winding speed as 30, 50, and 70 rpm. According to the Taguchi evaluation, the optimum spinning temperature was 110, the distance between the nozzle and winding rollers were 10 cm and the winding speed was 50 rpm. The experimental design analysis also showed that the distance showed the highest effect on mechanical strength, followed by the temperature and the winding speed. The predicted strength value was close to the verification experiment value. The ultimate tensile load increased by 69% with UV-treated MWCNTs addition. This increase in tensile strength was much more pronounced in meltspun filaments than in bulk composites due to the increased orientation and better alignment of CNTs by the drawing during fiber formation. No alignment could be observed which was attributed to the lower drawing ratio and shear rate. Besides, the influence of mechanical drawing during melt spinning on the crystallization of PE was investigated. The crystallinity of pure PE filaments was higher than bulk PE and it further increased with the addition of CNTs. Khan et al. (2011) incorporated MWNTs and UHMWPE into low density polyethylene (LDPE) filaments in order to improve its mechanical and thermal properties. LDPE was dry-mixed with MWNTs and UHMWPE and then meltspun into filaments. The loading of MWCNT and UHMWPE was 2 and 8 wt%, respectively. Increase in crystal size was observed with MWNT addition. While addition of MWNTs resulted in a decrease in melting temperature, addition of UHMWPE didn't affect melting temperature further. Crystallinity was improved with the addition of MWNTs. Incorporation of MWNT improved strength and modulus by about 23 and 57%, respectively, but reduced fracture strain by 45% and toughness by 36%. When UHMWPE was dispersed additionally as minor phase, the resulting LDPE/MWNT/UHMWPE filaments showed a 55% increase in toughness and 17% increase in elastic energy storage capacity without any loss of strength and improvements both in modulus and strain which was attributed to the increase in crystallinity, crystallite size, and sliding between the minor and major phases.

Li et al. (2006) compounded PET resin with CNTs using a twin-screw extruder and prepared composite pellets, filaments and fabric. The CNT content was changed as 0.5, 2, 4, 8, and 10 wt%. The composites of 4 wt% CNTs in PET had a volume electrical resistance of 10^3 Ωcm, which was 12 orders lower than pure PET. CNTs were well dispersed in PET matrix. Rheological behavior of CNTs/PET composites showed that the viscosity of CNTs/PET composites containing high nanotube loadings exhibited a large decrease with increasing shear frequency. Nucleating effect of CNTs in the cooling crystallization process of PET was confirmed by DSC analysis. While decrease was observed in crystallinity, the initial crystal temperature and crystal temperature increased with the increase in CNT content. No significant change was observed in melting temperature. Composite fiber was prepared using the conductive CNTs/PET composites and pure PET resin by composite spinning process. Fabric woven with the composite fiber and common terylene with the ratio 1:3 showed excellent antistatic electricity property and its charge surface density was only 0.25 μC/m^2. Mun et al. (2008) *in situ* synthesized PET in the presence of functionalized MWNTs and meltspun composite filaments with various functionalized MWNT contents and draw ratios. The CNT content was changed as 0.3, 0.5, 1, and 1.5 wt%. The addition of only a small amount of functionalized MWNTs was found to be sufficient to improve the properties of PET filaments. Agglomeration was observed with the increase in CNT content. The domain size of the dispersed MWNT phase decreased with increasing draw ratio. The maximum enhancement in the ultimate tensile strength was observed at MWNT content of 0.5 wt%. However, the initial modulus was found to increase linearly with increases in MWNT loading from 0 to 1.5 wt%. While the tensile strength of pure PET fibers increased, the tensile strength of composite fibers decreased as draw ratio increased. T_g and T_m increased with increase in CNT content. Thermal stability of the filaments was improved with CNT addition. The highest increase in initial degradation temperature was observed at CNT content of 0.5 wt%. The conductivities of the composite PET filaments were 5 orders of magnitude higher than those of pure PET fibers. Shen et al. (2008) prepared master batches of PET with four different kinds of functionalized MWNTs through mixing of MWNTs with PET (0.01:0.99 w/w) in trifluoroacetic acid/ dichloromethane mixed solvents (0.7:0.3 v/v) followed by the removal of the solvents in the mixture by flocculation. Pristine, acid treated and two different alkyl chain-grafted CNTs were used for reinforcement of PET filaments. Good dispersion of MWCTs was achieved. The reinforced fibers were fabricated by the melt spinning of PET chips with small amounts of the master batch and then further post-drawing. The optimal spinning conditions for the composite fibers were a 0.6 mm spinneret hole and a 250 m/ min wind-up speed. Among the four master batches, the fibers obtained from PET and acid treated CNTs had the highest improvement in mechanical properties. For a 0.02 wt% loading of acid-treated MWCT, the breaking strength of the composite fibers increased by 36.9%, and the initial modulus increased by 41.2%.

Another polymer that is widely used in research studies to produce melt-spun filaments with CNTs is PP. Kearns and Shambaugh (2002) produced PP filaments with SWNTs in order to improve the strength of the filaments. They used solvent processing to disperse SWNTs in PP. After the removal of the solvent, the composite polymer was meltspun and post-drawn into fibers. The SWNT content was changed as 0.5, 1, 1.5, and 2 wt%. Best mechanical properties were obtained at 1 wt% SWNT addition. At the 1 wt% loading of SWNTs, the fiber tensile strength and modulus increased 40% and 55%, respectively. Moore et al. (2004) added SWNTs (0.5, 1, and 2 wt%) to two different grades of PP which had different melt flow rates. While an increase was observed in tenacity with SWNT addition for PP, which had lower melt flow rate especially after post-drawing, a decrease was observed for PP with high melt flow rate. The changes in tensile properties were related to the grade of PP used. Li et al. (2004) used three differently produced CNTs to improve the antistatic properties of PP filaments. The filler content was varied as 0.5, 1, 2, and 3 wt%. The CNT concentration in the filler was 10 wt%. This type of fiber showed better antistatic properties than the fibers with only pure organic antistatic agent and conductive carbon black. Antistatic effects promoted by CNTs were observed to be enhanced with the decrease in the diameters and the curvatures of CNT walls. Jose et al. (2007) added 0.5 and 1 wt% CNTs to PP filaments and investigated the effect of the nanotubes on the crystallization, mechanical behavior and thermal properties of PP as well as the effect of draw ratio on the nanocomposite filament morphology and properties. Decrease in crystallinity was observed with the increase in CNT content. Three-fold increase in the modulus and a five-fold increase in the tensile strength was observed with the addition of CNTs. Further increase was observed in the strength after drawing which suggested that the CNTs were aligned further during drawing. Thermal stability of the fibers was also improved with CNT addition. McIntosh et al. (2006) used fluorinated SWNTs to reinforce PP filaments. The concentration of pristine and fluorinated SWNTs was changed as 2.5, 5, 7.5, 10 wt%. Partial defluorination occurred during melt processing by shear mixing providing the opportunity for *in situ* direct covalent bonding between the nanotubes and PP matrix, which resulted in better mechanical reinforcement of the composite. Maximum stress was observed at fluorinated CNT content of 10 wt%. Fluorinated SWNTs provided better dispersion, interfacial adhesion, and mechanical properties compared to pristine SWNTs. In another study, McIntosh et al. (2007) used benzoyl peroxide as initiator for the *in situ* functionalization of the SWNTs during PP filament spinning. The decomposition of benzoyl peroxide during the high shear and high temperature phase of processing resulted in the linkage of the SWNTs to the surrounding polypropylene matrix via a covalent bond. The SWNT content was varied as 2.5, 5, 7.5, and 10 wt%. Significant increases were obtained in tensile strength (173.1%) and tensile modulus (133.7%) at the SWNT content of 10 wt%. Choi et al. (2009) examined the thermal and mechanical properties of PP filaments reinforced with maleic anhydride polypropylene functionalized MWNTs. Composites with 0.1, 0.5, and 1 wt% MWNTs were prepared by melt compounding using a twin screw extruder which was then meltspun into composite filaments. The highest tensile strength was observed at the MWNT content of 0.1 wt%. No significant changes were observed in elongation. The crystallization temperature and crystallinity were slightly increased with MWNT addition. Skrifvars and Soroudi (2009) prepared PP/CNT composite filaments by melt spinning and characterized regarding morphology, conductivity, and thermal properties. CNT addition exhibited adverse effects on spinning, and appearance and surface quality of the fibers. While the conductivity of pure PP is lower than 1.0×10^{-10} S/cm, the conductivity of PP fibers with 1.5 wt% MWNT was 1.0×10^{-5} S/cm. The conductivity increased further with the increase in CNT content but both the spinning and the properties of the filaments were adversely affected. Thermal stability was improved with CNT addition. Soitong and Pumchusak (2011) used different dispersing agents such as 2-propanol, sodiumlauryl sulfate (SLS), and Triton X-100 to disperse the MWNTs in PP and then produced filaments of the PP/MWNT composites using melt spinning. MWNT content was changed as 1, 2, 3, 4, and 5 wt%. TEM images showed that the MWNTs were well dispersed and aligned. The use of 2-propanol as the dispersing agent resulted in the highest tensile strength and tensile modulus at the MWNT content of 1 wt%. The highest strength and modulus were observed at the MWNT content of 3 wt% with no variation in breaking elongation compared to pure PP filaments.

Some other polymers were also investigated for composite fiber production with CNTs. Haggenmueller et al. (2000) prepared PMMA composite fibers with SWNTs using solvent casting for SWNT dispersion and melt spinning for fiber production. The composite filaments of PMMA showed improved mechanical and electrical properties with excellent SWNT alignment. The dispersion of SWNTs in PMMA improved with the increased number of melt mixing cycles. Improvement was observed in elastic modulus and yield stress with nanotube loading and draw ratio. The conductivity increased from 0.118 to 11.5 S/m along the flow direction and from 0.078 to 7.0 S/m perpendicular to the flow direction as the SWNT content increased from 1.3 to 6.6 wt% in the composite films of PMMA/SWNT. The draw ratio of 4 was found to be sufficient to improve the conductivity of the composite filaments. Poetschke et al. (2005) produced PC/MWNT filaments containing 2 wt% MWNTs with different take-up velocities, up to 800 m/min and draw ratios up to 250. PC master batch with 15 wt% MWNTs was mixed with PC by melt mixing in an extruder. It was observed that the alignment of MWNTs increased and their curved shapes reduced as the draw ratio increased. While the volume resistivity of the composite material with 2 wt% MWNTs was measured as 550 Ω cm before melt spinning, the volume resistivities of the filaments were not in the measurable range of the device used. The elongation at break and tensile strength of the composite filaments were lower than pure PC filaments at low spinning speeds whereas higher elongation and comparable true stress at break were obtained at the highest take-up velocity of 800 m/min. Meng et al. (2008) prepared SMPU fibers with MWNTs using *in situ* polymerization and melt spinning. MWNT content was changed as 1, 3, 5, and 7 wt%. Functionalization was performed and ultrasonication was applied for better dispersion of the nanotubes. The alignment of MWNTs, and shape memory behavior of the fibers were investigated. MWNTs were found to be aligned in the fiber axial direction. The shape recovery ratio and recovery force were improved by the aligned MWNTs. The self-aligned MWNTs enabled the shape memory fibers to recover the original length more quickly that showed the possibility of fabricating smart instruments with higher sensitivity. Ferreira et al. (2011) prepared PLA/CNT monofilaments using melt compounding for dispersion of CNTs and melt spinning for fiber formation. CNT concentration was selected as 4 wt%. While the volume electrical resistivity of pure PLA was 10^{14} Ωm, it was measured as 0.14 Ωm for PLA composite filaments with 4 wt% CNTs. The resistivity values were observed to be changing with relative humidity, thus the composite filaments were suggested for use as humidity sensors.

Wet spinning has also been used to produce composite fibers/filaments with CNTs. The studies are mostly focused on the production of PAN and PVA filaments.

PAN is particularly popular because of its widespread applications as carbon fiber precursor (Zhou et al. 2011). Sreekumar et al. (2004) reported one-fold and 10-fold increase in tensile strength at room temperature and at 150°C, respectively, for PAN fibers containing 10% SWNTs. In addition, 40°C improvement in the glass transition temperature was observed. Chae et al. (2005) spun PAN/CNT composite fibers from solutions in dimethyl acetamide (DMAc), using SWNTs, double wall carbon nanotubes (DWNTs), MWNTs, and vapor grown carbon nanofibers (VGCNFs). Filler content was 5 wt%. Maximum increase in modulus and reduction in thermal shrinkage was observed for the

SWNT containing composite fibers while the maximum improvement in tensile strength, strain to failure, and work of rupture was observed for the MWNT containing composite fibers. The orientation of PAN macromolecules and PAN crystallite size in the composite fibers were found be larger than that of pure PAN, while the overall PAN crystallinity decreased slightly. Besides, CNT orientation was higher than PAN orientation. Mikołajczyk et al. (2010) produced PAN filaments using 4 different types of CNTs such as chemically modified MWNTs, SWNTs, long MWNTs, and short MWNTs. CNTs were dispersed in DMF using ultrasonication. The presence of CNTs in the PAN spinning solution improved its deformability during the drawing stage, which resulted in a higher tensile strength. The use of a three-stage drawing process resulted in a significant increase in the tensile strength of PAN fibers modified with MWNTs. Zhou et al. (2011) investigated the effects of mixing methods on the PAN/CNT filaments. MWNTs were mixed with PAN by *in situ* polymerization or by mechanical mixing and then the mixtures were wetspun into fibers. CNT content was changed as 3 and 5 wt%. *In situ* polymerization resulted in a thin layer of PAN molecules covering the surface of the CNTs, which increased their diameters. Raman spectroscopy indicated that the layer of PAN molecules was strongly attached to the CNTs through grafting polymerization. No chemical interactions could be observed in the fibers prepared by mechanical mixing. While both mixing methods provided production of composite filaments with increased tensile strength, the fibers prepared by *in situ* polymerization mixing were much stronger.

Mercader et al. (2012) reported the production of polyvinyl alcohol (PVA) composite fibers with high amount of carbon nanotubes (SWNTs and MWNTs) using a water-based spinning. CNTs were stabilized in the spinning solution with the use of surfactants. The concentration of SWNTS and MWNTs were about 12 and 17 wt%, respectively, according to TGA results. The Young's modulus, strength to failure, and energy to failure were higher for composite filaments and they increased further with hot-drawing. The resistance of annealed PVA/MWNT was measured as about 300 Ω/cm, which was found to be sufficient for applications such as heating, sensing, or antistatic textiles. Lai et al. (2015) used wet spinning process to produce PVA composite fibers with MWNTs. MWNTs were functionalized before use by covalent and noncovalent functionalization and ultrasonication was used to disperse 5, 10, and 20 wt% CNTs in water. The composite fibers were investigated for structural, thermal, and mechanical properties. While elongation decreased with the addition of CNTs, the maximum tensile strength and Young's modulus were observed at the CNT contents of 10 wt% and 20 wt%, respectively. The CNTs were well dispersed in PVA; however, some agglomerates were observed as the CNT content increased. Decrease was observed in degree of order, melting and crystallization temperatures and enthalpies with addition of CNTs. Jee et al. (2012a) produced PVA/MWNT composite filaments with a different method, which was called co-flowing wet spinning technique. Ultrasonication was used for the dispersion of the nanotubes in water in the presence of a surfactant. The homogeneous MWNT dispersion was injected into the center of the vertically aligned glass pipe in which PVA solution was flowing coaxially at room temperature and thus composite fibers were obtained. Structural, mechanical, and electrical properties of PVA/MWNT composite fibers were investigated as a function of draw ratio. The MWNT concentration was reported to be about 60 wt% depending on the TGA results. The alignment of CNTs, initial moduli, tensile strength, and electrical conductivity were improved with increase in the draw ratio. While the conductivity was measured as 4.2 S/cm for undrawn fibers, it increased to 41 S/cm with 30% drawing. In another study, Jee et al. (2012b) investigated the effects of varying the spinning geometry (inner diameter of glass pipe) on structural, mechanical, and electrical properties of the composite fibers. The alignment of MWNTs was improved along the composite fiber axis with increasing the aligning shear stress of the spinning process. As a result of the improved alignment, initial moduli and tensile strengths of the composite fibers were significantly increased. The conductivity of the composite fibers was measured about 6 S/cm and only slight increase in conductivity was observed with increased aligning shear stress.

UHMWPE was also investigated for the production of composite filaments with CNTs. Wang et al. (2005) prepared UHMWPE/CNT composite fibers by gel spinning. The CNTs were purified and functionalized by titanate coupling agents before use for better dispersion. The content of CNTs was changed as 0.25, 1, 2, and 3 wt%. Ultrasonication was used to disperse nanotubes in the polymer solution. The results showed that the CNTs were well dispersed in the polymer and a good interaction between CNTs and UHMWPE matrix was established. The CNTs contributed to the alignment of the polymer macromolecules. The tensile strength of composite fibers was improved up to the CNT content of 1 wt%. Further CNT addition resulted in a decrease in tensile strength. Thermal stability was improved with CNT addition.

It is seen that improvement in crystallinity, mechanical properties, thermal properties, thermal shrinkage, electrical properties could be observed depending on the CNT concentration, CNT type and properties, CNT pretreatment method, mixing method, production method and process parameters. As the key factors in the reinforcement ability of the carbon nanotubes, CNT dispersion, alignment, and adhesion to the polymer were also investigated in many of these studies.

The highest CNT content of 60 wt% was reached with co-flow wet spinning method, which resulted in a conductivity value of 6 S/cm (Jee et al. 2012a). In melt spinning, the highest loading of CNTs was 10 wt% (Sandler et al. 2004, Li et al. 2006, McIntosh et al. 2006, 2007) while in wet spinning, it was 20 wt% (Lai et al. 2015).

Composite fibers/filaments with silver nanoparticles

Textile fibers and fabrics normally have no inhibitory effects on bacteria. On the contrary, they provide suitable environment for the growth of bacteria that may result in color change, unpleasant odor release, fiber degradation, and potential health risks (Dastjerdi et al. 2009b). Among many methods used to produce antibacterial textiles, adding silver to the filament structure during fiber production provides the embedding of silver to the filament structure that results in a long lasting effect with a slower silver release (Yazdanshenas et al. 2012). Besides, it has been possible to develop semi conductive filaments with addition of silver to the filament structure (Kizildag and Ucar 2016a). A SEM image of composite filaments with AgNPs are presented in Figure 5 while a list of the studies about composite fibers/filaments with AgNPs is presented in Table 2.

Yeo and Jeong (2003) produced bicomponent sheath-core fibers using PP chips and AgNPs. PP/Ag master batches with two different Ag contents as 3 and 10 wt% were prepared by a conventional twin-screw extruder. A general conjugate spinning machine composed of two extruders and gear pumps was used in the production of the bicomponent filaments. PP/Ag master batches were added as either the core or the shell while pure PP formed the other layer. SEM images showed that the average diameter of the silver nanoparticles was approximately 30 nm and some particles were aggregated. The results of the DSC and XRD indicated that the crystallinity slightly decreased with the addition of silver nanoparticles. While the fibers which contained silver in the core part did not show antibacterial effects, fibers with silver in the sheath part exhibited excellent antibacterial effects.

Mikołajczyk et al. (2009) added AgNPs to PAN spinning solutions, and they investigated the effects of the amount of AgNPs on the rheological properties of the solutions and the structure and properties of the fibers produced. The AgNPs were dispersed in DMF by ultrasound for 30 min. Fibers were formed using wet spinning method. PAN solutions with a concentration of 22 wt% in DMF, containing different amounts of silver nanoparticles, were used as spinning solutions. The fibers were tested for supramolecular structure, porosity, thermal, and tensile strength properties. It was found that the addition of AgNPs in an amount of up to 1.5% does not cause a decrease in the susceptibility of the fiber matter to deformation at the drawing stage. The composite fibers were characterized by an increased total volume

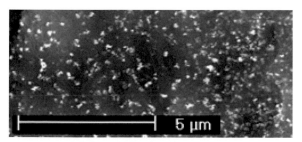

Figure 5. SEM image of composite PAN filament with 3 wt% AgNO$_3$ (after chemical reduction process).

Table 2. List of the studies about composite fibers/filaments with AgNPs.

Additive type & amount	Matrix material	Method of production	Focus of research	Reference
AgNPs (0.3, 0.9, 1.5 and 5 wt%)	PP	melt spinning	dispersion thermal properties crystallinity antibacterial activity	Yeo and Jeong 2003
AgNPs (1, 2, and 3 wt%)	PAN	wet spinning	supramolecular structure porosity thermal properties mechanical properties	Mikołajczyk et al. 2009
Ag-based antibacterial agent (0.1, 0.2, 0.4, 0.6 wt%)	PP	melt spinning	dispersion rheological properties morphology crystallinity shrinkage mechanical properties thermal properties antibacterial properties	Dastjerdi et al. 2009a
Ag/TiO$_2$ (0.2 and 0.5 wt%)	PP	melt spinning	rheological properties morphology crystallinity shrinkage mechanical properties thermal properties antibacterial activity	Dastjerdi et al. 2010
AgNPs	PA 6	melt spinning	mechanical properties antibacterial activity	Damerchely et al. 2011
AgNPs (0.5, 1 and 4 wt%)	PA 6	melt spinning	dispersion morphology thermal properties mechanical properties antibacterial activity	Yazdanshenas et al. 2012
AgNPs (0.05, 0.1, 0.5, 1 wt%)	PP	melt spinning	dispersion morphology crystallinity thermal properties mechanical properties antibacterial activity antistatic properties	Gawish et al. 2012
Encapsulated AgNPs (0.1, 0.5, 1, 20 wt%)	PET	melt spinning	dispersion mechanical properties antibacterial activity	Sun et al. 2014
AgNO$_3$ (1 and 3 wt%)	PAN	wet spinning	morphology chemical structure tensile properties crystallinity conductivity thermal properties silver ion release behavior antibacterial activity	Kizildag and Ucar 2016a

of pores of 0.35 cm^3/g and tenacity of more than 34 cN/tex. Dastjerdi et al. (2009a) prepared PP fibers containing 0.1, 0.2, 0.4, and 0.6 wt% silver based antibacterial agent. Pure PP granules and mixture of PP granules with master batches containing antibacterial agent were meltspun into fibers at the spinning temperature of 240°C and take-up speed of 2000 m/min. SEM images showed that the particles were dispersed relatively good in fiber structure. Although the crystallinity of the as-spun fibers and drawn fibers tended to decrease with the addition and increase in the content of antibacterial agent, generally, increase was observed in the tenacity of the as-spun and drawn fibers, especially up to additive content of 0.4 wt%. While the breaking elongation of the as-spun fibers decreased as the additive content increased, some increase in elongation could be observed depending on the additive content for the drawn fibers. The fabrics produced with the composite filaments showed antibacterial activity against S. Aureus. Even at the additive content of 0.2 wt%, the antibacterial efficiency was over 99%. Dastjerdi et al. (2010) prepared composite PP filaments with 0.2 and 0.5 wt% Ag/TiO$_2$ nanoadditive using melt spinning method with the aim of producing antibacterial PP fibers. The nanoadditives were premixed and melt blended with PP granule in a co-rotating screw extruder. Composite PP fibers were produced by a pilot plant melt-spinning machine. Physical and structural properties of as-spun and drawn yarns with constant and variable draw ratios were investigated. Pure PP and all other nanocomposite samples showed good spinnability at the spinning temperature of 240°C and take-up speed of 2000 m/min by a pilot plant melt-spinning machine. SEM images confirmed that the nanoparticles were dispersed quite well. Decrease was observed in the crystallinity of as-spun nanocomposite fibers, which was compensated with the drawing process. Tensile properties of drawn composite yarns with the variable draw ratio were higher than the pure PP, whereas the inverse observation was noticed in the case of constant draw ratio. The composite PP fibers with 0.2 and 0.5 wt% nanoadditive displayed antibacterial efficiency over 99%. Damerchely et al. (2011) investigated some physical, mechanical, and antibacterial properties of meltspun PA6/AgNP multifilament yarns. While the composite yarns with AgNPs showed antibacterial property, their tensile properties were not much affected by AgNP addition. Yazdanshenas et al. (2012) produced PA6 filaments doped with different amounts of AgNPs (0.5, 1, and 4 wt%) using melt spinning method and investigated their physical, mechanical and antimicrobial properties. The average particle size of AgNPs was about 60 nm, which implied that some agglomeration had occurred. The agglomeration increased as the additive content increased. AgNPs up to 1 wt% did not change the mechanical properties. Yarns with 0.5–1.0 wt% of AgNP content were found to display significant antimicrobial activity. The antibacterial activity was shown to be affected by the thickness of the filaments. Gawish et al. (2012) prepared PP fibers containing silver, zinc and silver exchanged zeolite nanoparticles. The additive content was varied as 0.05, 0.1, 0.5, and 1 wt%. The nanometals were generally uniformly dispersed in PP. Some aggregation was observed on fiber surface and in fiber cross-sections. PP/0.5% Ag yarn had the highest degree of crystallinity (44.7%), which was very close to that of the control sample. While the mechanical properties of composite fibers with silver were not very much affected at the silver content less than 0.1 wt%, fiber tenacity decreased as the amount of silver content increased to 1.0%. The spinning performance measured by the number of broken filaments worsened with the increase in the additive content. Extruded composite fibers with 0.72% silver and 0.60% zinc nanoparticles showed outstanding antibacterial efficiency against *E. coli* and *S. aureus*. Furthermore, the composite fibers with AgNPs had improved antistatic properties. Sun et al. (2014) designed and prepared AgNPs with hydrophobic surface of small size (< 10 nm), capped by 18 wt% aliphatic acid to improve their dispersibility and compatibility in the PET matrix. They produced composite PET filaments varying the additive content as 0.1, 0.5, 1 and 20 wt%. The AgNPs had a fine and homogeneous dispersion in the PET matrix as determined from TEM analysis. The PET nanocomposites with lower content (≤ 500 ppm) of Ag had good spinnability, and the stress of the resulting fibers remained higher than 2.5 cN/dtex under non-optimized conditions. The composite filaments showed good antibacterial activity (> 96%) against *S. aureus* and *E. coli*. Kizildag and Ucar (2016a) added silver nitrate to PAN filament structure and applied chemical reduction to obtain PAN filaments with AgNPs showing electrostatic dissipative and antibacterial properties. 1 and 3 wt% silver nitrate was added to DMF, and the dispersion was homogenized with ultrasonic tip for 15 minutes and ultrasonic bath for 45 minutes. 21 wt% PAN was added to the solution which was then stirred magnetically. A laboratory type wet spinning machine was used for the production of the composite filaments. The formation of nanoparticles in the filament structure was confirmed by scanning electron microscope and UV-visible spectroscopy. Breaking strength and breaking elongation increased at silver nitrate content of 1%. Composite filaments displayed improved thermal stability. Their conductivities were measured in the semiconductive range. Atomic absorption spectroscopy confirmed that necessary amounts of silver release for antibacterial activity occurred, while the antibacterial activity analysis showed that the composite filaments had excellent antibacterial activity. The composite filaments were suggested for electrostatic dissipative and antibacterial applications.

Small amounts of silver nanoparticles (less than 1 wt%) were found to be sufficient in providing the necessary bacteria inhibition effect to the filaments (Dastjerdi et al. 2010, Yazdanshenas et al. 2012, Kizildag and Ucar 2016a). The crystallinity, thermal properties, and mechanical properties were also affected by the addition of silver nanoparticles. While some improvement could be observed in fiber properties at the lower content of AgNPs, increase in the additive content adversely affected the properties and spinnability of the filaments. Research about production of polymeric filaments with AgNPs showed the possibility of producing antibacterial and semiconductive filaments with the addition of silver nanoparticles to the filament structure.

Composite fibers/filaments with polyaniline

Polyaniline is an inherently conductive polymer, which is relatively easy to synthesize, has high chemical and thermal stability in conductive form, and low cost of raw material. Although the processability has been a problem and limited its wide use for many years, attempts to make it soluble in common solvents (Cao et al. 1992a,b) and fusible (Laska et al. 1993) have led to the development of conductive filaments with PANI.

Conductive network formation is one of the key aspects of developing conductivity in insulative polymeric materials. While many conductive additives have been investigated for developing conductive polymeric materials, the high amount of the additives needed to obtain a certain level of conductivity distorts the mechanical properties of the polymers. In this regard, blending inherently conductive polymers with insulative polymers offers an efficient way. Recently, polyaniline is being used frequently in studies to form a conductive network in the insulative polymeric materials. Many factors, such as doping level, dopant type, and distribution in the matrix affect the properties of composite materials. There are also some studies about the production of fibers with PANI. Both melt spinning (Passiniemi et al. 1997, Fryczkowski et al. 2004, Kim et al. 2004, Cullen et al. 2008, Soroudi and Skrifvars 2011, 2012) and wet spinning (Zhang et al. 2001, 2002, Lee et al. 2003, Jianming et al. 2005) have been used to develop conductive fibers/filaments with polyaniline. A list of the studies about composite fibers/filaments with PANI is presented in Table 3.

PP is widely preferred in the melt spinning studies with its relatively lower melting temperature. Passiniemi et al. (1997) prepared meltspun binary fibers of polypropylene and a commercial polyaniline-complex (Panipol™). Panipol™ was produced first, mixed with PP in a twin-screw extruder and pellets of PP/PANI (85/15) were obtained which were then meltspun into fibers. The PP fibers with PANI showed a highly ordered PANI structure and a phase separated morphology with continuous fibrils of PANI. The continuous PANI phase resulted in improved conductivity of up to 10^{-3} S/cm. Fryczkowski et al. (2004) prepared meltspun PP fibers that contained PANI doped with dodecylbenzenesulfonic acid (DBSA). PANI content was kept constant as 5 wt% while the DBSA amount was changed as 5, 10, 15, 20, 25, 30 wt%. It was observed that the increase in DBSA content improved the homogeneity and miscibility of PANI and PP by acting as a compatibilizer. Depending on the UV-Visible spectroscopy results, 20 wt% was found to be the optimum amount for DBSA, as adding more acid to the solution did not affect the absorption of polaron band significantly. The highest conductivity was obtained with 25 wt% DBSA addition. It was about 10^{-6} S/cm and didn't increase significantly with the increase in DBSA amount. Kim et al. (2004) produced PP filaments with PANI using melt-spinning method and characterized the composite filaments in terms of electrical properties. The conductivity of the fibers were measured around 10^{-9} S/cm and showed no dependence on the PANI amount within the studied range of 1 wt% to 30 wt%. Cullen et al. (2008) prepared fibers consisting of a blend of PANI and PP as the backbone of encapsulated tissue-engineered neural-electrical relays. The composite fiber had a resistivity of 5.2 Ω cm. They observed a significant increase (greater than ten-fold) in the density of viable neurons on fiber surfaces by manipulating surrounding surface charges which resulted in robust neuritic extension and network formation directly along the fibers. Soroudi and Skrifvars (2011) blended melt-

Table 3. List of the studies about composite fibers/filaments with PANI.

Additive type & amount	Matrix material	Method of production	Focus of research	Reference
Panipol 15 wt%	PP	melt spinning	morphology electrical conductivity	Passiniemi et al. 1997
PANI 3, 8, 10, 16, 20 wt%	PA 11	wet spinning	morphology electrical conductivity	Zhang et al. 2001
PANI 8, 16, 20 wt%	PA 11	wet spinning	morphology mechanical properties electrical conductivity	Zhang et al. 2002
PANI 10, 20, 30, 40, 50 wt%	PAN	wet spinning	morphology mechanical properties electrical conductivity	Lee et al. 2003
PANI-DBSA 5 wt%	PP	melt spinning	morphology chemical structure electrical conductivity	Fryczkowski et al. 2004
PANI 1, 2, 5, 20, 30 wt%	PP	melt spinning	morphology electrical conductivity	Kim et al. 2004
PANI-DBSA 2.5, 5, 7, 10 wt%	PAN	wet spinning	morphology mechanical properties thermal stability electrical conductivity	Jianming et al. 2005
PANI 5, 11.5, 14.2, 20, 25 w/w%	PA 6	wet spinning	electrical conductivity electroactivity	Mirmohseni et al. 2006
Panipol	PP	melt spinning	electrical conductivity neurocompatibility neuronal adhesion and growth	Cullen et al. 2008
Panipol 20 wt%	PP	melt spinning	morphology electrical conductivity	Soroudi and Skrifvars 2011
Panipol 20 wt%	PP/PA	melt spinning	morphology mechanical properties electrical conductivity	Soroudi and Skrifvars 2012

processable polyaniline complex (Panipol CXL) with PP under different mixing conditions, meltspun into fiber filaments under different draw ratios and characterized the fibers regarding their morphology and conductivity. Panipol CXL polyaniline-complex and fiber grade polypropylene were melt mixed using a twin-screw compounder. Two sample series were prepared. First group was prepared at 200°C and with a screw rotation rate of 70 rpm for 12 min. The second group was processed at 220°C and with screw rotation rate of 70 rpm for 15 min. PANI content was 20 wt% in both of the samples. Fibers were melt spun at suitable processing conditions and drawn between two Godet rolls with draw ratios of 2, 3, 4, 5 and 6. Both the mixing conditions and the draw ratios had substantial effects on morphology and conductivity of the fibers. Morphology of fibers was found to be two phased, containing a continuous PP rich phase and a dispersed polyaniline rich phase. The size of the dispersed Panipol™ rich phase, which was controlled by melt blending conditions, and the draw ratio showed substantial effects on electrical resistance of composite fibers. Second group showed lower resistance due to the better dispersion of the polyaniline in fiber structure especially at the draw ratio of 2 and 3. Soroudi and Skrifvars (2012) produced melt spun fibers using a ternary blend of PP/PA 6/PANI complex. PP and PA6 were melt-mixed with the conductive PANI-complex using a twin-screw micro compounder. Two different blends (binary blend of PP/PANI (80/20) and ternary blend of PP/PA6/PANI (55/25/20)) were prepared. SEM images of the ternary blends displayed at least a three-phase morphology. 20% PANI addition to the insulating PP and PP/PA6 blends resulted in conductive fibers that possess conductivities in the range of 10^{-3}–10^{-4} S/cm. The conductivity of meltspun composite filaments was affected by the applied draw ratio. Ternary blend fibers showed higher conductivity and higher tensile strength only at the draw ratio of 5. Since both conductivity and tensile strength depended on the formation of fibrils from the core-shell dispersed phase of the PA6/PANI-complex, the authors concluded that the draw ratio is critical for the ternary blend fibers.

Zhang et al. (2001) produced PA 11/PANI filaments using wet spinning method. The spinning solution was prepared by blending PANI in the emeraldine base form with PA 11 in concentrated sulfuric acid. As-spun fibers were obtained by spinning into coagulation bath water or dilute acid while drawn fibers were obtained by drawing the as-spun fibers in warm drawing bath water. PANI content was varied as 3, 8, 10, 16, and 20%. SEM and TEM images showed that PA 11 and PANI were incompatible and PANI was in fibrillar form, which was valuable for producing conducting channels. While the electrical conductivity changed between 10^{-6} to 10^{-1} S/cm depending on the PANI content, the percolation threshold was observed to be about 5 wt% which is much lower compared to PET/carbon black samples which showed a percolation threshold of 18 wt%. They concluded that the morphology of the conductive component was of great importance regarding electrical conductivity. In another study, Zhang et al. (2002) investigated the effect of acid concentration in the coagulation bath and PANI percentage in the fibers on the microstructure of the as-spun fibers, the mechanical properties, and conductivity of the filaments. The microstructure of the fibers, mechanical properties, and drawability were affected by both the acid concentration of the coagulation bath and PANI content. Increasing the acid concentration in the coagulation bath led to the shrinkage of the void size of the as-spun fibers, improved the drawing process and increased mechanical properties. On the other hand, increasing PANI content resulted in decrease in the mechanical properties although decrease in the void size was observed with PANI addition. The conductivity of the fibers ranged from 10^{-5} S/cm with 5 wt% PANI to 10^{-1} S/cm with 20% PANI. Lee et al. (2003) prepared conductive PAN/sulfonated PANI (SPAN) fibers. PANI in emeraldine base oxidation state was synthesized using phosphoric acid and then sulfonated PANI (S/N ratio: 0.5) was obtained by the reaction of PANI powder with sulfuric acid. For the preparation of the spinning solutions, PAN and SPAN were dissolved in DMSO. SPAN ratio was changed as 10, 20, 30, 40 and 50 wt%. The fibers were produced using wet spinning method. The fibers were drawn in a hot water bath during the take-up process. XRD spectra showed that phase separation had occurred between the polymers. While the tenacity and elongation of the fibers decreased with the increase in PANI content, drawing resulted in some improvements in the mechanical properties. All the samples were conductive without additional doping process. Jianming et al. (2005) produced conductive composite fibers of PAN with dodecylbenzene sulfonic acid doped polyaniline (PANI-DBSA) and investigated the effects of PANI-DBSA amount on several properties such as the mechanical properties, electrical conductivity and thermal stability. While the total concentration of the polymer (PAN and PANI-DBSA) was kept as 17 wt%, the concentration of PANI-DBSA was varied as 2.5, 5, 7 and 10 wt%. Films were also prepared for comparison. SEM and TEM images showed that PANI-DBSA was well dispersed in PAN. The fiber with 7 wt% PANI-DBSA showed a conductivity of 10^{-3} S/cm which was lower than the corresponding film. The tensile strength first increased, reached the highest value at the PANI-DBSA content of 5 wt% and then decreased with the further increase of PANI-DBSA amount. On the other hand, the elongation at break reduced with the increase in PANI-DBSA content. The composite fiber displayed better thermal properties. Mirmohseni et al. (2006) blended PANI with PA6 in concentrated formic acid and spun fibers using wet spinning method. PANI content was varied between 5 and 25 wt/wt% and the electrical resistances of blend fibers were measured between 0.665 and 0.015 MΩ/cm. Cyclic voltammetric studies showed that the fibers with PANI were electroactive.

In the reviewed studies about polymer/PANI filaments, the focus was mainly on the dependence of conductivity on PANI concentration. PANI concentration was varied between 2.5 and 30 wt% and the conductivity values reported were between 10^{-9} and 10^{-1} S/cm. Better dispersion of PANI in filament structure resulted in higher conductivity values. The conductivity of the composite filaments was also affected by the draw ratio. Application of the optimum draw ratio was found to be necessary to be able to obtain the highest conductivity values. While some improvement could be observed in tensile strength with the addition of low amount of PANI, decrease was observed in elongation values.

Composite fibers containing CNTs, AgNPs and PANI in different combinations

Addition of two or more different nanostructures into a polymer has proven to be a promising technique in combining the desirable properties of different structures into one material or obtaining a synergistic effect with regards to a special property. This technique has attracted great interest recently and has been widely applied in composite film production (Zhang et al. 2010, Fortunati et al. 2011, Grinou et al. 2012, Mi and Xu 2012), nanofiber production (Kizildag and Ucar 2016b, Eren et al. 2016), and filament production (Soroudi and Skrifvars 2010, Rangari et al. 2010).

Soroudi and Skrifvars (2010) prepared PP/PANI (Panipol CXL)/MWNT blends and melt spun into filaments. Master batch of PP with 15 w% CNTs was used in compounding to eliminate the difficulties regarding CNT dispersion. Different blend combinations such as PP/Panipol CXL, PP/CNT and PP/Panipol CXL/CNT were prepared in twin-screw microcompounder. While the concentration of PANI was kept constant as 20 wt%, the concentration of CNTs was changed as 1.5 wt% and 7.5 wt%. Composite filaments were characterised regarding conductivity, morphology, and thermal properties. PP containing 20 wt% PANI modified with 7.5 wt% CNTs displayed the maximum conductivity of about 0.16 S/cm. Panipol CXL showed nucleating effect in neat PP, and adding CNT to the PP/Panipol blend increased the nucleating effects further. Comparison between neat PP and the sample containing Panipol CXL regarding thermal decomposition temperatures showed that Panipol CXL increased the onset of decomposition temperature of the blend. When the PP/Panipol CXL blend was additionally modified with 7.5 wt% CNT, the decomposition temperature increased further. The authors concluded that the results were promising regarding the possibility to prepare conductive polymer fibers suitable for smart and functional textiles. Rangari et al. (2010) prepared Ag-coated CNTs hybrid nanoparticles (Ag/CNTs) by ultrasonic irradiation of dimethylformamide (DMF) and silver (I) acetate precursors in the presence of CNTs and used them in the production of PA6 filaments. Composite fibers with Ag-coated CNTs displayed better antibacterial performance than PA6/Ag and PA6/CNT composite filaments. In addition, improvement was observed in Young's modulus, breaking stress, thermal stability, and crystallinity of the composite fibers at the Ag-coated CNT content of 1 wt%.

Conclusions

Nanomaterials and conductive polymers have been widely used in the production of composite fibers with the aim of either improving the existing properties or developing some new functionalities. Carbon nanotubes have attracted great interest both from industry and academy with their unique mechanical, thermal and electrical properties since their discovery. Especially, the high aspect ratio of the carbon nanotubes resulting in low percolation thresholds makes them outstanding conductive additives. Although great progress has been made in the studies regarding the processing and very promising results have been obtained, more work and research are necessary to be able to fully utilize the properties of carbon nanotubes. Silver nanoparticles have been reported to provide both antistatic ability and antibacterial property while polyaniline, which is an inherently conductive polymer, is used to improve conductivity in filaments by forming conductive paths in the filament structure. Studies in literature show the possibility of producing functional fibers with these additives. A relatively new subject in composite fiber production is the combined addition of the additives into fiber structure, which may enable to improve the existing properties further with synergistic effects or combine different functionalities in a fiber/filament. The results of the studies are very encouraging and show that there is a wealth of opportunities in nanocomposite fibers/filaments.

Acknowledgements

The support of The Scientific and Technological Research Council of Turkey (TUBITAK) through Project 112M877 is acknowledged.

References

Arbab, S., P. Noorpanah, N. Mohammadi and A. Zeinolebadi. 2011. Exploring the effects of non-solvent concentration, jet-stretching and hot-drawing on microstructure formation of poly(acrylonitrile) fibers during wet-spinning. J. Polym. Res. 18: 1343–1351.

Cao, Y., P. Smith and A.J. Heeger. 1992a. Counter-ion induced processability of conducting polyaniline and of conducting polyblends of polyaniline in bulk polymers. Synthetic Met. 48: 91–97.

Cao, Y. and A.J. Heeger. 1992b. Magnetic susceptibility of polyaniline in solution in non-polar organic solvents and in polyblends in poly(methyl methacrylate). Synthetic Met. 52: 193–200.

Carfagna, C. and P. Persico. 2006. Functional textiles based on polymer composites. Macromol. Symp. 245-246: 355–362.

Chae, H.G., T.V. Sreekumar, T. Uchida and S. Kumar. 2005. A comparison of reinforcement efficiency of various types of carbon nanotubes in polyacrylonitrile fiber. Polymer 46: 10925–10935.

Choi, S., Y. Jeong, G.W. Lee and D.H. Cho. 2009. Thermal and mechanical properties of polypropylene filaments reinforced with multiwalled carbon nanotubes via melt compounding. Fiber Polym. 10(4): 513–518.

Cruz-Estrada, R.H. and M.J. Folkes. 2000. *In situ* production of electrically conductive fibres in polyaniline-SBS blends. J. Mater. Sci. 35: 5065–5069.

Cullen, D.K., A.R. Patel, J.F. Doorish, D.H. Smith and B.J. Pfister. 2008. Developing a tissue-engineered neural-electrical relay using encapsulated neuronal constructs on conducting polymer fibers. J. Neural. Eng. 5(4): 374–384.

Damerchely, R., M.E. Yazdanshenas, A.S. Rashidi and R. Khajavi. 2011. Morphology and mechanical properties of antibacterial nylon 6/nano-silver nano-composite multifilament yarns. Text. Res. J. 81(16): 1694–1701.

Dastjerdi, R., M.R.M. Mojtahedi, A.M. Shoshtari, A. Khosroshahi and A.J. Moayed. 2009a. Fiber to fabric processability of silver/zinc-loaded nanocomposite yarns. Text. Res. J. 79(12): 1099–1107.

Dastjerdi, R., M.R.M. Mojtahedi and A.M. Shoshtari. 2009b. Comparing the effect of three processing methods for modification of filament yarns with inorganic nanocomposite filler and their bioactivity against Staphylococcus aureus. Macromol. Res. 17(6): 378–387.

Dastjerdi, R., M.R.M. Mojtahedi, A.M. Shoshtari and A. Khosroshahi. 2010. Investigating the production and properties of Ag/TiO$_2$/PP antibacterial nanocomposite filament yarns. J. Text. I. 101(3): 204–213.

Demirsoy, N., N. Ucar, A. Onen, I. Karacan, N. Kizildag, O. Eren et al. 2016. The effect of dispersion technique, silver particle loading, and reduction method on the properties of polyacrylonitrile–silver composite nanofiber. J. Ind. Text. 45(6): 1173–1187.

Devaux, E., C. Aubry, C. Campagne and M. Rochery. 2011. PLA/carbon nanotubes multifilament yarns for relative humidity textile sensor. J. Eng. Fiber Fabr. 6(3): 13–24.

Eren, O., N. Ucar, A. Onen, N. Kizildag and I. Karacan. 2016. Synergistic effect of polyaniline, nanosilver, and carbon nanotube mixtures on the structure and properties of polyacrylonitrile composite nanofiber. J. Compos. Mater. 50(15): 2073–2086.

Ferreira, A., F. Ferreira, M.C. Paiva, B. Oliveira and J.A. Covas. 2011. Monofilament Composites with Carbon Nanotubes for Textile Sensor Applications. Autex 2011, Mulhouse, France.

Ferreira, A., F. Ferreira and M.C. Paiva. 2013. Textile sensor applications with composite monofilaments of polymer/carbon nanotubes. Adv. Sci. Tech. 80: 65–70.

Fortunati, E., F. D'Angelo, S. Martino, A. Orlacchio, J.M. Kenny and I. Armentano. 2011. Carbon nanotubes and silver nanoparticles for multifunctional conductive biopolymer composites. Carbon. 49(7): 2370–2379.

Fryczkowski, R., W. Binias, J. Farana, B. Fryczkowska and A. Wlochowicz. 2004. Spectroscopic and morphological examination of polypropylene fibres modified with polyaniline. Synthetic Met. 145: 195–202.

Gao, J., M.E. Itkis, A. Yu, E. Bekyarova, B. Zhao and R.C. Haddon. 2005. Continuous spinning of a single-walled carbon nanotube-nylon composite fiber. J. Am. Chem. Soc. 127: 3847–3854.

Gawish, S.M., H. Avci, A.M. Ramadan, S. Mosleh, R. Monticello, F. Breidt et al. 2012. Properties of antibacterial polypropylene/nanometal composite fibers. J. Biomat. Sci.-Polym. E. 23: 43–61.

Grinou, A., H. Bak, Y.S. Yun and H.J. Jin. 2012. Polyaniline/silver nanoparticle-doped multiwalled carbon nanotube composites. J. Disper. Sci. Technol. 33(5): 750–755.

Haggenmueller, R., H.H. Gommans, A.G. Rinzler, J.E. Fisher and K.I. Winey. 2000. Aligned single-wall carbon nanotubes in composites by melt processing methods. Chem. Phys. Lett. 330: 219–225.

Iijima, S. 1991. Helical microtubulues of graphitic carbon. Nature 354: 56–58.

Jee, M.H., J.U. Choi, S.H. Park, Y.G. Jeong and D.H. Baik. 2012a. Influences of tensile drawing on structures, mechanical, and electrical properties of wet-spun multi-walled carbon nanotube composite fiber. Macromol. Res. 20(6): 650–657.

Jee, M.H., S.H. Park, J.U. Choi, Y.G. Jeong and D.H. Baik. 2012b. Effects of wet-spinning conditions on structures, mechanical and electrical properties of multi-walled carbon nanotube composite fibers. Fiber Polym. 13(4): 443–449.

Jianming, J., P. Wei, Y. Shenglin and L. Guang. 2005. Electrically conductive PANI-DBSA/Co-PAN composite fibers prepared by wet spinning. Synthetic Met. 149: 181–186.

Jose, M.V., D. Dean, J. Tyner, G. Price and E. Nyairo. 2007. Polypropylene/carbon nanotube nanocomposite fibers: Process–morphology–property relationships. J. Appl. Polym. Sci. 103: 3844–3850.

Kearns J.C. and R.L. Shambaugh. 2002. Polypropylene fibers reinforced with carbon nanotubes. J. Appl. Polym. Sci. 86: 2079–2084.

Khan, M.R., H. Mahfuz, T. Leventouri, V.K. Rangari and A. Kyriacou. 2011. Enhancing toughness of low-density polyethylene filaments through infusion of multiwalled carbon nanotubes and ultrahigh molecular weight polyethylene. Polym. Eng. Sci. 51: 654–662.

Kim, B., V. Koncar, E. Devaux, C. Dufour and P. Viallier. 2004. Electrical and morphological properties of PP and PET conductive polymer fibers. Synthetic Met. 146(2): 167–174.

Kizildag, N. and N. Ucar. 2016a. Nanocomposite polyacrylonitrile filaments with electrostatic dissipative and antibacterial properties. J. Compos. Mater. (in press).

Kizildag, N. and N. Ucar. 2016b. Investigation of the properties of PAN/f-MWCNTs/AgNPs composite nanofibers. J. Ind. Text. (in press).

Laska, J., M. Trzandel and A. Pron. 1993. Solubilisation and plastification of polyaniline in the protonated state. Mater. Sci. Forum. 122: 177–184.

Lai, D., Y. Wei, L. Zoun, Y. Xu and H. Lu. 2015. Wet spinning of PVA composite fibers with a large fraction of multi-walled carbon nanotubes. Prog. Nat. Sci. 25: 445–452.

Lee, S.J., H.J. Oh, H.A. Lee and K.S. Ryu. 2003. Fabrication and physical properties of conductive polyacrylonitrile-polyaniline derivative fibers. Synthetic Met. 135-135: 399–400.

Li, C., T. Liang, W. Lu, C. Tang, X. Hu, M. Cao et al. 2004. Improving the antistatic ability of polypropylene fibers by inner antistatic agent filled with carbon nanotubes. Compos. Sci. Technol. 64: 2089–2096.

Li, Z., G. Luo, F. Wei and Y. Huang. 2006. Microstructure of carbon nanotubes/PET conductive composites fibers and their properties. Compos. Sci. Technol. 66: 1022–1029.

McIntosh, D., V.N. Khabashesku and E.V. Barrera. 2006. Nanocomposite fiber systems processed from fluorinated single-walled carbon nanotubes and a polypropylene matrix. Chem. Mater. 18: 4561–4569.

McIntosh, D., V.N. Khabashesku and E.V. Barrera. 2007. Benzoyl peroxide initiated *in situ* functionalization, processing, and mechanical properties of single-walled carbon nanotube-polypropylene composite fibers. J. Phys. Chem. C. 111: 1592–1600.

Meng, Q. and J. Hu. 2008. Self-organizing alignment of carbon nanotube in shape memory segmented fiber prepared by *in situ* polymerization and melt spinning. Compos. Part A-Appl. S. 39: 314–321.

Meng, Q.J., Z.M. Wang, X.X. Zhang, X.C. Wang and S.H. Bai. 2010. Fabrication and properties of polyamide-6,6-functionalized carboxylic multi-walled carbon nanotube composite fibers. High Perform. Polym. 22: 848–862.

Mercader, C., V. Denis-Lutard, S. Jestin, M. Maugey, A. Derré, C. Zakri et al. 2012. Scalable process for the spinning of PVA–carbon nanotube composite Fibers. J. Appl. Polym. Sci. 125: E191–E196.

Mi, H. and Y. Xu. 2012. Ag-loaded polypyrrole/carbon nanotube: One-step *in situ* polymerization and improved capacitance. Adv. Mat. Res. 531: 35–38.

Mikołajczyk, T., G. Szparaga and G. Janowska. 2009. Influence of silver nano-additive amount on the supramolecular structure, porosity, and properties of polyacrylonitrile precursor fibers. Polym. Advan. Technol. 20: 1035–1043.

Mikołajczyk, T., G. Szparaga, M. Bogun, A. Fraczek-Szczypta and S. Blazewicz. 2010. Effect of spinning conditions on the mechanical properties of polyacrylonitrile fibers modified with carbon nanotubes. J. Appl. Polym. Sci. 115: 3628–3635.

Min, B.G., H.G. Chae, M.L. Minus and S. Kumar. 2009. Polymer/carbon nanotube composite fibers—an overview. pp. 43–73. *In*: K.P. Lee, A.I. Gopalan and F.D.S. Marquis (eds.). Functional Composites of Carbon Nanotubes and Applications. Transworld Research Network, Kerala, India.

Mirmohseni, A., D. Salari and R. Nabavi. 2006. Preparation of conducting polyaniline/nylon 6 blend fibre by wet spinning technique. Iran. Polym. J. 15(3): 259–264.

Moore, E.M., D.L. Ortiz, V.T. Marla, R.L. Shambaugh and B.P. Grady. 2004. Enhancing the strength of polypropylene fibers with carbon nanotubes. J. Appl. Polym. Sci. 93: 2926–2933.

Mun, S.J., Y.M. Jung, J.C. Kim and J.H. Chang. 2008. Poly(ethylene terephthalate) nanocomposite fibers with functionalized multiwalled carbon nanotubes via *in situ* polymerization. J. Appl. Polym. Sci. 109: 638–646.

Passiniemi, P., J. Laakso, H. Osterholm and M. Pohl. 1997. TEM and WAXS characterization of polyaniline/PP fibers. Synthetic Met. 84(1-3): 775–776.

Perrot, C., P.M. Piccione, C. Zakri, P. Gaillard and P. Poulin. 2009. Influence of the spinning conditions on the structure and properties of polyamide 12/carbon nanotube composite fibers. J. Appl. Polym. Sci. 114: 3515–3523.

Poetschke, P., H. Brünig, A. Janke, D. Fischer and D. Jehnichen. 2005. Orientation of multiwalled carbon nanotubes in composites with polycarbonate by melt spinning. Polymer 46: 10355–10363.

Prabhu, S. and E.K. Poulose. 2012. Silver nanoparticles: Mechanism of antimicrobial action, synthesis, medical applications, and toxicity effects. Int. Nano Lett. 32: 2–10.

Rangari, V.K., M. Yousuf, S. Jeelani, M.X. Pulikkathara and V.N. Khabashesku. 2008. Alignment of carbon nanotubes and reinforcing effects in nylon-6 polymer composite fibers. Nanotechnology 19(245703): 1–9.

Sandler, J.K.W., S. Pegel, M. Cadek, F. Gojny, M. van Es, J. Lohmar et al. 2004. A comparative study of melt spun polyamide–12 fibres reinforced with carbon nanotubes and nanofibers. Polymer. 45: 2001–2015.

Shen, L., X. Gao, Y. Tong, A. Yeh, R. Li and D. Wu. 2008. Influence of different functionalized multiwall carbon nanotubes on the mechanical properties of poly(ethylene terephthalate) fibers. J. Appl. Polym. Sci. 108: 2865–2871.

Skrifvars, M. and A. Soroudi. 2009. Melt spinning of carbon nanotube modified polypropylene for electrically conducting nanocomposite fibres. Sol. St. Phen. 151: 43–47.

Soitong, T. and J. Pumchusak. 2011. Morphology and tensile properties of polypropylene-multiwalled carbon nanotubes composite fibers. J. Appl. Polym. Sci. 119: 962–967.

Sondi, I. and B. Salopek-Sondi. 2004. Silver nanoparticles as antimicrobial agent: a case study on *E. coli* as a model for Gram-negative bacteria. J. Colloid Interf. Sci. 275: 177–182.

Soroudi, A. and M. Skrifvars. 2010. Melt blending of carbon nanotubes/polyaniline/polypropylene compounds and their melt spinning to conductive fibres. Synthetic Met. 160: 1143–1147.

Soroudi, A. and M. Skrifvars. 2011. Polyaniline–polypropylene melt-spun fiber filaments: the collaborative effects of blending conditions and fiber draw ratios on the electrical properties of fiber filaments. J. Appl. Polym. Sci. 119(1): 558–564.

Soroudi, A. and M. Skrifvars. 2012. Electroconductive polyblend fibers of polyamide-6/polypropylene/polyaniline: Electrical, morphological, and mechanical characteristics. Polym. Eng. Sci. 52(7): 1606–1612.

Sreekumar, T.V., T. Liu, B.G. Min, H.N. Guo, S. Kumar and R.H. Hauge. 2004. Polyacrylonitrile single-walled carbon nanotube composite fibers. Adv. Mater. 16: 58–61.

Sulong, A.B., J. Park, C.H. Azhari and K. Jusoff. 2011. Process optimization of melt spinning and mechanical strength enhancement of functionalized multi-walled carbon nanotubes reinforcing polyethylene fibers. Compos Part B-Eng. 42: 11–17.

Sun, B., B. Hua, X. Ji, Y. Shi, Z. Zhou, Q Wang et al. 2014. Preparation of silver nanoparticles with hydrophobic surface and their polyester based nanocomposite fibres with excellent antibacterial properties. Mater. Res. Innov. 18(4): 869–874.

Ucar, N., N. Demirsoy, A. Onen, I. Karacan, N. Kizildag, O. Eren et al. 2015. The effect of reduction methods and stabilizer (PVP) on the properties of polyacrylonitrile (PAN) composite nanofibers in the presence of nanosilver. J. Mater. Sci. 50: 1855–1864.

Wang, Y., R. Cheng, L. Liang and Y. Wang. 2005. Study on the preparation and characterization of ultra-high molecular weight polyethylene–carbon nanotubes composite fiber. Compos. Sci. Technol. 65: 793–797.

Yazdanshenas, M.E., R. Damerchely, A.S. Rashidi and R. Khajavi. 2012. Bioactive nano-composite multifilament yarns. J. Eng. Fiber. Fabr. 7(1): 69–78.

Yeo, S.Y. and S.H. Jeong. 2003. Preparation and characterization of polypropylene/silver nanocomposite fibers. Polym. Int. 52: 1053–1057.

Zhang, Q., H. Jin, X. Wang and X. Jing. 2001. Morphology of conductive blend fibers of polyaniline and polyamide-11. Synthetic Met. 123: 481–485.

Zhang, Q., X. Wang, D. Chen and X. Jing. 2002. Preparation and properties of conductive polyaniline/poly-ω-aminoundecanoyle fibers. J. Appl. Polym. Sci. 85: 1458–1464.

Zhang, W., W. Li, J. Wang, C. Qin and L. Dai. 2010. Composites of polyvinyl alcohol and carbon nanotubes decorated with silver nanoparticles. Fiber. Polym. 11(8): 1132–1136.

Zhang, X. 2014. Fundamentals of Fiber Science. DEStech Publications, Pennsylvania, USA.

Zhou, H., X. Tang, Y. Dong, L. Chen, L. Zhang, W. Wang et al. 2011. Multiwalled carbon nanotube/polyacrylonitrile composite fibers prepared by *in situ* polymerization. J. Appl. Polym. Sci. 120: 1385–1389.

15

Processing and Mechanical Properties of Carbon Nanotube Reinforced Metal Matrix Composites

Abdollah Hajalilou,[1,]* *Ebrahim Abouzari-lotf,*[2,3] *Katayoon Kalantari,*[4]
Hossein Lavvafi[5] *and Amalina Binti Muhammad Afifi*[4]

Introduction

Metal matrix composites, as the name implies, are composed of high strength and modulus components placed in a highly ductile and tough metal matrix. Such combination yields a unique property that in not attainable in either the reinforcing agents or the matrix, creating greater compression strength, shear and higher service temperature capabilities. Thus, their high performance make them a superior candidate for devices that need low weight and high strength such as aerospace, recreation, automotive industries, etc.

With further progress in technology, there is always a demand in design and development of novel materials in aerospace, automotive industries due to their requirement for materials with high strength and light weight. Efforts endeavored in this area are increasing day by day to discover replacements for the conventional macro/micro reinforcements. In this respect, nano-sized reinforcements have proven to be suitable candidates due to the increased interfacial contact area and smaller inter-particle spacing. However, nanoparticles of various materials are used as discontinuous reinforcement (Table 1) in metal, ceramic or polymeric matrix. Among them, carbon nanotubes (CNTs) have recently attracted more attention because of their amazing mechanical, thermal, electrical and other properties (Marikani 2009). For example, CNTs' great mechanical strength and stiffness are well-known due to the intrinsic strength of carbon-carbon sp^2 bond and their special structure.

[1] Faculty of Mechanical Engineering, Department of Materials Engineering, University of Tabriz, 51666, Tabriz, Iran.
[2] Advanced Materials Research Group, Center of Hydrogen Energy, Universiti Teknologi Malaysia, 54100, Kuala Lumpur, Malaysia.
[3] Department of Chemical Engineering, Universiti Teknologi Malaysia, 81310 Johor Bahru, Malaysia.
[4] Center of Advanced Material (CAM), Faculty of Mechanical Engineering, University of Malaya, 50603 Kuala Lumpur, Malaysia.
[5] University of Toledo Medical Center, Department of Radiation Oncology, 3000 Arlington Avenue, MS 1151, Toledo, OH 43614, Dana Cancer Center.
* Corresponding author: e.hajalilou@yahoo.com

Table 1. Discontinuous reinforcements characteristics (Khare and Bose 2005, Li and Langdon 1997, Spigarelli et al. 2002).

Reinforcement	Density (g/cm³)	Young's modulus (GPa)	Strength (MPa)	Thermal expansivity (10⁻⁶/°C)
CNT	1.2–2.6	1400	100000	−1.5 (SWCNT)
Al_2O_3	3.98	379 (1090°C)	221 (1090°C)	7.92
Al_3Ti	3.3	217	-	-
AlN	3.26	310 (1090°C)	2069	4.84
B_4C	2.52	448 (24°C)	2759 (24°C)	6.08
C	2.18	690	-	−1.44
MgO	3.58	317 (1090°C)	41 (1090°C)	11.61
Si	2.33	112	-	3.06
SiC	3.21	324 (1090°C)	-	5.4
Si_3N_4	3.18	207	-	1.44
SiO_2	2.66	73	-	< 1.08
TiB_2	4.5	414 (1090°C)	-	8.28
TiC	4.93	269 (24°C)	55 (1090°C)	7.6
WC	15.63	269 (24°C)	-	5.09
ZrB_2	6.09	503 (24°C)	-	8.28
ZrC	6.73	359 (24°C)	90 (1090°C)	6.66
ZrO_2	5.89	132 (1090°C)	83 (1090°C)	12.01

Carbon nanotube (CNT) basics

Carbon family

Until 1980s, graphite and diamond were the only known physical form of carbon where the atoms are covalently bonded. Graphite crystallizes in a layer or sheet-liked structure and the atoms are connected to each other in a hexagonal lattice structure by covalent bonding and layers by van der Walls forces (Figure 1a) (Marikani 2009). On the other hand, diamond crystallizes in cubic unit cell and has two

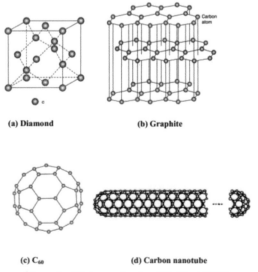

(a) Diamond   (b) Graphite

(c) C₆₀   (d) Carbon nanotube

Figure 1. Allotropy of carbon (Marikani 2009).

interpenetrating FCC lattice structure, where the atoms connect in similar manner with graphite (Figure 1b) (Marikani 2009).

Spherical shaped carbon liked as a ball with 32 faces was discovered in 1985 and was called C_{60} or bucker minster fullerence or buckyball (Figure 1c). Afterward, other allotropes of carbon with molecules like C36, C70, C76 were discovered, which are composed of pentagons and hexagons structure. Later on, cylinder shaped with unique geometric construction of carbon atoms in the bukyball shape were found that are known as carbon nanotubes (CNTs) (Figure 1d).

CNTs

Carbon nanotubes (CNTs) were first observed at the negative end of the electrode during the arc discharge production of fullerenes. Iijima (Iijima 1991) used TEM to observe the multi-wall carbon nanotubes (MWCNTs), identifying an important discovery. Consequently, the single-walled nanotubes were discovered in 1993 (Bethune et al. 1993, Iijima and Ichihashi 1993). The CNT atomic structure can be considered as a lay of graphene (a single atom thick graphite), rolling up into a seemless nano-sized cylinder. Rolling the graphene in a specific direction results in the formation of various types of CNTs such as chiral (n, m), zig-zag (n, 0), and armchair (n, n), as shown in Figure 2.

These nano-wieres include multi-walled (MW), double-walled (DW) and single-walled (SW) carbon nanotubes (CNTs). Indeed, they are made of a lay (in SCNTs), double lay (in DWCTNs) or more than two lays (MWCNTs) of graphite rolling up into a cylinder with the two ends capped. High resolution transmission electron microscopy (HRTEM) is required to observe these features.

Basically, CNTs are chemically bonded with sp^2 bonds, an extremely strong form of molecular interaction, which identify the extra-high strength and stiffness of the CNTs. The SWCNTs possess Young moduli and density ranging from 1 to 5 TPa and 1.8 g/cm^3, respectively, while the MWCNTs represent an average value of 1.8 TPa and 1.8 g/cm^3. These result in a high specific strength (about 55 GPa) and high specific modulus (about 555 GPa) of CNTs, making them a desirable candidate as reinforcement agents in composites (George et al. 2005). These materials in comparison with the conventional micron-sized reinforcement such as SiC_p and carbon fibers are elastic and have feasibility to bend, flatten, twist, and bukle or tangle without breaking. On the other hand, those traditional reinforcement components are stiff but brittle and behave as rigid rod. Thus, it can be concluded that CNTs can be flexible, and undergo arbitrary shapes in composite during bending (Ajayan and Tour 2007).

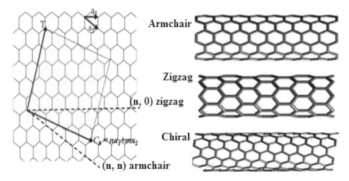

Figure 2. Various types of CNTs.

CNTs-reinforced metal matrix composites

There is always an attempt to develop advanced CNTs-based composites utilizing ceramic, polymeric and metallic matrix. Collected information by Bakshi et al. (Bakshi et al. 2010). Revealed that the CNTs-metallic composites are less studied materials compared to two others. This is quite astonishing that metals are the widely employed structural materials all over the world. The main reason may be associated with difficulty in their fabrication in comparison with the polymeric and ceramic matrix composites.

Table 2. Basic characteristics of metals used as matrix.

Atomic symbol	Ti	Mg	Al	Cu
Atomic Number	22	12	13	29
Density (g/cm^3)	4.506	1.738	2.6989	8.92
Atomic Weight	47.86	24.305	26.9815	63.546
Specific Heat Capacity (J/g.°C)	0.523	1.020	0.900	0.385
Thermal Conductivity (W/m.K)	21.9	156	210	401
Boiling Point (°C)	3287	1.091	2519	2.562
Melting Point (°C)	1670	650	660	1.085
Electron Configuration	[Ar] 3d^{2}4s^2	[Ne] 3s^2	3s^2 3p^1	[Ar] 3d^{10} 4s^1
Atomic Radius (pm)	147	145	143.1	128
Shear Modulus (GPa)	44	17	25	48
Elastic Modulus (GPa)	110		68	
Hardness, Vickers	830–3420	-	160–350	343–369

Among metals, many may be used any use as matrix, but the most widely used and applicable metals are listed in Table 2 along with their characteristics. In the following section, their corresponding composites are briefly described.

CNTs-Al matrix composite

From Table 2, it can be extracted that the main reasons of using Al and its alloys as matrix are their low density, low price, requirement in most applications, and high mechanical properties like workability, machinability, corrosion resistance, stiffness and strength.

To this point, several researches but not only limited to following references (Choi et al. 2009, Esawi and El Borady 2008, Esawi et al. 2009, Pérez-Bustamante et al. 2009) have developed CNTs reinforced Al composites. Despite advances in composite materials, only a few developments on CNTs-Al composites have been reported. The cost of fabrication seems to be the main reason, but it seems that a failure to understand the strengthening mechanism and processing difficulties are other contributing factors. Furthermore, most researchers focused on the room temperature static mechanical properties and not the fatigue resistance, high temperature tensile strength, hardness, etc.

CNTs-Ti matrix composite

Titanium and titanium alloys represent high strength to weight ratios and exhibit potential corrosion resistance but simultaneously these materials possess low Young's modulus, wear resistance, and exhibit lower heat resistance in comparison with carbon nanotubes. Due to a superior thermal, electrical, physical and mechanical behavior of CNTs, the CNTs-Ti matrix composites are appropriate for a wide range of engineering aspects such as automotive, aerospace, and industrial applications.

In the current era, the demand is to minimize the fuel consumptions for global sustainability. This requires developing lightweight, high strength and cost effective structural materials. Carbon nanotubes (CNTs) reinforced titanium metal matrix composites offer the prospective to replace unreinforced titanium alloys for various applications.

CNTs-Cu matrix composite

Copper, because of its superior electrical conductivity, has been employed as a conventional material for interconnects of ultra-large scale integrated circuits (ULSI) for several decades. However, by demanding

ULSI for nano-sized interconnects (Dubin et al. 2007), it becomes difficult for Cu to carry out for nano-interconnects owing to the grain boundary scattering and electron surface scattering effects from Cu electrodes. CNTs are considered as a perfect candidate to have the role of "incorporating element" in composite interconnects, due to its one-dimensional structure, as well as unique thermal, mechanical and electrical properties (Liu et al. 2008). The employed incorporation routes play a key role in the performance of CNTs/Cu composites.

CNTs-Mg matrix composite

Owing to their low density, high mechanical and physical properties of magnesium matrix composites, they are widely used in various applications in aerospace and defense industry. The improvement of mechanical properties such as wear, damping behavior, stiffness and specific strength, as well as creep and fatigue characteristics are noticeably affected by the introduction of reinforcing elements into the metallic matrix in comparison with the conventional engineering materials (Dey and Pandey 2015).

The reinforcement of CNTs gives rise to improvement of the tensile strength, bonding strength and wettability of Magnesium matrix composites. Yet, the introduction of CNTs to the matrix may lead to the basal plane texture weakening (Yu et al. 2000).

Processing techniques

Various parameters such as starting powder size selection, morphology of starting powders, choosing the appropriate volume fractions, processing techniques, dispersion mechanisms of CNTs in the matrix, interfacial reactions and stability of CNTs in the matrix, consolidation techniques, and sintering are always challengeable in composite production and have a key role in determining the final properties of the composites. Reinforcement of metal matrices with CNTs is challenging because of (i) uniform dispersion of CNTs in the matrix, in a way it causes an isotropic mechanical properties achievement (Akhtar 2008, Akhtar et al. 2007, Farid et al. 2007, Hussainova 2003). On the other hand, non-uniform distribution gives rise to agglomeration (Agarwal et al. 2016, Bakshi et al. 2010, Hertel et al. 1998, Jiang et al. 2006, Kondoh et al. 2009, Zeng et al. 2010). (ii) Structural damages to CNTs during processing; CNTs can be damaged during the consolidation process and ultimately lose their properties' assistances. (iii) Minimal CNTs-matrix reactions (intermetallic compounds (i.e., TiC) formation at high temperatures). These compounds may have positive or negative effect on the final properties depending on the aim and application. Therefore, considering such challenges, it is required to adopt a proper production route, to ensure the homogenous dispersion and minimal structural damage to CNTs.

Several process techniques (Figure 3) have been employed to fabricate CNTs-Metal Matrix Composites (CNTs-MMC). In all techniques, the challenges are to disperse CNTs homogeneously through the metal matrix (MM), to strengthen interfacial bond between CNTs and MM, and to keep CNTs structural and chemical integrity in the matrix throughout the process.

Powder metallurgy route

Integration of CNTs as reinforcement in metal matrices is highly dependent on not only the applied method but also its process parameters. These parameters would tend to control the microstructure and consequently the final properties of the composite. Powder metallurgy is proposed to be a proper processing technique for metal matrix composites for several reasons (Agarwal et al. 2016, Bakshi et al. 2010, Kondoh et al. 2009, Kuzumaki et al. 2000). One priority of employing this route over ingot metallurgy is that a wider range of compositions can be produced through this technique unlike ingot metallurgy, as it is not limited by the phase diagrams and the thermodynamics. Furthermore, in the case of ingot metallurgy, since CNTs have low density in comparison with metal matrices, CNTs are likely to get separated due to the buoyancy forces.

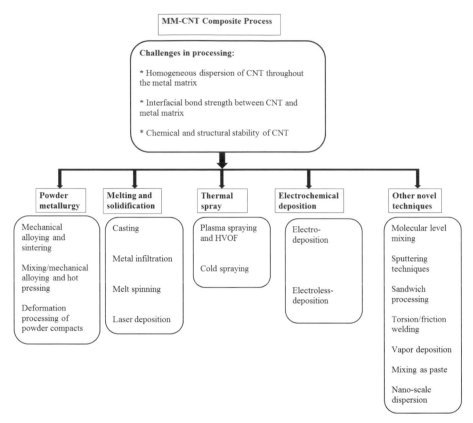

Figure 3. An illustration of various processes employed for CNTs-MMC fabrication (Bakshi et al. 2010).

The basic concept of powder metallurgy is to weigh the raw materials in terms of a molar ratio or atomic weight percentage. Then, the mixed powders are subjected to two types of milling regardless of using a wet or dry media: (i) mechanical alloying (MA) is operated in high-energy ball mill with the aim of mechanochemical purpose (Hajalilou et al. 2014a, 2015a, 2015b, Kuzumaki et al. 2000, Zeng et al. 2007, Liao and Tan 2011, Li et al. 2013) and (ii) mechanical mixing is carried out in normal mill and used for simple mixing of powders (Hajalilou et al. 2014b, 2018). Figure 4 shows a schematic representation of powder metallurgy route for the metal matrix composites development.

The mechanical alloying or mixing is sometimes not enough to provide enough energy to fabricate the desired composite, and is thus followed by (i) the post-sintering deformation process which is accomplished by compacting and shaping through extrusion, equi-channel angular, rolling, etc., and/

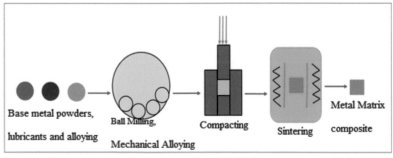

Figure 4. The powder metallurgy process.

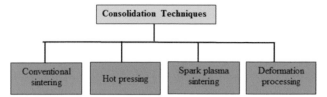

Figure 5. Different methods of consolidation of metal matrix composites.

or (ii) the sintering process such as spark plasma, hot/cold pressing and so on. These processes are considered as consolidation routes and briefly can be categorized as in Figure 5.

Conventional compaction of the mixed powder followed by the sintering is not an effective method to improve the CNTs dispersion in the metal matrix. This is mainly because of (i) any agglomeration formed in the composite during the mixing process would also occur in final composite after consolidation, and (ii) a less dense structure with porosity which would give rise to poor mechanical properties. Thus, researchers attempt to adopt other processing routes of consolidation such as hot pressing, spark plasma and deformation assisted sintering techniques.

Hot pressing is a suitable consolidation technique which is applied for the alloyed powder by given pressure and temperature in order to obtain dense compacts. In fact, heating during the compaction of powder mixtures in a carbon or graphite die can be incorporated by induction heating, electric resistance heating, and/or radiation. The grain growth of metal powder is prohibited by CNTs during the hot pressing because of their high thermal stability under elevated temperatures (Agarwal et al. 2016, Pang et al. 2007). The initial distribution of CNTs in the metal powder during the MA process plays a main role as it retains in the compact (Agarwal et al. 2016). A schematic representation of a hot pressing set-up is shown in Figure 6. The time required to attain composites with higher densities is 1 h in most researches carried out to date.

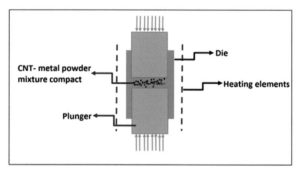

Figure 6. Schematic representation of a hot pressing set up.

Spark plasma sintering (SPS) is a modern consolidation route through a pass of direct pulse current from the die inclosing the alloyed powder (Agarwal et al. 2016, Bakshi et al. 2010). It is therefore possible to decrease the sintering time, which is very vital for CNTs-based-MM composites in order to prevent the interfacial reactions formation during the sintering (Hulbert et al. 2009). Furthermore, it is also possible to consolidate nano-powders without giving them sufficient time to incorporate grain growth (Bakshi et al. 2010). SPS is assisted by uniaxial press, punch electrodes, vacuum chamber, controlled atmosphere, DC pulse generator, temperature, and pressure measuring units (Kessel et al. 2010). A schematic representation of a spark plasma sintering set-up is shown in Figure 7. Researchers have successfully developed CNT-Ti matrix composites by this method (Kondoh et al. 2009, Li et al. 2013, Munir et al. 2006, Xue et al. 2010). It can be concluded that the sintering time of the alloyed powder reduces and the interfacial reactions miniaturizes by utilizing this technique, improving the CNT-MM composites mechanical properties.

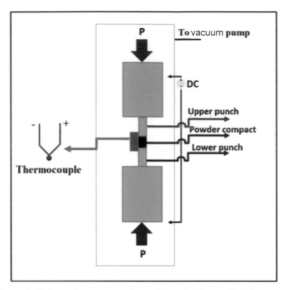

Figure 7. Schematic representation of a spark plasma sintering set-up.

Deformation processing is another consolidation technique that employed to break down the agglomerated CNTs structure with high shear forces (Agarwal et al. 2016). Better consolidated metal matrix composites structure with noticeable improvement in mechanical properties was observed by employing this method for composites productions (Deng et al. 2008, Kuzumaki et al. 1998, Kwon et al. 2009). This route includes hot rolling and hot extrusion which are considered to be effective processes to break down the CNT clustering into a consolidated composite structure (Agarwal et al. 2016). Researchers marked rolling a better processing technique than extrusion on the grounds to obtain better alignment of

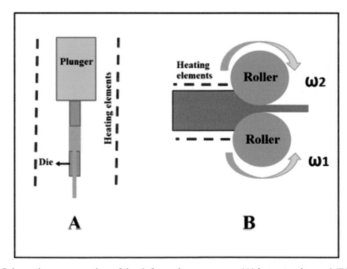

Figure 8. Schematic representation of the deformation processes: (A) hot extrusion and (B) hot rolling.

CNT clusters in the metal matrix (Dong et al. 2001, Kim et al. 2006), which results to the improvement in wear resistance properties of the composites due to a better alignment and dispersion of CNTs in the metal matrix (Bakshi et al. 2010). A schematic representation of hot extrusion and hot rolling set-ups is shown in Figure 8.

Melting and solidification route

As the name implies, a metal is melted and CNTs are scattered in the melted matrix and is finally solidified. This is the most commonly used conventional route for the CTNs-MMC fabrication. This technique is rarely used due to several limitations: firstly, the melting process needs high temperatures and consequently high temperature furnaces that are not economically suitable. Secondly, using high temperature may cause unwanted compounds in the CNTs-M interface or leads to CNTs damages. Thirdly, controlling the solidification process is another issue which strongly affects the composite mechanical properties. Finally, the suspended CNTs have a tendency to produce clusters because of tension forces. Consequently, this route is only desirable for those composites where their matrix has a low melting point such as magnesium. This process can be accomplished by using one or combination of the following process.

Casting

CNTs-reinforced Zr-based bulk metallic glass was the first product of this technique (Bian et al. 2002, Bian et al. 2004). CNTs were mixed with pre-alloyed products and were compacted into cylinders and subsequently melted and casted to produce 10% vol.-% CNT composite rods. ZrC was formed at CNT and matrix interface due to an increase in the matrix crystallinity, which results in the reduction of Zr from amorphous matrix. In spite of this, hardness of the composite increases due to CNT reinforcement. Besides, the manufactured composite represented superior acoustic wave absorption ability, which is associated with a large amount of interfaces induced by CNT reinforcement.

Lower melting point of Al makes it suitable candidate for casting. CNT-Al composite were synthesized by high pressure die casting (Ishikawa 2006). For this aim, 0.05 wt.% MWCNTs were wrapped in Al foil to produce a ball. As seen in Figure 2, four balls containing MWCNTs were set on the entrance of the die and then, the molten Al-alloy (239D) was squeezed into the die by the piston with a high velocity. The melt speeds up to 250 m/s due to the small cross section at the entrance, which results in a turbulent flow in the die. This will contribute to the CNTs dispersion (Ishikawa 2006).

Inclusion of 0.05 wt.% MWCNTs caused an increase of the elongation at fracture and tensile stress validated through tensile testing. Such result is because of the advanced mechanical properties of CNTs.

No further increase in the CNTs content was attempted may be because of the strong agglomeration of CNTs with enhancing fraction in the matrix, high viscosity of Al, and the poor CNTs-Al wettability.

Friction stir is another process that was employed to fabricate MWCNTs-Al composites (Shi et al. 2007). The SWCNT temperature stability was found to be up to 750°C in air and 2800°C in vacuum (Laha and Agarwal 2008). In the stir casting process, homogeneous distribution was not obtained once regularly tangled nanotubes were utilized as the base materials. For the better distribution of CNT in the matrix, multiple passes are needed.

Casting is also a suitable route for Mg/alloys due to their relatively low melting point. In some cases, to produce CNTs-Mg composites, the CNTs were Ni plated in order to improve their wettability with the matrix (Zhou et al. 2007). This led to improvement of mechanical characteristics with addition of only 0.67 wt.% CNTs. The cast billet were hot extruded (623 K) to achieve a better CNTs reinforced in the matrix. Increasing CNTs content results in the reduction of the cycle number to failure under fatigue. This is due to the existence of voids at the CNT matrix interface, which give rise to weaker reinforcement (Zhou et al. 2007).

Melt spinning

This route comprises pouring a molten alloy drop by drop on to a rotating Cu wheel. Owing to rapid solidification, the droplets are converted into amorphous ribbons. CNT-$Fe_{82}P_{18}$ bulk metallic glass composite ribbons of 40 µ thickness were produced by this route (Zhou et al. 2007). Even though there was no investigation to report on dispersion, reinforcement and the CNTs and matrix interface nature, amorphous nature of the composite and reduction undamaged CNTs were achieved.

Laser deposition

A few researchers have used this route to fabricate CNTs-MMCs. Ni-10 vol.-%CNT composite obtained is an example of this method, which is processed through laser deposition route after roller blending of Ni and CNT powders. Nonetheless, though the process is subject to high temperatures, CNTs were still retained. This is while the density of CNTs graphication and induced defects increased, which is normal for the high processing condition.

Metal infiltration

In this technique, first a porous solid structure with dispersed CNTs is made and then a liquid metal is infiltrated into the pores and solidified to fabricate a composite structure. However, it is possible to control an even distribution of CNTs in matrix, a proper filling up of the pores in order to produce high-densed composite which becomes a critical step. Furthermore, the CNTs' movement and consequently their agglomeration may happen because of the high pressure during infiltration (Bakshi et al. 2010). This technique is accompanied by squeeze casting and has proven to be a suitable way to produce CNT-based Al or Mg alloy composites with a proper distribution of CNTs and without pores (Uozumi et al. 2008).

Thermal spray

This technique can be described as the spraying of molten or semi-molten particles onto a substrate to produce a coating/deposit through high impact and rapid solidification. It presents the benefits of large cooling rates as high as 10^8 K/s during the solidification, leading to the nanocrystalline structure formation/retention in the coatings (Berndt and Lavernia 1998, Fauchais et al. 2004, Meyer 1996).

Electrochemical routes

The CNTs' electrochemical properties normally originate from their superior electrocatalytic behaviors and their high surface-to-volume ratio in comparison with other carbon materials that are mostly utilized as electrode materials (McCreery 1991). In this regard, the electrochemistry has been developed into a well-designed tool for the functionalization or modification of CNTs in a selective and controlled manner. Indeed, a CNT electrode is immersed in a solution containing a suitable reagent and a constant current (galvanostatic) or a constant potential (potentiostatic) is applied and results in a highly reactive (radical) species formation through electron transfer between the CNT and the reagent. Those species would tend to react with the initial reagent or to self-polymerize, leading to a polymer coating on the tubes.

Electrochemical and electroless deposition techniques are utilized to synthesise MM-CNT coatings and thin films. The composite structures produced through different deposition techniques are restricted in thicknesses less than 200 μm (Arai et al. 2008).

It is not feasible to synthesize freestanding MM-CNT composites of larger thickness for load-bearing structural applications through an electrochemical processing route. MM-CNT coatings, made by the electrode position techniques, can be divided broadly into three categories based on the fabrication method, as well as the morphology of the composite:

1. Coating on thin film synthesized by co-deposition of CNTs and metal-ions from the electrochemical or chemical bath on a substrate. Most of the research on MM-CNT composites by the electrode position has been carried out in this category.
2. Metal is deposited on a uniform, aligned array of CNTs or CNT film. A uniform and homogeneous distribution of CNT is achievable in the metal matrix through this process.
3. A single CNT is coated by the electrodeposition of metallic particles on it, also known as one-dimensional (1-D) composite. This is primarily used in different types of nano-sensors, electrodes, inter-connects, and magnetic recorder heads in computer applications and also as the precursor

powder for larger MM-CNT composite structures, processed through a powder metallurgy route (Dingsheng and Yingliang 2006).

The first category of electrodeposited MM-CNT composite films and coatings is the one fabricated through co-deposition by both electrochemical and electroless processes. Synthesis of MM-CNT composite through co-deposition route involves a similar approach for electrodeposition (Chen et al. 2001, Wang et al. 2003, Xue and Wu 2007). Electroless deposition route CNTs are thoroughly suspended in the electrolytic bath and simultaneously deposited on the substrate along with the metal ions, thus producing a composite layer.

The main challenge for electrochemical techniques is also the dispersion of CNTs in the metal matrix. The natural tendency of agglomeration of CNTs, due to their high surface energy, hinders uniform dispersion and good suspension in the electrolyte bath. This ultimately results in heterogeneous distribution of CNTs in the metal matrix coating. The most commonly used technique to improve CNT dispersion in the electrolyte bath includes agitation by means of mechanical stirrer, ultrasonication, and the magnetic forces. Ball milling of CNTs prior to mixing in the bath also contributes to a better dispersion of CNT. It helps in making the CNTs shorter, reducing the tendency of agglomeration, and thus resulting in better dispersability (Arai et al. 2004, Arai et al. 2006, Arai et al. 2007). Surface treatment of CNTs can also help in increasing the wettability to keep CNTs in the suspension and to modify their surface charge to reduce tendency of agglomeration. These surface treatments include acid treatment and sensitization, acid cleaning, adding surfactants (Chen et al. 2005, Shi et al. 2004), and pre-coating the CNTs with metallic materials (Dai et al. 2008, Sun et al. 2007). Modifying the bath by adding reagents (adding polyacrylic acid for Ni-CNT electro-deposition) (Arai et al. 2004) is another means to achieve better dispersion of CNTs in the bath.

For electrodepositon and electroless deposition techniques, the CNT content in the composite coating is not directly proportional to the amount of CNT added to the bath. It has been observed for both processing techniques that there is a critical concentration of CNT in the bath, which is related to deposition efficiency and hence CNT content in the composite coating. Electrodeposition and electroless deposition techniques being fundamentally very different methods, the optimum concentration for CNTs in the bath for best co-deposition is different for both techniques. The higher CNT content in the bath results in their agglomeration and hence, creates a decrease in the amount of dispersed and suspended CNTs available for co-deposition on substrate. The final concentration of CNTs in the deposited structure and coating is a combined effect of CNT concentration in the bath and process parameters of the respective techniques. Therefore, dependence on the multi-variable makes the analysis of CNT content more complicated. Research is being conducted to isolate the individual and cumulative effect of all the parameters on the CNT content in the composite coating and thin film deposited by the electrochemical techniques. The processing parameters that govern the nature and morphology of the deposited structure are also different in both electrochemical and electroless deposition techniques. Current density is an important factor in the case of electrodeposition. An increase in the current density helps in the deposition of CNTs due to an increase in the electrostatic force between the CNTs and the substrate. However, at the same time, the high electrical conductivity of CNTs causes them to be engulfed by metallic particles very rapidly, creating very large particles to be deposited on the surface. Thus, a critical current density is desired for an optimum CNT content, high deposition rate, smoothness, and uniformity of the MM-CNT deposited coating and thin films (Chen et al. 2002, Chen et al. 2001).

Apart from the current density, the type of power source (AC or DC) also affects the composite structure fabricated by the electrodeposition process (Yamamoto et al. 1998). Pulse current provides more nucleation sites for metal ions to deposit on the CNT surface. In addition, during the reverse cycle of pulse current, metal is selectively dissolved as it forms the top layer of the deposit, engulfing the CNTs. Thus, the CNTs content of the deposit increases and results in better mechanical properties, namely hardness and wear resistance for the composite coatings and thin films. The protruding parts of the deposited coating dissolve more during the reverse cycle, thus making the surface smoother for further deposition. This helps in uniform deposition and decreased porosity in the deposited composite coating. Increase in the pulse frequency and reverse ratio has also been reported to increase CNT content and smoothness of deposition, but more research is needed for the concluding evidence. Since electroless deposition is a thermo-chemical process, bath temperature and pH are two other important factors which determine the coating composition and morphology. These two parameters vary with respect to the composition of the bath and the target composition of the deposit (Arai et al. 2016, Goel et al. 2016, Uysal et al. 2016). A thorough study of these parameters on MM-CNT composite morphology, composition, and properties is yet to be carried out. Most studies carried out for this type of composite coating and film are on Ni-CNT system, copper (Arnaud et al. 2016) and zinc (Kim and Kim 2015) based systems. The second type of MM-CNT composite synthesized through the electrodeposition route is by depositing the metal into the voids between the CNT array of network. A CNT network on a non-metallic substrate (silicon, alumina, or glassy carbon) is prepared by two different techniques: (1) growing aligned CNTs on a substrate (Si or porous alumina) (Li et al. 2016a, Li et al. 2016b) and (2) pouring CNT solution on a substrate surface followed by drying (glassy carbon electrode) (Shahrokhian et al. 2015). Studies have been carried out on copper and nickel-based composites fabricated by electro-deposition of pre-arrayed CNTs. Prior distribution of CNTs on the surface ensures uniform distribution within the composite.

The third category of the MM-CNT composites is metal-coated freestanding CNTs, which can be synthesized by both electrodeposition (Baghayeri et al. 2016, Gioia and Casella 2016) and electroless deposition techniques (Choi et al. 2015, Hakamada et al. 2016, Lin et al. 2016). Such one-dimensional MM-CNT composites can be obtained by co-deposition of metal and CNTs from the bath on a substrate. Subsequently, the coated CNTs are separated from the substrate by ultrasonication. The coating deposited on the CNT surface is not very uniform and often results in the formation of metal particles. A better way of controlling the size and maintaining the uniformity of metallic coating on CNT through electrodeposition is by coating the vertically aligned array of CNTs inside a porous template substrate, which can be leached away further (Arai et al. 2004).

Electroless deposition of metal particles on CNTs is performed by suspending the CNTs in an electroless-plating bath. The reducing agent present in the bath reduces the metallic salts to release metal, which coats the CNTs. CNTs are functionalized prior to this process to create nucleation sites on the CNTs surface to aid metal deposition. Better surface treatment produces uniform metallic coating on CNTs (Agarwal et al. 2016). Reports are also available on one-dimensional coated CNT composites for metal particles such as Ni, Cu, Ag and Co with the coating thickness varying between 20 nm and 600 nm. Electrodeposition and electroless deposition techniques are very effective in producing thin MM-CNT composites and coatings with good dispersion on CNTs. Metal ions either co-deposit with CNTs or nucleate on CNT surfaces, producing a strong bonding. These factors are very important with respect to the properties exhibited by the composites and their performance in service condition. Deposition techniques are the most suitable for producing one-dimensional MM-CNT composites. However, there are two major limitations for electrochemical processing routes. First, these techniques are mostly developed for Ni and Ni-alloys, with a few investigations on Cu and Co. These techniques require more research in order to optimize the process for other metals including Cu and Co. Second, these processing methods can be applied mainly for synthesizing the MM-CNT composite in the form of thin coating or a film less than 200 μm. With the current knowledge, it is almost impossible to produce thick and freestanding MM-CNT composites for structural applications by electrochemical processing.

Other novel techniques

Conventional processing techniques for MM-CNT composites synthesis have been applied with partial success. However, due to the problems associated with the fabrication of CNT composites, several new strategies have to be designed. Some of these novel routes fall in the category of well-established techniques, but are modified according to the suitability for processing metal-matrix composites, while others are based on completely novel concepts. This section is a compendium of the scattered novel processing methods developed to overcome the main challenges in fabricating MM-CNT composites.

Molecular level mixing

Molecular level mixing is a composite powder preparation technique that was developed by Cha and co-workers in Korea Advanced Institute of Science and Technology (White et al. 2007). In this method, the dispersion of CNTs is achieved on the molecular level. The method consists of dispersing CNTs in copper acetate [$Cu(CH_3COO)_2.H_2O$] solution followed by drying the suspension while magnetically stirring it. Good dispersion requires the CNTs to be functionalized so the resultant salt powders are nicely dispersed. The salt is converted to CuO by calcination in air at 300°C. The CuO/CNT mixture is reduced with H_2 at 250°C to obtain Cu/CNT powders. Fabrication of Cu-CNT composite by SPS of these powders led to extraordinary strengthening. The compressive yield strength of Cu-5 vol.% CNT composite was measured to be 360 MPa, which is 2.3 times that of pure copper (White et al. 2007). The hardness and sliding wear resistance of Cu-10 vol.% CNT composite was improved by two and three times respectively, compared to copper (Chen et al. 2003). This study highlighted the significance of CNTs dispersion on the mechanical properties. In a variation of this method, Cu_2O-CNT particles were prepared by precipitation from suspension of $CuSO_4$-CNT by addition of NaOH. The oxide particles were subsequently reduced to Cu with H_2 at 400°C. This resulted in Cu powders containing excellent dispersion of CNTs. There are several other studies on preparation of metal-CNT composite powders that can be discussed in this category. Sn-CNT and $SnSb_{0.5}$-CNT composite powders have been produced by reduction of mixtures of $SnCl_2$ and $SnCl_3$ solutions with alkaline KBH_4 (Wang et al. 2005). Such composites were shown to have increased capacity for lithiation and de-lithiation compared to Sn and $SnSb_{0.5}$ without the CNTs and could serve as anodes for Li-ion battery applications. A similar method for reduction of $CuSO_4$ by alkaline KBH_4 has been carried out to obtain CNTs coated with Cu. CNTs were acid functionalized in order to facilitate the dispersion. These Cu-CNT mixtures showed superior catalytic performance in thermal decomposition of ammonium perchlorate due to the large surface area. Molecular level mixing is a promising technique for the production of metal powders containing dispersed CNTs. Since there are chances for introducing oxygen impurity due to incomplete reduction by H_2, the use of these powders may or may not be suitable.

Mechanisms of strengthening in CNTs-MM composites

CNTs are added into the metal matrix with the aim (i) to enhance the elastic modulus, and (ii) to improve the tensile strength of the composite. This is achievable due to a higher strength and stiffness of CNTs compared to metal matrix. The mechanism and how they contribute to the mechanical properties development is described in the following section.

Elastic modulus of CNTs-MM composites

Large modulus of CNTs about 350–970 GPa improves the composite elastic modulus (Yu et al. 2000). Most of the investigation on the CNT-polymer composites is also applicable for the metal CNT-metal matrix. Several micromechanical models have been introduced to evaluate the composite materials elastic modulus, which are also employed for the composites reinforced by CNTs (Coleman et al. 2006, Laborde-Lahoz et al. 2005, Seidel and Lagoudas 2006, Zalamea et al. 2007).

Widely used models are described as follows:

Cox model

Elastic modulus (E) based on this model is expressed by (Cox 1952, Laborde-Lahoz et al. 2005):

$$E = \frac{1}{5}\eta_l E_f V_f + E_m \ (1\text{-}V_f)$$

where $\eta_l = 1 - \dfrac{\tanh(\beta s)}{\beta s}$, $s = \dfrac{2l}{r}$ and $\beta = \dfrac{2\pi \, E_m}{E_f(1 + V_m)\ln(\frac{1}{v_f})}$, here in, l and r designate the length and radius of the reinforcement fiber.

Hashin-strikman model

This model determines the upper and lower bounds for the composite elastic modulus in terms of the variational principles (Hashin and Shtrikman 1962). It is independent of the particles' shape. It has been stated that the elastic modulus ranged between upper and lower bounds for CNTs-Al composites produced by PSF and HVOF (Laha et al. 2007).

Modified eshelby model

This model is employed to correlate the CNT-based composites to the volume fraction of CNTs and porosities (Bian et al. 2002).

The longitudinal elastic modulus value is expressed by

$$E_{11} = E_m \varepsilon_{11}^m (\varepsilon_{11}^m + V_f \varepsilon_{11}^{CNT})^{-1}.$$

This model proposes the values higher than the experimentally obtained. This is associated with a poor bonding between matrix and CNTs.

Halpin-tsai equations

These equations are used to attain the elastic modulus of randomly oriented fiber composite as given

$$E = \frac{3}{8}\left[\frac{1+(\frac{2l}{D})\eta_l V_f}{1-\eta_l V_f}\right] + \frac{5}{8}\left[\frac{1+2\eta_T V_f}{1-\eta_T V_f}\right]$$

where $\eta_T = \dfrac{\frac{E_f}{E_m}-1}{\frac{E_f}{E_m}+2}$, $\eta_l = \dfrac{\frac{E_f}{E_m}-1}{\frac{E_f}{E_m}+2l/D}$ and D and l designate the diameter and length of the CNT, accordingly.

These equations closely estimate mechanical properties in the case of small CNT concentration in metal or polymer matrix composites (Bakshi et al. 2008, Yeh et al. 2006).

Combined voigt-reuss model

The elastic modulus for randomly oriented fiber composite is expressed by

$$E = \frac{3}{8}E_\parallel + \frac{5}{8}E_\perp$$

where $E_\perp$ is the transverse modulus and is equal to $\dfrac{E_f E_m}{E_f(1 - V_f) + E_m E_f}$ and $E_\parallel$ is the longitudinal modulus and is equal to $V_f E_F + (1 - V_f)E_m$.

Dispersion based model

All above-mentioned equations assume that the CNTs are uniformly distributed in the matrix which is seldom, particularly at high concentrations. A developed model is proposed to consider clustering phenomena in CNT-based composites, computing the CNT clusters properties. The overall properties of the composites are achieved by taking into account it as a dilute suspension of the clusters (behaviors with subscript dsc) in the matrix

$$K_{dsc} = K_m + \frac{(K_{cluster}-K_m)C_c}{1+\frac{K_{cluster}-K_m}{K_m+4\mu_m}}$$

$$\mu_{dsc} = \mu_m \left[1-\frac{15(1\,V_m)(11\frac{\mu_{cluster}}{\mu_m})C_c}{7-5V_m+2(4-2V_m)\frac{\mu_{cluster}}{\mu_m}}\right]$$

where C_c designates the clusters volume fraction, which is attributed to the overall CNT fraction by $V_f = C_c C_f$, C_f is the CNT concentration in a cluster. More accurately, the values are obtained by this model than that of Cox model in the case of epoxy CNT composites (De Villoria and Miravete 2007).

Strengthening mechanisms in CNTs-based composites

CNTs as fibrous reinforcements are used to increase the elastic modulus and tensile strength of the composite. This is because the CNTs have a higher strength and stiffness compared to the metal matrix. Understanding the strengthening mechanism has been continually studied almost since five decades. Models employed such as shear lag (Ryu et al. 2003) for conventional fiber reinforced composites are also applicable for the CNTs-based composites. The stress is transferred to the fiber (σ_f) through the interface and is associated with the shear stress (τ_{mf}) between the matrix and fiber given by

$$\frac{\sigma_f}{D_f} = \frac{\sigma_f}{2\tau_{mf}}$$

where D_f and l_f are the CNTs diameter and length, accordingly. Large load transfer is achieved by a larger aspect ratio of CNTs which results in an efficient utilization of reinforcement. The value of σ_f is equal to the fracture strength of CNTs once there is a critical length l_c. For $l < l_c$, the fracture strength of composite is evaluated by

$$\sigma_c^{Frac} = V_f \sigma_f^{Frac}\left(\frac{l}{2l_c}\right) + V_m \sigma_m^{Frac}.$$

Generally, the mechanical properties, measurements of CNTs-MM composites such as Al-CNT composites are evaluated via tensile tests, and in the literature good tensile strength was found only with the low content of CNTs (less than 2wt.%) and it decreases by increasing the CNTs content. For instance, Al-CNT composites present reduction in tensile strength with addition of CNTs (> 0.5 wt.%) (Figure 9a). A similar behavior was observed in tensile elongation of the composite compared with the monolithic Al, as shown in Figure 9b. Evidence of the declined ductility of Al-CNT composite is stated in most reports (George et al. 2005, Kuzumaki et al. 1998, Kwon et al. 2009, Morsi et al. 2010). Figure 9 represents a summary of the reported tensile strength in terms of CNT content. Based on the geometry and physical properties of CNTs, three main strengthening mechanisms are proposed: (i) load transfer, (ii) Orowan mechanism, and (iii) thermal mismatch. It should be noted that the improved tensile strength

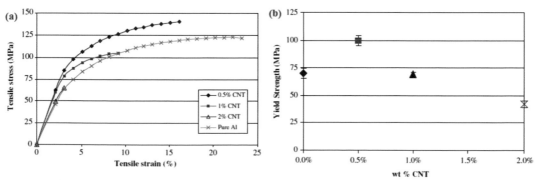

Figure 9. (a) Typical stress–strain curves for various wt.% CNT-Al composites, and (b) average strain-to-failure for the different wt.% CNT–Al indicating the decreasing ductility with increasing wt.% CNT (Esawi and El Borady 2008).

is presumably because of a simultaneous combination of several contributing factors. These mechanisms are associated with each other and in most cases, they operate simultaneously.

Load transfer

During loading CNTs-based composites, load partition is determined by the matrix/fiber interfacial shear stress (IFSS). A strong interface or good adhesion, would promote load transfer efficiency and as a result, it enhances the composite strength. The interfacial feature is critical to the mechanical properties of MMCs.

Generally speaking, three types of adhesion mechanisms between two surfaces are (i) chemical bonding, (ii) mechanical interlocking, and (iii) physical bonding (Liao and Tan 2012, Nalwa 2002).

Orowan mechanism

Normally, a composite material yield strength is the stress required to operate dislocation sources, and it is dominated by the presence and magnitude of the obstacles to hinder the dislocation movement. Orowan strengthening, derived by the resistance of closely spaced hard particles to the passing of dislocations, is an important phenomenon in CNTs-based metal matrix composites. It requires a small inter-particle spacing, hence it is not noticeable in the macro- and micro-sized particle reinforced composites. In contrast, it becomes more favorable in the nano-sized CNTs-reinforced composite, owing to the small filler size and small inter-particle spacing (Liao and Tan 2012, Zhang and Chen 2006).

Thermal mismatch

A thermal expansion coefficient of the CNTs is about 1×10^{-6} K^{-1}, which is almost lower than that of metal matrix. Thus, there is always thermal mismatch between the CNTs and metal matrix, especially once they are cooled from high temperature during the fabrication process. Because of the thermal mismatch, a high amount of dislocations will be generated near the reinforcement CNTs. In this case, the dislocation density (ρ) is given by

$$\rho = \frac{10 f \varepsilon}{b \,(1-f)} \frac{1}{D}$$

where ε, f, and D refer to misfit strain, CNTs volume fraction and CNTs diameter, respectively (Jinzhi 2012). Furthermore, there is a correlation between dislocation density and yield strength ($\Delta\sigma$) as follows:

$$\Delta\sigma = Gb\rho^{1/2}$$

where G is the shear modulus and b is the Burgers vector. It indicates that the yield strength and dislocation density have a direct relationship.

Interfacial phenomena in CNTs-MMCs

Chemical stability and interfacial phenomenon of the CNTs in the metal matrix composites are important factors for several reasons. The interfacial strength (Coleman et al. 2006) and fiber-matrix stress transfer (Dilandro et al. 1988) have a key role in strengthening. Indeed, a weak interface causes an inefficient utilization of fiber features and lower strength by facilitating pullout phenomena at low loads because of interface failure. On the other hand, the applied stress is transferred to the high strength fiber over the interfacial layer, in such a way that a strong interface would result in the very strong composite formation but at the expense of the composite ductility. Furthermore, the wetting conditions, i.e., angle, as well as interfacial reactions, i.e., interfacial phase(s) play a key role in strengthening of the CNTs-MM composites. In fact, interfacial reactions would result in interfacial phase(s) formation which improves the degree of the fiber wetting if the liquid has a lower contact angle with the produced phase. Degree

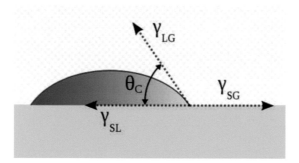

Figure 10. Schematic representation of wettability in terms of Young's equation.

of the reinforcement wetting by the liquid alloy is associated with the surface energies of the interacting species. As shown in Figure 10, the theoretical portrayal of contact derives from taking into account of a thermodynamic equilibrium between the three phases: gas or vapor (G or V), solid (S) and liquid (L). Surface energy between these phases can be written as $\gamma_{s/v}$, $\gamma_{s/l}$ and $\gamma_{l/v}$ and the equilibrium contact angle θ_C is given from the Young's equation and Young-Dupre's relation as follows

$$\gamma_{SV} - \gamma_{SL} - \gamma_{LV} \cos \theta_c = 0$$

$$\cos\theta_c = \frac{\gamma_{SV} - \gamma_{SL}}{\gamma_{LV}}$$

Then, the work of adhesion between the substrate and liquid form

$$\Delta W_{SLV} = \gamma_{LV}(1 + \cos \theta_c)$$

where ΔW_{SLV} is the solid–liquid adhesion energy per unit area when in the medium V.

Mechanical behaviors of various CNTs-MMCs

CNTs-Al matrix composite

Comparison of the 2009Al matrix and the CNTs/2009Al composites tensile curves is shown in Figure 11. It indicates the yield stress (YS) and the ultimate yield stress (UTS) enhanced by incorporating CNTs into the matrix; YS from 300 MPa to 350 MPa and UTS from 400 MPa to 450 MPa. On the other hand, the elongation declined from 18% to about 15% (Liu et al. 2011).

Substituting the obtained values (strength of matrix and composite) into the following equation

$$\sigma_c = \sigma_m(1 + RV_{CNTs})$$

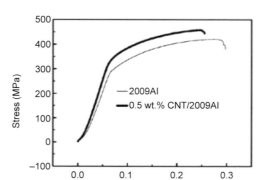

Figure 11. Stress and strain curves of as-forged 2009Al and CNTs/2009Al composite under natural ageing condition (Liu et al. 2011).

where R and V_{CNTs} refer to the CNTs strengthening efficiency and volume fraction, respectively. σ_c and σ_m are the composites and matrix alloy strength, respectively. The load transfer coefficient was found to be 26 which was very close to the value of 27 ~ 28 in uniformly distributed CNTs reinforced copper matrix composites produced by molecular mixing route (Kim et al. 2008). This reveals that the CNTs are highly efficient in transferring the applied load and the homogenous CNT distributions are favorable to strengthening.

CNTs-Ti matrix composite

Homogenous distribution of CNTs in the Ti matrix improves the Young's modulus and wear resistance of titanium and its alloy. In this respect, the role of employed technique for the fabrication process becomes important. Similarly, an increase in hardness of the composite in comparison with the unreinforced titanium alloy is associated with the TiC bonds formation during the mechanical alloying process (Kuzumaki et al. 2000). In fact, this process leads to the hardness and Young's modulus improvement at the expense of ductility in the final product.

Powder metallurgy route involving mechanical mixing followed by hot pressing resulted in a noticeable increase in Young's modulus and hardness of CNTs reinforced titanium as compared to pure titanium (Kuzumaki et al. 2000). The CNT reinforced Young's modulus of the composite obtained was about 198 GPa (comparable with steel, E D 200 GPa), which is 1.7 times greater than unreinforced titanium (ETi D 120 GPa). Likewise, the hardness increased up to 1216 (Vickers), which is 5.5 times the hardness of unreinforced titanium (221 Vickers). In another study (Li et al. 2013), CNTs-based Ti matrix composites were produced by adopting a powder metallurgy route accompanied by spark plasma sintering. Mechanical testing represented an 11.3% and 40.2% enhancement in tensile strength and yield strength, respectively.

A heterogeneous coacervation method was found to be another effective way to obtain a fine dispersion of CNTs in a Ti matrix (Xue et al. 2010). The fabricated composite exhibited a superior plastic behavior under high temperature loads.

Wet processing routes have also proven to be a desirable way to improve the mechanical properties of the CNTs-based Ti composites. Improvements in yield strength (48%), tensile strength (28%), and hardness (9%) in the final products were stated using 0.35 wt% CNTs (Fugetsu et al. 2005, Kondoh et al. 2009).

CNTs-Cu matrix composite

Improved yield strength and enhancing tensile properties and of CNT/Cu composites with increasing volume fraction of CNTs has been reported by Kim et al. (Kim et al. 2006). However, the same composite produced by powder metallurgy has not reached up to the expected values. This is because of an inhomogeneous dispersion of CNTs has not been perfectly solved in order to synthesize CNT/metal nanocomposites with minimized aggregation of CNTs. This issue was addressed by changing the applied technique to molecular-level mixing and consequently homogenous distribution of CNTs (Cha et al. 2005). Furthermore, Kim et al. (Kim et al. 2008) found that refining of matrix grains leads to the Cu-MM composites strengthening by dislocation motion and hardening mechanisms. As shown in Figure 12, the CNT/Cu nanocomposites yield strength increased by 27% from 360 to 460 MPa once the matrix strength increased from 150 to 190 MPa as the grain size reduced from 4 to 1.5 μ. Besides, the matrix hardening by grain refinement without more addition of CNTs results in additional strengthening of the composite. Their results indicated that the metallurgical treatment of the matrix is essential for the development of high-strength CNT/metal nanocomposites.

Friction coefficient as a mechanical property of Cu-MM composites is determined by the rule of mixtures (Van Trinh et al. 2010). The composite's friction coefficient decreases with an increase in CNTs mass fraction in the composite. 0.15 was found to be the appropriate value of the friction coefficient of the CNTs component in the Cu-CNTs composite.

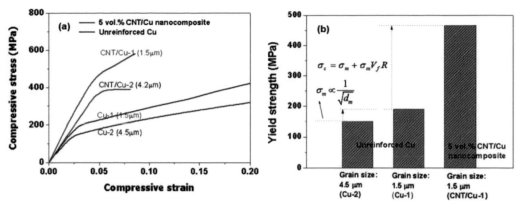

Figure 12. (a) Compressive stress-strain curves for 5 vol% CNT/Cu nanocomposites and unreinforced Cu with different grain sizes, and (b) Comparison of the yield strength of unreinforced Cu and 5 vol % CNT/Cu nanocomposites with different Cu matrix grain sizes (Kim et al. 2008).

Liu et al. (Liu et al. 2008) reported that the mechanical properties of Cu-CNTs composites are strongly dependent on the preparation technique. For example, the CNTs/Cu composite thin film prepared by electrophoresis and electroplating deposition techniques after annealing has noticeably better electrical properties than that of conventional pure Cu thin film. Moreover, the CNTs/Cu films electrical resistance declines faster than that of the pure Cu films, as the annealing temperature enhances.

Comparing the tribological property of Cu matrix composites reinforced by CNTs and pure Cu bulk, which were prepared by the powder metallurgy route and under the same consolidation processing condition, revealed that the wear rate and the friction coefficient of both samples increase with the increase in electrical current without exception, and the effect of electrical current is more evident on tribological behavior of pure Cu bulk than on that of CNTs/Cu composite. Arc erosion wear and plastic flow deformation are the dominant wear mechanisms, respectively. CNTs improve the tribological feature of the Cu matrix composites with electrical current (Wei et al. 2011).

In addition, Cu-10 vol.-% CNT composite fabricated by the spark plasma sintering route has improved the composite's hardness by 79% with a further improvement of up to 207% derived from rolling of the SPS composite. This improvement is associated with a better dispersion and reinforcement, resulted by SPS and rolling Super aligned carbon nanotube (SACNT) reinforced-(Kim et al. 2006, Kim et al. 2004) process and the mechanical features and transport properties of the composite with different SACNT content that was investigated. Their results revealed that the as-fabricated composites have a better comprehensive performance than pure copper. The simple rule of mixtures (ROM) is employed to evaluate the potential maximum behaviors of the Cu/SACNT composites. This composite is found to be a suitable candidate for electronics and communications applications (Shuai et al. 2016).

CNTs-Mg matrix composite

Catalytic chemical vapor deposition (CVD) was proposed to fabricate CNTs-based Mg matrix composites (Sun et al. 2013). The results indicated that as-obtained CNTs had an average size of 15 nm and were highly graphitizing and homogeneously dispersed in the Mg powder. Milling this composite powder for short duration resulted in embedding of CNTs into Mg, which result in grain refinement. An improvement in the CNTs-Mg composite tensile strength about 285 Mpa was observed to be about 45% higher than that of commercial pure Mg. Furthermore, the hardness of the composite enhanced by about 25%, up to the CNTs mass fraction of about 2.4%. Utilizing Mg as a catalyst carrier to fabricate CNT/Mg composite powders by CVD gives rise to high oxidization of the Mg matrix and the Ni catalyst reaction with Mg at high temperature, which results in activity loss (Kang et al. 2011).

A vacuum hot pressure method followed by extrusion for processing the CNTs-based Mg alloy composite reinforced with 1 wt.% CNTs showed an improvement in the Young's modulus, tensile strength and yield strength of about 23% (Shimizu et al. 2008).

The effects of CNTs on the microstructure of hot-extruded Mg–2.0Zn magnesium matrix composites reinforced with 1.0 wt% CNTs were investigated by Zeng et al. (Zeng et al. 2010). Later on, Li et al. (Li et al. 2014) employed the same method and studied the effect of CNTs on the microstructure and mechanical properties of Mg–6Zn matrix composite. The significant results from their study were: (i) homogenous distribution of CNTs in the Mg matrix without any phase or compound formation in the interfacial regions of Mg and CNTs, (ii) CNTs contents enhancement caused mainly due to grain size refinement that resulted in the mechanical properties' improvement such as yield strength, hardness, ultimate tensile strength, elastic modulus and yield strength. The interface study of the Mg–6 wt% Al alloy matrix composites reinforced with CNTs revealed presence of Al2MgC2 needle-like ternary carbides in the Mg and CNTs interface (Fukuda et al. 2011).

In another study (Rashad et al. 2014), the effect of two types carbon, namely, nanoplatelets (GNPs) and multi-walled carbon nanotube (MW-CNTs), on mechanical properties of pure magnesium was investigated. These samples were produced by semi-powder metallurgy route and their microstructures are shown in Figure 13.

The room temperature tensile and compressive for those samples are compared and shown in Figure 14. Once the CNTs were utilized as reinforcement (Mg–1Al–0.60CNTs), noticeable improvement in tensile properties (+ 22% E, +35% TYS, and +42% UTS) as well as in failure strain (+44%) was obtained. On the other hand, the addition of GNPs (Mg–1Al–0.60GNPs) gives rise to nominal improvement in tensile strength (+ 34% E; +31% TYS, and + 31% UTS) with a reduction in ductility.

The compressive strength of Mg–1Al–0.60CNTs and Mg–1Al–0.60GNPs composites improved by about (+34% E; + 137% CYS; and + 12% UCS) and (+ 52% E; + 130% CYS; and + 7% UCS), respectively. Even though the improvement in compressive strength (+34% E, +67% CYS, and + 5% UCS) of Mg–1Al–0.60 (CNT + GNPs) composite is lower than that of Mg–1Al–0.60GNPs or Mg–1Al–0.60CNTs composites, the compressive failure strain (%) of synthesized composites is higher than pure Mg and lower than Mg–1Al alloy. Similarly, tensile failure strain (%), the compression failure strain (%) of Mg–1Al–0.60 (CNT + GNPs) is higher than those of Mg–1Al–0.60GNPs and Mg–1Al–0.60CNTs composites.

In another study (Yoo et al. 2012), a new method was introduced to fabricate CNT-reinforced Mg composites by using accumulative roll bonding to magnesium sheets coated with ball-milled Al-CNT powder. The result represented a homogenous distribution of CNTs in the Mg matrix and introduction of small amount of CNTs remarkably improved the composite's mechanical properties.

Summary

The available reports on CNTs reinforced metal matrix composites fabricated by various fabrication methods were reviewed. The strengthening mechanisms that contribute to the CNTs-based metal matrix composites, as well as CNT content and evolution during the composite fabrication process were addressed. Finally, the challenges which ensue in realizing CNTs as reinforced in metal matrix composites were highlighted from the point of processing, cost considerations and safety considerations. The following conclusions can be made:

- The procedure for the CNTs-based metal matrix composites fabrication is a multistage process. The final quality of the resultant composite is determined by an appropriate combination of volume content of the reinforcement, the CNTs and metal powder mixing technique, the subsequent consolidation and secondary processing, etc. The expected strengthening mechanisms can be manipulated via choosing a proper combination of the affecting factors.
- There are various strengthening mechanisms that may take place in CNT reinforced metal matrix composites, including load transfer, Orowan mechanism and thermal mismatch. Some of these mechanisms are related to each other and in practice, they operate simultaneously.
- The challenges of the CNTs-based composite lie on the (i) dispersion degree, (ii) load transfer (interfacial bonding), (iii) volume content, (iv) CNT alignment, and (v) processing technique. Up till now, the CNT-based metal matrix composites such as CNTs-Al, CNTs-Mg, CNTs-Ti, CNTs-Cu and so on are still at a research phase and their full potential is yet to be realized.

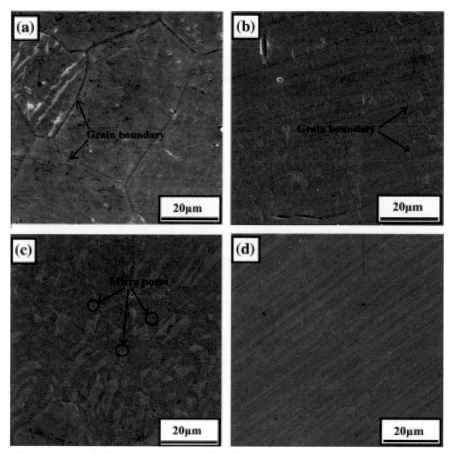

Figure 13. Microstructures of: (a) pure Mg, (b) Mg–1Al–0.6GNPs (c) Mg–1Al–0.6CNTs, and (d) Mg–1Al–0.6 (CNT + GNP) composites (Rashad et al. 2014).

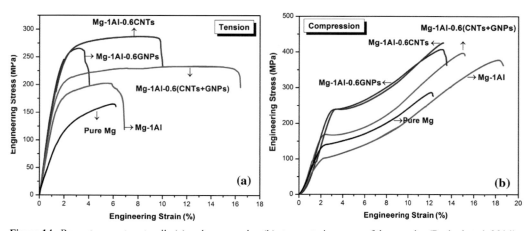

Figure 14. Room temperature tensile (a) and compression (b) stress–strain curves of the samples (Rashad et al. 2014).

References

Agarwal, A., S.R. Bakshi and D. Lahiri. 2016. Carbon Nanotubes: Reinforced Metal Matrix Composites. CRC press.

Ajayan, P.M. and J.M. Tour. 2007. Materials science: nanotube composites. Nature 447: 1066–1068.

Akhtar, F. 2008. An investigation on the solid state sintering of mechanically alloyed nano-structured 90W–Ni–Fe tungsten heavy alloy. Int. J. Refract. Met. Hard. Mater. 26: 145–151.

Akhtar, F., I.S. Humail, S. Askari, J. Tian and G. Shiju. 2007. Effect of WC particle size on the microstructure, mechanical properties and fracture behavior of WC–(W, Ti, Ta) C–6wt% Co cemented carbides. Int. J. Refract. Met. Hard. Mater. 25: 405–410.

Arai, S., M. Endo and N. Kaneko. 2004. Ni-deposited multi-walled carbon nanotubes by electrodeposition. Carbon 42: 641–644.

Arai, S., M. Endo, T. Sato and A. Koide. 2006. Fabrication of nickel–multiwalled carbon nanotube composite films with excellent thermal conductivity by an electrodeposition technique. Electrochem. Solid-State Lett. 9: C131–C133.

Arai, S., T. Saito and M. Endo. 2007. Low-internal-stress nickel multiwalled carbon nanotube composite electrodeposited from a sulfamate bath. J. Electrochem. Soc. 154: D530–D533.

Arai, S., A. Fujimori, M. Murai and M. Endo. 2008. Excellent solid lubrication of electrodeposited nickel-multiwalled carbon nanotube composite films. Mater Lett. 62: 3545–3548.

Arai, S., T. Osaki, M. Hirota and M. Uejima. 2016. Fabrication of copper/single-walled carbon nanotube composite film with homogeneously dispersed nanotubes by electroless deposition. Mater. Today Commun. 7: 101–107.

Arnaud, C., F. Lecouturier, D. Mesguich, N. Ferreira, G. Chevallier, C. Estournès et al. 2016. High strength–High conductivity double-walled carbon nanotube-copper composite wires. Carbon 96: 212–215.

Baghayeri, M., M.B. Tehrani, A. Amiri, B. Maleki and S. Farhadi. 2016. A novel way for detection of antiparkinsonism drug entacapone via electrodeposition of silver nanoparticles/functionalized multi-walled carbon nanotubes as an amperometric sensor. Mater. Sci. Eng. 66: 77–83.

Bakshi, S., D. Lahiri and A. Agarwal. 2010. Carbon nanotube reinforced metal matrix composites—a review. Int. Mater Rev. 55: 41–64.

Bakshi, S.R., V. Singh, K. Balani, D.G. McCartney, S. Seal and A. Agarwal. 2008. Carbon nanotube reinforced aluminum composite coating via cold spraying. Surf. Coat. Technol. 202: 5162–5169.

Berndt, C. and E. Lavernia. 1998. Thermal spray processing of nanoscale materials—a conference report with extended abstract. J. Therm. Spray. Technol. 7: 411–440.

Bethune, D., C. Klang, M. De Vries, G. Gorman, R. Savoy and J. Vazquez. 1993. Cobalt-catalysed growth of carbon nanotubes with single-atomic-layer walls. Nature 363: 605–607.

Bian, Z., M.X. Pan, Y. Zhang and W.H. Wang. 2002. Carbon-nanotube-reinforced Zr 52.5 Cu 17.9 Ni 14.6 Al 10 Ti 5 bulk metallic glass composites. Appl. Phys. Lett. 81: 4739–4741.

Bian, Z., R.J. Wang, W.H. Wang, T. Zhang and A. Inoue. 2004. Carbon-nanotube-reinforced Zr-based bulk metallic glass composites and their properties. Adv. Funct. Mater. 14: 55–63.

Cha, S.I., K.T. Kim, S.N. Arshad, C.B. Mo and S.H. Hong. 2005. Extraordinary strengthening effect of carbon nanotubes in metal-matrix nanocomposites processed by molecular-level mixing. Adv. Mater. 17: 1377–1381.

Chen, X., J. Peng, X. Li, F. Deng, J. Wang and W. Li. 2001. Tribological behavior of carbon nanotubes—reinforced nickel matrix composite coatings. J. Mater. Sci. Lett. 20: 2057–2060.

Chen, X., F. Cheng, S. Li, L.P. Zhou and D. Li. 2002. Electrodeposited nickel composites containing carbon nanotubes. Surf. Coat. Technol. 155: 274–278.

Chen, W., J. Tu, L. Wang, H. Gan, Z. Xu and X. Zhang. 2003. Tribological application of carbon nanotubes in a metal-based composite coating and composites. Carbon 41: 215–222.

Chen, X., C. Chen, H. Xiao, F. Cheng, G. Zhang and G. Yi. 2005. Corrosion behavior of carbon nanotubes–Ni composite coating. Surf. Coat. Technol. 191: 351–356.

Choi, H., J. Shin, B. Min, J. Park and D. Bae. 2009. Reinforcing effects of carbon nanotubes in structural aluminum matrix nanocomposites. J. Mater. Res. 24: 2610–2616.

Choi, J.-R., K.-Y. Rhee and S.-J. Park. 2015. Influence of electrolessly silver-plated multi-walled carbon nanotubes on thermal conductivity of epoxy matrix nanocomposites. Composites, Part B. 80: 379–384.

Coleman, J.N., U. Khan, W.J. Blau and Y.K. Gun'ko. 2006. Small but strong: a review of the mechanical properties of carbon nanotube–polymer composites. Carbon 44: 1624–1652.

Cox, H. 1952. The elasticity and strength of paper and other fibrous materials. Br. J. Appl. Phys. 3: 72.

Dai, P.-Q., W.-C. Xu and Q.-Y. Huang. 2008. Mechanical properties and microstructure of nanocrystalline nickel-carbon nanotube composites produced by electrodeposition. Mater. Sci. Eng.: A. 483: 172–174.

De Villoria, R.G. and A. Miravete. 2007. Mechanical model to evaluate the effect of the dispersion in nanocomposites. Acta Mater. 55: 3025–3031.

Deng, C., Y. Ma, P. Zhang, X. Zhang and D. Wang. 2008. Thermal expansion behaviors of aluminum composite reinforced with carbon nanotubes. Mater. Lett. 62: 2301–2303.

Dey, A. and K.M. Pandey. 2015. Magnesium metal matrix composites—a review. Rev. Adv. Mater. Sci. 42: 58–67.

Dilandro, L., A. Dibenedetto and J. Groeger. 1988. The effect of fiber-matrix stress transfer on the strength of fiber-reinforced composite materials. Polym. Compos. 9: 209–221.

Dingsheng, Y. and L. Yingliang. 2006. Electroless deposition of Cu on multiwalled carbon nanotubes. Rare Metals 25: 237–240.

Dong, S., J. Tu and X. Zhang. 2001. An investigation of the sliding wear behavior of Cu-matrix composite reinforced by carbon nanotubes. Materials Science and Engineering: A. 313: 83–87.

Dubin, V., R. Akolkar, C. Cheng, R. Chebiam, A. Fajardo and F. Gstrein. 2007. Electrochemical materials and processes in Si integrated circuit technology. Electrochim. Acta 52: 2891–2897.

Esawi, A.M. and M.A. El Borady. 2008. Carbon nanotube-reinforced aluminium strips. Compos. Sci. Technol. 68: 486–492.

Esawi, A.M., K. Morsi, A. Sayed, A.A. Gawad and P. Borah. 2009. Fabrication and properties of dispersed carbon nanotube–aluminum composites. Mater. Sci. Eng. A. 508: 167–173.

Farid, A., S. Guo, F.E. Cui, P. Feng and T. Lin. 2007. TiB 2 and TiC stainless steel matrix composites. Mater. Lett. 61: 189–191.

Fauchais, P., A. Vardelle, M. Vardelle and M. Fukumoto. 2004. Knowledge concerning splat formation: an invited review. J. Therm. Spray. Technol. 13: 337–360.

Fugetsu, B., W. Han, N. Endo, Y. Kamiya and T. Okuhara. 2005. Disassembling single-walled carbon nanotube bundles by dipole/dipole electrostatic interactions. Chem. Lett. 34: 1218–1219.

Fukuda, H., K. Kondoh, J. Umeda and B. Fugetsu. 2011. Interfacial analysis between Mg matrix and carbon nanotubes in Mg–6wt.% Al alloy matrix composites reinforced with carbon nanotubes. Compos. Sci. Technol. 71: 705–709.

George, R., K. Kashyap, R. Rahul and S. Yamdagni. 2005. Strengthening in carbon nanotube/aluminium (CNT/Al) composites. Scripta Mater. 53: 1159–1163.

Gioia, D. and I.G. Casella. 2016. Pulsed electrodeposition of palladium nano-particles on coated multi-walled carbon nanotubes/nafion composite substrates: Electrocatalytic oxidation of hydrazine and propranolol in acid conditions. Sensors Actuators B: Chem. 237: 400–407.

Goel, V., P. Anderson, J. Hall, F. Robinson and S. Bohm. 2016. Electroless Co–P-Carbon Nanotube composite coating to enhance magnetic properties of grain-oriented electrical steel. J. Magn. Magn. Mater. 407: 42–45.

Hajalilou, Abdollah, Abbas Kianvash, Hossein Lavvafi and Kamyar Shameli. 2017. Nanostructured soft magnetic materials synthesized via mechanical alloying: a review. J. Mater Sci. Mater. Electron. (http://dx.doi.org/10.1007/s10854-017-8082-0).

Hajalilou, A., M. Hashim, R. Ebrahimi-Kahrizsangi, H. Mohamed Kamari and N. Sarami. 2014. Synthesis and structural characterization of nano-sized nickel ferrite obtained by mechanochemical process. Ceram. Int. 40: 5881–87.

Hajalilou, A., M. Hashim, R. Ebrahimi-Kahrizsangi and Halimah Mohamed Kamari. 2015. Influence of evolving microstructure on electrical and magnetic characteristics in mechanically synthesized polycrystalline Ni-ferrite nanoparticles. J. Alloys. Compd. 633: 306–16. (http://dx.doi.org/10.1016/j.jallcom.2015.02.061).

Hajalilou, Abdollah, Mansor Hashim, Mehrdad Abbasi, Halimah Mohamed Kamari and Hassan Azimi. 2015. A comparative study on the effects of different milling atmospheres and sintering temperatures on the synthesis and magnetic behavior of spinel single phase Ni0.64Zn0.36Fe2O4 nanocrystals. J. Mater. Sci. Mater. Electron. 26(10): 7468–83. Retrieved (http://link.springer.com/10.1007/s10854-015-3381-9).

Hajalilou, Abdollah, Mansor Hashim, Halimah Mohamed and Javadi Samikannu. 2014. Synthesis of ZrC Nanoparticles in the ZrO2–Mg–C–Fe system through mechanically activated self-propagating high-temperature synthesis. Acta Metall Sin. (Engl. Lett.) 27(6): 1144–1151. DOI:10.1007/s40195-014-0152-1.

Hakamada, M., T. Abe and M. Mabuchi. 2016. Electrodes from carbon nanotubes/NiO nanocomposites synthesized in modified Watts bath for supercapacitors. J. Power Sources 325: 670–674.

Hashin, Z. and S. Shtrikman. 1962. On some variational principles in anisotropic and nonhomogeneous elasticity. J. Mech. Phys. Solids 10: 335–342.

Hertel, T., R. Martel and P. Avouris. 1998. Manipulation of individual carbon nanotubes and their interaction with surfaces. J. Phys. Chem. B. 102: 910–915.

Hulbert, D.M., A. Anders, J. Andersson, E.J. Lavernia and A.K. Mukherjee. 2009. A discussion on the absence of plasma in spark plasma sintering. Scripta Mater. 60: 835–838.

Hussainova, I. 2003. Effect of microstructure on the erosive wear of titanium carbide-based cermets. Wear. 255: 121–128.

Iijima, S. 1991. Synthesis of carbon nanotubes. Nature 354: 56–58.

Iijima, S. and T. Ichihashi. 1993. Single-shell carbon nanotubes of 1-nm diameter. Nature 363: 603–605.

Ishikawa, T. 2006. Overview of trends in advanced composite research and applications in Japan. Adv. Compos. Mater. 15: 3–37.

Jiang, L.Y., Y. Huang, H. Jiang, G. Ravichandran, H. Gao, K. Hwang et al. 2006. A cohesive law for carbon nanotube/polymer interfaces based on the van der Waals force. J. Mech. Phys. Solids 54: 2436–2452.

Kang, J., J. Li, N. Zhao, P. Nash, C. Shi and R. Sun. 2011. Study of Mg powder as catalyst carrier for the carbon nanotube growth by CVD. J. Nanomater. 2011: 7.

Kessel, H., J. Hennicke, R. Kirchner and T. Kessel. 2010. Rapid sintering of novel materials by FAST/SPS—Further development to the point of an industrial production process with high cost efficiency. FCT Systeme GmbH 96528.

Khare, R. and S. Bose. 2005. Carbon nanotube based composites—a review. Journal of Minerals and Materials Characterization and Engineering 4: 31.

Kim, C.H. and B.-H. Kim. 2015. Zinc oxide/activated carbon nanofiber composites for high-performance supercapacitor electrodes. J. Power Sources 274: 512–520.

Kim, K., J. Eckert, S. Menzel, T. Gemming and S. Hong. 2008. Grain refinement assisted strengthening of carbon nanotube reinforced copper matrix nanocomposites. Appl. Phys. Lett. 92: 121901.

Kim, K.T., K.H. Lee, S.I. Cha, C.-B. Mo and S.H. Hong. 2004. Characterization of carbon nanotubes/Cu nanocomposites processed by using nano-sized Cu powders. MRS Online Proceedings Library Archive 821.

Kim, K.T., S.I. Cha, S.H. Hong and S.H. Hong. 2006. Microstructures and tensile behavior of carbon nanotube reinforced Cu matrix nanocomposites. Mater. Sci. Eng: A. 430: 27–33.

Kondoh, K., T. Threrujirapapong, H. Imai, J. Umeda and B. Fugetsu. 2009. Characteristics of powder metallurgy pure titanium matrix composite reinforced with multi-wall carbon nanotubes. Compos. Sci. Technol. 69: 1077–1081.

Kuzumaki, T., K. Miyazawa, H. Ichinose and K. Ito. 1998. Processing of carbon nanotube reinforced aluminum composite. J. Mater. Res. 13: 2445–2449.

Kuzumaki, T., O. Ujiie, H. Ichinose and K. Ito. 2000. Mechanical characteristics and preparation of carbon nanotube fiber-reinforced Ti composite. Adv. Eng. Mater. 2: 416–418.

Kwon, H., M. Estili, K. Takagi, T. Miyazaki and A. Kawasaki. 2009. Combination of hot extrusion and spark plasma sintering for producing carbon nanotube reinforced aluminum matrix composites. Carbon 47: 570–577.

Laborde-Lahoz, P., W. Maser, T. Martinez, A. Benito, T. Seeger, P. Cano et al. 2005. Mechanical characterization of carbon nanotube composite materials. Mechanics of Advanced Materials and Structures 12: 13–19.

Laha, T., Y. Liu and A. Agarwal. 2007. Carbon nanotube reinforced aluminum nanocomposite via plasma and high velocity oxy-fuel spray forming. J. Nanosci. Nanotechnol. 7: 515–524.

Laha, T. and A. Agarwal. 2008. Effect of sintering on thermally sprayed carbon nanotube reinforced aluminum nanocomposite. Mater. Sci. Eng.: A. 480: 323–332.

Li, C., X. Wang, W. Liu, K. Wu, H. Shi, C. Ding et al. 2014. Microstructure and strengthening mechanism of carbon nanotubes reinforced magnesium matrix composite. Mater. Sci. Eng.: A. 597: 264–269.

Li, S., B. Sun, H. Imai, T. Mimoto and K. Kondoh. 2013. Powder metallurgy titanium metal matrix composites reinforced with carbon nanotubes and graphite. Compos. Part A-Appl. S. 48: 57–66.

Li, S., Y. Su, Q. Ouyang and D. Zhang. 2016a. *In situ* carbon nanotube-covered silicon carbide particle reinforced aluminum matrix composites fabricated by powder metallurgy. Mater. Lett. 167: 118–121.

Li, S., Y. Su, X. Zhu, H. Jin, Q. Ouyang and D. Zhang. 2016b. Enhanced mechanical behavior and fabrication of silicon carbide particles covered by *in situ* carbon nanotube reinforced 6061 aluminum matrix composites. Materials & Design 107: 130–138.

Li, Y. and T.G. Langdon. 1997. Creep behavior of an Al-6061 metal matrix composite reinforced with alumina particulates. Acta Mater. 45: 4797–4806.

Liao, J. and M.-J. Tan. 2011. Mixing of carbon nanotubes (CNTs) and aluminum powder for powder metallurgy use. Powder Technol. 208: 42–48.

Liao, J.Z. and M.J. Tan. 2012. Improved tensile strength of carbon nanotube reinforced aluminum composites processed by powder metallurgy. Paper Presented at the Advanced Materials Research.

Lin, C.-Y., J.-L. Pan, C.-C. Wu and W.-C. Chang. 2016. Study of nickel catalysts deposited by using the electroless plating method and growth of the multiwall carbon nanotubes. Paper presented at the Proceedings of the 3rd International Conference on Intelligent Technologies and Engineering Systems (ICITES2014).

Liu, P., D. Xu, Z. Li, B. Zhao, E.S.-W. Kong and Y. Zhang et al. 2008. Fabrication of CNTs/Cu composite thin films for interconnects application. Microelectron Eng. 85: 1984–1987.

Liu, Z., B. Xiao and Z. Ma. Fabrication of CNTs/Al composite with enhanced dispersion pre-treatment.

McCreery, R.L. 1991. Electroanalytical Chemistry. Marcel Dekker, Inc., New York 17: 221–374.

Meyer, W. 1996. Metal spraying in the United States: a jtst historical paper. J. Therm. Spray Technol. 5: 79–83.

Morsi, K., A. Esawi, P. Borah, S. Lanka and A. Sayed. 2010. Characterization and spark plasma sintering of mechanically milled aluminum-carbon nanotube (CNT) composite powders. J. Compos. Mater. 44: 1991–2003.

Munir, Z., U. Anselmi-Tamburini and M. Ohyanagi. 2006. The effect of electric field and pressure on the synthesis and consolidation of materials: a review of the spark plasma sintering method. J. Mater. Sci. 41: 763–777.

Nalwa, H.S. 2002. Handbook of Thin Film Materials: Nanomaterials and Magnetic Thin Films. Vol. 5: Academic Press, New York.

Pang, L., K. Sun, S. Ren, C. Sun and J. Bi. 2007. Microstructure, hardness, and bending strength of carbon nanotube—iron aluminide composites. J. Compos. Mater. 41: 2025–2031.

Pérez-Bustamante, R., C. Gómez-Esparza, I. Estrada-Guel, M. Miki-Yoshida, L. Licea-Jiménez, S. Pérez-García et al. 2009. Microstructural and mechanical characterization of Al–MWCNT composites produced by mechanical milling. Mater. Sci. Eng.: A. 502: 159–163.

Rashad, M., F. Pan, A. Tang, M. Asif and M. Aamir. 2014. Synergetic effect of graphene nanoplatelets (GNPs) and multi-walled carbon nanotube (MW-CNTs) on mechanical properties of pure magnesium. J. Alloys Compd. 603: 111–118.

Ryu, H.J., S.I. Cha and S.H. Hong. 2003. Generalized shear-lag model for load transfer in SiC/Al metal-matrix composites. J. Mater. Res. 18: 2851–2858.

Seidel, G.D. and D.C. Lagoudas. 2006. Micromechanical analysis of the effective elastic properties of carbon nanotube reinforced composites. Mech. Mater. 38: 884–907.

Shahrokhian, S., M. Azimzadeh and M.K. Amini. 2015. Modification of glassy carbon electrode with a bilayer of multiwalled carbon nanotube/tiron-doped polypyrrole: application to sensitive voltammetric determination of acyclovir. Materials Science and Engineering: C. 53: 134–141.

Shi, X., H. Yang, G. Shao, X. Duan, L. Yan, Z. Xiong et al. 2007. Fabrication and properties of W–Cu alloy reinforced by multi-walled carbon nanotubes. Mater. Sci. Eng.: A. 457: 18–23.

Shi, Y., Z. Yang, M. Li, H. Xu and H. Li. 2004. Electroplated synthesis of Ni–P–UFD, Ni–P–CNTs, and Ni–P–UFD–CNTs composite coatings as hydrogen evolution electrodes. Mater. Chem. Phys. 87: 154–161.

Shimizu, Y., S. Miki, T. Soga, I. Itoh, H. Todoroki, T. Hosono et al. 2008. Multi-walled carbon nanotube-reinforced magnesium alloy composites. Scripta Mater. 58: 267–270.

Shuai, J., L. Xiong, L. Zhu and W. Li. 2016. Enhanced strength and excellent transport properties of a superaligned carbon nanotubes reinforced copper matrix laminar composite. Compos. Part A-Appl. S. 88: 148–155.

Spigarelli, S., M. Cabibbo, E. Evangelista and T.G. Langdon. 2002. Creep properties of an Al–2024 composite reinforced with SiC particulates. Mater. Sci. Eng.: A. 328: 39–47.

Sun, F., C. Shi, K.Y. Rhee and N. Zhao. 2013. *In situ* synthesis of CNTs in Mg powder at low temperature for fabricating reinforced Mg composites. J. Alloys Compd. 551: 496–501.

Sun, Y., J. Sun, M. Liu and Q. Chen. 2007. Mechanical strength of carbon nanotube–nickel nanocomposites. Nanotechnology 18: 505704.

Uozumi, H., K. Kobayashi, K. Nakanishi, T. Matsunaga, K. Shinozaki, H. Sakamoto et al. 2008. Fabrication process of carbon nanotube/light metal matrix composites by squeeze casting. Mater. Sci. Eng.: A. 495: 282–287.

Uysal, M., H. Akbulut, M. Tokur, H. Algül and T. Çetinkaya. 2016. Structural and sliding wear properties of Ag/Graphene/WC hybrid nanocomposites produced by electroless co-deposition. J. Alloys Compd. 654: 185–195.

Van Trinh, P., T.B. Trung, N.B. Thang, B.H. Thang, T.X. Tinh, D.D. Phuong et al. 2010. Calculation of the friction coefficient of Cu matrix composite reinforced by carbon nanotubes. Comp. Mater. Sci. 49: S239–S241.

Wang, F., S. Arai and M. Endo. 2005. The preparation of multi-walled carbon nanotubes with a Ni–P coating by an electroless deposition process. Carbon 43: 1716–1721.

Wang, L.Y., J. Tu, W. Chen, Y. Wang, X. Liu, C. Olk et al. 2003. Friction and wear behavior of electroless Ni-based CNT composite coatings. Wear. 254: 1289–1293.

Wei, X., H. Rui, J.-S. Li and H.-Z. Fu. 2011. Effect of electrical current on tribological property of Cu matrix composite reinforced by carbon nanotubes. Transactions of Nonferrous Metals Society of China 21: 2237–2241.

White, A.A., S.M. Best and I.A. Kinloch. 2007. Hydroxyapatite–carbon nanotube composites for biomedical applications: a review. International Journal of Applied Ceramic Technology 4: 1–13.

Xue, F., S. Jiehe, F. Yan and C. Wei. 2010. Preparation and elevated temperature compressive properties of multi-walled carbon nanotube reinforced Ti composites. Mater. Sci. Eng.: A. 527: 1586–1589.

Xue, R.-J. and Y.-C. Wu. 2007. Mechanism and microstructure of electroless Ni-Fe-P plating on CNTs. J. China Univ. of Mining & Tech. 17: 424–427.

Yamamoto, K., S. Akita and Y. Nakayama. 1998. Orientation and purification of carbon nanotubes using ac electrophoresis. J. Phys. D.: Appl. Phys. 31: L34.

Yeh, M.-K., N.-H. Tai and J.-H. Liu. 2006. Mechanical behavior of phenolic-based composites reinforced with multi-walled carbon nanotubes. Carbon 44: 1–9.

Yoo, S., S. Han and W. Kim. 2012. Magnesium matrix composites fabricated by using accumulative roll bonding of magnesium sheets coated with carbon-nanotube-containing aluminum powders. Scripta Mater. 67: 129–132.

Yu, M.-F., O. Lourie, M.J. Dyer, K. Moloni, T.F. Kelly and R.S. Ruoff. 2000. Strength and breaking mechanism of multiwalled carbon nanotubes under tensile load. Science 287: 637–640.

Zalamea, L., H. Kim and R.B. Pipes. 2007. Stress transfer in multi-walled carbon nanotubes. Compos. Sci. Technol. 67: 3425–3433.

Zeng, Q., J. Luna, Y. Bayazitoglu, K. Wilson, A.M. Imam and E.V. Barrera. 2007. Metal coated functionalized single-walled carbon nanotubes for composites application. Paper Presented at the Mater Sci. Forum.

Zeng, X., G. Zhou, Q. Xu, Y. Xiong, C. Luo and J. Wu. 2010. A new technique for dispersion of carbon nanotube in a metal melt. Materials Science and Engineering: A. 527: 5335–5340.

Zhang, Z. and D. Chen. 2006. Consideration of Orowan strengthening effect in particulate-reinforced metal matrix nanocomposites: a model for predicting their yield strength. Scripta Mater. 54: 1321–1326.

Zhou, S.-m., X.-b. Zhang, Z.-p. Ding, C.-y. Min, G.-l. Xu and W.-m. Zhu. 2007. Fabrication and tribological properties of carbon nanotubes reinforced Al composites prepared by pressureless infiltration technique. Compos. Part A-Appl. S. 38: 301–306.

16

Fabrications of Graphene Based Nanocomposites for Electrochemical Sensing of Drug Molecules

Tien Song Hiep Pham, Peter J. Mahon and *Aimin Yu**

Introduction

Drugs are an important part of the modern world and are used in various areas to improve the quality of daily life. For example, different types of drugs have been employed in medicine to treat people with diseases (Thapliyal et al. 2015), or to treat bacterial infections (da Silva et al. 2015), and in agriculture to protect crops against attacks from pathogens (Li et al. 2016). However, environment and health related issues caused by the side effects of drugs have become a matter of concern with incidental exposure being a major risk factor. A recent scientific report estimates that approximately 2.5 million tons of pesticides were used for agriculture around the world annually (Qiu et al. 2015). This widespread use raises serious concerns to human and animal communities due to the potential persistence in the aquatic and soil environments resulting in pollution.

Also, it has been reported that antibiotics generally exhibit activity against different types of bacteria but their adverse effects can be a serious problem to the health of patients. For instance, levofloxacin, moxifloxacin, and ciprofloxacin cause nausea, headache, trouble sleeping; dizziness, sudden pain, psychotic reactions, swelling; and tearing, permanent nerve damage, respectively (Thapliyal et al. 2015). In several cases, the side effects of drugs result in serious health threats. Particularly, amikacin causes acute kidney injury, damage to the nervous system leading to permanent hearing loss, and balance problems (Modongo et al. 2014, Downes et al. 2015). Hence, developing methods for the sensitive detection of trace amount of drugs in different experimental and practical conditions has become extremely important in recent times.

Over the years, great efforts have been made by using a number of analytical methods to detect different types of drug molecules. Specifically, high performance liquid chromatography (HPLC), HPLC with fluorescence detection, LC–mass spectrometry analysis, LC coupled with tandem mass spectrometry, capillary electrophoresis, and ultraviolet-visible spectrophotometry have been used to determine fungicide

Swinburne University of Technology, Department of Chemistry and Biotechnology, John Street, Hawthorn, Victoria, Australia, 3122.
  Emails: tspham@swin.edu.au; pmahon@swin.edu.au
* Corresponding author: aiminyu@swin.edu.au

residues (Wang et al. 2016), fluoroquinolone (Arroyo-Manzanares et al. 2014) and sulphonamides (He and Blaney 2015), multiple environmental antibiotics (Yi et al. 2015), nonsteroidal anti-inflammatory drugs (Kang et al. 2014), paracetamol (Sultan et al. 2012), and lovastatin (a hypocholesterolemic drug) (Li et al. 2014c), respectively. However, those techniques are prone to several drawbacks such as their expense, the need for skilled operators, multi-step sample preparations and complicated analysis procedures. This has consequently led to the limitation in their application for the determination of drug molecules.

In contrast, electrochemical techniques have been considered by many scientists as powerful and versatile analytical methods due to their unique advantages including high sensitivity with reasonably fast responses, outstanding selectivity, ability to analyze within various matrices such as sweat, urine, serum, etc., simple operation (time saving) and low cost (Sanghavi et al. 2015). For this reason, these methods are widely applied for electrochemical detections of many inorganic and organic compounds, especially drugs. However, many drug molecules have showed poor electrochemical responses at commonly used bare working electrodes (Li et al. 2014b, Bai et al. 2015, Er et al. 2015). Such disadvantage results in the ineffective determination of drug molecules in trace amount. Fortunately, this limitation can be improved by modifying electrode surface with effective catalysts. This deliberate strategy becomes a very important driving force to stimulate the design and synthesis of materials used to enhance electrochemical signals. Among the materials of interest, much recent attention has been focussed on graphene based nanocomposites due to their unique structures and extraordinary physical and electrocatalytic properties. These materials have been considered as promising candidates for electrode modification. This chapter highlights the recent advance of graphene based nanocomposites in electrochemical sensing of drug molecules. Specifically, the first part is an overview of graphene and graphene based nanocomposites. The next few sections are focused on the fabrications of graphene based nanocomposites as well as their uses in electrode surface modification. Finally, their recent practical applications in electrochemical sensing of drug molecules are reviewed and future directions of this research field are discussed.

Graphene and graphene oxide—a special type of carbon material

Graphene has been widely known as one of the greatest discoveries of the 21st century with the ability to produce valuable applications in the fields of materials science, condensed matter physics, and applied electrochemistry (Novoselov et al. 2004). This material is generally recognized as the basic building block of all graphitic forms of carbon material. It is constructed by a single layer of sp^2-hybridized carbon atoms in a closely packed honeycomb two-dimensional (2D) lattice (Allen et al. 2009). To date, it is known as the thinnest (Meng et al. 2015) and the strongest material ever tested (Kuila et al. 2012). In referring to graphene fabrication, it is commonly prepared by "top-down" technique using mechanical cleavage and liquid phase exfoliation from graphite, or "bottom-up" technique from chemical vapour deposition, epitaxial growth on silicon carbide, etc. (Young et al. 2012). With special regards to the very unique structure and a wide range of remarkable physicochemical properties such as high levels of stiffness and strength (1060 GPa), high thermal conductivity range from 1500 to 5000 W $m^{-1}K^{-1}$ (Fugallo et al. 2014), high carrier mobility at room temperature (~ 10000 cm^2 V^{-1} s^{-1}) (Mayorov et al. 2011), and excellent optical transmittance ($\sim 97.7\%$) (Zhu et al. 2010b), it has become one of the most extensively studied nanomaterials over the past ten years. Figure 1 below shows the number of publications relating to graphene in comparison with that of graphene oxide and reduced graphene oxide in the period of 2004 to 2013.

There are many scientific fields to which this potential candidate has been successfully applied such as nanocomposite materials (Tian et al. 2014), capacitors (Hu et al. 2014, Mahmood et al. 2014, Zhu et al. 2014), catalysts (Jiao et al. 2014), conductors (Du et al. 2014, Vlassiouk et al. 2015), devices (Avouris 2010, Chen et al. 2015), sensors (Xing et al. 2014, Cui et al. 2015) or biosensors (Ruecha et al. 2014, ul Hasan et al. 2015), and so on. Remarkably, other interesting properties of graphene including large surface area (~ 2630 m^2g^{-1}) (Sun and Wang 2014), high electrical conductivity (Sanghavi et al. 2015), and very efficient electrocatalytic activity (Salavagione et al. 2014) enable it to improve electron transfer effectively when used to modify bare working electrode surface. This feature makes graphene an ideal material for use in electrochemical analysis, especially in the drug detection area.

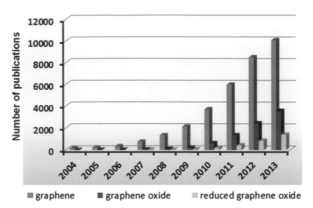

Figure 1. Number of publications related to graphene and its derivatives from 2004 to 2013 according to Scopus. The numbers were obtained using the key words of graphene, graphene oxide and reduced graphene oxide. Reprinted with permission from (Nguyen and Zhao 2014). Copyright@Royal Society of Chemistry.

Also, graphene has been used as an excellent supporting conductive surface for the attachment of electrocatalytic nanoparticles for electrochemical sensing and detection purposes (Pumera et al. 2010). However, pristine or synthetic graphene often exhibits serious problems of agglomeration or restacking caused by the weak attractive van der Waals forces amongst its adjacent nanosheets (Gong et al. 2014). Because of this disadvantage, this special carbon material possesses a very low solubility in water which reduces the effective surface area. These very defects consequently lead to a huge challenge in the production of monolayer graphene nanosheets and the limitation of its future applications. In order to overcome this obstacle, various techniques have been effectively employed to modify the surface of graphene using chemical reagents. This idea not only helps maximize the intrinsic value of graphene but also is the right direction for synthesizing graphene based nanocomposites with novel physicochemical properties.

Graphene oxide (GO) is the oxidized form of graphene. As one of the common derivatives of graphene, GO shares several similar typical properties with graphene such as original, isolated forms and true monolayers (Dreyer et al. 2014). However, there are still differences in the composition of their carbon structures that influence their physicochemical properties. In particular, graphene shows excellent electrical conductivity ($\sigma = 64$ mS cm^{-1}) (Martín and Escarpa 2014) while GO is a non-conductor (Compton and Nguyen 2010) due to the numerous defects that disrupt the sp^2 network. Additionally, as previously mentioned, graphene is constructed with sp^2 hybridized carbon atoms, whereas GO is mainly built by high level of sp^3 hybridization with a lot of electrochemically reducible oxygen containing groups such as alcohols, epoxides, and carboxylic acid. Due to the appearance of these hydrophilic functional groups on the surface, GO possesses higher solubility in aqueous environment in comparison with graphene. Irrespective of these basic differences, it is reasonable to consider GO as an essential precursor for making graphene. Therefore, the graphene synthesis from the subsequent reduction of exfoliated GO (reduced GO or RGO) has been intensively studied by many material scientists and chemists in recent times. There are three popular ways for the effective reduction of GO that include thermal, chemical and electrochemical methods (Pumera 2013). Thermally reduced graphene shows similar structure and electrochemical properties with amorphous carbon (Wong et al. 2012), while the chemical reduction technique with various reducing compounds allows graphene to maintain its inherent electrochemical nature or even being able to undergo further reduction (Chng and Pumera 2011). The authors also mentioned that electrochemically reduced graphene exhibits stable and wide electrochemical windows, which can be used in electrochemical analysis. The structures of graphene, GO and RGO can be viewed in Figure 2.

Electrochemical behaviour of graphene

Despite the great physical properties of graphene as discussed above, it is worthy to exploit its basic electrochemistry for relevant electrochemical sensing applications. There has been considerable effort from research communities into studying the electrochemical behaviour of this fundamental 2D carbon

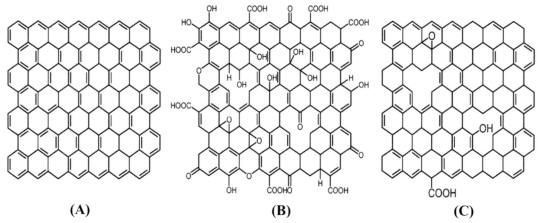

Figure 2. The chemical structures of graphene (A), GO (B), and RGO (C). Reprinted with permission from (Perreault et al. 2015). Copyright@Royal Society of Chemistry.

material since its discovery in 2004 (Novoselov et al. 2004). Several important aspects including the electron transfer rate and reduction-oxidation (redox) potential are usually taken into account when studying graphene's electrochemical nature. For example, it was reported that graphene modified glassy carbon electrode (graphene/GCE) possesses lower charge transfer resistance (R_{ct}) in comparison to that of GO modified GCE and bare GCE (Zhou et al. 2009). In other words, graphene/GCE shows the highest electrochemical efficiency although all the electrodes have similar electrochemical potential windows, which has a great impact on analytical applications. Similarly, other research in this area stated that graphene/GCE showed faster electron transfer rate constants than bare GCE with well-defined redox peaks observed with various redox-species systems such as $[Fe(CN)_6]^{3-/4-}$, $[Ru(NH_3)_6]^{2+/3+}$ and $Fe^{2+/3+}$ using cyclic voltammetry (CV) (Tang et al. 2009). Also, it has been mentioned that the peak-to-peak potential separations (ΔE_p) for most one-electron-transfer redox couples tested with graphene modified electrode using CV are close to the reversible limit (~ 59 mV) (McCreery 2008, Shang et al. 2008). Moreover, graphene obtained from the reduction of GO showed a better electrochemical response in comparison with that of graphite or glassy carbon (Pumera et al. 2010). In addition, the existence of a large amount of edges in the structure permits graphene to have higher heterogeneous electron transfer rates in contrast with carbon nanotubes (Pumera 2010). It is noticeable that graphene has the ability to enhance redox processes without the need of high over-potentials for reduction or oxidation of those species (Martín and Escarpa 2014). For such reasons, there is no doubt that graphene is a great potential candidate for fabricating excellent electroactive nanocomposites to serve in the field of electrochemical detection for different drug molecules.

Fabrication of graphene based nanocomposites

The synthesis of graphene based composites in the 1–100 nm size range has become one of the most exciting topics in electrochemistry and material science research recently. The introduction of noble metal nanoparticles, conductive polymers and non-conductive materials into graphene nanosheets not only helps reduce agglomerations on graphene surface but also creates novel hybrid nanocomposites. In general, graphene based nanocomposites can be classified into four common categories including graphene–noble metal nanoparticles, graphene–conductive polymers, graphene–non-conductive materials, and graphene–noble metal nanoparticles–conductive polymers (or non-conductive materials). Although there have already been attempts for the fabrication of new and outstanding graphene based nanocomposites with multiple functionalities for electrochemical sensing applications, challenging tasks still remain in this area. Over the years, a number of different fabrication methods have been effectively used to synthesize these types of nanocomposites. However, only a few of the more common methods will be reviewed in the following section and this includes solution mixing, *in situ* polymerization, chemical reduction and microwave assisted reduction.

Solution mixing

Solution mixing has been used as one of the most common methods for preparing graphene–polymer nanocomposites and has resulted in improved properties for electrical conductivity, thermal and chemical stability. This method is very simple, without the requirement of special equipment, but provides effective synthesis and allows production on a large scale. In general, the solution mixing method involves dispersing graphene in a solvent which is followed by mixing with a desired polymer solution (Fan et al. 2010). However, this process can sometimes be carried out in the opposite way by adding the polymer solution into the graphene dispersion. Additionally, magnetic agitation or high-energy sonication is employed to generate uniformity for the solution mixture in either water or organic solvents. The next step is the removal of solvent by evaporation to obtain a composite. For example, Yin et al. (2010a) successfully prepared a graphene–chitosan composite film by mixing 1.5 mg of graphene with 1 mL of 0.5% chitosan followed by 2 h sonication to create a homogeneous solution. This fabrication method was also used to generate a graphene–polystyrene nanocomposite (Qi et al. 2011), where the authors reported that an increase in the graphene content from ~ 0.11 to ~ 1.1 vol% resulted in improved electrical conductivity. Other nanocomposites such as graphene–nafion (Yin et al. 2010b), graphene–polypyrrole (or graphene–PPy) (Xu et al. 2011), and graphene–polyaniline (or graphene–PANI) (Ruecha et al. 2015) were also prepared using the same method. In the mixing solution technique, the solubility of graphene in the polymer solution needs to be taken into account due to the fact that pristine or synthetic graphene can aggregate on the polymer surface. For this reason, it is necessary to use surfactants to functionalize the graphene surface in order to overcome this problem. One typical example is when Stankovich et al. (2006) used phenyl isocyanate to functionalize the precursor of GO. The graphene sheets created from the reduction of GO during the synthesis could bind to the polystyrene matrix and the final graphene–polystyrene nanocomposite suspension was soluble. However, the use of a stabilizer could affect the properties of the final composite. Additionally, organic solvents are able to strongly adsorb on graphene surfaces and removing organic solvent from resulting nanocomposites remains a critical issue when using solution mixing as the fabrication method.

In situ polymerization

In situ polymerization is another popular method to synthesize graphene–polymer nanocomposites. Generally, in this method, a graphene suspension is mixed with a monomer or pre-polymer solution (Kuilla et al. 2010). The next stage is to initiate the polymerization process with the resulting polymer dependent on parameters including time and temperature. The advantage of *in situ* polymerization technique is that graphene is homogeneously dispersed and will strongly interact with the polymer matrix via atom transfer radical polymerization (Verdejo et al. 2011). This typical interaction consequently leads to improved electrical and thermal conductivity in the final composites. Over the years, there have been attempts of using this method for fabricating various types of graphene–polymer composites. For instance, a graphene–epoxy composite was successfully prepared by using the *in situ* polymerization technique (Liang et al. 2009). In addition, PANI is another common polymer that can be prepared by this polymerization approach, with the polymerization to form PANI being an oxidative process (Yan et al. 2010). Particularly, the graphene acts as an electron acceptor while the aniline monomer is an electron donor and the aniline monomers can adsorb onto the graphene sheet surface due to electrostatic attraction. An oxidant, namely ammonium persulfate, was employed to facilitate polymerization of aniline to produce the graphene–PANI nanocomposite. In a similar manner, PPy was used to prepare a graphene–PPy nanocomposite using *in situ* polymerization (Xu et al. 2011). Specifically, the pyrrole monomers adsorbed and bonded with graphene sheets surface via the strong π–π interactions, hydrogen bonds and Van der Waals forces. Next, the addition of the oxidant, ammonium persulfate solution, was added to initiate the formation of PPy layers via a chemical oxidation polymerization process and a graphene–PPy nanocomposite was successfully made with a rough surface. Likewise, another scheme for fabricating a graphene–PPy composite is shown in Figure 3. This fabrication method has also been utilized to prepare hybrid graphene–polyethylene (Shehzad et al. 2016) and graphene–polyurethane (Kim et al. 2010) with improved electrical conductivity. However, the presence of solvents in polymerization reaction conditions could be a drawback for solution processing and the viscosity of the polymerization reaction mixture needs to be considered because an increase in viscosity could limit manipulation and the loading fraction.

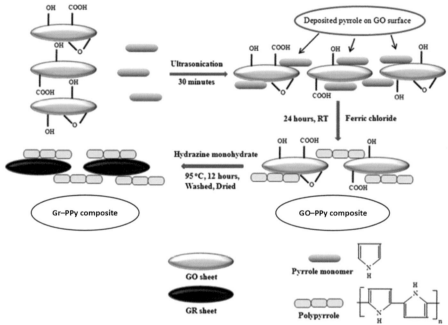

Figure 3. The preparation of graphene–PPy composite by *in situ* polymerization method. Reprinted with permission from (Bose et al. 2010). Copyright@Elsevier.

Chemical reduction

In recent times, chemical reduction has been used to synthesize graphene based nanocomposites. It is a very straightforward approach to effectively prepare graphene–noble metal nanoparticles nanocomposites or similar nanostructured materials based on the direct chemical reduction of noble metal precursors (normally their satls) in the presence of GO suspension by using chemical agents. Factors including reaction time, reaction temperature, type of reducing agent and pH of the reaction can be controlled to achieve the optimum synthesis condition when using this method. In general, chemically RGO nanosheets with residual oxygen containing groups act as attractive nucleation sites for the decoration of noble metal nanoparticles. For example, Er et al. (2015) used this facile method to successfully synthesize a graphene–gold nanoparticles–nafion (or graphene–AuNPs–nafion) nanocomposite. In their work, the $HAuCl_4$ solution was reduced to form AuNPs by using $NaBH_4$ and the GO was reduced to graphene with the help of tri-sodium citrate. The newly synthesized AuNPs combine with nafion to functionalize the surface of graphene nanosheets. The as-prepared graphene–AuNPs–nafion composites were used to modify the working electrode surface to enhance the electrochemical sensitivity. The schematic diagram in Figure 4 shows a simple synthesis process for graphene–AuNPs from the precursors of GO and $HAuCl_4$ via a single-step reduction. Besides $NaBH_4$, graphene–Au NPs nanocomposites can also be made by the utilization of other reducing and/or stabilizing agents such as strong base like NaOH (Tian et al. 2012), hydrazine under alkaline conditions (Shervedani and Amini 2014), sodium citrate (Goncalves et al. 2009) and ascorbic acid (Iliut et al. 2013).

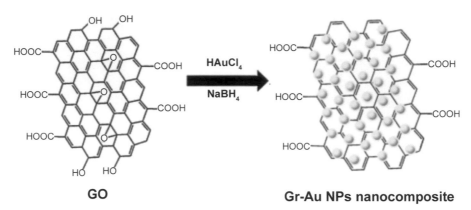

GO   **Gr-Au NPs nanocomposite**

Figure 4. The synthesis of graphene–Au NPs nanocomposite from GO and HAuCl$_4$ via a single-step reduction. Reprinted with permission from (Er et al. 2015). Copyright@Elsevier.

Furthermore, graphene–noble metal nanoparticles nanocomposites made by chemical reduction can be mediated with the addition of polyelectrolytes into the fabrication procedure. The crucial roles of such polymers are to support the entire reduction of GO nanosheets and to keep nanocomposites stable in the aqueous synthetic environment. Recently, researchers have used polyethyleneimine (Cai et al. 2012), polyacrilic acid (Tai et al. 2012), poly(diallyldimethylammonium chloride) (Zhu et al. 2013) and many others in order to synthesize graphene–silver nanoparticles (or graphene–Ag NPs) nanocomposites.

Microwave assisted reduction

Among the reduction methods, microwave assisted reduction has been recognized as one of the useful and facile methods to make graphene–noble metal nanoparticles nanocomposites by providing energy for chemical reactions. In this technique, the main mechanism is the simultaneous reduction of GO precursor and metal ions solution under microwave irradiation (Hassan et al. 2009). As a result, noble metal nanoparticles will be rapidly formed and decorated on the surface of reduced GO. The synthesis of graphene–Au NPs nanocomposite based on the microwave assisted reduction method has been one of the most intriguing topics in the nanomaterial related research field. For example, Figure 5 shows a scheme for fabricating a graphene–Au NPs nanocomposite using this reduction method (Sharma et al. 2013). In particular, the GO suspension was stirred with HAuCl$_4$ solution for 30 min, followed by the addition of ascorbic acid solution and water. Next, the whole mixture was treated with 5 min of microwave irradiation and stirred for 15 min. The graphene–Au NPs precipitate was finally obtained after removing excess of the treatment solution. Likewise, other researchers have generated graphene–Au NPs nanocomposite using hydrazine hydrate as the reducing agent (Hu et al. 2012) and polyethyleneimine

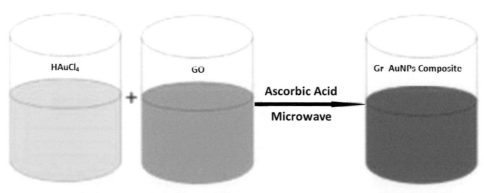

Figure 5. The schematic description of graphene–Au NPs nanocomposite synthesis using microwave assisted reduction method. Reprinted with permission from (Sharma et al. 2013). Copyright@Elsevier.

(Ponnusamy et al. 2014) for the reduction process or even in the absence of reducing (or stabilizing) agent (Jasuja et al. 2010). Apart from Au NPs, microwave assisted reduction has been applied to make Ag NPs (Liu et al. 2011c) and palladium nanoparticles (or Pd NPs) (Zhang et al. 2013) as examples. This method is very straightforward with simple preparation and is suitable for large-scale production with high efficiency. However, controlling the size uniformity and the distribution of noble metal nanoparticles on the graphene nanosheet surface are still significant challenges.

Methods to immobilize graphene based nanocomposites onto electrode surface

As was mentioned earlier, bare electrodes can have poor electrochemical responses in drug determination. For this reason, electrode surface modification is of obvious importance. It has been reported that electrochemical sensitivities of glassy carbon electrodes can be improved by modifications using nanomaterials (Sanghavi et al. 2014b). Most materials used for electrode surface modification are generally electroactive substances, which can exchange electrons with the electrode (can be either oxidized or reduced). The introduction of such materials onto bare electrodes enables an increase in electron transfer rate between the electrode and targeting molecules being studied (Karimi-Maleh et al. 2015), which has benefits for the enhancement of electrochemical performances of modified electrodes. Therefore, modifying the electrode substrate is an extremely important approach to deliberately create novel properties for bare electrode to serve in drugs detection applications by using electrochemical methods. Many graphene based nanocomposites modified working electrodes have been developed and used in different ways for electrochemical measurements. Among effective modification methods, chemical bonding, physical adsorption, electrochemical deposition and layer by layer self-assembly have received considerable interest due to their simplicity and effectiveness in electrode modification.

Chemical bonding

Chemical bonding is considered as one of the most effective ways for electrode surface modification. In general, some species seem to be more attracted to the electrode surface than bulk solution, which leads to the spontaneous attachment to the surface. For example, organic substances, with hydrophobic properties due to double bonds in their structures, are able to strongly attach to carbon materials based surface from aqueous solutions via either electrostatic or non-electrostatic interactions (Moreno-Castilla 2004). With a similar mechanism, graphene nanosheets electrostatically interact with noble metal nanoparticles and this bonding can be exploited as a key factor for creating graphene–noble metal nanoparticles modified electrodes. It follows that graphene can effectively and rapidly attach to a GCE surface based on the large number of double bonds in its structure. Li et al. (2013a) recently prepared sulfonated graphene–Au NPs/ GCE based on the formation of chemical bonds. Specifically, 5 µL of a sulfonated graphene dispersion was cast on to the surface of a polished GCE to form sulfonated graphene/GCE. The modified GCE was then placed in an Au NPs solution and sulfonated graphene–Au NPs/GCE was obtained after a period of 30 min due to electrostatic interaction between the graphene and AuNPs. The as-prepared modified electrode showed excellent electrochemical performance based on the increased electrocatalytic activity of Au NPs and high conductivity of graphene. In another study, Raj and John (2015) successfully made graphene–Au NPs/GCE based on the combination of chemical bonding and the spontaneous reduction with 1,6-hexanediamine (HDA) as the linker. The schematic description for making the modified electrode can be seen in Figure 6. Particularly, a cleaned GCE was first dipped into HDA solution for 8 h to create HDA modified GCE via electrostatic bonding between –NH$_2$ groups of HDA molecules and the electrode surface. The –NH$_2$ in HDA electrostatically reacted with the oxygen containing groups on the GO surface (–OH, C–O–C on the plane and –COOH at the edges) to obtain a GO–HDA modified GCE. The spontaneous reductions of GO to graphene and of HAuCl$_4$ solution to form Au NPs was carried out in the potential range 0 to –1.4 V in phosphate buffer solution (PBS) by CV after 15 cycles. The NPs strongly attached to graphene surface via electrostatic forces and the desired modified GCE was successfully fabricated. This method permits the formation of a thin monolayer of modifying material on an electrode surface.

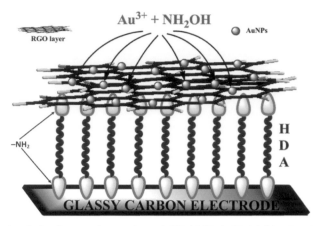

Figure 6. The schematic description for preparing graphene–AuNPs/GCE via HDA bridge. Reprinted with permission from (Raj and John 2015). Copyright@Royal Society of Chemistry.

Physical adsorption

It has been mentioned that electrodes coated with carbon based materials significantly increase the effective surface area for electron transfer in comparison with that of bare electrodes (Aregueta-Robles et al. 2015). For this reason, depositing graphene based nanocomposites onto electrode surface has become one of the most popular approaches for modifying electrodes. The technique allows a thin layer of such materials to form on the electrode surface when the solvent evaporates. For example, Guo (2011b) and co-workers used this method to successfully deposit a graphene–β-CD composite solution onto a well-polished GCE to obtain graphene–β-CD/GCE. Particularly, the bare GCE was firstly polished with 1 μm alumina slurry and was followed by rinsing with distilled water. This step was repeated with 0.3 and 0.05 μm alumina slurry and rinsing. The aim of this polishing step was to create mirror like surface for the working GCE (Gao et al. 2016). The next step was washing the polished GCE with nitric acid and acetone successively (with the ratio of 1:1), followed by ultrasonication in distilled water and air drying. This approach was carried out to remove the adsorbed residual alumina particles on the GCE surface (Ponnusamy et al. 2014). Finally, 5 μL of graphene–β-CD composite was dropped on to the polished GCE and dried in air to obtain graphene–β-CD/GCE. The as-prepared modified GCE showed excellent electrochemical properties for ultrasensitive detection of carbendazim. In a similar approach, Er and colleagues (2015) successfully prepared a graphene–Au NPs nanocomposite modified GCE for highly sensitive detection of silodosin using electrochemical technique. The modified GCE process for the determination of the targeted drug can be viewed in Figure 7. Sometimes, the drying process can be carried out using an IR lamp (Tian et al. 2012), N_2 gas (Ye et al. 2013) or in a vacuum oven (Liu et al. 2013); ethanol (Pruneanu et al. 2011) or methanol (You et al. 2013) are utilized in washing process when

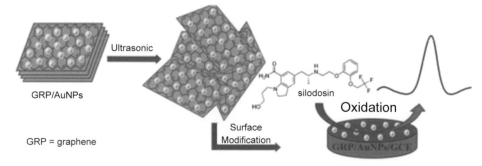

Figure 7. The process of modifying GCE by graphene–Au NPs nanocomposite for the detection of silodosin. Reprinted with permission from (Er et al. 2015). Copyright@Elsevier.

preparing graphene based composite modified electrode. The physical adsorption method saves time and provides fast and effective modification with resulting satisfactory performance. This method is suitable to coat any type of graphene based nanocomposite on to bare working electrodes.

Electrochemical deposition

Electrochemical deposition is popularly known as a controllable and robust technique to directly attach materials of interest on to working electrode surfaces. Usually, an electric current is applied to reduce dissolved metal cations in a GO solution resulting in a coherent composite coating on the surface of electrode. Many researchers have used this method to successfully prepare electroactive materials like graphene based nanocomposites modified electrodes for electrochemical analysis over the past decade. As an example, Liu et al. (2011a) made graphene–Au NPs nanocomposite modified GCE using this method. In their work, CV reduction was carried out in the deposition mixture solution of GO and HAuCl$_4$; a three-electrode system (working, auxiliary, reference electrodes) was applied with a potential scan in the range 0.6 to –1.5 V. Five potential cycles were applied to control the deposit loading and the working electrode was washed twice with distilled water when the deposition was completed. Similarly, Figure 8 shows the schematic description of fabricating RGO–Ag NPs nanocomposite on a modified indium tin oxide electrode via electrochemical deposition based on the precursor of GO in AgNO$_3$ solution. The as-modified electrode showed excellent electrochemical performance due to the remarkable electrocatalytic properties of RGO–Ag NPs as the electrode modifier. Also, other synthetic nanocomposites have been produced including RGO–Pd NPs/GCE (Konda and Chen 2015), graphene–platinum nanoparticles–nafion/GCE (or graphene–Pt NPs–nafion/GCE) (Kalambate et al. 2015) using this convenient electrochemical deposition method.

Parameters such as the applied potential, the time of electrodeposition and the analyte concentration should be considered for the control of size, density and morphology of the generated NPs in the final composites when using this deposition technique (Bosch-Navarro et al. 2016). For instance, Sundaram

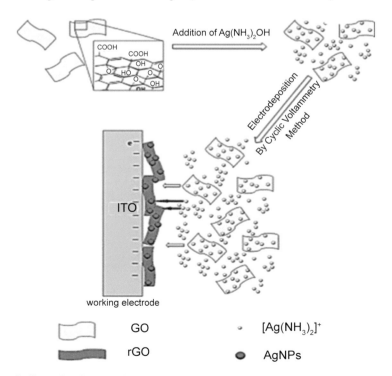

Figure 8. Schematic illustration for preparing a RGO–Ag NPs nanocomposite modified electrode using electrochemical deposition technique. Reprinted with permission from (Moradi Golsheikh et al. 2013). Copyright@Elsevier.

et al. (2008) stated that applying a more negative potential led to the thicker distribution of Pd NPs around the graphene surface. This consequently contributed to the appearance of a thick graphene–Pd NPs composite film on the electrode surface. It can be said that electrochemical deposition provides the advantage of making homogeneous and rigid composite films on a working electrode surface without the need for transferring any post-synthetic materials. Also, it is a rapid and environmental friendly technique. However, the drawback is that it can only be applied on conductive surfaces.

Layer by layer self-assembly

Layer by layer self-assembly (LBL) has been shown to be one of the most versatile and inexpensive ways for surface modification as well as to fabricate large scale assembled composite films on a substrate. LBL is simple but effective in monitoring the thickness of both micro and nanoscale structured thin films (Xi et al. 2012). The main process of this method is achieved based on alternately depositing complementary materials with opposite charges on a substrate or a surface via electrostatic, hydrogen or convalent bonding interactions (Quinn et al. 2007). Common materials used in this technique are charged species including polyelectrolytes, noble metal nanoparticles, graphene and its derivatives. One of the initial uses of this method was to deposit oppositely charged polymers on a surface (Decher et al. 1992). However, over the last 20 years, the LBL method has been employed to make highly ordered nanosized graphene based composites modified electrodes. For example, Gao and co-workers (2013) successfully utilized LBL to successfully prepare a graphene–PANI nanofiber composite modified electrode, which showed greatly improved electrochemical performance. Similarly, another research group made a graphene–PANI bilayer film modified indium tin oxide electrode (Sarker and Hong 2012). It was reported that the composite film was created based on electrostatic, hydrogen, π–π interactions between graphene and PANI. In addition, a three dimensional hybrid nanostructured graphene–Pt NPs multilayer composite film modified electrode was constructed via LBL using an ionic liquid as a linker (Zhu et al. 2010a). The authors reported that the indium tin oxide electrode as the substrate of interest was first treated to create a negatively charged surface, which was later immersed in a graphene–ionic liquid solution for 30 min. This surface was then washed and dipped in a Pt NPs solution for another 30 min. The depositing procedure was alternately repeated until the desired number of layers was achieved. Figure 9 depicts the process of making the modified electrode.

To summarise, LBL is a simple but versatile method, which can create highly tunable thin films on a particular substrate. It is a suitable method for creating graphene based nanocomposites on an electrode surface with controllable film layer thicknesses by adjusting the charge compensation mechanism. The optimum control in this technique is achieved by taking into account a few important factors such as temperature of adsorption, deposition time, cycle number, concentration and pH (or ionic strength) of materials being deposited.

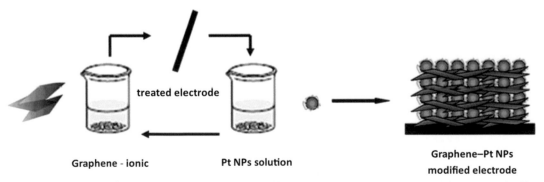

Figure 9. The process of preparing a three dimensional hybrid nanostructured graphene–Pt NPs multilayer composite film modified electrode using LBL method. Reprinted with permission from (Zhu et al. 2010a). Copyright@American Chemical Society.

Electrochemical characterization of graphene based nanocomposites modified electrodes

The electrochemical characterization of modified electrodes is one of the important stages in studying electrode surface modification and properties of modifiers. Among the effective characterization methods, electrochemical impedance spectroscopy (EIS) and CV have been commonly used in the characterization of graphene based nanocomposite modified electrodes in recent years. Particularly, EIS is used as a technique to study the impedance changes of an electrode surface. This important information enables better understanding about the change in properties of the electrode surface before and after the modification procedure. For example, Sanghavi et al. (2014a) reported that the charge transfer resistance, R_{ct} decreased from 900 Ω (at GCE) to 370 Ω (at graphene–Au NPs/GCE). In other words, the graphene–Au NPs composite on the GCE substrate greatly improved the critical properties that include higher electrical conductivity and increased surface area which enhances charge transfer sensitivity. Also, EIS was employed to investigate a RGO–Au NPs–nafion modified GCE compared to other electrodes. Figure 10A shows the Nyquist plots for EIS measurements of 1 mM $[Fe(CN)_6]^{3-/4-}$ at different electrodes. In the impedance spectra, the semicircle part at high frequency is electron transfer limited (Cao et al. 2009) and the semicircle diameter represents R_{ct} (Zhang et al. 2014a). The linear part at low frequency refers to the existence of Warburg diffusion resistance (Yang et al. 2011). It can be clearly seen that there was a remarkable decrease in the semicircle diameter in the impedance spectra from (i) to (iv). This observation matches with a decrease in R_{ct} values from 510.12 Ω (bare GCE) to 370.24 Ω (Au NPs–nafion/GCE), 250.21 Ω (RGO–nafion/GCE), and 90.16 Ω (RGO–Au NPs–nafion/GCE). The explanation for such phenomenon is that the RGO and Au NPs components of the RGO–Au NPs–nafion composite impart high electrocatalytic activity leading to the improvement in charge transfer at the modified electrode compared to that of bare and other GCEs. This result confirms that the RGO–Au NPs–nafion/GCE possessed the highest electron transfer efficiency among the electrodes. Similarly, a research group reported that a graphene–PANI/Pt electrode showed the lowest R_{ct} value compared to the other electrodes being studied (Cong et al. 2013). This is mainly based on the good electrical and outstanding ion diffusion properties of the flexible graphene–PANI composite as active modifiers on the Pt electrode surface.

Graphene based nanocomposites modified electrodes can also be characterized in redox system like $K_3[Fe(CN)_6]/K_4[Fe(CN)_6]$ solution using CV method. For example, Figure 10B depicts the electrochemical behaviour of bare and variously modified GCEs in 10 mM $[Fe(CN)_6]^{3-/4-}$ solution containing 0.1 M KCl. The well-defined and quasi-reversible peaks can be clearly seen in the cyclic voltammograms of every GCE. However, there was an increase in redox peak currents from the curve a to curve d. Moreover, the authors stated that ΔE_p (at bare GCE) was measured as 304 mV and this value was reduced to 152 mV (at Au NPs/GCE) and continued to drop (at graphene nanofibers/GCE) and reached the minimum of 117 mV (at graphene nanofibers–Au NPs/GCE). The obtained results can be explained by the remarkable

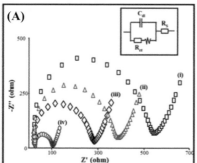
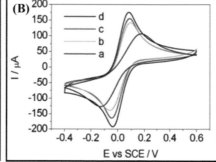

Figure 10. (A) Nyquist plots of EIS for bare GCE (i), Au NPs–nafion/GCE (ii); RGO–nafion/GCE (iii), and RGO–Au NPs–nafion/GCE (iv) in 1 mM $[Fe(CN)_6]^{3-/4-}$ solution; (B) Cyclic voltammograms of bare GCE (a), Au NPs/GCE (b), graphene nanofibers/GCE (c) and graphene nanofibers–Au NPs/GCE (d) in 10 mM $[Fe(CN)_6]^{3-/4-}$ solution containing 0.1 M KCl. Reprinted with permission from (Sanghavi et al. 2014b) and from (Niu et al. 2013), respectively. Copyright@Elsevier.

contribution of graphene nanofibers and Au NPs on the GCE surface, with greater electrical conductivity, larger surface area and improved electrocatalytic properties which has effectively enhanced electron transfer between the modified GCE and $[Fe(CN)_6]^{3-/4-}$ (Sanghavi et al. 2014a). In another work, Li and colleagues (2009) also tested the as-prepared nafion/GCE and graphene–nafion/GCE in 1 mM $K_3Fe(CN)_6$ solution containing 1 M KCl using CV method. They reported that nafion as a negatively charged polymer blocked the $Fe(CN)_6^{3-}$ diffusion into the electrode surface. For this reason, the electron transfer was limited and no redox peaks were observed in the cyclic voltammogram of nafion/GCE. The addition of graphene as a part of the composite acted as an electrically conductive bridge to enhance the electron transfer between $Fe(CN)_6^{3-}$ and graphene–nafion/GCE.

In CV, the electroactive surface area of the electrode (for a reversible process) can be calculated according to the Randles-Sevcik equation (Shahrokhian et al. 2016):

$$I_p = 2.69 \times 10^5 \, AD^{1/2}n^{3/2}v^{1/2}C \tag{1}$$

where I_p, A, D, n, v, and C are the peak current, area of the electrode, diffusion coefficient of the electroactive species, number of electrons transfer in the redox reaction, scan rate and concentration, respectively. Based on the differences in surface area between bare and modified electrodes, the electrode surface modification can be confirmed. The larger surface area of an electrode importantly contributes to larger currents for targeted molecules at an electrode surface and greater sensitivity. Qian et al. (2014) used Eq. (1) to calculate the surface area of their prepared electrodes and found that the surface areas of the bare GCE, graphene/GCE, and graphene–Au NPs–PPy/GCE were estimated as 0.067, 0.105, and 0.143 cm^2, respectively. This gradual increase in surface area was due to the large surface area of graphene and Au NPs in the composite. Therefore, it can be said that the GCE was successfully modified by the graphene–Au NPs–PPy composite.

Graphene based nanocomposites in drug sensing

As mentioned above, electrochemical methods have recently become one of the most important and effective tools for drug detection due to their simplicity, fast response, and relative cheapness (Sanghavi et al. 2014a, Jiang et al. 2016). In addition, graphene based nanocomposites are electroactive and able to actively take part in redox chemical reactions as electron donors or acceptors. In other words, they can act as highly selective, tuneable and effective electrocatalysts when applied in such analytical techniques for drugs sensing. This consequently leads to enhancement in the sensitivity, stability and reproducibility for measurements. Recognizing this important role of graphene based nanocomposites, great effort has been made to successfully prepare various electrochemical sensing platforms, which are suitable for the drugs detection. The following section will describe the utilization of different graphene based composites in the drug sensing area.

Graphene–noble metal nanoparticles nanocomposites in drug sensing

Noble metal nanoparticles are a class of very promising nanocatalysts that includes Au, Ag, Pt, and Pd. They have gained tremendous attention due to their unique size and shape dependent electrical properties as well as good electrocatalytic activities for many chemical reactions. Also, these materials show high surface-to-volume ratio, strong surface reaction and adsorption abilities (Lian et al. 2012). Therefore, it is believed that the combination of graphene and these materials has good prospects for the formation of graphene–noble metal nanoparticles nanocomposites for the redox reduction of drug molecules at electrode surfaces. Such composites can be employed as electrode surface modifiers with the aim of enhancing electrochemical signals between an electrode and drugs in terms of high selectivity and sensitivity.

For example, a research group prepared graphene–Au NPs and graphene–Ag NPs for modifying Pt electrodes to serve in the detection of carbamazepine using linear sweep voltammetry (LSV) (Pruneanu et al. 2013). The authors reported that graphene possesses high electrical conductivity and high surface area with a lot of edge-plane sites on its surface that mainly contributed to electron transfer in the oxidation of the drug. Additionally, Au and Ag NPs have great electrocatalytic behaviour that partly supported the chemical reaction. For this reason, the electrochemical signal between the modified Pt electrode improved, compared to that of bare electrode. The limit of detection (LOD) was found as 2.75×10^{-5} and 2.92×10^{-5} M for graphene–Au NPs/Pt electrode and graphene–Ag NPs/Pt electrodes, respectively. Jiang et al. (2012) successfully fabricated graphene–Au NPs/screen printed carbon electrodes to electrochemically detect cocaine via catalytic redox-cycling amplification. The composite of graphene–Au NPs with outstanding electrocatalytic activity and high conductivity efficiently promoted electron transfer while cocaine molecules were concentrated at the modified electrode surface by sulfur–Au interactions. Differential pulse voltammetry (DPV) was employed to sensitively detect the drug with LOD of 1 nM in the linear range of 1–500 nM.

Similar to graphene–Au NPs and graphene–Ag NPs nanocomposites, graphene–Pt NPs nanocomposites are expected to greatly contribute to the electrochemical sensing of drugs. For instance, a sensor was fabricated based on graphene–Pt NPs nanocomposites for selectively detecting rutin (Yu and Zhao 2011). Specifically, graphene as a part of the synthetic composite with large surface area enabled capturing the drug molecules at the modified GCE surface, where the Pt NPs on the graphene nanosheets played a crucial role of enhancing electron transfer for the oxidation of the drug molecules. Therefore, the composite was responsible for the current response improvement and consequently led to more sensitive detection. Moreover, their scan rates' effect study was carried out in the range 10–250 mV/s using CV method and the results showed that forward and reverse peak potential moved towards positive and negative directions, respectively, suggesting an adsorption–controlled process for electrode reaction of the drug. Also, the DPV results of this study showed the LOD of 6.7 nM and the peaks current increased in the linear drug concentration range between 2.0×10^{-8} M and 8.0×10^{-5} M. Another attempt at using a graphene–Pt NPs–carbon nanosphere nanocomposite as an electrode surface material was made for the successful detection of cefepime using LSV method (Shahrokhian et al. 2014). With enhanced electron transfer capability, large electrochemically active surface area due to excellent physicochemical properties of graphene and Pt NPs as constructing components, the as-synthesized composite modified electrode was able to attract and capture targeted drug molecules to provide better current responses than bare electrode. The LOD was found to be 1.2 nM with a wide linear range of 8 to 6000 nM. Also, Atta and co-workers (2014) prepared graphene–Pd NPs hybrid composite modified GCE with rapid electron transfer and great catalytic capabilities for sensitive determination of morphine by DPV with the LOD of 12.95 nM. The modified electrodes were applied to successfully detect this drug in diluted human urine with satisfactory reproducibility, sensitivity and recovery.

Following the same approach, graphene–Pd NPs material have recently been fabricated to effectively modify a working electrode to serve in the separate detection of rutin and isoquercitrin molecules at concentrations as low as 3.0×10^{-10} M and 1.6×10^{-12} M, respectively (Liu et al. 2016). Generally speaking, Pd NPs exhibit better abundance (Li et al. 2014b), lower cost and toxicity (Leng et al. 2015) compared to Au and Pt NPs. For this reason, they are more popularly employed in the synthesis of graphene based nanocomposites for electrochemical drug sensing related studies than other noble metal nanoparticles.

Graphene-conductive polymers nanocomposites in drug sensing

Conductive polymers have gained special attention from many researchers in the past few decades. With unique structures and good electrical, electronic, magnetic and optical properties, they have been utilized as one of the key materials in a wide range of practical applications in many research areas including biomedical (Dong et al. 2015), electronic devices (Michinobu 2015), supercapacitors (Zeigler et al. 2015) and electrochemical sensors (Li et al. 2015b). Over the years, rapid growth of interest is observed in the integration of conductive polymers with the capability of effectively transferring electric charges (Ahuja et al. 2007) and graphene for hybrid nanocomposites production to modify working electrode surface, aimed at improving the electrochemical performance of electrodes for highly sensitive and selective detection of analytes. This idea has been widely exploited for the construction of various novel and reliable drug sensing platforms.

For instance, Li et al. (2015a) successfully prepared an electrode functionalized with PPy grafted nitrogen doped graphene composite for electrochemical sensing of the toxic herbicide, paraquat. As previously mentioned, graphene has high electrical conductivity and large surface area. PPy also has high electrical conductivity and good redox reversibility (Qian et al. 2014). In addition, the amine functional group on the ring of its structure possibly increases the ability for sensing biomolecules (Ulubay and Dursun 2010). For this reason, the as-prepared graphene–PPy/GCE with outstanding electrocatalytic activity showed great electrochemical performance in detecting the targeted drug molecules with a detection limit of 41 nM with a linear range from 50 to 2000 nM.

In addition, graphene–PANI nanocomposites have recently been utilized in electrochemical sensing platforms for the effective detection of antibiotics. For example, researchers created a novel biosensor for successfully sensing the antimalarial drug, artesunate (Radhapyari et al. 2013). The robust hybrid graphene–PANI nanocomposite acted as an ideal supporting matrix with remarkable electrochemical properties and stability for the attachment of horseradish peroxidase. The as-synthesized composite effectively captured the drug molecules as well as stimulated the electron transfer resulting in enhanced current responses. The sensitivity of the detection method was found as 0.15 µA ng/mL in the range from 0.05 to 0.40 ng/mL. Also, the authors mentioned that the biosensor was tested with different practical conditions with satisfactory recovery and selectivity. Likewise, graphene–PANI/SPCE was made for the detection of sulphonamide residues (Thammasoontaree et al. 2014).

Graphene-non-conductive materials nanocomposite in drug sensing

Unlike conductive polymers, non-conductive materials with unique physicochemical properties have contributed to the fabrication of graphene based nanocomposites in their own ways. These materials are normally used to functionalize graphene surfaces by non-covalently binding to functional groups on the basal plane and edges of graphene. As a result, the restacking of graphene's adjacent nanosheets is effectively minimized. Nanocomposites synthesized based on graphene and non-conductive materials can also be applied in electrochemical methods for the detection of drug molecules.

For example, graphene and β-CD were used as starting materials for the creation of a new type of nanocomposite to serve in modifying the bare GCE surface for the determination of the fungicide, carbendazim (Guo et al. 2011b). As a part of the prepared nanocomposite, β-CD molecules with hydrophobic interior cavities and hydrophilic exterior sides with many hydroxyl groups (Li et al. 2014d) were able to form host-guest complexes with carbendazim. For this reason, β-CD molecules could concentrate the drug molecules at the graphene–β-CD/GCE substrate. In addition, as a highly conductive component of the nanocomposite, graphene effectively accelerated the electron transfer between the drug molecules and the modified GCE. As a result, the peak currents obtained at graphene–β-CD/GCE were much higher compared to that of graphene/GCE and bare GCE using the CV method. Also, the increase in scan rates from 50 to 500 mV/s led to the gradual shift of redox potentials, which can be seen in Figure 11A. The

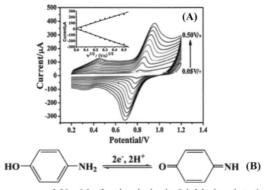

Figure 11. (A) Cyclic voltammograms of 50 µM of carbendazim in 0.1 M phosphate buffer solution (PBS) (pH 7) at graphene–β-CD/GCE with various scan rates; (B) the mechanism of ciprofloxacin oxidation. Reprinted with permission from (Guo et al. 2011b) and from (Yin et al. 2010a), respectively. Copyright@Elsevier.

observed redox peaks indicated that carbendazim undergoes a quasi-reversible redox process. Moreover, the peak current varied linearly with the square root of the scan rate suggesting a diffusion-controlled process. In order to detect the trace amounts of carbendazim, DPV was employed instead of CV. The reason is that DPV is highly sensitive in quantitative analysis with the detection range being μM to nM or even lower by effectively inhibiting the background current (Li et al. 2014b), whereas CV is known as a type of potential-dynamic electrochemical measurement method widely used to study mechanisms and rates of redox process occurring at various potentials (Casado et al. 2016). The DPV results exhibited a LOD of 2 nM and the peak currents increased when the carbendazim concentration was linearly increased from 5 nM to 0.45 μM. Based on the as-prepared graphene–β-CD nanocomposite, the authors made a sensor to electrochemically test carbendazim in spiked water with satisfactory stability, selectivity and reproducibility. In addition, the as-prepared sensor was applied to separately detect doxorubicin and methotrexate with concentrations as low as 0.1 nM and 20 nM, respectively (Guo et al. 2011a).

Similarly, Wu et al. (2011) created a graphene–β-CD nanocomposite based sensor to determine one of the nitroaromatic organophosphate pesticides, namely methyl parathion. It was reported that the drug highly adsorbed onto graphene–β-CD/GCE via strong π–π interaction and the composite facilitated electron transfer leading to the fast and sensitive signal responses. The LOD of drug was found as 0.05 parts per billion (ppb) using the DPV method. Also, the optimal experimental conditions to achieve best drug sensing were reported as 5 min (for time accumulation), 0.6 μg of graphene–β-CD composite and pH of 7.0.

A graphene–chitosan composite film was prepared as a GCE modifier for the determination of acetaminophen (Zheng et al. 2013). The attachment of graphene–chitosan onto a GCE resulted in an increase in the active surface area for the modified electrode. The catalytic effect of the composite film allowed higher rates of electron transfer between the drug molecules and the modified GCE. For this reason, there was a change in ΔE_p from 245 mV (bare GCE) to 190 mV (modified GCE) when measured with CV. Further analysis where pH was varied produced a measured value (62.1 mV pH^{-1}), which compares favourably with the Nernstian theoretical value (59 mV pH^{-1}) confirming that the loss of electrons and protons were the same and equal to 2 (Manjunatha et al. 2011). The DPV results showed the LOD of 3.0×10^{-7} M is in the range of 1.0×10^{-6}–1.0×10^{-4} M. Likewise, Yin and colleagues (2010a) successfully applied a graphene–chitosan composite in electrochemically detecting 4-aminophenol by DPV with a LOD of 5.7×10^{-8} M. The electrochemical redox reaction mechanism of the drug can be viewed in Figure 11B.

Graphene-noble metal nanoparticles–conductive polymer (or non-conductive materials) nanocomposites in drug sensing

Recently, researchers have combined Au NPs with graphene and nafion to successfully synthesize an excellent composite for sensitively detecting sumatriptan using a voltammetric method (Sanghavi et al. 2014a). In this work, Au NPs were added to the graphene surface to prevent agglomeration of graphene nanosheets. Additionally, Au NPs are not only able to reduce the over-potentials of electrochemical reactions but also improve the reversibility of redox reactions (Saha et al. 2012). It has been reported that excellent electrical conductivity of Au NPs enable them to increase the electron transfer rate between electrodes and electroactive species (Ahmad et al. 2015) while nafion with superior chemical and thermal stability enables the binding substrate to effectively adhere to other modifiers on an electrode surface (Lue et al. 2015). Therefore, the stable graphene–Au NPs–nafion nanocomposite with high electrical conductivity and large surface area helped facilitate electron transfer between drug molecules and the composite modified GCE. Consequently, the electrochemical signal was effectively improved resulting in the successful detection of the drug with a very low LOD of 7.03×10^{-10} M.

A similar trend was described for the fabrication of an electrochemical sensor prepared by Er and colleagues (2015) using graphene–Au NPs–nafion composite to sensitively detect silodosin. The authors stated that the –NH$_2$ groups of silodosin interacted with oxygen containing groups like –COOH at the edges of the composite resulting in the adsorption of the drug molecules to the modified GCE. Additionally, the synergistic effects from the combination of conductive graphene, AuNPs and nafion in the composite helped facilitate electron transfer. Thus, the graphene–Au NPs–nafion/GCE displayed the strongest electrochemical response with the drug compared to the bare and other modified GCE in this study. Also, the authors used CV to investigate the scan rate effect of the electrochemical reaction for the drug at the modified GCE in the range from 10 to 500 mV/s. The results indicated that silodosin underwent an irreversible electrochemical reaction based on the shift of the oxidation peak to positive potential and the peak currents were proportional to the scan rate implying an adsorption controlled process. Remarkably, 4 electrons are involved in the electrode reaction as calculated based on the Eq. (2) below (Zhao et al. 2014):

$$E_{pa} = E^{\theta} + \left(\frac{RT}{\alpha nF}\right)\ln\left(\frac{RTk^{\theta}}{\alpha nF}\right) + \left(\frac{RT}{\alpha nF}\right)\ln\nu \tag{2}$$

where E_{pa}, α, k^{θ}, n, ν, E^{θ}, R, T, F are the peak potential, transfer coefficient, standard rate constant of the reaction, electron transfer number, scan rate, formal redox potential, gas constant, absolute temperature and Faraday constant, respectively. Figure 12A exhibits the DPV results for silodosin detection in the study.

A graphene–Pt NPs–nafion nanocomposite was employed as a modified electrode for electrochemically determining paracetamol (Kalambate et al. 2015). The CV results suggested that the redox reaction of paracetamol was considered as a quasi-reversible process with the exchange of 2 electrons and 2 protons. The adsorptive stripping square wave voltammetry (SWV) results showed sensitive and selective detection as low as 0.106 nM in the linear range of 8.2 x 10^{-6}–1.6 x 10^{-9} M. Figure 12B shows one of the SWV results of the study.

The combined graphene–Au NPs–PANI nanocomposite was employed with an important thiazine dye, thionines, to fabricate an electrochemical aptasensor for successfully sensing kanamycin (Li et al. 2014a). The excellent synergistic effects from such integration provided a lot of active sites for effective electron transfer between the modified electrode and the drug being studied. The testing with DPV exhibited the LOD of 8.6 nM in the linear range of 1.0 x 10^{-8}–2.0 x 10^{-7} M. Using a similar approach, researchers have fabricated electrochemical sensors based on graphene–Ag NPs–poly(methylene blue), graphene–Au NPs–chitosan–Pt NPs and graphene–Au NPs–chitosan–plant esterase nanocomposites to successfully detect rutin, erythromycin and malathion with LODs of 1.0 x 10^{-8} M (Yang et al. 2014), 2.3 x 10^{-8} M (Lian et al. 2012), and 1.51 x 10^{-9} M (Bao et al. 2015), respectively.

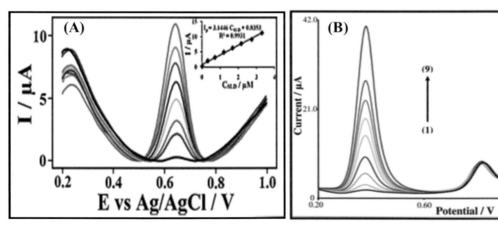

Figure 12. (A) DP voltammograms of silodosin at increasing concentration in pH 4.0 Britton–Robinson buffer solution (PBS) at graphene–Au NPs–nafion/GCE with the inset showing the linear range of silodosin; and (B) SW voltammograms of paracetamol at concentrations from 1.57 nM (1) to 8.26 µM (9) at the graphene–Pt NPs–nafion/GCE in the presence of 19.5 µM domperidone. Reprinted with permission from (Er et al. 2015) and from (Kalambate et al. 2015), respectively. Copyright@Elsevier.

Graphene based nanocomposites for drug sensing in real samples

For many years, graphene based nanocomposites have played an important role as effective electrode modifiers to serve in practical applications of drugs detection. For example, Li et al. (2014a) successfully tested for kanamycin with graphene–PANI–Au NPs/GCE in diluted milk and the drug concentration recoveries were estimated to be from 97.2 to 104.9%. Another research group studied the antimalarial drug, artesunate, in biological fluids such as urine, serum, and plasma samples with graphene–PANI/GCE and obtained impressive recovery percentages in the range 98.23–100.3% (Radhapyari et al. 2013). Likewise, Lian and co-workers (2012) made an attempt to effectively detect erythromycin in real spiked samples including milk and honey using a graphene–Au NPs–chitosan–Pt NPs/Au electrode. They reported that the recovery of the proposed sensor was estimated to be about 95.8 to 103%. The group of Zheng (2013) effectively determined acetaminophen in clinical samples and pharmaceutical industry. They stated that the peak current deviation was about 3.6% with a recovery range of 92–107% and their fabricated sensor exhibited excellent selectivity for this particular drug. Similarly, Er et al. (2015) created a novel silodosin sensor from graphene–Au NPs–nafion nanocomposite with high sensitivity and accuracy (99% recovery) without any interference effect. Bao et al. (2015) tested organophosphate pesticides contaminated carrot and apple samples with graphene–Au NPs–chitosan–plant esterase/GCE with good recovery from 92 ± 3.85 to $109.5 \pm 5.26\%$. In addition, Yin and colleagues (2010a) prepared a graphene–chitosan based sensor to selectively detect 4-aminophenol in water samples with the recovery between 97.2% and 103.8%. The test results show that this method can be applied to determine this particular drug in environmental and pharmaceutical samples. Apart from that, Guo and co-workers (2011b) fabricated an ultrasensitive sensing platform from graphene–β-CD to efficiently detect carbendazim and the sensor revealed excellent stability with only 3.2% loss when being stored around 25°C after 15 days. All of these sensing platforms showed good stability, reproducibility as well as high sensitivity in drug detection in the presence of different interfering substances such as glucose, sucrose, citric acid, ascorbic acid, uric acid, dopamine, Na^+, K^+, Mg^{2+}, Zn^{2+}, Al^{3+}, SO_4^{2-}, NH_4^+, CO_3^{2-} and Cl^- (Bao et al. 2015, Er et al. 2015, Li et al. 2015a).

Summary and future directions

The rapid development of material science has a great impact on analytical electrochemistry. The electrochemical detection of drug molecules is a typical example to prove that point. This chapter presents a comprehensive insight on the fabrications of graphene based nanocomposites and their application in electrochemical sensing of drug molecules. With the capacity to increase the effective surface area, improve the conductivity and electrocatalytic activity, there is no doubt that graphene is one of the preferred electrode surface modifiers. Such materials could also effectively attract drug molecules onto electrode surface. For electroactive drug molecules that contain functional groups (–N, –S, –OH, etc.), graphene composites could accelerate the electrons transfer at the electrode interface resulting in enhanced electrochemical signals. As a result, the detection limits of drug molecules are improved. For electrochemical techniques, CV provides clear insight about redox reaction between target drugs and graphene based nanocomposite modified electrode surface while pulse voltammetry such as DPV and SWV can further increase the sensitivity based on the fact that these methods are able to limit the background current, as mentioned previously. Compared to SWV, DPV is more popularly used for drug detection. To summarize, the successful utilization of graphene based nanocomposites in effective detection of drug molecules has become the driving force for widely applying these nanosized materials to determine inorganic, organic, and biological molecules in various conditions.

A challenge for the future research work might be using graphene based materials in the detection of non-electroactive drug molecules, which inactively take part in electrochemical reactions. However, it is believed that the continuous development of modern technology in the next decades or even centuries will be able to overcome these limitations. Also, metal oxides such as TiO_2, Co_3O_4, Fe_3O_4, SnO_2, Cu_2O, CuO and MnO_2 have been used to reduce agglomeration phenomenon among adjacent nanosheets on graphene surface for the production of graphene based nanocomposites. Nevertheless, practical applications of such materials have been popularly focused on inorganic molecules sensing, drug delivery as well as energy storage. Therefore, the importance of metal oxides in graphene based nanocomposites for electrochemically detecting drug molecules, like the previously mentioned noble metal nanoparticles and conductive polymers, should deserve much more attention from electrochemical scientists. This idea might be a novel and innovative pathway for many researchers to follow in the near future.

Table 1. Electrochemical detection of various drugs with graphene based nanocomposite modified electrodes using voltametric techniques.

Modified electrode	Drug name	Detection method	Linear range (in M)	Detection of limit (in M)	References
Graphene–β-CD/GCE	Carbendazim	DPV	5.0×10^{-9}–4.5×10^{-7}	2.0×10^{-9}	(Guo et al. 2011b)
RGO–β-CD polymer/GCE	Imidacloprid	DPV	5.0×10^{-8}–1.5×10^{-5} and 2.0×10^{-5}–1.5×10^{-4}	2.0×10^{-8}	(Chen et al. 2013)
Graphene–nafion–β-CD/GCE	Rutin	DPV	6.0×10^{-9}–1.0×10^{-5}	2.0×10^{-9}	(Liu et al. 2011b)
Graphene–chitosan/GCE	4-aminophenol	DPV	2.0×10^{-7}–5.5×10^{-4}	5.7×10^{-8}	(Yin et al. 2010a)
Graphene–Au NPs–PANI/GCE	Kanamycin	DPV	1.0×10^{-8}–2.0×10^{-7}	8.6×10^{-9}	(Li et al. 2014a)
Graphene–Au NPs–PPy/GCE	Levofloxacin	DPV	1.0×10^{-8}–1.0×10^{-6}	0.53×10^{-6}	(Wang et al. 2014)
Ni(OH)$_2$ NPs–RGO–GO/GCE	Rifampicin	LSV	0.06×10^{-7}–1.0×10^{-5}	4.16×10^{-9}	(Rastgar and Shahrokhian 2014)
Graphene–Au NPs–nafion/GCE	Silodosin	DPV	0.01×10^{-6}–3.3×10^{-6}	3.8×10^{-9}	(Er et al. 2015)
Graphene–nafion/GCE	Codeine	SWV	5.0×10^{-8}–3.0×10^{-5}	1.5×10^{-8}	(Li et al. 2013b)
NiFe$_2$O$_4$ NPs–graphene/carbon paste electrode	Tramadol	SWV	1.0×10^{-8}–9.0×10^{-6}	3.0×10^{-9}	(Afkhami et al. 2014)
Graphene–Au NPs–nafion/GCE	Sumatriptan	DPV	2.14×10^{-9}–4.12×10^{-5}	7.03×10^{-10}	(Sanghavi et al. 2014a)
Graphene–Pt NPs–nafion/GCE	Paracetamol	SWV	8.2×10^{-6}–1.6×10^{-9}	1.06×10^{-10}	(Kalambate et al. 2015)
Nitrogen doped graphene–Au NPs/GCE	Chloramphenicol	LSV	2.0×10^{-6}–8.0×10^{-5}	5.9×10^{-7}	(Borowiec et al. 2013)
Sulfonated graphene–Ag NPs/GCE	Chloramphenicol and Metronidazole	DPV	2.0×10^{-8}–2.0×10^{-5} and 1.0×10^{-7}–2.0×10^{-5}	1.0×10^{-8} and 5.0×10^{-8}	(Zhai et al. 2015)
Graphene–molecularly imprinted polymer/GCE	Trimethoprim	SWV	1.0×10^{-6}–1.0×10^{-4}	1.3×10^{-7}	(da Silva et al. 2014)
Graphene–poly(alizarin red)/GCE	Ciprofloxacin	DPV	4.0×10^{-8}–1.2×10^{-4}	1.0×10^{-8}	(Zhang et al. 2014b)
Graphene–Pd NPs–β-CD/GCE	Rutin and Isoquercitrin	DPV	1.0×10^{-9}–3.0×10^{-5} and 5.0×10^{-12}–5.0×10^{-6}	3.0×10^{-10} and 1.6×10^{-12}	(Liu et al. 2016)
Graphene–Pt NPs–carbon nanospheres/GCE	Cefepime	LSV	8.0×10^{-9}–6.0×10^{-6}	1.2×10^{-9}	(Shahrokhian et al. 2014)
Nitrogen-doped graphene–PPy/GCE	Paraquat	DPV	5.0×10^{-8}–2.0×10^{-6}	4.1×10^{-8}	(Li et al. 2015a)
Graphene–Au NPs–chitosan–Pt NPs/Au electrode	Erythromycin	CV	7.0×10^{-8}–9.0×10^{-5}	2.3×10^{-8}	(Lian et al. 2012)

References

Afkhami, A., H. Khoshsafar, H. Bagheri and T. Madrakian. 2014. Preparation of $NiFe_2O_4$/graphene nanocomposite and its application as a modifier for the fabrication of an electrochemical sensor for the simultaneous determination of tramadol and acetaminophen. Anal. Chim. Acta 831: 50–59.

Ahmad, R., N. Griffete, A. Lamouri, N. Felidj, M.M. Chehimi and C. Mangeney. 2015. Nanocomposites of gold nanoparticles @ molecularly imprinted polymers: chemistry, processing, and applications in sensors. Chem. Mater. 27: 5464–5478.

Ahuja, T., I.A. Mir, D. Kumar and Rajesh. 2007. Biomolecular immobilization on conducting polymers for biosensing applications. Biomaterials 28: 791–805.

Allen, M.J., V.C. Tung and R.B. Kaner. 2009. Honeycomb carbon: a review of graphene. Chem. Rev. 110: 132–145.

Aregueta-Robles, U.A., A.J. Woolley, L.A. Poole-Warren, N.H. Lovell and R.A. Green. 2015. Organic electrode coatings for next-generation neural interfaces. Front. Neuroeng. 7: 57–74.

Arroyo-Manzanares, N., L. Gámiz-Gracia and A.M. García-Campaña. 2014. Alternative sample treatments for the determination of sulfonamides in milk by HPLC with fluorescence detection. Food Chem. 143: 459–464.

Atta, N.F., H.K. Hassan and A. Galal. 2014. Rapid and simple electrochemical detection of morphine on graphene–palladium-hybrid-modified glassy carbon electrode. Anal. Bioanal. Chem. 406: 6933–6942.

Avouris, P. 2010. Graphene: electronic and photonic properties and devices. Nano Lett. 10: 4285–4294.

Bai, H., C. Wang, J. Chen, J. Peng and Q. Cao. 2015. A novel sensitive electrochemical sensor based on *in situ* polymerized molecularly imprinted membranes at graphene modified electrode for artemisinin determination. Biosens. Bioelectron. 64: 352–358.

Bao, J., C. Hou, M. Chen, J. Li, D. Huo, M. Yang et al. 2015. Plant esterase–chitosan/gold nanoparticles–graphene nanosheet composite-based biosensor for the ultrasensitive detection of organophosphate pesticides. J. Agric. Food Chem. 63: 10319–10326.

Borowiec, J., R. Wang, L. Zhu and J. Zhang. 2013. Synthesis of nitrogen-doped graphene nanosheets decorated with gold nanoparticles as an improved sensor for electrochemical determination of chloramphenicol. Electrochimi. Acta 99: 138–144.

Bosch-Navarro, C., J. Rourke and N. Wilson. 2016. Controlled electrochemical and electroless deposition of noble metal nanoparticles on graphene. RSC Adv. 6: 73790–73796.

Bose, S., T. Kuila, M.E. Uddin, N.H. Kim, A.K.T. Lau and J.H. Lee. 2010. *In situ* synthesis and characterization of electrically conductive polypyrrole/graphene nanocomposites. Polymer 51: 5921–5928.

Cai, X., M. Lin, S. Tan, W. Mai, Y. Zhang, Z. Liang et al. 2012. The use of polyethyleneimine-modified reduced graphene oxide as a substrate for silver nanoparticles to produce a material with lower cytotoxicity and long-term antibacterial activity. Carbon 50: 3407–3415.

Cao, Q., H. Zhao, L. Zeng, J. Wang, R. Wang, X. Qiu et al. 2009. Electrochemical determination of melamine using oligonucleotides modified gold electrodes. Talanta 80: 484–488.

Casado, N., G. Hernández, H. Sardon and D. Mecerreyes. 2016. Current trends in redox polymers for energy and medicine. Prog. Polym. Sci. 52: 107–135.

Chen, K., S. Song, F. Liu and D. Xue. 2015. Structural design of graphene for use in electrochemical energy storage devices. Chem. Soc. Rev. 44: 6230–6257.

Chen, M., Y. Meng, W. Zhang, J. Zhou, J. Xie and G. Diao. 2013. β-Cyclodextrin polymer functionalized reduced-graphene oxide: application for electrochemical determination imidacloprid. Electrochimi. Acta 108: 1–9.

Chng, E.L.K. and M. Pumera. 2011. Solid-state electrochemistry of graphene oxides: absolute quantification of reducible groups using voltammetry. Chem. Asian J. 6: 2899–2901.

Compton, O.C. and S.T. Nguyen. 2010. Graphene oxide, highly reduced graphene oxide, and graphene: versatile building blocks for carbon-based materials. Small 6: 711–723.

Cong, H.P., X.C. Ren, P. Wang and S.H. Yu. 2013. Flexible graphene–polyaniline composite paper for high-performance supercapacitor. Energy Enviro. Sci. 6: 1185–1191.

Cui, M., J. Huang, Y. Wang, Y. Wu and X. Luo. 2015. Molecularly imprinted electrochemical sensor for propyl gallate based on PtAu bimetallic nanoparticles modified graphene–carbon nanotube composites. Biosens. Bioelectron. 68: 563–569.

da Silva, H., J.G. Pacheco, J. McS Magalhães, S. Viswanathan and C. Delerue-Matos. 2014. MIP-graphene-modified glassy carbon electrode for the determination of trimethoprim. Biosens. Bioelectron. 52: 56–61.

da Silva, H., J. Pacheco, J. Silva, S. Viswanathan and C. Delerue-Matos. 2015. Molecularly imprinted sensor for voltammetric detection of norfloxacin. Sens. Actuat. B Chem. 219: 301–307.

Decher, G., J. Hong and J. Schmitt. 1992. Buildup of ultrathin multilayer films by a self-assembly process: III. Consecutively alternating adsorption of anionic and cationic polyelectrolytes on charged surfaces. Thin Solid Films 210: 831–835.

Dong, R., Y. Zhou, X. Huang, X. Zhu, Y. Lu and J. Shen. 2015. Functional supramolecular polymers for biomedical applications. Adv. Mater. 27: 498–526.

Downes, K.J., N.R. Patil, M.B. Rao, R. Koralkar, W.T. Harris, J.P. Clancy et al. 2015. Risk factors for acute kidney injury during aminoglycoside therapy in patients with cystic fibrosis. Pediatr. Nephrol. 30: 1879–1888.

Dreyer, D.R., A.D. Todd and C.W. Bielawski. 2014. Harnessing the chemistry of graphene oxide. Chem. Soc. Rev. 43: 5288–5301.

Du, J., S. Pei, L. Ma and H.M. Cheng. 2014. 25th anniversary article: carbon nanotube- and graphene-based transparent conductive films for optoelectronic devices. Adv. Mater. 26: 1958–1991.

Er, E., H. Çelikkan, N. Erk and M.L. Aksu. 2015. A new generation electrochemical sensor based on graphene nanosheets/gold nanoparticles/nafion nanocomposite for determination of silodosin. Electrochimi. Acta 157: 252–257.

Fan, H., L. Wang, K. Zhao, N. Li, Z. Shi, Z. Ge et al. 2010. Fabrication, mechanical properties, and biocompatibility of graphene-reinforced chitosan composites. Biomacromolecules 11: 2345–2351.

Fugallo, G., A. Cepellotti, L. Paulatto, M. Lazzeri, N. Marzari and F. Mauri. 2014. Thermal conductivity of graphene and graphite: collective excitations and mean free paths. Nano Lett. 14: 6109–6114.

Gao, J., S. Zhang, M. Liu, Y. Tai, X. Song, Y. Qian et al. 2016. Synergistic combination of cyclodextrin edge-functionalized graphene and multiwall carbon nanotubes as conductive bridges toward enhanced sensing response of supramolecular recognition. Electrochimi. Acta 187: 364–374.

Gao, Z., W. Yang, J. Wang, H. Yan, Y. Yao, J. Ma et al. 2013. Electrochemical synthesis of layer-by-layer reduced graphene oxide sheets/polyaniline nanofibers composite and its electrochemical performance. Electrochimi. Acta 91: 185–194.

Goncalves, G., P.A. Marques, C.M. Granadeiro, H.I. Nogueira, M. Singh and J. Gracio. 2009. Surface modification of graphene nanosheets with gold nanoparticles: the role of oxygen moieties at graphene surface on gold nucleation and growth. Chem. Mater. 21: 4796–4802.

Gong, Y., S. Yang, L. Zhan, L. Ma, R. Vajtai and P.M. Ajayan. 2014. A bottom-up approach to build 3D architectures from nanosheets for superior lithium storage. Adv. Funct. Mater. 24: 125–130.

Guo, Y., Y. Chen, Q. Zhao, S. Shuang and C. Dong. 2011a. Electrochemical sensor for ultrasensitive determination of doxorubicin and methotrexate based on cyclodextrin-graphene hybrid nanosheets. Electroanalysis 23: 2400–2407.

Guo, Y., S. Guo, J. Li, E. Wang and S. Dong. 2011b. Cyclodextrin–graphene hybrid nanosheets as enhanced sensing platform for ultrasensitive determination of carbendazim. Talanta 84: 60–64.

Hassan, H.M., V. Abdelsayed, S.K. Abd El Rahman, K.M. AbouZeid, J. Terner, M.S. El-Shall et al. 2009. Microwave synthesis of graphene sheets supporting metal nanocrystals in aqueous and organic media. J. Mater. Chem. 19: 3832–3837.

He, K. and L. Blaney. 2015. Systematic optimization of an SPE with HPLC-FLD method for fluoroquinolone detection in wastewater. J. Hazard. Mater. 282: 96–105.

Hu, H., X. Wang, C. Xu, J. Wang, L. Wan, M. Zhang et al. 2012. Microwave-assisted synthesis of graphene nanosheets–gold nanocomposites with enhancing electrochemical response. Fuller. Nanotub. Car. N. 20: 31–40.

Hu, R., W. Sun, Y. Chen, M. Zeng and M. Zhu. 2014. Silicon/graphene based nanocomposite anode: large-scale production and stable high capacity for lithium ion batteries. J. Mater. Chem. A 2: 9118–9125.

Iliut, M., C. Leordean, V. Canpean, C.M. Teodorescu and S. Astilean. 2013. A new green, ascorbic acid-assisted method for versatile synthesis of Au–graphene hybrids as efficient surface-enhanced Raman scattering platforms. J. Mater. Chem. C 1: 4094–4104.

Jasuja, K., J. Linn, S. Melton and V. Berry. 2010. Microwave-reduced uncapped metal nanoparticles on graphene: Tuning catalytic, electrical, and raman properties. J. of Phys. Chem. Lett. 1: 1853–1860.

Jiang, B., M. Wang, Y. Chen, J. Xie and Y. Xiang. 2012. Highly sensitive electrochemical detection of cocaine on graphene/AuNP modified electrode via catalytic redox-recycling amplification. Biosens. Bioelectron. 32: 305–308.

Jiang, Z., G. Li and M. Zhang. 2016. Electrochemical sensor based on electro-polymerization of β-cyclodextrin and reduced-graphene oxide on glassy carbon electrode for determination of gatifloxacin. Sens. Actuat. B Chem. 228: 59–65.

Jiao, Y., Y. Zheng, M. Jaroniec and S.Z. Qiao. 2014. Origin of the electrocatalytic oxygen reduction activity of graphene-based catalysts: a roadmap to achieve the best performance. J. Am. Chem. Soc. 136: 4394–4403.

Kalambate, P.K., B.J. Sanghavi, S.P. Karna and A.K. Srivastava. 2015. Simultaneous voltammetric determination of paracetamol and domperidone based on a graphene/platinum nanoparticles/nafion composite modified glassy carbon electrode. Sens. Actuat. B Chem. 213: 285–294.

Kang, J., S.J. Park, H.C. Park, V. Gedi, B. So and K.J. Lee. 2014. Multiresidue determination of ten nonsteroidal anti-inflammatory drugs in bovine, porcine, and chicken liver tissues by HPLC-MS/MS. Appl. Biochem. Biotechnol. 174: 1–5.

Karimi-Maleh, H., F. Tahernejad-Javazmi, N. Atar, M.L. t. Yola, V.K. Gupta and A.A. Ensafi. 2015. A novel DNA biosensor based on a pencil graphite electrode modified with polypyrrole/functionalized multiwalled carbon nanotubes for determination of 6-mercaptopurine anticancer drug. Ind. Eng. Chem. Res. 54: 3634–3639.

Kim, H., Y. Miura and C.W. Macosko. 2010. Graphene/polyurethane nanocomposites for improved gas barrier and electrical conductivity. Chem. Mater. 22: 3441–3450.

Konda, S.K. and A. Chen. 2015. One-step synthesis of Pd and reduced graphene oxide nanocomposites for enhanced hydrogen sorption and storage. Electrochem. Commun. 60: 148–152.

Kuila, T., S. Bose, A.K. Mishra, P. Khanra, N.H. Kim and J.H. Lee. 2012. Chemical functionalization of graphene and its applications. Prog. Mater. Sci. 57: 1061–1105.

Kuilla, T., S. Bhadra, D. Yao, N.H. Kim, S. Bose and J.H. Lee. 2010. Recent advances in graphene based polymer composites. Prog. Poly. Sci. 35: 1350–1375.

Leng, J., P. Li, L. Bai, Y. Peng, Y. Yu and L. Lu. 2015. Facile synthesis of Pd nanoparticles-graphene oxide hybrid and its application to the electrochemical determination of rutin. Int. J. Electrochem. Sci. 10: 8522–8530.

Li, F., Y. Guo, X. Sun and X. Wang. 2014a. Aptasensor based on thionine, graphene–polyaniline composite film, and gold nanoparticles for kanamycin detection. Eur. Food Res. Technol. 239: 227–236.

Li, J., S. Guo, Y. Zhai and E. Wang. 2009. High-sensitivity determination of lead and cadmium based on the Nafion-graphene composite film. Anal. Chim. Acta 649: 196–201.

Li, J., J. Liu, G. Tan, J. Jiang, S. Peng, M. Deng et al. 2014b. High-sensitivity paracetamol sensor based on Pd/graphene oxide nanocomposite as an enhanced electrochemical sensing platform. Biosens. Bioelectron. 54: 468–475.

Li, J., W. Lei, Y. Xu, Y. Zhang, M. Xia and F. Wang. 2015a. Fabrication of polypyrrole-grafted nitrogen-doped graphene and its application for electrochemical detection of paraquat. Electrochimi. Acta 174: 464–471.

Li, S.-J., G.-Y. Zhao, R.X. Zhang, Y.L. Hou, L. Liu and H. Pang. 2013a. A sensitive and selective nitrite sensor based on a glassy carbon electrode modified with gold nanoparticles and sulfonated graphene. Microchimi. Acta 180: 821–827.

Li, S.-W., H.P. Song and Y. Leng. 2014c. Rapid determination of lovastatin in the fermentation broth of aspergillus terreus using dual-wavelength UV spectrophotometry. Pharm. Biol. 52: 129–135.

Li, S., X. Wu, Q. Zhang and P. Li. 2016. Synergetic dual recognition and separation of the fungicide carbendazim by using magnetic nanoparticles carrying a molecularly imprinted polymer and immobilized β-cyclodextrin. Microchimi. Acta 183: 1–7.

Li, Y., K. Li, G. Song, J. Liu, K. Zhang and B. Ye. 2013b. Electrochemical behavior of codeine and its sensitive determination on graphene-based modified electrode. Sens. Actuat. B Chem. 182: 401–407.

Li, Y., Y. Liu, J. Liu, J. Liu, H. Tang, C. Cao et al. 2015b. Molecularly imprinted polymer decorated nanoporous gold for highly selective and sensitive electrochemical sensors. Sci. Rep. 5: 7699–7706.

Li, Z., L. Zhang, X. Huang, L. Ye and S. Lin. 2014d. Shape-controlled synthesis of Pt nanoparticles via integration of graphene and β-cyclodextrin and using as a noval electrocatalyst for methanol oxidation. Electrochimi. Acta 121: 215–222.

Lian, W., S. Liu, J. Yu, X. Xing, J. Li, M. Cui et al. 2012. Electrochemical sensor based on gold nanoparticles fabricated molecularly imprinted polymer film at chitosan–platinum nanoparticles/graphene–gold nanoparticles double nanocomposites modified electrode for detection of erythromycin. Biosens. Bioelectron. 38: 163–169.

Liang, J., Y. Wang, Y. Huang, Y. Ma, Z. Liu, J. Cai et al. 2009. Electromagnetic interference shielding of graphene/epoxy composites. Carbon 47: 922–925.

Liu, C., K. Wang, S. Luo, Y. Tang and L. Chen. 2011a. Direct electrodeposition of graphene enabling the one-step synthesis of graphene–metal nanocomposite films. Small 7: 1203–1206.

Liu, K., J. Wei and C. Wang. 2011b. Sensitive detection of rutin based on β-cyclodextrin@ chemically reduced graphene/nafion composite film. Electrochimi. Acta 56: 5189–5194.

Liu, S., J. Tian, L. Wang and X. Sun. 2011c. Microwave-assisted rapid synthesis of Ag nanoparticles/graphene nanosheet composites and their application for hydrogen peroxide detection. J. Nanopart. Res. 13: 4539 4548.

Liu, Y., H. Wang, J. Zhou, L. Bian, E. Zhu, J. Hai et al. 2013. Graphene/polypyrrole intercalating nanocomposites as supercapacitors electrode. Electrochimi. Acta 112: 44–52.

Liu, Z., Q. Xue and Y. Guo. 2016. Sensitive electrochemical detection of rutin and isoquercitrin based on SH-β-cyclodextrin functionalized graphene-palladium nanoparticles. Biosens. Bioelectron. (in press).

Lue, S.J., Y.L. Pai, C.M. Shih, M.C. Wu and S.M. Lai. 2015. Novel bilayer well-aligned nafion/graphene oxide composite membranes prepared using spin coating method for direct liquid fuel cells. J. Membr. Sci. 493: 212–223.

Mahmood, N., C. Zhang, H. Yin and Y. Hou. 2014. Graphene-based nanocomposites for energy storage and conversion in lithium batteries, supercapacitors and fuel cells. J. Mater. Chem. A 2: 15–32.

Manjunatha, R., D.H. Nagaraju, G.S. Suresh, J.S. Melo, S.F. D'Souza and T.V. Venkatesha. 2011. Electrochemical detection of acetaminophen on the functionalized MWCNTs modified electrode using layer-by-layer technique. Electrochimi. Acta 56: 6619–6627.

Martín, A. and A. Escarpa. 2014. Graphene: the cutting–edge interaction between chemistry and electrochemistry. TrAC Trends Anal. Chem. 56: 13–26.

Mayorov, A.S., R.V. Gorbachev, S.V. Morozov, L. Britnell, R. Jalil, L.A. Ponomarenko et al. 2011. Micrometer-scale ballistic transport in encapsulated graphene at room temperature. Nano Lett. 11: 2396–2399.

McCreery, R.L. 2008. Advanced carbon electrode materials for molecular electrochemistry. Chem. Rev. 108: 2646–2687.

Meng, F., W. Lu, Q. Li, J.H. Byun, Y. Oh and T.W. Chou. 2015. Graphene-based fibers: a review. Adv. Mater. 27: 5113–5131.

Michinobu, T. 2015. Click functionalization of aromatic polymers for organic electronic device applications. Macromol. Chem. Phys. 216: 1387–1395.

Modongo, C., R.S. Sobota, B. Kesenogile, R. Ncube, G. Sirugo, S.M. Williams et al. 2014. Successful MDR-TB treatment regimens including amikacin are associated with high rates of hearing loss. BMC Infect. Dis. 14: 542.

Moradi Golsheikh, A., N.M. Huang, H.N. Lim, R. Zakaria and C.Y. Yin. 2013. One-step electrodeposition synthesis of silver-nanoparticle-decorated graphene on indium-tin-oxide for enzymeless hydrogen peroxide detection. Carbon 62: 405–412.

Moreno-Castilla, C. 2004. Adsorption of organic molecules from aqueous solutions on carbon materials. Carbon 42: 83–94.

Nguyen, K.T. and Y. Zhao. 2014. Integrated graphene/nanoparticle hybrids for biological and electronic applications. Nanoscale 6: 6245–6266.

Niu, X., W. Yang, G. Wang, J. Ren, H. Guo and J. Gao. 2013. A novel electrochemical sensor of bisphenol: a based on stacked graphene nanofibers/gold nanoparticles composite modified glassy carbon electrode. Electrochimi. Acta 98: 167–175.

Novoselov, K.S., A.K. Geim, S. Morozov, D. Jiang, Y. Zhang, S. Dubonos et al. 2004. Electric field effect in atomically thin carbon films. Science 306: 666–669.

Perreault, F., A.F. de Faria and M. Elimelech. 2015. Environmental applications of graphene-based nanomaterials. Chem. Soc. Rev. 44: 5861–5896.

Ponnusamy, V.K., V. Mani, S.M. Chen, W.T. Huang and J.F. Jen. 2014. Rapid microwave assisted synthesis of graphene nanosheets/polyethyleneimine/gold nanoparticle composite and its application to the selective electrochemical determination of dopamine. Talanta 120: 148–157.

Pruneanu, S., F. Pogacean, A.R. Biris, S. Ardelean, V. Canpean, G. Blanita et al. 2011. Novel graphene-gold nanoparticle modified electrodes for the high sensitivity electrochemical spectroscopy detection and analysis of carbamazepine. J. Phys. Chem. C 115: 23387–23394.

Pruneanu, S., F. Pogacean, A.R. Biris, M. Coros, F. Watanabe, E. Dervishi et al. 2013. Electro-catalytic properties of graphene composites containing gold or silver nanoparticles. Electrochimi. Acta 89: 246–252.

Pumera, M. 2010. Graphene-based nanomaterials and their electrochemistry. Chem. Soc. Rev. 39: 4146–4157.

Pumera, M., A. Ambrosi, A. Bonanni, E.L.K. Chng and H.L. Poh. 2010. Graphene for electrochemical sensing and biosensing. TrAC Trends Anal. l Chem. 29: 954–965.

Pumera, M. 2013. Electrochemistry of graphene, graphene oxide and other graphenoids: review. Electrochem. Commun. 36: 14–18.

Qi, X.Y., D. Yan, Z. Jiang, Y.K. Cao, Z.Z. Yu, F. Yavari et al. 2011. Enhanced electrical conductivity in polystyrene nanocomposites at ultra-low graphene content. ACS Appl. Mater. Interfaces 3: 3130–3133.

Qian, T., C. Yu, X. Zhou, S. Wu and J. Shen. 2014. Au nanoparticles decorated polypyrrole/reduced graphene oxide hybrid sheets for ultrasensitive dopamine detection. Sens. Actuat. B Chem. 193: 759–763.

Qiu, X., W. Zeng, W. Yu, Y. Xue, Y. Pang, X. Li et al. 2015. Alkyl chain cross-linked sulfobutylated lignosulfonate: a highly efficient dispersant for carbendazim suspension concentrate. ACS Sustain. Chem. Eng. 3: 1551–1557.

Quinn, J.F., A.P. Johnston, G.K. Such, A.N. Zelikin and F. Caruso. 2007. Next generation, sequentially assembled ultrathin films: beyond electrostatics. Chem. Soc. Rev. 36: 707–718.

Radhapyari, K., P. Kotoky, M.R. Das and R. Khan. 2013. Graphene–polyaniline nanocomposite based biosensor for detection of antimalarial drug artesunate in pharmaceutical formulation and biological fluids. Talanta 111: 47–53.

Raj, M.A. and S.A. John. 2015. Assembly of gold nanoparticles on graphene film via electroless deposition: spontaneous reduction of Au 3+ ions by graphene film. RSC Adv. 5: 4964–4971.

Rastgar, S. and S. Shahrokhian. 2014. Nickel hydroxide nanoparticles-reduced graphene oxide nanosheets film: layer-by-layer electrochemical preparation, characterization and rifampicin sensory application. Talanta 119: 156–163.

Ruecha, N., R. Rangkupan, N. Rodthongkum and O. Chailapakul. 2014. Novel paper-based cholesterol biosensor using graphene/polyvinylpyrrolidone/polyaniline nanocomposite. Biosens. Bioelectron. 52: 13–19.

Ruecha, N., N. Rodthongkum, D.M. Cate, J. Volckens, O. Chailapakul and C.S. Henry. 2015. Sensitive electrochemical sensor using a graphene–polyaniline nanocomposite for simultaneous detection of Zn (II), Cd (II), and Pb (II). Anal. Chimi. Acta 874: 40–48.

Saha, K., S.S. Agasti, C. Kim, X. Li and V.M. Rotello. 2012. Gold nanoparticles in chemical and biological sensing. Chem. Rev. 112: 2739–2779.

Salavagione, H.J., A.M. Díez-Pascual, E. Lázaro, S. Vera and M.A. Gómez-Fatou. 2014. Chemical sensors based on polymer composites with carbon nanotubes and graphene: the role of the polymer. J. Mater. Chem. A 2: 14289–14328.

Sanghavi, B.J., P.K. Kalambate, S.P. Karna and A.K. Srivastava. 2014a. Voltammetric determination of sumatriptan based on a graphene/gold nanoparticles/nafion composite modified glassy carbon electrode. Talanta 120: 1–9.

Sanghavi, B.J., W. Varhue, J.L. Chávez, C.F. Chou and N.S. Swami. 2014b. Electrokinetic preconcentration and detection of neuropeptides at patterned graphene-modified electrodes in a nanochannel. Anal. Chem. 86: 4120–4125.

Sanghavi, B.J., O.S. Wolfbeis, T. Hirsch and N.S. Swami. 2015. Nanomaterial-based electrochemical sensing of neurological drugs and neurotransmitters. Microchimi. Acta 182: 1–41.

Sarker, A.K. and J.D. Hong. 2012. Layer-by-layer self-assembled multilayer films composed of graphene/polyaniline bilayers: high-energy electrode materials for supercapacitors. Langmuir 28: 12637–12646.

Shahrokhian, S., N. Hosseini-Nassab and M. Ghalkhani. 2014. Construction of Pt nanoparticle-decorated graphene nanosheets and carbon nanospheres nanocomposite-modified electrodes: application to ultrasensitive electrochemical determination of cefepime. RSC Adv. 4: 7786–7794.

Shahrokhian, S., L. Naderi and M. Ghalkhani. 2016. Modified glassy carbon electrodes based on carbon nanostructures for ultrasensitive electrochemical determination of furazolidone. Mater. Sci. Eng. C 61: 842–850.

Shang, N.G., P. Papakonstantinou, M. McMullan, M. Chu, A. Stamboulis, A. Potenza et al. 2008. Catalyst-free efficient growth, orientation and biosensing properties of multilayer graphene nanoflake films with sharp edge planes. Adv. Funct. Mater. 18: 3506–3514.

Sharma, P., G. Darabdhara, T.M. Reddy, A. Borah, P. Bezboruah, P. Gogoi et al. 2013. Synthesis, characterization and catalytic application of Au NPs-reduced graphene oxide composites material: an eco-friendly approach. Catal. Commun. 40: 139–144.

Shehzad, F., M. Daud and M.A. Al-Harthi. 2016. Synthesis, characterization and crystallization kinetics of nanocomposites prepared by *in situ* polymerization of ethylene and graphene. J. Therm. Anal. Calorim. 123: 1501–1511.

Shervedani, R.K. and A. Amini. 2014. Novel graphene-gold hybrid nanostructures constructed via sulfur modified graphene: preparation and characterization by surface and electrochemical techniques. Electrochimi. Acta 121: 376–385.

Stankovich, S., D.A. Dikin, G.H. Dommett, K.M. Kohlhaas, E.J. Zimney, E.A. Stach et al. 2006. Graphene-based composite materials. Nature 442: 282–286.

Sultan, M.A., H.M. Maher, N.Z. Alzoman, M.M. Alshehri, M.S. Rizk, M.S. Elshahed et al. 2012. Capillary electrophoretic determination of antimigraine formulations containing caffeine, ergotamine, paracetamol and domperidone or metoclopramide. J. Chromatogr. Sci. 51: 502–510.

Sun, W. and Y. Wang. 2014. Graphene-based nanocomposite anodes for lithium-ion batteries. Nanoscale 6: 11528–11552.

Sundaram, R.S., C. Gómez-Navarro, K. Balasubramanian, M. Burghard and K. Kern. 2008. Electrochemical modification of graphene. Adv. Mater. 20: 3050–3053.

Tai, Z., H. Ma, B. Liu, X. Yan and Q. Xue. 2012. Facile synthesis of Ag/GNS-g-PAA nanohybrids for antimicrobial applications. Colloids Surf. B 89: 147–151.

Tang, L., Y. Wang, Y. Li, H. Feng, J. Lu and J. Li. 2009. Preparation, structure, and electrochemical properties of reduced graphene sheet films. Adv. Funct. Mater. 19: 2782–2789.

Thammasoontaree, N., P. Rattanarat, N. Ruecha, W. Siangproh, N. Rodthongkum and O. Chailapakul. 2014. Ultra-performance liquid chromatography coupled with graphene/polyaniline nanocomposite modified electrode for the determination of sulfonamide residues. Talanta 123: 115–121.

Thapliyal, N., R.V. Karpoormath and R.N. Goyal. 2015. Electroanalysis of antitubercular drugs in pharmaceutical dosage forms and biological fluids: a review. Anal. Chimi. Acta 853: 59–76.

Tian, T., X. Shi, L. Cheng, Y. Luo, Z. Dong, H. Gong et al. 2014. Graphene-based nanocomposite as an effective, multifunctional, and recyclable antibacterial agent. ACS Appl. Mater. Interfaces 6: 8542–8548.

Tian, X., C. Cheng, H. Yuan, J. Du, D. Xiao, S. Xie et al. 2012. Simultaneous determination of l-ascorbic acid, dopamine and uric acid with gold nanoparticles–β-cyclodextrin–graphene-modified electrode by square wave voltammetry. Talanta 93: 79–85.

ul Hasan, K., M.H. Asif, M.U. Hassan, M.O. Sandberg, O. Nur, M. Willander et al. 2015. A miniature graphene-based biosensor for intracellular glucose measurements. Electrochimi. Acta 174: 574–580.

Ulubay, Ş. and Z. Dursun. 2010. Cu nanoparticles incorporated polypyrrole modified GCE for sensitive simultaneous determination of dopamine and uric acid. Talanta 80: 1461–1466.

Verdejo, R., M.M. Bernal, L.J. Romasanta and M.A. Lopez-Manchado. 2011. Graphene filled polymer nanocomposites. J. Mater. Chem. 21: 3301–3310.

Vlassiouk, I.V., G. Polizos, R. Cooper, I.N. Ivanov, J.K. Keum, F. Paulauskas et al. 2015. Strong and electrically conductive graphene based composite fibers and laminates. ACS Appl. Mater. Interfaces 7: 10702–10709.

Wang, F., L. Zhu and J. Zhang. 2014. Electrochemical sensor for levofloxacin based on molecularly imprinted polypyrrole–graphene–gold nanoparticles modified electrode. Sens. Actuat. B Chem. 192: 642–647.

Wang, L., M. Zhang, D. Zhang and L. Zhang. 2016. New approach for the simultaneous determination fungicide residues in food samples by using carbon nanofiber packed microcolumn coupled with HPLC. Food Control 60: 1–6.

Wong, C.H.A., A. Ambrosi and M. Pumera. 2012. Thermally reduced graphenes exhibiting a close relationship to amorphous carbon. Nanoscale 4: 4972–4977.

Wu, S., X. Lan, L. Cui, L. Zhang, S. Tao, H. Wang et al. 2011. Application of graphene for preconcentration and highly sensitive stripping voltammetric analysis of organophosphate pesticide. Anal. Chimi. Acta 699: 170–176.

Xi, Q., X. Chen, D.G. Evans and W. Yang. 2012. Gold nanoparticle-embedded porous graphene thin films fabricated via layer-by-layer self-assembly and subsequent thermal annealing for electrochemical sensing. Langmuir 28: 9885–9892.

Xing, F., G.X. Meng, Q. Zhang, L.T. Pan, P. Wang, Z.B. Liu et al. 2014. Ultrasensitive flow sensing of a single cell using graphene-based optical sensors. Nano Lett. 14: 3563–3569.

Xu, C., J. Sun and L. Gao. 2011. Synthesis of novel hierarchical graphene/polypyrrole nanosheet composites and their superior electrochemical performance. J. Mater. Chem. 21: 11253–11258.

Yan, J., T. Wei, B. Shao, Z. Fan, W. Qian, M. Zhang et al. 2010. Preparation of a graphene nanosheet/polyaniline composite with high specific capacitance. Carbon 48: 487–493.

Yang, S., C. Shen, Y. Liang, H. Tong, W. He, X. Shi et al. 2011. Graphene nanosheets-polypyrrole hybrid material as a highly active catalyst support for formic acid electro-oxidation. Nanoscale 3: 3277–3284.

Yang, S., G. Li, J. Zhao, H. Zhu and L. Qu. 2014. Electrochemical preparation of Ag nanoparticles/poly(methylene blue) functionalized graphene nanocomposite film modified electrode for sensitive determination of rutin. J. Electroanal. Chem. 717-718: 225–230.

Ye, X., Y. Du, D. Lu and C. Wang. 2013. Fabrication of β-cyclodextrin-coated poly(diallyldimethylammonium chloride)-functionalized graphene composite film modified glassy carbon-rotating disk electrode and its application for simultaneous electrochemical determination colorants of sunset yellow and tartrazine. Anal. Chimi. Acta 779: 22–34.

Yi, X., S. Bayen, B.C. Kelly, X. Li and Z. Zhou. 2015. Improved detection of multiple environmental antibiotics through an optimized sample extraction strategy in liquid chromatography-mass spectrometry analysis. Anal. Bioanal. Chem. 407: 9071–9083.

Yin, H., Q. Ma, Y. Zhou, S. Ai and L. Zhu. 2010a. Electrochemical behavior and voltammetric determination of 4-aminophenol based on graphene–chitosan composite film modified glassy carbon electrode. Electrochimi. Acta 55: 7102–7108.

Yin, H., Y. Zhou, Q. Ma, S. Ai, P. Ju, L. Zhu et al. 2010b. Electrochemical oxidation behavior of guanine and adenine on graphene–nafion composite film modified glassy carbon electrode and the simultaneous determination. Process Biochem. 45: 1707–1712.

You, J.M., D. Kim, S.K. Kim, M.S. Kim, H.S. Han and S. Jeon. 2013. Novel determination of hydrogen peroxide by electrochemically reduced graphene oxide grafted with aminothiophenol–Pd nanoparticles. Sens. Actuat. B Chem. 178: 450–457.

Young, R.J., I.A. Kinloch, L. Gong and K.S. Novoselov. 2012. The mechanics of graphene nanocomposites: a review. Compos. Sci. Technol. 72: 1459–1476.

Yu, S.H. and G.C. Zhao. 2011. Preparation of platinum nanoparticles-graphene modified electrode and selective determination of rutin. Int. J. Electrochem. 2012: 1–6.

Zeigler, D.F., S.L. Candelaria, K.A. Mazzio, T.R. Martin, E. Uchaker, S.-L. Suraru et al. 2015. N-type hyperbranched polymers for supercapacitor cathodes with variable porosity and excellent electrochemical stability. Macromolecules 48: 5196–5203.

Zhai, H., Z. Liang, Z. Chen, H. Wang, Z. Liu, Z. Su et al. 2015. Simultaneous detection of metronidazole and chloramphenicol by differential pulse stripping voltammetry using a silver nanoparticles/sulfonate functionalized graphene modified glassy carbon electrode. Electrochimi. Acta 171: 105–113.

Zhang, M., J. Xie, Q. Sun, Z. Yan, M. Chen, J. Jing et al. 2013. *In situ* synthesis of palladium nanoparticle on functionalized graphene sheets at improved performance for ethanol oxidation in alkaline media. Electrochimi. Acta 111: 855–861.

Zhang, X., Y. Wei and Y. Ding. 2014a. Electrocatalytic oxidation and voltammetric determination of ciprofloxacin employing poly (alizarin red)/graphene composite film in the presence of ascorbic acid, uric acid and dopamine. Anal. Chimi. Acta 835: 29–36.

Zhao, Y., Y. Du, D. Lu, L. Wang, D. Ma, T. Ju et al. 2014. Sensitive determination of vanillin based on an arginine functionalized graphene film. Anal. Methods 6: 1753–1758.

Zheng, M., F. Gao, Q. Wang, X. Cai, S. Jiang, L. Huang et al. 2013. Electrocatalytical oxidation and sensitive determination of acetaminophen on glassy carbon electrode modified with graphene–chitosan composite. Mater. Sci. Eng. C 33: 1514–1520.

Zhou, M., Y. Zhai and S. Dong. 2009. Electrochemical sensing and biosensing platform based on chemically reduced graphene oxide. Anal. Chem. 81: 5603–5613.

Zhu, C., S. Guo, Y. Zhai and S. Dong. 2010a. Layer-by-layer self-assembly for constructing a graphene/platinum nanoparticle three-dimensional hybrid nanostructure using ionic liquid as a linker. Langmuir 26: 7614–7618.

Zhu, J., D. Yang, Z. Yin, Q. Yan and H. Zhang. 2014. Graphene and graphene-based materials for energy storage applications. Small 10: 3480–3498.

Zhu, Y., S. Murali, W. Cai, X. Li, J.W. Suk, J.R. Potts et al. 2010b. Graphene and graphene oxide: synthesis, properties, and applications. Adv. Mater. 22: 3906–3924.

Zhu, Z., M. Su, L. Ma, L. Ma, D. Liu and Z. Wang. 2013. Preparation of graphene oxide–silver nanoparticle nanohybrids with highly antibacterial capability. Talanta 117: 449–455.

17

Towards CNTs; Functionalization and Their Sensing Applications

Nada F. Atta, * *Hagar K. Hassan* and *Ahmed Galal*

Introduction

CNTs were discovered in 1991 by a Japanese scientist, Sumio Iijima (Iijima 1991). Iijima noticed CNTs formation in the soot of the arc discharge. The morphology of CNTs was then confirmed using transmission electron microscope (TEM). However, the initial history of CNT started in 1970 by Morinobu Endo through his preparation for planned carbon filaments but unfortunately, the required measurements for completion his work were not available in his time. Hence, the first complete discovery of CNT could be credited to Sumio Iijima (Chen et al. 2011). In 1993, CNTs began to attract the attention of researchers due to their unique structural, mechanical, electrical and electrochemical properties. Generally, CNT can be considered as a rolled-up graphene sheet and according to the number of graphene sheets, CNT is classified into single-walled, double-walled and multiple-walled CNT (Chen et al. 2011). CNTs have the dimension of outer diameter as 2–20 nm and 1 nm for its inner diameter. CNTs can reach tens of microns in length and 340 nm inter-tubular distances that is slightly higher than the inter-planar distance in graphite (Endo et al. 1993). CNTs have three different structures based on the position of their carbon-carbon bonds with respect to the tube axis (chiral axis) that is denoted by chiral indices (n, m) see Figure 1 (Choudhary and Gupta 2011). For example, if CNT has (n, 0) structure, i.e., carbon-carbon bonds are parallel to the tube axis in a zig-zag pattern, it is called zig-zag CNT. If carbon-carbon bonds are perpendicular to the chiral axis in (n, n) structure, CNT will have an armchair structure. Both zigzag and armchair structures do not have a mirror-image whereas if it has a mirror image, this will be a chiral CNT with chiral indices (n, m) where, n ≠ m and not equal zero (Balasubramanian and Burghard 2005).

Mainly, there are three common methods for CNT preparation including chemical vapor deposition (CVD), laser ablation and arch-discharge method. During arc discharge method, two graphite rods are put close to each other and a high potential difference is applied between them. This results in the formation of CNT on the cathode and consuming of the positive electrode (anode).

The disadvantage of the arc-discharge method is the formation of carbon soot impurities, which contain amorphous carbon, anions and fullerenes. Additionally, this method is not suitable for the large-scale production of CNT. Another method of CNT preparation is the laser ablation method. It is based on

Department of Chemistry, Faculty of Science, Cairo University, Giza 12613, Egypt.
* Corresponding author: nada_fah1@yahoo.com

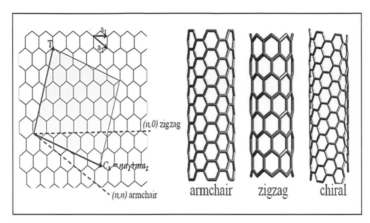

Figure 1. Rolling of graphene sheets to form different types of CNT. (Reprinted with permission from Choudhary and Gupta 2011. Copyright © In Tech.)

applying an intense laser beam to a carbonaceous target in the presence of a small amount of metal catalysts such as nickel and cobalt in an inert atmosphere. This method usually leads to the formation of SWCNTs that can be collected on a cold finger. The most widely used method for CNT production is chemical vapor deposition (CVD). In this method, a mixture of hydrocarbons, metal catalyst, and an inert gas are introduced into the reaction chamber while the decomposition of hydrocarbons occurs at temperatures 700–900°C, resulting in the deposition of CNT on the substrate (Choudhary and Gupta 2011).

Covalent and non-covalent functionalization of CNTs

CNT, in principal, is chemically inert and has very poor solubility as well as high tendency to form aggregates. Hence, further functionalization of CNT is of prime priority in order to increase its potential in several applications in science and technology. On the other hand, understanding the core of functionalization and selecting the suitable functional group as well as the suitable method of functionalization is very important to achieve the desired goal. Most of the researchers classified CNT functionalization processes into covalent and non-covalent based on whether a true covalent bond is formed between CNT and the functional entities or not. Such covalent functionalization can be performed through many routes including thermal, photochemical or electrochemical methods (Balasubramanian and Burghard 2005). Electrochemical functionalization can be achieved when a constant current (galvanostatic) or a constant potential (potentiostatic) is applied to CNT. For example, the electrochemical coupling of aromatic diazonium salts and their attachments to SWCNTs was successfully achieved by Bahr et al. (Bahr et al. 2001). Photochemical functionalization is usually accomplished by using photo-irradiation to generate a reactive species such as nitrenes that in their turns attach to the sidewalls of CNT, as mentioned elsewhere (Moghaddam et al. 2004).

On the other side, non-covalent functionalization of CNT attracts much attention because it does not induce any structural transformations or damages and is much simpler than covalent functionalization (Chen et al. 2011). Non covalent functionalization can be carried out through weaker interactions such as van der Waals forces, hydrogen bonding or π–π interaction. Polymer wrapping, metal binding and biomolecule binding (Meng et al. 2009) are examples of non-covalent functionalization of CNT. Generally, the chemical functionalization of CNT is energetically favored compared to the regular planner graphene sheet due to the presence of a curvature in the former structure that facilitates the transformation from trigonal-planner geometry into a tetrahedral structure. This transformation is accompanied by the transformation of sp^2 to sp^3 hybridized carbon atom as a result of the addition reactions. The presence of defects in CNT makes the addition reaction more favored. Stone-Wales defect, the two-pairs of five-membered and seven-membered rings 7-5-5-7 defect, leads to deformation in the graphitic sidewall of CNT and thereby increases the curvature and facilitates its chemical functionalization (Balasubramanian and Burghard 2005).

Covalent functionalization of CNTs and adding functional groups

Acid functionalization of CNTs is an example of covalent functionalization. The purpose of this method is to generate defects on the sidewalls and tube tips; those defects represent sites for further functionalization (Li and Xing 2007). Ultrasonic treatment in a mixture of concentrated nitric and sulfuric acid leads to opening of the tube caps and formation of holes in the sidewalls as well as the formation of short tubes in the range of 100–300 nm. While under a less vigorous condition such as refluxing in nitric acid, shortening of the tubes is minimized (Chen et al. 1998). Chiang et al. (Chiang et al. 2011) stated that carboxylic group attachment to CNT can be achieved by the treatment of CNT with the acid mixture for two days. Furthermore, the treatment of CNT with only nitric acid causes the tube caps to open while maintaining the pristine mechanical and electronic properties of CNT (Chen et al. 2011). It is worth mentioning that introduction of carboxyl groups into CNT is considered as the first step to further functionalization of the tubes with ester or amide groups. Carboxyl-functionalized CNT can also be used as a raw-material for doping of nucleic acids, enzymes, metal nanoparticles, semiconductors and polymeric materials. A schematic diagram showing the further functionalization of carboxylic group-functionalized CNT is shown in Figure 2 (Balasubramanian and Burghard 2005). Another advantage of carboxylic group functionalization is to separate the individual CNT by decreasing van der Walls interaction.

Moreover, a variety of solution-based chemistry can be carried out on carboxyl-functionalized CNT by increasing the solubility of CNT in aqueous media (Balasubramanian and Burghard 2005). However, such acid treatment usually reduces the electrical conductivity of CNT. So, it is important to develop more effective functionalization methods for CNT without sacrificing its electrical and mechanical properties which represents the major challenge in CNT applications. By using bifunctional reagent such as α-ω diamine, the nanotubes can be covalently cross-linked with each other (Stevens et al. 2003). Another example for covalent functionalization is introducing a hydroxyl group into CNT. This can be performed by several routes such as one-step hydroxylation of CNT that can be performed by free radical addition of 4,4'-azobis(4-cyanopentanol) (ACP) in water (Yang et al. 2010). Treating SWCNT with KOH or NaOH leads to introducing hydroxyl groups onto the surface of SWCNT. Also, such base treatment leads to increasing of the surface area of CNT. Raymundo-Piñero et al. (Raymundo-Piñero et al. 2005) showed that NaOH is effective only for the disordered materials while KOH is effective with the ordered ones and leads to distortion of the disordered materials.

Several chemical reactions can be carried out on the surface of the sidewalls of CNT. Some are substitution reactions starting with carboxylic group addition followed by a substitution reaction as

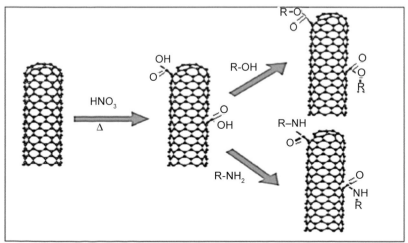

Figure 2. Chemical modification of nanotubes through thermal oxidation followed by subsequent esterification or amidization of the carboxyl groups. (Reprinted with permission from Balasubramanian and Burghard 2005. Copyright @ John Wiely and Sons.)

illustrated in Figure 2. A variety of functional groups can be successfully added to CNT through addition reactions. Balasubramanian and Burghard summarized in their review article (Balasubramanian and

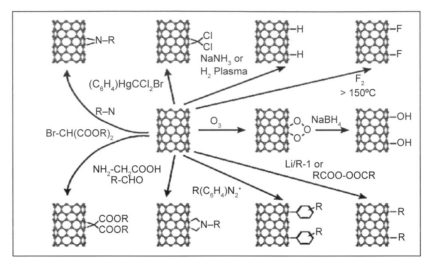

Figure 3. Overview of possible addition reactions for the functionalization of the nanotube sidewall. (Reprinted with permission from Balasubramanian and Burghard 2005. Copyright @John Wiely and Sons.)

Burghard 2005) the various function groups that can be introduced to CNT through the addition reactions as depicted in Figure 3.

Polymeric functionalization of CNT

As mentioned before, polymers are functional entities that can be attached to CNT through either covalent or non-covalent functionalization. Choudhary and Gupta (Choudhary and Gupta 2011) summarized the disadvantages of both methods in only one statement "covalent functionalization often leads to tube fragmentation and the non-covalent functionalization results in poor exfoliation".

Chemical functionalization of CNT with polymeric material usually proceeds by two techniques, either "grafting from-" or "grafting to-" techniques. Before this step, an initial functionalization of CNT by adding some functional groups such as -OH, -COOH or -NH$_2$ groups are required. These groups, polymer or monomer species, can attach to the surface of CNT. The basic concept of "grafting from" technique is the initial immobilization of initiator on the surface of CNT followed by *in situ* polymerization step. This technique leads to high grafting density and is suitable for grafting of poly(methyl-methacrylate) (PMMA) and the related polymers to CNT. Grafting-to approach is through attaching a functionalized polymer to a functionalized carbon nanotube via a chemical reaction. An example of grafting-to approach is that published by Fu et al. (Fu et al. 2001) that describes attachment of a carboxylic acid group to CNT surface followed by refluxing with thionyl chloride in order to convert the carboxylic group into acyl chloride that can be further reacted with the hydroxyl group of dendritic poly(ethylene glycol) (PEG) polymers via esterification reactions (Balasubramanian and Burghard 2005). Attachment of CNT to conducting polymers such as polyaniline, polypyrrole, polythiophene, ethylenedioxy-thiophene (EDOT), etc. attracted considerable interest. Such composites are considered as promising materials in many applications including electrochemical sensors and biosensors due to the presence of various oxidation structures. One method to enhance the solubility of CNT is by its functionalization with amphiphilic molecules such as homo-polymers block copolymers, lipids and/or surfactants (Park et al. 2007). Functionalization of CNT with amphiphilic copolymers (PEtOz-PCL) not only enhances the solubility of CNT but also provides a possibility for further controlled assembly and can be used as a template for the formation of Au nanoparticles on the surface of CNT. Interaction of CNT with epoxies can be facilitated by the initial introduction of the amine group to MWCNT (Li et al. 2008). The introduced amine group may react with the epoxy group of epoxy-prepolymer; hence, the interaction between MWCNTs and epoxy is improved.

Metal binding

The research on transition metal functionalized CNTs was initiated by Planeix and coworkers in 1994 (Planeix et al. 1994). Their work is based on the preparation of Ru/nanotubes catalyst. Thus, Ru nanoparticles were loaded on SWCNTs with 0.2% and with a uniform distribution of sizes. The Ru/nanotubes catalyst was used in the hydrogenation of cinnamic aldehyde.

Basically, CNT functionalizations with transition metals are divided into three main categories: the first is based on *in situ* reduction of metal precursors that were mixed or loaded onto CNTs; the second is non-covalent functionalization of CNTs with the already prepared metal nanoparticles and finally, covalent functionalization of CNTs with metal nanoparticles. For the *in situ* reduction of metal ion loaded to CNT, noble or rare metal nanoparticles such as Pt, Au, Pd, Ag, Rh, Ru and Ni can be directly deposited onto the surface of CNTs by the aid of van der Waals force. Kim and Sigmund (Kim and Sigmund 2004) also used layer-by-layer self-assembly method to prepare metal nanoparticles-functionalized CNTs. Another method to incorporate the metal ions to CNT is by using metal-organic frameworks (MOFs) (Zheng et al. 2003a,b). It is based on the coordination chemistry between metal ion and SWCNTs, which provides an alternative type of template. MOFs attracted a great interest due to their potential applications in many areas including hydrogen storage and DNA sensing. The work of Meng et al. (Meng et al. 2009) is concentrated on the incorporation of Zn to SWCNTs through MOF approach. A wide range of transition metals could be incorporated into SWCNT systems using the same approach. It is worth mentioning that the properties of MOF-CNT can be easily tuned via changing the ligand type (Meng et al. 2009).

CNT-biomolecule binding

CNT can cross the cell membrane easily, so it can be used in drug delivery or in the introducing of proteins, nucleic acids or peptides to the cell (Chen et al. 2011). Zheng et al. (Zheng et al. 2003b) first reported that bundled SWCNT could be effectively dispersed in water by its sonication in the presence of single-stranded DNA (ssDNA). They found that short oligonucleotides having repeating sequences of guanines and thymines $(dGdT)_n$ (n = 10–45) could wrap in a helical manner around CNT with a periodic pitch. They also found that wrapping of CNTs by ssDNA was sequence-dependent. Proteins and polypeptides can also bind with CNT through the hydroxyl, sulfhydryl, and amino groups present in the proteins (Meng et al. 2009). Devis et al. (Devis et al. 2003) found that metallo-proteins, cytochrome C, and ferritin, could be easily adsorbed onto CNTs and forming pH and ionic strength independent complexes. CNT-carbohydrates complexes have been also studied through molecular dynamic simulation by Xie and Soh (Xie and Soh 2005). They found that van der Waals forces were the dominant interaction force in carbohydrate/CNT complexes. On the other hand, cyclodextrins as well as, gum Arabic are used for CNT functionalization (Kong et al. 2007, He et al. 2006, Wang et al. 2005). Cellulose, linear polymeric D-glucose units, cross-linked together through 1,4-glucoside bonds, were used as a promising modifier for CNTs. The resulting cellulose-CNT composite is water soluble and stable at room temperature for a year and shows good biocompatibility (Meng et al. 2009). Moreover, Gea et al. (Gea et al. 2011) studied the binding of SWCNTs with human blood serum and they found that the competitive bindings of blood proteins to SWCNT can greatly affect their cellular interaction pathways and reduce the cytotoxicity for these protein-coated SWCNTs. They studied the binding of blood proteins as fibrinogen, immunoglobulin, albumin, transferrin, and ferritin to CNT, experimentally, and via molecular dynamic simulation. They also found that SWCNTs bind differently with blood-proteins and the resulting complexes showed different cytotoxicity. Atashbar et al. (Atashbar et al. 2006) used the phenomenon of the charge carrier concentration reduction and consequently the decrease in the current level upon non-covalent binding between CNT and biomolecule to construct a conductometric and gravimetric sensor for Streptavidin and mouse monoclonal immunoglobulin G (IgG). On the other hand, Singh et al. (Singh et al. 2005) reported that functionalization of SWCNTs with ammonium promotes the penetration of the human cells and facilitates DNA delivery.

Functionalized CNT in sensing applications

Because of the high surface area of CNT as well as high electrical conductivity and high mechanical strength, CNT has been widely used in many applications including electrochemical actuators (Baughman et al. 1999), electrochemical capacitors (Niu et al. 1997), lithium ion batteries (Ahan et al. 2012), catalysis (Gao et al. 1999, Disma et al. 1997), hydrogen storage (Silambarasan et al. 2014)and as sensors and biosensors. The scope of this chapter is to highlight the use of functionalized CNT in sensing and biosensing applications.

Functionalized CNT as sensor for drugs and biologically active compounds

Raoof et al. (Raoof et al. 2009) fabricated modified 1,2-naphthoquinone-4-sulfonic acid sodium (Nq)-SWNT through electrodeposition technique to fabricate a sensor for cysteamine determination. The functionalization process was carried out through two successive steps. Initially, SWCNT was acid treated in 1:3 ratio (HNO_3:H_2SO_4) to get more carboxyl groups on CNT surface. The acid functionalized SWCNT was then sonicated in dimethyl-formamide (DMF) to produce SWCNT dispersion. The resulted dispersion was pipetted on GCE surface and was subjected to electropolymerization of Nq. This sensor provided good selectivity and sensitivity toward cysteamine with a detection limit of 3.0×10^{-6} M in the linear range from 5×10^{-6} to 2.7×10^{-4} M. Additionally, SWCNT-Nq sensor provided a shift in the potential to 710 mV, which was more negative compared to GCE as well as an increase in the anodic peak current. They attributed the increase in the peak current to the interaction between cysteamine and the oxidized form of Nq through what is known by "Michael addition" reaction that facilitates further oxidation of the product.

Non-covalent functionalization of CNT with the covalently functionalized poly ethyleneimine with dopamine (PEI-DA) allows a stable and high performance with respect to nicotinamide adenine dinucleotide (NADH). This composite overcame the problem of surface fouling that usually accompanied NADH determination as well as it showed a very good electrochemical performance compared to bare CNT or PEI-DA. The obtained limit of detection is 3.0×10^{-6} M in the linear range of 1.0×10^{-5} to 1.0×10^{-4} M and an oxidation potential as low as –25 mV vs Ag/AgCl (3M KCl). The hydrophilic nature of the previous composite arises from the π–π interaction between the aromatic region of DA attached to PEI and hydrophobic region of CNT. This π–π interaction between CNT and PEI results in wrapping the PEI around CNT. Owing to the polymer wrapping, amine groups extend on the surface leading to more hydrophilic nature and consequently improve the dispersion of the composite in aqueous medium (Gasnier et al. 2012). One of the most common methods for covalent functionalization of MWCNT is the amidation method. This process is carried out via three successive steps. Firstly, CNT is oxidized or acid treated followed by reaction with thionyl chloride, $SOCl_2$ (MWCNT-COCl was obtained) and finally the designated amine-containing compound can be added (Kim et al. 2012, Xie et al. 2012). As we stated before, this method allows the binding of any bio-molecule containing amine group to CNT. Cãnete-Rosales et al. (Cãnete-Rosales et al. 2014) used this method to construct a new sensor for double strand DNA (ds-DNA) based on the functionalization of MWCNT with ethylenediamine (eda) as illustrated in Figure 4. The electroanalytical method is based on the direct adsorption of ds-DNA on eda-CNT modified GCE. They claimed that eda functionalization of CNT allows its dispersion in phosphate buffer pH 7.0 that in turn promotes ds-DNA electrostatic adsorption. This sensor provided a limit of detection 0.971 ppm after 10 minutes of accumulation in the linear range from 5 to 60 ppm with a sensitivity of 0.0315 ± 0.0003 µA mg^{-1} L.

Using the previous procedure of amidation of CNT, the simultaneous determination of folic acid (FA) and methotrexate (MTX) was achieved using quaternary amine-functionalized CNT modified GCE by Zhu et al. (Zhu et al. 2013). The attached amine groups were quaternized by stirring overnight in iodo-methane in the presence of 18-crown-6 ether as a catalyst. The used method of analysis is the ion chromatography

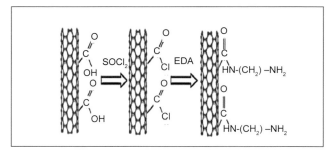

Figure 4. The preparation of ethylendiamine-functionalized MWCNTs (Reprinted with permission from Cãnete-Rosales et al. 2014. Copyright @ Elsevier).

with electrochemical detection (IC-ECD) that provides high sensitivity and good separation of FA and MTX. The obtained detection limits (LODs) for MTX and FA were 0.2 and 0.4 μg L^{-1} (S/N = 3), respectively, in the linear range from 0.01 to 20 mg L^{-1} for both MTX and FA.

On the other hand, a biosensor based on the functionalization of CNT with ferrocene and biotin has been developed for the electrochemical determination of avidin (Fabre et al. 2012). The synthesis of Fc-Biot-MWCNTs is described in Figure 5. In this work, biotin is used as a molecular receptor for avidin while CNT acts as an efficient electrical communication between the immobilized center and the electrode surface. The role of ferrocene is to act as an electrochemical transducer of biotin-avidin interaction (Hauquier et al. 2004).

They observed that there was a decrease in both anodic and cathodic peak currents upon binding of avidin with biotin. As described by Liu and Goodling (Liu and Gooding 2009), this decrease in the peak current reflects the change in the microenvironment of ferrocene units arising from formation of biotin-avidin complex that consequently affects the counterion mobility, electron pathway and electron transfer.

Attachment of polymeric materials to CNT has attracted a great attention especially in sensing applications. Wang and Bi used poly(Tyrosine) functionalized CNT for the electrochemical determination of uric acid (UA), ascorbic acid (AA) and dopamine (DA) (Wang and Bi 2013). This sensor provided a peak separation of 160 and 150 mV between AA-DA and DA-UA, respectively, as well as a very high peak current in pH 7.4. The sensor also showed a detection limit of 2.0 μM, 0.02 μM and 0.30 μM for

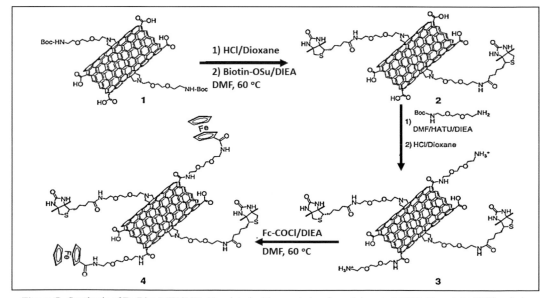

Figure 5. Synthesis of Fc-Biot-MWCNTs (Reprinted with permission from Fabre et al. 2012. Copyright @ Elsevier).

AA, DA and UA, respectively. Functionalization of MWCNT with poly(Tyrosine) was performed via electropolymerization of tyrosine on the surface of MWCNT-COOH (obtained by acid treatment). The electropolymerization of tyrosine was carried out by cyclic voltammetry from −0.8 to 1.8 V vs. SCE at 100 mV s^{-1} for sixteen cycles in pH 6.0 phosphate buffer solution containing 1.0×10^{-3} M Tyrosine and they used Nafion for immobilization of the composite on the surface of the working electrode. Lin et al. used poly(Xanthurenic Acid) (poly Xa) to functionalize the acid treated SWCNT and MWCNT. They found that poly Xa/MWCNTs showed a higher response for the electrochemical determination of AA, DA and UA with a detection limit of 10 µM, 1 µM, and 5 µM using amperometric technique (Lin et al. 2015). Filik et al. (Filik et al. 2016) constructed another sensor for the simultaneous determination of AA, DA, UA and tryptophan (Trp) by functionalization of CNT with Azure A (AzA) dye and Au nanoparticles as well as nafion. The functionalization process was carried out through sonication of AzA dye with acid treated MWCNT in order to get a well adsorbed AzA/MWCNT. The composite was added to gold nanoparticles (AuNPs) dispersion with vigorous agitation for 8 hours. They reported that the electrostatic interaction between AzA$^+$ and negatively charged AuNPs leads to an increase in the density of the adsorbed AuNPs on the surface of MWCNT. The resulting AuNPs/AzA/MWCNT electrode was sonicated with nafion solution and drop casted onto GCE. AuNPs/AZA/MWCNT/nafion showed a detection limit of 16 µM, 0.014 µM, 0.028 µM and 0.56 µM for AA, DA, UA, and Trp, respectively. Moreover, the previous sensor showed very good peak separation between AA, DA, UA and Trp.

On the other hand, the electropolymerized poly(basic red 9) (BR9) on MWCNT-COOH was used as a promising sensor for neurotransmitters such as serotonin and epinephrine (Ep) (Li et al. 2014). They stated that the presence of functionalized MWCNT improves the surface coverage and the stability of electropolymerized BR9. The calculated detection limits for EP and serotonin was 7.0 µM and 9.0 µM. The sensor also provided an excellent long-term stability where only 6% decrease in the peak current was observed after 30 days.

An electrochemical sensor for bromhexine, a drug used for the treatment of respiratory disorders, was constructed by Kutluay and Aslanoglu based on the functionalization of CNT with metal nanoparticles (Kutluay and Aslanoglu 2014). Metal nanoparticles binding to CNTs has attracted great attention due to their excellent long-term stability and sensitivity, fast response time, high resistance to surface fouling, low overpotential and limits of detection, high conductivity and large surface area (Goyal et al. 2008, Karuwan et al. 2009, Lijun et al. 2009, Goyal et al. 2011, Huang et al. 2011).

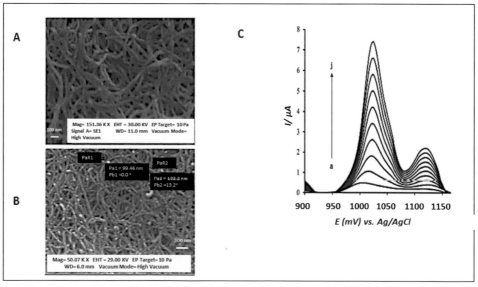

Figure 6. SEM of MWCNT/Pt (A), NiNPs/MWCNT/Pt (B) and square wave voltammograms (SWVs) of various concentrations of bromhexine at the surface of NiNPs/MWCNT/Pt electrode (C). (Reprinted with permission from Kutluay and Aslanoglu 2014. Copyright @ Elsevier.)

As mentioned before, there are different routes to bind metal nanoparticles to CNT surface. One of these routes is via the sonication of metal precursors with MWCNTs in the presence of a reducing agent. This method provides good distribution of metal nanoparticles on the surface of MWCNT and increases the adsorption affinity to metal nanoparticles, as shown in Figures 6A and 6B. CNT should be acidified in an acid mixture to provide more oxygenated groups on MWCNT surface before mixing with metal precursors. The previous sensor provided improved voltammetric response, long-term stability, excellent reproducibility and a detection limit as low as 3.0 μM. Figure 6C shows square wave voltammograms of various concentrations of bromhexine at the surface of NiNPs/MWCNT/Pt electrode.

Yola et al. (Yola et al. 2014) presented a promising sensor for the antibiotic Cefixime by a combination of the molecular imprinting technique with 2-aminoethanethiol-functionalized CNT and metal nanoparticles (Fe@Au NPs in core-shell structure). This is a little bit complicated sensor as it provided very low detection limit for Cefixime as low as 2.2×10^{-11} M in the linear range from 1.0×10^{-10} to 1.0×10^{-8} M in a human plasma sample. This sensor was validated according to the international conference of harmonization (ICH) guidelines. The procedure of the sensor construction is summarized in Figure 7. The Molecular imprinting technique provides high selectivity and sensitivity towards the target analyte. It includes the polymerization of the monomers in the presence of the target template followed by extraction of the template molecules leaving a cavity specific for this target molecule. The previous sensor allows new β-lactam antibiotics recovery with high sensitivity and selectivity.

Metal oxides have been also used for CNT functionalization and have provided a good response in the field of electrochemical sensors. TiO_2 nanoparticles have specific properties that enable it to be good candidates for electrochemical sensors and biosensors. Some of these properties are well biocompatibility, nontoxicity, low cost, high conductivity, chemical durability and high efficient photocatalytic properties

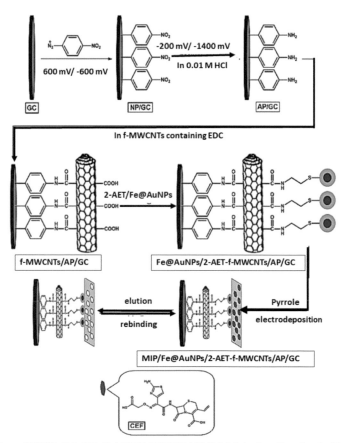

Figure 7. The procedure of MIP/Fe@AuNPs/2-AET-f-MWCNTs/AP/GC fabrication. (Reprinted with permission from Yola et al. 2014, Copyright @ Elsevier.)

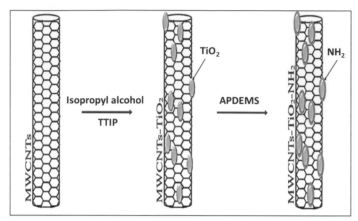

Figure 8. Preparation of NH$_2$-TiO$_2$-MWCNT/GCE where, TTIP is titanium (IV) isopropoxide and APDEMS is 3-aminopropyl-(diethoxy)-methylsilane (Reprinted with permission from Arvand and Palizkar 2013. Copyright @ Elsevier).

(Yola et al. 2014, Benvenuto et al. 2009, Zhu et al. 2009, Wang et al. 2008, Jiang et al. 2009, Lu et al. 2009, Tang et al. 2010). The combination of CNT and TiO$_2$ enhances the photocatalytic activity of TiO$_2$. A sensor for Olanzapine based on TiO$_2$-MWCNT has been constructed by Arvand and Palizkar (Arvand and Palizkar 2013). The preparation of the electrode surface is based on the sol-gel technique. The hydroxyl group of TiO$_2$ in solution can be easily attached to 3-aminopropyl (diethoxy) methylsilane forming a suitable amine functional group at the surface and NH$_2$-TiO$_2$-MWCNT was prepared as shown in Figure 8. The proposed NH$_2$-TiO$_2$-MWCNT/GCE is applied also for the determination of olanzapine in human blood serum and in commercial tablet.

The detection limit of Olanzapine at NH$_2$-TiO$_2$-MWCNT/GCE surface was 0.09 µM. TiO$_2$-MWCNT was also used in combination with ionic liquid (IL) and benzofuran derivative for the electrochemical determination of isoproterenol and serotonin (Mazloum-Ardakani et al. 2014). They used 9-(1,3-dithiolan-2-yl)-6,7 dihydr-oxy-3,3-dimethyl-3,4-dihydro-dibenzo[b,d]furan(2H)-one (DDF) as benzofuran derivative and 1-butyl-3-methylimidazolium-tetrafluoroborate as IL. TiO$_2$ nanoparticles are doped on the surface of CNT through ultrasonication of titanium isopropoxide with CNT in a mass ratio of 1:5 followed by autoclave heating at 270ºC for two hours. The composite was then filtered and dried in a vacuum oven. The DDF-CNT-TiO$_2$ composites were prepared by mixing of DDF with CNT-TiO$_2$ in DMF under stirring for 48 h at room temperature, then IL was added to the resultant composite which was then subjected to ultrasonication. DDF-CNT-TiO$_2$/IL composite was drop-casted on the surface of GC electrode and used directly for electrochemical determination. DDF-CNT-TiO$_2$/IL/GCE showed very good peak separation between serotonin and isoproterenol with a low detection limit of 154 µM and 28 ± 2 nM, respectively, and a good recovery was obtained in the human blood serum sample. See Table 1 for more functionalized CNT-based sensors.

Functionalized CNT as metal sensors

Exposure to high level of heavy metals has a detrimental effect on both human health and the environment. Additionally, heavy metals themselves are not biodegradable and remain in the ecological systems and in the food chains (Seiler et al. 1998, Rassaei et al. 2011). Therefore, constructing a novel sensor to determine the traces of heavy metals in the human fluids, soils and water is crucial. Carbon nanotubes possess a great potential as a superior adsorbent for removing various divalent metal ions from aqueous solutions. Also, they are widely used as a promising material for heavy metal ions determination (Wang et al. 2012, Kang et al. 2006, Zhad et al. 2013, Liu et al. 2005). The physicochemical properties of CNT can be tuned easily by the surface functionalization to achieve a certain property based on the desired application. Many attempts to functionalize carbon nanotube with various functional groups or molecules to be used as metal sensors are covered in the present section.

Table 1. Comparison between the electroanalytical properties of different functionalized CNT electrodes towards some drugs and biologically active compounds.

Electrode	Analyte	Limit of detection (μM)	Linear range	Sensitivity ($\mu A/\mu M$)	Method of determination	Reference
SWNT-Nq	Cysteamine	3.0	5 to 270 μM	58.07	potentiometry	Raoof et al. 2009
MWCNT/PEI-DA/GC	NADH	3.0	10 to 100 μM	1.9×10^4 ($\mu A/mol/dm^3$)	Amperometry	Gasnier et al. 2012
Eda-CNT	Ds-DNA	0.971 ppm	5 to 60 ppm	0.0315 ± 0.0003 μA mg^{-1} L	potentiometry	Cãnete-Rosales et al. 2014
Quaternary amine-functionalized CNT	Folic acid and methotrexate	0.4 and 0.2 μg L^{-1}	0.01 to 20 mg L^{-1}	11.63 and 16.34 (nA min/mg/l)	ion chromatography with electrochemical detection (IC-ECD)	Zhu et al. 2013
Fc-Biot-MWCNTs	Avidin	----	0.9–20 nM	----	potentiometry	Fabre et al. 2012
Poly(Tyr)/MWCNTs-COOH/GCE	AA, UA and DA	2.0, 0.3 and 0.02	50–1000, 1.0–350 0.1–30 μM	2.322, 0.3531 and 0.1215	potentiometry	Wang and Bi 2013
Poly Xa/MWCNTs	AA, UA and DA	10, 1.0 and 5.0	10–2.3, 5–1650 and 5–165 μM	10–1000, 10–320 and 10–1000 μM	potentiometry	Lin et al. 2015
AuNPs/AzA/MWCNT/Nafion	AA, UA, DA, Trp	16, 0.028 0.014 and 0.56	300–10,000, 0.5–50, 0.5–50 and 1.0–100 μM	2.8862, 1.4151, 3.5062 (0.5 to 10 μM) and 0.8299	potentiometry	Filik et al. 2016
Poly(basic red 9) (BR9) on M-WCNT-COOH	Serotonin and epinephrine	9 and 7	----	----	potentiometry	Li et al. 2014
Fe@AuNPs/2-AET-f-MWCNTs/AP/GC	Cefixime	2.2×10^{-5}	$1.0–100 \times 10^{-10}$ M	0.962 $\mu A/nM$	Potentiometry	Yola et al. 2014
NH$_2$-TiO$_2$-MWCNT/GCE	Olanzapine	0.09	0.12 to 33 μM	0.098	Potentiometry	Arvand and Palizkar 2013
DDF-CNT-TiO$_2$/IL/GCE	Isoproterenol and serotonin	$28 \pm 2 \times 10^{-3}$, 154	0.1–60 and 1.0–650 μM	0.21 and 0.039	Potentiometry	Mazloum-Ardakani et al. 2014

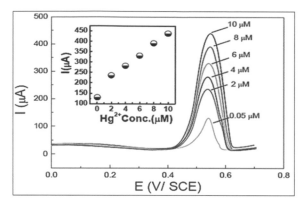

Figure 9. SWV curves recorded on PDAN/CNT-modified electrode for various Hg^{2+} concentrations and corresponding calibration curve (Reprinted with permission from Naguyan et al. 2011. Copyright @ Elsevier).

Liang et al. used phytate to non-covalently functionalize CNT to be used as a promising Cu^{2+} sensor (Liang et al. 2014). Phytic acid and its salts are present in seeds, all *eukaryotic* cells and in the grains (Maga 1982, Raboy et al. 2003). Phytate (IP_6) is usually used as an oral cleansing agent, food additive and in water treatment (Yang et al. 2004). IP6 can capture many metal ions and form stable complexes with them (Pang et al. 2006, Crea et al. 2008). The stability constant of the formed complexes depends on the type of metal ions. As reported by Tamim and Angel (Tamim and Angel 2003), the stability of metal–phytate complexes are in the order of Cu(II) > Zn(II) > Co(II) > Mn(II) > Fe(II) > Ca(II). Therefore, phytate can be used as a promising sensor for Cu(II) ions with no or little interference from other metal ions. The functionalization process occurred via sonication of the acid treated MWCNTs with IP6 for five hours. The composite was kept for adsorption on indium tin oxide surface (ITO) for another three hours to form IP6-MWCNTs-ITO electrode. IP6-MWCNTs-ITO exhibited very good sensitivity and selectivity toward Cu(II) ions with no interference from Ni(II), Mg(II), Ca(II), Pb(II), Cd(II), Mn(II), Zn(II) and Fe(II, III) as well as some anionic and organic species. The detection limit obtained was 2.5×10^{-9} M in the linear range between 0.01 to 0.8 µM. A negative shift in the oxidation potential from 0.4 and 0.72 V (two oxidation peaks were are observed) on bare ITO to 0.29 V (only one oxidation peak was obtained) on IP6-MWCNTs-ITO was obtained.

Conducting polymers/CNT nanocomposites were used also as effective metal sensors with high sensitivity. Naguyan et al. (Naguyan et al. 2011) used a mixture of the electropolymerized polydiamino-phthalene (PDAN) and CNT on integrated arrays for the electrochemical determination of Hg^{2+}. The electrochemical determination method was performed via stripping voltammetry. The preconcentration step was performed at open circuit potential (OCP) followed by square wave voltammetry (SWV). Figure 9 showed square wave voltammograms (SWVs) as well as the calibration curve of Hg^{2+} using PDAN/CNT electrode. The same research group used P1, 5DAN/CNT with nafion for the electrochemical determination of Pd^{2+} and Cd^{2+} using square wave anodic stripping voltammetry (SWASV). Detection limits of 3.2 and 2.1 µg L^{-1} were obtained for Cd^{2+} and Pd^{2+}, respectively, in the linear range of 4–150 µg L^{-1} (Vu et al. 2015). The sensor did not show any interference from Na^+, Ca^{2+}, Zn^{2+}, Fe^{2+}, Al^{3+}, Cu^{2+} (each 1.0 mg L^{-1}) and Cl^-, Br^-, SO_4^{2-} (each 10.0 mg L^{-1}) to 100-µg L^{-1} Pb^{2+} acetate buffer solution. Moreover, PANI was also used to functionalize CNT and to detect Pb^{2+} ions by square wave anodic stripping voltammetry (SWASV) by Wang et al. (Wang et al. 2011). Initially, CNT was oxidized to CNT-COOH and then mixed with PANI. The solution used to disperse PANI-CNT was 0.5 M H_2SO_4. Many other solvents were used to disperse PANI-CNT composite in their study but 0.5 M H_2SO_4 showed the highest electrochemical response as shown in Figure 10.

Salmanipour and Taher (Salmanipour and Taher 2011) functionalized GC electrode with a mixture of multi-walled carbon nanotubes (MWCNTs) and 2-(5-bromo-2-pyridylazo)-5-diethylaminophenol (5-Br-PADAP). The sensor was employed to detect Pb^{2+} by ASV in the range from 1 to 115 µg L^{-1}. Polypyrrole/CNT composite was used to detect lead, mercury, and iron ions by surface plasmon technique

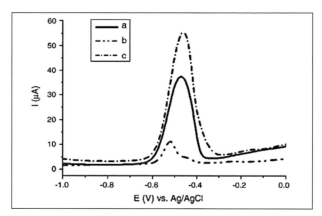

Figure 10. Voltammograms of different electrodes in (a) ethanol solution containing MWCNT-COOH and PANI, (b) ethanol only and (c) sulfuric acid solution containing MWCNT-COOH and PANI. (Reprinted with permission from Wang et al. 2011. Copyright @ Elsevier.)

(Sadrolhosseini et al. 2014). The authors found that Hg^{2+} ions bonded to the sensing layer more strongly than did by Pb^{2+} and Fe^{3+} ions. The limitation of the sensor was calculated to be about 0.1 ppm, which produced an angle shift in the region of 0.3° to 0.6° (Sadrolhosseini et al. 2014).

A highly selective and sensitive sensor based on CNT functionalization for the simultaneous determination of Cd^{2+}, Pd^{2+} and Hg^{2+} was constructed by Shaban and Galaly (Shaban and Galaly 2016). Porous anodic alumina (PAA) membrane was functionalized with $CoFe_2O_4$ nanoparticles and used as a substrate for the growing of very long helical-structured CNTs. The diameter of CNTs is less than 20 nm. The determination method is based on the *in situ* surface enhanced Raman spectroscopy (SERS). The authors showed that PAA membranes of pore diameters greater than 50 nm are used for the growth of highly ordered CNTs. These membranes have the advantages of the ease of preparation, high thermal stability, low cost and high controllability (Wang et al. 2001).

Functionalized CNT as gas sensors

The level of pollutants and greenhouse gases such as CO_2, CH_4, N_2O, CO, SO_2, NO_x and NH_3 has increased in the atmosphere as a result of industrial development and human activity. Increasing the emission of these gases causes climate changes, affects the plant and animal growth and causes health problem to the human beings. For example, CO may replace oxygen in the blood and if its concentration reaches 35 ppm, it causes headache and dizziness and may lead to death (Trung et al. 2014).

Also, a high concentration of SO_2 leads to eye irritation, respiratory distress and death (Boudiba et al. 2012). Hence, it is very important to develop devices that can detect very low concentrations of these gases. These devices are known as "gas sensor".

A gas sensor is the device whose one or more physical properties change upon exposure to gas molecules (Jesús et al. 2013). It should have some characteristics to be applicable such as high selectivity for different gases, high sensitivity to very low concentration of gases, low cost, low power consumption, and fast and better response if they operate at room temperature (Jesús et al. 2013).

Pristine CNTs are chemically inactive due to strong sp^2 bonding in a hexagonal network that prevents their reaction with surface molecules. However, functionalization of CNTs can improve their chemical reactivity. The added functional groups are chosen according to the target molecule. For example, the carboxylic functionalization are used for base molecules, while amine functionalization for acid molecules, and aromatic groups (such as thiols) for large organic molecules (Afrin and Shah 2015). The chemical functionalization of CNTs leads to the formation of defects on the surface of CNT. These defects facilitate the interaction with the surrounding gases leading to either chemical or physical adsorption. The adsorbed molecules themselves act as dopants for CNT leading to a change in Fermi-level position depending on the nature of these adsorbed molecules (Goldoni et al. 2010). If the adsorbed gas molecules have an oxidizing property, this leads to an electron withdrawing from CNT leaving holes that in turn causes a shift in the Fermi-level down closer to the valence band. The opposite process happens if the adsorbed gas has a reducing property, which leads to more electrons and consequently shifts the Fermi-level up toward the conduction band. These processes influence the electronic properties of CNTs and consequently lead to a change in their electrical resistance (Goldoni et al. 2010). For example, NO_2, an oxidizing gas, takes electrons from CNTs and produces holes. As a result, the resistance decreases after the interaction. Since NH_3 is a reducing gas that donates electrons, it increases the resistance of CNTs after the interaction.

Afrin et al. (Afrin et al. 2015) prepared CNT backypaper functionalized with carboxylic groups (CNT-CBP) and thiol groups (CNT-TBP) to be used as NH_3 and CO gas sensor. Backypaper CNT can be synthesized by using a cross-linker such as 4-4'-diaminobenzophenon in an appropriate ratio. Their work was based on the change in the resistance of the functionalized CNT upon exposure to CO (electron acceptor) or NH_3 (electron donor). Both gas molecules affect the Fermi-level as discussed before. Afrin and co-workers showed that CNT-TBP has a higher sensitivity toward CO and NH_3 than the corresponding CNT-CBP (Afrin et al. 2015). On the other hand, liquid phase electrochemical deposition of Pd on CNT was used as a hydrogen sensor and showed a high sensitivity toward a very low concentration of H_2 gas, as low as 0.01% (Suehiro et al. 2007). CNT was trapped onto a microelectrode by positive dielectrophoresis, then Pd^{2+} was reduced and deposited onto CNT by a simple liquid phase electrochemical deposition. Additionally, Dhall et al. (Dhall et al. 2013) could detect 0.05% H_2 at room temperature using acid functionalized CNT gas sensor. The acid functionalization this time was carried out by an acid mixture of H_2SO_4, HNO_3, and H_2O_2. They found that the current carrying capacity of acid-functionalized CNT increased to 35 mA (while the original value was 49 µA) at a low scan rate and the recovery time of 0.05% H_2 decreased to 100 s compared to 190 s using the pristine CNT. Moreover, DNA with metal nanoparticles was also used for CNT functionalization and employed in gas sensing (Su et al. 2013). In this work, DNA was used as a dispersing agent for SWCNT and as a metal ion chelating center for nanoparticles formation. Various metal nanoparticles such as Pt, Pd and Au nanoparticles were used to functionalize SWCNTs that were previously dispersed in DNA. The resultant sensor was used for the determination of H_2, NH_3, NO_2 and H_2S gases. Additionally, Pd/DNA-SWCNT has the highest sensitivity toward H_2, NH_3 and H_2S compared to other electrodes. Pt/DNA-SWCNT showed the highest response towards NO_2 (Su et al. 2013). Another gas sensor based on metal nanoparticles/CNT is that presented by Abdelhalim et al. (Abdelhalim et al. 2015). Their sensor was designed for the determination of CO, CO_2, ethanol and NH_3. The construction of the sensor is based on the sensor array technique. Sensor array is a device that contains several sensing materials and is designed to enhance the selectivity of the gas sensor in general (Abdelhalim et al. 2015). Each sensing element has its own functionalization that provides a specific interaction with the gas molecules in varying degrees.

In the work by Abdehalim et al., spray-depositing technique was used to deposit layers of CNT and the thermal evaporation technique was used for CNT functionalization with Au, Pd, Ag, Ti, Cr and Al nanoparticles. The CNT thickness and nanoparticles loading were optimized. They found that the best combination of the metallic functionalization was Au at a load of 1.5 nm, Cr at a load of 1.0 nm and Pd at a load of 0.2 nm. This gas sensor arrays provided a short response time as low as 100 s. Lu et al. (Lu et al. 2006) could construct a gas sensor array based on CNT containing 32 sensing elements. The sensor array contained a pristine CNT, metal clusters CNT and a CNT-polymer composite. CNT sensor arrays were also used to detect lung cancer biomarkers, as that performed by Peng et al. (Peng et al. 2008). Their sensor is based on CNT-nonpolymeric organic compounds and could discriminate between lung-cancer volatile organic compounds (VOCs) present in the patient's breath in relation to healthy persons. The determination method is based on gas-chromatography linked with mass-spectroscopy (GC-MS) analysis of a real exhaled breath. On the other hand, the functionalization of CNT with polyaniline (PANI) results in a higher sensitivity towards H_2 gas as well as a reduced response time compared with PANI (Srivastava et al. 2009). The increased performance of PANI/CNT was attributed to the hollow structure of CNT as well as its high surface area that provides extra sites for hydrogen adsorption. In addition to PANI, aminosilane polymers enhanced the sensitivity of SWCNT towards the target gas (Tran et al. 2008). Aminosilanes contain amino group that react with SWCNT and lead to an increase in the electron density on SWCNT surface. When the oxidizing gas interacts with SWCNT-aminosilane composite, it accepts the electrons from SWCNT and results in an increase in the electric conductivity. Field-effect transistor based on SWCNT-π-conjugated poly(9,9-dioctylfluorene-co-bithiophene) (F8T2) composite was used as NO_2 gas sensor (Septiania and Yuliarto 2016). The determination method is based on monitoring the change in the current-voltage (I-V) characteristics of the sensor when exposed to various concentrations of NO_2. This device could be used to study the adsorption isotherm as well as the kinetics of gas adsorption in details. Other functionalized CNT-based gas sensors are tabulated in Table 2 with their sensitivity, sensing and recovery times.

Functionalized CNT-based sensors in health

A unique stretchable and wearable sensor made of CNT and gum was constructed by Darabi et al. (Darabi et al. 2015) to be used as bodily motion sensors. The sensor showed a high sensitivity and capability of tracing slow breathing. Gum sensor has the ability to monitor humidity changes with a high sensitivity and fast resistance response leads to monitoring the human breathing. The sensor also could detect humidity changes, a feature that could be used to track breathing, which releases water vapor with every exhale. A schematic diagram of the preparation method for the MWCNT/gum membrane as well as the application of using MWCNT/gum sensor to detect breathing is shown in Figures 11 and 12.

Additionally, chemists from MIT have devised an inexpensive, portable sensor that can detect gases emitted by rotten meat. This allows consumers to determine whether the meat in their grocery store or refrigerator is safe to eat or not. The sensor is based on the functionalization of CNT with cobalt porphyrin. Biogenic amines such as putrescine and cadaverine are produced by decaying meat, can bind with Co-porphyrin/CNT and causes an increase in its resistance that can be easily measured (Liu et al. 2015).

Also, researchers from Stanford University have devised a transparent skin-like stretchable pressure sensor based on CNT (Lipomi et al. 2011). The sensor uses a transparent film of single-walled carbon

Table 2. CNT-based gas sensors with their sensitivity, sensing and recovery times.

Electrode	Target gas	Response time	Sensitivity	Recovery time	Reference
Pd/DNA/SWCNTs	H_2	10.6 min	0.5%	1.6 min	Su et al. 2013
Acid functionalized MWCNTs		----	0.8%	100 s	Dhall et al. 2013
Pd/SWCNT		5–10 s	---	400 s	Kong et al. 2001
F-MWCNTs/SnO_2		---	2.8%	9 to 30 s	Jaggi and Dhal 2014
Pt/DNA/SWNTs	NH_3	12 min	0.8%	10 min	Su et al. 2013
PANI-SWNT		Few minutes	2.44%	Few hours	Zhang et al. 2006
Ag NC–MWCNTs		7 s	9%	Several minutes	Cui et al. 2012
Au/DNA/SWNTs	NO_2	10 min	–6%	12.6 min	Su et al. 2013
SnO_2/MWCNT		4 min			Leghrib et al. 2010
MWCNTs-COOH		230 s (100 ppm)	26.88%	288 s	Pisal et al. 2014
CNT-CBP (Ratio 1:7)	CO	7.7 s	1.01%	08.16 s	Afrin et al. 2015
Acid Functionalized CNTs		16 s	0.30%	16.02 s	Afrin et al. 2015
CNT-TBP (Ratio 1:5)		22 s	4.78%	33 s	Afrin et al. 2015
MWCNTs/SnO_2		Less than 5 min	----	Less than 5 min	Hieu et al. 2008
CoOOH/MWCNT Using complementary metal oxide semiconductor (CMOS) process		23 s (200 ppm, CO)	0.19 mV/ ppm	34 s	Dai et al. 2010
SnO_2/MWCNT/Ru sol gel method (MWCNT:SnO_2 is 1:50)	Isobutane	5–7 s (at 200°C) 1–2 (at 250°C)	-----	60 s 10 s	Aroutiounian et al. 2013

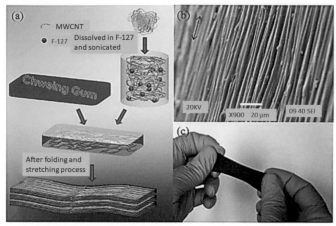

Figure 11. A schematic diagram of the preparation method for the MWCNT/gum membrane, (a) SEM image of the MWCNT/gum membrane (the arrow shows the stretch direction) (b) and Optical image of MWCNT/gum membrane (6 wt% of MWCNT) (c). (Reprinted with permission from Darabi et al. 2015. Copyright @ American Chemical Society, ACS.)

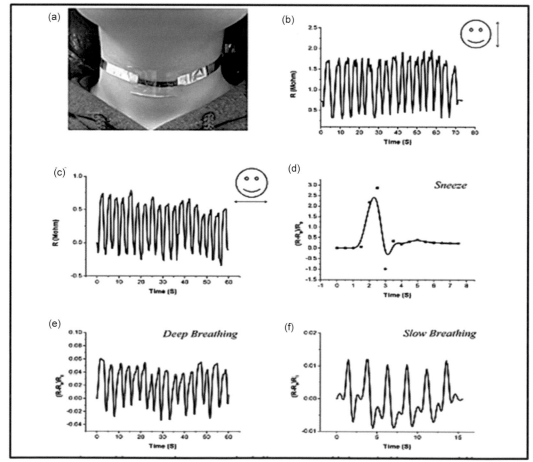

Figure 12. Applications of the gum sensor (6 wt % MWCNT) as bodily motion sensors. (a) Photograph of the gum sensor sealed between two PDMS films attached to the throat. Responses of the gum sensor attached to the throat, (b) when moving head up and down repeatedly, (c) when moving head left and right repeatedly, and (d) with a sneeze. Resistance traces showing (e) deep breathing and (f) slow breathing. (Reprinted with permission from Darabi et al. 2015. Copyright @ American Chemical Society, ACS.)

nanotubes that act as tiny springs, enabling the sensor to measure the force on it. The transparent film of the carbon "nano-springs" is created by spraying nanotubes in a liquid suspension onto a thin layer of silicone, which is then stretched. When the silicone is stretched, some of the "nano-bundles" get pulled into alignment in the direction of the stretching. When the silicone is released, it rebounds back to its original dimensions. Then the nanotubes buckle and form little nanostructures that look like springs. The second stretching of the nanotube-coated silicone, in the direction perpendicular to the first direction, causes some of the other nanotube bundles to align in the second direction. That makes the sensor completely stretchable in all directions, with total rebounding afterward.

References

Abdelhalim, A., M. Winkler, F. Loghin, C. Zeiser, P. Lugli and A. Abdellah. 2015. Highly sensitive and selective carbon nanotube-based gas sensor arrays functionalized with different metallic nanoparticles. Sens. Actuat. B-Chem. 220: 1288–1296.

Afrin, R. and N.A. Shah. 2015. Room temperature gas sensors based on carboxyl and thiol functionalized carbon nanotubes buckypapers. Diam. Relat. Mater. 60: 42–49.

Ahn, D., X. Xiao, Y. Li, A.K. Sachdev, H.W. Park, A. Yu et al. 2012. Applying functionalized carbon nanotubes to enhance electrochemical performances of tin oxide composite electrodes for Li-ion battery. J. Power Source 212: 66–72.

Aroutiounian, V.M., A.Z. Adamyan, E.A. Khachaturyan, Z.N. Adamyan, K. Hernadi, Z. Pallai et al. 2013. Study of the surface-ruthenated SnO_2/MWCNTs nanocomposite thick-film gas sensors. Sens. Actuat. B-Chem. 177: 308–315.

Arvand, M. and B. Palizkar. 2013. Development of a modified electrode with amine-functionalized TiO_2/multi-walled carbon nanocomposite for electrochemical sensing of the atypical neuroleptic drug olanzapine. Mat. Sci. Eng. C 33: 4876–4883.

Atashbar, M.Z., B.E. Bejcek, and Srikanth Singamaneni. 2006. Carbon Nanotube Network-Based Biomolecule Detection, IEEE Sens. J. 6: 524–528.

Bahr, J.L., J. Yang, D.V. Kosynkin, M.J. Bronikowski, R.E. Smalley and J.M. Tour. 2001. Functionalization of carbon nanotubes by electrochemical reduction of aryl diazonium salts: a bucky paper electrode. J. Am. Chem. Soc. 123: 6536–6542.

Balasubramanian, K. and M. Burghard. 2005. Chemically functionalized carbon nanotubes. Small 1: 180–192.

Baughman, R.H., C. Cui, A.A. Zakhidov, Z. Iqbal, J.N. Barisci, G.M. Spinks et al. 1999. Carbon nanotube actuators. Science 284: 1340.

Benvenuto, P., A.K.M. Kafi and A. Chen. 2009. High performance glucose biosensor based on the immobilization of glucose oxidase onto modified titania nanotube arrays. J. Electroanal. Chem. 627: 76–81.

Boudiba, A., C. Zhang, C. Bittencourt, P. Umek, O. Marie-Georges, R. Snyders et al. 2012. SO_2 gas sensors based on WO_3 nanostructures with different morphologies. Procedia Engineering 47: 1033–1036.

Cânete-Rosales, P., A. Alvarez-Lueje and S. Bollo. 2014. Ethylendiamine-functionalized multi-walled carbon nanotubes prevent cationic dispersant use in the electrochemical detection of dsDNA. Sens. Actuat. B 191: 688–694.

Chen, J., M.A. Hamon, H. Hu, Y. Chen, A.M. Rao, P.C. Eklund et al. 1998. Solution properties of single-walled carbon nanotubes. Science 282: 95–98.

Chen, L., H. Xie and W. Yu. 2011. Functionalization methods of carbon nanotubes and its applications, functionalization methods of carbon nanotubes and its applications. *In*: Jose Mauricio Marulanda (ed.). Carbon Nanotubes Applications on Electron Devices. InTech. DOI:10.5772/18547.

Chiang, Y.C., W.H. Lin and Y.C. Chang. 2011. The influence of treatment duration on multiwalled carbon nanotubes functionalized by H_2SO_4/HNO_3 oxidation. Appl. Surf. Sci. 257: 2401–2410.

Choudhary, V. and A. Gupta. 2011. Polymer/carbon nanotube nanocomposites. *In*: Siva Yellampalli (ed.). Carbon Nanotubes - Polymer Nanocomposites. InTech. DOI:10.5772/18423.

Crea, F., C.D. Stefano, D. Milea and S. Sammartano. 2008. Formation and stability of phytate complexes in solution. Coordin. Chem. Rev. 252: 1108.

Cui, S., H. Pu, G. Lu, Z. Wen, E.C. Mattson, C. Hirschmugl et al. 2012. Fast and selective room-temperature ammonia sensors using silver nanocrystal-functionalized carbon nanotubes. ACS Appl. Mater. Interfaces 4: 4898–4904.

Dai, C.L., Y.C. Chen, C.C. Wu and C.F. Kuo. 2010. Cobalt oxide nanosheet and CNT micro carbon monoxide sensor integrated with readout circuit on chip. Sensors 10(3): 1753–1764.

Darabi, M.A., A. Khosrozadeh, Q. Wang and M. Xing. 2015. Gum sensor: A stretchable, wearable, and foldable sensor based on carbon nanotube/chewing gum membrane. ACS Appl. Mater. Interfac. DOI:10.1021/acsami.5b08276.

Davis, J.J., K.S. Coleman, B.R. Azamian, B.C. Bagshaw and M.L.H. Green. 2003. Chemical and biochemical sensing with modified single walled carbon nanotubes. Chem. Eur. J. 9: 3732–3739.

Dhall, S., N. Jaggi and R. Nathawat. 2013. Functionalized multiwalled carbon nanotubes based hydrogen gas sensor. Sens. Actuat. A Phys. 201: 321–327.

Disma, F., C. Lenain, B. Beaudoin, L. Aymard and J.-M. Tarascon. 1997. Unique effect of mechanical milling on the lithium intercalation properties of different carbons. Solid State Ionics 98: 145–158.

Endo, N., K. Takeuchi, S. Igarashi, K. Kobori, M. Shiraishi and H.W. Kroto. 1993. The production and structure of pyrolytic carbon nanotubes (PCNTs). J. Phys. Chem. Solid 54: 1841–1848.

Fabre, B., C. Samorì and A. Bianco. 2012. Immobilization of double functionalized carbon nanotubes on glassy carbon electrodes for the electrochemical sensing of the biotin–avidin affinity. J. Electroanal. Chem. 665: 90–94.

Filik, H., A.A. Avan and S. Aydar. 2016. Simultaneous detection of ascorbic acid, dopamine, uric acid and tryptophan with Azure A-interlinked multi-walled carbon nanotube/gold nanoparticles composite modified electrode. Arab. J. Chem. 9: 471–480.

Fu, K., W. Huang, Y. Lin, L.A. Riddle, D.L. Carroll and Y.-P. Sun. 2001. Defunctionalization of functionalized carbon nanotubes. Nano Lett. 1: 439–441.

Gao, B., A. Kleinhammes, X.P. Tang, Bower, L. Fleming, Y. Wu and O. Zhou. 1999. Electrochemical intercalation of single-walled carbon nanotubes with lithium. Chem. Phys. Lett. 307: 153–157.

Gasnier, A., M.L. Pedano, F. Gutierrez, P. Labbe, G.A. Rivas and M.D. Rubianes. 2012. Glassy carbon electrodes modified with a dispersion of multi-wall carbon nanotubes in dopamine-functionalized polyethylenimine: Characterization and analytical applications for nicotinamide adenine dinucleotide quantification. Electrochim. Acta 71: 73–81.

Gea, C., J. Dua, L. Zhao, L. Wang, Y. Liu, D. Li et al. 2011. Binding of blood proteins to carbon nanotubes educes cytotoxicity. PNAS 108: 16969–16973.

Goldoni, A., L. Petaccia, S. Lizzit and R. Larciprete. 2010. Sensing gases with carbon nanotubes: a review of the actual situation. J. Phys. Condens. Matter. 22: 013001.

Goyal, R.N., D. Kaur, S.P. Singh and A.K. Pandey. 2008. Effect of graphite and metallic impurities of C60 fullerene on determination of salbutamol in biological fluids. Talanta 75: 63–69.

Goyal, R.N., S. Bishnoi and B. Agrawal. 2011. Single-walled-carbon-nanotube-modified pyrolytic graphite electrode used as a simple sensor for the determination of salbutamol in Urine. 2011. Int. J. Electrochem. doi:10.4061/2011/373498.

Hauquier, F., G. Pastorin, P. Hapiot, M. Prato, A. Bianco and B. Fabre. 2006. Carbon nanotube-functionalized silicon surfaces with efficient redox communication. Chem. Commun. 21: 4536–8.

He, J.L., Y. Yang, X. Yang, X.Y.-L. Liu, Z.-H. Liu, G.-L. Shen et al. 2006. Beta-cyclodextrin incorporated carbon nanotube-modified electrode as an electrochemical sensor for rutin. Sens. Actuat. B Chem. 114: 94–100.

Hieu, V., Nguyen, L.T. Thuy and N.C. Chien. 2008. Highly sensitive thin film NH$_3$ gas sensor operating at room temperature based on SnO$_2$/MWCNTs composite. Sens. Actuat. B-Chem. 129: 888–895.

Huang, J., Q. Lin, X. Zhang, X. He, X. Xing, W. Lian et al. 2011. Electrochemical immunosensor based on polyaniline/poly(acrylic acid) and Au-hybrid graphene nanocomposite for sensitivity enhanced detection of salbutamol. Food Res. Int. 44: 92–97.

Iijima, S. 1991. Helical microtubules of graphitic carbon. Nature 354(6348): 56.

Jaggi, N. and S. Dhall. 2014. Hydrogen gas sensing properties of multiwalled carbon nanotubes network partially coated with SnO$_2$ nanoparticles at room temperature. International Scholarly and Scientific Research & Innovation 8: 1295–1298.

Jesús, E.C., J. Li and C.R. Cabrera. 2013. Latest advances in modified/functionalized carbon nanotube-based gas sensors, syntheses and applications of carbon nanotubes and their composites. Dr. Satoru Suzuki (ed.). InTech. DOI:10.5772/52173.

Jiang, L.C. and W.D. Zhang. 2009. Electrodeposition of TiO$_2$ nanoparticles on multiwalled carbon nanotube arrays for hydrogen peroxide sensing. Electroanal. 21: 988–993.

Kang, S.Z., Z. Cui and J. Mu. 2006. High sensitivity to Cu^{2+} ions of electrodes coated with ethylenediamine-modified multi-walled carbon nanotubes. Nanotechnology 17: 4825.

Karuwan, C., A. Wisitsoraat, T. Maturos, D. Phokharatkul, A. Sappat, K. Jaruwon-grungsee et al. 2009. Flow injection based microfluidic device with carbon nanotube electrode for rapid salbutamol detection. Talanta 79: 995–1000.

Kim, B. and W.M. Sigmund. 2004. Functionalized multiwall carbon nanotube/gold nanoparticle composites. Langmuir 20: 8239–8242.

Kim, S.W., T. Kim, Y.S. Kim, H.S. Choi, H.J. Lim, S.J. Yang et al. 2012. Surface modifications for the effective dispersion of carbon nanotubes in solvents and polymers. Carbon 50: 3–33.

Kong, J., M.G. Chapline and H. Dai. 2001. Functionalized carbon nanotube for molecular hydrogen sensing, functionalized carbon nanotubes for molecular hydrogen sensors. Adv. Mater. 13: 1384–1386.

Kong, B., T.L. Yin, X.Y. Liu and W. Wei. 2007. Voltammetric determination of hydroquinone using beta cyclodextrin/poly(N-acetylaniline)/carbon nanotube composite modified glassy carbon electrode. Anal. Lett. 40: 2141–50.

Kutluay, A. and M. Aslanoglu. 2014. Nickel nanoparticles functionalized multi-walled carbon nanotubes at platinum electrodes for the detection of bromhexine. Sens. Actuat. B 192: 720–724.

Leghrib, R., R. Pavelko, A. Felten, A. Vasiliev, C. Cané, L. Gracia, I. et al. 2010. Gas sensors based on multiwall carbon nanotubes decorated with tin oxide nanoclusters. Sens. Actuat. B-Chem. 145: 411–416.

Li, H., C. Wen, Y. Zhang, D. Wu, S.-L. Zhang and Z.-J. Qiu. 2016. Accelerating gas adsorption on 3D percolating carbon nanotubes. Scientific Reports | 6: 21313 | DOI:10.1038/srep21313.

Li, L. and Y. Xing. 2007. Pt-Ru nanoparticles supported on carbon nanotubes as methanol fuel cell catalysts. J. Phys. Chem. C 111: 2803–2808.

Li, X., S.Y. Wong, W.C. Tjiu, B.P. Lyons, S.A. Oh and C.B. He. 2008. Non-covalent functionalization of multi walled carbon nanotubes and their application for conductive composites. Carbon 46: 829–831.

Li, Y., M.A. Ali, S.-M. Chen, S.Y. Yang, B.-S. Lou and F.M.A. Al-Hemaid. 2014. Poly(basic red 9) doped functionalized multi-walled carbon nanotubes as composite films for neurotransmitters biosensors. Colloid Surface. B 118: 133–139.

Liang, Y., Y. Liu, X. Guo, P. Ye, Y. Wen and H. Yang. 2014. Phytate functionalized multi walled carbon nanotubes modified electrode for determining trace Cu(II) usin differential normal pulse anodic stripping voltammetry. Sens. Actuat. 201: 107–113.

Lijun, L., Y. Laibo, C. Hao, C. Qifeng, W. Fengmin, C. Tian et al. 2008. The determination of salbutamol sulfat based on a flow-injection coupling irreversible biamperometry at poly(aminosulfonic acid)-modified glassy carbon electrode. Anal. Lett. 40: 3290–3308.

Lin, K.-C., Y.-S. Li and S.-M. Chen. 2015. Carboxy-functionalized multi-walled carbon nanotubes hybridized with poly(Xanthurenic Acid) enhance the electrocatalytic oxidation of ascorbic acid, dopamine, and uric. Acid. Int. J. Electrochem. Sci. 10: 2764–2775.

Lipomi, D.J., M. Vosgueritchian, B.C.-K. Tee, S.L. Hellstrom, J.A. Lee, C.H. Fox et al. 2011. Skin-like pressure and strain sensors based on transparent elastic films of carbon nanotubes. Nature Nanotechnology, Nature Nanotech. 6: 788–792.

Liu, G.,Y. Lin, Y. Tu and Z. Ren. 2005. Ultrasensitive voltammetric detection of trace heavy metal ions using carbon nanotube nanoelectrode array. Analyst 130: 1098.

Liu, G. and J.J. Gooding. 2009. Towards the fabrication of label-free amperometric immunosensors using SWNTs. Electrochem. Commun. 11: 1982–1985.

Liu, S., F. Petty, R. Alexander, S. Graham, T. Swager and M. Timothy. 2015. Single-walled carbon nanotube/metalloporphyrin composites for the chemiresistive detection of amines and meat spoilage. Angew. Chem. 54: 6554–6557.

Lu, Y., C. Partridge, M. Meyyappan and J. Li. 2005. A carbon nanotube sensor array for sensitive gas discrimination using principal component analysis. J. Electroanal. Chem. 593: 105–110.

Luo, Y., Y. Tian, A. Zhu, Q. Rui and H. Liu. 2009. Direct electron transfer of superoxide dismutase promoted by high conductive TiO$_2$ nanoneedles. Electrochem. Commun. 11: 174–176.

Maga, J.A. 1982. Phytate: its chemistry, occurrence, food interactions, nutritional significance, and met hods of analysis. J. Agr. Food Chem. 30: 1–9.

Mazloum-Ardakani, M. and A. Khoshroo. 2014. Electrocatalytic properties of functionalized carbon nanotubes with titanium dioxide and benzofuran derivative/ionic liquid for simultaneous determination of isoproterenol and serotonin. Electrochim. Acta 130: 634–641.

Meng, L., C. Fu and Q. Lu. 2009. Advanced technology for functionalization of carbon nanotubes. Prog. Nat. Sci. 19: 801–810.

Moghaddam, M.J., S. Taylor, M. Gao, S.M. Huang, L.M. Dai and M.J. McCall. 2004. Highly efficient binding of DNA on the sidewalls and tips of carbon nanotubes using photochemistry. Nano Lett. 4: 89–93.

Nguyen, T.D., L.D. Tran, H.L. Nguyen, B.H. Nguyen and V.H. Nguyen. 2011. Modified interdigitated arrays by novel poly(1,8-diaminonaphthalene)/carbon nanotubes composite for selective detection of mercury(II). Talanta 85: 2445–2450.

Niu, C., E.K. Sickel, R. Hoch, D. Moy and H. Tennent. 1997. Appl. Phys. Lett. 70: 1480.

Pang, Y. and T.J. Applegate. 2006. Effects of copper source and concentration on *in vitro* phytate phosphorus hydrolysis by phytase. J. Agr. Food Chem. 54: 1792.

Park, C., S. Lee, J.H. Lee, J. Lim, S.C. Lee, M. Park et al. 2007. Controlled assembly of carbon nanotubes encapsulated with amphiphilic block copolymer. Carbon 45: 2072–2078.

Peng, G., E. Trock and H. Haick. 2008. Detecting simulated patterns of lung cancer biomarkers by random network of single-walled carbon nanotubes coated with nonpolymeric organic materials. Nano Lett. 8: 3631–3635.

Pisal, S.H., N.S. Harale, T.S. Bhat, H.P. Deshmukh and P.S. Patil. 2014. Functionalized multi-walled carbon nanotubes for nitrogen sensor. J. Appl. Chem. 7: 49–52.

Planeix, J.M., N. Coustel, J. Coq, V. Brotons, P.S. Kumbhar, R. Dutartre et al. 1994. Application of carbon nanotubes as supports in heterogeneous catalysis. J. Am. Chem. Soc. 116: 7935–7936.

Raboy, V. 2003. Molecules of interest myo-inositol-1,2,3,4,5,6-hexakisphosphate. Phytochemistry 64: 1033–1043.

Raoof, J.B., R. Ojani and F. Chekin. 2009. Fabrication of functionalized carbon nanotube modified glassy carbon electrode and its application for selective oxidation and voltammetric determination of cysteamine. J. Electroanal. Chem. 633: 187–192.

Rassaei, L., M. Amiri, C.M. Cirtiu, M. Sillanpaa, F. Marken and M. Sillanpaa. 2011. Nanoparticles in electrochemical sensors for environmental monitoring. Trend. Anal. Chem. 30: 1704–1715.

Raymundo-Piñero, E., P. Azaïs, T. Cacciaguerra, S.D. Cazorla-Amoró, A. Linares-Solano and F. Béguin. 2005. KOH and NaOH activation mechanisms of multiwalled carbon nanotubes with different structural organization. Carbon 43: 786–795.

Sadrolhosseini, A.R., A.S.M. Noor, Afarin Bahrami, H.N. Lim, Z.A. Talib and M.A. Mahdi. 2014. Application of polypyrrole multi-walled carbon nanotube composite layer for detection of mercury. Lead and Iron Ions Using Surface Plasmon Resonance Technique. PlOS ONE 9: e93962.

Salmanipour, A. and M.A. Taher. 2011. An electrochemical sensor for stripping analysis of Pb(II) based on multiwalled carbon nanotube functionalized with 5-Br-PADAP. J. Solid State Electrochem. 15: 2695–2702.

Seiler, H.G., A. Sigel and H. Sigel. 1998. Handbook on Toxicity of Inorganic Compounds. Marcel-Dekker, New York.

Septiania, L.W. and B. Yuliarto. 2016. The development of gas sensor based on carbon nanotubes Ni. J. Electrochem. Soc. 163: B97–B106.

Shaban, M. and A.R. Galaly. 2016. Highly sensitive and selective *in situ* SERS detection of Pb^{2+}, Hg^{2+}, and Cd^{2+} using nanoporous membrane functionalized with CNTs. Sci. Rep. 6: 25307–25307.

Silambarasan, D., V. Vasu, K. Iyakutti, V.J. Surya and T.R. Ravindran. 2014. Reversible hydrogen storage in functionalized single-walled carbon nanotubes. Physica E: Low-dimensional. Systems and Nanostructures 60: 75–79.

Singh, R., D. Pantarotto, D. McCarthy, O. Chaloin, J. Hoebeke, C.D. Partidos et al. 2005. Binding and condensation of plasmid DNA onto functionalized carbon nanotubes; toward the construction of nanotube-based gene delivery vectors. J. Am. Chem. Soc. 127: 4388–4396.

Srivastava, S., S. Sharma, S. Kumar, S. Agrawal, M. Singh and Y. Vijay. 2009. Characterization of gas sensing behavior of multi walled carbon nanotube polyaniline composite films. Int. J. Hydrogen Energy 34: 8444–8450.

Stevens, J.L., A.Y. Huang, H. Peng, I.W. Chiang, V.N. Khabashesku and J.L. Margrave. 2003. Sidewall amino-functionalization of single-walled carbon nanotubes through fluorination and subsequent reactions with terminal diamines. Nano Letters 3(3): 331–336.

Su, H.C., M. Zhang, W. Bosze, J.-H. Lim and N.V Myung. 2013. Metal nanoparticles and DNA co-functionalized single-walled carbon nanotube gas sensors. Nanotechnology 24: 505502–5055013.

Suehiro, J., S. Yamane and K. Imasaka. 2007. Carbon nanotube-based hydrogen gas sensor, electrochemically functionalized with palladium. Presented in "Sensors IEEE" conference, Atlanta, pp. 554–557.

Tamin, N.M. and R. Angel. 2003. Phytate phosphorus hydrolysis as influenced by dietary calcium and micro-mineral source in broiler diets. J. Agri. Food Chem. 51: 4687–4693.

Tang, H., F. Yan, Q. Tai and H.L.W. Chan. 2010. The improvement of glucose bioelectrocatalytic properties of platinum electrodes modified with electrospun TiO_2 nanofibers. Biosens. Bioelectron. 25: 1646–1651.

Tran, T.H., J.-W. Lee, K. Lee, Y.D. Lee and B.-K. Ju. 2008. The gas sensing properties of single-walled carbon nanotubes deposited on an aminosilane monolayer. Sens. Actuat. B Chem. 129: 67–71.

Trung, D.D., N.D. Hoa, P.V. Tong, N.V. Duy, T. Dao, H. Chung et al. 2014. Effective decoration of Pd nanoparticles on the surface of SnO_2 nanowires for enhancement of CO gas-sensing performance. J. Hazard. Mater. 265: 124–132.

Vu, H.D., H.L. Nguyen, T.D. Nguyen, H.B. Nguyen, T.L. Nguyen and D.L. Tran. 2015. Anodic stripping voltammetric determination of Cd^{2+} and Pb^{2+} using interpenetrated MWCNT/P1,5-DAN as an enhanced sensing interface. Ionics 21: 571–578.

Wang, G.Y., X.L. Liu, G.A. Luo and Z. Wang. 2005. Alpha-Cyclodextrin incorporated carbon nanotube-coated electrode for the simultaneous determination of dopamine and epinephrine. Chin. J. Chem. 23: 297–302.

Wang, L., X. Wang, G. Shi, C. Peng and Y. Ding. 2012. Thiacalixarene covalently functionalized multiwalled carbon nanotubes as chemically modified electrode material for detection of ultratrace Pb^{2+} ions. Anal. Chem. 84: 10560–10567.

Wang, Q.H., M. Yan and R.P.H. Chang. 2001. Flat panel display prototype using gated carbon nanotube field emitters. Appl. Phys. Lett. 78: 1294–1296.

Wang, S., L.J. Ji, B. Wu, Q. Gong, Y. Zhu and J. Liang. 2008. Influence of surface treatment on preparing nanosized TiO_2 supported on carbon nanotubes. Appl. Surf. Sci. 225: 3263–3266.

Wang, Y. and C. Bi. 2013. Simultaneous electrochemical determination of ascorbic acid, dopamine and uric acid using poly (tyrosine)/functionalized multi-walled carbon nanotubes composite film modified electrode. J. Mol. Liq. 177: 26–31.

Wang, Z., E. Liu, D. Gu and Y. Wang. 2011. Glassy carbon electrode coated with polyaniline-functionalized carbon nanotubes for detection of trace lead in acetate solution. Thin Solid Films 519: 5280–5284.

Xie, H., C. Sheng, X. Chen, X. Wang, Z. Li and J. Zhou. 2012. Multi-wall carbon nanotube gas sensors modified with amino-group to detect low concentration of formaldehyde. Sens. Actuat. B Chem. 168: 34–38.

Xie, Y.H. and A.K. Soh. 2005. Investigation of non-covalent association of single walled carbon nanotube with amylose by molecular dynamics simulation. Mater. Lett. 59: 971–975.

Yang, H.F., J. Feng, Y.L. Liu, Y. Yang, Z.R. Zhang, G.L. Shen et al. 2004. Electrochemical and surface enhanced raman scattering spectroelectrochemical study of phytic acid on the silver electrode. J. Phys. Chem. B. 108: 17412–17417.

Yang, Y.K., S.Q. Qiu, C.G. He, W.J. He, L.J. Yu and X.L. Xie. 2010. Green chemical functionalization of multiwalled carbon nanotubes with polyin ionic liquids. App. Sur. Sci. 257: 1010–1014.

Yola, M.L., T. Eren and N. Atar. 2014. Molecularly imprinted electrochemical biosensor based on Fe@Au nanoparticles involved in 2-aminoethanethiol functionalized multi-walled carbon nanotubes for sensitive determination of cefexime in human plasma. Biosens. Bioelectron. 60: 277–285.

Zhad, H.R.L.Z., F. Aboufazeli, V. Amani, E. Najafi and O. Sadeghi. 2013. Modification of multiwalled carbon nanotubes by dipyridile amine for potentiometric determination of lead(II) ions in environmental samples. J. Chem. 2013: 414375–414382.

Zhang, T., M.B. Nix, B.-Y. Yoo, M.A. Deshusses and N.V. Myung. 2006. Electrochemically functionalized single-walled carbon nanotube gas sensor. Electroanalysis 18: 1153–1158.

Zheng, M., A. Jagota, M.S. Strano, Adelina, P. Santos, P. Barone, S.G. Chou et al. 2003a. Structure-based carbon nanotube sorting by sequence-dependent DNA assembly. Science 302: 1545–1548.

Zheng, M., A. Jagota, E.D. Semke, B.A. Diner, R.S. Mclean, S.R. Lustig et al. 2003b. DNA-assisted dispersion and separation of carbon nanotubes. Nat. Mater. 2: 338–342.

Zhu, Y., H. Cao, L. Tang, X. Yang and C. Li. 2009. Immobilization of horseradish peroxidase in three-dimensional macroporous TiO_2 matrices for biosensor applications. Electrochim. Acta 54: 2823–2827.

Zhu, Z., H. Wu, S. Wu, Z. Huang, Y. Zhu and L. Xi. 2013. Determination of methotrexate and folic acid by ion chromatography with electrochemical detection on a functionalized multi-wall carbon nanotube modified electrode. J. Chromatogr. A 1283: 62–67.

18

CNT-Based Bio-Nanocomposite as Electrochemical Sensors

*S.K. Suja** and G. *Jayanthi Kalaivani*

Introduction

"Nano", being the hot buzz word, is currently revolutionizing the whole world. It refers to one billionth (10^{-9}) of a material. Nanostructures are those material systems whose electrons are confined to nanoscale dimensions (1–100 nm) but are free to move in other dimensions. Their properties are therefore unique and different compared to the bulk materials. Several phenomena like quantum confinement, quantum coherence and surface/interface effects tend to affect the properties of nanostructured materials. According to Siegel, nanostructured materials are classified as zero dimension (0 D), one dimension (1 D), two dimension (2 D) and three dimension (3 D) nanostructures (Moriarty 2001, Hari 2002). The understanding and exploitation of the materials around us was started in the 20th century when the revolutionary physicist Richard Feynman who, in his lecture "There is plenty of room at the bottom" delivered in the year 1959, talked about nanoscience and nanotechnology (Feynman 1961).

Carbon nanostructures

The only element in the periodic table that occurs in different allotropic forms ranging from 0 D to 3 D is carbon owing to its different hybridization capabilities (Saito et al. 1998). Carbon nanostructures were considered to be artificially made structures that are composed of carbon atoms within nanometer range. The study of carbon nanostructures began in the late 1950s when Roger Bacon at Union carbide observed the formation of straight hollow tubes of carbon consisting of layers of graphite. In the 1970s, Morinobu Endo also found these tubes to be consisting of a single layer of graphite sheet being rolled into a tube. But the major breakthrough in the research on 1 D nanostructures occurred when the Japanese electron microscopist, Sumi Iijima in the year 1991, observed multiwalled carbon nanotubes (MWCNTs) produced by carbon-arc discharge technique (Iijima 1991). About two years later, in the year 1993, Iijima and Ichihashi, and Donald Bethune and his team, observed independently, the formation of single-walled carbon nanotubes (SWCNTs). CNTs can be broadly classified as SWCNTs and MWCNTs (Wang 2005).

Department of Chemistry, Lady Doak College, Madurai-625002, Tamil Nadu, India.
  Email: jayanthikvani@yahoo.com
* Corresponding author: senthija2007@gmail.com

SWCNTs were found to possess a cylindrical nanostructure with high aspect ratio (length to diameter ratio) which is in the order of 10^4 to 10^5 and their diameter may range from 0.7 to 1.4 nm which depends on the temperature at which these nanostructures have been synthesized. When the temperature is high, vibrational entropy will favor hexagonal rings compared to pentagonal rings thereby making the diameter larger. MWCNTs comprise of several layers of graphite cylinders with an interlayer spacing of about 3.4 Å and their diameter may be up to 100 nm (Saito et al. 1998, Wang 2005). CNTs can be prepared by employing chemical vapor deposition (CVD), carbon arc discharge methods, or laser evaporation, and high pressure carbon monoxide (HiPCO) techniques (Kannan and Burghard 2006, Simon et al. 2015). They tend to possess interesting properties (Figure 1) viz. chemical, electrical, mechanical, optical and thermal properties that make them unique and they have opened a new avenue for a wide variety of applications.

Despite these exhilarating properties, the major concern is with regard to their purity. The as-produced CNTs contain some metal nanoparticles and amorphous carbon as their major impurities. Purification can be done by treatment with acids which may lead to the removal of amorphous carbon and to oxidation of metals. But aggressive purification may end up with the shortening of CNTs and the production of oxidized CNTs. Also, CNTs without any prior treatment were found to have large surface area, high aspect ratio and tend to align themselves into huge bundles due to their strong van der Waals attraction. The formation of strong aggregates leads to a highly complex networking pattern which prevents their homogeneous dispersion thereby making them practically insoluble in any of the solvents (Mitchell et al. 2002) and may lead to high toxicity. In order to effectively utilize the properties and to reduce toxicity, CNTs need to be properly modified. Suitable modification leads to the formation of a large number of biocompatible functional groups which may prevent the toxicity of CNTs (Firme and Bandaru 2010). Modification of CNTs therefore becomes an important area of research.

There are several approaches for the modification of CNTs (Figure 2) (Tasis et al. 2006), the most important of which are physical and chemical approaches (Mittal 2011). Chemical approach or covalent modification includes sidewall and defect modification. Sidewall modification is based on change in hybridization of carbon from sp^2 to sp^3 while defect modification is based on defect transformation. Both the type cause damage to CNTs and lead to reagglomeration of the tubes. Physical approach or non-covalent modification includes polymer wrapping, surfactant adsorption and endohedral modification. Among them, both polymer wrapping and surfactant adsorption cause no damage to CNTs and prevent agglomeration of the tubes while endohedral modification is difficult and causes damage to CNTs. On the whole, non-covalent modification of CNTs protects the extended π-conjugation of the nanotubes thereby preserving the integrity of the tubes and consequently retains the structural, mechanical electronic and thermal properties of the CNTs. The advantages and the disadvantages of the different modification

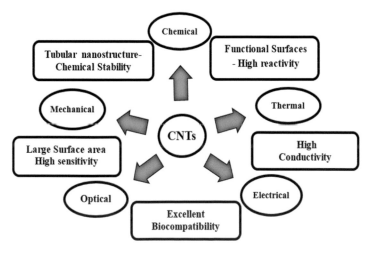

Figure 1. Properties of CNTs.

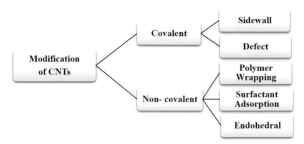

Figure 2. Methods of modification of CNTs.

methods have been tabulated (Table 1) (Ma et al. 2010). In a broader sense, both the physical and the chemical modification of CNTs will help in improving the dispersing ability of CNTs in solvents.

Along with solubility, stability should also be considered while modifying CNTs. For improving the stability, polymers play a vital role. In the present scenario, terms like "nano-biocomposites" or "biopolymer nanocomposites" are most frequently used in environmentally benign research studies. Synthetic polymers have been widely used in various applications of nanocomposites but they form a major source of waste after use owing to their poor biodegradability. Also, most of them show no biocompatibility under *in vivo* and *in vitro* environments. The modification of CNTs using biopolymers has been found to be the most promising route to preserve the properties of CNTs.

Therefore, scientists have turned their attention towards biopolymers as biodegradable materials so that several groups of natural biopolymers, such as polysaccharides, proteins, and nucleic acids have emerged in various applications (Moridi et al. 2011).

Among them, polysaccharides or carbohydrate biopolymers play a vital role in various applications and are economically feasible. They are considered as the most abundant natural biopolymers being widely distributed in nature and can be found in plants, animals and microorganisms (Poli et al. 2011). Those that are obtained from plants are found to be non-toxic, biocompatible, biodegradable and water soluble. Hence, they are suitable for different pharmaceutical, electrochemical and biomedical purposes.

Table 1. Advantage and disadvantage of modification of CNTs (Ma et al. 2010).

Method		Principle	Possible damage to CNTs	Easy to use	Re-agglomeration of CNTs
Chemical (Covalent)	Side wall	Hybridization of sp^2 to sp^3	Yes	No	Yes
	Defect	Defect transformation	Yes	Yes	Yes
Physical (Non-Covalent)	Polymer wrapping	van der Waals force, π–π stacking interactions	No	Yes	No
	Surfactant adsorption	Physical adsorption	No	Yes	No
	Endohedral	Capillary effect	No	No	Yes

Dispersion and stabilization of CNTs assisted by carbohydrate biopolymers

The dispersion and stabilization of CNTs can be simply achieved by using varieties of carbohydrate biopolymers such as starch (Star et al. 2002, Lucia et al. 2011), cellulose (Yun and Kim 2011, Moridi et al. 2011), chitosan (Zhai et al. 2006, Zhang et al. 2007, Yan et al. 2008, Laura et al. 2012), amylose (Kim et al. 2003), pullulan (Wuxu et al. 2011), gellan gum (Panhuis et al. 2007), phospholipid-dextran (Goodwin et al. 2009), alginate (Sui et al. 2012), hydraulic acid (Moulton et al. 2007), etc. The source and the structure of the different polysaccharides that can be used for the modification of CNTs were given in Table 2. Yang et al., in the year 2006, has proved that the physical purification of CNTs by chitosan modification was

Table 2. Source and structure of polysaccharides.

Polysaccharide	Composition	Source	Structure	Monomer unit(s)	Reference
Starch	Branched structure consists of short chains of α-(1→4)-linked D-glucosyl units interconnected through α-(1→6)-linkages	Potatoes, rice, corn, wheat, etc.		Amylase and amylopectin	(Varatharajan and Bertoft 2014)
Cellulose	Linear chain of β-(1→4) glucopyranose units	Green plants, algae, oomycetes, etc.		Glucose	(Johnsy and Sabapathi 2015)
Amylose	Helical polymer with α-(1→4) linked α-D-glucose units	Grain crops, potatoes, bananas, etc.		Glucose	(Nelson and Cox 2008)
Chitosan	Randomly distributed β-(1-4)-linked D-glucosamine and N-acetyl-D-glucosamine	Shrimps and other crustaceans		D-glucosamine and N-acetyl-D-glucosamine	(Islem and Rinaudo 2015)

Pectin	Linear chain of α-(1→4)-linked D-galacturonic acid	Citrus fruits, Apple pomace, Sugar beet, Sunflower, etc.		D-galactose, L-arabinose and D-xylose	(Pranati and Malviya 2011)
Inulin	The fructose units in inulin are joined by β (2→1) glycosidic bonds of various length, terminated generally by a single glucose unit	Onion, leek, garlic, banana, asparagus, chicory, and Jerusalem artichoke		Glucose and fructose	(Robertfroid 2007)
Pullulan	Linear-D-glucan with α-(1→6) linked polymer of maltotriose subunits	Strains of Aureobasidium		Maltotriose	(Leathers 2003)

Table 2 contd. ....

...Table 2 contd.

Polysaccharide	Composition	Source	Structure	Monomer unit(s)	Reference
Guar gum or Guaran	Linear chain of $(1\rightarrow 4)$-linked β-D-mannopyranose main-chain with a branched α-D-galactopyranose unit at 6 position	Endosperm of guar beans		Galactose and mannose	(Prasad et al. 2009)
Xanthan gum	$(\beta\text{-}(1\rightarrow 4)\text{-glucan})$ with trisaccharide side chains attached to alternate glucose units in the main-chain	Bacterium *Xanthomonas campestris*		Glucose, mannose, and glucuronic acid in the molar ratio 2:2:1	(Garcia-Ochoa et al. 2000, Izawa et al. 2009)
Gellan gum	O-5-acetyl and O-2-glyceryl groups on the $(1\rightarrow 3)$-linked glucose residue	Microorganism, *Sphingomonas elodea* that lives on an aquatic plant, *Elodea Canadensis*		Glucose	(Mahdi et al. 2015)

Alginate o:- alginic acid or a:ign	Linear copolymer with homopolymeric blocks of (1-4)-linked β-D-mannuronate (M) and its C-5 epimer α-L-guluronate (G) residues	Sea weeds		Mannuraonic and Guluronic acid	(Kuen and Mooney 2012)
Hyaluronic acid or Hyaluronan	Alternating units of D-glucuronic acid and N-acetyl-D-glucosamine, linked together via alternating β-1,4 and β-1,3 glycosidic bonds	Leafy greens, root vegetables and soy products		Glucuronic acid and Glucosamine	(Amir and Berkland 2013)

found to be easy and effective. The composites of CNTs with these biopolymers were found to exhibit excellent sensing performance as they were also found to be electroactive. These biopolymers tend to introduce hydrophilic groups onto the surface of CNTs rendering them biocompatible and biodegradable and making them suitable for a variety of medical applications (Bianco et al. 2011). In addition, they can help in attaching almost any desired chemical species to them, which not only enhances the solubility and biocompatibility of the carbon nanotubes but also reduces the toxicity of CNTs (Xiaoming et al. 2012, Sheva et al. 2013).

The dispersion and stabilization of CNTs in water may be understood by studying the interaction between CNTs and water through nonpolar-polar interfacial induced force of attraction. Pristine CNTs have a stronger tendency to aggregate in water making their dispersion in water difficult without hard surface treatment (Walther et al. 2001). Currently, a lot of research work is going on mainly to understand the mechanism of aggregation of CNTs in water. The orientation of water molecules around CNT surface was found to be higher than that in bulk water which causes a rise in energy of the molecules around the surface of CNTs and force them to aggregate into bundles in order to prevent or minimize the overall rise in the energy of the system. Most of the hydrophilic biopolymers have large number of hydroxyl, carboxyl, amino groups, etc., and when they wrap around the CNTs, hydrogen bonding interactions can occur between the hydrophilic groups of them and water molecules. This interaction greatly enhances the dispersion ability of CNTs and in turn improves the mechanical properties of the biopolymers. Thus, the formation of bio-nanocomposite of CNTs with the carbohydrate biopolymers leads to synergistic reinforcement in the properties of both of them.

Fabrication of bio-nanocomposite

The most common method for the fabrication of bio-nanocomposite of CNTs with biopolymers is solution processing. The general protocol (Figure 3) of it includes the dispersion of CNT powder in water by vigorous stirring followed by ultrasonication, and then mixing the CNT dispersion with a biopolymer solution followed by evaporating the solvent in a controlled way. Ultrasonication is very important as it breaks the aggregates of CNTs (Coleman et al. 2006).

Such non-covalent modification with biopolymers is more advantageous as it not only generates a new class of bioactive CNTs but also help in preserving the remarkable properties exhibited by them compared to covalent modification. Large aspect ratios of CNTs can enhance the electrical conductivity of bio-nanocomposite. However, formation of stable dispersion of CNTs in a biopolymer matrix is very difficult as they have very large surface areas and strong van der Waals force, which results in the

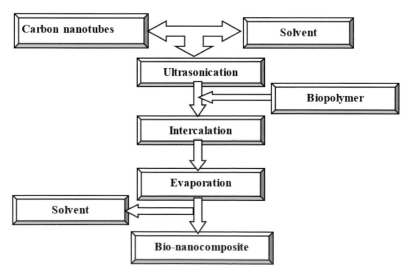

Figure 3. Schematic representation for fabrication of nanocomposites.

aggregation. Therefore, well-dispersed CNTs with enhanced affinity for the biopolymers and also greater stability of the CNT-based bio-nanocomposite are required for pertinent application (Gouvea 2011).

The stability of the bio-nanocomposite can be improved by incorporating metal nanoparticles (Au, Ag, Ni, Pt, Pd, Cu, etc.) (Kannan and Burghard 2006) or semiconducting nanoparticles (SiO_2, TiO_2, Fe_2O_3, Al_2O_3, ZrO_2, ZnO, CdS, ZnS, etc.) (Mahbubur et al. 2010) as they improve the performance properties of the individual materials.

CNT-based bio-nanocomposite as electrochemical sensors

A sensor is a device that consists of a transducer and a detector as its main component to convert physical or chemical changes of an analyte into a quantifiable electrical signal. It can be classified based on the type of energy transfer as thermal, electromagnetic, mechanical and electrochemical. Among them, electrochemical sensor is a promising tool owing to its high degree of selectivity and sensitivity (Saleh Ahammad et al. 2009). Biosensors are analytical devices that combine biological components such as enzymes, nucleic acids, proteins, etc. Electrochemical biosensors are currently the most popular type of biosensors. Conventional electrochemical biosensors are based on either glassy carbon electrodes (GCE) or metal electrodes (Au, Pt or Cu). These electrodes have poor sensitivity, less stability, low reproducibility, large response times and a high over potential for electron transfer reactions. Carbon materials have been used as components in electrochemical biosensors for over a decade. CNTs are promising materials for sensing applications due to several intriguing properties. CNTs can overcome most of these disadvantages due to their ability to undergo fast electron transfer, provide wide potential window, flexible surface chemistry, good biocompatibility and the resistance of CNT-modified electrodes to surface fouling (Kannan and Burghard 2006).

Nowadays, enzyme based electrodes that combine the specificity of enzymes with the analytical power of electrochemical devices are extremely useful for clinical diagnostics or environmental monitoring. Enzymes like glucose oxidase (GOx), urease, cholesterol oxidase (ChOx) and tyrosinase, etc., were immobilized on suitable supports for better stability and reusability and associated with transducers for biosensor monitoring of clinical metabolites or biomarkers such as glucose, urea, cholesterol, creatinine, etc. The significant milestones in the development of the different biosensors is shown in Figure 4.

A biosensor is said to be efficient based on 3S concept. They are sensitivity, selectivity and stability. For example, a glucose biosensor should be sensitive in the concentration range from few μM to 15 mM as 6 mM glucose concentration in blood is considered to be normal whereas 7 mM or higher indicates diabetic condition. Similarly, a cholesterol biosensor is said to be efficient only when it exhibits sensitivity in the concentration range 2.5 mm to 10 mM. If the level of cholesterol in blood is < 5 mM, it is said to be risk-free while a level > 6 mM is dangerous and referred to as hypercholesterolemia. Under normal conditions, the urea concentration in the blood lies between 2.5–6.7 mM (15–40 mg/dl) while pathophysiological range covers 30–150 mM (180–900 mg/dl). The normal level of creatinine concentration in blood is found to be 0.7–1.4 mg/dL. If the level exceeds, it leads to chronic kidney disease. Thus, accurate monitoring of all these biomarkers is important in early diagnosis of diseases.

Another parameter is the selectivity of the biosensor. Normally, the biological fluids under investigation viz. blood, urine, saliva, etc., have more than one component. The biosensor has to be selective in the analysis of the desired analyte. The next important parameter is the stability of the biosensor. The biosensor incorporating the enzyme needs to stable for a longer time so that reliable and reproducible results can be obtained.

The sensitivity and selectivity of CNT-based electrodes can be improved by adopting suitable method for the immobilization of the enzyme. Non covalent method viz. physical, adsorption, electrochemical entrapment and encapsulation were found to be better as it retains the activity of the enzyme. The various parameters that can affect the activity of the enzyme are surface area, hydrophilic character of the immobilizing matrix, environmental conditions and the immobilization method.

In CNT-based electrodes, CNTs can serve as the transducer communicating the signal effectively from the redox site of the enzyme to the substrate. The CNT-based bio-nanocomposite that incorporates biopolymers will help in introducing hydrophilic groups on the surface of CNT thereby increasing the

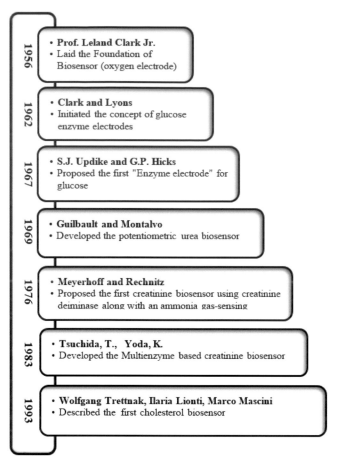

Figure 4. Milestones in the development of biosensors.

surface area. Hence, they provide a suitable platform for immobilizing large amount of the enzyme and can prevent leaching of the enzyme. In recent years, composite materials based on biopolymers, redox mediators (CNTs), metal or semiconducting nanoparticles that synergize the interesting features of the individual components have been used for the fabrication of the biosensor.

Glucose sensing

Diabetes mellitus has become the growing concern as it has gained the status of a potential epidemic all over the world. According to an estimate by International Diabetes Federation (IDF), more than 387 million adults were living with diabetes in 2014 and the number of people with the disease may reach beyond 592 million in 2035 (Joao et al. 2016). Consequently, quantization of the glucose (diabetes biomarker) is of extreme importance. Most of the conventional methods used for measuring blood glucose involve complex procedures, low reproducibility and stability. Therefore, several studies have been undertaken to develop simple, stable, selective, sensitive, cost-effective, reliable and fast enzymatic/ nonenzymatic glucose sensors (Mahbubur et al. 2010).

Detection of glucose is one of the most frequently performed routine analyses in medicine. Glucose sensors normally incorporate glucose oxidase (GOx), an enzyme which catalyses the oxidation of β-D-glucose to D-glucono-1,5-lactone, using oxygen (O_2) as electron acceptor (Figure 5). The generated hydrogen peroxide (H_2O_2) is then electrochemically detected at an appropriate electrode. This is the mechanism involved in the first generation glucose biosensor. The major drawback with this biosensor is

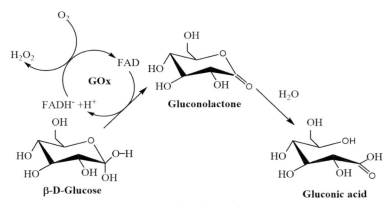

Figure 5. Mechanism of oxidation of glucose by GOx.

that there will be fluctuations in the oxygen tension leading to fluctuations in the concentrations of H_2O_2 generated, thereby producing inaccurate results.

In the second generation biosensor, a redox mediator that is dissolved in the electrolyte solution is utilized so that electron shuttling between the mediator and the enzyme will be made possible thereby overcoming the problem of oxygen fluctuations as found in the first generation biosensors. The current produced due to the oxidation of the mediator is proportional to the concentration of glucose. The introduction of the third generation biosensors helped scientists to achieve direct electron transfer between the active site of the enzyme and the electrode by co-immobilizing the mediator and the enzyme. Electrode modification with nanomaterials like carbon nanotubes has shown a great promise in the study of electrochemical biosensing (William and Ronkainen 2013).

Considering the aspects of stability and commercialization of the electronic device for glucose analysis, the fourth generation biosensors that are based on non-enzymatic analysis of glucose with the help of metal nanoparticles have emerged (Prakash et al. 2013, Yaovi et al. 2017). The various generations

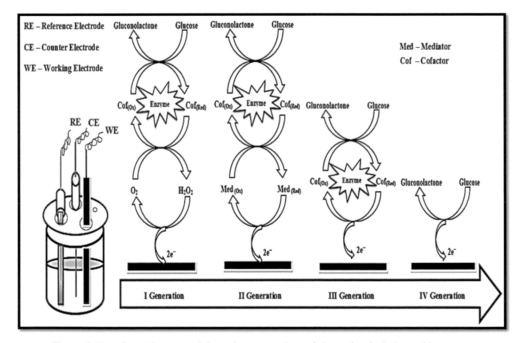

Figure 6. Experimental set up and the various generations of electrochemical glucose biosensors.

of biosensors along with the experimental set up to carry out electrochemical studies are shown in Figure 6.

The pioneering work on enzyme based glucose sensor was done by Clark and Lyons in the year 1962 and the first glucose sensor was commenced by Updike and Hicks in 1967. Since then, exploding researches on glucose oxidase (GOx) based biosensors have been made.

More than 80,000 research articles related to various biosensors have been published since the discovery of the first enzyme based electrode by Clark and Lyons, out of which ~ 60% of the articles are related to enzymatic glucose biosensors. Recent interest in nanomaterials is evident from the fact that ~ 80% of the reported glucose biosensors research exploits the properties of nanomaterials for improved biosensing. Among the various techniques used, electrochemical glucose biosensors are the most widely studied which is due to the ease of fabrication and the cost effectiveness of the biosensor. Table 3 below shows the performance characteristics of a few glucose biosensors.

Table 3. The performance characteristics of glucose biosensor.

Type of support for GOx	Biopolymer	Linear range	Sensitivity $(\mu AmM^{-1}cm^{-2})$	Detection limit	Stability	Reference
CNTs	Chitosan	-	0.52	-	-	(Liu et al. 2005)
Au–PtNPs-MWCNTs	Chitosan	0.001 mM–7.0 mM	8.53	0.2 μM		(Xiaoyang et al. 2007)
Cu/CNT	Chitosan	0.05 mM–12 mM	-	0.02 mM		(Ying et al. 2008)
MWCNT-PtNP-SiO$_2$	Chitosan	1 μM–23 mM	58.9	1 μM	-	(Zou et al. 2008)
MWCNTs	Cellulose	0.05 mM–1.0 mM	6.57	-	10 days (89%)	(Wu et al. 2009)
AgNps/CNTs	Chitosan	0.5 μM–5.0 μM	135.9	0.1 μM	9% decrease in 10 days	(Lin et al. 2009)
MWCNT	Chitosan	1 mM–10 mM	0.45	21 μM	-	(Bao et al. 2009)
AuNPs/CNT	Chitosan	5 mM–6 μM	-	3 μM	-	(Ying et al. 2009)
PTFE/GOx/ MWCNTs/PTH/ GCE	Chitosan	0.04 mM–2.5 mM	2.80	5 μM	3 weeks	(Wenwei et al. 2014)
Graphene-MWCNTs/AuNPs	Pectin	(i) 10 μM–2 mM (ii) 2 mM–5.2 mM	(i) 0.695 (ii) 0.2380	(i) 4.1 μM (ii) 0.95 mM	94.37% Two months	(Rajkumar et al. 2015)

PTH: polythionine; PTFE: polytetrafluoroethylene; NP: Nanoparticle.

Cholesterol sensing

Cholesterol, an essential lipid produced by liver, plays a vital role for normal body functioning. It is important in cell membranes as it regulates the membrane over a range of physiological temperatures. However, an excessive amount of cholesterol in the blood can significantly raise the risk of arterial disease (Zhang et al. 2012). Public concerns about the risks of high cholesterol levels in blood began to rise in the 1980s. The normal level of cholesterol concentration in blood is 0.2–0.1 g dl^{-1} (Ikonen 2008). The high levels of cholesterol in the blood serum lead to coronary heart diseases, hypertension, atherosclerosis and dysfunction of the lipid metabolism (Zhang et al. 2013).

Therefore, rapid and accurate determination of cholesterol levels is important for clinical diagnosis, and several biosensors have been proposed for this purpose. Cholesterol oxidase (ChOx) is the most commonly used enzyme in the fabrication of cholesterol biosensor. ChOx catalyzes the oxidation of cholesterol to H$_2$O$_2$ and cholest-4-en-3-one (Figure 7). The current associated with the production of

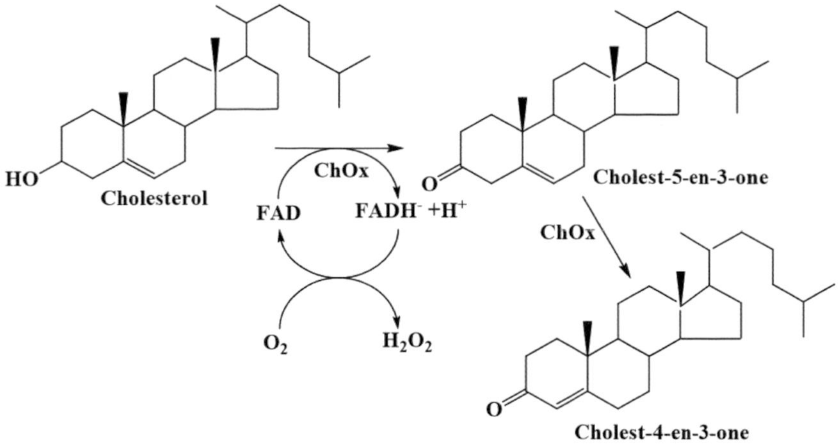

Figure 7. Mechanism of oxidation of cholesterol by ChOx.

hydrogen peroxide can be detected at a suitable potential and can be correlated with the concentration of cholesterol.

Carbon nanotubes (CNTs) and their composite with nanoparticles, owing to low potential detection of cholesterol, high sensitivity and fast response, have gained much interest for application to cholesterol biosensor (Guo et al. 2004, Yang et al. 2006, Li et al. 2005).

Li et al. (2005) have fabricated a cholesterol biosensor using MWCNTs modified carbon electrodes. They have shown that for cholesterol estimation in 100–400 mg/dl, the presence of MWCNTs promotes the electron transfer and increases the sensitivity from 0.0032 to 0.0059 μA (mg/dl)$^{-1}$.

A new type of amperometric cholesterol biosensor based on sol-gel chitosan/silica and MWCNTs organic–inorganic hybrid composite material has been developed. The hybrid composite film is used to immobilize cholesterol oxidase on the surface of Prussian blue modified glass carbon electrode. This sensor has been used to estimate free cholesterol concentration in real human blood samples. The sensitivity of the biosensor was found to be 1.55 mA/mM cm^{-2}. The linear range for cholesterol determination is 4.06 to 7.06 nM with a response time of < 20 s. (Tan et al. 2005).

Ashutosh and Gong (2008) have fabricated a composite electrode made of chitosan-SiO$_2$-multiwall carbon nanotube coated on the indium-tin oxide glass substrate and then cholesterol oxidase (ChOx) was covalently immobilized. The fabricated bioelectrode exhibited a sensitivity of 3.4 nA/mg dl^{-1} in the linear range 50–650 mg dl^{-1} with a response time of 5 s and was stable for about six months.

A biosensor for cholesterol has been fabricated by immobilizing ChOx onto MWCNT-Chitosan-Pt. The biosensor response was found to exhibit a sensitivity of 0.044 AM^{-1} cm^{-2} and a response time of about 8 s. Further, it retained the initial activity of 60% after 7 days of storage (Tsai et al. 2008).

Barik et al. (2010) have immobilized cholesterol oxidase (ChOx) covalently onto polyaniline-carboxymethyl cellulose (PANI-CMC) nanocomposite film. It was then deposited onto indium tin oxide (ITO) coated glass plate using glutaraldehyde as a cross-linker for the detection of cholesterol.

A cholesterol biosensor of ChOx/MWCNTs has been successfully coated on glassy carbon electrode (GCE). It exhibited excellent performance for the detection of cholesterol with a sensitivity of 1261.74 μA mM^{-1} cm^{-2}, a linear range of 4.68×10^{-5}–2.79×10^{-4} M, and a detection limit of 4.68×10^{-5} M compared to ChOx/GCE, the ChOx/MWCNTs/GCE (Jeng et al. 2011).

Since serum cholesterol is present in the form of ester, apart from ChOx other enzymes are also required. In a recently developed method for the determination of total cholesterol in serum samples, three enzymes viz. cholesterol esterase, ChOx and peroxidase were covalently immobilized onto gold nanoparticles and carboxylated multiwalled carbon nanotubes. The biosensor exhibited low response time of 20 seconds and a wide linear range from 0.01 mM to 5.83 mM with a detection limit of 0.01 mM. The biosensor was found to be stable for about 60 days, reusable for more than 45 times and selective towards cholesterol determination without any significant interference (Lata et al. 2016).

Urea sensing

Urea is one of the final products of protein metabolism. Urea is an omnipresent compound present in blood and various other fluids like urine. Determination of urea has been performed regularly in the medical field to study the proper functioning of the kidney. Kidney related disorders are ranked third among the other life-threatening diseases, next to cancer and cardiac ailments. Hence, accurate monitoring of urea in the biological samples becomes important (Luo and Do 2004). The primary function of the kidney is to remove wastes from the body. When the kidney malfunctions, such substances begin to accumulate. Over time, progressive kidney failure can result in uremia. Amongst the various methods, detection of urea through electrochemical mode is highly adopted and versatile. This method involves the use of electrochemical urea biosensor. In the development of electrochemical urea biosensors, immobilization of urease (EC 3.5.1.5) over modified electrodes is the key parameter which decides the sensitivity and reproducibility of the sensor. In urea biosensors, the urease enzyme immobilized to the surface of the electrode catalyzes the hydrolysis of the urea, in an overall reaction leading to the formation of ammonium, bicarbonate and hydroxide ions as shown in Figure 8.

The ionic products of the above reaction change the electronic properties of the biosensor. The commercial biosensors that are available suffer the drawbacks of high cost, complicated construction, and electrical interferences. Therefore, the development of cost effective and disposable biosensors for the detection of clinically important metabolites is of great importance. The pioneering work on potentiometric urea biosensor was developed by Guilbault, Smith and Montalvo based on a cation-selective glass electrode for the determination of the ammonium ion (Guilbault et al. 1969), after which a number of research papers on urea biosensors have been published. A few latest research papers were highlighted.

Azam and Sara (2013) have prepared polyaniline–MWCNTs composite as the matrix for urease enzyme entrapment and monitored its response amperometrically. The biosensor has shown good sensitivity of 12×10^{-5} AmM^{-1}cm^{-2} within the linear range 10^{-2} to 10^{-5} M concentration of urea with a low detection limit of 0.04 mM and has retained 50% of its original activity after 15 days.

Recently, Esma et al. have reported the incorporation of ferrocene-poly(amidoamine) (Fc-PAMAM) dendrimers into MWCNTs for the fabrication of new electrochemical urea biosensor. Amperometric studies were carried out to show the performance of the bioelectrode in urea detection and was found to exhibit a wide linear range of 0.2–1.8 mM, low detection limit of 0.05 mM, fast response of 3 s, and good sensitivity of 1.085 μA cm^{-2} mM^{-1}. The electrochemical biosensor was assessed based on real sample application performed in healthy human blood sample and excellent performance in terms of selectivity and sensitivity was observed (Esma et al. 2017).

A novel amperometric urea biosensor was fabricated by immobilizing urease onto self-assembled polyamidoamine grafted multiwalled carbon nanotube dendrimers over gold electrode. The fabricated biosensor was found to exhibit a wide linear range of 1–20 mM with a detection limit of 0.4 mM, response time of 3 s and sensitivity of 6.6 nA/mM. The biosensor was applied for the real time analysis of urea in human serum samples (Muamer et al. 2017).

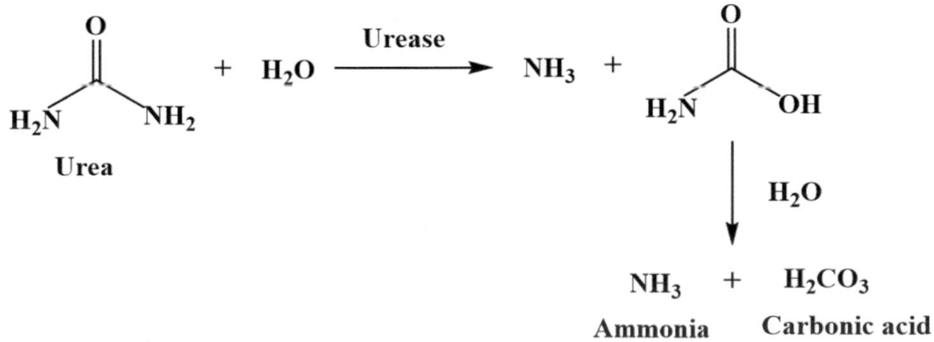

Figure 8. Mechanism of hydrolysis of urea by urease.

A non-enzymatic nickel-metal organic framework/MWCNT composite based biosensor has been recently proposed to exhibit remarkably high sensitivity (685 μAmM^{-1} cm^{-2}), low detection limit (3 μM) with a response time of 10 s, good stability (30 days) and feasibility for the analysis of urea in urine samples (Thao et al. 2017).

Creatinine sensing

Chronic end-stage kidney failure affects many patients in the world. The consequence of this disease is the significant increase in the level of metabolic waste products in blood including urea and creatinine. Creatinine plays vital role in kidney dialysis process. Since the creatinine concentration in serum and urinary excretion relative to that of urea is less affected by sepsis, trauma, fever, or dietary changing factors, its level gives a more sensitive and specific index for evaluating glomerular filtration rate and for assessing renal, thyroid, and muscular functions. Hence, there is a considerable interest in the design of biosensors specific for creatinine sensing.

The first creatinine biosensor was proposed by Meyerhoff and Rechnitz, who used creatinine deiminase along with an ammonia gas-sensing electrode (Meyerhoff and Rechnitz 1976). Enzymatic methods use a three-enzyme system viz. creatinine amidohydrolase (CA; EC 3.5.2.10), Creatine amidinohydrolase (CI; EC 3.5.3.3) and Sarcosine oxidase (SO; 1.5.3.1) that catalyze the conversion of creatinine to amperometrically measurable hydrogen peroxide.

Creatinine amidohydrolase (creatininase) obtained from *Pseudomonas putida* has a molecular weight of 1,75,000 by ultracentrifugal analysis and the molecular weight of the subunit estimated by SDS-polyacrylamide gel electrophoresis was 23,000, suggesting that the enzyme is composed of eight monomeric subunits. The enzyme was found to contain about one gram atom of zinc per monomer subunit and it catalyzes the hydrolysis of creatinine to creatine (Rikitake et al. 1979). Creatine amidinohydrolase (creatinase) is a homodimer with the subunit molecular weight of approximately 45 kDa. The two active sites of this protein are at the interface of the monomers being shared by each monomer and only the dimer is active. The enzyme catalyzes the hydrolysis of creatine (Figure 9). Sarcosine oxidase from the Arthrobacter sp. is a monomer with a molecular weight of 43 kDa. The monomeric sarcosine oxidases are flavine proteins that contain 1 mol of flavine adenine dinucleotide (FAD) that is covalently linked to the enzyme by a cysteine residue. It catalyzes the oxidative demethylation of sarcosine (N-methylglycine) and forms equimolar amounts of formaldehyde, glycine, and hydrogen peroxide (Berberich et al. 2005a, Berberich et al. 2005b). The complexity of the three-enzyme system makes the process of commercialization slow.

A new ZnONPs/Chitosan/MWCNT-COOH/PANI composite film has been synthesized on Pt electrode. Three enzymes viz. CA, CI and SO, were immobilized onto the composite film. The fabricated biosensor has shown 0.5 mM LOD at a signal to noise ratio of 3 within 10 s in the linear range from 10–650 mM and sensitivity of 0.030 mA mM^{-1} cm^{-2} (Yadav et al. 2011a).

A creatinine biosensor was fabricated using enzyme/c-MWCNT/PANI/Pt as working electrode, Ag/AgCl as reference electrode, and Pt wire as auxiliary electrode connected through potentiostat. The

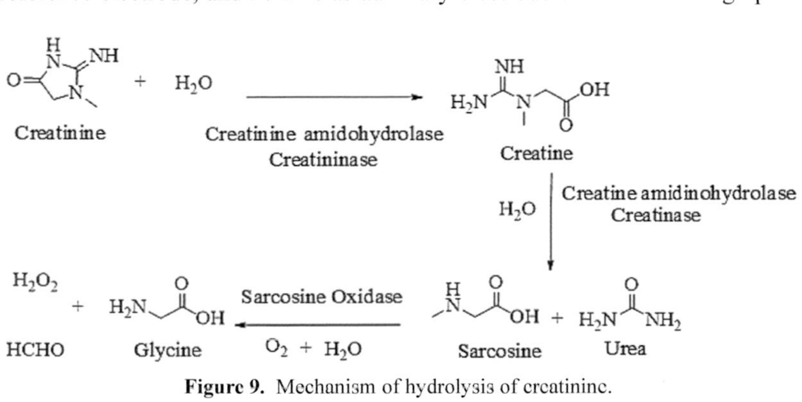

Figure 9. Mechanism of hydrolysis of creatinine.

biosensor detected creatinine levels as low as 0.1 μM were estimated at a (S/N = 3), within 5s at pH 7.5 and 35°C. The optimized biosensor showed a linear response range of 10 to 750 μM creatinine with sensitivity of 40 $\mu AmM^{-1} cm^{-2}$. The fabricated biosensor was successfully employed for determination of creatinine in human serum. The biosensor showed only 15% loss in its initial response after 180 days when stored at 4°C (Yadav et al. 2011b).

Recently in the year 2015, Utkarsh, Narang and Chauhan reported a creatinine biosensor fabricated using CA, CI and SO enzymes immobilized on AuNPs/CHIT/c-MWCNT modified GC electrode where the sensor exhibited optimum response of 4 s at 25°C in pH 7.5 and a linear dependence on creatinine concentration ranging from 0.5 to 1000 μM with a detection limit of 0.5 μM. The biosensor also exhibited long term storage stability.

Final thoughts

The research on CNT based carbohydrate biopolymer nanocomposite has been progressing rapidly over the past few years. Such nanocomposites possess the exhilarating features of both CNTs and the carbohydrate biopolymers get utilized in wide variety of applications especially electrochemical sensing applications. This chapter provided a general overview about the significance of enhancing the hydrophilicity of MWCNTs by modifying those using environmentally benign polysaccharides and their ability to sense clinically significant biomarkers viz. glucose, cholesterol, urea and creatinine. It is evident that the modification of MWCNTs using natural biopolymers will definitely pave way for exploring a new dimension in the field of electrochemical sensing of biomolecules.

Acknowledgements

The authors greatly acknowledge all those who helped us while writing the chapter. Our sincere thanks are due to the Principal, Management and Head of the Department of Chemistry for their support and encouragement.

References

Amir, F. and C. Berkland. 2013. Applications and emerging trends of hyaluronic acid in tissue engineering, as a dermal filler, and in Osteoarthritis Treatment. Acta. Biomater. 9: 7081–7092.

Ashutosh, T. and S. Gong. 2008. Electrochemical study of chitosan-SiO_2-MWNT composite. Electrodes for the fabrication of cholesterol biosensors. Electroanal. 20: 2119–2126.

Bao, -Y.W., S.-H. Hou, M. Yu, X. Qin, S. Li and Q. Chen. 2009. Layer-by-layer assemblies of chitosan/multi-wall carbon nanotubes and glucose oxidase for amperometric glucose biosensor applications. Mat. Sci. Eng. C. 29: 346–349.

Barik, A., P.R. Solanki and A. Kaushik. 2010. Polyaniline carboxymethyl cellulose nanocomposite for cholesterol detection. J. Nanosci. Nanotech. 10: 6479–6488.

Berberich, J.A., L.W. Yang, J. Madura, I. Bahar and A.J. Russell. 2005a. A stable three-enzyme creatinine biosensor.1. Impact of structure, function and environment on PEGylated and immobilized sarcosine oxidase. Acta. Biomater. 1: 173–181.

Berberich, J.A., L.W. Yang, I. Bahar and A.J. Russell. 2005b. A stable three enzyme creatinine biosensor. 2. Analysis of the impact of silver ions on creatine amidinohydrolase. Acta. Biomater. 1: 183–191.

Bethune, D.S., C.H. Kiang, M.S. Devries, G. Gorman, R. Savoy and J. Vazquez. 1993. Cobalt-catalysed growth of carbon nanotubes with single-atomic-layer walls. Nature. 363: 605–607.

Bianco, A., K. Kostarelos and M. Prato. 2011. Making carbon nanotubes biocompatible and biodegradable. Chem. Commun. 47: 10182–10188.

Clark, L.C. 1956. Monitor and control of blood and tissue oxygenation. Trans. Am. Soc. Artif. Intern. Organs. 2: 41–48.

Clark, L.C. Jr and C. Lyons. 1962. Electrode systems for continuous monitoring in cardiovascular surgery. Ann. N. Y. Acad. Sci. 102: 29–45.

Coleman, J.N., U. Khan, W.J. Blau and Y.K. Gun'ko. 2006. Small but strong: a review of the mechanical properties of carbon nanotube-polymer composites. Carbon. 44: 1624–1652.

Esma, D., M. Dervisevic, J.N. Nyangwebah and M. Şenel. 2017. Development of novel amperometric urea biosensor based on Fc-PAMAM and MWCNT bio-nanocomposite film. Sens. Actuators, B. 246: 920–926.

Feynman, R.P. 1961. There's plenty of room at the bottom. pp. 282–296. *In*: Gilbert, H.D. (ed.). Miniaturization. Reinhold Publishing Corporation. New York.

Firme, C.P. and P. R. Bandaru. 2010. Toxicity issues in the application of carbon nanotubes to biological systems. Nanomedicine: NBM 6: 245–256.

García-Ochoa, F., V.E. Santos, J.A. Casas and E. Gómez. 2000. Xanthan gum: production, recovery, and properties. Biotechnol. Adv. 18: 549–579.

Goodwin, A.P., S.M. Tabakman, K. Welsher, S.P. Sherlock, G. Prencipe and H. Dai. 2009. Phospholipid-dextran with a single coupling point: A useful amphiphile for functionalization of nanomaterials. J. Am. Chem. Soc. 131: 289–296.

Guilbauit, G.G. and J. Montalvo. 1969. A urea specific enzyme electrode. J. Am. Chem. Soc. 91: 2164–2169.

Guilbault, G.G., R.K. Smith and J.G. Montalvo Jr. 1969. Use of ion selective electrodes in enzymic analysis. Cation electrodes for deaminase enzyme systems. Anal. Chem. 41: 600–605.

Guo, M., J. Chen, J. Li, L. Nie and S. Yao. 2004. Carbon nanotubes-based amperometric cholesterol biosensor fabricated through layer-by-layer technique. Electroanal. 16: 1992–1998.

Gouvea, C. 2011. Biosensors for health applications. Biosensors for Health, Environment and Biosecurity. Prof. Pier Andrea Serra (Ed.), InTech. ISBN: 978-953-307-443-6.

Hari Singh Nalwa (ed.). 2002. Nanostructured Materials and Nanotechnology. Academic Press, London.

Iijima, S. 1991. Helical microtubules of graphitic carbon. Nature. 354: 56–58.

Iijima, S. and T. Ichihashi. 1993. Single-Shell carbon nanotubes of 1-nm diameter. Nature 363: 603–605.

Ikonen, E. 2008. Cellular cholesterol trafficking and compartmentalization. Nat. Rev. Mol. Cell Biol. 9: 125–138.

Islam, Y. and M. Rinaudo. 2015. Chitin and chitosan preparation from marine sources. Structure, Properties and Applications. Mar. Drugs 13: 1133–1174.

Izawa, H., Y. Kaneko and J. Kadokawa. 2009. Unique gel of xanthan gum with ionic liquid and its conversion into high performance hydrogel. J. Mater. Chem. 19: 6969–6972.

Joao da, R.F., K. Ogurtsova, U. Linnenkamp, L. Guariguata, T. Seuring, P. Zhang et al. 2016. IDF diabetes atlas estimates of 2014 global health expenditures on diabetes. Diabetes Res. Clin. Pract. 117: 48–54.

Jeng, Y.Y., Y. Li, S.M. Chen and K.C. Lin. 2011. Fabrication of a cholesterol biosensor based on cholesterol oxidase and multiwall carbon nanotube hybrid composites. Int. J. Electrochem. Sci. 6: 2223–2234.

Johnsy, G. and S.N. Sabapathi. 2015. Cellulose nanocrystals: synthesis, functional properties, and applications. Nanotechnol. Sci. Appl. 8: 45–54.

Kannan, B. and M. Burghard. 2006. Biosensors based on carbon nanotubes. Anal. Bioanal. Chem. 385: 452–468.

Kim, O.K., J. Je, J.W. Baldwin, S. Kooi, P.E. Pehrsson and L.J. Buckley. 2003. Solubilization of single-wall carbon nanotubes by supramolecular encapsulation of helical amylose. J. Am. Chem. Soc. 125: 4426–4427.

Kuen, Y.L. and D.J. Mooney. 2012. Alginate: properties and biomedical application. Prog. Polym. Sci. 37: 106–126.

Lata, K., V. Dhull and V. Hooda. 2016. Fabrication and optimization of ChE/ChO/HRp-AuNPs/c-MWCNTs based silver electrode for determining total cholesterol in serum. Biochem. Res. Int. 2016: 1545206.

Laura, C., K. Hibbert, F. Akindoju, C. Johnson, M. Stewart, C. K-. Brown et al. 2012. Synthesis, characterization and stability of chitosan and poly(methyl methacrylate) grafted carbon nanotubes. Spectrochim. Acta PartA. 96: 380–386.

Leathers, T.D. 2003. Biotechnological production and applications of pullulan. Appl. Microbiol. Biotechnol. 62: 468–473.

Li, G.J., M. Liao, G.Q. Hu, N.Z. Ma and P.J. Wu. 2005. Study of carbon nanotube modified biosensor for monitoring total cholesterol in blood. Biosens. Bioelectron. 20: 2140–2144.

Lin, J., C. He, Y. Zhao and S. Zhang. 2009. One-step synthesis of silver nanoparticles/carbon nanotubes/chitosan film and its application in glucose biosensor. Sens. Actuators, B. 137: 768–73.

Liu, Y., M. Wang, F. Zhao, Z. Xu and S. Dong. 2005. The direct electron transfer of glucose oxidase and glucose biosensor based on carbon nanotubes/chitosan matrix. Biosens. Bioelectron. 21: 984–988.

Lucia, M.F., V. Pettarin, S. Goyanes and C. Bernal. 2011. Starch/multi-walled carbon nanotubes composites with improved mechanical properties. Carbohydr. Poly. 83: 1226–1231.

Luo, Y.C. and J.S. Do. 2004. Urea biosensor base on PANi(urease)-Nafion(R)/Au composite electrode. Biosens. Bioelectron. 20: 15–23.

Ma, P.C., N.A. Siddiqui, G. Marom and J.K. Kim. 2010. Dispersion and functionalization of carbon nanotubes for polymer-based nanocomposites: a review. Composites, Part A. 41: 1345–1367.

Mahbubur, R., A.J. Saleh Ahammad, J.-H. Jin, S.J. Ahn and J-. Joon. 2010. A comprehensive review of glucose biosensors based on nanostructured metal-oxides. Sensors 10: 4855–4886.

Mahdi, M.H., B.R. Conway and A.M. Smith. 2015. Development of mucoadhesive sprayable gellan gum fluid gels. Int. J. Pharm. 488: 12–19.

Meyerhoff, M. and G.A. Rechnitz. 1976. Activated enzyme electrode for creatinine. Anal. Chim. Acta. 85: 277–285.

Mitchell, D.T., S.B. Lee, L. Trofin, N. Li, T.K. Nevanen, H. Söderlund and C.R. Martin. 2002. Smart nanotubes for bio separations and biocatalysis. J. Am. Chem. Soc. 124: 11864–11865.

Moriarty, P. 2001. Nanostructured materials. Rep. Prog. Phys. 64: 297.

Moridi, Z., V. Mottaghitalab and A.K. Haghi. 2011. A detailed review of recent progress in carbon nanotube/chitosan nanocomposites. Cellulose Chem. Technol. 45: 549–563.

Moulton, S.E., M. Maugey, P. Poulin and G.G. Wallace. 2007. Liquid crystal behavior of single-walled carbon nanotubes dispersed in biological hyaluronic acid solutions. J. Am. Chem. Soc. 129: 9452–9457.

Muamer, D., E. Dervisevic and M. Şenel. 2017. Design of amperometric urea biosensor based on self-assembled monolayer of cystamine/PAMAM-grafted MWCNT/urease. Sens. Actuators, B. (In press).

Nelson, David and M.M. Cox. 2008. Principles of Biochemistry. 5th ed. New York: W. H. Freeman and Company.

Panhuis, M., A. Heurtematte, W.R. Small and V.N. Paunov. 2007. Inkjet printed water sensitive transparent films from natural gum-carbon nanotube composites. Soft Matter. 3: 840–843.

Poli, A., G. Anzelmo, G. Fiorentino, B. Nicolaus, G. Tommonaro and P. Di Donato. 2011. Polysaccharides from wastes of vegetable industrial processing: New opportunities for their eco-friendly re-use. *In*: Magdy Elnashar (ed.). Biotechnology of Biopolymers.

Prakash, S., T. Chakrabarty, A.K. Singh and V.K. Shahi. 2013. Polymer thin films embedded with metal nanoparticles for electrochemical biosensors applications. Biosens. Bioelectron. 41: 43–53.

Pranati, S. and R. Malviya. 2011. Sources of Pectin, extraction and its applications in pharmaceutical industry. Ind. J. Nat. Prod. Resources 2: 10–18.

Prasad, K., H. Izawa, Y. Kaneko and J. Kadokawa. 2009. Preparation of temperature-induced shapeable film material from guar gum-based gel with an ionic liquid. J. Mater. Chem. 19: 4088–4090.

Rajkumar, D., V. Mani, S.-M. Chen, S.-T. Huang, T.-T. Huang, C.-M. Lin et al. 2015. Glucose biosensor based on glucose oxidase immobilized at gold nanoparticles decorated graphene-carbon nanotubes. Enzyme Microb. Technol. 78: 40–45.

Rikitake, K., I. Oka, M. Ando, T. Yoshimoto and D. Tsuru. 1979. Creatinine amidohydrolase (Creatininase) from *Pseudomonas putida* purification and some properties. J. Biochem. 86: 1109–1117.

Robertfroid, M.B. 2007. Inulin-type fructans: functional food ingredients. J. Nutrition. 137: 2493–2502.

Saito, R., G. Dresselhaus and M.S. Dresselhaus. 1998. Physical Properties of Carbon Nanotubes. London: Imperial College Press. 279.

Saleh Ahammad, A.J., J.-J. Lee and Md. A. Rahman. 2009. Electrochemical Sensors Based on Carbon Nanotubes. Sensors. 9: 2289–2319.

Sheva, N., M. Jafari, F. Edalat, K. Raymond, A. Khademhosseini and P. Chen. 2013. Biocompatibility of engineered nanoparticles for drug delivery. J. Controlled Release. 166: 182–194.

Simon, J.H., J.V. Anguita and S.R.P. Silva. 2015. Synthesis of carbon nanotubes. Encyclopedia of Nanotechnology 1–9.

Star, A., D.W. Steueran, J.R. Heath and J.F. Stoddar. 2002. Starched carbon nanotubes. Angew. Chem. 41: 2508–2512.

Sui, K., Y. Li, R. Liu, Y. Zhang, X. Zhao, H. Liang et al. 2012. Biocomposite fiber of calcium alginate/multi-walled carbon nanotubes with enhanced adsorption properties for ionic dyes. Carbohydr. Polym. 90: 399–406.

Tan, X., M. Li, P. Cai, L. Luo and X. Zou. 2005. An amperometric cholesterol biosensor based on multiwalled carbon nanotubes and organically modified sol-gel/chitosan hybrid composite film. Anal. Biochem. 337: 111–120.

Tasis, D., N. Tagmatarchis, A. Bianco and M. Prato. 2006. Chemistry of carbon nanotubes. Chem. Rev. 106: 1105–1136.

Thao, Q.N.T., G. Das and H.H. Yoon. 2017. Nickel-metal organic framework/MWCNT composite electrode for non-enzymatic urea detection. Sens. Actuators, B. 243: 78–83.

Tsai, Y.-C., S.-Y. Chen and C.-A. Lee. 2008. Amperometric cholesterol biosensors based on carbon nanotube-chitosan-platinum-cholesterol oxidase nanobiocomposite. Sens. Actuators, B. 135: 96–101.

Tsuchida, T. and K. Yoda. 1983. Multi-enzyme membrane electrodes for determination of creatinine and creatine in serum. Clin Chem. 29: 51–5.

Updike, S.J. and G.P. Hicks. 1967. The enzyme electrode. Nature 214: 986–988.

Utkarsh, J., J. Narang and N. Chauhan. 2015. Creatinine biosensing by immobilizing creatininase, creatinase and sarcosine oxidase on nanohybrid interface. Int. J. Adv. Res. 3: 1482–1497.

Varatharajan, V. and E. Bertoft. 2014. Structure-function relationships of starch components. Starch. 66: 1–14.

Vikas Mittal (ed.). 2011. Surface Modification of Nanotube Fillers. Technology and Engineering. John Wiley and Sons. Germany.

Walther, J.H., R. Jaffe, T. Halicioglu and P. Koumoutsakos. 2001. Carbon nanotubes in water: structural characteristics and energetics. J. Phys. Chem. B. 105: 9980–9987.

Wang, J. 2005. Carbon-nanotube based electrochemical biosensors: a review. Electroanal. 17: 7–14.

Wenwei, T., L. Li, L. Wu, J. Gong and X. Zeng. 2014. Glucose biosensor based on a glassy carbon electrode modified with polythionine and multiwalled carbon nanotubes. PLoS One. 9(5): e95030.

William, P. and N.J. Ronkainen. 2013. Immobilization techniques in the fabrication of nanomaterial-based electrochemical biosensors: a review. Sensors 13: 4811–4840.

Wolfgang, T., I. Lionti and M. Mascini. 1993. Cholesterol Biosensors Prepared by Electropolymerization of Pyrrole 5: 753–763.

Wu, X., F. Zhao, J.R. Varcoe, A.E. Thumser, C. Avignone-Rossa and R.C.T. Slade. 2009. Direct electron transfer of glucose oxidase immobilized in an ionic liquid reconstituted cellulose–carbon nanotube matrix. Bioelectrochem. 77: 64–68.

Wuxu, Z., Z. Zhang and Y. Zhang. 2011. The application of carbon nanotubes in target drug delivery systems for cancer therapies. Nano. Res. Lett. 6: 555–576.

Xiaoyang, K., Z. Mai, X. Zou, P. Cai and J. Mo. 2007. A novel glucose biosensor based on immobilization of glucose oxidase in chitosan on a glassy carbon electrode modified with gold–platinum alloy nanoparticles/multiwall carbon nanotubes. Anal. Biochem. 369: 71–79.

Xiaoming, L., L. Wang, Y. Fan, Q. Feng and F.-Z. Cui. 2012. Biocompatibility and toxicity of nanoparticles and nanotubes. J. Nanomater. 2012: 1–19.

Yadav, S., R. Devi, A. Kumar and C.S. Pundir. 2011a. Tri-enzyme functionalized ZnO-NPs/CHIT/c-MWCNT/PANI composite film for amperometric determination of creatinine. Biosens. Bioelectron. 28: 64–70.

Yadav, S., A. Kumar and C.S. Pundir. 2011b. Amperometric creatinine biosensor based on covalently immobilized enzymes onto carboxylated multiwalled carbon nanotubes/polyaniline composite film. Anal. Biochem. 419: 277–283.

Yaovi, H., S. Tingry, K. Servat, T.W. Napporn, D. Kornu and K.B. Kokoh. 2017. Nanostructured inorganic materials at work in electrochemical sensing and biofuel cells. Catalysts 7: 1–42.

Yan, L.Y., Y.F. Poon, M.B. Chan-Park, Y. Chen and Q. Zhang. 2008. Individually dispersing single-walled carbon nanotubes with novel neutral pH water-soluble chitosan derivatives. J. Phys. Chem. C. 112: 7579–7587.

Yang, M., Y. Yang, H. Yang, G. Shen and R. Yu. 2006. Layer-by-layer self-assembled multilayer films of carbon nanotubes and platinum nanoparticles with polyelectrolyte for the fabrication of biosensors. Biomaterials 27: 246–255.

Ying, W., W. Wei, J. Zeng, J.X. Liu and X. Zeng. 2008. Fabrication of a copper nanoparticle/chitosan/carbon nanotube-modified glassy carbon electrode for electrochemical sensing of hydrogen peroxide and glucose. Microchim. Acta. 160: 253.

Ying, W., W. Wei, X. Liu and X. Zeng. 2009. Carbon nanotube/chitosan/gold nanoparticle-based glucose biosensor prepared by a layer-by-layer technique. Mater. Sci. Eng. C. 29: 50–54.

Yun, S. and J. Kim. 2011. Mechanical, electrical, piezoelectric and electroactive behaviour of aligned multiwalled carbon nanotube/cellulose composites. Carbon. 49: 518–527.

Zhai, X., W. Wei, J. Zeng, S. Gong and J. Yin. 2006. Layer-by-layer assembled film based on chitosan/carbon nanotubes, and its application to electrocatalytic oxidation of NADH. Microchim. Acta. 154: 315–320.

Zhang, J.P., Q. Wang, L. Wang and A. Wang. 2007. Manipulated dispersion of carbon nanotubes with derivatives of chitosan. Carbon. 45: 1911–1920.

Zhang, H.F., R.X. Liu and J.B. Zheng. 2012. Selective determination of cholesterol based on cholesterol oxidase-alkaline phosphatase bioenzyme electrode. Analyst. 137: 5363–5367.

Zhang, Y.H., T. An, R.C. Zhang, Q. Zhou, Y. Huang and J. Zhang. 2013. Very high fructose intake increases serum LDL-cholesterol and total cholesterol: a meta-analysis of controlled feeding trials. J. Nutr. 143: 1391–1398.

Zou,Y., C. Xiang, L.-X. Sun and F. Xu. 2008. Glucose biosensor based on electrodeposition of platinum nanoparticles onto carbon nanotubes and immobilizing enzyme with chitosan-SiO_2 sol-gel. Biosens. Bioelectron. 237: 1010–1016.

19

Polypropylene Nanocomposites Containing Multi-walled Carbon Nanotubes and Graphene Oxide by Ziegler-Natta Catalysis

Jin-Yong Dong[1,2,*] and *Yawei Qin*[1,2]

Introduction

Nanotechnology is widely considered as a major area that will make great technological progress in the 21st century. In materials category, polymer nanocomposites, hailed as a 'radical alternative to conventional filled polymers or polymer blends' (see Vaia and Wagner 2004) and 'a revolutionary new class of materials that have demonstrated vastly improved properties compared to those of conventional composites' (see Prashantha et al. 2009), are experiencing an unprecedented rapid development. For instance, polymer nanocomposites containing carbon nanotubes, graphenes, and layered silicates have been shown with significant improvement in various material properties including mechanical, thermal, electrical, and gas barrier properties (see Byrne and Gun'ko 2010, Garces et al. 2000, Leuteritz et al. 2003, Manias et al. 2001, Spitalsky et al. 2010, Stankovich et al. 2006a, Vaia and Maguire 2007, Winey and Vara 2007). It is believed that once some key issues regarding their fabrications and scaling up are resolved, polymer nanocomposites will have a promising future by playing an essential role in polymeric materials applications (see Coleman et al. 2006).

Although invented more than half a century ago and has been a commodity polymer for decades, polypropylene, abbreviated as PP, still shows very strong vitality with an annual global production as well as consumption volume well exceeding 40 million tons in the past few years. It remains one of the fastest growing types in the whole thermoplastic polymer regime, with an average annual growth rate over 5% second only to polycarbonate. With the many highly appreciated material properties in mechanical, thermal, and processing respects, in conjunction with its inherent low density and excellent durability and recyclability characteristics, PP is now being seen as an effective solution to greenization of polymer

[1] CAS Key Laboratory of Engineering Plastics, Institute of Chemistry, Chinese Academy of Sciences, Beijing 100190, China.
[2] University of Chinese Academy of Sciences, Beijing 100049, China.
* Corresponding author: jydong@iccas.ac.cn

materials application by being able to replace most of the conventional polymer types made of fossil fuel that not only consume the non renewable resource but cause severe long-term geo-pollution due to their non-degrading drawbacks. Moreover, the application arena of PP itself has never stopped expanding; most notably, such high profile articles like bank notes in certain countries have been made of PP. It is quite convincing that PP is experiencing a rejuvenating age, which demands that its performance excels or at least keeps pace with the ever-growing, widespread property expectations through, say, forming polymer nanocomposites (see Pasquini 2005).

Polymer nanocomposites can generally be accessed through two routes: direct polymer/nano filler combination and *in situ* monomer polymerization on the surfaces of the nano fillers (see Kim et al. 2010, Okamoto 2004, Spitalsky et al. 2010). The former is usually carried out by solution or melt blending. The latter, however, is incorporated into the polymer formation process and is thus free of those energy and environmental concerns associated with solution and melt polymer blending processes, a highly desired advantage that will be more prominent during materials' mass production. In particular, the *in situ* approach is interesting for fabricating polyolefins [PP, PE, etc.]—the largest entity of thermoplastic polymers—nanocomposites, for the chemically inert, often crystalline polymers with very low surface energies are usually thermodynamically prevented from being exhaustively dispersed by nano fillers with very high surface energies (see Frankowski et al. 2007). However, by embedding olefin polymerization catalyst species immediately on the surfaces of nano fillers' nano entities (e.g., nano silicate layer for clay, individual nanotube for carbon nanotube, graphene oxide sheet for graphite oxide, etc.), and allowing polyolefins chains to freshly grow on the nano filler substrate, the *in situ* polymerization approach takes a detour to successfully avoid the thermodynamic barrier in polyolefin nanocomposites fabrication, which, in practice, has been adequately proved of its effectiveness (see Guo et al. 2007, He et al. 2004, Huang et al. 2010, Kim et al. 2009, Koval'chuk et al. 2008a,b, Park et al. 2008, Sun et al. 2009, Toti et al. 2008, Yang et al. 2007). In this context, for polyolefin nanocomposites' preparation, the *in situ* polymerization approach is not only "green" but essential.

For *in situ* polymerization to polyolefin nanocomposites, two systems of catalyst are currently available in a rather mature state in the field of polyolefin catalysis, namely, the traditional Ziegler-Natta catalysts and the metallocene catalysts. Between the two types of catalyst system, certainly the latter is more advanced than the former in terms of its catalytic functionality. Despite this fact, however, presently, or even in the foreseeable future, industrial PP processes will still be dominated by traditional Ziegler-Natta catalysts that are mostly with the supported $MgCl_2/TiCl_4$ family (see Galli et al. 2001). This is a well-accepted vision due both to economic and performance reasons. This much more economical catalyst system is nothing less as compared to metallocene system with regards to polymerization efficiency and polymer properties, especially when only PP homopolymers or copolymers with only slight incorporations of comonomer is to be produced. Moreover, it is generally viewed that the replacement of highly expensive metallocene catalyst system by more facile Ziegler-Natta catalyst system is crucial for practically applying the *in situ* polymerization technology to PP-based nanocomposites preparations.

PP/Multi-walled carbon nanotubes nanocomposites

General preparation

Carbon nanotubes (CNTs, including multi-walled and single-walled, MWCNTs and SWCNTs) possess excellent mechanical properties with a Young's modulus as high as 1.2 TPa and a tensile strength of 50–200 GPa (see Ma and Kim 2009). The combination of these exceptional mechanical properties with low density, high aspect ratio, and high surface area make CNTs ideal reinforcement candidates for polymers. Moreover, remarkably advantageous to carbon fibers which are brittle and easily fracture during composite processing, CNTs are highly flexible and thus their composites are well processable without significant loss of aspect ratios. Still with excellent electrically and thermally conductive properties, CNTs have become the choice of filler for many polymer composite materials targeting various performance improvements (see Byrne 2010).

PP is a large-volume thermoplastic polymer of extremely high commercial importance (see Pasquini 2005). The incorporation of CNTs into PP making PP/CNTs nanocomposites is highly desirable, as simply no more than a few percent of CNTs loading may greatly enhance many of the mechanical properties of PP, namely, elastic modulus, tensile strength, impact resistance and fracture toughness, without deteriorating its light weight characteristic (see Prashantha et al. 2009). Tasis et al. summarized the literature reports concerning the reinforcing effect of CNTs on PP, where one can find that, depending on CNTs types and processing and testing methods, as high as ~ 400% increase in tensile strength and ~ 300% increase in Young's modulus can be achieved with generally less than 10 weight percent of CNTs loadings (see Spitalsky et al. 2010). In addition, nucleation effect is frequently reported with both MWCNTs and SWCNTs in PP (see Valentini et al. 2003, Wu et al. 2008), which also contributes to the improvements of PP's mechanical and processing properties. Besides conventional reinforcement, CNTs can also impart its electrical conductivity to PP, transforming the typical insulating polymer into antistatic or even semi-conductive materials (see Bao et al. 2008, Deng et al. 2009).

For the fabrication of PP/CNTs nanocomposites, three major approaches are currently at work, which are melt compounding (see Andrews et al. 2002, Barber et al. 2004, Hong et al. 2007), solution/latex mixing (see Lee et al. 2009, Lu et al. 2008), and *in situ* polymerization (see Funck et al. 2007, Kaminsky et al. 2006a,b, Wiemann et al. 2005, Koval'chuk et al. 2008a,b). The former two are essentially polymer mixing methods which, though direct in processing, are challenging as regards CNTs' dispersion due to the vast surface energy gap between the chemically inert PP matrix and the nano-sized CNTs that have extremely large surface area and strong intra-van der Waals attractions. In contrast, the latter accesses PP/CNTs nanocomposite via surface-polymerization of monomeric propylene on CNTs, which leads to a synchronous formation of PP matrix that exempts the nanocomposite fabrication from the polymer mixing dilemma. In consequence, this *in situ* approach gives more readiness for unraveling CNTs' entangled bundles in PP, and thus is widely considered as the most effective method for PP/CNTs nanocomposites preparation. To successfully practice such an approach, as reflected in many of the reports, the key is to generate isospecific polymerization active sites on CNTs surface. In their pioneering work, Kaminsky et al. achieved this goal by basic heterogenization of metallocene-type catalyst system with CNTs (see Kaminsky et al. 2007). Metallocenes with typically C_2-symmetric structure such as *rac*-Me$_2$Si(2-Me-4-Ph-Ind)$_2$ZrCl$_2$ coupled with methylaluminoxane (MAO) form excellent catalyst system for isospecific propylene polymerization (see Kaminsky et al. 2006a). With MAO heterogenized by CNTs, polymerization active sites are formed on the surface of CNTs. With the *in situ* polymerization of monomeric propylene, PP/CNTs nanocomposites are afforded. In recent years, the preparation of PP/CNTs nanocomposites by *in situ* polymerization technique has been generally conducted following Kaminsky's approach, with the use of metallocene catalyst system. For instance, Koval'chuk et al. reported the preparation of PP/MWCNTs nanocomposite using a similar approach to Kaminsky's but carrying out the polymerization reaction in propylene bulk, by which they claimed that they had made a step forward toward the industrial realization of the *in situ* preparation of PP/CNTs nanocomposites (see Koval'chuk et al. 2008a,b).

In situ synthesis via Ziegler-Natta catalysis

For *in situ* preparation of PP/CNTs nanocomposites, there have been few examples of using the more convenient Ziegler-Natta catalyst system to undertake the task of generation of isospecific polymerization active sites on CNTs surface. The challenge lies in the fact that most of the MgCl$_2$/TiCl$_4$ Ziegler-Natta catalysts with good isospecificity control are heterogeneous catalysts which are difficult to be homogeneously combined with CNTs. In contrast, without the prerequisite of high stereospecificity, Ziegler-Natta catalyst system has been successfully applied in the preparation of PE/SWCNTs nanocomposites with *in situ* polymerization technology (see Tong et al. 2004).

This issue is resolved by, instead of struggling with the well-established 4th or 5th–generation Ziegler-Natta catalysts and their preparation processes, looking for new ways of Ziegler-Natta catalyst formation that render intimate, basically individual tubes-based CNTs/catalyst combinations. This goal is achieved by covalently anchoring Mg-Cl bond on hydroxyl-functionalized MWCNTs (MWCNTs-OH) via the efficient Grignard reagent-hydroxyl reaction, which exhaustively transforms original MWCNTs-

OH into Mg-Cl bond-covered MWCNTs that are able to support $TiCl_4$ to form MWCNTs-supported, propylene polymerization-efficient Mg-Ti complex interconnected via double Cl bridges (**I**), as illustrated in Scheme 1 (see Wang et al. 2012). Similar to other Ziegler-Natta catalysts, this catalyst set also allows the incorporation of Lewis bases such as diether compounds (e.g., 9, 9-bis(methoxymethyl)fluorine, BMMF) as internal electron donor to boost its catalyst performance (**II**).

The use of Grignard reagent and $TiCl_4$ to generate active Ti species on the surface of hydroxyl-containing heterogeneous carriers such as silica was first documented by Nowlin et al. who prepared highly active Ziegler-Natta catalyst for linear low density polyethylene (LLDPE) by interaction of the Grignard compound ethylmagnesium chloride (EtMgCl) with silica followed by further reaction with $TiCl_4$ (see Nowlin et al. 1988). Munox-Escalona et al. later also committed *n*-BuMgCl as Grignard reagent and $TiCl_4$ in the preparation of a silica-supported Ziegler-Natta catalyst for copolymerization of ethylene and 1-hexene (see Munoz-Esealona 1990). It was authentically demonstrated that silica-surfaced hydroxyls, especially those in isolated states, quantitatively reacted with Grignard reagents with the formation of surface–O-Mg-Cl moieties. Further treatment with $TiCl_4$ caused the complexation of $TiCl_4$ on the surface –O-Mg-Cl moieties. Activated by alkylaluminum cocatalysts such as $AlEt_3$, the so-formed supported catalysts were able to promote ethylene homo- and co-polymerizations. Nevertheless, no trial on propylene polymerization has ever been conducted. Therefore, no information on the characteristics of propylene polymerization such as catalyst activity and stereospecificity was currently available.

MWCNTs-OH (Figure 1) containing 3.06 wt.% of surface hydroxyl groups (O-H stretching band at $3400\ cm^{-1}$, inset FTIR spectrum in Figure 1a), is taken as the raw material. This MWCNTs-OH has an average outer diameter of 10–20 nm and average length of *ca.* 50 μm. B.E.T. surface area measurement gave a specific surface area of 164.5 m^2/g, with which the surface concentration of hydroxyl groups was roughly estimated to be 11 per cm^2. From these estimations, it may seem reasonable to deem the hydroxyl groups as isolated from each other. *n*-BuMgCl is taken as the Grignard reagent, and its reaction with MWCNTs-OH was carried out in THF, with *n*-BuMgCl in much excess of MWCNTs-OH as regards its hydroxyl population. Characterizations of the MWCNTs-OH-*n*-BuMgCl reaction product (MWCNTs/Mg) are presented in Figure 2. In generally less than 2 h, the hydroxyl groups on MWCNTs-OH had been undetectable by FTIR spectroscopy. Instead, XPS examination revealed the presence of well discernible Mg and Cl atoms. The Mg content in the so-formed MWCNTs/Mg was determined to be 1700 μmol/g via titration, which is roughly in consistence with the original hydroxyl concentration on MWCNTs-OH.

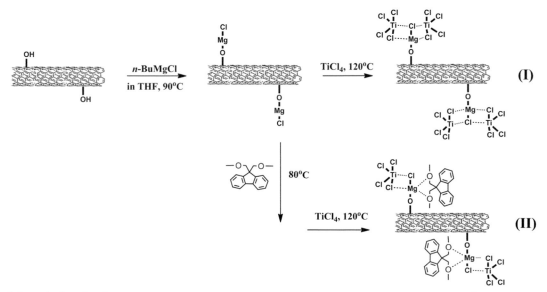

Scheme 1. Functionalization of MWCNTs with Ziegler-Natta catalytic species. Reprinted with permission from Wang, N., Y.W. Qin, Y.J. Huang and J.Y. Dong. 2012. Appl. Catal. A Chem. 435-436: 107. Copyright@Elsevier.

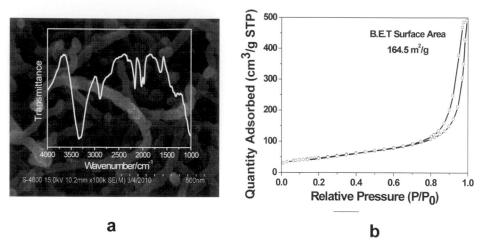

a   b

Figure 1. Characterizations of pristine hydroxyl-functionalized MWCNTs, MWCNTs-OH. (a) SEM image with inset FTIR spectrum; (b) N_2 sorption profile. Reprinted with permission from Wang, N., Y.W. Qin, Y.J. Huang and J.Y. Dong. 2012. Appl. Catal. A Chem. 435-436: 107. Copyright@Elsevier.

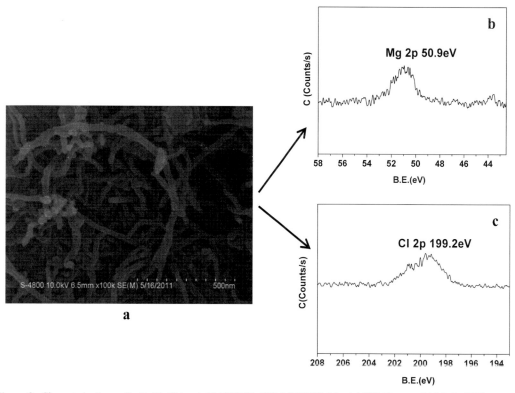

Figure 2. Characterizations of *n*-BuMgCl-treated MWCNTs-OH, MWCNTs/Mg. (a) SEM image; (b) Mg 2p XPS spectra.; (c) Cl 2p XPS spectra. Reprinted with permission from Wang, N., Y.W. Qin, Y.J. Huang and J.Y. Dong. 2012. Appl. Catal. A Chem. 435-436: 107. Copyright@Elsevier.

With Mg-modified MWCNTs-OH, MWCNTs/Mg, ready, its reaction with $TiCl_4$ to form Ziegler-Natta catalyst-functionalized MWCNTs, herein referred as MWCNTs/Mg/Ti, was completed by soaking MWCNTs/Mg in $TiCl_4$ at elevated temperature. Characterizations of MWCNTs/Mg/Ti are summarized in Figure 3. While Mg content was still determined by titration method which gave roughly an unchanged value of 1.7 mmol/g, the Ti content in MWCNTs/Mg/Ti was measured to be 1.0 mmol/g by a spectrophotometer

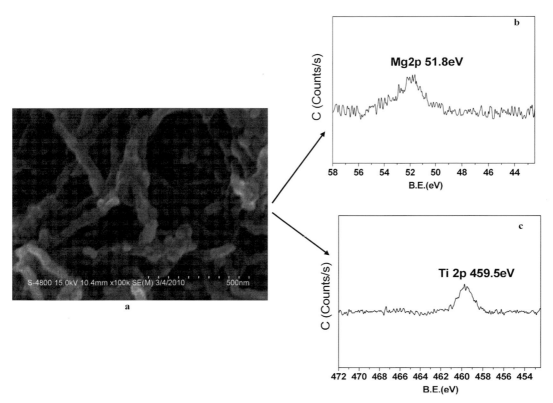

Figure 3. Characterizations of MWCNTs/Mg/Ti. (a) SEM image; (b) Mg 2p XPS spectrum and (c) Ti 2p XPS spectrum. Reprinted with permission from Wang, N., Y.W. Qin, Y.J. Huang and J.Y. Dong. 2012. Appl. Catal. A Chem. 435-436: 107. Copyright@Elsevier.

operated at the wavelength of 410 nm. The molar ratio between Ti and Mg was estimated to be 1.7, slightly lower than the expected ratio of 2.

The Ziegler-Natta catalyst-functionalized MWCNTs (**I**), MWCNTs/Mg/Ti, was then charged in propylene polymerization to test its catalytic performance. Typical Ziegler-Natta polymerization conditions were applied, including using AlEt$_3$ as cocatalyst and DDS as external electron donor. In Table 1 are summarized the detailed polymerization conditions and results. The polymerization kinetic profiles at some representative temperatures, i.e., 40°C, 60°C, and 70°C, are plotted in Figure 4. Basically, MWCNTs/Mg/Ti is fairly effective for propylene polymerization, with its kinetic profiles of polymerizations being untypical decay-type as the instantaneous polymerization rates kept reducing as polymerizations proceeded at all three temperatures (Figure 4). With the initial use of 1/10 equiv of DDS (with respect to AlEt$_3$, runs A-1 ~ 3), which is typical for Ziegler-Natta catalyst system, the yields were given at 69.5 ~ 89 g PP per gram catalyst (MWCNTs included) per hour, from which the catalytic activities were derived to be in the range of 6.6 × 10^4 ~ 8.4 × 10^4 g PP/mol Ti.h. By extracting the neat PP polymers from their MWCNTs-containing raw products, the PP polymers are revealed with relatively high molecular weights (> 300 kg/mol). Though able to render propylene polymerization, MWCNTs/Mg/Ti was found to be only of medium stereospecificity, with the isotacticity indexes (I.I.) of the PP polymers measured at no more than 81%. Increment in DDS equivalent in the catalyst system from 1/10 to 1/2 did affect the increment of I.I. to ~ 90%; however, the activity of polymerization was significantly reduced by nearly 1/3. Further increasing DDS equivalent, pitifully, didn't result in higher I.I., only to cause continued sacrifice of catalyst activity.

For improving catalytic performances of MgCl$_2$/TiCl$_4$ as well as other Ziegler-Natta catalysts, incorporation of Lewis base compounds as internal electron donors has been well known. Among the many Lewis base compounds that have proved their efficacies, diethers such as 9, 9-bis(methoxymethyl) fluorine (BMMF) have been successfully used in the 5th-generation commercial Ziegler-Natta catalysts. By

Table 1. A summary of conditions[a] and results of propylene polymerization with the two MWCNTs-immobilized Mg-Ti Ziegler-Natta catalysts, MWCNTs/Mg/Ti and MWCNTs/Mg/BMMF/Ti. Reprinted with permission from Wang, N., Y.W. Qin, Y.J. Huang and J.Y. Dong. 2012. Appl. Catal. A Chem. 435-436: 107. Copyright@Elsevier.

Run	T_p (°C)	[DDS]/ [AlEt$_3$] (mol/mol)	Yield (g PP/g cat.h)	Catalyst Activity [b]	MWCNTs content (wt.%)	I.I. [c] (%)	M_η [d] (× 10^5 g/ mol)	T_m [e] (°C)	ΔH_f [e] (J/g)
A-1	40	0.1	73.5	0.70	0.68	76.4	3.88	157.6	50.8
A-2	60	0.1	89.0	0.84	0.56	80.6	3.23	155.8	45.1
A-3	70	0.1	69.5	0.66	0.72	81.3	3.02	158.6	55.0
A-4	60	0.5	56.0	0.53	0.89	90.0	5.10	158.5	69.2
A-5	60	1.0	26.3	0.25	1.90	91.0	6.77	159.3	69.0
A-6	60	2.0	24.9	0.23	2.00	90.8	8.71	159.3	72.5
A-7	60	8.0	8.1	0.08	5.95	90.9	8.78	160.4	75.3
B-1	40	0	111.0	1.76	0.45	92.5	4.33	157.4	55.7
B-2	60	0	131.1	2.10	0.38	92.2	2.56	158.7	60.8
B-3	70	0	55.4	0.88	0.90	90.2	2.09	159.3	68.0
B-4	60	0.1	74.3	1.19	0.67	95.4	5.24	159.2	66.1
B-5	60	0.5	49.5	0.79	1.01	98.1	5.88	159.9	65.1
B-6	60	1.0	23.2	0.37	2.15	97.5	5.60	160.4	66.8
B-7	60	2.0	15.7	0.25	3.17	97.7	10.50	161.9	64.3

[a] General conditions: slurry polymerization, 100 mL hexane as solvent, 1.0 mL TEA (1.8 M) as cocatalyst, 3 atm propylene, 2 h; runs A-1 ~ A-7 use MWCNTs/Mg/Ti as catalyst with a Ti content of 5.0 wt%, runs B-1 ~ B-7 use MWCNT/Mg/BMMF/ Ti as catalyst with a Ti content of 3.0 wt%; [b] Catalyst activity, × 10^5 g PP/mol Ti.h; [c] Isotacticity index, determined by boiling heptanes extraction (12 h) as the percentage of insoluble portion; [d] Viscosity-average molecular weight; [e] Determined by DSC, data collected from the first heating scan.

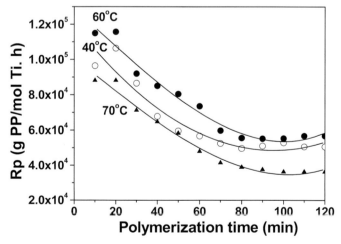

Figure 4. Kinetic profiles of propylene polymerization with MWCNTs/Mg/Ti–AlEt$_3$/DDS at 40°C, 60°C, and 70°C. (Polymerization conditions: 100 mL hexane as solvent, 1.0 mL TEA (1.8 M) as cocatalyst, 3 atm propylene, [DDS]/[AlEt$_3$] = 1/10.) Reprinted with permission from Wang, N., Y.W. Qin, Y.J. Huang and J.Y. Dong. 2012. Appl. Catal. A Chem. 435-436: 107. Copyright@Elsevier.

complexing with Mg atom through the lone electron pairs of the two O atoms, BMMF helps increase the peripheral electron density of Ti atom which is interconnected with Mg through the double Cl bridges as well as confine its spatial geometry so as to restrain the coordination and insertion variations of incoming propylene monomers, thus significantly improving the activity and stereospecificity of the catalysts. BMMF was incorporated into MWCNTs/Mg/Ti by reacting MWCNTs/Mg with BMMF prior to with TiC_4. With otherwise identical reaction conditions and procedures, this led to a significant alteration of the product composition, with not only BMMF being incorporated (4.0 wt.%) but Ti content significantly reduced from 5.0 wt.% to 3.0 wt.%. Apparently, BMMF occupied about half of the complexation vacancies on Mg atoms previously filled by $TiCl_4$. Results of XPS examination of the BMMF-incorporated MWCNTs/Mg/Ti, MWCNTs/Mg/BMMF/Ti, are shown in Figure 5. Both Mg 2p (52.1 ev) and Ti 2p (460.1 ev) signals shift noticeably from in MWCNTs/Mg/Ti (51.8 and 459.5 ev, respectively), manifesting the effect of BMMF coordination.

MWCNTs/Mg/BMMF/Ti was re-engaged in propylene polymerization, with $AlEt_3$ still being used as the cocatalyst but no DDS added at first in order to clarify the effect of BMMF incorporation (runs B-1 ~ 3). The polymerization results are also summarized in Table 1. In Figure 6 are plotted the kinetic profiles of polymerizations at 40°C, 60°C, and 70°C, respectively.

Comparing runs B-1 ~ 3 with A-1 ~ 3, one can see that, even without the use of DDS, the BMMF-containing MWCNTs/Mg/BMMF/Ti rendered PP of isotacticity indexes already exceeding 90%, suggesting its much higher stereospecificity. Moreover, its catalytic activities as measured per mole of Ti were also increased from the prototypical MWCNTs/Mg/Ti, particularly at polymerization temperatures of 40°C and 60°C (runs B-1 and B-2) when those increments could reach almost 2.5 folds. With the inherent high stereospecificity and activity, the use of DDS during polymerization further boosted the isotacticity indexes

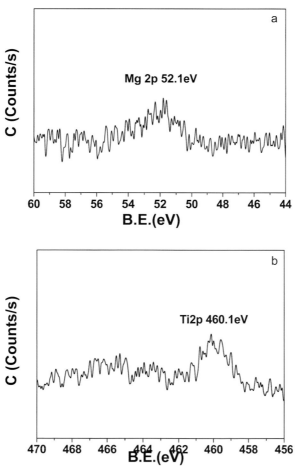

Figure 5. (a) Mg 2p and (b) Ti 2p XPS spectra of MWCNTs/Mg/BMMF/Ti. Reprinted with permission from Wang, N., Y.W. Qin, Y.J. Huang and J.Y. Dong. 2012. Appl. Catal. A Chem. 435-436: 107. Copyright@Elsevier.

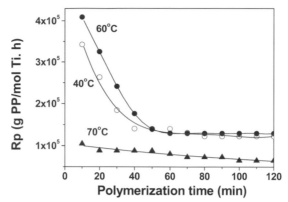

Figure 6. Kinetic profiles of propylene polymerization with MWCNTs/Mg/BMMF/Ti–AlEt$_3$ at 40°C, 60°C, and 70°C. (Polymerization conditions: 100 mL hexane as solvent, 1.0 mL TEA (1.8 M) as cocatalyst, 3 atm propylene, no DDS added.) Reprinted with permission from Wang, N., Y.W. Qin, Y.J. Huang and J.Y. Dong. 2012. Appl. Catal. A Chem. 435-436: 107. Copyright@Elsevier.

of the resultant PP polymers, with, for example, PP of 98.1% of isotacticity index being afforded when 0.5 equiv of DDS were added with respect to AlEt$_3$ (run B-5). The kinetic profiles of propylene polymerization with MWCNTs/Mg/BMMF/Ti were of typical decay-type at polymerization temperatures of 40°C and 60°C (Figure 6), with the instantaneous polymerization rates becoming gradually steady after 50 min of polymerization at ~ 1.3 × 10^5 gPP/molTi.h. This relatively high polymerization rate could be maintained at least within 2 hours period, thus allowing control of the relative amounts of PP and MWCNTs in the polymerization products so as to achieve PP/MWCNTs nanocomposites of different MWCNTs loadings.

The PP/MWCNTs nanocomposites samples prepared with both MWCNTs/Mg/Ti and MWCNTs/Mg/BMMF/Ti were determined by their chemical compositions and physical properties in addition to I.I., including MWCNTs loadings, PP viscosity-average molecular weights (M_η), melting temperatures (T_m) and thermal enthalpies during melting (ΔH). MWCNTs loadings were determined by TGA as the percentage of 600°C char residues. PP/MWCNTs nanocomposites with MWCNTs loadings ranging from 0.38 to 5.95 wt.% were obtained, controlled mainly by the yields of polymerizations. M_η of PP were in the range of 209 k to 1050 k, depending mainly on polymerization temperatures and DDS equiv, with no significant distinctions between MWCNTs/Mg/Ti and MWCNTs/Mg/BMMF/Ti. T_m and ΔH were determined by DSC from the first heating scan, the values of which were of typical PP polymers (157.6–161.9°C) and no clear line could be drawn between polymers from MWCNTs/Mg/Ti and from MWCNTs/Mg/BMMF/Ti. The DSC thermograms (Figure 7) showed generally smooth curves with relatively sharp endothermic peaks, reflecting the overall homogeneity of the matrix PP. This implied that the catalyst-functionalized MWCNTs (both MWCNTs/Mg/Ti and MWCNTs/Mg/BMMF/Ti) were homogeneous in terms of active sites distributions in the MWCNTs as a whole.

Both MWCNTs/Mg/Ti and MWCNTs/Mg/BMMF/Ti resulted in PP/MWCNTs nanocomposites with MWCNTs being homogeneously dispersed in PP matrices primarily as individual tubes, which should be thankful to the intimacy of Zielger-Natta catalyst species and MWCNTs in these catalyst-functionalized MWCNTs that induced a wrapping-like polymerization around each individual tube. Figure 8 shows typical SEM images of the PP/MWCNTs nanocomposites retrieved directly from polymerization reactor. The wrapping-like growth of PP on MWCNTs is not a phenomenon difficult to discern.

Figure 9 shows TEM images of two PP/MWCNTs nanocomposites polymerized respectively with the two Ziegler-Natta catalyst-functionalized MWCNTs. The observation specimens were also prepared from the as-retrieved polymerization products. In fact, well dispersion of MWCNTs, irrespective of their loadings, was general in all the PP/MWCNTs nanocomposites samples. With the catalyst-functionalization of MWCNTs being as exhaustive as that of all individual nanotubes containing their shares of catalyst moieties with sufficiently high activity, highly appreciable nano-scale composite morphology should be expected in thus-resulted PP/MWCNTs nanocomposites.

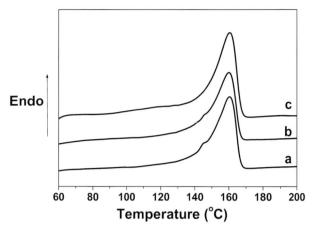

Figure 7. DSC thermograms of PP/MWCNTs nanocomposites containing (a) 1.01 wt.% (run B-5 in Table 1), (b) 2.15 wt.% (run B-6 in Table 1), and (c) 5.95 wt.% (run A-7 in Table 1) of MWCNTs. Reprinted with permission from Wang, N., Y.W. Qin, Y.J. Huang and J.Y. Dong. 2012. Appl. Catal. A Chem. 435-436: 107. Copyright@Elsevier.

Figure 8. SEM images of PP/MWCNTs nanocomposites containing (a) 2.15 wt.% (run B-6 in Table 1) and (b) 5.95 wt.% (run A-7 in Table 1) of MWCNTs. Reprinted with permission from Wang, N., Y.W. Qin, Y.J. Huang and J.Y. Dong. 2012. Appl. Catal. A Chem. 435-436: 107. Copyright@Elsevier.

PP/Graphene oxide nanocomposites

General preparation

Graphene oxide (GO), basically graphene bearing generally hydroxyl and carboxyl functional groups, is a layered carbon nanomaterial produced by the oxidation of natural graphite flakes followed by ultrasonication treatment of its aqueous solution (see Boehm et al. 1968, Hummers et al. 1958, Lerf et al. 1998, Nakajima et al. l988, Shen et al. 2009). Since GO is chemically similar to carbon nanotube and structurally analogous to layered clay (see de Heer et al. 2007, Geim et al. 2007, Katsnelson 2007, Thostenson et al. 2005), it has a great potential to simultaneously improve not only the mechanical and barrier properties but also functional properties like electrical and thermal conductivities of polymers (see Causin et al. 2006, Kalaitzidou et al. 2007a,b, Nguyen et al. 2009, Salavagione et al. 2009, Stankovich et al. 2006b, Szabo et al. 2005, Xu et al. 2002). Despite the potential advantages, the synthesis of GO-reinforced polymer nanocomposites was challenging in obtaining well dispersed GO sheets in polymer matrix, especially when the polymer is selected from nonpolar polymer category typically represented by a polyolefin like PE or PP. Generally, solution mixing is one ideal way to polymer/GO nanocomposites

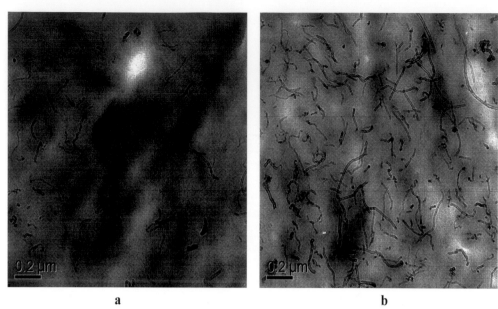

a   b

Figure 9. TEM images of PP/MWCNTs nanocomposites containing (a) 2.15 wt.% (run B-6 in Table 1) and (b) 5.95 wt.% (run A-7 in Table 1) of MWCNTs. Reprinted with permission from Wang, N., Y.W. Qin, Y.J. Huang and J.Y. Dong. 2012. Appl. Catal. A Chem. 435-436: 107. Copyright@Elsevier.

provided the polymer is readily soluble in common organic solvents. Such polar polymers as PMMA, PAN, PAA and polyesters have been successfully applied in nanocomposite fabrication with GO using this art (see Debelak and Lafdi 2007, Kim and Macosko 2008, Ramanathan et al. 2008). This approach was recently further boosted via GO surface modification with reagents like alkylchlorosilanes, alkylamine, isocyanates, etc. (see Matsuo et al. 2003, 2004, 2005, 2007a,b). For instance, GO was isocyanate-treated in DMF and mixed with polystyrene to form polystyrene/GO nanocomposites (see Stankovich et al. 2006b). Poly(vinyl alcohol) was mixed in DMSO with esterificated GO to fabricate PVA-based GO nanocomposites (see Salavagione et al. 2009). Most recently, the solution method was also applied to prepare thermoplastic polyurethane (TPU)/GO nanocomposites (see Nguyen et al. 2009). At the same time, melt mixing is also adoptable for polymer/GO nanocomposite fabrication, which in fact is applicable to not only those solvent-resolvable polar polymers but nonpolar polymers like polyolefins as well. For example, Macosko et al. employed a melt mixing method to successfully prepare poly(ethylene-2,6-naphthalate)/GO nanocomposites (see Kim and Macosko 2008). As for nanocomposites based on polyolefins, several groups started from the prototype of GO—graphite—to prepare PP-based nanocomposite (see Causin et al. 2006, Kalaitzidou et al. 2007a,b), which, due to a poor exfoliation of graphite, resulted in only limited degree of property improvement. The graphite exhibited significant impacts on the mechanical properties of PP only at loadings higher than 5 vol.% (see Kalaitzidou et al. 2007a,b) and on the thermal conductivity at contents over 20 wt.% (see Causin et al. 2006). Only until recently has Torkelson and co-workers adopted a special method (solid-state shear pulverization, SSSP) to improve the exfoliation of graphite and produced PP/graphite nanocomposites with well-dispersed nano graphite sheets (graphene) (see Stankovich et al. 2006b, Xu et al. 2008, Wakabayashi et al. 2008).

In situ synthesis via Ziegler-Natta catalysis

In situ synthesis of PP/GO nanocomposites containing well exfoliated and dispersed GO sheets by Ziegler-Natta catalysis is based on a reaction scheme illustrated in Scheme 2 (see Huang et al. 2012). First, a GO-intercalated Ziegler-Natta catalyst is synthesized by the reaction of a Grignard reagent, *n*-BuMgCl, with GO in THF followed by excess $TiCl_4$ treatment to generate Mg/Ti catalyst species on individual GO sheet surfaces. Thus-obtained GO-intercalated Ziegler-Natta catalyst thenceforth initiates efficient and iso-specific propylene polymerization that concurrently makes the PP matrix and renders exfoliation and dispersion of GO and results in the formation of PP/GO nanocomposites.

GO was prepared from natural graphite by the Hummers method. The as-prepared powdery GO contained 57.84 wt% of C, 1.82 wt% of H by Elemental Analysis measurement. As high as 40.44 wt% of O could be calculated based on the contents of C and H in GO, indicating that the degree of oxidation was very high. Morphology of GO was studied with Scanning Electron Microscopy (SEM) (Figure 10a). Obviously, layered structure of GO was retained. To further characterize the structure of GO, TEM analysis of cast film samples at the concentration of 1.0 mg/mL in THF was conducted. The TEM image of GO (Figure 10b) indicates that GO was fully exfoliated to nanosheets with micrometer-long wrinkles by ultrasonic treatment, exhibiting clearly a flake-like shape of the individual graphene oxide sheets. The inset is the measured electron diffraction pattern of the nano sheets which shows many diffraction dots, suggesting these graphene oxide sheets are highly crystalline.

Grignard reagent (e.g., n-BuMgCl, EtMgCl) had been reported to be able to react with hydroxyl groups of thermally treated SiO_2 forming RMgCl-modified SiO_2, which was taken as a support to further react with $TiCl_4$ for the synthesis of SiO_2-supported Ziegler-Natta catalyst. However, in the midst of GO reacting with n-BuMgCl the Grignard reagent, it was found, intriguingly, that the amount of n-BuMgCl

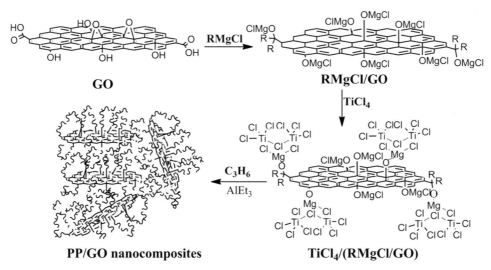

Scheme 2. Synthetic scheme for PP/GO nanocomposites. Reprinted with permission from Huang, Y.J., Y.W. Qin, Y. Zhou, H. Niu, Z.Z. Yu and J.Y. Dong. 2010. Chem. Mater. 22: 4096. Copyright@American Chemical Society.

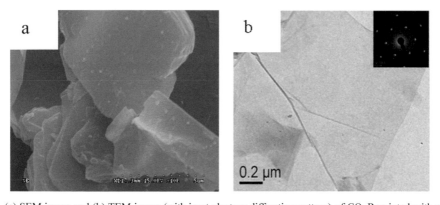

Figure 10. (a) SEM image and (b) TEM image (with inset electron diffraction pattern) of GO. Reprinted with permission from Huang, Y.J., Y.W. Qin, Y. Zhou, H. Niu, Z.Z. Yu and J.Y. Dong. 2010. Chem. Mater 22: 4096. Copyright@American Chemical Society.

used substantially affects the structure of GO that is derived. In the experiment, three dosages of the Grignard reagent, *n*-BuMgCl, were used to treat GO in THF, which, in reference to the total amount of O functional groups, were 1, 3, and 6 equivalents. The products (rGO-Mg-1~3), along with pristine GO, were thoroughly characterized by ^{13}C MAS NMR, XPS, Raman spectroscopy, WAXD, and powdery electrical conductivity measurement. Clearly, *n*-BuMgCl causes the reduction of GO. For instance, shown in Figure 11 is the ^{13}C MAS NMR spectrum of GO before and after *n*-BuMgCl treatment. Both the C-OH (70 ppm) and C-O-C (60 ppm) resonances become significantly weakened at the treatment of *n*-BuMgCl at 1 and 3 equiv. Increasing *n*-BuMgCl to 6 equiv., both resonances are hardly detectable, and the sp^2-carbon resonance continuously up-field shifting with *n*-BuMgCl becomes dominant in the NMR spectrum.

The most extensively reduced GO (rGO-Mg-3) with the highest conductivity was then continued to proceed with TiCl$_4$ immobilization. The reaction was carried out in pure TiCl$_4$ at 120°C under nitrogen atmosphere. The product was exhaustively washed with anhydrous hexane. The impregnated Ti content was determined to be 0.88 wt.% using a spectrophotometer. Other characterizations of the catalytically active Ti-bearing GO (rGO-Mg-Ti-3) were focused on a structural impact of TiCl$_4$ treatment on GO, and the results are summarized in Figure 12. Matter-of-factly, except for absorbing TiCl$_4$ on its sheets surfaces, the structure of GO was hardly affected by this inflictive process. As evidenced in Figure 12, the sp^2-carbon resonance stands as the sole signal in the ^{13}C MAS NMR spectrum (Figure 12a), with a noticeably down-field-shifted peak at 129 ppm (from 119 ppm) probably due to the immobilization of TiCl$_4$ generating many peripheral Cl atoms. Raman spectrum (Figure 12b) gives a similar D/G intensity ratio at 2.0 (versus 2.3 for rGO-Mg-3). Interestingly, besides showing the majority of C atoms are sp^2-hybridized, the C1s XPS spectrum (Figure 12c) is detected with C-O, C-O-Mg, and even the π → π* shake-up satellite peak, a characteristic of aromatic or conjugated systems. WAXD (Figure 12d) showing no (002) diffraction peak indicates the reduced GO remained exfoliated.

The reduced GO-supported Mg-Ti (rGO-Mg-Ti-3) was then combined with AlEt$_3$ to catalyze propylene polymerization in slurry conditions. The results are summarized in Table 2. In general, the combination of rGO-Mg-Ti-3 and AlEt$_3$ affords a fairly high efficiency and iso-specificity catalyst system for propylene polymerization. Thus, by varying polymerization conditions such as propylene pressures (monomer concentrations) and polymerization durations, a series of PP polymers containing 0.09–10.2 volume per cent (vol.%) of reduced GO (actually graphene sheets) had been materialized. To explore the dispersion quality of the graphene sheets in these composite materials, SEM and TEM techniques were employed and applied to 190°C hot-press-processed samples. As expected, the carbon sheets are found well dispersed in the PP matrix.

The conducting feature of graphene motivated the conductivity measurement of the PP nanocomposites even though GO was reported to have a much lower electrical conductivity than graphene. The polymerized samples were compressed to be films with thickness around 0.3–0.5 mm at 190°C for electrical conductivity measurement. Under the processing temperature, GO cannot be thermally reduced to graphene. An in-plane direct current electrical conductivity was measured at ambient temperature for each specimen and recorded

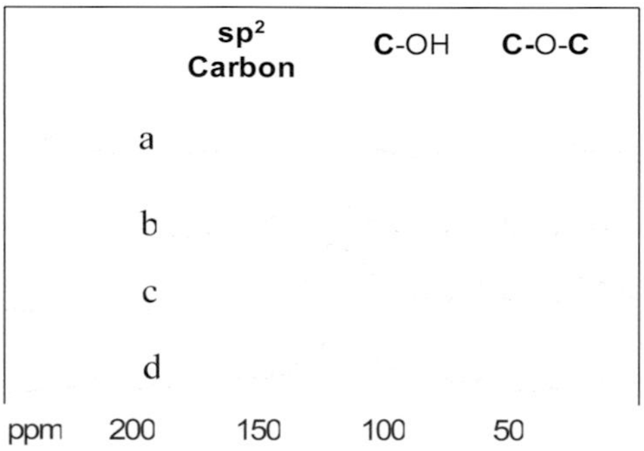

Figure 11. ^{13}C MAS NMR spectra of (a) GO, (b) GO treated with 1 equiv. of *n*-BuMgCl, rGO-Mg-1, (c) GO treated with 3 equiv. of *n*-BuMgCl, rGO-Mg-2, and (d) GO treated with 6 equiv. of *n*-BuMgCl, rGO-Mg-3. Reprinted with permission from Huang, Y.J., Y.W. Qin, Y. Zhou, H. Niu, J.Y. Dong, J.P. Hu et al. 2012. Macromol. Chem. Phys. 213: 720. Copyright@Wiley.

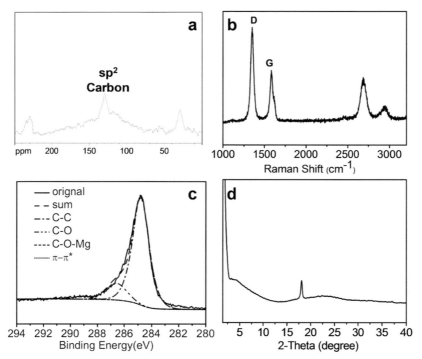

Figure 12. Characterization of rGO-Mg-3. (a) ^{13}C MAS NMR spectrum; (b) Raman spectrum; (c) C 1S XPS spectrum; (d) WAXD pattern. Reprinted with permission from Huang, Y.J., Y.W. Qin, Y. Zhou, H. Niu, J.Y. Dong, J.P. Hu et al. 2012. Macromol. Chem. Phys. 213: 720. Copyright@Wiley.

Table 2. Results of *in situ* propylene polymerization catalyzed by rGO-Mg-Ti-3. Reprinted with permission from Huang, Y.J., Y.W. Qin, Y. Zhou, H. Niu, J.Y. Dong, J.P. Hu et al. 2012. Macromol. Chem. Phys. 213: 720. Copyright@Wiley.

Run no.	Polymerization conditions [a]				Yield (g)	filler content [b] (vol. %)	M_w [c] ($\times 10^4$ g/mol)	PDI	T_m [d] (°C)	ΔH_m [d] (J/g)
	Cat. Feed (g)	T (°C)	P (MPa)	t (min)						
1	0.025	60	0.5	60	11.7	0.09	33.1	6.6	160.9	81.0
2	0.029	60	0.5	10	4.1	0.3	44.1	5.2	161.3	82.6
3	0.050	40	0.5	10	3.5	0.6	45.4	8.4	161.7	82.8
4	0.053	60	0.3	30	2.0	1.2	63.7	5.5	161.8	77.3
5	0.051	60	0.5	8	0.6	4.1	48.4	9.1	162.0	79.4
6	0.056	40	0.1	20	0.3	10.2	n.d.	n.d.	161.2	50.0

[a] General conditions: Slurry polymerization, 50 mL hexane as solvent, 2.0 mL TEA (1.8 M) as cocatalyst, [TEA]/[DDS] = 20/1. [b] The graphene sheets contents in PP composites are expressed by volume fraction v (vol %) transformed from mass fraction ω (wt. %) via the following equation: $v = w\rho_p/[w\rho_p + (1-w)\rho_g]$, where v and w are the volume and mass fractions of graphene sheets. ρ_p and ρ_g represent the density of the PP matrix and graphene nanosheets, which can be taken as 0.9 g/cm^3 and 2.2 g/cm^3, respectively. [c] Determined by GPC. [d] Melting temperature and thermal enthalpy determined by DSC.

for establishing a relationship between graphene sheets volume per cent loading and composite electrical conductivity. PP is a known insulating material, having an electrical conductivity as low as < 10^{-13} S·m^{-1} (Figure 13A). With 0.09 vol.% of the reduced graphene sheets present, the composite is measured with an electrical conductivity of 2.51 × 10^{-8} S·m^{-1}. Further increasing the loading to 0.20 vol.%, the obtained conductivity, 2.88 × 10^{-6} S·m^{-1}, has well surpassed the antistatic criterion value (10^{-6} S·m^{-1}) for thin films. Over the loading span from 0.12 vol.% to 1.2 vol.%, a rapid, 6 orders of magnitude increase of electrical conductivity is observed, which is followed by a more gradual increment that, nonetheless, still gives such high electrical conductivities as 3.92 S·m^{-1} at 1.2 vol.%, 28.5 S·m^{-1} at 4.1 vol.%, and 163.1 S·m^{-1} at 10.2 vol.%, respectively (Figure 13A). A bond percolation model is adopted to describe the overall conductivity behavior of the PP/graphene nanocomposites. The conductivity, σ_c, above the percolation threshold is treated with a power law: $\sigma_c = \sigma_f[(\phi-\phi_c)/(1-\phi_c)]^t$, where σ_f is the conductivity of the graphene sheets, ϕ the graphene sheets volume fraction, ϕ_c the percolation threshold and t the "universal critical exponent". Plot of logarithmic σ_c vs. logarithmic $(\phi-\phi_c)$ (Figure 13B) using the electrical conductivity data at different graphene sheets loadings renders the estimation of t to be 2.38 ± 0.21, $\sigma_f 10^{-1.98 \pm 0.11}$ S·m^{-1}, and ϕ_c 0.2 vol.%. The low percolation threshold (0.2 vol.%), together with the high absolute electrical conductivities, is deemed to be the result of the overall structural superiorities of the *in situ*-prepared nanocomposites that excel not only in graphene sheets dispersion but in sp^2-carbon network's completeness in the graphene sheets themselves.

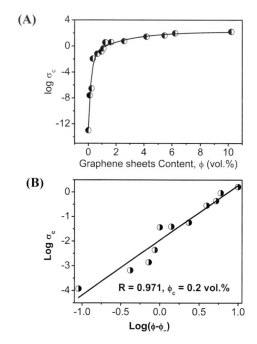

Figure 13. Electrical conductivity of the PP/graphene composites as a function of graphene sheets volume fraction. (A) Composite conductivity, σc (S m–1), is logarithmized, and plotted against filler volume fraction, φ; (B) Logσc plotted against log(φ–φc), where φc is the percolation threshold. Inset in A: the four-probe setup for DC conductivity measurement. Reprinted with permission from Huang, Y.J., Y.W. Qin, Y. Zhou, H. Niu, J.Y. Dong, J.P. Hu et al. 2012. Macromol. Chem. Phys. 213: 720. Copyright@Wiley.

Conclusions

In this chapter, we summarized the synthesis of PP-based nanocomposites containing multi-walled carbon nanotubes (MWCNTs) and graphene oxide (GO) as nanofillers using the technically challenging yet economically favorable *in situ* polymerization approach. Particularly, we introduced in detail our initiative attempt to apply Ziegler-Natta catalyst system to the *in situ* preparations of these two types of nanocomposites. We found that by simply treating surface hydroxyl-containing MWCNTs, MWCNTs-OH, with *n*-BuMgCl and TiCl$_4$, a Mg-Ti Ziegler-Natta catalyst species could be formed on MWCNTs, only that it was of moderate catalyst activity as well as stereospecificity. To resolve this issue, we resorted to an electron-donating reagent, BMMF, and incorporated it into the Mg-Ti-functionalized MWCNTs. The BMMF-incorporated Mg-Ti-functionalized MWCNTs, MWCNTs/Mg/BMMF/Ti, was found to be able to produce PP/MWCNTs nanocomposites with high PP isotacticity indexes (up to 98.0%). Together with the ability to afford highly appreciable nano-scale composite morphology, the newly set

BMMF-boosted Mg-Ti Ziegler-Natta catalyst on MWCNTs could serve as an eligible alternative to the much more expensive metallocene catalysts that currently prevail in the *in situ* preparation of PP/CNTs nanocomposites.

On the other hand, intercalation of Ziegler-Natta catalyst in GO was conducted by treating GO with the Grignard reagent of *n*-BuMgCl and $TiCl_4$ successively. It was found that treatment of GO with *n*-BuMgCl the Grignard reagent entailed triple effects on GO structure simultaneously: anchoring Mg-Cl species on the graphene oxide sheet surfaces, enlarging sheet-sheet inter-distances, and reducing the highly functionalized (oxidized) sheets back to sp^2-carbon-dominant graphene sheets. With some excess amount of *n*-BuMgCl (e.g., 6 equiv. of *n*-BuMgCl to O functionalities in GO), the three effects could be reconciled to render a Mg-Cl-functionalized graphene material ready to immobilize $TiCl_4$ and to further undergo *in situ* olefin polymerization with concomitant graphene sheets exfoliation and dispersion to prepare actually polyolefin/graphene nanocomposites. Featured by well-restored sp^2-carbon-networked graphene sheets, these PP/GO nanocomposites thus prepared possessed a rather low electrical percolation threshold (~ 0.2 vol.%) and showed high electrical conductivities.

For future studies of *in situ* preparation of PP nanocomposites which certainly includes the above two, we suggest that engineers follow suit to come up with plans that allow petrochemical industry to accept the *in situ* approach and practice the nanocomposite preparation in industrial PP processes. We believe that only by then will this scientifically interesting chemistry become a practically meaningful technology.

References

Andrews, R., D. Jacques, D.L. Qian and T. Rantell. 2002. Multiwall carbon nanotubes: Synthesis and application. Acc. Chem. Res. 35: 1008–1017.

Bao, H.D., Z.X. Guo and J. Yu. 2008. Effect of electrically inert particulate filler on electrical resistivity of polymer/multi-walled carbon nanotube composites. Polymer 49: 3826–3831.

Barber, A.H., S.R. Cohen, S. Kenig and H.D. Wagner. 2004. Interfacial fracture energy measurements for multi-walled carbon nanotubes pulled from a polymer matrix. Compos Sci. Tech. 64: 2283–2289.

Boehm, H.P. and W. Scholz. 1968. New results on chemistry of graphite oxide. Carbon 6: 226–227.

Byrne, M.T. and Y.K. Gun'ko. 2010. Recent advances in research on carbon nanotube-polymer composites. Adv. Mater. 22: 1672–1688.

Causin, V., C. Marega, A. Marigo, G. Ferrara and A. Ferraro. 2006. Morphological and structural characterization of polypropylene/conductive graphite nanocomposites. Eur. Polym. J. 42: 3153–3161.

Coleman, J.N., U. Khan and Y.K. Gun'ko. 2006. Mechanical reinforcement of polymers using carbon nanotubes. Adv. Mater. 18: 689–706.

Debelak, B. and K. Lafdi. 2007. Use of exfoliated graphite filler to enhance polymer physical properties. Carbon 45: 1727–1734.

de Heer, W.A., C. Bergera, X. Wu, P.N. First, E.H. Conrad, X. Li et al. 2007. Epitaxial graphene. Solid State Commun. 143: 92–100.

Deng, H., R. Zhang, E. Bilotti, J. Loos and T. Pejis. 2009. Conductive polymer tape containing highly oriented carbon nanofillers. J. Appl. Polym. Sci. 113: 742–751.

Frankowski, D.J., S.A. Khan and R.J. Spontak. 2007. Chain-scission-induced intercalation as a facile route to polymer nanocomposites. Adv. Mater. 19: 1286.

Funck, A. and W. Kaminsky. 2007. Polypropylene carbon nanotube composites by *in situ* polymerization. Compos. Sci. Tech. 67: 906–915.

Galli, P. and G. Vecellio. 2001. Technology: driving force behind innovation and growth of polyolefins. Prog. Polym. Sci. 26: 1287–1336.

Garces, J.M., D.J. Moll, J. Bicerano, R. Fibiger and D.G. McLeod. 2000. Polymeric nanocomposites for automotive applications. Adv. Mater. 12: 1835–1839.

Geim, A.K. and K.S. Novoselov. 2007. The rise of graphene. Nat. Mater. 6: 183–191.

Guo, N., S.A. DiBenedetto, D.–K. Kwon, L. Wang, M.T. Russel, M.T. Lanagan et al. 2007. Supported metallocene catalysis for *in situ* synthesis of high energy density metal oxide nanocomposites. J. Am. Chem. Soc. 129: 766–767.

He, A.H., H.Q. Hu, Y.J. Huang, J.Y. Dong and C.C. Han. 2004. Isotactic poly(propylene)monoalkylimidazolium-modified montmorillonite nanocomposites: Preparation by intercalative polymerization and thermal stability study. Macromol. Rapid Commun. 25: 2008–2013.

Hong, C.E., J.H. Lee, K. Prashantha, G.H. Yoo and S.G. Advani. 2007. Effects of oxidative conditions on properties of multi-walled carbon nanotubes in polymer nanocomposites. Compos Sci. Tech. 67: 1027–1034.

Huang, Y.J., Y.W. Qin, Y. Zhou, H. Niu, Z.Z. Yu and J.Y. Dong. 2010. Polypropylene/graphene oxide nanocomposites prepared by *in situ* Ziegler-Natta polymerization. Chem. Mater. 22: 4096–4102.

Huang, Y.J., Y.W. Qin, Y. Zhou, H. Niu, J.Y. Dong, J.P. Hu et al. 2012. Reduction of graphite oxide with a grignard reagent for facile *in situ* preparation of electrically conductive polyolefin/graphene nanocomposites. Macromol. Chem. Phys. 213: 720–728.

Hummers, W. and R. Offeman. 1958. Preparation of graphitic oxide. J. Am. Chem. Soc. 80: 1339–1339.

Kalaitzidou, K., H. Fukushima and L.T. Drzal. 2007a. Mechanical properties and morphological characterization of exfoliated graphite-polypropylene nanocomposites. Composites: Part A 38: 1675–1682.

Kalaitzidou, K., H. Fukushima and L.T. Drzal. 2007b. Multifunctional polypropylene composites produced by incorporation of exfoliated graphite nanoplatelets. Carbon 45: 1446–1452.

Kaminsky, W., A. Funck and K. Wiemann. 2006a. Nanocomposites by *in situ* polymerization of olefins with metallocene catalysts. Macromol. Symp. 239: 1–6.

Kaminsky, W. and K. Wiemann. 2006b. Polypropene nanocomposites by metallocene/MAO catalysts. Compos. Interf. 13: 365–375.

Katsnelson, M.I. 2007. Graphene: Carbon in two dimensions. Mater Today 10: 1–8.

Kim, H. and C.W. Macosko. 2008. Morphology and properties of polyester/exfoliated graphite nanocomposites. Macromolecules 41: 3317–3327.

Kim, H., A.A. Abdala and C.W. Macosko. 2010. Graphene/polymer nanocomposites. Macromolecules 43: 6515–6530.

Kim, J., S.M. Hong, S. Kwak and Y. Seo. 2009. Physical properties of nanocomposites prepared by *in situ* polymerization of high-density polyethylene on multiwalled carbon nanotubes. Phys. Chem. Chem. Phys. 11: 10851–10859.

Koval'chuk, A.A., A.N. Shchegolikhin, V.G. Shevchenko, P.M. Nedorezova, A.N. Klyamkina and A.M. Aladyshev. 2008a. Synthesis and properties of polypropylene/multiwall carbon nanotube composites. Macromolecules 41: 3149–3156.

Koval'chuk, A.A., V.G. Shevchenko, A.N. Shchegolikhin, P.M. Nedorezova, A.N. Klyamkina and A.M. Aladyshev. 2008b. Effect of carbon nanotube functionalization on the structural and mechanical properties of polypropylene/MWCNT composites. Macromolecules 41: 7536–7542.

Lee, J.-I., S.-B. Yang and H.-T. Jung. 2009. Carbon nanotubes-polypropylene nanocomposites for electrostatic discharge applications. Macromolecules 42: 8328–8334.

Lerf, A., H. He, M. Forster and J. Klinowski. 1998. Structure of graphite oxide revisited. J. Phys. Chem. B 102: 4477–4482.

Leuteritz, A., D. Pospiech, B. Kretzschmar, M. Willeke, D. Jehnichen, U. Jentzsch et al. 2003. Progress in polypropylene nanocomposite development. Adv. Eng. Mater. 5: 678–681.

Lu, K.B., N. Grossiord, C.E. Koning, H.E. Miltner, B.V. Mele and J. Loos. 2008. Carbon nanotube/isotactic polypropylene composites prepared by latex technology: Morphology analysis of CNT-induced nucleation. Macromolecules 41: 8081–8085.

Ma, P.C. and J.-K. Kim. 2009. Carbon Nanotubes for Polymer Reinforcement, CRC Press.

Manias, E., A. Touny, L. Wu, K. Strawhecker, B. Lu and T.C. Chung. 2001. Polypropylene/montmorillonite nanocomposites. Review of the synthetic routes and materials properties. Chem. Mater. 13: 3516–3523.

Matsuo, Y., K. Watanabe, T. Fukutsuka and Y. Sugie. 2003. Characterization of n-hexadecylalkylamine-intercalated graphite oxides as sorbents. Carbon 41: 1545–1550.

Matsuo, Y., T. Fukunaga, T. Fukutsuka and Y. Sugie. 2004. Silylation of graphite oxide. Carbon 42: 2117–2119.

Matsuo, Y., T. Tabata, T. Fukunaga, T. Fukutsuka and Y. Sugie. 2005. Preparation and characterization of silylated graphite oxide. Carbon 43: 2875–2882.

Matsuo, Y., T. Miyabe, T. Fukutsuka and Y. Sugie. 2007a. Preparation and characterization of alkylamine-intercalated graphite oxides. Carbon 45: 1005–1012.

Matsuo, Y., Y. Nishino, T. Fukutsuka and Y. Sugie. 2007b. Introduction of amino groups into the interlayer space of graphite oxide using 3-aminopropylethoxysilanes. Carbon 45: 1384–1390.

Munoz-Esealona, A., A. Fuentes, J. Liscano, A. Albornoz, K. Tominaga and S. Kazuo. 1990. High active Ziegler-Natta catalysts for homo- and copolymerization of ethylene by supporting a Grignard compound and $TiCl_4$ on SiO_2. Stud. Surf. Sci. Catal. 56: 377–404.

Nakajima, T., A. Mabuchi and R. Hagiwara. 1988. A new structure model of graphite oxide. Carbon 26: 357–361.

Nguyen, D.A., Y.R. Lee, A.V. Raghu, H.M. Jeong, C.M. Shinb and B.K. Kim. 2009. Morphological and physical properties of a thermoplastic polyurethane reinforced with functionalized graphene sheet. Polym. Int. 58: 412–417.

Nowlin, T.E., J.V. Kissin and K.P. Wagner. 1988. High-activity Ziegler-Natta catalysts for the preparation of ethylene copolymers. J. Polym. Sci. Polym. Chem. 26: 755–764.

Okamoto, M. 2004. Polymer/Clay Nanocomposites, American Scientific Publishers, Stevenson Ranch, California.

Park, S., S.W. Yoon, H. Choi, J.S. Lee, W.K. Cho, J. Kim et al. 2008. Pristine multiwalled carbon nanotube/polyethylene nanocomposites by immobilized catalysts. Chem. Mater. 20: 4588–4594.

Pasquini, N. 2005. Polypropylene Handbook, 2nd Edition, Hanser, Munchen, Germany.

Prashantha, K., J. Soulestin, M.F. Lacrampe and P. Krawczak. 2009. Present status and key challenges of carbon nanotubes reinforced polyolefins: a review on nanocomposites manufacturing and performance issues. Polym. Polym. Compos. 17: 205–245.

Ramanathan, T., A.A. Abdala, S. Stankovich, D.A. Dikin, M. Herrera-alonso, R.D. Piner et al. 2008. Functionalized graphene sheets for polymer nanocomposites. Nat. Nanotechnol. 3: 327–331.

Shen, J., Y. Hu, M. Shi, X. Lu, C. Qin, C. Li et al. 2009. Fast and facile preparation of graphene oxide and reduced graphene oxide nanoplatelets. Chem. Mater. 21: 3514–3520.

Salavagione, H.J., M.A. Gomez and G. Martinez. 2009. Polymeric modification of graphene through esterification of graphite oxide and poly(vinyl alcohol). Macromolecules 42: 6331–6334.

Spitalsky, Z., D. Tasis, K. Papagelis and C. Galiotis. 2010. Carbon nanotube-polymer composites: Chemistry, processing, mechanical and electrical properties. Prog. Polym. Sci. 35: 357–401.

Stankovich, S., D.A. Dikin, G.H.B. Dommett, K.M. Kohlhaas, E.J. Zimney, E.A. Stach et al. 2006a. Graphene-based composite materials. Nature 442: 282–286.

Stankovich, S., R.D. Piner, S.T. Nguyen and R.S. Ruoff. 2006b. Synthesis and exfoliation of isocyanate-treated graphene oxide nanoplatelets. Carbon 44: 3342–3347.

Sun, L., J. Liu, S.R. Kirumakki, E.D. Schwerdtfeger, R.J. Howell, K. Al-Bahily et al. 2009. Polypropylene nanocomposites based on designed synthetic nanoplatelets. Chem. Mater. 21: 1154–1161.

Thostenson, E.T., C. Li and T.W. Chou. 2005. Nanocomposites in context. Compos. Sci. Technol. 65: 491–516.

Tong, X., C. Liu, H.-M. Cheng, H.C. Zhao, F. Yang and X. Q. Zhang. 2004. Surface modification of single-walled carbon nanotubes with polyethylene via *in situ* Ziegler-Natta polymerization. J. Appl. Polym. Sci. 92: 3697–3700.

Toti, A., G. Giambastiani, C. Bianchini, A. Meli, S. Bredeau, P. Dubois et al. 2008. Tandem action of early-late transition metal catalysts for the surface coating of multiwalled carbon nanotubes with linear low-density polyethylene. Chem. Mater. 20: 3092–3098.

Vaia, R.A. and H.D. Wagner. 2004. Framework for nanocomposites. Mater Today 7: 32–37.

Vaia, R.A. and J.F. Maguire. 2007. Polymer nanocomposites with prescribed morphology: Going beyond nanoparticle-filled polymers. Chem. Mater. 19: 2736–2751.

Valentini, L., J. Biagiotti, J.M. Kenny and S. Sautucci. 2003. Effects of single-walled carbon nanotubes on the crystallization behavior of polypropylene. J. Appl. Polym. Sci. 87: 708–713.

Wakabayashi, K., C. Pierre, D.A. Dikin, R.S. Ruoff, T. Ramanathan, L.C. Brinson et al. 2008. Polymer-graphite nanocomposites: Effective dispersion and major property enhancement via solid-state shear pulverization. Macromolecules 41: 1905–1908.

Wang, N., Y.W. Qin, Y.J. Huang and J.Y. Dong. 2012. Functionalized multi-walled carbon nanotubes with stereospecific Ziegler-Natta catalyst species: Towards facile *in situ* preparation of polypropylene nanocomposites. Appl. Catal. A Chem. 435-436: 107–114.

Wiemann, K., W. Kaminsky, F.H. Gojny and K. Schulte. 2005. Synthesis and properties of syndiotactic poly(propylene)/carbon nanofiber and nanotube composites prepared by *in situ* polymerization with metallocene/MAO catalysts. Macromol. Chem. Phys. 206: 1472–1478.

Winey, K. and R.A. Vara. 2007. Polymer Nanocomposites. MRS Bulletin. Materials Research Society, Pittsburgh.

Wu, D.F., Y.R. Sun, L. Wu and M. Zhang. 2008. Linear viscoelastic properties and crystallization behavior of multi-walled carbon nanotube/polypropylene composites. J. Appl. Polym. Sci. 108: 1506–1513.

Xu, C., X. Wu, J. Zhu and X. Wang. 2008. Synthesis of amphiphilic graphite oxide. Carbon 46: 386–389.

Xu, J., Y. Hu, L. Song, Q. Wang, W. Fan and Z. Chen. 2002. Increasing the electromagnetic interference shielding effectiveness of carbon fiber polymer-matrix composite by using activated carbon fibers. Carbon 40: 445–467.

Yang, K.F., Y.J. Huang and J.Y. Dong. 2007. Efficient preparation of isotactic polypropylene/montmorillonite nanocomposites by *in situ* polymerization technique via a combined use of functional surfactant and metallocene catalysis. Polymer 48: 6254–6261.

20

Recent Advances in Graphene Metal Oxide Based Nanocomposite for Energy Harvesting/Thermoelectric Application

Abhijit Dey

Introduction

Thermoelectric stealth/energy harvesting material

The consumption of power in 2007 due to worldwide civilization was 495 quadrillion British thermal units (BTUs) (Leila et al. 2012). This is comparable to an average power of 16.6 Terawatts (TW). Due to increase in population and the rapid development of economy, the energy need of the planet is continuously growing. The Energy Information Administration (EIA) of Department of Energy, USA expects that the average global energy spending will rise by 49% (or yearly 1.4%), from ~ 495 quadrillions BTUs in 2007 to ~ 750 quadrillion BTUs by 2035. These numbers are a reason for alarm, not only because it will be a task to supply energy on this level, but also the majority (85%) of energy is currently produced by burning of fossil fuels (Heremans et al. 2002, Tech report 2010). There are severe concerns about the penalty of the burning of fossil fuels along with the query about the long-term sustainability of non-renewable fuels. The combustion of fossil fuels creates a large amount of greenhouse gas, carbon dioxide (CO_2) that plays vital role to the phenomenon of global warming (He et al. 2015). Alternative energy sources without carbon need to be adopted in coming future on a huge scale to mitigate the crisis of climate change by stabilizing the level of CO_2 at reasonable target values. The technical analyses of the literature (Hoffert et al. 1998, Caldeira et al. 2003) point out that 10–30 TW of carbon-free primary power technology will need to be positioned by 2050 to achieve modest carbon dioxide stabilization goals. Thus, the majority or even the entire global energy consumption is required to be supplied by sources that are free from CO_2

High Energy Materials Research Laboratory, Defence Research & Development Organization, Sutarwadi, Pune, India-411 021.
Email: abhidey_bkn @yahoo.com

emission. Many solutions are available to meet up this challenge in a sustainable fashion (Hoffert et al. 1998). Nuclear fission technology is established and has the prospect to play a partial role for achieving the terawatt task. However, in addition to severe concerns related to nuclear waste, long plant start-up times, and development of nuclear weapons, energy from fission may be limited on this scale by the profusion of suitable nuclear fuel (Zhang et al. 2013). We need to make a way to other forms of energy to execute at least 10 TW of carbon-free energy by 2050.

Enormous amount (120000 TW) of energy is coming to the earth's surface round the clock in the form of solar energy that can be handy to satisfy the global energy demand (Tech report 2010, Zhang et al. 2013). Solar energy is one of the most promising renewable energy resources. As a first step, to circumvent today's energy demands, resources based on renewable energy need to be anticipated. To cultivate the renewable energy using solar energy, three different steps are essential like capture, transformation, and storage (Suman et al. 2015). Development of highly efficient thermoelectric materials (with high figures of merit, ZT) is one of the provocative area of research for the utilization of solar energy. A comparative data of the amount of energy (TW) produced from different forms of energy resources have been summarized in Table 1 (Sundarraj et al. 2014).

Table 1. Comparison of different forms of energy sources.

S. NO.	Energy resources	Amount of energy (TW)
1.	Hydroelectric resources	≤ 0.5
2.	All tides and ocean currents in the world	≤ 2.0
3.	Wind power (Globally extractable)	2–4
4.	Solar Energy	120000

On the other hand, the increasing need for the surrogacy and handy sources of energy has motivated considerable effort to develop new forms of energy-conversion and storage devices. Ambient energy is mostly leading compared to the mechanical and thermal forms and their transfiguration into electrical energy may play a vital role in upcoming technologies such as remotely accessible electronics, self-powered sensors, and other medical related devices. There is increasing necessity for self-sufficient power sources for wireless sensors and electronics which can spread device performance beyond what is accessible from conservative batteries. Substantial attention has been given on thermal energy as a potential source of energy that is widely obtainable in the surroundings and can be rehabilitated to electrical energy by employing TE or pyroelectric (PE) modules (Hoffert et al. 2002, Yun et al. 2014). A large number of TE devices have functioned under temperature gradient conditions (> 50K). Thermal gradients in the surroundings are directly rehabilitated to electrical energy through the TE effect. Evaluation of the coefficient of performance (COP) related to the thermoelectric figure of merit, the characterization of the cooling capacity has been pointed out by Enescu et al. in the form of review (Enescu et al. 2014). TE module consisting of n- and p-type materials are electrically connected at two ends having a variation of temperatures. The developed voltage and power corresponding to the temperature difference is termed as Seebeck coefficient (S) of the TE materials. Large thermal gradients are vital to producing realistic voltage and required power.

Thermoelectric materials are very useful for generation of electricity from unwanted heat, i.e., conversion of heat into electricity. The major challenge lies in the realistic approach that how a small variation of temperature on the environmental temperature can be effectively harvested. The transformation of waste heat to electricity by the TE system has been depicted schematically in Figure 1. The low efficiency of the existing TE devices is the remarkable shortcoming. If the efficiency of the TE using ZT can be amplified, these devices can be a viable solution of the TW energy problems. Therefore, development of efficiency or figure of merit is a vital issue in the allied areas of research. Hence, in this introduction chapter, our attempt is to offer a consolidated review of recent work on state-of-the-art thermoelectric composites.

Several generalized literature (Snyder et al. 2008, Gautois et al. 2013, Zhang et al. 2014, Toberer et al. 2010, Kleinke et al. 2010, Kanatzidis et al. 2010, Wang et al. 2014, Gangopadhyay et al. 2000, Zebarjadi et al. 2012, Ashiska et al. 2014) of various composite thermoelectric materials are already available, but in this chapter we primarily focus on graphene-based polymeric composites as upcoming/new generation

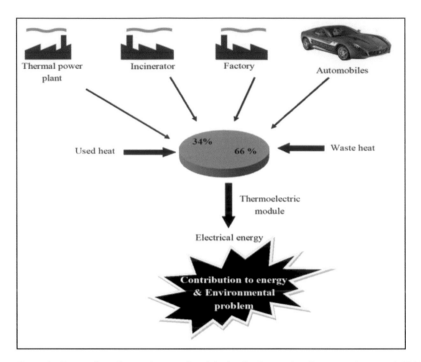

Figure 1. Conversion of waste heat to electricity by the thermoelectric system (Dey et al. 2015).

TE materials. The partial supply of natural fuels and adversative public opinion on nuclear energy led to search for an alternative resource of energy and subsequently this has turned into an important research subject in a global scenario. TE system is very promising for transformation of heat into electricity, i.e., harvesting of electricity from unwanted heat with a small gradient of temperature with respect to the surrounding temperature (Snyder et al. 2008, Tritt et al. 2006) and such system is highly dependable on reduced level of noise and maintenance-free long term operation due to the absence of moving parts and simpler leg type structure (Ono et al. 1998, Yan et al. 2001).

The concept of thermoelectric involves three major fundamental effects viz., (a) Peltier effect (b) Seebeck effect (c) Thomson effect. In the case of Peltier effect, temperature gradients generate from electrical energy, whereas in Seebeck effect, electricity generates due to temperature gradient. For Thomson effect, the heat is changed to the temperature gradient in a single conductor while passing through an electric current. These effects are basically related to the characteristics of TE (Hereman et al. 2008, Bell et al. 2008, Kjelstrup et al. 2013), and it is demonstrated in Figure 2. TE module is an association of 'p' type (electron deficient) and 'n' type (electron rich) semiconductors with high accuracy and precession (Ashiska et al. 2007, Lim et al. 2012). Presently, the use of thermoelectric devices is restricted to a niche application due to mediocre efficiency than the conventional generators. Although these three effects were discovered independently, correlation (Bubhove et al. 2012, Heremans et al. 2001, Lampinen et al. 1991) can be made through Kelvin relationship that describes the basics of thermoelectric behavior as follows (Eqs. 1 and 2). The coefficient comprised in the Kelvin relation correlates electric (α) and heat (λ) current variation with a temperature gradient and electric field (see Sondheimer et al. 1956)

$$\vec{i} = \sigma(\vec{E} - \alpha \vec{\nabla} T) \tag{1}$$

$$\vec{q} = \alpha T \vec{i} - \lambda \vec{\nabla} T \tag{2}$$

where q, i, E, σ, α, λ, and T denote electric current density, heat current density, electric field, electrical conductivity, Seebeck coefficient, thermal conductivity and absolute temperature, respectively. S is the key factor for electric current to generate the Thomson and Peltier effect in the electrical circuit (Zhang et al. 2008).

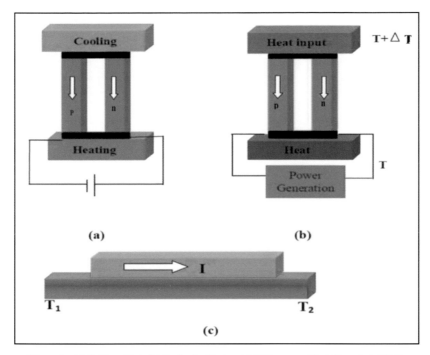

Figure 2. (a) Peltier effect, (b) Seebeck effect and (c) Thomson effect (Dey et al. 2015).

Figure 3 indicates a thermoelectric module showing the direction of charge flow for both cooling and power generation (Tritt et al. 2006). The efficiency of TE materials is expressed as figure of Merit (ZT), a dimensionless parameter (Elsheikh et al. 2014) (Eq. 3).

$$ZT = \frac{\sigma S^2 T}{\lambda} \qquad (3)$$

where, σ, S, λ and T are electrical conductivity, Seebeck coefficient, thermal conductivity, and absolute temperature, respectively. Thermoelectric power factor (TPF) depends on σ and S. It can be formulated as TPF $= S^2\sigma$. TE modules with a ZT of 1 work at 10% Carnot efficiency; however, a simple generator works with 30% Carnot efficiency. An equal amount of efficiency could be achieved by a TE module (Disalvo 1999) with a ZT $\sim$ 4. Therefore, to escalate the TE efficiency, a combination of low κ with high σ is indispensable. Such parameters are inter-related to each other and determined by the electronic structure like band shape, band gap and band degeneracy near the Fermi level and scattering of charge carriers (electrons and holes) (Chung et al. 2000). The best TE materials conveyed till today for devices that works at room temperature and have ZT value $\sim$ 2.

The interest regarding TE materials is escalating day by day. This statement can be validated by counting the number of scientific publication over time (Figure 4) on the research area of TE materials. It clearly describes that TE materials are acquiring significant interest day by day which is reflected by the hits of SciFinder research publications.

Nano inorganic thermoelectric materials have shown comparatively higher ZT values in recent times. But expensive raw material and high production cost as well as environmental pollution by adulteration of heavy metal and low process ability restricts their application (Yan et al. 2001, Li et al. 2010, Yao et al. 2010, Toshima et al. 2002). The low κ property of a polymer gives them a considerable advantage over regular TE materials regarding efficiency that is represented by ZT. For example, with the augmentation of carrier concentration, σ can be enhanced; however, it decreases the S. Accordingly, the improvement of TPF gets stalled.

An enhancement of carrier mobility is the most effective way to improve both the parameter σ and S (Hiroshige et al. 2006, Lee et al. 2012, Yao et al. 2005, Makala et al. 2003, Yamashita et al. 1999). The

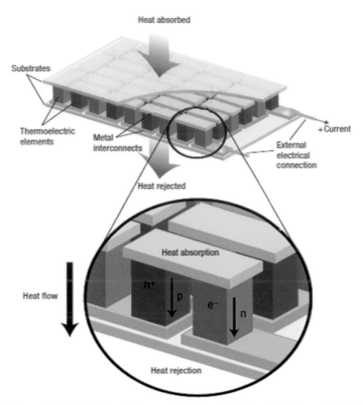

Figure 3. Thermoelectric module. Reprinted with the permission from Ref. (Snyder 2008) copyright 2008 nature publishing group.

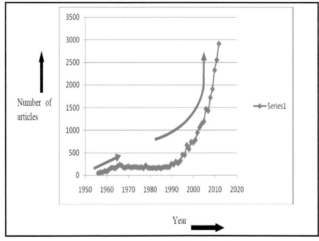

Figure 4. The number of articles on thermoelectric materials published as a function of year from 1955 to 2012 (Dey et al. 2015) copyright 2015 Elsevier.

movement of the carriers is mainly organized by the intra-chain and inter-chain hopping processes and transport mechanism based on variable range hopping model (VRH) (Long et al. 2004). Conformation and arrangement of the polymer chain are some of the important parameter that control carrier mobility. The expanded chain conformation and aligned arrangement decrease the hindrance of both intra-chain and inter-chain hopping and thus, improve carrier mobility (Hiroshige et al. 2006, Lee et al. 2012, Yao et al. 2005, Macdamid et al. 1995).

How to enhance TE efficiency?

From Eq. 3, it is clear that ZT can be improved by enhancing σ and *S* and by decreasing κ. Since these three parameters are interrelated, they form a magic triangle. These parameters need to be cautiously optimized in the high-performance thermoelectric material. The enhanced density of state due to quantum confinement effects enhances *S* without diminishing σ. Boundary scattering at interfaces can diminish κ more than it does for *σ*. From Eq. 3, it has been expressed that improvement in ZT is tricky due to the interconnectivity of σ, S and κ with each other and all are strongly dependent on the electronic structure and crystal structure of material and carrier concentration. A good TE material must have a high figure of merit over a wide range of operational temperatures with good thermal and mechanical characteristics. Low-dimensional physics offers an added control to resolve this paradox.

Thermoelectric properties: polymer materials vs. current thermoelectric

The schematic assessment between polymer composite based thermoelectric materials and current thermoelectric materials has been illustrated in Figure 5. The current researches on TE materials mainly related with crystalline nano-structured semiconductor material (Zhao et al. 2006, Chen et al. 2003, Pennelli et al. 2014).

These semiconductor materials are enormously doped to generate the large value of the figure of merit. Bi_2Te_3-Sb_2O_3 based alloy is a commercially available TE material with a room temperature ZT (Chen et al. 2003, Minnich et al. 2009) of around 1. Both the classes have their advantages as well as disadvantages. Improvement in the figure of merit for polymer based nano-composites may be of interest due to their non toxicity, low cost, malleability and manufacturability nature (LeBlanc et al. 2014). Although the common TE materials have the benefit of high figure of merit, but their rigid, expensive and toxic nature causes less appeal.

Graphene filled polymer composite: An efficient futuristic thermoelectric material

Graphene is a unique class of carbon having a 2D sheet which is just one atom thick. It acquires great attention in present time due to special features like optical, electrical, mechanical and catalytic properties.

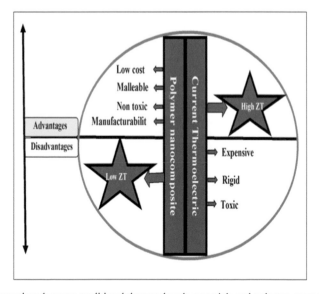

Figure 5. Schematic comparison between traditional thermoelectric materials and polymer composites. Dey et al. 2015(a) copyright 2015 Elsevier.

The nanosheets of graphene show very high σ at room temperature due to the high mobility of electrons in it. Graphene was invented in 2004 by Konstantin Novoselov and Andre Geim. They obtained graphene sheet by splitting graphite crystal into more and more thin units until individual atomic planes were reached. This notable invention was felicitated by Noble Prize (discipline: physics 2010) and led to a rapid increase of research interest in graphene (Enoki et al. 2009, Liang et al. 2009).

Graphene is the origin for all the graphitic forms. It is a building block for carbon nano structures of all other dimensionalities. For example, 0D Bucky balls, 1D nanotubes and 3D graphite (Figure 6).

It has several similarities to CNT like structure-properties. It has high aspect ratio (i.e., lateral size/thickness), good mechanical properties, large surface area and rich electronic state. It has ample scope in various areas where CNT have been already explored. Graphene is a superior electrode material with respect to CNT. The 2D planar geometry of graphene sheet helps to enhance electron transport. Table 2 highlights some relevant properties of graphene as compared with CNT.

Due to its outstanding properties with large surface area, graphene nanosheets (GNs) and 2D geometry have been considered as the latest class of promising materials for potential applications in solar cells, actuators, sensors, field-effect transistors, field-emission devices, batteries and supercapacitors

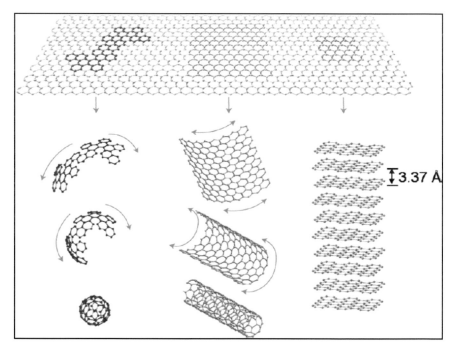

Figure 6. Graphene origin of different graphitic forms is a two dimensional building material for carbon materials of all different dimensionalities and can be wrapped up into zero dimensional bucky balls, rolled into one dimensional nanotube or stacked into three dimensional graphite (Dey et al. 2015a) copyright 2015 Elsevier.

Table 2. Comparison of important properties of graphene with CNT.

Properties	Graphene	CNT
Fracture strength (GPa)	124 (Modulus: 1100 GPa)	45
Density (g cm^{-3})	> 1	1.33
Thermal conductivity (W/mK)	5000	3000
Electrical conductivity (S cm^{-1})	10^6	5000
Charge mobility (cm^{2}V^{-1} s^{-1})	200000	100000
Specific surface area	2630	400

(Enoki et al. 2009, Liang et al. 2009, Wang et al. 2008a, Yoo et al. 2008, Vivekchand et al. 2008, Wang et al. 2008b, Eda et al. 2008, Yang et al. 2010, Weng et al. 2011).

Li et al. have reported a simplistic approach to produce graphene materials (Li et al. 2008). It opens up a remarkable opportunity for different technological applications. Graphene is composed of sp^2 hybridized carbon atoms with honeycomb structure (Novoselov et al. 2004, Geim et al. 2007, Li et al. 2008, Shao et al. 2010). The maximum S value of graphene was mentioned to be about 80 mV K^{-1}, and the ZT (Bolotin et al. 2008, Lee et al. 2008, Zuev et al. 2009, Wei et al. 2009) value was estimated to be around 0.006 at 300 K. SWCNT can look like as rolled sheet of graphene at the molecular level. There may be some supplementary coaxial tubes around the SWCNT formed multiwall carbon nanotube (Seol et al. 2010, Kateb et al. 2010, Yu et al. 2007). A methodical review by Noorden (Noorden 2011) on various carbon nanostructure concluded that research area on fullerene and carbon nanotube are nearly saturated, but the enhanced research interest may be anticipated for graphene in the coming decade (Figure 6). Two important concepts have been developed to lead the TE research from the last twenty years (Dresselhaus et al. 2007, Zhao et al. 2014). First one was suggested by Hicks and Dresselhaus in 1993. They suggested enhancing ZT by going down to lower dimensions (Hicks et al. 1993, Hicks et al. 1993). On the basis of this concept, substantial enhancement in ZT ($\sim$ 1) values was observed in thin film (Venkatasubramanian et al. 2001) and quantum dot super lattice (Harman et al. 2002). The other concept proven by two crucial experiments (Hochbaum et al. 2008, Boukai et al. 2008) in 2008 suggests improvement of ZT by nano-structuring. The trial shows a 100 times enhancement in ZT in SiNWs over the bulk substrate that generally helps to diminish κ by boundary scattering.

These two concepts are logically combined in nanostructure of graphene and can be beneficial for thermoelectric research and applications. To get an enhanced value of ZT in graphene system, two key drawbacks should be conquered, i.e., (i) high value of κ for graphene (ii) low value of S due to zero gap band structure. Boundary scattering can potentially decrease κ largely. Theoretical calculation predicts that disordered edge GNR (Sevincli et al. 2010, Xu et al. 2010) and graphene quantum dots (Xu et al. 2010) may maintain large ZT. In addition, the introduction of loosely coupled interface can resist thermal conduction efficiently. CNT also experience large κ, but theoretical calculation predicted especially small κ of 0.13–0.2 Wm^{-1} K^{-1} when they formed random network (Estrada et al. 2011, Behnam et al. 2012, Timmermans et al. 2012, Gupta et al. 2012, Gupta et al. 2013). It may be due to junction formation between CNTs (Prasher et al. 2009). These concepts are quite relevant in the case of graphene to lower down κ value. Generally, the enhancement of ZT can be attained by diminishing κ value while keeping σ intact. Thus, more research attention is required for framing approaches to adequately decouple electron and phonons. Some current inventions regarding graphene composite as the TE application is highlighted as follows:

PANI/Graphene filled composite

It was mentioned for the first time by Du et al. (Du et al. 2012) that S and σ of PANI-GNs nanocomposite followed an escalating trend with an enhancement of GNs concentration (up to a ratio of 1:1). Four dissimilar compositions were processed with different ratio, i.e., PANI-GNs = 4:1, 3:1, 2:1 and 1:1, respectively. Maximum σ and S was observed at a concentration of PANI-GNs = 1:1. It can be explained by considering the reality that carrier mobility is enhanced significantly. The efficiency, i.e., power factor of the pallets and film increased from 0.64 to 5.6 μWm^{-1}K^{-2} and 0.05 to 1.47 μWm^{-1}K^{-2} at 50 wt.% GNs, respectively.

Xiang et al. (Xiang et al. 2012) synthesized PANI-Graphene composite by *in situ* polymerization of aniline monomer in the presence of GNP. PANI and GNP both have strong п-п bond interaction, producing a consistent coating on GNs. Authors produced a paper-like composite. S of product varies with the initial level of aniline concentration in the mix as well as with ratio of polyaniline. Its value raises upto 33 μV/K for composite with 40 wt.% concentration of PANI through a protonation ratio of 0.2. Due to the occurrence of GNP, electrical conductivity, σ of the composite gets enhanced to 59 S/cm. Hence, ZT was increased by two orders of magnitude, larger than that of either graphene or PANI. For enhanced performance, higher protonation is not essential owing to the contrary relationship between σ and S. Additionally, graphene is also incorporated as a template in the polymer matrix to elongate the polymeric chains of the PANI leading to enhancement in carrier mobility. The collective effect increases PF and suppresses the augment of κ. This is beneficial in achieving a significant gain in performance.

Lu et al. also synthesized PANI-GN composite by *in situ* process. During this experiment, PANI (Lu et al. 2012) was soaked on the surface of the graphene and modification in morphology of PANI (twist structure to extended structure) was also observed after the introduction of GNs. The authors have analysed the TE properties in the range of 323K to 453K. The ZT value of the graphene composite containing 30 wt.% of graphene was seventy times higher at 453K than that of the neat PANI.

PEDOT:PSS-graphene filled composite

It is already reported that carbon nanotubes can increase the TE properties of poly(3,4-ethylene dioxythiophene): poly(styrene sulfonate) (PEDOT: PSS). Literatures (Lu et al. 2012, Kim et al. 2012, Hwang et al. 2013a, Hwang et al. 2013b, Kong et al. 2013, Li et al. 2013, Nardecchia et al. 2013, Xiang et al. 2012, Wang et al. 2012) suggests that ten times enhancement in the ZT was noticed by the incorporation of 2–3 wt.% of graphene. This value is comparatively higher than that of CNT-based composites. The reason is strong π–π interaction facilitated by superior dispersion. These interactions are well supported by the Raman spectroscopy. When graphene dispersed uniformly, it increases the interface area by 2–10 times more in comparison to CNT with the same weight.

Superior carrier mobility in graphene increases the σ but the κ of (PEDOT: PSS): graphene is moderately smaller than PEDOT: PSS thin film contains 35 wt.% of SWCNT (Dresselhaus et al. 2007). This abnormality of the two parameter was well demonstrated by the author, i.e., firstly, well-defined porous multilayer structure works as a centre for phonon scattering and secondly, thermal conductivity, κ of such materials is dominated by phonon. Hence, lattice thermal conductivity (kl) is more dominant over electronic thermal conductivity (ke). The highest ZT (i.e., ZT-2.1 × 10^{-2}) has been achieved in PEDOT: PSS composite at 2 wt.% of graphene at temperature 300K. The acquired value was 10 times higher than that of pure PEDOT: PSS. Composite fabrication addition of graphene in PEDOT: PSS matrix is an effective and noble way for augmentation of TE performance.

In recent times, fullerene (Zhang et al. 2013) has been employed to improvise TE properties of graphene-PEDOT: PSS. In this system, fullerene interacted with functionalized graphene with by *p-p* stacking in an interface of liquid-liquid medium and added into poly(3,4-ethylenedioxythiophene) poly(styrenesulfonate) matrix. The specialty of this formulation is that graphene enhances σ while fullerene increases *S* and diminishes κ. These synergistic effects help to raise the thermoelectric properties. By addition of nano-hybrids, the σ and κ is enhanced by upto 70000 S/m and 2 W/mK, respectively. In addition, a fourfold enhancement in *S* was achieved. Hence, nano-hybrid incorporated polymer composite experiences ZT, i.e., figure of merit as high as 6.7 × 10^{-2} that is one order superior than the single phase filler based polymer composite.

Yoo et al. (2014) have also synthesized PEDOT: PSS/graphene composite and evaluated its application in harvesting of energy from waste heat. The composite consists of 3 wt.% of graphene exhibiting higher σ, *S* and power factor. Fascinatingly, very recent synthesis of poly(3,4-ethylenedioxythiophene)/reduced graphene oxide (PEDOT/rGO) nanocomposites displayed higher thermoelectric performance relative to those of the corresponding neat PEDOT. Authors have developed three various *in situ* chemical oxidative polymerization method to attain PEDOT/rGO nanocomposites, viz. (1) spin-coating and successive liquid layer polymerization, (2) spin-coating followed by vapour phase polymerization and (3) *in situ* polymerization and subsequent post treatment by ethylene glycol (EG) immersing (Xu et al. 2015).

Graphene nanoribbons (GNRs)

Graphene nanoribbons (GNRs) are basically the narrow stripes of graphene having a lattice of hexagonal carbon with quasi-one-dimensional geometry. Because of the different edge structures, material with low-dimension are likely to demonstrate outstanding TE effects (Dresselhaus et al. 1999, Venkatasubramaniam et al. 2001, Bouki et al. 2008, Hochbaum et al. 2008, Nakada et al. 1996). In the first GNR TE effect report, Ouyang et al. (2009) have discussed the consequences of vacancies and edge roughness on TE properties. It has been revealed that vacancies and edge roughness could diminish the thermal conductivity and thus enhance the Seebeck coefficient. However, the ZT is still dormant by the reduced electrical conductivity that suggests the employment of kinked GNRs (Chen et al. 2010, Huang et al. 2011, Liang et al. 2012, Xie et al. 2012, Yang et al. 2012). The first synthesis (Cai et al. 2010) of kinked GNRs was basically a simple, surface-based bottom-up chemical process. Experimentalists first used the especially designed halogen-substituted bianthryl monomers and performed the planned two-step thermolysis on gold and silver surfaces in a vacuum to synthesize different GNRs. The widespread thermoelectric phenomenon can be observed from theoretical inputs but is tricky to be considered in practical. Table 3 represents thermoelectric properties of some graphene based polymer nanocomposite.

Scope and objectives

Some number of literatures has been focused to cover synthesis of graphene-based metal oxide for various applications such as a catalyst, anode material for lithium ion battery, solar cell, fuel cell, etc. But there is no systematically reported literature on the thermoelectric effect of graphene-metal oxide nanocomposite (GMNC) for energy harvesting application. Hence, there is a huge scope for the development of simple, ecofriendly synthetic route of the preparation of GMNC and its application in thermoelectric/energy harvesting. By taking the advantage of unique properties of graphene, several metal oxides like TiO_2, ZnO, SnO_2, MnO_2, Co_3O_4, Fe_3O_4, Fe_2O_3, NiO, Cu_2O, etc., can be placed over the graphene surface for thermoelectric application. Nano metal oxide particle can be placed by the *in situ* or *ex situ* method. These nano metal oxide particle helps to reduce the Van der Wall force between the graphene layers, hence minimizing the stacking nature. The agglomerating tendency of the nano particles is reduced drastically due to decoration over the graphene that helps to increase thermoelectric efficiency. In thermoelectric, GMNC can be used in various polymer matrices like PVAc, PEDOT:PSS, PANI, polypyrrole, and P3HT. Primarily, polymer matrix acts as a bonding agent that helps to bind filler material. Non-conducting polymer like PVAc can be useful for it low thermal conductivity. PEDOT:PSS is a conducting polymer that may lead to increased electrical conductivity and reduced thermal conductivity simultaneously. However, the filler loading capability of PEDOT:PSS is relatively lesser compared to PVAc. The combination of these two polymers can be suitable for the development of flexible thermoelectric material. Other polymers like PANI, polypyrrole and P3HT can also be used for making the composite. All the above mentioned GMNC can be synthesized by means of ecofriendly, simple way and subsequently they can be deployed for both catalysis and as efficient thermoelectric material. Therefore, keeping all these in mind, the main objectives of the investigation have been set as follows:

The chapter is mainly focused on the synthesis of graphene iron oxide nanocomposite, i.e., GINC, graphene titanium dioxide nanocomposite, i.e., GTNC and graphene zinc oxide nanocomposite, i.e., GZnNC by an ecofriendly route. Metal oxides have been chosen on the basis of electronic configuration. Iron oxide, titanium dioxide and zinc oxide have electronic configuration d^5, d^0 and d^{10}, respectively. To process the nanocomposite, ecofriendly techniques like ultrasonication and microwave irradiation have been employed to achieve following objectives:

- Synthesis of GINC, GTNC and GZnNC by nano metal oxide impregnation over graphene surface.
- Evaluation of thermoelectric properties of GINC, GTNC, and GZnNC filled conducting (PEDOT/PSS)/nonconducting (PVAc) polymer matrix for thermoelectric/energy harvesting application.

Synthesis of graphene metal oxide nanocomposite (GMNCs)

Synthesis of graphene metal oxide nanocomposites (GMNC) is very simple. Ecofriendly techniques like ultrasonication and microwave irradiation have been used to synthesize GMNCs. The detailed procedures for synthesis of various ingredients and nanocomposites are as follows:

Table 3. Reported thermoelectric properties of graphene based polymer nanocomposite.

Thermoelectric materials	σ, S/m	S, μV/k	κ, W/mK	PF or ZT
Graphene (Weng et al. 2011)	10^6	5000	--	ZT = 0.006 at 300 K
PANi/Graphene composite (Du et al. 2012)				
PANi	10^3	14	--	PF = 0.2
Graphene	2×10^4	15	--	PF = 8
Pallet:			--	PF = 0.7
PANi:Graphene:: 4:1	1.4×10^3	20	--	PF = 2.0
PANi:Graphene:: 3:1	3.0×10^3	26	--	PF = 4.0
PANi:Graphene:: 2:1	4.0×10^3	28	--	PF = 5.6
PANi:Graphene:: 1:1	5.0×10^3	30	--	PF = 0.04
Film			--	PF = 0.12
PANi:Graphene:: 4:1	20	27	--	PF = 0.20
PANi:Graphene:: 3:1	80	38	--	PF = 1.2
PANi:Graphene:: 2:1	150	39		
PANi:Graphene:: 1:1	700	41		
PANi+GNP (*In situ* polymerization with protonation ratio-0.2) (Xiang et al. 2012)	150	7	0.6	ZT (300K) = 3.68
Neat PANi	2×10^4	5	74	$\times 10^{-6}$
Neat GNP	5900	33	13	3.04×10^{-6}
PANi/GNP (50 mM, as made)	1.74×10^4	19	15	1.51×10^{-4}
PANi/GNP (50 mM, reprotonated)				1.26×10^{-4}
PANI + 30% Graphene (Lu et al. 2012) (*In situ* polymerization)	5×10^3 at 323K	12	--	ZT = 1.95×10^{-3} at 453 K (70 times higher than PANi)
PANI at 420K	500	13		PF = 0.1
PANI + 5% Graphene	700	28	0.6	PF = 0.4
PANI + 15% Graphene	1000	32		PF = 0.8
PANI + 20% Graphene	3600	20		PF = 1.6
PANI + 30% Graphene (Lu et al. 2012)	4.0×10^3	26		PF = 2.6
PANI+HClO$_4$ + Graphite (50 wt.%) (Wang et al. 2011)	1.2×10^4	19	1.2	PF = 1.2
PEDOT:PSS + GNP (2–3%) (Zhao et al. 2006)				at 300 K
PEDOT:PSS + GN	74	165.8	0.24	2×10^{-3}
100 1	1469	46.92	0.19	4.6×10^{-3}
99 2	3213	58.77	0.14	2.1×10^{-2}
98 3	3170	44.75	0.30	5.3×10^{-3}
97 4				
PEDOT:PSS (Pristine PEDOT:PSS synthesized without graphene)	45300	23.1	--	PF = 24.173
	4.73	0.52	--	PF = 0.00013
Pristine reduced graphene oxide (rGO)	52800	24.375	--	PF = 31.375
PEDOT:PSS/Graphene (1%)	54800	24.750	--	PF = 33.568
PEDOT:PSS/Graphene (2%)	63700	26.778	--	PF = 45.677
PEDOT:PSS/Graphene (3%)	48200	23.250	--	PF = 26.055
PEDOT:PSS/Graphene (3%) with simple mixing	55600	24.715	--	PF = 33.962
PEDOT:PSS/Graphene (4%)	55900	21.750	--	PF = 26.444
PEDOT:PSS/Graphene (5%) (Yoo et al. 2014)				

Synthesis of nano metal oxide

i) Nano Fe_2O_3

To make nanocomposite, graphene nano sheet and nano iron oxide were prepared independently. Nano iron oxide was synthesized from iron nitrate, i.e., Fe $(NO_3)_3$ in a two step method. In the first step, preparation of citrate polymeric gel was carried out and calcinations of gel were carried out in second step.

ii) Nano TiO_2

Nano TiO_2 (Anatase) have been prepared by using titanium tetrachloride ($TiCl_4$) and myristic acid. The detailed synthesis process for nano TiO_2 is based on the literature methodology published by Dey et al. 2015b. Where produced nano TiO_2 powder was calcined at 500°C to enhance its crystalline anatase-TiO_2.

iii) Nano ZnO

The synthesis was carried out by preparing the alkali solution of zinc nitrate by dissolving 0.5M zinc nitrate [Zn $(NO_3)_2.6H_2O$] and 1M NaOH, in 100 ml distilled water. The solution was continuously mixed at 75°C, maintaining at pH 12. Polyvinylpyrrolidone (PVP, 0.2 gm) was then mixed to the reaction mixture which was mixed for 5–6 hrs at 75°C to get white suspended ZnO nanoparticles. The suspension was separated through centrifugation followed by washing with distilled water and ethanol. Hence, white precipitate was obtained. After that, this white precipitate was annealed at 400°C for two hours in a muffle furnace. After annealing, highly nano-crystalline ZnO nanoparticles were obtained.

Synthesis of graphene from graphite flake

In the first step, graphite oxide was synthesized from graphitic flakes by Hummers method. In the second step, thermally expanded graphene oxide (TEGO) was synthesized by thermal expansion/exfoliation at 1050°C (Ar, 30s). Finally, graphene nanosheets (GNS) were produced by hydrogenation of TEGO at 400°C for 2 hrs.

Synthesis of graphene iron oxide nanocomposite (GINC)

During GINC synthesis, 50 mg of graphene was first distributed in absolute ethanol by ultrasonication for 40 min, following which iron oxide nano particles was mixed into the graphene dispersion with ultrasonication continued for 2 hrs. The resulting dispersion was kept in ambient condition for drying. After complete evaporation of ethanol, nanocomposite was deposited over the Petri dish. The nanocomposite was positioned into the microwave reactor for 2–3 min for improved exfoliation and on consequent cooling; the GINC was collected in to the sample vial.

Synthesis of graphene titanium dioxide nanocomposite (GTNC)

GNTC was synthesized by dispersing 50 mg graphene nanosheets in ethanol medium by ultrasonication. In second step, pre-dispersed titanium dioxide nanoparticles were added to the graphene dispersion. The so-developed dispersion was ultrasonicated further for 2 hrs. Drying of composite was carried out at room temperature to take out solvent. After complete drying, GTNC was deposited over a Petri Dish and subjected to microwave irradiation for 2 min to get improved exfoliation.

Synthesis of graphene zinc oxide nanocomposite (GZnNC)

GZnNC was also synthesized by dispersing 50 mg graphene in ethanol by ultrasonication and by adding well dispersed nano zinc oxide to the dispersed graphene solution. The dispersion was additionally ultrasonicated for 2 hrs followed by drying at room temperature to eliminate solvent. After drying,

GZnNC was deposited over a petridish and positioned to microwave irradiation for 2 min to make better exfoliation.

Characterization of graphene metal oxide nanocomposites (GMNCs)

Graphene iron oxide nanocomposite (GINC)

GINC has been characterized by several sophisticated techniques like HRTEM, FTIR, RAMAN, XRD, FESEM and UV-Vis spectrometer. The detailed characterization has been given below.

HRTEM analysis

Selected area electron diffraction (SAED) pattern and high-resolution transmission electron microscopy (HRTEM) were carried out to examine the quality of nano iron oxide decoration over graphene layer. HRTEM micrographs with an SAED pattern of graphene, nano iron oxide and GINC are shown in

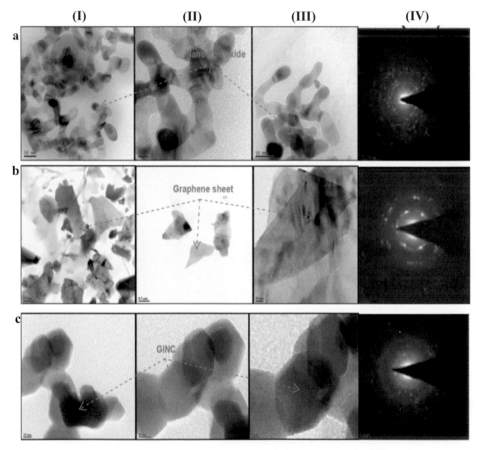

Figure 7. HR-TEM image (I, II, III) and SAED pattern (IV) of (a) nano iron oxide, (b) graphene, (c) graphene-iron oxide nanocomposite (GINC). Copyright (Dey et al. 2015a) copyright 2015 Royal Society of Chemistry.

Figure 7 which confirms the size of nano Fe_2O_3 and graphene. Graphene sheets were evidently observable, and crystalline nature nano iron oxide and graphene is confirmed by SAED pattern. Iron oxides nano particles were located over the graphene sheet and formed nano composite (Figure 7c) which displayed arrangement of lattice fringes of GINC.

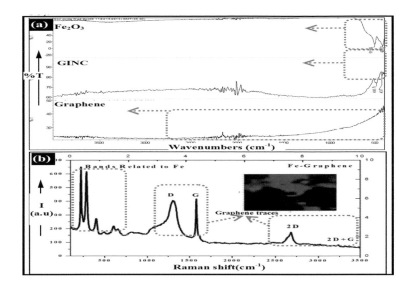

Figure 8. (a) FTIR traces confirm low defect content in graphene. (b) Characteristic raman signature. Copyright (Dey et al. 2015a) copyright 2015 Royal Society of Chemistry.

FTIR and RAMAN analysis

FTIR spectrum in Figure 8a confirms the nonattendance of functional group and attendance of a low defect on top of the graphene. The observed Raman traces of Fe_2O_3 suggest the eminence of graphene sheet after and before the formation of nanocomposite employing techniques as mentioned. The most prominent Raman traces observed were D band at 1310 cm^{-1} equivalent to defect and G band at 1575 cm^{-1} that equivalent to in-plane vibration of sp^2 hybridized carbon. 2D band at 2627 cm^{-1} as a consequence of two-phonon double resonance process has also been observed. As per the spectra recorded in Figure 8b, D band of graphite was found to be weak while graphene demonstrates D band of small intensity. The occurrence of a lower defect on flakes of graphene is revealed by the lower intensity of D band. Coleman et al. recommended that the defects are primarily in attendance at the edges of the flakes, and the basal plane is found to be free from any defect. The ratio I (D)/I (G) of the GINC is augmented by two folds (0.993) with regard to the corresponding ratio of 0.497 for pure graphene. During the formation of the nanocomposite, more than a few defects with sp^2 domain were formed with characteristic Raman signature (Figure 8b). Raman mapping was employed to confirm the decoration of nano iron oxide above graphene substrate (Figure 8b).

XRD analysis

The adornment of nano Fe_2O_3 over graphene substrate is established by Raman mapping tool. The representative XRD pattern of nano iron oxide, graphene, and GINC from angle expanse of 0–100 degree is depicted in Figure 9a,b,c. Each and every peak is assigned to corresponding crystallographic phases. Figure 9c illustrates the XRD pattern of GINC, graphene peaks as well as nano Fe_2O_3 peaks. The crystallographic phases of graphene and iron oxide remain integral during processing. The premeditated crystallite dimension of graphene nano sheet (GNS) and nano Fe_2O_3 were 28 nm and 38 nm, respectively, by Scherrer equation.

Polymer GINC composite: Characterization

Environmental scanning electron microscopy (ESEM) images with various magnification (800X and inset, 3000X) have been presented in Figure 18. Figure 18 consists of various micrographs of different

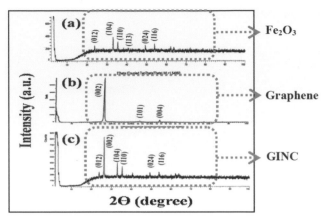

Figure 9. XRD traces of (a) iron oxide, (b) graphene (c) graphene iron oxide nanocomposite (GINC). Copyright (Dey et al. 2015a) copyright 2014 Royal Society of Chemistry.

cellulose polymer GINC nanocomposites with concentration variation of polymer GINC nanocomposites. The micrograph shows that cellulosic fibre pores were filled with polymer-GINC composite. It shows the consistent dispersion of PEDOT:PSS/GINC over the cellulosic film. Inset of the micrographs highlighted the consistent network formation between polymer GINC composite and cellulosic fibre. As concentration increases, coating ability of polymer GINC over cellulosic fibre enhances drastically. Therefore, efficient conductive network has been formed. Hence, efficiency gets enhanced. Figure 10j represents the simple cellulosic film. In this micrograph, small pores are visible clearly.

UV-Vis spectroscopy

To examine optical band gap of the produced composites, optical diffuse reflectance extents were performed on finely grounded powders at ambient conditions. The spectra have been documented at the range of 200 nm to 800 nm by employing a UV-Vis spectrometer of Cary 5000. The absorption (α/Λ) information was extracted from reflectance data using Kubelka-Munk equation: $\alpha/\Lambda = (1-R)^2/(2R)$, where α and Λ are the absorptions and scattering coefficients, respectively, and R is referred as reflectance. Lastly, the band gaps of energy were obtained from the plot α/Λ vs. E (eV). The detailed graphical representation is depicted in Figure 11. The detailed assignments of those are highlighted in Table 4.

In general, the thermoelectric properties viz. electrical conductivity and Seebeck coefficient increase with the decrease of a band gap. In the above table, insignificant variations were observed in band gap calculation. Therefore, it is very difficult to establish any connection between band gap and electrical conductivity of studied composites.

Graphene titanium dioxide nanocomposite (GTNC)

GTNC has been characterized by several sophisticated techniques like XRD, FTIR, RAMAN, HRTEM, FESEM. The detailed characterization has been given below.

XRD analysis

XRD scans have been documented for the samples to confirm the existence of anatase TiO_2 nanoparticles as well as graphene in GTNC (Figure 12a). The TiO_2 nanoparticles obtained by sol-gel manner illustrated the crystalline scenery with 2θ peaks at 25.25 (101), 47.9 (200), 37.8 (004), 62.36 (204) and 53.59 (105). The lane peak at 2 = 25.8° in GTNC points out an indiscriminate stuffing of graphene sheets which matches the graphite (002) plane. The FTIR band of the GTNC has not disclosed any data regarding the

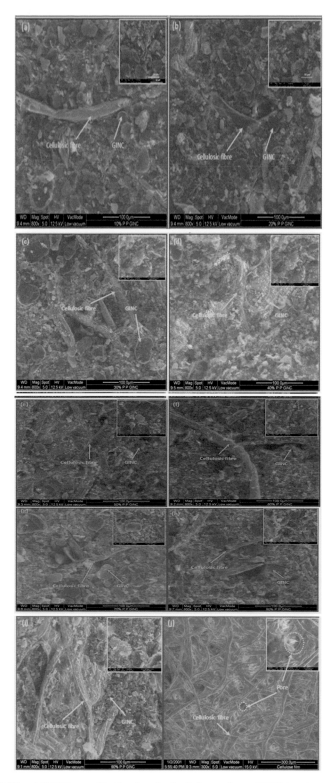

Figure 10. ESEM at 800X and 3000X magnification (inset) of cellulose polymer GINC based composite with (a) 10% (b) 20% (c) 30% (d) 40% (e) 50% (f) 60% (g) 70% (h) 80% (i) 90% PEDOT:PSS solution respectively and (j) cellulosic film (Dey et al. 2016) copyright 2016 Royal Society of Chemistry.

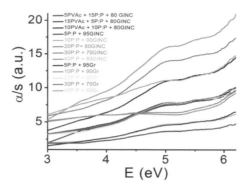

Figure 11. (i) Graphical representation of energy band gap derived from α/Λ vs. *E* (eV) plots (Dey et al. 2016) copyright 2016 Royal Society of Chemistry.

Table 4. Optical band gaps for various compositions.

Sr. No.	Sample name	Band gap (eV)
1	10P:P + 90Gr	3.04
2	30P:P + 70Gr	3.06
3	5P:P + 95Gr	3.12
4	10PVAc + 10P:P + 80GINC	3.13
5	20P:P + 80Gr	3.20
6	15PVAc + 5P:P + 80GINC	3.11
7	30P:P + 70GINC	3.07
8	20P: P + 80GINC	3.25
9	40P:P + 60GINC	3.31
10	10P:P + 90GINC	3.27
11	5P:P + 95GINC	3.26
12	5PVAc + 15P:P + 80 GINC	3.32
13	40P:P + 60Gr	3.02

existence of organic components. On the other hand, it was inveterate that the peaks for pristine graphene are present, and the peaks for graphene oxide were not present (Figure 12b). The Raman bands were used to examine the eminence of sheets of graphene before and after formation of the nanocomposite. The Raman traces of nano TiO_2, graphene, and GTNC are provided in Figure 12c.

RAMAN and FTIR analysis

The mainly distinct Raman traces (Figure 12c and Figure 12d) are found as D band at 1310 cm⁻¹ match to defect, and G band 1575 cm⁻¹ match to in-plane vibration of sp² carbon. The two-dimensional band at 2727.4 cm⁻¹ is engendered because of two-phonon dual resonance progression. The inferior intensity D band corresponds to the existence of a minute quantity of defects lying on graphene slivers. The ratio between I (D) and I (G) of GTNC is augmented by 1.5 folds with regard to pristine graphene. More than a few defects using sp² province were shaped throughout nanocomposite preparation.

HRTEM analysis

HRTEM results of the GTNC reveal the existence of TiO_2 nanoparticles within graphene (Figure 13a,b). It has inveterate the dimension of TiO_2 nanoparticles and nanometer graphene regime leading to 20 nm

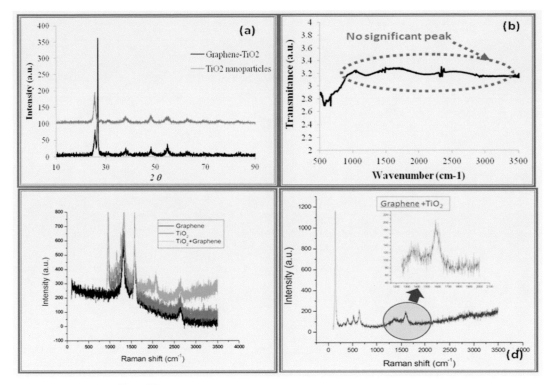

Figure 12. (a) XRD profiles of TiO$_2$ nanoparticles and GTNC, (b) FTIR spectrum of GTNC (c) Raman traces of graphene, TiO$_2$ and GTNC (d) Raman traces of GTNC (Dey et al. 2015c) copyright 2015 Royal Society of Chemistry.

TiO$_2$ nanoparticles. Clear visibility of skinny graphene sheets in the images suggested its presence and acted as a substrate for TiO$_2$ nanoparticles. Likewise, FESEM images (Figure 13c,d) depict homogeneous spreading of TiO$_2$ nanoparticles above graphene sheets in aggregated form. The exfoliated structure of graphene layers and nano-sized spherical of TiO$_2$ ensures the gigantic amount of surface on GTNC, which is a vital parameter for catalytic action.

Graphene zinc oxide nanocomposite (GZnNC)

GZnNC has been characterized by several sophisticated techniques like XRD, FTIR, RAMAN, HRTEM, FESEM. The detailed characterizations have been given below.

XRD and FTIR analysis

The presence of ZnO nanoparticles and graphene in the GZnNC was confirmed by X-Ray diffraction scan of the samples (Figure 14).

The XRD spectrum of ZnO nanoparticles obtained by sol-gel process displayed somewhat wide peaks owing to its nano-sized nature. The 2θ peaks present at 25.25° (101), 37.8° (004), 47.9° (200), 53.59° (105) and 62.36° (204) confirmed the wurtzite crystal structure of ZnO. The XRD spectra of GZnNC sample showed a broad peak at $2 = 25.8°$ which indicates a haphazard packing of graphene layers and relates to the graphite (002) plane. The FT-IR data of GZnNC (Figure 15d) has not disclosed any evidence concerning the existence of organic components in the product and the corresponding peaks for graphene oxide were preoccupied. However, peaks were present at 3400 cm^{-1} corresponding to –OH group due to moisture and a peak at 1675 cm^{-1} due to skeletal vibration in sheets of graphene. The sharp peak at 2355 cm^{-1} could be due to a presence of adsorbed air-borne CO$_2$ molecules. The GZnNC sample did show a peak at around

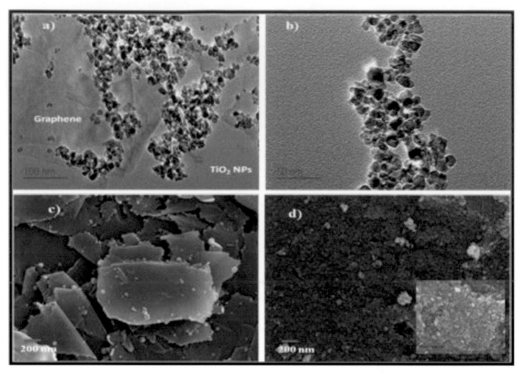

Figure 13. HR-TEM micrographs of (a) GTNC and (b) TiO$_2$ nanoparticles. FE-SEM images (c) GTNC and (d) TiO$_2$ nanoparticles. The images reveal that the spherical shaped nano TiO$_2$ are nicely dispersed in the layers of graphene (Dey et al. 2015c) copyright 2015 Royal Society of Chemistry.

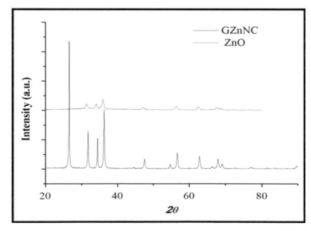

Figure 14. XRD profiles of graphene zinc oxide nano-composite (GZnNC) and ZnO nanoparticles.

1380 cm^{-1} that could be due to the carbon moiety carried by the ZnO nanoparticles. This carbon moiety can be formed after the sintering process of ZnO nanoparticles which resulted in decomposition of PVP capping, producing carbon residue in the sample.

Raman analysis

Figure 15a,b and c represents Raman touches of graphene, nano ZnO, and GZnNC. The Raman spectrum of pristine graphene showed the characteristic in-plane vibration peak (band G at 1580 cm^{-1}) which originates from sp^2 carbon atoms along with primary and second order connotation of diverse in-plane vibration peaks D band (1350 cm⁻¹) and 2D (2690 cm⁻¹). It was observed that the pristine graphene was composed of multi-stacked layers as the intensity of 2D band peak was less than the G band peak. The calculated ratio for I_{2D}/I_G (~ 2.5) for GZnNC sample was less as compared to the pristine graphene sample (~ 1.5). Thus, the stacking of graphene sheets decreased in the nanocomposite sample that could be a result of the presence of ZnO nanoparticles between the graphene sheets. Also, the ratio of I_G/I_D was 2.6 for pristine graphene and more than 5 for GZnNC sample. Such high values for I_G/I_D ratio accounts for high disorders and defect density in the samples. The defect intensity increased in the GZnNC sample as compared to pristine graphene, making the composite material more nanocrystalline. It is well

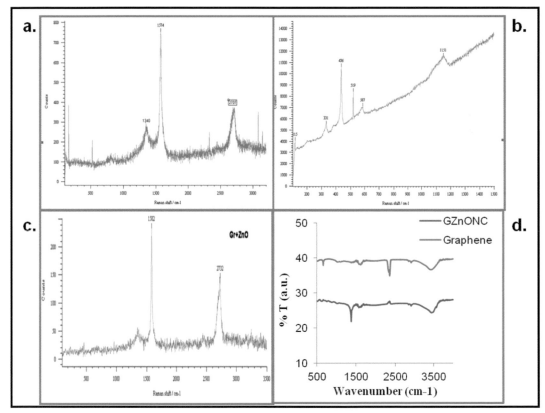

Figure 15. Raman traces of (a) graphene sheets, (b) ZnO nanoparticles; (c) GZnNC sample and (d) FT-IR spectrum of GZnNC and graphene sample.

documented that nanocrystalline graphite phases in the graphene samplers lead to higher electron-phonon scattering, increasing the I_D/I_G ratio in the process. Figure 15d represents the FTIR spectrum of graphene and graphene zinc oxide nanocomposite (GZnNC). Graphene oxide traces were absent in graphene as well as GZnNC.

FESEM and HRTEM analysis

Field emission scanning electron microscopy (FESEM) images of samples at different magnification (Figure 16a and 16b) highlighted the presence of exfoliated graphene sheet. The FESEM images of GZnNC samples indicated the uniform presence of ZnO nanoparticles over and around layers of graphene. It has been clearly observed that ZnO nanoparticles are existent in combined shape in GZnNC. The exfoliated nature of graphene sheets in conjunction with nano-sized distorted shaped ZnO will ensure

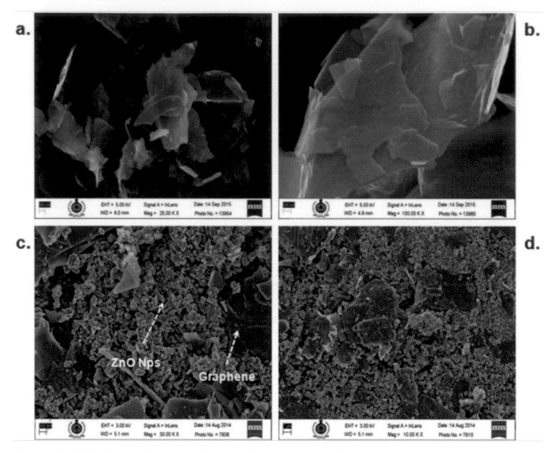

Figure 16. FE-SEM images of graphene sheets (a) with a magnification of 25 KX; (b) with a magnification of 100 KX; (c) FESEM images of GZnNC nanocomposite with a magnification of 30 KX and (d) with a magnification of 10 KX. The uniform presence of ZnO nanoparticles and graphene sheets in the nanocomposite would be advantageous for its use as a burn-rate enhancer and thermoelectric material.

enormous surface availability of the GZnNC. The enormous surface area of GZnNC could be vivacious for desired catalytic nature.

HRTEM of the GZnNC studies were conducted to identify the shape and size of ZnO nanoparticles in GZnNC (Figure 17).

The distorted shaped ZnO nanoparticles were confirmed by TEM images as various pebble shaped ZnO nanoparticles with a particle size in the expanse of 20–100 nm were formed after sintering. The exfoliation of graphene was also evident for the pristine graphene samples (Figure 17a) which helped in providing a platform for a large number of ZnO nanoparticles to assemble on its surface (Figure 17b). It was interesting to observe that distorted rod-like, hammer-like ZnO nanoparticles were also present on the graphene sheets in GZnNC sample (Figure 17c). The presence of such shaped ZnO nanoparticles would facilitate a significant improvement in surface area in GZnNC sample which, as mentioned above, can play a vital role in catalysis applications. The clear visibility of lattice fringes with spacings d = 0.281 nm (corresponding to 100 crystal plane) not only confirmed the presence of hexagonal ZnO but revealed its nanocrystalline nature (Figure 17d and 17e). The EDAX quantification of the GZnNC sample obtained from HRTEM further confirmed the presence of ZnO when the electron beam was projected on the observed rod cum hammer like nanoparticles (Figure 17f).

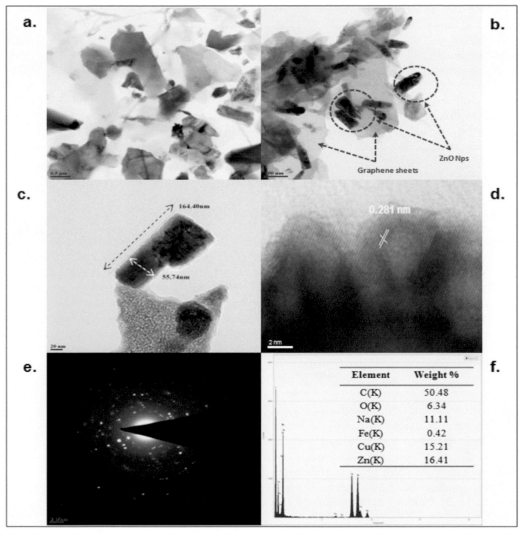

Element	Weight %
C(K)	50.48
O(K)	6.34
Na(K)	11.11
Fe(K)	0.42
Cu(K)	15.21
Zn(K)	16.41

Figure 17. HR-TEM images of (a) graphene sheets with a scale bar of 500 nm; (b) graphene-zinc oxide nano-composite (GZnNC) with a scale bar of 200 nm; (c) rod-like distorted ZnO nano particles present in the GZnNC nanocomposite (scale bar is 20 nm); (d) lattice fringes of ZnO nanoparticles with d = 0.281 nm corresponding to (100) plane of wurtzite ZnO; (e) SAED image of GZnNC nanocomposite showing nanocrystalline nature of the sample and (f) EDAX quantification of GZnNC nanocomposite from HRTEM images.

Seebeck coefficient/thermoelectric power measurements

To measure the thermopower using temperature, samples (Dey et al.) have been prepared with a dimension of 30 mm (l) × 6 mm (w) × 1 mm (t) from polymer nanocomposite film and positioned on a thermally insulated fiber glass. At one termination of the sample, a thermally conductive epoxy (electrical insulating 2763 Stycast) has been positioned, whereas at the other end, a copper piece (sink of heat) made a connection with the Peltier cooling site. The voltage fall and temperature gradient through the film were dignified using thermocouples organized in series using a couple of copper wires. To make unquestionable that the voltage fall and thermal gradient were being dignified at the same location, two minor copper films were devoted to the polymer-GZnNC film with electrically or thermally conducting silver epoxy. The voltage wires and the thermocouple were devoted to copper films. The thermoelectric energies were scrutinized regarding temperature alteration using Keithley (2182A). Peltier cooling module has been employed to change the base temperature. The TE power was dogged via two self-determining resources: (1) After accomplishment of a stable state through a smeared current to the heater. (2) By fitting the linear correlation for V vs. ΔT rejoinder to a heating pulse. The deviance among two methods and between diverse conducting tests was always smaller than 5%.

Owing to the high electrical conductive nature of the composite, delta mode four probe methods was used to measure electrical resistivity. The lowest conceivable current was obtained (100 mA) by Keithley 6220 and the corresponding voltage was scrutinized with nanovoltmeter of Keithley (2182A). The tiniest conceivable current was used to evade heating of the sample at small temperatures. Polymer nanocomposite sample using dimensions 8 mm × 3 mm × 1 mm have been prepared and subjected to measure electrical conductivity.

Thermoelectric/energy harvesting application

Polymer based graphene iron oxide nanocomposite (GINC)

Figure 18a shows that the electrical conductivity of PVAc-GINC composite enhances up to 4–5 order with respect to PVAc graphene composite. Raw PVAc shows an electrical conductivity of approximately 10^{-13} S/m. The electrical conductivity was measured at room temperature. After two months, sample showed identical results. This indicates excellent stability of the nanocomposite within a period. The Seebeck coefficient (see Figure 18b) also exhibits increasing trend and reaches a maximum value with filler concentration, 80 wt.% then decreases. Figure 18c exhibits the variation of power factor (PF) as a function of filler concentration. According to Figure 18c, PF enhances and achieves a very high value, i.e., 32 $\mu W\ m^{-1}K^{-2}$ at 80 wt.% filler concentration. The achieved value is found to be maximum ever reported in the literature for polyvinyl acetate based system without conducting polymer. The enhancement of electrical conductivity was following the percolation law of the composite that predicts an increase in electrical conductivity up to a critical concentration level of filler. These phenomena work when two unlike materials with a large difference in electrical conductivity are mixed.

During the evaluation of thermoelectric properties, PVAc-GINC composite offered two times enhancement in Seebeck coefficient and ten-fold augmentation in electrical conductivity were observed with respect to the PVAc-graphene composite with same filler loading (80 wt%). Hence, the computed power factor for PVAc/GINC composite (density: 1.47 g/cc) enhances up to 27 times with respect to PVAc-graphene composite (density: 1.32 g/cc). Thermal conductivity is found to be 3.21 W/mK. Therefore, ZT reaches to 0.0031 for PVAc-GINC composite. This is one of the noteworthy findings. In GINC, iron oxide nano particles were ornamented over two dimensional graphene sheet. The occurrence of iron oxide nano particle helps to crack the thermally conductive network, but electrical network remains undamaged. When GINC is used as conducting filler, it not only diminishes the interdependency of σ and S but also increases both the parameter concurrently. The augmentation of Seebeck coefficient is insignificant incomparing electrical conductivity of PVAc-GINC composite. We are demonstrating this feature 1st time for GINC. This shows the novelty of the experiment. Figure 22 represents the schematic diagram of thermally disconnected but electrically connected network in PVAc/GINC composite.

The extraordinary increase in electrical conductivity (6.7×10^4 S/m) of PEDOT:PSS/GINC composite has been achieved (Figure 20d). Thermal conductivity and Seebeck coefficient have deviated by a small degree. Virgin PVAc has an electrical conductivity of 10^{-13} S/m and the electrical conductivity was measured at room temperature. Reproducible results show superiorstability of the nanocomposite with a period even after two months. The Seebeck coefficient (see Figure 20b) also shows attractive tendency with preliminary decrease and then the final enhancement to arrive at a maximum with 20 wt.% PEDOT:PSS concentration. Figure 20c shows the extent of change of power factor (PF) as a function of filler concentration. According to Figure 20c, PF enhances and achieves a very high value, 34.17 $\mu W\ m^{-1}K^{-2}$ at filler concentration of 20 wt.%. Likewise, ZT, i.e., thermoelectric figure of merit, reached the upper limit, i.e., 0.003.

In Figure 21, five dissimilar compositions (CP1–CP5) consist of various concentrations of filler and PEDOT:PSS, i.e., GINC/graphene have been evaluated using thermal conductivity, electrical conductivity, Seebeck coefficient, power factor and ZT. In graphics of Figure 21a, the thermal conductivity

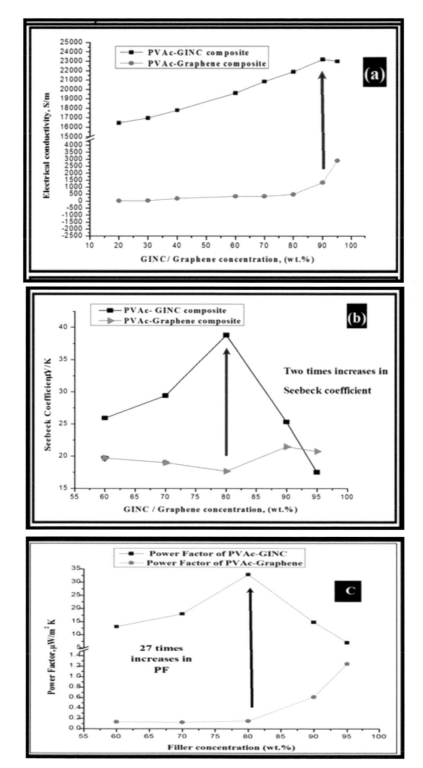

Figure 18. Thermoelectric properties like (a) Electrical conductivity (b) Seebeck coefficient (c) power factor with respect to filler concentration (graphene, GINC) at room temperature (300K). The arrow indicates the enhancement of all the three properties, i.e., Seebeck coefficient, electrical conductivity and power factor at a concentration level of 80–90 wt.% (Dey et al. 2015d) copyright 2014 Royal Society of Chemistry.

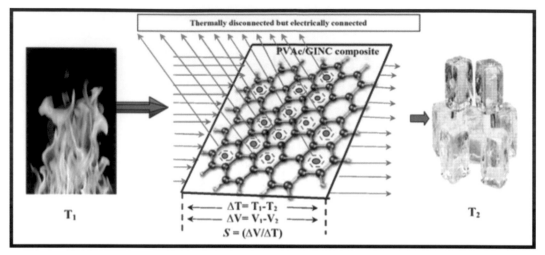

Figure 19. Schematic representation of augmentation of electrical conductivity and Seebeck coefficient simultaneously. (⟶) symbolize electrical conduction and (⟶) thermal conduction (Dey et al. 2015d) copyright 2015 Royal Society of Chemistry.

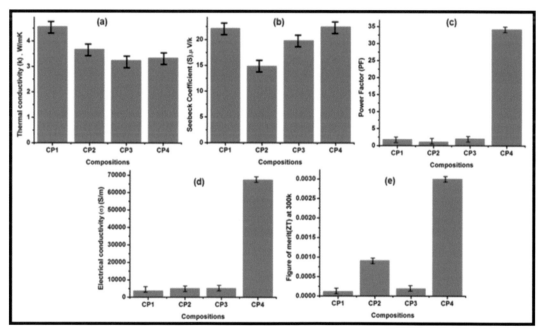

Figure 20. (a) Thermal conductivity (b) Seebeck coefficient (c) power factor (d) Electrical conductivity and (e) ZT with the variation of PVAc:PEDOT:PSS concentration at room temperature (300K). CP1: 15% PVAc+ 5% PEDOT:PSS solution +80% GINC, CP2: 10% PVAc+ 10% PEDOT:PSS solution +80% GINC, CP3: 5% PVAc + 15% PEDOT:PSS solution +80% GINC, CP4: 20% PEDOT:PSS solution +80% GINC (Dey et al. 2016) copyright 2016 Royal Society of Chemistry.

of PEDOT:PSS-graphene composite is much higher compared to the PEDOT:PSS-GINC composite that is the main drawback for graphene-based composite. In another way, Seebeck coefficient is also higher for PEDOT:PSS-GINC composite compared to PEDOT:PSS graphene composite. In the same way, the electrical conductivity of PEDOT:PSS-GINC composite has been found to be much higher than PEDOT:PSS graphene composite. Hence, a higher value of PF and ZT has been achieved in the case of PEDOT:PSS-GINC composite. While optimization of PEDOT:PSS and GINC concentration in the composite, composition with 5 wt.% PEDOT:PSS solution and 95 wt.% GINC shows very high PF and

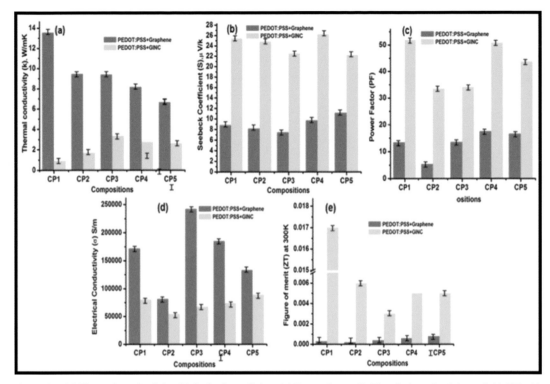

Figure 21. (a) Thermal conductivity (b) Seebeck coefficient (c) Power factor (d) Electrical conductivity and (e) ZT with respect to PEDOT:PSS concentration at room temperature (300 K). CP1 = 5% PEDOT:PSS solution +95% Graphene/GINC, CP2 = 10% PEDOT:PSS solution +90% Graphene/GINC CP3 = 20% PEDOT:PSS solution +80% Graphene/GINC CP4 = 30% PEDOT:PSS solution +70% Graphene/GINC, CP5 = 40% PEDOT:PSS solution +60% Graphene/GINC (Dey et al. 2016) copyright 2016 Royal Society of Chemistry.

ZT value, i.e., 51.93 μW m^{-1}K^{-2} and 0.017 respectively. This value is found to be the highest ever reported in the literature for PEDOT:PSS-based system. The improvement of electrical conductivity was following the percolation law of the composite that predicts an augmentation of electrical conductivity up to a certain concentration level of filler. These phenomena come into play when two dissimilar materials with a large difference in electrical conductivity are mixed.

Figure 22 represents the bar diagram of electrical conductivity, Seebeck coefficient and power factor with concentration variation of PEDOT:PSS (5 wt.%)/GINC (95 wt.%) composite in the cellulose matrix. Although the material became more mechanically robust, power factor value is found to be very less compared to the bare composite.

During the evaluation of thermoelectric properties, PEDOT:PSS/GINC composite experiences approximately 50 fold enhancement in ZT and four times enhancement in power factor were found with respect to the PEDOT:PSS-graphene composite with an equal filler loading (95 wt%). We have an implicit atypical mechanism of PEDOT:PSS in the existence of GINC. PEDOT:PSS is a polar conductive polymer. PEDOT:PSS is well-suited with GINC. During preparation of polymer nanocomposite, PEDOT:PSS is simply coated over GINC. Therefore, junctions present in the interlayer were modulated in such a way that diminish thermal conductivity but enhances electrical conductivity, and Seebeck coefficient which, therefore, enhances the power factor. Also, phonons are liable to thermal conductivity. Phonone gets spread during conduction which diminishes the thermal conductivity. In PVAc, though its thermal conductivity is the comparatively low but the efficiency to decreases the thermal conductivity is also relatively poor. PVAc, being a non-conducting polymer, is not adequately compatible with GINC. Hence, modulation with such a type of polymer to the interlayer junctions becomes tricky. Besides this, Phonon scattering is also not efficient for PVAc, with respect to PEDOT:PSS.

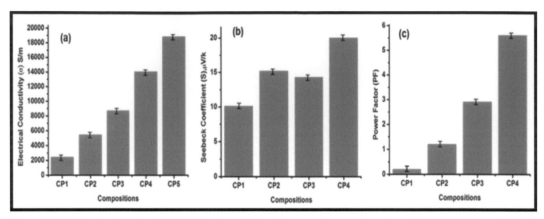

Figure 22. (a) Electrical conductivity (b) Seebeck coefficient (c) power factor as a function of PEDOT:PSS (5 wt.%)/GINC (95 wt.%) composite. The concentration at room temperature (300 K). CP1: Cellulose paper + PEDOT:PSS-GINC composite (10 wt.%), CP2: Cellulose paper + PEDOT:PSS-GINC composite (15 wt.%), CP3: Cellulose paper + PEDOT:PSS-GINC composite (20 wt.%), CP4: Cellulose paper + PEDOT:PSS-GINC composite (30 wt.%), CP5: Cellulose paper + PEDOT:PSS-GINC composite (40 wt.%) (Dey et al. 2016) copyright 2016 Royal Society of Chemistry.

Polymer based graphene titanium dioxide nanocomposite

Seebeck coefficient/thermopower, power factor and electrical conductivity are the three major parameters of thermoelectric properties. Electrical conductivity and thermopower were measured to calculate power factor (PF) from experimental values. Seebeck coefficient, electrical conductivity and power factor using filler concentration at 300 K in a non-conducting polymer matrix, i.e., PVAc are revealed in Figure 23a–c. Likewise, electrical conductivity, variation of Seebeck coefficient and power factor using filler concentration at 300 K for conducting polymer matrix like PEDOT:PSS are shown in Figure 23d–f.

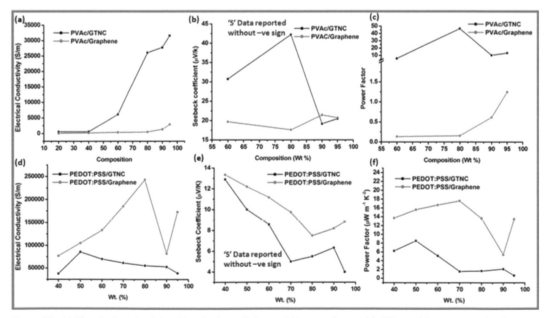

Figure 23. (a) Electrical conductivity, (b) seebeck coefficient and (c) power factor with different filler concentration for PVAc based composite. (d) electrical conductivity, (e) seebeck coefficient and (f) power factor with different filler concentration for PEDOT:PSS-based composite (Dey et al. 2015c) copyright 2015 Royal Society of Chemistry.

Figure 23a to 23c designates the augmentation of Seebeck coefficient, electrical conductivity and power factor at a filler quantity level of 80–90 wt% in PVAc matrix. Figure 23d–f suggests the increase in electrical conductivity and power factor with the variation of filler concentration in PEDOT:PSS matrix (reaches a maximum at 80% wt. filler) but reduces the value of Seebeck coefficient to a certain level, then enhances. Additionally, we have processed several compositions like CP-1 to CP-6. Figure 23 represents comparative graphs of (a) electrical conductivity (b) Seebeck coefficient and (c) power factor of various compositions from CP-1 to CP-6. It has been found that CP-1, i.e., PVAc (20%) and GTNC (80%) exhibit the highest power factor, i.e., 46.54 $\mu W\ m^{-1}K^{-2}$ with elevated Seebeck coefficient and moderate electrical conductivity that is superior than the thermoelectric material based on other compositions. The elaborated assignments have been highlighted in Table 5 to 8. The combination of conducting (PEDOT:PSS) and non-conducting (PVAc) polymer has been engaged as polymer matrix for GTNC nanocomposite preparation. The PF values are found to be very low compared to CP-1 composition, i.e., PVAc (20%) and GTNC (80%). The bar diagram has been given in Figure 24 and detail assignments highlighted in Table 9.

Table 5. Thermoelectric properties of PVAc/GTNC composite with concentration.

Sr. No.	Composition	Electrical conductivity (σ) S/m, (20°C)	Seebeck coefficient (*S*), μV/k	Power factor (PF) μW m⁻¹K⁻²
1.	PVAc (80%) + GTNC (20%)	538	--[a]	--[a]
2.	PVAc (60%) + GTNC (40%)	564.3	--[a]	--[a]
3.	PVAc (40%) + GTNC (60%)	6157	−30.71	5.806
4.	PVAc (20%) + GTNC (80%)	26096	−42.23	46.54
5.	PVAc (10%) + GTNC (90%)	27777	−19.18	10.21
6.	PVAc (5%) + GTNC (95%)	31605	−20.48	13.26

Table 6. Thermoelectric properties of PVAc/graphene composite with concentrations.

Sr. No.	Composition	Electrical conductivity (σ) S/m (20°C)	Seebeck coefficient (*S*), μV/k	Power factor (PF) μW m⁻¹K⁻²
1.	PVAc (80%) + Graphene (20%)	21.47	--	--
2.	PVAc (60%) + Graphene (40%)	186.48	--	--
3.	PVAc (40%) + Graphene (60%)	346.32	−19.69	0.134
4.	PVAc (20%) + Graphene (80%)	484.84	−17.64	0.154
5.	PVAc (10%) + Graphene (90%)	1333.33	−21.44	0.612
6.	PVAc (5%) + Graphene (95%)	2898.55	−20.73	1.245

Table 7. Thermoelectric properties of PEDOT:PSS/GTNC composite with concentrations.

Sr. No.	Composition	Electrical conductivity (σ) S/m (20°C)	Seebeck coefficient (*S*), μV/k	Power factor (PF) μW m⁻¹K⁻²
1.	PEDOT:PSS (60%) + GTNC (40%)	37514	−12.9	6.24
2.	PEDOT:PSS (50%) + GTNC (50%)	85530	−10.0	8.55
3.	PEDOT:PSS (40%) + GTNC (60%)	69930	−8.58	5.14
4.	PEDOT:PSS (30%) + GTNC (70%)	61236	−5.0	1.53
5.	PEDOT:PSS (20%) + GTNC (80%)	55634	−5.4	1.62
6.	PEDOT:PSS (10%) + GTNC (90%)	52671.5	−6.33	2.11
7.	PEDOT:PSS (5%) + GTNC (95%)	38461.5	−4.00	0.61

Table 8. Thermoelectric properties of PEDOT:PSS/GTNC composite with concentrations.

Sr. No.	Composition	Electrical conductivity (σ) S/m (20°C)	Seebeck coefficient (S), μV/k	Power factor (PF) μW m⁻¹K⁻²
1.	PEDOT:PSS (60%) + Graphene (40%)	76987	−13.34	13.7
2.	PEDOT:PSS (50%) + Graphene (50%)	105067	−12.2	15.6
3.	PEDOT:PSS (40%) + Graphene (60%)	133333	−11.17	16.65
4.	PEDOT:PSS (30%) + Graphene (70%)	185125	−9.75	17.6
5.	PEDOT:PSS (20%) + Graphene (80%)	242541	−7.5	13.64
6.	PEDOT:PSS (10%) + Graphene (90%)	81726	−8.2	5.4
7.	PEDOT:PSS (5%) + Graphene (95%)	172057	−8.85	13.48

Table 9. Thermoelectric properties of PEDOT:PSS/GTNC composite with concentration.

Sr. No.	Composition	Electrical conductivity (σ) S/m, (20°C)	Seebeck coefficient (S), μV/k	Power factor (PF) μW m⁻¹K⁻²
CP-1	PVAc (20%) + GTNC (80%)	26096	−42.23	46.54
CP-2	PVAc (15%) + PEDOT:PSS (5%) + GTNC (80%)	6555	−16.21	1.72
CP-3	PVAc (10%) + PEDOT:PSS (10%) + GTNC (80%)	6405	−12.5	1.01
CP-4	PVAc (5%) + PEDOT:PSS (15%) +GTNC (80%)	19503	−20.0	7.80
CP-5	PEDOT:PSS (20%) + GTNC (80%)	55634	−5.4	1.62
CP-6	PEDOT:PSS sheet	27972	−28	21.93

For CP-1, thermal conductivity reached 2.9 W/mK and the associated ZT moved towards 0.0048. These results are more hopeful than our earlier research regarding PVAc-GINC composite. In GTNC, nano TiO_2 is decorated over 2D graphene sheet, and its survival assists to wipe out thermally conducted network whereas electrical network remains intact. The inclusion of GTNC as conductive filler decouples S and σ as well as augments both the parameters simultaneously. Such enhancement of Seebeck coefficient is insufficient about electrical conductivity in the case of PVAc-GTNC. The key fact of diminishing the thermal conductivity of the material can be acquired by thermal padding nature of the PVAc. According to the scheduled reference in Table 10, PVAc composite exhibits PF value up to 12. By the addition of PEDOT:PSS, values enhances to more than 30. In our earlier work, we have demonstrated PF value is 32.9 μW m⁻¹K⁻² and associated ZT is 0.0031 for PVAc-GINC composite. In our current study, we have

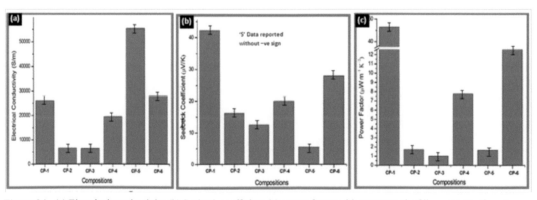

Figure 24. (a) Electrical conductivity (b) Seebeck coefficient (c) power factor with respect to the filler concentration at room temperature (300K) CP-1: PVAc (20%), GTNC (80%), CP-2: PVAc (15%), PEDOT:PSS (5%), GTNC (80%), CP-3: PVAc (10%), PEDOT:PSS (10%), GTNC (80%)CP-4: PVAc (5%), PEDOT:PSS (15%), GTNC (80%), CP-5: PEDOT:PSS (20%), GTNC (80%), CP-6: PEDOT sheet (Dey et al. 2015c) copyright 2015 Royal Society of Chemistry.

Table 10. Thermoelectric properties of PVAc/GZnNC composite with concentrations.

Sr. No.	Composition	σ, S/m (20°C)	S, μV/k, (– ve)	PF μW m⁻¹K⁻²
1.	PVAc (80%) + GZnNC (20%)	0.0029	23.8	1.61×10^{-6}
2.	PVAc (60%) + GZnNC (40%)	2.34	24.2	1.3×10^{-3}
3.	PVAc (40%) + GZnNC (60%)	82.5	26.3	5.7×10^{-3}
4.	PVAc (30%) + GZnNC (70%)	917.4	25.76	0.608
5.	PVAc (20%) + GZnNC (80%)	1478.2	23.7	0.831
6.	PVAc (10%) + GZnNC (90%)	2043.7	28.40	0.144
7.	PVAc (5%) + GZnNC (95%)	5078.2	32.6	5.39

raised PF value up to 47 and associated ZT is 0.0048 at ambient condition. According to literature survey, no one has reported such exciting feature in the literature till now.

The major area of thermoelectric properties is the interconnection of three parameter like electrical conductivity, Seebeck coefficient and thermal conductivity with each other. If one parameter is enhanced, other parameter will also be enhanced. Generally, graphene is not an outstanding thermoelectric material due to its high thermal conductivity. It has very high electrical conductivity which makes it an effective TE material by some means. In the current study, nano TiO_2 particles are positioned in between the graphene sheets during processing, which diminishes the stacking nature of the graphene. Besides, it helps in diminishing its thermal conductivity. Hence, thermal conductivity and electrical conductivity decouples and enhances power factor value.

Polymer based graphene zinc oxide nanocomposite (GZnNC)

PVAc/GZnNC composite

Figure 25 represents the comparative bar diagram of Seebeck coefficient, electrical conductivity and power factor at different PVAc/GZnNC composition (C1 to C7). Electrical conductivity and power factor increases abruptly, but an insignificant upsurge in Seebeck coefficient was perceived. From composition C1 to C7, GZnNC concentration increases systematically to find out the effect of filler. Composition C7 shows maximum PF value. The detailed composition is given in figure caption.

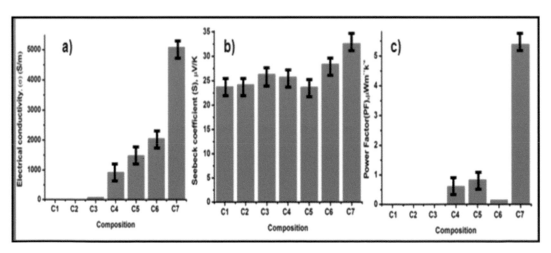

Figure 25. Comparative bar diagram of (a) Electrical conductivity, (b) Seebeck coefficient, (c) Power factor (PF) with different composition based on PVAc and GZnNC. C1: PVAc (80%) + GZnNC (20%), C2: PVAc (60%) + GZnNC (40%), C3: PVAc (40%) + GZnNC (60%), C4: PVAc (30%) + GZnNC (70%), C5: PVAc (20%) + GZnNC (80%), C6: PVAc (10%) + GZnNC (90%), C7: PVAc (5%) + GZnNC (95%).

Table 11. Thermoelectric properties of PVAc/graphene composite with concentrations.

Sr. No.	Composition	σ, S/m (20°C)	S, μV/k (– ve)	PF μW m⁻¹K⁻²
1.	PVAc (80%) + Graphene (20%)	21.47	--	--
2.	PVAc (60%) + Graphene (40%)	186.48	--	--
3.	PVAc (40%) + Graphene (60%)	346.32	19.69	0.134
4.	PVAc (20%) + Graphene (80%)	484.84	17.64	0.154
5.	PVAc (10%) + Graphene (90%)	1333.33	21.44	0.612
6.	PVAc (5%) + Graphene (95%)	2898.55	20.73	1.245

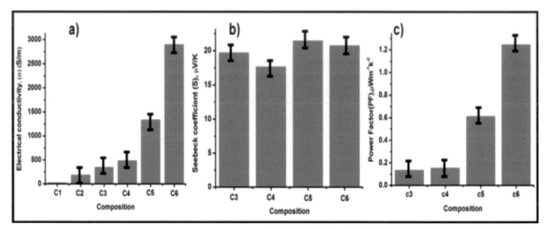

Figure 26. Comparative bar diagram of (a) Electrical conductivity, (b) Seebeck coefficient, (c) Power Factor (PF) with different composition based on PVAc and graphene. C1: PVAc (80%) + Graphene (20%), C2: PVAc (60%) + Graphene (40%), C3: PVAc (40%) + Graphene (60%), C4: PVAc (20%) + Graphene (80%), C5: PVAc (10%) + Graphene (90%), C6: PVAc (5%) + Graphene (95%).

PVAc/graphene composite

Figure represents the comparative bar diagram of Seebeck coefficient, electrical conductivity and power factor at different PVAc/graphene composition (C1 to C6). The detailed composition, bar diagram, and assignment are given below. Here, graphene was used as filler in place of GZnNC. Due to conducting nature of graphene, electrical conductivity increases but a marginal change was observed in Seebeck coefficient. Hence, calculated PF increases with filler concentration. But the overall PF value is relatively low.

PEDOT:PSS/GZnNC composite

Figure 27 depicts the disparity of electrical conductivity, Seebeck coefficient and power factor in PEDOT:PSS/GZnNC composites. To identify the optimum composition, several compositions have been made and evaluated. The comparative chart indicates C2 composition is the optimized composition of these systems which experiences maximum Seebeck coefficient, electrical conductivity and power factor values (see Table 12).

PEDOT:PSS/graphene composite

On the contrary, to find out the effect of Graphene on PEDOT:PSS matrix, seven compositions have been made. In these compositions, electrical conductivity increases to a certain level, then decreases.

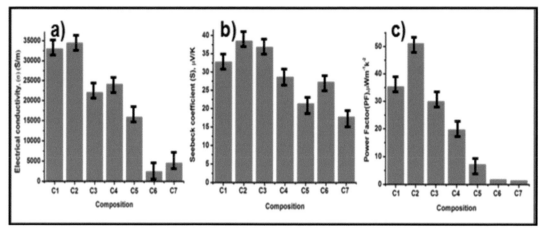

Figure 27. Comparative bar diagram of (a) Electrical conductivity, (b) Seebeck coefficient, (c) Power factor (PF) with different composition based on PEDOT:PSS and GZnNC. C1: PEDOT:PSS (60%) + GZnNC (40%), C2: PEDOT:PSS (50%) + GZnNC (50%), C3: PEDOT:PSS (40%) + GZnNC (60%), C4: PEDOT:PSS (30%) + GZnNC (70%), C5: PEDOT:PSS (20%) + GZnNC (80%), C6: PEDOT:PSS (10%) + GZnNC (90%), C7: PEDOT:PSS (5%) + GZnNC (95%).

Table 12. Thermoelectric properties of PEDOT:PSS/GZnNC composite with concentrations.

Sr. No.	Composition	σ, S/m (20°C)	S, µV/k (– ve)	PF µW m⁻¹K⁻²
1.	PEDOT:PSS (60%) + GZnNC (40%)	32980.2	32.7	35.3
2.	PEDOT:PSS (50%) + GZnNC (50%)	34465.2	38.5	51.1
3.	PEDOT:PSS (40%) + GZnNC (60%)	22026.4	36.8	29.9
4.	PEDOT:PSS (30%) + GZnNC (70%)	24063.0	28.6	19.7
5.	PEDOT:PSS (20%) + GZnNC (80%)	15847.8	21.3	7.2
6.	PEDOT:PSS (10%) + GZnNC (90%)	2421.3	27.2	1.8
7.	PEDOT:PSS (5%) + GZnNC (95%)	4545.4	17.7	1.4

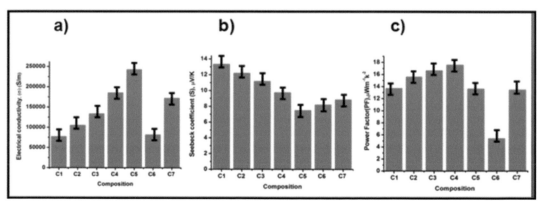

Figure 28. Comparative bar diagram of (a) Electrical conductivity, (b) Seebeck coefficient, (c) PF with different compositions based on PEDOT:PSS/GZnNC. C1: PEDOT:PSS (60%) + Graphene (40%), C2: PEDOT:PSS (50%) + Graphene (50%), C3: PEDOT:PSS (40%) + Graphene (60%), C4: PEDOT:PSS (30%) + Graphene (70%), C5: PEDOT:PSS (20%) + Graphene (80%), C6: PEDOT:PSS (10%) + Graphene (90%), C7: PEDOT:PSS (5%) + Graphene (95%).

Seebeck coefficient decreases with graphene concentration, but PF follows slightly increasing trends than decreasing trends (see Figure 28, Table 13). When compositions have been made in a combination of PVAc/PEDOT:PSS, all the three properties were found to be inferior in compare to PEDOT:PSS-based composition (Figure 29, Table 14).

Table 13. Thermoelectric properties of PEDOT:PSS/GZnNC composite with concentrations.

Sr. No.	Composition	σ, S/m (20°C)	S, μV/k (– ve)	PF μW m⁻¹K⁻²
1.	PEDOT:PSS (60%) + Graphene (40%)	76987	13.34	13.7
2.	PEDOT:PSS (50%) + Graphene (50%)	105067	12.2	15.6
3.	PEDOT:PSS (40%) + Graphene (60%)	133333	11.17	16.65
4.	PEDOT:PSS (30%) + Graphene (70%)	185125	9.75	17.6
5.	PEDOT:PSS (20%) + Graphene (80%)	242541	7.5	13.64
6.	PEDOT:PSS (10%) + Graphene (90%)	81726	8.2	5.4
7.	PEDOT:PSS (5%) + Graphene (95%)	172057	8.85	13.48

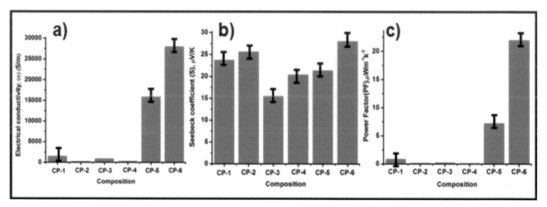

Figure 29. Comparative bar diagram of (a) Electrical conductivity, (b) Seebeck coefficient, (c) Power factor (PF) with different compositions based on PVAc/PEDOT:PSS and GZnNC. CP-1: PVAc (20%) + GZnNC (80%), CP-2: PVAc (15%) + PEDOT:PSS (5%) + GZnNC (80%), CP-3: PVAc (10%) + PEDOT:PSS (10%) + GZnNC (80%), CP-4: PVAc (5%) + PEDOT:PSS (15%) + GZnNC (80%), CP-5: PEDOT:PSS (20%) + GZnNC (80%), CP-6: PEDOT:PSS sheet.

Table 14. Thermoelectric properties of PEDOT:PSS/GZnNC composite with concentration.

Sr. No.	Composition	σ, S/m (20°C)	S, μV/k	PF μW m⁻¹K⁻²
CP-1	PVAc (20%)+GZnNC (80%)	1478.2	23.7	0.831
CP-2	PVAc (15%) + PEDOT:PSS (5%) + GZnNC (80%)	202.02	25.58	0.132
CP-3	PVAc (10%) + PEDOT:PSS (10%) + GZnNC (80%)	860.58	15.44	0.205
CP-4	PVAc (5%) + PEDOT:PSS (15%) + GZnNC (80%)	274.72	20.35	0.113
CP-5	PEDOT:PSS (20%) + GZnNC (80%)	15847.8	21.3	7.2
CP-6	PEDOT:PSS sheet	27972	28	21.93

PVAc/PEDOT:PSS/GZnNC composite

To optimize the ratio of polymer concentrations, i.e., PVAc and PEDOT: PSS, several compositions based on both the polymers have been formulated and evaluated for thermoelectric properties. Figure 29 represents the comparative bar diagram of electrical conductivity, Seebeck coefficient and power factor. The detailed assignments have been consolidated in Table 14.

In the course of studying the thermoelectric properties, PEDOT:PSS/GTNC composite displayed 12 fold upsurge in electrical conductivity and a two-fold upsurge in Seebeck coefficient was perceived as compared to a composite of PVAc-graphene. The premeditated power factor for PEDOT:PSS/GZnNC composite increased to 50 folds than a composite of PVAc-graphene. Thermal conductivity is instituted to be 3.01 W/mK, leading to ZT of 0.005. Since nano zinc oxide was decorated over 2D graphene sheet in

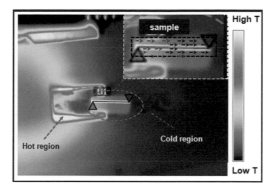

Scheme 30. High-resolution IR camera during testing of sample for Seebeck coefficient measurement.

GZnNC, attendance of nano zinc oxide particle helps to abolish thermally conductive linkage and keeps intact the electrical network. When GZnNC is engaged as conducting filler, it not only decouples σ and S but also improves both the parameter instantaneously which is a very significant finding though the Seebeck coefficient enhancement is minimally about electrical conductivity for PEDOT:PSS/GZnNC composite.

Scheme 30 represents images from high-resolution IR camera during testing of a sample for Seebeck coefficient measurement. The temperature gradient is developed at the ends. The red portion gives an indication of the hotter part whereas blue portion indicates the collar part. The arrow indicates the movement of 'e' and the voltage was developed between the two ends of the sample. IR images help to measure the actual temperature at a particular probes location.

Summary and conclusions

This section discusses the goals achieved during this thesis work, limitations, and scope for future work. Anticipated futuristic developmental study of this research work has also been discussed.

Before making the graphene-metal oxide nanocomposites (GMNC), individual nano metal oxides viz. iron oxide (Fe_2O_3), titanium dioxide (TiO_2) and zinc oxides (ZnO) have been prepared separately. Few-layer graphene has been prepared from graphite flakes by following Hammers method. Three oxides of d-block elements viz. Fe^{3+} (d^5 half filled d orbital), Ti^{4+} (d^0, vacant d orbital) and Zn^{2+} (d^{10}, completely filled d orbital) have been selected and subjected to make graphene iron oxide nanocomposite (GINC), graphene titanium dioxide nanocomposite (GTNC) and graphene zinc oxide nanocomposite (GZnNC), respectively. All the graphene-metal oxide nanocomposites have been prepared by ultrasonication followed by microwave irradiation. Such method of nanocomposite preparation is found to be a green/ecofriendly method. These GMNCs have been characterized by FTIR, RAMAN, SEM, FESEM, TEM, XRD, DSC, and TGA. Also, thermoelectric properties measurement instrument has been fabricated and validated with a standard sample. After validation, thermoelectric properties like Seebeck coefficient, electrical conductivity, and thermal conductivity have been evaluated and computed its power factor and thermoelectric figure of merit. The catalytic behavior of all the metal oxides enhanced significantly in the presence of graphene.

We have attempted to exploit the designed GMNCs as efficient thermoelectric candidates. Polymer like PVAc, a non-conducting and PEDOT:PSS, a conducting polymer has been employed to process polymer nanocomposite along with added advantage of efficacy. In brief, PEDOT:PSS-based GINC and GZnNC show enhanced power factor (PF) value, i.e., > 50 μW $m^{-1}K^{-2}$ at room temperature. But PVAc based GTNC attributed PF value to up to 47 μW $m^{-1}K^{-2}$. The elaborate discussion on the experimental findings ultimately leads to the following conclusions:

Also, highly conductive PVAc/graphene composites and PVAc/GINC have been processed through hot compaction in the direction of their thermoelectric solicitations. The concentrations of graphene and GINC were altered prudently and optimized to achieve desirable Seebeck coefficient and high electrical conductivity. The contrived PVAc/GINC film displayed a conductivity of 2.18 x 10^4 S m^{-1} and Seebeck coefficient of 38.8 μV K^{-1}, respectively. As a result, the power factor (PF) extends to a value of

32.90 μW m^{-1} K^{-2}, which is 27 fold more than PVAc based graphene composite thermoelectric material. Such power factor is considered to be the uppermost ever testified in the absence of conducting polymer in the literature. We have extended our study with PEDOT:PSS, a conducting polymer, to enhance the power factor (PF) and figure of merit (ZT) of the system further. The evaluation suggested that the PEDOT:PSS/ GINC composite (5:95) upsurges thermoelectric possessions viz. the electrical conductivity of 8.0×10^4 S/m, Seebeck coefficient of 25.42 μV/K and thermal conductivity of 0.90 W/mK remarkably. Therefore, PF and ZT reach to 51.93 μW/mK^2 and 0.017, respectively, when we incorporate PEDOT:PSS. In this composite, cellulose fibre has been appended to improve the mechanical forte of the resultant polymer composite. Though the addition of cellulose fibre increases the mechanical strength of the composite, it has reduced the PF to 5.6, which is smaller than the PEDOT:PSS/GINC composite by ten folds.

To evaluate GTNC as a thermoelectric material, several conductive composites like PEDOT:PSS-GTNC, PVAc-GTNC, PEDOT:PSS-graphene, PVAc-graphene have been made by ultrasonication and hot compaction. The concentration of Graphene, GTNC and polymer matrix were sensibly mottled in the mixture and optimized in the direction of reaching electrical conductivity and Seebeck coefficient that tends to higher power factor (PF) value. The PVAc based composite, i.e., PVAc (20%) and GTNC (80%) was found to be the most promising material with an electrical conductivity of 2.6×10^4 S/m and Seebeck coefficient of -42 μV/K. As a result, the PF reaches to 47 μW m^{-1}K^{-2} that is approximately 37 times more than PVAc-graphene based composite and five times more than PEDOT:PSS-based composite. The enhancement of thermoelectric efficiency of GTNC composite originates from the synergistic effect of TiO$_2$ nanoparticle and graphene nanosheet. This PF value is found to be maximized, ever reported in the absence of conducting polymer.

Such GZnNC nano-composites are also executed as an effective energy harvesting or thermoelectric material. Throughout the study of thermoelectric possessions, PEDOT:PSS/GTNC composite displayed 12 fold increase in electrical conductivity and a two-fold increase in Seebeck coefficient than PVAc-graphene composite. The resultant power factor for PEDOT:PSS/GZnNC composite increases by 50 folds than PVAc-graphene composite. Thermal conductivity is found to be 3.01 W/mK along with the ZT of 0.005. In GZnNC, nano zinc oxide was decorated over 2D graphene sheet that not only decouples the σ and S but also enhances both the parameters all together.

Overall, it has been observed that graphene and the type of the metal oxide play a vital role for both the applications. In the presence of graphene, the thermoelectric effect of the metal oxide enhanced ~ 50% drastically. Also, the presence of graphene enhances the conductivities, but metal oxide helps to retard the phonon transport without affecting the electron transport. Hence, thermoelectric PF and ZT increase significantly. In the presence of PEDOT:PSS, PF, and ZT value enhance markedly for GINC and GZnNC. But GTNC administered more PF and ZT value in the presence of PVAc.

Contributions of the present work

A green/ecofriendly method has been developed for the preparation of graphene-metal oxide nanocomposite preparation. Three d-block elements viz. Fe^{3+} (d^5 half-filled d orbital), Ti^{4+} (d^0, vacant d orbital) and Zn^{2+} (d^{10}, filled d orbital) have been chosen on the basis of their electronic configuration. Vacant, half filled and full filled orbital's impart extra stability to the oxidation state of the metal oxide even after decoration on graphene nanosheets. Other d-block elements also show catalytic and thermoelectric properties, but they reduce very fast which is not acceptable. Graphene iron oxide nanocomposite (GINC), graphene titanium dioxide nanocomposite (GTNC) and graphene zinc oxide nanocomposite (GZnNC) have been made by ultrasonication followed by microwave irradiation. All GMNCs were used as burn rate enhancer for composite propellant. Results show that decomposition temperature of AP decreases up to 150°C and heat release increases in the presence of each GMNC. Activation energy also reduces significantly, which reflects the catalytic nature of the composites. In the case of propellant, burning rate increases significantly in the presence of GMNC. Between these GMNCs, GINC shows comparatively high catalytic efficiency due to the presence of half filled d orbital that imparts better stability to the

Fe^{2+} ion. Overall, catalytic efficiency enhances approximately 50% in the presence of GINC, GTNC or GZnNC. Rest of the d-block elements shows unstable variable oxidation state and hence, catalytic activity goes down very fast during aging. Additionally, thermoelectric properties of these composites have been evaluated at room temperature. Among GINC, GTNC, and GZnNC, GINC and GZnNC shows better thermoelectric properties due to stable oxidation states d^5 and d^{10}, respectively. Power factor value of PEDOT:PSS/GINC and PEDOT:PSS/GZnNC polymer nanocomposites reaches to significantly higher power factor value, i.e., PF > 50 $\mu W\ m^{-1}K^{-2}$, which is quite inspiring in nature. The decoration of metal oxide over graphene surface helps to obstruct the phonon transport but allows electronic transport leading to increase in the thermoelectric performance. The addition of polymer matrix improved the electronic transport further by modulating the junction of the graphene layer. Thus, the present thesis contributed to developing a multi-functional nanocomposites material along with their applications in two major areas.

Limitations and scope for future work

To the best of our knowledge, this thesis for the first time presents a systematic study on synthesis and characterization of graphene-based metal oxide nanocomposites. As an extension of this research, the structure-property relationship of polymer/GMNC composites can be quantified by doing analytical or by empirical modeling. The acquired knowhow can be helpful for further optimization of GMNC and polymer/GMNC composite towards their actual application. Outstanding steps in the development of this research should entail comprehensive studies on the simplest processing techniques of GMNCs and polymer/GMNC composites. Several other composites can be made by following above mentioned methodology. The role of the nano metal oxide decorated over graphene surface as well as the actual mechanistic pathway for obstruction of phonon without affecting the electron transport can be studied in detail. The study regarding the catalytic and thermoelectric effect at high temperature for GMNCs and polymer/GMNCs was untouched. This study lacks experimental results and ideal analytical solutions to predict systematic catalytic and thermoelectric properties of GMNCs. All the measured properties depend on type and loading of metal oxides and GMNCs. Also, polymers played a vital role for thermoelectric application to enhance PF and ZT value. Among the tested applications, the nanocomposites appear to be more suitable for catalysis and as a thermoelectric material. There is a huge scope for the better use of these nanocomposites as radar absorbing material, sensors, microelectronics, optoelectronics, super capacitor and lithium ion battery. These could not be completed either due to the time limitations or due to the unavailability of experimental facilities; however, they can be extended as scopes for future research.

Further, the results obtained in this thesis may be extended for other d block metal oxide systems in general. There is a tremendous scope for the GMNCs in some applications, where improvements in properties especially the catalytic as well as thermoelectric do play a major role.

References

Ahiska, R. and J. Turk. 2007. New method for investigation of dynamic parameters of thermo-electric modules. Elec. Engin. 15: 51–65.

Ahiska, R. and H. Mamur. 2014. A review: Thermoelectric generators in renewable energy. Int. J. Renew. Ener. Res. 4: 128–136.

Bakker, F.L., A. Slachter, J.P. Adam and B.J. Vanwees. 2010. Interplay of peltier and seebeck effects in nanoscale nonlocal spin valves. Phys. Rev. Lett. 105: 136601–136605.

Beeby, S.P., M.J. Tudor and N.M. White. 2006. Energy harvesting vibration sources for microsystems applications. Meas. Sci. Technol. 17: R175–R195.

Behnam, A.V.K., X. Sangwan, F. Zhong, D. Lian, D. Estrada, A. Jariwala et al. 2012. High-field transport and thermal reliability of sorted carbon nanotube network devices. ACS Nano 7: 482–490.

Bell, L.E. 2008. Cooling, heating, generating power, and recovering waste heat with thermoelectric systems. Science 32: 1457–1461.

Bolotin, K.I., K.J. Sikes, Z. Jiang, M. Klima, G. Fudenberg and J. Hone. 2008. Ultrahigh electron mobility in suspended graphene. Solid State Commun. 146: 351–355.

Boukai, A.I., Y. Bunimovich, J. Tahir-Kheli, J.K. Yu, W.A. Goddard Iii and J.R. Heath. 2008. Silicon nanowires as efficient thermoelectric materials. Nature 451: 168–171.

Bubhove, O. and X. Cripson. 2012. Towards polymer-based organic thermoelectric generators. Energy Environ. Sci. 5: 9345–9362.

Cai, J., P. Ru_eux, R. Jaafar, M. Bieri, T. Braun, S. Blankenburg et al. 2010. Atomically precise bottom-up fabrication of graphene nanoribbons. Nature 466: 470–473.

Caldeira, K., A. Jain and M. Hoffert. 2003. Climate sensitivity uncertainty and the need for energy without CO_2 emission. Science 299: 2052–2054.

Chen, G., M.S. Dresselhaus, G. Dresselhaus, J.P. Flexurial and T. Eaillat. 2003. Recent developments in thermoelectric materials. Inter. Mat. Rev. 48: 45–66.

Chen, Y., T. Jayasekera, A. Calzolari, K.W. Kim and B.M. Nardelli. 2010. Thermoelectric properties of graphene nanoribbons, junctions and superlattices. Journal of Physics: Condensed Matter. 22: 372202–372205.

Chung, D.Y., T. Hogan, P. Brazis, M. Rocci-Lane, C. Kannewurf and M. Bastea. 2000. $CsBi_4Te_6$: A high performance thermoelectric material for low-temperature applications. Science 287: 1024-.

Dey, A., J. Athar, P. Varma, H. Prasanth, A.K. Sikder and S. Chattopadhyay. 2015a. Graphene-iron oxide nanocomposite (GINC): an efficient catalyst for ammonium perchlorate (AP) decomposition and burn rate enhancer for AP based composite propellant. RSC Adv. 5: 1950–1960.

Dey, A., O.P. Bajpai, A.K. Sikder, S. Chattopadhyay and M.A.S. Khan. 2015b. Recent advances in CNT/graphene based thermoelectric polymer nanocomposite: A proficient move towards waste energy harvesting. Renew. Sust. Energ. Rev. 53: 19248–255.

Dey, A., S. hadawale, S. Khan, P. More, P.K. Khanna, A.K. Sikder et al. 2015c. Graphene/Titanium dioxide nanocomposite (GTNC): An emerging ecofriendly and efficient thermoelectric material. Dalton trans. 44: 1950–1960.

Dey, A., S. Panja, A.K. Sikder and S. Chattopadhyay. 2015d. One pot green synthesis of graphene-iron oxide nanocomposite (GINC): An efficient material for enhancement of thermoelectric performance. RSC Adv. 5: 10358–10364.

Dey, A., S. Hadawale, M.A.S. Khan, A.K. Sikder, and S. Chattopadhyay. 2016. PVAc/PEDOT:PSS/Graphene-iron oxide nanocomposite (GINC): an efficient thermoelectric material. RSC Adv., 6, 22453–22460

DiSalvo, F. J. 1999. Thermoelectric cooling and power generation. Science 285: 703–706.

Dresselhaus, M.S., G. Dresselhaus, X. Sun, Z. Zhang, S.B. Cronin, T. Koga et al. 1999. The promise of low-dimensional thermoelectric materials. Microscale Thermophysical Engineering 3: 89–100.

Dresselhaus, M.S., G. Chen, M.Y. Tang, R. Yang, H. Lee, D. Wang, Z. Ren et al. 2007. New directions for low-dimensional thermoelectric materials. Adv. Mater. 19: 1043–1053.

Du, Y., S.Z. Shen, W. Yang, R. Donelson, K. Cai and P.S. Casey. 2012. Simultaneous increase in conductivity and seebeck coefficient in a polyaniline/graphene nanosheets thermoelectric nanocomposite. Synth. Met. 161: 2688–2692.

Eda, G., G. Fanchini and M. Chhowalla. 2008. Large-area ultrathin films of reduced graphene oxide as a transparent and flexible electronic material. Nat. Nanotechnol. 3: 270–274.

Elsheikh, M.H., D.A. Shnawah, M.F.M. Sabri, S.B.M. Said, M.H. Hassan, M.B.A. Bashir et al. 2014. Review on thermoelectric renewable energy: Principle parameters that affect their performance. Renewable and Sustainable Energy Reviews 30: 337–355.

Energy information administration, international Energy Outlook. 2010. Tech. Rep., Department of Energy, US, 2010.

Enescu, D. and E.O. Virjoghe. 2014. A review on thermoelectric cooling parameters and performance. Renewable and Sustainable Energy Reviews 38: 903–916.

Enoki, T., K. Takai, V. Osipov, M. Baidakova and A. Vul. 2009. Nanographene and nanodiamond; new members in the nanocarbon family. Chem. Asian J. 4: 796–804.

Estrada, D. and E. Pop. 2011. Imaging dissipation and hot spots in carbon nanotube network transistors. Appl. Phys. Lett. 98: 073102–073103.

Gangopadhyay, R. and A. De. 2000. Conducting polymer nanocomposites: a brief overview. Chem. Mater. 12: 608–622.

Gautois, M.W., T.D. Sparks, C.K.H. Borg, R. Seshadri, W.D. Boificio and D.R. Clarke. 2013. Data-driven review of thermoelectric materials: performance and resource considerations. Chem. Mater. 25: 2911–2920.

Geim, A.K. and K.S. Novoselov. 2007. The rise of graphene. Nat. Mater. 6: 183–191.

Gupta, M.P., L. Chen, D. Estrada, A. Behnam, E. Pop and S.J. Kumar. 2012. Impact of thermal boundary conductances on power dissipation and electrical breakdown of carbon nano-tube network transistors. Appl. Phys. 112: 124506–124507.

Gupta, M.P., A. Behnam, F. Lian, D. Estrada, E. Pop and S. Kumar. 2013. High field breakdown characteristics of carbon nanotube thin film transistors. Nanotechnology 24: 405204–405213.

Hereman, J.P., C.M. Trush and D.T. Morelli. 2001. Geometrical magnetothermopower in semiconductors. Phys. Rev. Lett. 86: 2098–2101.

Hereman, J.P., C.M. Thrush, D.T. Morelli and M.C. Wu. 2002. Thermoelectric power of bismuth nanocomposites. Phys. Rev. Lett. 88: 4361–4365.

Hereman, J.P., V. Jovovic, E.S. Toberer, A. Saramat, K. Kurosaki and A. Charo–enphakdeem. 2008. Enhancement of thermoelectric efficiency in PbTe by distortion of the electronic density of states. Science 321: 554–557.

Harman, T.C., P.J. Taylor, M.P. Walsh and B.E. LaForge. 2002. Quantum dot super lattice thermoelectric materials and devices. Science 297: 2229–2232.

He, W., G. Zhang, X. Zhang, J. Ji, G. Li and X. Zhao. 2015. Recent development and application of thermoelectric generator and cooler. Applied Energy 143: 1–25.

Hicks, L. and M.S. Dresselhaus. 1993. Effect of quantum-well structures on the thermoelectric figure of merit. Phys. Rev. B. 47: 12727–12731.

Hicks, L. and M.S. Dresselhaus. 1993. Reflection on thermoelectric. Phys. Rev. B. 47: 16631–16634.

Hiroshige, Y., M. Ookawa and N. Toshima. 2006. High thermoelectric performance of poly(2,5-dimethoxy phenylenevinylene) and its derivatives. Synyh. Met. 156: 1341–1347.

Hochbaum, A., R. Chen, R. Delgado, W. Liang, E. Garnett, M. Najarian et al. 2008. Enhanced thermoelectric performance of rough silicon nanowires. Nature 451: 163–167.

Hoffert, M., K. Caldeira, A. Jain, E. Haites, L. Harvey and S. Potter. 1998. Energy implications of future stabilization of atmospheric CO_2 content. Nature 395: 881–884.

Hoffert, M., K. Caldeira, G. Benford, D. Criswell, C. Green and H. Herzog. 2002. Advanced technology paths to global climate stability: energy for a greenhouse planet. Science 298: 981–987.

Huang, W., J.S. Wang and G. Liang. 2011. Theoretical study on thermoelectric properties of kinked graphene nanoribbons. Physical Review B 84: 045410–045417.

Hwang, S.H., H.W. Park and Y.B. Park. 2013a. Piezoresistive behavior and multi-directional strain sensing ability of carbon nanotube–graphene nanoplatelet hybrid sheets. Smart Materials and Structures 22: 015013–015022.

Hwang, S.H., H.W. Park, Y.B. Park, M.K. Um, J.H. Byun and S. Kwon. 2013b. Polymeric nanofibre composites with aligned ZnO nanorods, Science and Technology 89: 9–17.

Kanatzidis, M.G. 2010. Nanostructured thermoelectrics: The new paradigm? Chem. Mater. 22: 648 659.

Kateb, B., V. Yamamoto, D. Alizadeh, L. Zhang, H.M. Manohara, M.J. Bronikowski et al. 2010. Multiwalled carbon nanotube (MWCNT) synthesis, preparation, labeling, and functionalization. Methods Mol. Biol. 651: 307–317.

Kim, G.H., D.H. Hwang and S.I. Woo. 2012. Thermoelectric properties of nanocomposite thin films prepared with poly(3,4-ethylenedioxythiophene) poly(styrenesulfonate) and graphene. Phys. Chem. Chem. Phys. 14: 3530–3536.

Kjelstrup, S., P.J.S. Vie, L. Akyalcin, P. Zefaniya, J.G. Pharoah and O.S. Burheim. 2013. The seebeck coefficient and the peltier effect in a polymer electrolyte membrane cell with two hydrogen electrodes. Electrochimica. Acta 99: 166–175.

Kleinke, H. 2010. New bulk materials for thermoelectric power generation: clathrates and complex antimonides. Chem. Mater. 22: 604–611.

Kong, H.X. 2013. Hybrids of carbon nanotubes and graphene/graphene oxide. Curr. Opin. Solid State Mater. Sci. 17: 31–37.

Lampinen, M.J. 1991. Thermodynamic analysis of thermoelectric generator. J. Appl. Phys. 69: 4318–4323.

Lee, C., X. Wei, J.W. Kysar and J. Hone. 2008. Measurement of the elastic properties and intrinsic strength of monolayer graphene. Science 321: 385–388.

Lee, J.K., M.W. Oh, S.D. Park, S. Kim, B.K. Min, M.H. Kim et al. 2012. Improvement of thermoelectric properties through controlling the carrier concentration of $AgPb_{18}SbTe_{20}$ alloys by Sb addition. Elec. Mater. Lett. 8: 659–663.

Leila, S. and Y. Masoud. 2012. An empirical study of the usefulness of SARFIMA models in energy science. Inter. J. Ener. Sci. 2: 59–63.

LeBlanc, S., S.K. Yee, M.L. Scullin, C. Dames and K.E. Goodson. 2014. Material and manufacturing cost considerations for thermoelectric. Renewable and Sustainable Energy Reviews 32: 313–327.

Liang, L., E. Cruz-Silva, E.C. Girão and V. Meunier. 2012. Enhanced thermoelectric figure of merit in assembled graphene nanoribbons. Physical Review B 86: 115438 (1–8).

Liang, J., Y. Xu, Y. Huang, L. Zhang, Y. Wang and Y. Ma. 2009. Infrared-triggered actuators from graphene based nanocomposites. J. Phy. Chem. C 113: 9921–9927.

Lim, C.H., S.M. Choi, W.S. Seo and H.H. Park. 2012. A power-generation test for oxide- based thermoelectric modules using p-Type $Ca_3Co_4O_9$ and n-Type $Ca_{0.9}Nd_{0.1}MnO_3$ legs. J. Elec. Mater. 41: 1247–1255.

Li, D., M.B. Muller, S. Gilje, R.B. Kaner and G.G. Wallace. 2008. Processable aqueous dispersions of graphene nanosheets. Nat. Nanotechnol. 3: 101–105.

Li, J.J., X.F. Tang, H. Li, Y.G. Yan and Q.J. Zhang. 2010. Synthesis and thermoelectric properties of hydrochloric acid-doped polyaniline. Synth. Met. 160: 1153–1158.

Li, W., D. He and J. Bai. 2013. Hybrids of carbon nanotubes and graphene/graphene oxide. Composites Part A 54: 28–34.

Li, X., G. Zhang, X. Bai, X. Sun, X. Wang and E. Wang. 2008. Highly conducting graphene sheets and Langmuir–Blodgett films. Nat. Nanotechnol. 3: 538–542.

Long, Y., Z. Chen, X. Zhang, J. Zhang and Z. Liu. 2004. Synthesis and electrical properties of carbon nanotube polyaniline composites. Appl. Phys. Lett. 85: 1796–1798.

Lu, Y., Y. Song and F. Wang. 2012. Thermoelectric properties of graphene nanosheets-modified polyaniline hybrid nanocomposites by an *in situ* chemical polymerization. Materials Chemistry and Physics 53: 4202–4210.

MacDiamid, A.G. and A.J. Epstein. 1995. Secondary doping in polyaniline. Synth. Met. 69: 85–92.

Makala, R.S., K. Jagannadham and B.C.J. Sales. 2003. Pulsed laser deposition of Bi_2Te_3-based thermoelectric thin films. J. Appl. Phy. 94: 3907–3919.

Minnich, A.J., M.S. Dresselhaus and Z. Ren. 2009. Bulk nanostructured thermoelectric materials: current research and future prospects. Chem. Energy Environ. Sci. 2: 466–479.

Nakada, K., M. Fujita, G. Dresselhaus and M.S. Dresselhaus. 1996. Edge state in graphene ribbons: Nanometer size effect and edge shape dependence. Physical Review B 54: 17954–17961.

Nardecchia, S., D. Carriazo, M.L. Ferrer, M.C. Gutierrez and F. del Monte. 2013. Three dimensional macroporous architectures and aerogels built of carbon nanotubes and/or graphene: synthesis and applications. Chem. Soc. Rev. 42: 794–830.

Noorden, R.V. 2011. Chemistry: The trials of new carbon. Nature 469: 14–16.

Novoselov, K.S., A.K. Geim, S.V. Morozov, D. Jiang, Y. Zhang and S.V. Dubonos. 2004. Electric field effect in atomically thin carbon films. Science 306: 666–669.

Ono, K. and R.O. Suzuki. 1998. Thermoelectric power generation: converting low-grade heat into electricity. Met. Mater. Soc. 40: 49–51.

Ouyang, Y. and J. Guo. 2009. A theoretical study on thermoelectric properties of graphene nanoribbons, Applied Physics Letters 94: 263107 (1–3).

Pennelli, G.J. 2014. Review of nanostructured devices for thermoelectric applications. Beilstein J. Nano technol. 5: 1268–1284.

Prasher, R.S., X. Hu, Y. Chalopin, N. Mingo, K. Lofgreen, S. Volz et al. 2009. Turning carbon nanotubes from exceptional heat conductors into insulators. Phys. Rev. Lett. 102: 105901–105905.

Seol, J.H., I. Jo, A.L. Moore, L. Lindsay, Z.H. Aitken and M.T. Pettes. 2010. Two-dimensional phonon transport in supported graphene. Science 328: 213–216.

Sevinçli, H. and G. Cuniberti. 2010. Enhanced thermoelectric figure of merit in edge-disordered zigzag graphene. Phys. Rev. B 81: 113401–113404.

Shao, Y., J. Wang, H. Wu, J. Liu, I.A. Aksay and Y. Lin. 2010. Graphene based electrochemical sensors and biosensors: a review. Electroanalysis 22: 1027–1036.

Snyder, G.J. and E.S. Toberer. 2008. Complex thermoelectric materials. Nat. Mater. 7: 105–114.

Sondheimer, E.H. 1956. The kelvin relations in thermo-electricity. R. Soc. London, Ser. A234: 391–398.

Suman, S., M.K. Khan and M. Pathak. 2015. Performance enhancement of solar collectors—A review. Renewable and Sustable Energy Reviews 49: 192–210.

Wang, Q.H., D.O. Bellisario, L.W. Drahushuk, R.M.S. Jain Kruss, M.P. Landry, S.G.E. Mahajan et al. 2014a. Low dimensional carbon materials for applications in mass and energy transport. Chem. Mater. 26: 172–183.

Wang, Q., Q. Yao, J. Chang and L. Chen. 2012b. Enhanced thermoelectric properties of CNT/PANI composite nanofibres by highly orienting the arrangement of polymer chains. J. Mater. Chem. 22: 17612–17618.

Wang, X., L. Zhi and K. Mullen. 2008a. Conductive graphene electrodes for dye-sensitized solar cells. Nano Lett. 8: 323–327.

Wang, X., L. Zhi, N. Tsao, Ž. Tomović, J. Li and K. Müllen. 2008b. Transparent carbon films as electrodes in organic solar cells. Angew Chem. Int. Ed. 47: 2990–2992.

Wei, P., W. Bao, Y. Pu, C.N. Lau and J. Shi. 2009. Anomalous thermoelectric transport of Dirac particles in graphene. Phys. Rev. Lett. 102: 166808–166812.

Weng, Z., Y. Su, D.W. Wang, F. Li, J. Du and H.M. Cheng. 2011. Graphene–cellulose paper flexible supercapacitors. Adv. Energy Mater. 1: 917–922.

Xiang, J. and T.D. Lawrence. 2012. Templated growth of polyaniline on exfoliated graphene nanoplatelets (GNP) and its thermoelectric properties. Polymer 53: 4202–4210.

Xie, Z.X., L.M. Tang, C.N. Pan, K.M. Li, K.Q. Chen and W. Duan. 2012. Enhancement of thermoelectric properties in graphene nanoribbons modulated with stub structures. Applied Physics Letters 100: 073105 (1–4).

Xu, K., G. Chen and D. Qiu. 2015. *In situ* chemical oxidative polymerization preparation of poly(3,4-ethylenedioxythiophene)/graphene nanocomposites with enhanced thermoelectric performance. Chemistry-An Asian Journal 10: 1225–1231.

Xu, Y., X. Chen, J.S. Wang, B.L. Gu and W. Duan. 2010. Thermal transport in graphene junctions and quantum dots. Phys. Rev. B 81: 195425–195431.

Yamashita, O. and N. Sadatomi. 1999. Dependence of seebeck coefficient on carrier concentration in heavily b- and p-doped Si1-xGex (x0.05) system. J. Appl. Phys. 38: 6394–6400.

Yan, H., T. Ohta and N. Toshima. 2001. Stretched polyaniline films doped by (l)-10 camphor sulfonic acid: anisotropy and improvement of thermoelectric properties. Macromol. Mater. Eng. 286: 139–142.

Yang, K., Y. Chen, R.D. Agosta, Y. Xie, J. Zhong and A. Rubio. 2012. Enhanced thermoelectric properties in hybrid graphene/boron nitride nanoribbons. Physical Review B 86: 045425–045428.

Yang, W., K.R. Ratinac, S.P. Ringer, P. Thordarson, J.J. Gooding and F. Brat. 2010. Carbon nanomaterials in biosensors: should you use nanotubes or graphene? Angew Chem. Int. Ed. 49: 2114–2138.

Yao, Q., L.D. Chen, X.C. Xu and C.F. Wang. 2005. The high thermoelectric properties of conducting polyaniline with special submicron-fibre structure. Chem. Lett. 34: 522–523.

Yao, Q., L.D. Chen, W.Q. Zhang, S.C. Liufu and X.H. Chen. 2010. Enhanced thermoelectric performance of single-walled carbon nanotubes/polyaniline hybrid nanocomposites. ACS Nano 4: 2445–2451.

Yoo, D., J. Kim and J.H. Kim. 2014. Direct synthesis of highly conductive PEDOT:PSS/graphene composites and their applications in energy harvesting systems. Nano Research 7: 717–730.

Yoo, E.J., J. Kim, E. Hosono, H. Zhou, T. Kudo and I. Honma. 2008. Large reversible Li storage of graphene nanosheet families for use in rechargeable lithium ion batteries. Nano Lett. 8: 2277–2282.

Yu, D. and F. Liu. 2007. Synthesis of carbon nanotubes by rolling up patterned graphene nanoribbons using selective atomic adsorption. Nano Lett. 7: 3046–3050.

Yun, J. and S.S. Lee. 2014. Human movement detection and identification using pyroelectric infrared sensors. Sensors 14: 8057–8081.

Zebarjadi, M., K. Esfarjani, M.S. Dresselhaus, Z.F. Ren and G. Chen. 2012. Perspectives on thermoelectrics: from fundamentals to device applications. Energy Environ. Sci. 5: 5147–5162.

Zhang, J.C. 2008. Recent advances on thermoelectric materials. Front Phy. 3: 269–279.

Zhang, Y. and G.D. Stucky. 2014. Heterostructured approaches to efficient thermoelectric materials. Chem. Mater. 26: 837–848.

Zhang, K., Y. Zhang and S. Wang. 2013. Enhancing thermoelectric properties of organic composites through hierarchical nanostructures. Scientific Reports 3: 3448 (1–7).

Zhao, H., J. Sui, Z. Tang, Y. Lan, Q. Jie, D. Kraemer et al. 2014. High thermoelectric performance of MgAgSb-based materials. Nano Energy 7: 97–103.

Zhao, L., T. Lu, M. Yosef, M. Steinhart, M. Zacharias, U. Gösele and S. Schlecht. 2006. Single-crystalline CdSe nanostructures: from primary grains to oriented nanowires. Chem. Mater. 18: 6094–6096.

Zuev, Y.M., W. Chang and P. Kim. 2009. Thermoelectric and magneto thermoelectric transport measurements of graphene. Phys. Rev. Lett. 102: 096802–096807.

21

Synthesis and Properties of Carbon Nanotube/Alumina Nanocomposites

Soumya Sarkar* and *Probal Kumar Das*

Introduction

Carbon nanotubes, being one of the outstanding sp^2 hybridized allotropes of the element carbon, offer excellent properties in several aspects. Such properties coupled with low density (1.3–2.1 g/cc) eventually render outstanding specific properties to CNTs as well (Sarkar and Das 2014a). Consequently, just after its discovery, R&D work on assessing the real-life applicability of various types of CNT not only as individual but also as a promising nano-filler in various metallic, polymeric and ceramic matrices have been initiated globally (Bogue 2011, Chen et al. 2011, Iijima 1991, Rahmat and Hubert 2011, Umma et al. 2012). Beside the lucrative properties, another advantage of using CNT as a superior reinforcing phase over conventional microscopic fillers is the requirement of very low filler concentration due to its nano-dimension and very high specific surface area. On the contrary, issues related to heterogeneous dispersion of CNT due to its clustering tendency and hydrophobicity, inadequacy of the interface region, phase segregation and morphological damage of CNT at high T_{sin} and pressure, especially, during densification of ceramic based nanocomposites are still under active consideration prior to full phased real-life utilization of components based on CNT reinforced nanocomposites (Bogue 2011, Camargo et al. 2009, Estili and Sakka 2014, Sarkar and Das 2014a). Although several ceramic matrices have been reinforced with CNT, e.g., Al_2O_3, ZrO_2, Si_3N_4, and SiC, reports on CNT/Al_2O_3 nanocomposites starting from processing details, microstructure and interface studies to property evaluation are the maximum (Belmonte et al. 2014, Clark et al. 2012, Melk et al. 2015, Michálek et al. 2014, Sarkar et al. 2016, Sarkar and Das 2012b,c,d, 2014b,c, 2015). Availability of different grades of Al_2O_3 in any quantity at an affordable price and its versatility for effective utilization in both commercial and strategic sectors are the two prime factors that always attract researchers for further property enhancement through compositing of Al_2O_3 ceramics with suitable fillers (Mata-Osoro et al. 2012, MD et al. 2014, Medvedovski 2002, Silva et al. 2014, Tang et al. 2012). The major limitations of Al_2O_3 ceramics that can be evaded through proper reinforcement are its extensive brittleness, inadequate toughness, electrical insulating nature and poor thermal conductivity. Thus, beside microscopic reinforcements, researchers have also used CNT as a nano-filler in Al_2O_3 to

CSIR-Central Glass and Ceramic Research Institute, 196 Raja S.C. Mullick Road, Kolkata-700032, India.
  Email: probalkdas@rediffmail.com
* Corresponding author: soumya@cgcri.res.in

get superior properties in the ultimate product over pure Al_2O_3 (Estili and Sakka 2014, Sarkar and Das 2014a). Significant improvements in properties of CNT/Al_2O_3 nanocomposites over monolithic Al_2O_3 suitable for different high-performance applications are discussed. ~ 21% higher HV over pure Al_2O_3 was obtained at ~ 4 wt.% CNT loading through uniform dispersion of CNT in the matrix phase and adequate densification of nanocomposite to promote effective load sharing between the matrix and filler through interface region (An and Lim 2002, An et al. 2003). About 13% higher HV over pure Al_2O_3 for a hot-pressed 2 wt.% MWCNT/Al_2O_3 nanocomposite (*HV* = 18 GPa) was also achieved (Ahmad et al. 2010). However, improvement in HV of CNT reinforced nanocomposite, especially at higher filler loading, is not an obvious event. This is mainly due to the fact that CNT being extremely elastic and softer than Al_2O_3, generally, undergoes higher deformation during indentation loading cycle causing higher deformation of the nearby Al_2O_3 matrix. During the unloading cycle, although, the elastically deformed CNTs can easily return to their initial position, the plastically deformed matrix eventually yield lower HV values for the nanocomposite, especially at higher CNT loading compared to monolithic Al_2O_3 (Sarkar et al. 2016, Sikder et al. 2016). On the contrary, improvements in toughness and flexural strength are quite common. Although 20–100% improvement in K_{IC} values of CNT/Al_2O_3 nanocomposites over pure Al_2O_3 was found, toughness increment beyond 100% was also reported (Estili and Sakka 2014, Sarkar and Das 2014a). A SPS processed 10 vol.% SWCNT/Al_2O_3 nanocomposite offered ~ 195% higher K_{IC} over pure Al_2O_3 (K_{IC} ≈ 3.3 MPa-m$^{0.5}$) (Jiang et al. 2007). ~ 150% higher K_{IC} was obtained at 2.48 wt.% MWCNT loading in SPS processed Al_2O_3 (K_{IC} of pure Al_2O_3 ≈ 4.5 MPa-m$^{0.5}$) (Lee et al. 2012). Furthermore, ~ 120% higher K_{IC} over pure Al_2O_3 (K_{IC} ≈ 3.56 MPa-m$^{0.5}$) was achieved for a hot-pressed 1.5 wt.% nano-ZrO_2 coated MWCNT containing nanocomposite (Zhu et al. 2008). Depending on the type of CNT used and its content, processing history, microstructure, and interface, improvement in σ_{FS} values generally ranges from 10–70% (Estili and Sakka 2014, Sarkar and Das 2014a). Improvements in electrical conductivity of CNT/Al_2O_3 nanocomposites have also been reported. While a SPS processed 10 vol.% SWCNT/Al_2O_3 nanocomposite offered electrical conductivity of ~ 3340 S/m (Zhan and Mukherjee 2004), conductivity as high as ~ 3330 and ~ 4810 S/m were also reported at 19.1 wt.% and 20 vol.% MWCNT loading, respectively, in SPS processed pure Al_2O_3 (Estili et al. 2012, Kumari et al. 2009). In the present chapter, the basic material properties of PLS processed MWCNT/Al_2O_3 are presented to elucidate the adequacy of PLS technique towards fabricating such nanocomposites having improved performance over pure Al_2O_3.

Experimentation

Nanocomposites were prepared using simple wet mixing of as-received commercial grade MWCNT and polycrystalline Al_2O_3 powder followed by pressureless sintering at 1500, 1600 and 1700°C with 2 hrs dwell at each of the T_{sin}'s under static Argon atmosphere. Composition, specimen nomenclature and TD values of the studied batches are given in Table 1. The sintered specimens were characterized in terms of physical properties, e.g., BD, AP and LS values, mechanical, thermal and electrical properties. HV values were evaluated at 4.9, 9.8, and 19.6 N with 10 s dwell at each of the test loads. K_R and K_{IC} values were evaluated using DCM and SENB technique, respectively (Liang et al. 1990, Niihara et al. 1982a,b, Sarkar and Das 2012c). Thermal properties were evaluated using laser flash technique at RT and 300°C and electrical conductivity was measured using standard 2-probe method. Microstructure related studies

Table 1. Details of the studied compositions.

Composition	Specimen id	Theoretical density (*TD*)
Pure Al_2O_3	*R*	3.970
Al_2O_3 + 0.15 vol.% MWCNT	*R1*	3.967
Al_2O_3 + 0.30 vol.% MWCNT	*R2*	3.963
Al_2O_3 + 0.60 vol.% MWCNT	*R3*	3.957
Al_2O_3 + 1.20 vol.% MWCNT	*R4*	3.944
Al_2O_3 + 2.40 vol.% MWCNT	*R5*	3.917

were carried out using scanning and transmission electron microscopy. Details of raw materials used as well as processing and characterizations of present specimens are available in other publications by the present authors (Sarkar and Das 2012b,c,d, 2014c).

Results and discussion

Morphology of MWCNT

Considering the structural integrity aspect of CNT in sintered nanocomposites, morphological stability of present nanotubes during high temperature exposure in inert atmosphere was studied prior to the nanocomposite fabrication. Figure 1a–b illustrate the qualitative and quantitative tube diameter distribution of as-received MWCNT, respectively. A few metallic nano-impurities were noticed within hollow channels and at end-caps of the nanotubes (Figure 1c–d) (Sarkar et al. 2011, Sarkar and Das 2012a). Figure 2a depicts unaffected CNTs having well separated internal bamboos in 600°C heat-treated specimen. It can be noted that the as-received MWCNT itself had internal bamboo morphology which was retained in 600°C heat-treated specimen because formation of internal bamboos by metal catalyst removal at only 600°C was not feasible. Additionally, heat-treatment at 600°C was also found to be inadequate for aligning the graphene layers toward the tube axis (Figure 2b). CNTs exposed at 1200°C in Argon also possessed structurally survived CNTs of different diameter (Figure 3) (Sarkar and Das 2013). Since higher temperature ensured better impurity removal, population of internal bamboos observed in 1200°C heat-treated specimen was found to be higher than that observed in as-received or 600°C heat-treated CNTs (Figure 3) (Wang et al. 2009). While Figure 4a illustrates presence of unaffected CNTs in 1500°C heat-treated specimen, Figure 4b shows clear internal channel and perfectly aligned graphene layers toward the tube axis in the same specimen. This suggested that heat-treatment at 1500°C in Argon

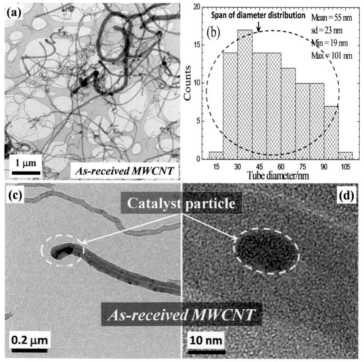

Figure 1. (a) Bright field TEM image of as-received MWCNTs showing wide distribution of tube diameter (b) Histogram of tube diameter distribution of as-received MWCNTs, (c–d) Bright field TEM images of as-received MWCNTs showing entrapped catalyst particles at the end-cap and within hollow channel, respectively.

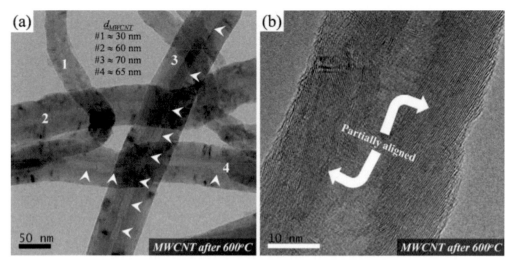

Figure 2. (a) Bright field TEM image of MWCNTs treated at 600°C in Argon atmosphere showing presence of CNTs of different diameters and internal bamboo morphology (pointed arrows), (b) HRTEM image of 600°C treated CNT showing partially aligned graphene layers.

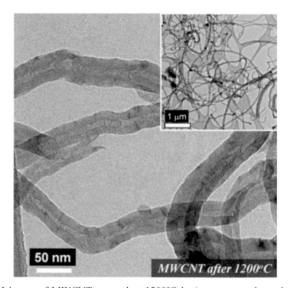

Figure 3. Bright field TEM image of MWCNTs treated at 1200°C in Argon atmosphere showing presence of CNTs of different diameters (Inset; scale bar: 1 μm) and internal bamboo morphology. Reprinted with permission from Sarkar, S. and P.K. Das. 2013. Mater. Res. Bull. 48: 41. Copyright@Elsevier.

was favorable to align the layers without damaging the tubular morphology. Beside existence of well-spaced internal compartments, the black spots and steps observed on the external wall of the 1700°C heat-treated tubes were primarily aroused from irregular overlapping of graphene layers and bending/twisting of the tubes (Figure 5a). Figure 5b confirmed that heat-treatment at 1700°C rendered better alignment of graphene layers in thick CNTs without creating any notable defect on the tube wall. Figure 6a shows structurally survived CNTs after heat-treatment at 1800°C in Argon. On the contrary, splitting of strained/defective graphene layers and subsequent loop formation were also noticed in 1800°C heat-treated nanotubes (Figure 6b–c) (Sarkar et al. 2011, Sarkar and Das 2013). Figure 6d shows formation of ribbon- and onion-like graphene structures in 1800°C heat-treated CNT specimen that corroborated other reports (Zhang et al. 1999).

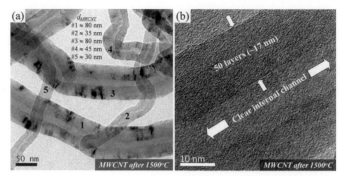

Figure 4. (a) Bright field TEM image of MWCNTs treated at 1500°C in Argon atmosphere showing presence of CNTs of different diameters, (b) HRTEM image of 1500°C treated CNT showing perfectly aligned graphene layers and impurity free internal channel.

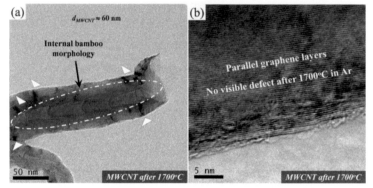

Figure 5. (a) Bright field TEM image of MWCNTs treated at 1700°C in Argon atmosphere showing structurally survived CNT having internal bamboo morphology and few dark spots (solid triangles), (b) HRTEM image of 1700°C treated CNT showing perfectly aligned graphene layers with no traceable morphological defect.

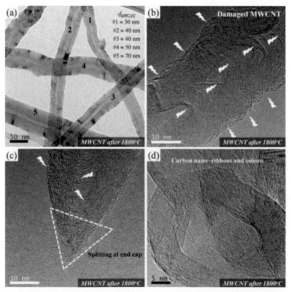

Figure 6. (a) Bright field TEM image of MWCNTs treated at 1800°C in Argon atmosphere showing presence of morphologically stable CNTs of different diameters and internal bamboo morphology, HRTEM image of 1800°C treated (b) comparatively thin CNT showing extent of damage to the graphene sheets, (c) CNT showing splitting of graphene layers at the sharp bent of end-cap and internal and external nano-scale corrugations, (d) CNT specimen showing formation of various carbon nano-structures during thermal treatment.

Physical properties

Figure 7a–b illustrate the homogeneous dispersion of nanotubes in Al_2O_3 matrix at 0.15 vol.% filler loading. On the contrary, since the total energy supplied during the dispersion process was kept constant, at increased CNT content, extent of homogeneous dispersion of CNT decreased. At 0.6 vol.% filler loading, population of dispersed and clustered CNTs was found to be almost the same (Figure 7c). Finally, at the highest CNT loading (i.e., 2.4 vol.%), large CNT clusters were noticed in the powder mixture (Figure 7d). Figure 8a–c show that except pure Al_2O_3, nanocomposites sintered at 1500°C were < 90% dense having moderate to high amount of open porosity (10–23%) and low LS (6–10%). Beside monolithic Al_2O_3, sintering at 1600°C offered significant increase in densification of the nanocomposites, especially up to 0.6 vol.% CNT loading (Figure 8a). Finally, sintering at 1700°C resulted in almost similar RD values (~ 100% of TD) for pure Al_2O_3 and nanocomposites containing up to 0.6 vol.% CNT containing negligible open pores and 12–13% LS (Figure 8a–c). Contrary to that, *R4* (1.2 vol.% CNT) and *R5* (2.4 vol.% CNT) batches remained inadequately dense (≤ 90%) even after sintering at 1700°C (Figure 8a–c). Decreased densification at increased CNT content observed at each of the T_{sin}'s was primarily caused by the incorporation of light weight CNT in Al_2O_3 matrix and porous rope-like structure, extensive agglomeration at higher loading and chemical inertness of CNT (Ahmad et al. 2010, Fan et al. 2006, Yamamoto et al. 2008a,b). Figure 8a–c further indicate that up to 0.6 vol.% CNT loading, only T_{sin} played the major role in controlling densification of the studied specimens. Beyond that concentration, CNT played an unfavorable role towards densification by limiting the pore removal process and material transport through grain boundary regions (Figure 8a–c) (Sarkar and Das 2012b,c,d, 2014c).

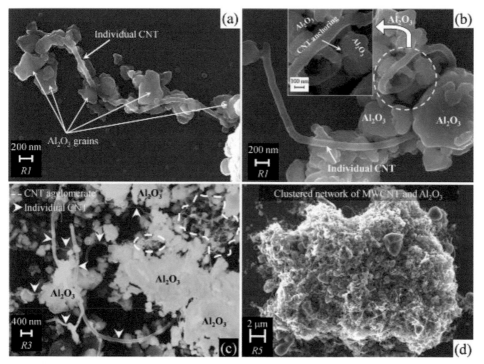

Figure 7. FESEM images of (a–b) 0.15 vol.% MWCNT/Al_2O_3 powder mixture showing presence of well-dispersed long CNTs attached with several matrix grains and nano-anchors formed by the dispersed CNTs, respectively, (c) 0.6 vol.% MWCNT/Al_2O_3 powder mixture showing concurrent presence of individual (pointed arrows) and clustered (dotted regions) CNTs within the matrix phase, (d) 2.4 vol.% MWCNT/Al_2O_3 powder mixture showing clustered network of Al_2O_3 and CNT.

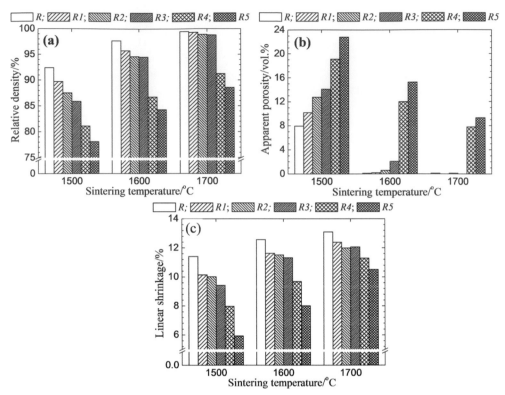

Figure 8. Physical properties of the studied specimens after sintering at three different temperatures (a) relative density plot, (b) apparent porosity plot, and (c) linear shrinkage plot.

Microstructure

Figure 9a shows porous microstructure of pure Al_2O_3 containing small under developed matrix grains (~2 μm) after sintering at 1500°C. Similar observations were made for the 1500°C sintered nanocomposites containing up to 0.6 vol.% CNT. Beyond 0.6 vol.% CNT loading, nanocomposites sintered at 1500°C showed formation of large surface cracks (Figure 9b). At higher CNT loading, those cracks primarily appeared due to the presence of randomly distributed extremely fluffy CNT clusters squeezed within the matrix grains. During heating, clustered CNTs acting as pores of equivalent diameter not only restricted densification of matrix but their volume expansion resulted in the formation of large cracks in the specimens. Alternatively, the CNT clusters in high nanotube containing batches facilitated localized densification of the matrix phase that formed a segregated microstructure and eventually, produced the macroscopic cracks. Presence of dense regions in 1.2 vol.% MWCNT/Al_2O_3 nanocomposite after sintering at 1500°C reinforced the above statement of differential sintering (Figure 9c). From microstructural observations and RD values, it can be stated that the samples underwent the intermediate stage of sintering at 1500°C, where channel/open pores did exist and densification prevailed the grain growth (Boch and Leriche 2007). Sintering at 1600°C of pure Al_2O_3 resulted in the formation of well developed (~ 4.48 ± 2.21 μm) and nearly equiaxed Al_2O_3 grains (Figure 10a). Presence of inter-granular closed pores mainly at quadruple grain junctions (Figure 10a) and grain size increment suggested that the system had entered the final stage of sintering at 1600°C. At this stage, coalescence of smaller particles with convex boundaries resulted in observed grain growth and concave grains having similar shape grew in size (Boch and Leriche 2007). Similar microstructural changes of the nanocomposites containing up to 0.6 vol.% CNT confirmed that

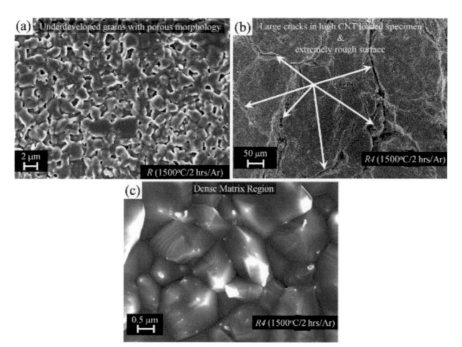

Figure 9. FESEM images of specimens sintered at 1500ºC for 2 hr in Argon (a) pure Al$_2$O$_3$, (b) 1.2 vol.% MWCNT/Al$_2$O$_3$ nanocomposite showing large surface cracks, and (c) relatively dense region in 1.2 vol.% MWCNT/Al$_2$O$_3$ nanocomposite.

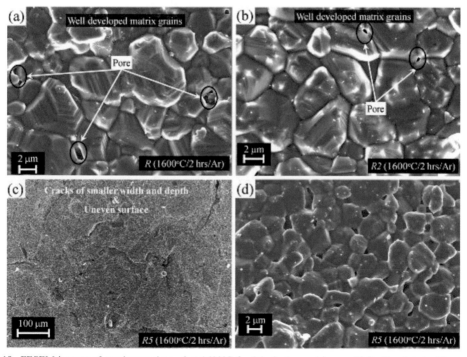

Figure 10. FESEM images of specimens sintered at 1600ºC for 2 hr in Argon (a) pure Al$_2$O$_3$, Reprinted with permission from Sarkar, S. and P.K. Das. 2012. Ceram. Int. 38: 423. Copyright@Elsevier. (b) 0.3 vol.% MWCNT/Al$_2$O$_3$ nanocomposite, (c) 2.4 vol.% MWCNT/Al$_2$O$_3$ nanocomposite showing surface cracks, and (d) dense region in 2.4 vol.% MWCNT/Al$_2$O$_3$ nanocomposite.

these specimens also underwent the final stage of sintering at 1600°C (Figure 10b). On the contrary, complete elimination of surface cracks, pore reduction and grain growth were still not observed in 1.2 and 2.4 vol.% MWCNT/Al$_2$O$_3$ nanocomposites even after sintering at 1600°C (Figure 10c–d). Sintering at 1700°C resulted in further grain growth of pure Al$_2$O$_3$ and nanocomposites containing up to 0.6 vol.% CNT. Grain size of pure Al$_2$O$_3$ after sintering at 1700°C was found to be 9.87 ± 4.67 µm (Figure 11a) that was ~ 2.2 times higher than that obtained after 1600°C (Sarkar and Das 2012b). Grain size values of the nanocomposites containing 0.15, 0.3 and 0.6 vol.% CNT after sintering at 1700°C were found to be ~ 8.2, ~ 6.7 and ~ 6.1 µm, respectively (Figure 11a–b) (Sarkar and Das 2012b). Figure 11c indicates that the severity of surface cracks in 1700°C sintered 2.4 vol.% MWCNT/Al$_2$O$_3$ nanocomposite was lower compared to that observed after 1600°C sintering (Figure 10c). Increase in grain size (~ 2 times) of 1700°C sintered 1.2 vol.% MWCNT/Al$_2$O$_3$ nanocomposite (Figure 11d) over 1600°C sintered *R4* specimen (~ 2.7 µm) was found to be almost the same as obtained for pure Al$_2$O$_3$. Thus, microstructural observations also confirmed that high CNT loaded nanocomposites required higher T$_{sin}$ to achieve adequate densification and grain growth. The gradual and continuous decrease in matrix grain size on CNT addition indicated *grain refining effect* of CNT which in turn could help in enhancing mechanical properties of the resulting nanocomposites. Different CNT/Al$_2$O$_3$ interactions were found during microstructural observations of the sintered nanocomposites, e.g., bridging of grains by individual CNTs and CNT Y-junctions, CNTs within *intergranular* pore, clustered CNT in high CNT loaded specimen and CNT entrapped within Al$_2$O$_3$ grains (Figure 12a–d).

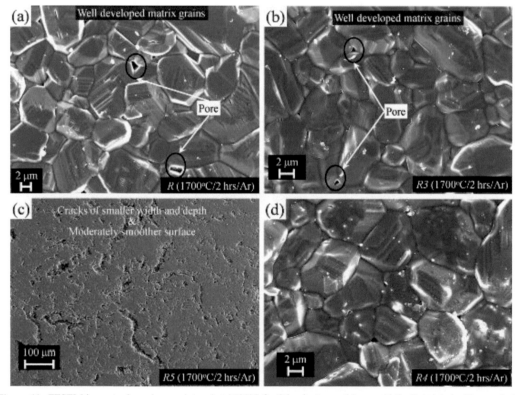

Figure 11. FESEM images of specimens sintered at 1700°C for 2 hrs in Argon (a) pure Al$_2$O$_3$, Reprinted with permission from Sarkar, S. and P.K. Das. 2012. Ceram. Int. 38: 423. Copyright@Elsevier. (b) 0.6 vol.% MWCNT/Al$_2$O$_3$ nanocomposite, (c) 2.4 vol.% MWCNT/Al$_2$O$_3$ nanocomposite showing surface cracks, and (d) dense region in 1.2 vol.% MWCNT/Al$_2$O$_3$ nanocomposite.

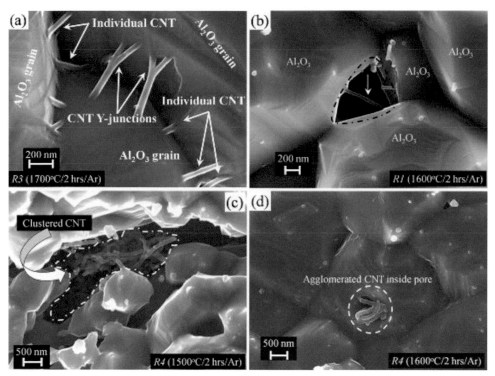

Figure 12. FESEM images of sintered specimens showing bridging of matrix grains by (a) individual CNTs and CNT Y-junctions. Reprinted with permission from Sarkar, S. and P.K. Das. 2012. Mater. Chem. Phys. 137: 511. Copyright[@] Elsevier. (b) isolated CNT within *intergranular* pore, (c) agglomerated CNTs along grain boundary regions, and (d) CNT cluster entrapped within growing matrix grain.

TEM study

Till date, limited work has been done on interface related issues of CNT/Al_2O_3 nanocomposites and thus, it requires much attention to precisely predict the structure-property relation in such nanocomposites (Ahmad et al. 2010, Wei et al. 2008a,b). This section describes the sequential changes observed at the interface region of the studied nanocomposites as a function of sintering temperature. HRTEM image of 1500°C sintered 0.6 vol.% MWCNT/Al_2O_3 nanocomposite revealed formation of an amorphous interface region of $\leq$ 5 nm thickness (Figure 13a). On the other hand, 1700°C sintered 2.4 vol.% MWCNT/Al_2O_3 nanocomposite showed existence of a definite reaction zone between the matrix and filler (Figure 13b). This indicated that sintering at 1700°C most likely helped in the formation of a well-defined engineered interface in present CNT/Al_2O_3 nanocomposite that could help in mechanical property enhancement. To further confirm this, nanocomposite powder having 35 wt.% CNT loading was sintered under identical conditions and observed using TEM. Formation of a few nanometer thick well-defined graphene layer coating on the Al_2O_3 grains was found to be the characteristic of the 1600°C sintered 35 wt.% CNT/Al_2O_3 powder (Figure 13c). Finally, observation of the 1700°C sintered nanocomposite powder confirmed existence of the same thin graphene layer encapsulation on the Al_2O_3 grains through which CNTs were attached to the matrix grains (Figure 13d). Since no detrimental reaction product (e.g., Al–O–C phases) was formed at the interface, chances of structural collapse of MWCNTs through extensive reaction between Al_2O_3 and carbon even at 1700°C in Argon were successfully evaded (Sarkar and Das 2012b, 2014c).

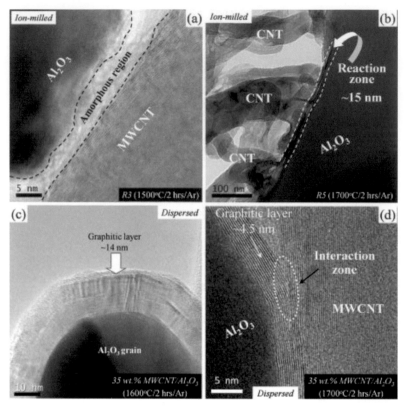

Figure 13. TEM image of interface region in (a) 1500°C sintered 0.6 vol.% MWCNT/Al₂O₃ nanocomposite, Reprinted with permission from Sarkar, S. and P.K. Das. 2014. Ceram. Int. 40: 2723. Copyright@Elsevier. (b) 1700°C sintered 2.4 vol.% MWCNT/Al₂O₃ nanocomposite, Reprinted with permission from Sarkar, S. and P.K. Das. 2014. Ceram. Int. 40: 2723. Copyright@Elsevier. HRTEM images of 35 wt.% MWCNT/Al₂O₃ nanocomposite powder showing (c) presence of a few nanometer thick graphene layer encapsulation on matrix grains after 1600°C, and (d) attachment between CNT and matrix grain through the thin graphene layer after sintering at 1700°C. Reprinted with permission from Sarkar, S. and P.K. Das. 2014. Ceram. Int. 40: 2723. Copyright@ Elsevier.

Micro-vickers hardness

^{avg}HV, SD and ^{ch}HV values of all specimens sintered at 1500°C are shown in Figure 14a. ^{ch}HV values suitable for designing purpose were evaluated using the 2-parameter *Weibull* statistics (Girolamo et al. 2011, Nevarez-Rascon et al. 2011). The specimens sintered at 1500°C possessed ample porosity, small under developed grains and amorphous interface with insufficient load bearing ability. Despite the above facts, the exceptional flexibility coupled with very high energy absorption capacity of CNT resulted in 7–20% improvement in HV values even in these partially dense nanocomposites at least up to 0.6 vol.% CNT loading over pure Al₂O₃ (Figure 14a). Irrespective of indentation load, specimens sintered at 1600°C offered better hardness and lower scatter of experimental data primarily due to improved densification and interface formation to ensure effective load sharing between the constituents (Figure 14b). Hardness of nanocomposites sintered at 1600°C increased up to 0.3 vol.% CNT loading. The highest HV value of ~ 21 GPa at 1 kgf was obtained for 0.3 vol.% CNT/Al₂O₃ nanocomposite which was ~ 22% higher than that of pure Al₂O₃ (Figure 14b). Specimens containing more than 0.3 vol.% CNT offered lower enhancement in HV values due to the presence of clustered and non-uniformly dispersed CNTs that acted as defects with no load carrying capacity. Furthermore, during indentation loading cycle, indentation dimension gradually increased due to extreme flexible nature and lower hardness of CNT compared to pure Al₂O₃ which allowed progressively higher depth of penetration, especially, at higher CNT loading (Hanzel et al. 2014, Morales–Rodríguez et al. 2014). During the unloading cycle, although the elastically deformed CNTs returned to their initial position, the adjacent Al_2O_3 matrix which was deformed plastically resulted in relatively larger indentation diagonals and thus, lower HV values of the nanocomposites at higher CNT loading (Sarkar and Das 2012b). Eventually, the 1.2 vol.% MWCNT/Al_2O_3 nanocomposite sintered at 1600°C offered ~ 14% increase in HV_1 than that of pure Al_2O_3 (Figure 14b). Similarly, among 1700°C sintered samples, the highest ^{avg}HV (21.4 GPa) was obtained for 0.3 vol.% MWCNT/Al_2O_3 nanocomposite which was nearly 23% higher over pure Al_2O_3, at 1 kgf load (Figure 14c). For 1.2 vol.% MWCNT/Al_2O_3 nanocomposite sintered at 1700°C, HV_1 increased a little with less scatter of experimental data than that obtained after 1600°C (Figure 14c). Eventually, the 1700°C sintered 1.2 vol.% MWCNT/Al_2O_3 nanocomposite offered ~ 16% higher hardness over pure Al_2O_3. Extents of hardness improvement of present nanocomposites were found to lie on the higher side of literature data (Sarkar and Das 2014a). This was probably achieved through uniform dispersion of CNTs at low loading, especially, up to 0.3 vol.%, large aspect ratio of present CNTs that effectively bridged multiple Al_2O_3 grains and rendered improved load sharing between Al_2O_3 and CNT in dense nanocomposites. Due to the lowest densification and presence of large clustered nanotubes in 2.4 vol.% MWCNT/Al_2O_3 nanocomposite, it was not possible to achieve the required surface finish of this particular specimen for HV measurement. *Weibull* moduli of HV data at the three loads for specimens sintered at 1500°C are shown in Figure 15a. It may be seen from the bar graphs

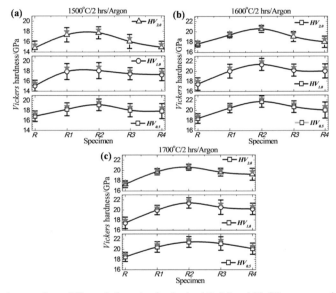

Figure 14. *Vickers* hardness at three different indentation loads viz. 4.9, 9.8 and 19.6N versus specimen composition of (a) 1500, (b) 1600, and (c) 1700°C sintered specimens. Open symbols represent average HV values and solid stars represent corresponding characteristic values obtained from *Weibull* analyses.

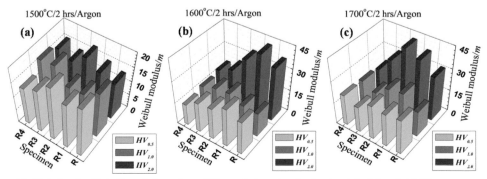

Figure 15. Variation in *Weibull* moduli at three different indentation loads viz. 4.9, 9.8 and 19.6N versus specimen composition of (a) 1500, (b) 1600, and (c) 1700°C sintered specimens.

that irrespective of test load and specimen type, '*m*' values remained almost the same. This suggested that in these partially dense specimens, defect population was such that the probability of interacting with the *weakest* flaw or overall averaging effect of randomly distributed flaws was comparable. On the contrary, specimens sintered either at 1600 or 1700°C offered notable increase in *Weibull* modulus at increased indentation load (Figure 15b–c). This indicated better reliability of HV data at higher T_{sin} and applied load. Such changes in '*m*' with applied load have also been reported for other materials (Dey et al. 2009, Girolamo et al. 2011). However, 1.2 vol.% MWCNT/Al_2O_3 nanocomposite sintered either at 1600 or 1700°C exhibited almost similar consistency of HV data at each of the test loads. Presence of clustered CNTs, porous microstructure and inadequate interface in this high CNT loaded nanocomposite resulted in much higher defect population. Eventually, irrespective of test load, probability of interacting with the *weakest* flaw or overall averaging effect due to randomly distributed flaws was similar which resulted in the observed steady '*m*' for 1.2 vol.% MWCNT/Al_2O_3 nanocomposite (Figure 15b–c). For 1700°C sintered specimens, the best reliability of HV data was obtained for 0.3 vol.% MWCNT/Al_2O_3 nanocomposite which was ~ 40% better compared to those obtained for pure Al_2O_3 (Figure 15c) (Sarkar and Das 2012b).

Fracture toughness

As far as changes in fracture resistance/toughness of the nanocomposites over pure Al_2O_3 are concerned, Figure 16a indicates that irrespective of the evaluation technique, all the nanocomposites excluding 1.2 vol.% MWCNT/Al_2O_3 nanocomposite offered higher K_{IC} than pure Al_2O_3 after sintering at 1500°C. Sintering at 1600°C offered additional enhancement in K_{IC} with better reproducibility indicating uniform toughening effect in nanocomposites (Figure 16b). Among the 1600°C sintered specimens, the highest K_{IC} was obtained for 0.3 vol.% MWCNT/Al_2O_3 nanocomposite (K_{IC} = 4.5 ± 0.2 MPa-m$^{0.5}$) which was ~ 26% higher than pure Al_2O_3 (Figure 16b). Similar observations were made in specimens sintered at 1700°C (Figure 16c). The highest values of K_R and K_{IC} were also obtained for the 0.3 vol.% MWCNT/Al_2O_3 nanocomposite which were around 34% higher than pure Al_2O_3 (Figure 16c). However, analogous to the hardness data, extent of toughness enhancement reduced at higher CNT content. Eventually, toughness of 1.2 vol.% MWCNT/Al_2O_3 nanocomposite (K_R = 4.23 ± 0.14 MPa-m$^{0.5}$; K_{IC} = 4.06 ± 0.21 MPa-m$^{0.5}$) was found

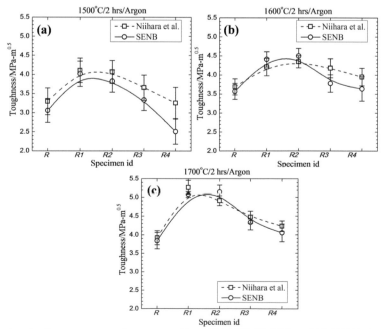

Figure 16. Fracture toughness versus specimen composition of (a) 1500, (b) 1600, and (c) 1700°C sintered specimens.

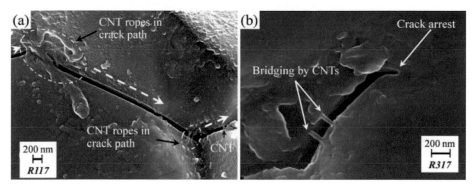

Figure 17. FESEM images of crack extension paths in sintered nanocomposite specimens showing (a) crack deflection and bridging by dispersed CNTs. Reprinted with permission from Sarkar, S. and P.K. Das. 2012. Mater. Chem. Phys. 137: 511. Copyright@Elsevier and (b) early crack arrest by bridging CNTs.

to be only 6–8% higher than those obtained for pure Al_2O_3 (Figure 16c). Crack path analyses of indented nanocomposites was carried out to reveal the crack growth nature and toughening mechanisms offered by CNTs that resulted in improved toughness over pure Al_2O_3. Figure 17a shows the toughening mechanisms, i.e., crack deflection and bridging involved in the nanocomposites that rendered enhanced fracture energy dissipation. Presence of perpendicularly oriented CNTs to the crack extension path also hindered crack width expansion and early crack arrest (Figure 17b). Although extent of toughness improvement (6–34%) of present nanocomposites matched well with literature data, in some cases present toughness values were found to be lower than achieved, particularly in hot-pressed or spark plasma sintered nanocomposites (Estili and Sakka 2014, Sarkar and Das 2014a). The observed discrepancy in the extent of toughness improvement in CNT/Al_2O_3 nanocomposites is quite common and yet ambiguous. Basically, several factors may contribute in this including purity, aspect ratio, and orientation of CNTs, dispersion of nanotubes, surface characteristics of filler and matrix, sintering technique, interface nature, bridging characteristics, residual stress field around filler and matrix and matrix grain size, etc.

Thermal and electrical properties

Thermal conductivity of 1500 and 1700°C sintered specimens at RT and 300°C are shown in Figure 18a–b. For a particular composition, κ increased with increasing T_{sin} and decreased with increasing T_{exp}. It has been demonstrated in earlier sections that the 1500°C sintered specimens were porous in nature and possessed amorphous interface (for nanocomposites). Higher pore volume resulted in a reduction of overall κ of the specimens because heat transfer through pores is usually slow and inefficient due to

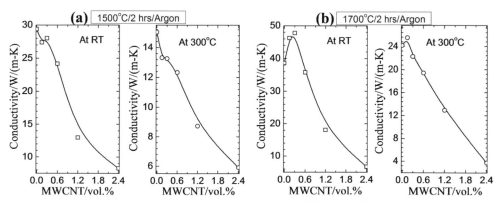

Figure 18. Variation in thermal conductivity with CNT content in (a) 1500 and (b) 1700°C sintered specimens at room temperature and 300°C.

higher phonon scattering. In addition, the interfacial *Kapitza* resistance at the amorphous interface was higher because amorphous materials with highly disordered and irregular structure have higher phonon scattering ability than crystalline phases (Callister 2007, Tritt 2004). On the contrary, for 1700°C sintered specimens, the unfavorable effects of porosity and *Kapitza* resistance were lowered due to formation of dense microstructure and thin graphene-like interface. On the other hand, increasing T_{exp} from RT to 300°C resulted in enhanced phonon-phonon scattering process, i.e., *Umklapp* scattering (Tritt 2004) and eventually, κ of the specimens decreased (Figure 18a–b). It may be seen from Figure 18a that among 1500°C sintered specimens, the highest conductivity values (κ_{RT} = 29.39 W/m-K; κ_{300} = 15.04 W/m-K) were obtained for pure Al_2O_3 having the highest RD among the studied specimens (Figure 8a). However, owing to improved densification and adequate interface formation, the 0.3 vol.% MWCNT/ Al_2O_3 nanocomposite (κ_{RT} = 47.78 W/m-K) offered ~ 24% higher thermal conductivity over pure Al_2O_3 when sintering was performed at 1700°C (Figure 18b). Furthermore, between 1500 and 1700°C sintered pure Al_2O_3, the RT thermal conductivity ($^{1500}\kappa_{Al2O3}$ ≈ 29 W/m-K; $^{1700}\kappa_{Al2O3}$ ≈ 39 W/m-K) was increased by ~ 30%. On the other hand, among 1500 and 1700°C sintered 0.15 and 0.3 vol.% MWCNT/Al_2O_3 specimens, the extent of increase in thermal conductivity values were found to be ~ 70% ($^{1500}\kappa_{R1}$ ≈ 27 W/m-K and $^{1700}\kappa_{R1}$ ≈ 46 W/m-K, $^{1500}\kappa_{R2}$ ≈ 28 W/m-K and $^{1700}\kappa_{R2}$ ≈ 48 W/m-K). Finally, Figure 18a–b also show that irrespective of T_{sin} and T_{exp}, nanocomposites having CNT loading ≥ 1.2 vol.% always had significantly lower conductivity values than rest of the specimens. Even after sintering at 1700°C, conductivity of 1.2 vol.% MWCNT/Al_2O_3 nanocomposite was found to be only around 18 W/m-K which was ~ 50% lower than that obtained for pure Al_2O_3 (Figure 18b). In a similar fashion, thermal properties of 2.4 vol.% CNT/Al_2O_3 nanocomposite ($^{1700}\kappa_{R5}$ ≈ 7 W/m-K) was found to be ~ 80% lower than that obtained for pure Al_2O_3 (Figure 18b).

It is known that pure Al_2O_3 is an insulator (resistivity range: 10^{10}–10^{12} Ω-m) that has limited its applications. Thus, the outstanding electrical conductivity of MWCNT (> 10^3 to 10^5 S/m) can be exploited in making conductive CNT/Al_2O_3 nanocomposites coupled with acceptable mechanical properties (Mondal et al. 2008, Thess et al. 1996). σ_{DC}'s of the studied specimens sintered at the three temperatures are shown in Figure 19a–c. σ_{DC} of pure Al_2O_3 was taken as 10^{-12} S/m (Zhan and Mukherjee 2004). Increasing CNT content in highly resistive Al_2O_3 rendered significant enhancement in σ_{DC}, especially at CNT loading ≥ 0.6 vol.%. It may be further visualized from the figures that with increase in T_{sin}, the *percolation threshold* was decreased from 1.2 vol.% MWCNT loading for 1500°C sintered specimens to 0.6 vol.% MWCNT for 1600 and 1700°C sintered nanocomposites. Such lowering of *percolation threshold* at increased T_{sin} was aroused due to the fact that at low T_{sin}, shrinkage of nanocomposites was less and thus, desired interconnected network of highly conducting CNTs was only achieved for specimens containing 1.2 or 2.4 vol.% CNT. Since, 1.2 and 2.4 vol.% MWCNT/Al_2O_3 nanocomposites characteristically contained agglomerated CNTs along the matrix grain boundaries, even at low shrinkage, these specimens exhibited highly conducting nature. At higher T_{sin} (i.e., 1600 or 1700°C), further shrinkage of the nanocomposites containing ≥ 0.6 vol.% CNT resulted in the formation of desired percolating CNT network by reducing the spacing among the tubes and ensured enhanced tunneling effect of electrons through individual CNTs or bundles. While an abrupt increase in σ_{DC} from ~ 10^{-12} S/m for pure Al_2O_3 to 10^{-4} S/m for 1.2 vol.% MWCNT/Al_2O_3 nanocomposite was observed among the 1500°C sintered specimens, among 1600 and 1700°C sintered specimens, *percolation* was noticed at 0.6 vol.% CNT loading and σ_{DC} changed from ~ 10^{-12} S/m for pure Al_2O_3 to ~ 10^{-3} S/m for the nanocomposite (Figure 19a–c). The low percolation concentration (i.e., ≥ 0.6 vol.% CNT) obtained for present CNT/Al_2O_3 nanocomposites was consistent with previous reports on CNT/ceramic nanocomposites (Sarkar and Das 2014a). Among all the specimens, the highest σ_{DC} was obtained for 2.4 vol.% MWCNT/Al_2O_3 nanocomposite that ranged from ~ 0.02 S/m after 1500°C sintering to ~ 21 S/m after 1700°C sintering (Figure 19a–c) (Sarkar and Das 2014c).

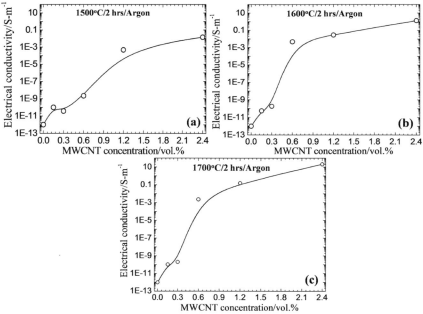

Figure 19. Variation in electrical conductivity with CNT content in (a) 1500, (b) 1600, and (c) 1700°C sintered specimens.

Conclusions

a. Heat treatment of MWCNTs up to 1800°C in inert atmosphere can be an effective method for impurity removal, graphene layer alignment, and internal bamboo formation. However, considering the morphological stability of CNTs, thermal exposure up to 1700°C is found to be the optimum.

b. It is possible to fabricate MWCNT/Al_2O_3 nanocomposites by simple wet mixing of as-received raw materials followed by pressureless sintering.

c. While low sintering temperature (e.g., 1500°C) forms an amorphous interface, high sintering temperature (e.g., 1700°C) ensures formation of a few nanometer thick prominent interface of graphene layers between CNT and alumina in the pressureless sintered nanocomposites.

d. In general, nanocomposites containing up to 0.6 vol.% CNT offer better reinforcing effects in structural Al_2O_3 because of adequate dispersion of nanotubes, improved densification, matrix grain refinement, and better interface performance.

e. Among 1700°C sintered samples, the 0.3 vol.% MWCNT/Al_2O_3 nanocomposite offers the highest hardness (avg. 21.4 GPa) and reliability of HV data which are ~ 23% and ~ 40% higher, respectively, over pure Al_2O_3. The same nanocomposite also offers the highest toughness, i.e., ~ 34% higher K_{IC} than that obtained for pure Al_2O_3 ($K_{IC} \approx 3.84$ MPa-m$^{0.5}$).

f. The 0.3 vol.% MWCNT/Al_2O_3 nanocomposite also offers the highest thermal conductivity (~ 48 W/m-K), i.e., ~ 24% higher than that of pure Al_2O_3. Among all the specimens, the highest σ_{DC} values are obtained for 2.4 vol.% MWCNT/Al_2O_3 nanocomposite ranging from ~ 0.02 S/m after 1500°C sintering to ~ 21 S/m after 1700°C sintering.

Acknowledgements

The authors express their sincere gratitude to the Director, CSIR–Central Glass and Ceramic Research Institute (CSIR–CGCRI), Kolkata, India for his interest in the research work. The authors are also grateful to the staff members of Analytical Facility Division and Materials Characterization Unit of CSIR–CGCRI for their extensive help in carrying out TEM analyses. Both the authors sincerely thank all the members of NOCCD, CSIR–CGCRI for their extensive help and support offered during the execution of the entire work. The first author acknowledges the financial support of the Council of Scientific and Industrial Research (CSIR), India (CSIR–P–81).

References

Ahmad, I., M. Unwin, H. Cao, H. Chen, H. Zhao, A. Kennedy et al. 2010. Multi-walled carbon nanotubes reinforced Al_2O_3 nanocomposites: Mechanical properties and interfacial investigations. Compos. Sci. Technol. 70: 1199–1206.

An, J.-W. and D.-S. Lim. 2002. Effect of carbon nanotube additions on the microstructure of hot-pressed alumina. J. Ceram. Process. Res. 3: 201–204.

An, J.-W., D.-H. You and D.-S. Lim. 2003. Tribological properties of hot-pressed alumina-CNT composites. Wear 255: 677–681.

Belmonte, M., S.M. Vega-Díaz, A. Morelos-Gómez, P. Miranzo, M.I. Osendi and M. Terrones. 2014. Nitrogen-doped-CNTs/Si_3N_4 nanocomposites with high electrical conductivity. J. Eur. Ceram. Soc. 34: 1097–1104.

Boch, P. and A. Leriche (eds.). 2007. Chapter 3. Sintering and Microstructure of Ceramics, First ed., pp. 55–92, ISTE, California.

Bogue, R. 2011. Nanocomposites: a review of technology and applications. Assembly Autom. 31: 106–112.

Callister, W.D. 2007. Chapter 19: Thermal properties. pp. 722–739. *In*: Materials Science and Engineering: An Introduction. edited, John Wiley & Sons, Inc, New York.

Camargo, P.H.C., K.G. Satyanarayana and F. Wypych. 2009. Nanocomposites: Synthesis, structure, properties and new application opportunities. Mater. Res. 12: 1–39.

Chen, Y.L., B. Liu, Y. Huang and K.C. Hwang. 2011. Fracture toughness of carbon nanotube-reinforced metal- and ceramic-matrix composites. J. Nano Mat. 2011: 746029.

Clark, M.D., L.S. Walker, V.G. Hadjiev, V. Khabashesku, E.L. Corral and R. Krishnamoorti. 2012. Polymer precursor-based preparation of carbon nanotube–silicon carbide nanocomposites. J. Am. Ceram. Soc. 95: 328–337.

Dey, A., A.K. Mukhopadhyay, S. Gangadharan, M.K. Sinha and D. Basu. 2009. Weibull modulus of nano-hardness and elastic modulus of hydroxyapatite coating. J. Mater. Sci. 44: 4911–4918.

Estili, M., A. Kawasaki and Y. Sakka. 2012. Highly concentrated 3D macrostructure of individual carbon nanotubes in a ceramic environment. Adv. Mater. 24: 4322–4326.

Estili, M. and Y. Sakka. 2014. Recent advances in understanding the reinforcing ability and mechanism of carbon nanotubes in ceramic matrix composites. Sci. Tech. Adv. Mater. 15: 064902.

Fan, J., D. Zhao, M. Wu, Z. Xu and J. Song. 2006. Preparation and microstructure of multi-wall carbon nanotubes-toughened Al_2O_3 composite. J. Am. Ceram. Soc. 89: 750–753.

Girolamo, G.D., F. Marra, C. Blasi, E. Serra and T. Valente. 2011. Microstructure, mechanical properties and thermal shock resistance of plasma sprayed nanostructured zirconia coatings. Ceram. Int. 37: 2711–2717.

Hanzel, O., J. Sedláček and P. Sajgalík. 2014. New approach for distribution of carbon nanotubes in alumina matrix. J. Eur. Ceram. Soc. 34: 1845–1851.

Iijima, S. 1991. Helical microtubules of graphitic carbon. Nature 354: 56–58.

Jiang, D., K. Thomson, J.D. Kuntz, J.W. Ager and A.K. Mukherjee. 2007. Effect of sintering temperature on a single-wall carbon nanotube-toughened alumina-based nanocomposite. Scripta Mater. 56: 959–962.

Kumari, L., T. Zhang, G.H. Du, W.Z. Li, Q.W. Wang, A. Datye et al. 2009. Synthesis, microstructure and electrical conductivity of carbon nanotube–alumina nanocomposites. Ceram. Int. 35: 1775–1781.

Lee, K., C.B. Mo, S.B. Park and S.H. Hong. 2012. Mechanical and electrical properties of multiwalled CNT-alumina nanocomposites prepared by a sequential two-step processing of ultrasonic spray pyrolysis and spark plasma sintering. J. Am. Ceram. Soc. 94: 3774–3779.

Liang, K.M., G. Orange and G. Fantozzi. 1990. Evaluation by indentation of fracture toughness of ceramic materials. J. Mater. Sci. 25: 207–214.

Mata-Osoro, G., J.S. Moya and C. Pecharroman. 2012. Transparent alumina by vacuum sintering. J. Eur. Ceram. Soc. 32: 2925–2933.

MD, J.A.D.A., W.N.C. MD and M.N. BS. 2014. High survivorship with a titanium-encased alumina ceramic bearing for total hip arthroplasty. Clin. Orthop. Relat. Res. 472: 611–616.

Medvedovski, E. 2002. Alumina ceramics for ballistic protection Part 2. Am. Ceram. Soc. Bull. 81: 45–50.

Melk, L., J.J.R. Rovira, M.-L. Antti and M. Anglada. 2015. Coefficient of friction and wear resistance of zirconia–MWCNTs composites. Ceram. Int. 41: 459–468.

Michálek, M., J. Sedláček, M. Parchoviansky, M. Michálková and D. Galusek. 2014. Mechanical properties and electrical conductivity of alumina/MWCNT and alumina/zirconia/MWCNT composite. Ceram. Int. 40: 1289–1295.

Mondal, K.C., A.M. Strydom, R.M. Erasmus, J.M. Keartland and N.J. Coville. 2008. Physical properties of CVD boron-doped multiwalled carbon nanotubes. Mater. Chem. Phy. 111: 386–390.

Morales–Rodríguez, A., A. Gallardo–López, A. Fernández–Serrano, R. Poyato, A. Muñoz and A. Domínguez–Rodríguez. 2014. Improvement of Vickers hardness measurement on SWNT/Al$_2$O$_3$ composites consolidated by spark plasma sintering. J. Eur. Ceram. Soc. 34: 3801–3809.

Nevarez-Rascon, A., A. Aguilar-Elguezabal, E. Orrantia and M.H. Bocanegra-Bernal. 2011. Compressive strength, hardness and fracture toughness of Al$_2$O$_3$ whiskers reinforced ZTA and ATZ nanocomposites: Weibull analysis. Int. J. Refract. Metals. Hard. Mater. 29: 333–340.

Niihara, K., R. Morena and D.P.H. Hasselman. 1982a. Further reply to comment on elastic/plastic indentation damage in ceramics: the median/radial crack system. J. Am. Ceram. Soc. 65: C116–C118.

Niihara, K., R. Morena and D.P.H. Hasselman. 1982b. Evaluation of K$_{IC}$ of brittle solid by the indentation method with low crack-to-indent ratios. J. Mater. Sci. Lett. 1: 13–16.

Rahmat, M. and P. Hubert. 2011. Carbon nanotube–polymer interactions in nanocomposites: a review. Compos. Sci. Technol. 72: 72–84.

Sarkar, K., S. Sarkar and P.K. Das. 2016. Spark plasma sintered multiwalled carbon nanotube/silicon carbide composites: densification, microstructure, and tribo-mechanical characterization. J. Mater. Sci. 51: 6697–6710.

Sarkar, S., P.K. Das and S. Bysakh. 2011. Effect of heat treatment on morphology and thermal decomposition kinetics of multiwalled carbon nanotubes. Mater. Chem. Phy. 125: 161–167.

Sarkar, S. and P.K. Das. 2012a. Non-isothermal oxidation kinetics of single- and multi-walled carbon nanotubes up to 1273 K in ambient. J. Therm. Anal. Calorim. 107: 1093–1103.

Sarkar, S. and P.K. Das. 2012b. Statistical analysis of mechanical properties of pressureless sintered multiwalled carbon nanotube/alumina nanocomposites. Mater. Chem. Phy. 137: 511–518.

Sarkar, S. and P.K. Das. 2012c. Temperature and load dependent mechanical properties of pressureless sintered carbon nanotube/alumina nanocomposites. Mater. Sci. Eng. A 531: 61–69.

Sarkar, S. and P.K. Das. 2012d. Microstructure and physicomechanical properties of pressureless sintered multiwalled carbon nanotube/alumina nanocomposites. Ceram. Int. 38: 423–432.

Sarkar, S. and P.K. Das. 2013. Thermal and structural stability of single and multiwalled carbon nanotubes up to 1800°C in Argon studied by Raman spectroscopy and transmission electron microscopy. Mater. Res. Bull. 48: 41–47.

Sarkar, S. and P.K. Das. 2014a. Processing and properties of carbon nanotube/alumina nanocomposites: a review. Rev. Adv. Mater. Sci. 37: 53–82.

Sarkar, S. and P.K. Das. 2014b. Effect of sintering temperature and nanotube concentration on microstructure and properties of carbon nanotube/alumina nanocomposites. Ceram. Int. 40: 7449–7458.

Sarkar, S. and P.K. Das. 2014c. Role of interface on electrical conductivity of carbon nanotube/alumina nanocomposite. Ceram. Int. 40: 2723–2729.

Sarkar, S. and P.K. Das. 2015. Indentation fracture resistance vs. conventional fracture toughness of carbon nanotube/alumina nanocomposites. Metall. Mater. Trans. A 46A: 5072–5079.

Sikder, P., S. Sarkar, K.G. Biswas, S. Das, S. Basu and P.K. Das. 2016. Improved densification and mechanical properties of spark plasma sintered carbon nanotube reinforced alumina ceramics. Mater. Chem. Phy. 170: 99–107.

Silva, M.V., D. Stainer, H.A. Al-Qureshi, O.R.K. Montedo and D. Hotza. 2014. Alumina-based ceramics for armor application: Mechanical characterization and ballistic testing. J. Ceram. 2014: 618154.

Tang, D., H.-B. Lim, K.-J. Lee, C.-H. Lee and W.-S. Cho. 2012. Evaluation of mechanical reliability of zirconia-toughened alumina composites for dental implants. Ceram. Int. 38: 2429–2436.

Thess, A., R. Lee, P. Nikolaev, H. Dai, P. Petit, J. Robert et al. 1996. Crystalline ropes of metallic carbon nanotubes. Science 273: 483–487.

Tritt, T.M. (ed.). 2004. Thermal Conductivity: Theory, Properties and Applications. Kluwer Academic/Plenum Publishers, New York.

Umma, A., M.A. Maleque, I.Y. Iskandar and Y.A. Mohammed. 2012. Carbon nano tube reinforced aluminium matrix nano-composite: a critical review. Aust. J. Basic. Appl. Sci. 6: 69–75.

Wang, C.-Y., C.-P. Liu and C.B. Boothroyd. 2009. Thermal stability of catalytically grown multi-walled carbon nanotubes observed in transmission electron microscopy. Appl. Phys. A. 94: 247–251.

Wei, T., Z. Fan, G. Luo and F. Wei. 2008a. A new structure for multi-walled carbon nanotubes reinforced alumina nanocomposite with high strength and toughness. Mater. Lett. 62: 641–644.

Wei, T., Z. Fan, G. Luo, F. Wei, D. Zhao and J. Fan. 2008b. The effect of carbon nanotubes microstructures on reinforcing properties of SWNTs/alumina composite. Mater. Res. Bull. 43: 2806–2809.

Yamamoto, G., M. Omori, T. Hashida and H. Kimura. 2008a. A novel structure for carbon nanotube reinforced alumina composites with improved mechanical properties. Nanotechnol. 19: 315708.

Yamamoto, G., M. Omori, K. Yokomizo and T. Hashida. 2008b. Mechanical properties and structural characterization of carbon nanotube/alumina composites prepared by precursor method. Diam. Rel. Mater. 17: 1554–1557.

Zhan, G.-D. and A.K. Mukherjee. 2004. Carbon nanotube reinforced alumina-based ceramics with novel mechanical, electrical, and thermal properties. Int. J. Appl. Ceram. Technol. 1: 161–171.

Zhang, M., D.W. He, L. Ji, B.Q. Wei, D.H. Wu, X.Y. Zhang et al. 1999. Microstructural changes in carbon nanotubes induced by annealing at high pressure. Carbon 37: 657–662.

Zhu, Y.-F., L. Shi, J. Liang, D. Hui and K.-t. Lau. 2008. Synthesis of zirconia nanoparticles on carbon nanotubes and their potential for enhancing the fracture toughness of alumina ceramics. Compos. Part B: Eng. 39: 1136–1141.

List of abbreviation(s)

Name	Abbreviation
Alumina	Al_2O_3
Aluminum carbide	Al_4C_3
Apparent porosity	AP
Average Vickers hardness	$^{avg.}HV$
Bulk density	BD
Carbon nanotube	CNT
Characteristic Vickers hardness	^{ch}HV
Direct crack measurement	DCM
Electrical conductivity	σ_{DC}
Field emission scanning electron microscope	FESEM
Flexural strength	σ_{FS}
Fracture resistance	K_R
Fracture toughness	K_{IC}
Linear shrinkage	LS
Multiwalled carbon nanotube	MWCNT
Pressureless sintering	PLS
Relative density	RD
Room temperature	RT
Silica	SiO_2
Silicon carbide	SiC
Silicon nitride	Si_3N_4
Single edge notched beam	SENB
Singlewalled carbon nanotube	SWCNT
Sintering temperature	T_{sin}
Spark plasma sintering	SPS
Standard deviation	SD
Theoretical density	TD
Thermal conductivity	κ
Transmission electron microscope	TEM
Vickers hardness	HV
Zirconia	ZrO_2

Index